Lecture Notes in Mechanical Engineering

Lecture Notes in Mechanical Engineering (LNME) publishes the latest developments in Mechanical Engineering—quickly, informally and with high quality. Original research or contributions reported in proceedings and post-proceedings represents the core of LNME. Volumes published in LNME embrace all aspects, subfields and new challenges of mechanical engineering.

To submit a proposal or request further information, please contact the Springer Editor of your location:

Europe, USA, Africa: Leontina Di Cecco at Leontina.dicecco@springer.com
China: Ella Zhang at ella.zhang@cn.springernature.com
India, Rest of Asia, Australia, New Zealand: Swati Meherishi at swati.meherishi@springer.com

Topics in the series include:

- Engineering Design
- Machinery and Machine Elements
- Mechanical Structures and Stress Analysis
- Automotive Engineering
- Engine Technology
- Aerospace Technology and Astronautics
- Nanotechnology and Microengineering
- Control, Robotics, Mechatronics
- MEMS
- Theoretical and Applied Mechanics
- Dynamical Systems, Control
- Fluid Mechanics
- Engineering Thermodynamics, Heat and Mass Transfer
- Manufacturing Engineering and Smart Manufacturing
- Precision Engineering, Instrumentation, Measurement
- Materials Engineering
- Tribology and Surface Technology

Indexed by SCOPUS, EI Compendex, and INSPEC.

All books published in the series are evaluated by Web of Science for the Conference Proceedings Citation Index (CPCI).

To submit a proposal for a monograph, please check our Springer Tracts in Mechanical Engineering at https://link.springer.com/bookseries/11693.

Sai Sidhardh · S. Suriya Prakash ·
Ratna Kumar Annabattula · Phani Mylavarapu
Editors

Advances in Structural Integrity for Mechanical, Civil, and Aerospace Applications

Proceedings of SICE 2022, Volume 1

Springer

Editors
Sai Sidhardh
Department of Mechanical and Aerospace Engineering
Indian Institute of Technology Hyderabad
Sangareddy, Telangana, India

Ratna Kumar Annabattula
Department of Mechanical Engineering
Indian Institute of Technology Madras
Chennai, Tamil Nadu, India

S. Suriya Prakash
Department of Civil Engineering
Indian Institute of Technology Hyderabad
Sangareddy, Telangana, India

Phani Mylavarapu
Defence Metallurgical Research Laboratory
Hyderabad, Telangana, India

ISSN 2195-4356 ISSN 2195-4364 (electronic)
Lecture Notes in Mechanical Engineering
ISBN 978-981-97-6366-5 ISBN 978-981-97-6367-2 (eBook)
https://doi.org/10.1007/978-981-97-6367-2

This Springer imprint is published by the registered company Springer Nature Singapore Pte Ltd.
The registered company address is: 152 Beach Road, #21-01/04 Gateway East, Singapore 189721, Singapore

Contents

About the Editors

Dr. Sai Sidhardh is an Assistant Professor in the Department of Mechanical and Aerospace Engineering at Indian Institute of Technology Hyderabad. Before joining IIT Hyderabad in 2021, he was a Post-Doctoral Research Associate at the Ray W. Herrick laboratories, Purdue University, USA. His research interests include developing computational methods and solvers (mesh and meshless) for solid mechanics, artificial intelligence and machine learning techniques for applications in solid mechanics, multiphysics (coupled with solid mechanics) modeling and simulations, nonlocal mechanics, and applied fractional calculus. He has published close to 30 research papers in internationally reputed journals. He also serves as a Reviewer for various journals in numerical methods, fracture, composite structures, vibrations, among other general areas of solid mechanics.

Dr. S. Suriya Prakash is a Professor in the Department of Civil Engineering at IIT Hyderabad. He received a prestigious Ramanujan fellowship from the Government of India. Before joining IIT Hyderabad, he was a design engineer at Structural Group Inc., Baltimore, USA. He received his Ph.D. from Missouri University of Science and Technology, USA, in 2009. He obtained his MS degree in Structural Engineering from IIT Madras in 2005. His research interests include using advanced composite materials for civil infrastructure, precast systems for affordable housing, and repairing and rehabilitating structures. He has fifteen years of research and teaching experience, guided 8 Ph.D.s published more than 85 research papers in internationally reputed journals and presented more than 100 papers at national and international conferences. He is also the associate editor for the *ASCE Journal of Bridge Engineering* and a member of ACI, ASCE, ICI and ASTR. He is also an executive member of the Indian Institution of Bridge Engineers. He also serves on the editorial board of the *ASCE Journal of Composites for Construction.*

Dr. Ratna Kumar Annabattula is a Professor in the Department of Mechanical Engineering at Indian Institute of Technology Madras. His areas of interests are stimuli-responsive soft matter, granular materials and coupled problems in mechanics. He has more than 85 peer-reviewed international journal publications and has been a reviewer for several journals in the general area of mechanics and soft matter. He has guided 7 PhDs, 8 MS by research and 47 M.Tech. students. He is a member of two centres of excellence at IIT Madras namely, Centre of Excellence in Soft and Biological Matter, Centre for Additive Manufacturing. He is currently leading a group of 2 postdocs, 11 Ph.D. students and 9 Master's students.

Dr. Phani Mylavarapu is currently working as Scientist "E" in Non-Destructive Testing Group at Defence Metallurgical Research Laboratory, DRDO, Hyderabad. His current research interests include reliability assessment of NDT techniques for aerospace applications, modelling of NDT techniques, surface treatment techniques for inducing beneficial residual stresses in aeroengine components, applications of residual stresses in life revision studies of aeroengine components and characterization of powder metallurgy based ceramic composite materials used for advanced armor and structural applications using advanced NDT Techniques, etc. He has over 50 publications including various international journal papers, conference proceedings and presentations. He also serves as a reviewer for various journals in the area of surface engineering and non-destructive characterization of materials.

Modelling Bearing Capacity of Bored Piles Under Vertical Eccentric Load Within Python Framework

Pratyusha Bandaru and Hemaraju Pollayi

Abstract This paper deals with the modelling of load bearing capacity of bored-pile groups under the vertical eccentric load using Python. In many practical problems, the pile foundations of civil engineering structures (such as bridges, retaining walls, offshore structures) are frequently subjected to eccentric loads. There are two main tasks considered in the present work: (1) a group of equally spaced, identical piles subjected to an external moment vector, and (2) general case of a group of unevenly distributed and dissimilar piles. It is observed that the pile group subjected to vertical eccentric load failed by a cap rotation about the head of a pile and all piles within the group achieve the ultimate axial capacity in compression under the collapse mechanism. Finally, a wind turbine (with 95 m height) having a circular raft (with 16 bored piles, 22 m long, 0.8 m diameter) is modelled in the present work considering typical sub-soil profile (consisting of a deep layer of over consolidated, inorganic silty clay of high plasticity up to a depth of about 15 m with the ground water table 2.0 m from the ground level) and estimated the undrained shear strength in the intermediate layer and in the lower stiff clay. The authors will optimise the pile group under eccentric loading for the cost–benefit analysis.

Keywords Bored-pile group · Vertical eccentric load · Collapse mechanism · Sub-soil profile · Wind turbine

Nomenclature

(x, y)	Coordinates of the pile
α	Adhesion factor

P. Bandaru (✉) · H. Pollayi
Department of Civil Engineering, GITAM University, Hyderabad, Telangana 502329, India
e-mail: pbandaru2@gitam.edu

H. Pollayi
e-mail: dpollayi@gitam.edu

S. Sidhardh et al. (eds.), *Advances in Structural Integrity for Mechanical, Civil, and Aerospace Applications*, Lecture Notes in Mechanical Engineering,
https://doi.org/10.1007/978-981-97-6367-2_1

p	Number of piles in a group
P_k	Characteristic value of pile base resistance

1 Introduction

Typically, during analysis and design, super-structures are assumed to be fixed at the base. However, in reality, the super structure building's soil-foundation system has an influence on the resulting structure. Soil-structure interaction is the phenomenon of the superstructure's response to the soil-foundation system and the soil-response foundation's to the superstructure (SSI). In a more recent study, employed limit analysis theorems to create interaction diagrams for pile groups subjected to vertical eccentric loads in order to forecast their bearing capacity. They noted that 'cap rotation' was the most frequent failure mechanism [1]. Three studies examined the nonlinear-coupled response to multi-directional loads, which has previously been investigated in terms of linear elasticity. The pile group response to uni-directional stresses in the vertical, horizontal, or torsional directions has mostly been assessed using nonlinear simplified models [2–4]. Tall buildings are frequently exposed to moments from wind or other lateral forces, which can result in the entire structure to tilt and proposed a technique to measure the footing's effective width while being loaded eccentrically [5]. SSI impacts may change how damage is seen as well as how well structures with stand disasters; hence, they incorporated these effects into consideration [6]. The behaviour of pile groups under moderate-to-large eccentric vertical loads is the main subject of this paper. The capacity of a piling group is defined by current design methodologies as the eccentric vertical load corresponding to the accomplishment of the axial capacity of the outermost pile. The pile group is still able to support more external loads by drawing on the system's ductility reserves; therefore, this is not the beginning of yielding rather than the foundation failing [7].

2 Interaction Diagrams from Limit Analysis Theorems

The following presumptions are taken in order to calculate the collapse load of a pile group under vertical eccentric loads: (a) pile heads are connected through a rigid cap with infinite strength, (b) piles' connections to the cap are modelled as hinges, so that internal forces on piles only include axial loads, and (c) piles are modelled as rigid-plastic independent uniaxial elements with two yielding loads, one in compression (N_u), and one ($-S_u$), intended as the capacities of the single pile within the group (pinned piles). This work describes the findings of recent research based on numerical modelling, and it is intended to explore the parameters

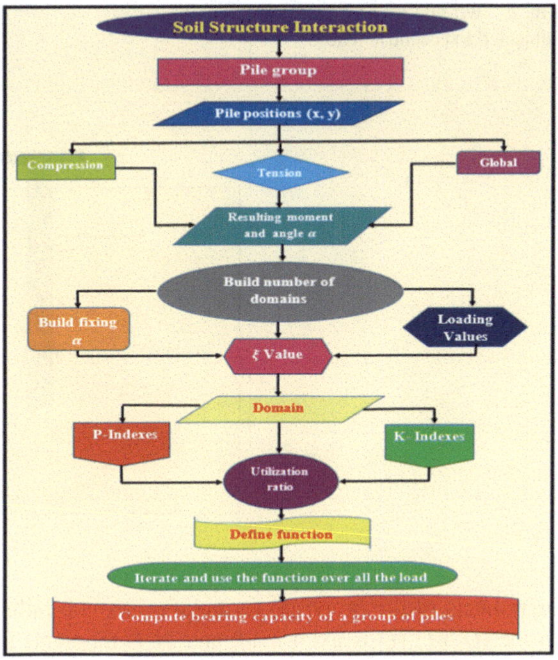

Fig. 1 Flowchart of soil-structure interaction of bearing capacity modelling of pile group

affecting the interaction between raft, soil, and piles while approaching failure. It begins by evaluating the knowledge currently available on the bearing capacity of piled rafts. Figure 1 shows flowchart of soil structure interaction (SSI) of bearing capacity modelling of pile group.

2.1 Arithmetic Framework: The Interaction Factor Concept

The proposed formulation is based on the interaction factor method and a stepwise incremental approach to account for nonlinear behaviour of piles. The main assumptions made to derive the mathematical framework are: (a) piles modelled uniaxial, nonlinear elements, with load–settlement behaviour in compression not necessarily equal to that in uplift, (b) irreversibility of the piles' behaviour, (c) piles' connecting cap idealised as a rigid body, clear from the soil, (d) pile-to-pile interaction effects modelled using superposition through the approximate solution [8], (e) piles' connections to the raft idealised as hinges. The assumption of rigid cap (c) is a reasonable hypothesis in case of 'small' piled rafts as originally defined [9]. Notably, for 'large' piled rafts, the load at foundation level is not remarkably eccentric and the response to combined moment-axial loads is a problem of minor concern. The same holds for rotational stiffness of individual piles, since this favourable contribution is small compared to the rocking stiffness associated with the axial capacity of piles. The basic

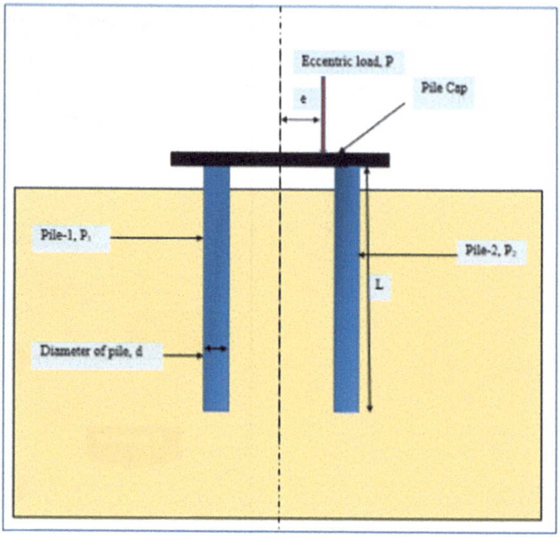

Fig. 2 Two-pile group subjected to eccentric load

ingredients of the proposed formulation are the failure envelope of the pile group in the (Q, M) (Q is axial load, and M is moment) plane and the load–settlement relationship of the piles. In the most general case of unevenly distributed, dissimilar piles, the interaction diagram coming from this approach is an irregular polygon.

2.2 Two-Pile Group

Consider a free-standing two-pile group with piles of length, L, and diameter, d, with spacing's subjected to vertical load, P, and an eccentricity, e, from the centre. For estimating pile settlement and pile-to-pile interaction, the ground is modelled as an elastic half space with deformation modulus, E_s, and Poisson's ratio, v_s. Figure 2 depicts the two-pile group's free body diagram and kinematics. Pile 1 represents the pile on the eccentricity side, and Pile 2 represents the pile on the opposite side. Piles 1 and 2 are compressed and settle by δ_1 and δ_2, respectively. Pile 1 experiences greater displacement than Pile 2, as expected. Due to the eccentric vertical load, P, the difference in displacements of both piles causes the two-pile group to rotate through an angle, $\theta = \frac{\delta_1 - \delta_2}{S}$.

2.3 Row of Equally Spaced Identical Piles

First, a set of evenly spaced, identical piles that are being external moment vector, M, that is inclined by α_M with respect to the y-axis are looked at Fig. 3. Because piles are designed to the centre of the capacity in compression, denoted by C_N, and in uplift, denoted by C_T, are identical. The origin of the reference system (X, Y) is considered coincident with C_N for simplicity's sake (or C_T). As seen in Fig. 3, a new reference system (n, g) can be established, with the g-axis aligned with the direction of the resulting moment vector M, as shown in Fig. 3. – 2p points make up the interaction diagram, where p is the quantity of piles in the group. As before, the vertices' coordinates can be expressed as follows:

$$\begin{cases} Q_{ui} = (i-1)N_u - (p-i+1)S_u \\ \quad i = 1, \ldots, p \\ M_{ui} = -N_u \sum_{j=1}^{i-1} \xi_j + S_u \sum_{j=i}^{p} \xi_j \end{cases} \tag{1}$$

and

$$\begin{cases} Q_{uk} = -(k-1)N_u + (p-i+1)S_u \\ \quad k = 1, \ldots, p \\ M_{uk} = S_u \sum_{j=1}^{i-1} \xi_j - N_u \sum_{j=k}^{p} \xi_j, \end{cases} \tag{2}$$

where

$$\xi_j = x_j \cos \alpha_M - y_j \sin \alpha_M. \tag{3}$$

The existence of conjunction lines between these points as an exact solution can be easily confirmed. The conjunction lines between two successive points are defined by the same equations as for the row of piles in the specific case of a rectangular group of identical piles with constant spacing s along both the x and y directions and

$$\alpha_M = 0$$

$$M_u = M_{ui} + (Q_u - Q_{ui})s/2 + (p - 2i + 1) \quad \text{for } i = 1 \ldots, 2p. \tag{4}$$

2.4 Row of Unevenly Distributed Dissimilar Piles

The aforementioned strategy can be simply expanded to the general case of a collection of disparate, unevenly distributed piles:

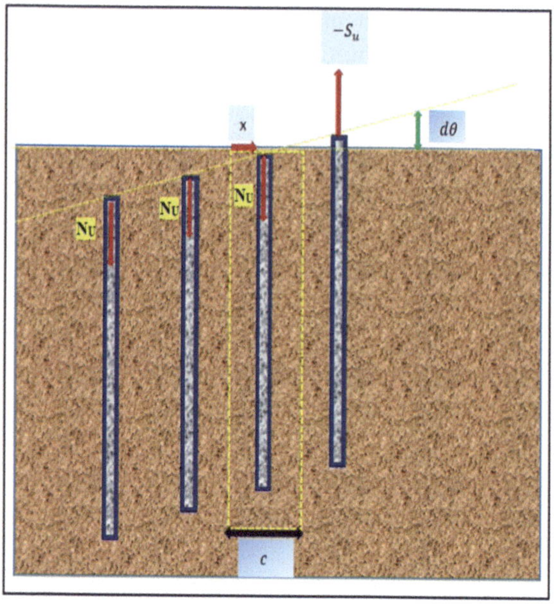

Fig. 3 Rotation about a point between two piles acts as the collapse mechanism for a row of equally spaced, identical pile

$$\begin{cases} Q_{ui} = \sum_{j=1}^{i-1} N_{uj} - \sum_{j=i}^{p} S_{uj} \\ i = 1\ldots, p \end{cases} \tag{5}$$

$$\begin{cases} M_{ui} = -\sum_{j=1}^{i-1} N_{uj}\xi_j + \sum_{j=i}^{p} S_{uj}\xi_j \end{cases} \tag{6}$$

$$\begin{cases} Q_{u(k+p)} = \sum_{j=k}^{p} N_{uj} - \sum_{j=1}^{k-1} S_{uj} \\ k = 1\ldots, p \end{cases} \tag{7}$$

$$\begin{cases} M_{u(k+p)} = \sum_{j=1}^{k-1} S_{uj}\xi_j - \sum_{j=k}^{p} N_{uj}\xi_j \end{cases} \tag{8}$$

Figure 4 shows the group of unevenly distributed dissimilar piles. Each of which has a different failure load in both compression and tension. In addition, the domain's maximum and minimum values of the ultimate axial load are connected to a nonzero ultimate moment. The domain is characterised by a polar symmetry that does not coincide with the origin of the reference system.

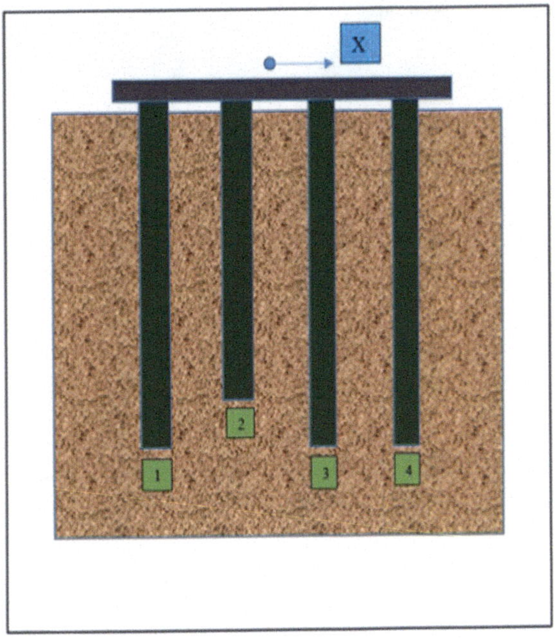

Fig. 4 Row of four piles, each of them having different failure loads in both compression and tension

3 Approximation of the Yielding Moment Incidence at the Pile Head

To benefit from the added strength provided by yielding moments occurring at the pile heads, the connections of the piles to the cap can be idealised as rigid-plastic internal fixities. Applying limit analysis theorems in this situation results in:

$$\begin{cases} \frac{Q_{ui}}{N_u} = i - 1 - \frac{S_u}{N_u} - (n - i + 1) \\ \quad i = 1 \ldots, n + 1 \\ \frac{M_{ui}}{sN_u} = \frac{1}{2}\left(1 + \frac{S_u}{N_u}\right)(i - 1)(n - i + 1) + (n - i + 1)m_{yt} + (i - 1)m_{yc} \end{cases} \tag{9}$$

$$\begin{cases} \frac{Q_{u(k+n+1)}}{N_u} = -\frac{S_u}{N_u}(k - 1) + (n - k + 1) \\ \quad k = 1 \ldots, n + 1 \\ \frac{M_{u(k+n+1)}}{sN_u} = -\frac{1}{2}\left(1 + \frac{S_u}{N_u}\right)(k - 1)(n - k + 1) - (n - k + 1)m_{yt} - (k - 1)m_{yc} \end{cases} \tag{10}$$

where

$$m_{yt} = \frac{M_{yt}}{sN_u} \tag{11a}$$

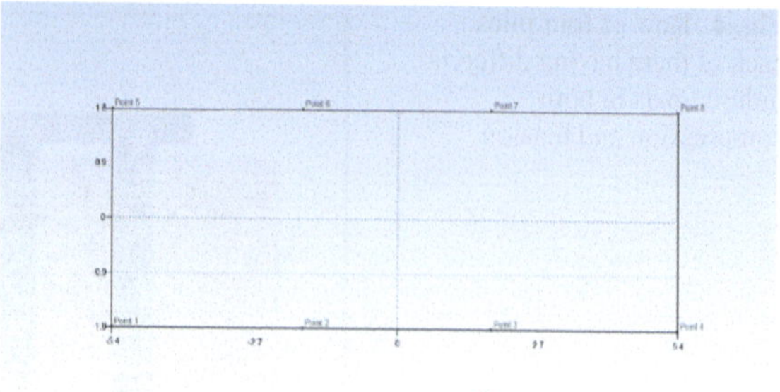

Fig. 5 Pile positions

$$m_{\mathrm{yc}} = \frac{M_{\mathrm{yc}}}{sN_{\mathrm{u}}} \quad (11b)$$

and M_{yt} and M_{yc} yield bending moments corresponding to $(-S_{\mathrm{u}})$ and N_{u}, respectively. In this case, the domain has $(2n+2)$ vertexes. Noticeably, $m_{\mathrm{yt}} = m_{\mathrm{yc}} = 0$ (Fig. 5).

$$Q_{\mathrm{u}i} = \sum_{j=1}^{i-1} N_{\mathrm{u}j} - \sum_{j=i}^{p} S_{\mathrm{u}j} \quad (12)$$

$$M_{\mathrm{u}i} = \sum_{j=1}^{i-1} \left(M_{\mathrm{yc}j} - N_{\mathrm{u}j}\xi_j\right) + \sum_{j=i}^{p} \left(M_{\mathrm{yt}j} + S_{\mathrm{u}j}\xi_j\right) \quad i = 1\ldots, p+1 \quad (13)$$

4 Case Study Application

This article applies a case study to illustrate the potential positive impact of the suggested approach. In this paper, a wind farm is the case being analysed. Includes 23 wind turbines built in southern Italy. Each wind turbine has a height of 95 m and is supported by 16 cast in situ drilled piles that are each 22 m long and 0.8 m in diameter. The total length of the piles, *nL*, which is approximately proportionate to the price of the piles, is 8096 m. A wind turbine's cross section and plan view are shown in Fig. 6. The typical sub-soil profile found at the site under investigation is also depicted in this figure. It consists primarily of a deep layer of over consolidated, inorganic silty clay with a high plasticity up to a depth of about 15 m, which is underlain by a very stiff, inorganic clay layer with a medium plasticity, and which is covered by a layer of silty sand. The ground water table was discovered at 2.0 m. While laboratory experiments mostly comprised unconsolidated and undrained tri axial tests, site investigations also included boreholes, conventional penetration tests, and down-hole tests. Therefore, it was calculated that the undrained shear strength, S_{u}, was 196 kPa in the intermediate layer and 232 kPa in the bottom stiff clay.

The following equations were used to evaluate unit base and shaft strength under undrained conditions in terms of total stress:

$$p_k = \sigma_{VL} + (N_C S_u), \tag{14}$$

$$s_k = \alpha(S_u), \tag{15}$$

where σ_{VL} is the total overburden stress at the depth L of the pile base, $N_c = 9$, and α is a coefficient depending on the pile installation technique and the undrained shear strength. A can be assumed to be equal to 0.35 [10] for replacement piles and $S_u - C$ 70 kPa. Finally, the following expression [11] was used to estimate the design value of the single pile capacity in compression (Fig. 7):

$$N_u = \frac{P_k}{\gamma_b} + \frac{S_k}{\gamma_s} = \frac{\pi d^2}{4\xi_3 \gamma_b} + \left(\sigma_{VL} + \frac{N_C S_u}{\gamma_{su}}\right) + \frac{\pi d}{\xi_3 \gamma_s \gamma_{su}} \int_0^L s_k dz. \tag{16}$$

C_S_u is the partial resistance factor for undrained shear strength, c_b and c_s are partial resistance factors for pile base and shaft capacity, and n_3 is the correlation factor to determine characteristic value from ground test data. Given that the shaft capacity under compression and the uplift capacity of a bored pile in clay are projected to be identical in the near future [12, 13].

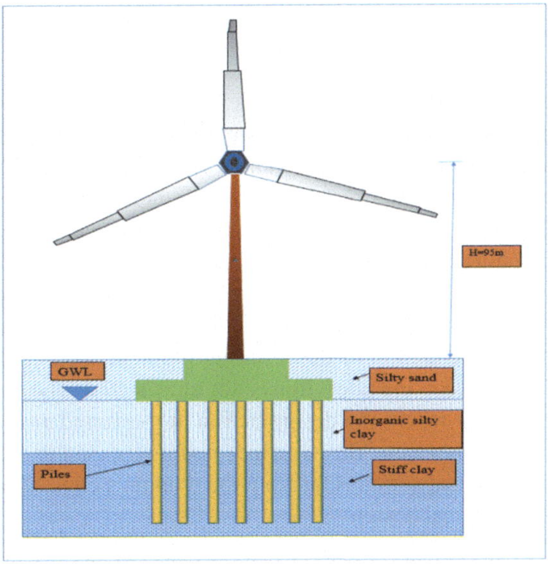

Fig. 6 Cross section of the piled foundation of the wind turbine

Fig. 7 Plan of the piled foundation of the wind turbine

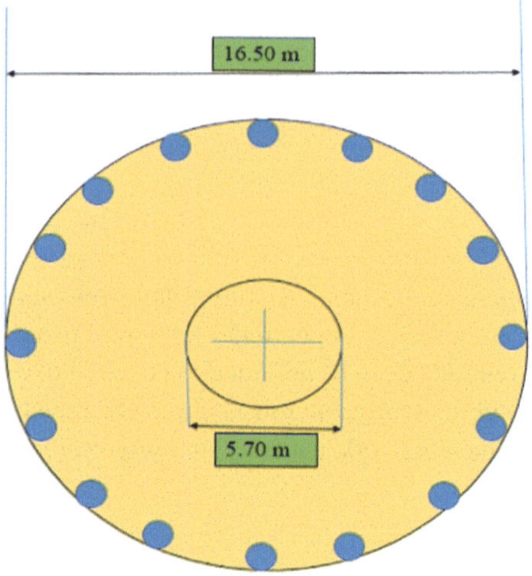

$$S_u = \frac{S_k}{\gamma_{st}} = \frac{\pi d}{\xi_3 \gamma_{st} \gamma_{su}} \int_0^L s_k dz, \tag{17}$$

where γ_{st} is the shaft capacity's partial factor in uplift. $S_u = 2614$ kN and $N_u = 2855$ kN were obtained using Design Approach 2 of Euro code 7 and partial factors from the Italian national appendix

$$(\gamma_b = 1.35, \gamma_s = 1.15, \gamma_{st} = 1.25, \gamma_{su} = 1, \xi_3 = 1.45).$$

As is expected for thin piles in clay, the axial capacity in uplift is thus extremely close to the one in compression. The group's two capacities were assessed under the assumption that its effectiveness was equal to one. The interaction diagrams based on the traditional and suggested methodologies are shown in Fig. 8. To examine the impact of wind direction on the ultimate axial capacity of the foundation, both domains were analysed for two values of the inclination *aL* of the applied moment vector *M*: $\alpha_M = 0$ and $\alpha_M = \pi/n$, with *n* number of piles supporting a single wind turbine. Evidently, the size and shape of the interaction diagrams are little affected by the inclination of the resulting moment vector. Given the close angular separation between the piles, this is not unexpected. The conventional domain is near the boundary where the design action at the foundation level equivalent to high wind conditions occurs. Based on the maximum capacity of the entire foundation, however, the identical action is far from the domain's edge. Consequently, there is space for a more logical design.

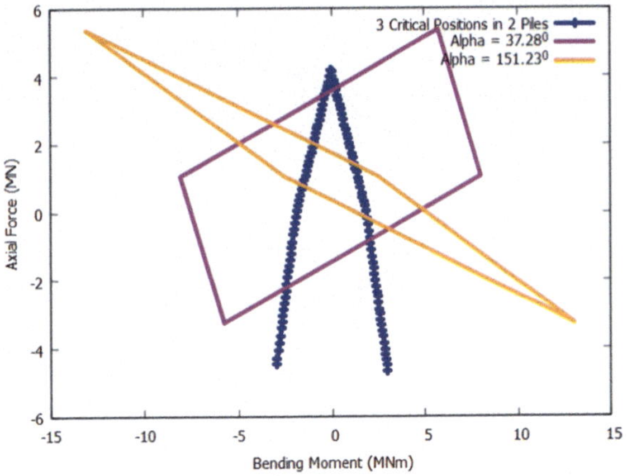

Fig. 8 Variation of the moment generated with respect to the load on the two-pile cap from the superstructure by changing the adhesion factor (α)

5 Results and Discussion

To validate the present work, the authors selected three different problems and observed the moment-resisting capacity of the pile group by changing the adhesion factor (α), which affects the behaviour of the bored piles under the vertical eccentric load. In the first example, a two-pile system is considered and the variation of the moment and load with respect to the adhesion factor (α) is plotted as shown in Fig. 8. It is observed that the ultimate moment capacity of the pile group is increased to $M_{\text{ult}} = 13.26$ MNm for the adhesion factor $\propto = 151.23°$. In the second example, for the four-pile system, the adhesion factor (α) is varied and considered $\propto = 37.28°$, $\propto = 151.23°$, $\propto = 197.73°$, $\propto = 351.24°$. It is observed that the ultimate moment capacity of the pile group is increased to be $M_{\text{ult}} = 31.2$ MNm for the adhesion factor $\propto = 351.24°$ (Fig. 9).

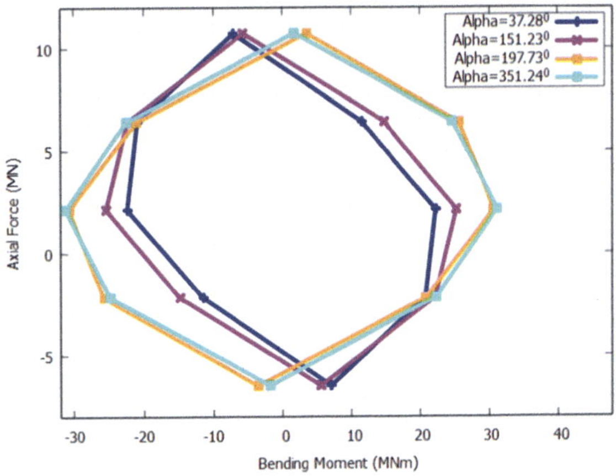

Fig. 9 Variation of the moment generated with respect to the load on the four-pile cap from the superstructure by changing the adhesion factor (α)

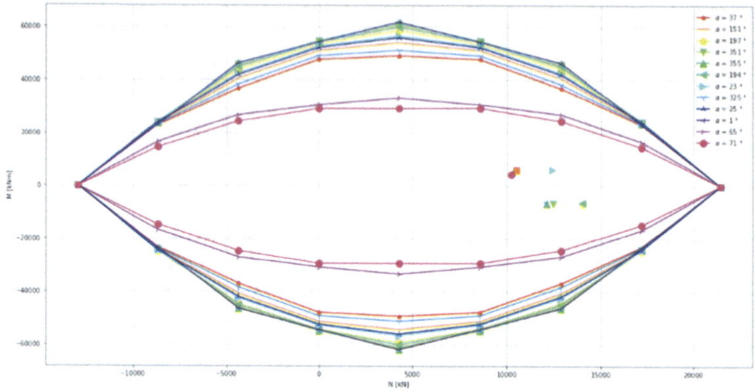

Fig. 10 Interaction diagram of the foundation of the wind turbine

Finally, to highlight the potential beneficial effect of the proposed methodology, a wind turbine is modelled and estimated the undrained shear strength. Figure 10 shows the interaction diagram of the foundation of the wind turbine. It is evident that all the piles achieve their ultimate axial strength and the group of piles loaded eccentrically will fail by a cap rotation about the head of a pile.

6 Conclusions

The present approach is employed to access the bearing capacity of pile groups under vertical eccentric loads assumes as ultimate strength that corresponding to the achievement of the axial capacity on the outermost pile. To identify an alternative, the more rational approach is employing the ultimate moment-axial force interaction diagrams. To analyse the problem under examination consists of a rigid cap, clear from the ground, surmounting a group of unevenly distributed, dissimilar piles, each of them having specified values of ultimate load in compression and uplift. If the external load corresponds to a vertex, failure may occur according to a cap rotation about any arbitrary point in between two consecutive piles in this case. This work is aimed at providing merely a theoretical framework regarding the bearing capacity of pile groups under vertical eccentric loads. All the piles achieve their ultimate strength.

References

1. Di Laora R, de Sanctis L, Aversa S (2019) Bearing capacity of pile groups under vertical eccentric load. Acta Geotech 14(1):193–205
2. Zhang Q, Liu S, Zhang S, Zhang J, Wang K (2016) Simplified non-linear approaches for response of a single pile and pile groups considering progressive deformation of pile–soil system. Soils Found 56:473–484
3. Ooi PS, Chang BK, Wang S (2004) Simplified lateral load analyses of fixed-head piles and pile groups. J Geotech Geoenviron Eng 130:1140–1151
4. Chen S, Kong L, Zhang L (2016) Analysis of pile groups subjected to torsional loading. Comput Geotech 71:115–123
5. Meyerhof GG (1953) The bearing capacity of foundations under eccentric and inclined loads. In: Proceedings of 3rd international conference on soil mechanics and foundation engineering, vol 1, pp 440–445
6. Hayashi Y, Takahashi I (2004) Soil structure interaction effects on building response in recent earthquakes. In: Proceedings of the 3rd UJNR workshop on soil–structure interaction, California, USA, pp 1–8
7. AASHTO L (2012) AASHTO LRFD bridge design specifications
8. Dobry R, Gazetas G (1988) Simple method for dynamic stiffness and damping of floating pile groups. Geotechnique 38(4):557–574
9. Russo G, Viggiani C (1998) Factors controlling soil-structure interaction for piled rafts. In: International conference on soil structure interaction in urban civil engineering, Darmstadt geotechnics, vol 4, pp 297–322
10. Viggiani C (1993) Further experience with auger piles in Naples area. In: Inpe WF (ed) Proceedings of the second international geotechnical seminar on deep foundations on bored and auger piles, BAP II. Balkema, Rotterdam, pp 445–458
11. CEN (pr)EN 1997-1 (2003) Euro code 7: geotechnical design-part I: general rules. European Committee for Standardization Technical Committee 250, Brussels, Belgium
12. Fleming WGK, Weltman AJ, Randolph MF, Elson WK (1992) Piling engineering, 2nd edn. Blackie Academic and Professional, Glasgow
13. St John HD, Randolph MF, Avoy RP, Gallagher KA (1983) The design of piles for tethered platforms. In: Proceedings of the conference on the design and construction of offshore structures. Institution of Civil Engineers, London, pp 53–64

An Improved Methodology for Precise Estimation of Fracture Process Zone Size

Mansi, Tutika Kavya, and Sonali Bhowmik

Abstract Concrete is widely used as a construction material; due to its heterogeneous behavior, the service life of the structures becomes crucial so as to avoid the sudden brittle failures. Fracture in concrete is accompanied by emergence and evolution of an inelastic zone, fracture process zone (FPZ), around the crack tip. Inside this FPZ, many mechanisms occur: crack blunting, aggregate bridging, and microcracking, which prevent brittle failure. The softening of stress–strain curve in post-peak region and the size-effect phenomena is attributed due to the existence of this FPZ. As the size of FPZ influences the toughness of concrete—larger the FPZ, higher will be toughness; it becomes necessary to quantify the length and width of FPZ. Till now, the research was primarily focused on the length of fully developed FPZ. However, for better insight into the crack growth, an accurate prediction of FPZ size is necessary. Hence, precise quantification of width is also important. In this work, an attempt has been made to estimate the width of fracture process zone using *v*-displacement data of digital image correlation (DIC). The study aims to propose a methodology to find the FPZ width in plain concrete members. The maximum FPZ width for monotonic loading is estimated for geometrically similar beam samples. The proposed methodology has been verified against the existing literature on other non-contact techniques, and a good agreement has been found between them.

Keywords Fracture process zone · Digital image correlation · Plain concrete · Microcrack

Mansi (✉) · T. Kavya · S. Bhowmik
Department of Civil Engineering, National Institute of Technology Rourkela, Rourkela, Odisha, India
e-mail: 521ce1012@nitrkl.ac.in

S. Sidhardh et al. (eds.), *Advances in Structural Integrity for Mechanical, Civil, and Aerospace Applications*, Lecture Notes in Mechanical Engineering,
https://doi.org/10.1007/978-981-97-6367-2_2

1 Introduction

In quasi-brittle materials such as concrete, rock, and ceramics, a zone of microcracking predates the fracture. This nonlinear zone is termed as fracture process zone (FPZ) which forms and evolves around the propagating crack tip and is of finite size. In this zone, nucleation of many microcracks gives rise to a major macrocrack. Thus, FPZ can be defined as nonlinear zone of distinct volume where coalescence of multiple microcracks happens to form a macrocrack, by grouping of microcracks in the vicinity of crack tip. A schematic diagram representing the development of FPZ is shown in Fig. 1. FPZ is also responsible for the nonlinear behavior of quasi-brittle materials and deviation from Linear Elastic Fracture Mechanics (LEFM). It has been seen that inside FPZ, fracture is characterized by several toughening mechanisms like aggregate interlocking, crack-tip blunting, microcrack shielding, and crack bridging which determines the ultimate crack trajectory. Thus, FPZ serves as a key to understanding fracture mechanisms in quasi-brittle materials. Precise observation of FPZ size and properties helps in predicting the material response of structure under the action of various loads, in turn bridging the gap between design life and service life of the structure. Hence, FPZ grabbed the attention of many researchers over the years.

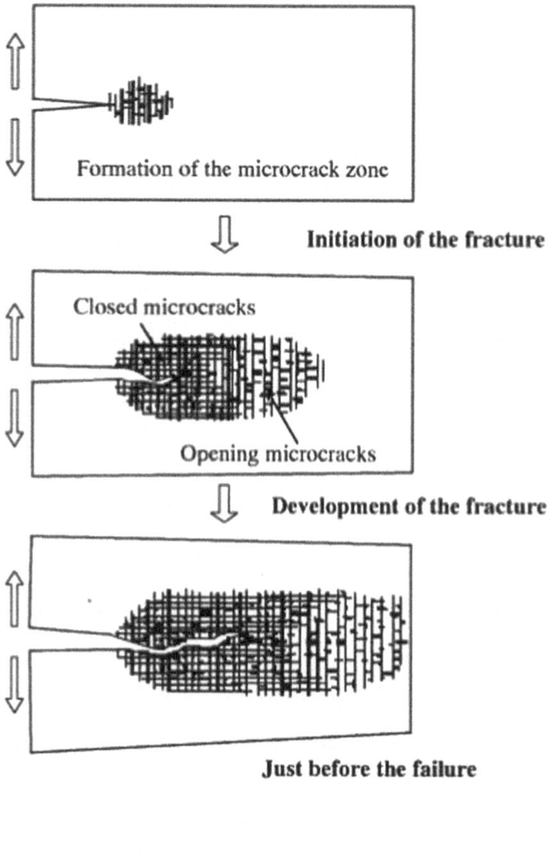

Fig. 1 Development of fracture process zone [1]

Earliest works on the existence of FPZ can be dated back to Hillerborg's fictitious crack length model [1]. Labuz et al. [2] proved the existence of FPZ in granite using Acoustic Emission (AE) and ultrasound probing. Castro-Montero et al. [3] using holographic setup and image analysis of strain field showed the existence of FPZ in mortar plate specimens with a notch at the center. In this study, deviation in experimentally observed strain fields and LEFM-based strain fields was regarded as the FPZ. Zhang and Wu [4] conducted AE tests on concrete beams with three-point bending while varying notch depths. This study quantified the development of the FPZ and shows that its length is independent of material properties. Following the AE analysis, Otsuka and Date [5] provided a clear understanding of FPZ in concrete at different stages of failure.

For a thorough understanding of FPZ, works have been dedicated to find its length and width using analytical concepts and experimental data. Bazant and Kazemi [6] correlated the length of FPZ with the brittleness number for rock and concrete samples. Bažant and Oh [7] showed width of FPZ to be independent of specimen size and totally a function of material parameter. For concrete, it was taken as three times the maximum size of aggregate. Saucedo et al. [8] determined the size of fully developed FPZ through an analytical cohesive model. Grassl et al. [9] studied the effect of size of structure on FPZ size in notched and un-notched concrete specimens through mesoscale approach. Relation between beam size and FPZ width was highlighted in this study. It was shown that for notched beam specimen, FPZ width was independent of beam size, and for un-notched specimen, width increases with beam size. Fakhimi and Tarokh [10] used Bazant's size-effect law to propose an analytical model. In their model, the size of FPZ was represented as a function of size of specimen and length-to-height ratio of specimen. Width of FPZ was calculated in regard to specimen size and was seen to increase with an increase in specimen size. Hu et al. [11] proposed an analytical model to estimate length and width of FPZ in terms of microstructure size for different quasi-brittle materials. The study highlighted the role of crack-tip blunting with respect to FPZ width at peak load. Ohno et al. [12] correlated the width of FPZ with the width of AE cluster. In their findings, it was highlighted that as the width of AE cluster increases, FPZ width was also found to be expanding. Saliba et al. [13] found out length and width of FPZ using AE event clusters. It was mentioned that width of FPZ increases with a decrease in relative notch depth. Despite all these observations, a question about the precision of estimated size of FPZ remains. Ambiguity over the accuracy of observations remained a debatable topic among researchers.

Recently, other non-contact techniques such as digital image correlation (DIC) are being used by researchers to estimate the size of FPZ. With the advancement in computer-based technologies, there has been rapid growth in the application of DIC in various civil engineering purposes which can be briefly summarized by Sutton and Hild [14]. In this line, Shah and Chandra Kishen [15] performed experiments on concrete–concrete bi-material interfaces to determine various fracture properties. Wu et al. [16] investigated FPZ length in plain concrete using DIC. They found that length of FPZ increases up to the point of its complete development and decreases subsequently. Alam et al. [17] investigated the size effect in concrete using DIC.

They also combinedly used DIC and AE to find out the size of FPZ [18]. Skarżyński and Tejchman [19] used DIC to characterize the size of FPZ in plain and reinforced concrete under the effect of shape, size, and volume of aggregate. Le et al. [20] applied DIC on rocks to determine the FPZ length under monotonic and cyclic loading. In their study, it was reported that length of FPZ in case of cyclic loading was observed to be 60% greater than in monotonic loading. Erarslan [21] studied the FPZ development using DIC on Mode I fracture toughness of rocks. Dong et al. [22] used DIC to study FPZ in rock–concrete interface. Bhowmik and Ray [23] developed a new methodology to quantize the length of FPZ in plain concrete under the action of monotonic as well as fatigue loading. Li et al. [24] used DIC and Bazant crack band model to determine the position of crack tip and FPZ length in concrete.

In a nutshell, it can be seen from the literature that several techniques can be found for the length estimation. However, due to the common perception of length being greater than width, very few literature can be found for FPZ width estimation. The knowledge of width is significant in explaining the toughening mechanisms inside FPZ such as crack bridging and crack-tip blunting. The estimation of FPZ width using DIC data has not been attempted yet. In this line, a basic attempt has been made to find the width of FPZ using DIC v-displacement data under the action of monotonic loading.

In this study, the existing methodologies [16, 20–25] for calculating the size of FPZ using DIC have been improved and modified to estimate the width of FPZ. The existing methods mainly use u-displacement data and its variation across the span to find the FPZ length. The tip of FPZ is determined either by the location of merged contours or by finding displacement jumps. Latter is more commonly used for precise estimation of FPZ length in quasi-brittle materials. Here, an attempt has been made to modify the gradient-based displacement jump methodology for estimating the width of FPZ. For this, DIC data of v-displacement has been used in place of u-displacement. The properties of v-displacement are still unexplored, and an attempt has been made to study FPZ using v-displacement. The variation of v-displacement across the cross section can be used to locate the boundary of FPZ. Following this, an improved methodology has been proposed for finding the FPZ width, which is further discussed in Sect. 3.1.

2 Digital Image Correlation (DIC)

Digital image correlation is a technique that uses image analysis tools and algorithms to estimate surface displacements and strain fields in any given sample. Digital images before and after application of load are compared and analyzed to find the displacement and strain fields. DIC works on the principle that light reflected from the deformed and undeformed specimens has a unique intensity and is stored and processed as grayscale images. Images are captured using camera with Charged Couple Device (CCD)-sensor array which has a high pixel count and can record thousands of pixels per line. The images are then digitized as an array of dark (0) intensity and light (100) intensity pixels and then convoluted into grayscale images.

Two-dimensional and three-dimensional analysis techniques can be used to conduct DIC studies. The key distinction between the 2-D and 3-D DIC approaches is that while the 2-D analysis calls for a single camera to take a series of photographs of a deformed planar object, the 3-D approach calls for a stereo vision system with two cameras to take a series of pair images. While two-dimensional DIC typically can only detect displacements in the image plane, the three-dimensional variation makes use of an additional camera to locate all three (spatial) components of the displacement vector. A 2-D DIC analysis can be utilized to measure in-plane surface displacements with a single camera if out-of-plane displacements are minimal [26]. The specimen in this situation needs to be parallel to the sensor plane and should be planned [27]. Few researchers have suggested using the 3-D DIC method since it is challenging to apply rigid body displacements to specimens for 2-D DIC analysis in order to evaluate the DIC algorithms. By putting more space between the camera and the object while using a conventional lens, the impact of the object's out-of-plane motion on in-plane measurements can be reduced [28].

A manually chosen area of interest (AOI) is divided into an equally spaced virtual grid in order to undergo DIC analysis. The virtual grid consists of a group of pixels called subsets which are used to track deformation. The displacements are computed with respect to each subset to obtain full-field deformations. Using specified correlation functions, such as cross-correlation or normal cross-correlation, the subgroup is tracked. In order to have sufficient distinct and distinguishable characteristics, subset size is taken such that it is large enough to include at least two or three speckles throughout its width and height and small enough to satisfy the basic assumption of image correlation that the deformation is homogeneous inside the subset. Subset spacing determines the distance between data points. Small spacing means that the point grid will be denser. As a result, the analysis will take longer to run, but the shape will be smoother, and the measured deformation will be more local.

In DIC, the sum of squared difference (SSD) correlation criteria or the cross-correlation criterion is typically used to represent the similarity pattern between the reference and deformed subsets. Two different cross-correlation criteria are employed in DIC algorithm to measure the deformation of subset. One is for determining initial guess and other one is for the subsequent refinement of the guess. Normal cross-correlation (NCC) is used for finding initial guess as shown in Eq. (1).

$$C_{\mathrm{NSD}} = \sum_{i=-m}^{m} \sum_{i=-m}^{m} \left[\frac{f(x_i, y_j)}{\overline{f}} - \frac{f\left(x_i^*, y_j^*\right)}{\overline{g}} \right]^2, \quad (1)$$

where $\overline{f}$ and $\overline{g}$ can be expressed as below:

$$\overline{f} = \sqrt{\sum_{i=-m}^{m} \sum_{i=-m}^{m} |f(x_i, y_j)|^2}, \tag{2}$$

$$\overline{g} = \sqrt{\sum_{i=-m}^{m} \sum_{i=-m}^{m} \left|g\left(x_i^*, y_j^*\right)\right|^2}. \tag{3}$$

For the present DIC analysis, a subset size of 29 × 29 pixels with subset spacing value of 1 has been selected to ensure the accuracy of displacement measured. For strain analysis, a strain radius of 35, with subset spacing of 1 has been chosen and maintained through the analysis. The data obtained from DIC is in pixels, and a conversion factor of 0.082, 0.115, and 0.19 has been used for small, medium, and large beam samples, respectively. The data extracted from DIC has been used in such a way that an estimation of width of fracture process zone can be made.

3 Experimental Details

The image data for static analysis has been taken from the experiments of Bhowmik and Ray [23] tested under 250 kN servo-hydraulic MTS machine. They experimented on geometrically similar beams of small, medium, and large sizes under the span-to-depth ratio of 4 and a constant width of 50 mm under the action of static and fatigue loads. Notch-to-depth ratio is maintained at 0.2 for all beam sizes. Details of specimen size are summarized in Table 1.

In this study, only the static image dataset has been analyzed using the modified version of the NCORR code [29]. From the code, strain and displacement plots are generated, which are suitably used for the interpretation of fracture process zone width by further analysis. The detailed methodology for the aforementioned task is explained in the following section.

Table 1 Specimen dimensions [23]

Specimen designation	Span (*S*) (mm)	Length (*L*) (mm)	Depth (*D*) (mm)	Width (*B*) (mm)	Notch size (*a*0) (mm)
Small	200	300	50	50	10
Medium	400	550	100	50	20
Large	800	1000	200	50	40

3.1 Methodology

The non-contact DIC technique has been used to monitor the fracture process of the beam. The DIC analysis was carried out at peak load and in post-peak regions at 90% of peak load, 70% of peak load, 50% of peak load, 30% of peak load, and 15% of peak load. Figure 2 shows the load–CMOD curve highlighting the analyzed post-peak regions.

DIC gives two types of displacements: opening displacement or *u*-displacement along the *X*-direction and sliding displacement or *v*-displacement along the *Y*-direction. In DIC, *u*-displacement has been used by researchers to find the length of FPZ [22–25]. Nevertheless, the estimation of FPZ width using DIC data has not been attempted yet. Here, the width of FPZ is estimated using DIC *v*-displacement data.

The objective is to find the threshold layer that separates the regions of major and minor microcracking which can be identified by differentiating the area with higher displacements from areas with lesser or no displacement. Minor microcracking is due to heterogeneity of concrete, and it does not contribute to the nonlinearity of concrete. The region with major microcracking is considered for estimating the width of FPZ. The methodology used for estimating the width of FPZ is represented in Fig. 3 and is explained as follows:

1. The entire AOI is covered by a series of parallel reference lines in the direction of span and height. These reference lines are arbitrarily chosen, and displacement data is plotted for the set of reference lines in the direction of span.
2. Reference lines in the direction of height are chosen in such a way that it traces the coordinates of the threshold layer. This set of values gives an estimate of the shape and size of FPZ.
3. The difference between the values corresponding to the left and right of notch gives the width of FPZ at that depth. The maximum value represents the maximum depth of FPZ in the direction of span and is reported in the results.

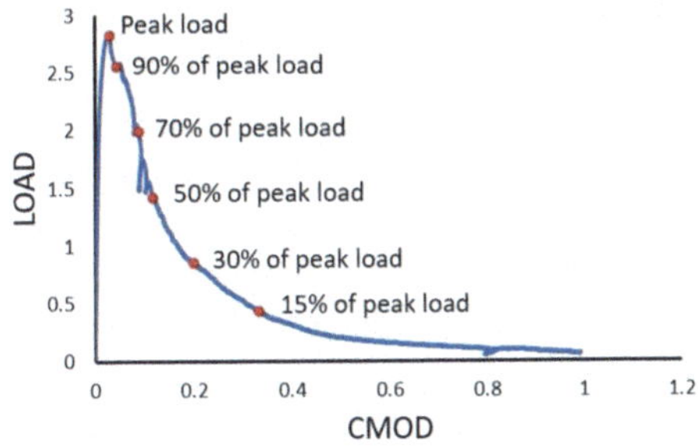

Fig. 2 Load–CMOD curve [23]

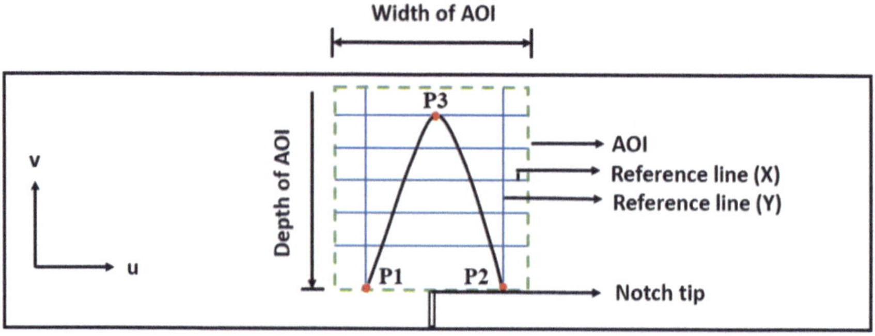

Fig. 3 Schematic diagram showing proposed methodology

Fig. 4 Geometrically similar beams [23]

The area of interest (AOI) has been chosen in accordance with the ratio of the depth of beams for the analysis. Figure 4 shows the geometrically similar beams. The value of AOI for small, medium, and large beam samples is reported as 505.38 mm^2, 2021.52 mm^2, and 8086.11 mm^2, respectively, as shown in Fig. 5.

4 Result and Discussion

Displacement contours obtained for different post-peak regions for medium-size beam are shown in Fig. 6. It can be seen that with progression in post-peak region, the plots are getting more distinct, and the value of maximum displacement is also increasing. It can be noted that the yellow-colored contour serves as the threshold layer in all these plots. Hence, the coordinates of this contour are used to find the maximum width of FPZ at 15% of post-peak load. Hence, for small and large beam samples, only displacement contours at 15% of post-peak load are reported. Figure 7 shows the displacement plots for small, medium, and large beam samples.

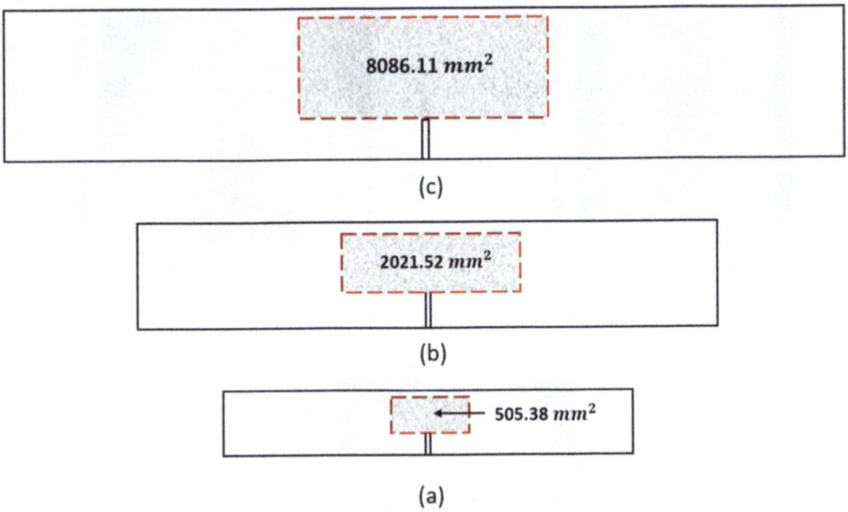

Fig. 5 Schematic diagram of geometrically similar beams showing AOI

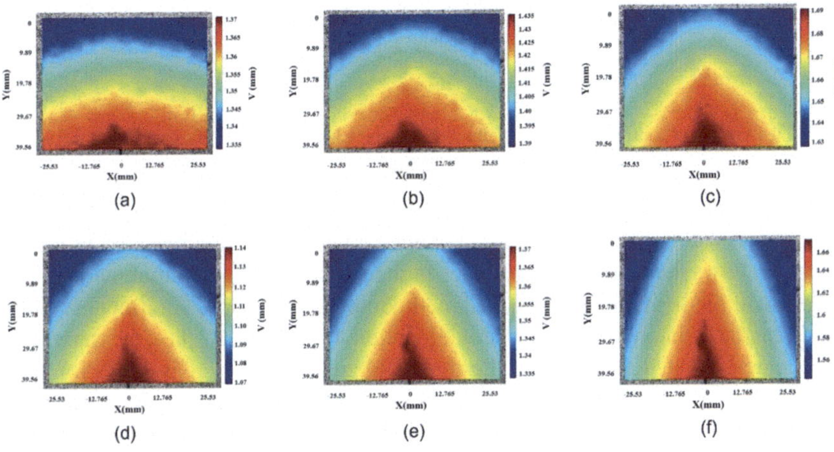

Fig. 6 Displacement contours for medium beam specimen at different post-peak regions at **a** peak load, **b** 90% of peak load, **c** 70% of peak load, **d** 50% of peak load, **e** 30% of peak load, **f** 15% of peak load

While plotting the displacement data at peak load and 15% of post-peak load in a single graph as shown in Fig. 8, a significant rise in displacement is observed. The vertical displacement at peak load is almost negligible in comparison to the displacement corresponding to 15% of post-peak load. Hence, it can be inferred that at peak load, the FPZ has not been formed and the microcracks are still insufficient to generate a macrocrack. The coordinates for the threshold region are extracted from

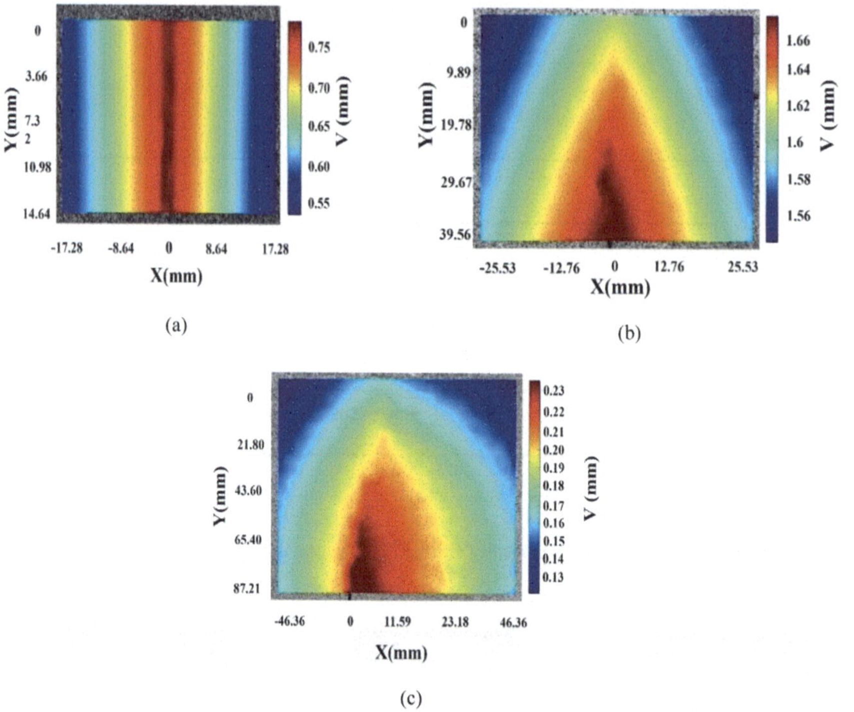

Fig. 7 Displacement contours for **a** small, **b** medium, and **c** large beam samples at 15% of post-peak load

the DIC data using the methodology explained above and are represented in Fig. 9. Figure 9a shows the extreme coordinates in the *v*-displacement contour, and Fig. 9b shows the implementation of proposed methodology. From these coordinates, width of FPZ can be estimated at that particular location. From the DIC plot, the maximum displacement located at the tip of the threshold zone is reported as 1.618 mm and plotted as tip displacement point at P3 in Fig. 9b. Coordinates of threshold zone and the estimated width are mentioned in Table 2. The maximum width of FPZ is reported as 31.51 mm for medium beam sample.

Large-size beam samples behave in a similar way as medium-size samples. The threshold region is forming under the yellow line contour, which encompasses majority of the displacement. However, for a small beam sample, there is no contour even for 15% of the post-peak load. This indicates that the fracture process zone has not been formed in the small beam samples. This leads to the size-effect phenomena in concrete that can be further explored as the future scope of this work. For now, it can be stated that FPZ width also changes with the change in size of sample.

The width of the FPZ is often regarded as the characteristic length of microstructure and may be considered as a material property. Numerous experimental findings indicate that the process zone has a semi-elliptical (narrow-band) shape [5, 16,

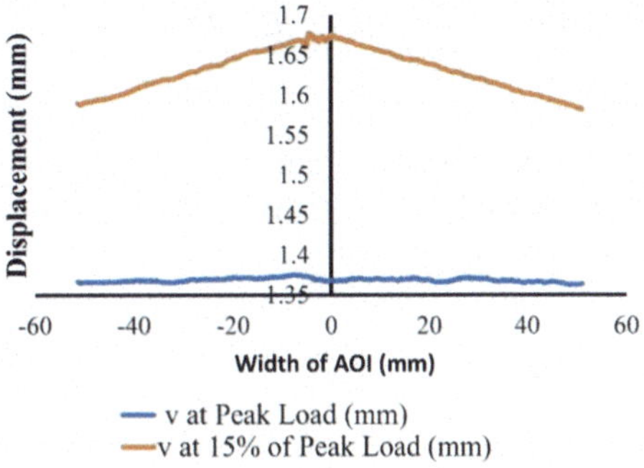

Fig. 8 *v*-displacement curves for medium size beam sample at peak load and 15% of peak load in post-peak region

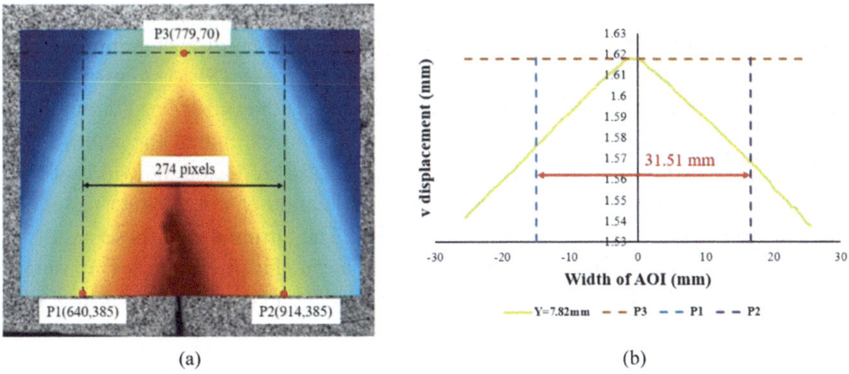

Fig. 9 **a** Schematic diagram representing the selected points with respective coordinates for width estimation, **b** figure estimating the width of FPZ

Table 2 Estimated width value

Point	Distance from notch (pixel)	Distance from notch (mm)
P1	130	14.95
P2	144	16.56
P3	70	36.225
Width		31.51

* Notch is taken as center

30]. This shape is well supported by the strip-yield (Dugdale–Barenblatt) model's assumption of a cohesion-based process zone along the crack plane [31, 32]. Bažant and Oh [7] put forward the crack band theory, in which the width of FPZ was assumed to be a constant value. The material fracture toughness was given by:

$$G_f = h_c\left(1 + \frac{E}{E_t}\right)\frac{f_t^2}{2E}, \quad (4)$$

where G_f is product of area under stress strain curve, E is modulus of elasticity, E_t is strain-softening model, h_c is width of crack band, and f_t is tensile strength of material.

To calculate the value of h_c, an approximate function was given in Bazant's crack band model as $h_c = n_a d_a$, where d_a is maximum aggregate size and n_a is coefficient relating width of FPZ to aggregate maximum size. The value of n_a was approximated to be 3 for concrete and 5 for rock as mentioned earlier. Zhou et al. [33] modeled the mesoscale evolution of fracture process zone in concrete and pointed out that the width of FPZ is less than 3 times of the maximum size of coarse aggregate. On the other hand, Skarżyński et al. [30] stated that instead of coarse aggregate dimension, FPZ width is related to size of specimen and mixing characteristics of concrete. The authors reported that the FPZ width varies between 3.5 and 5.5 mm for gravel concrete (with $d_{max} = 8$ mm) and sand concrete (with $d_{max} = 3$ mm). Later on, various researchers used non-contact damage quantification techniques such as Acoustic Emission (AE) and Scanning Electron Microscope (SEM) to estimate the width of FPZ. Mihashi and Nomura [34] reported the width of FPZ as 70 mm using AE. Otsuka and Date [5] also used AE technique and estimated the width to be between 20 and 45 mm. Similarly, Hadjab et al. [35, 36] reported the width of FPZ as 33 and 40 mm using SEM. Again, in this study, another non-contact technique (DIC) has been used to estimate the FPZ width. The summary of these results is provided in Fig. 10. From the figure, it can be seen that the value of n_a ranges between 2.25 and 3, which is in agreement with crack band model. The width of FPZ as determined from the improved methodology is also reported to be in accordance with the existing literature.

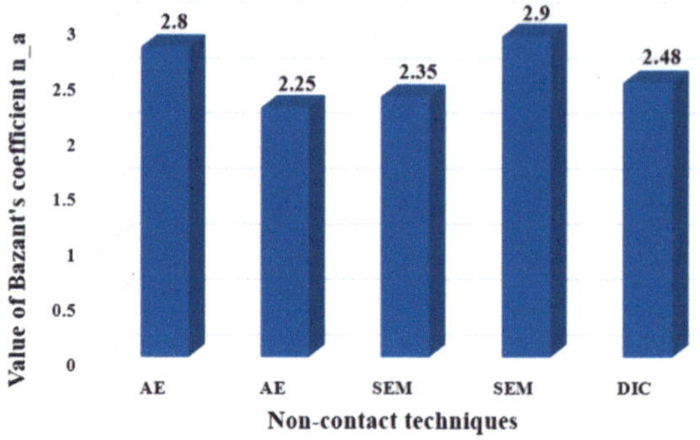

Fig. 10 Comparison of FPZ width values with existing experimental results

5 Conclusion

The study of fracture process zone (FPZ) is important to understand the fracture behavior of concrete. FPZ size has a significant role play in explaining the various toughening mechanisms. So far, various methods to estimate the length of fracture process zone have been developed. But for accurately predicting the fracture process zone size, width is also important. However, very few studies can be found to estimate the width of fracture process zone. In this regard, digital image correlation technique has been explored here to estimate the width of fracture process zone using vertical displacement data. A new methodology is presented in this study which can estimate the fracture process zone width. A threshold layer which separates the major and minor displacements is identified from the v plots as a yellow-colored contour, and its coordinates are estimated from the proposed methodology. The maximum fracture process zone width is reported. This methodology can also be used for estimating FPZ width for large beam samples. However, it was seen that fracture process zone is not fully developed for small beam samples, indicating evidence of size effect. Effect of size on width of fracture process zone can be further explored for the geometrically similar beams. The proposed methodology can be extended for cyclic fracture process zone as a future scope of this work.

References

1. Mihashi H (1987) States of the art and a view of the fracture mechanics of concrete. J JCI 25(2):14–25. Hillerborg A, Modéer M, Petersson P-E (1976) Analysis of crack formation and crack growth in concrete by means of fracture mechanics and finite elements. Cem Concr Res 6(6):773–781
2. Labuz JF, Shah SP, Dowding CH (1987) The fracture process zone in granite: evidence and effect. Int J Rock Mech Min Sci Geomech Abstr 24(4):235–246
3. Castro-Montero A, Shah SP, Miller RA (1990) Strain field measurement in fracture process zone. J Eng Mech 116(11):2463–2484
4. Zhang D, Wu K (1999) Fracture process zone of notched three-point-bending concrete beams. Cem Concr Res 29(12):1887–1892
5. Otsuka K, Date H (2000) Fracture process zone in concrete tension specimen. Eng Fract Mech 65(2–3):111–131
6. Bazant ZP, Kazemi MT (1990) Determination of fracture energy, process zone length and brittleness number from size effect, with application to rock and concrete. Int J Fract 44(2):111–131
7. Bažant ZP, Oh BH (1983) Crack band theory for fracture of concrete. Mat Constr 16:155–177
8. Saucedo L, Yu RC, Ruiz G (2012) Fully-developed FPZ length in quasi-brittle materials. Int J Fract 178(1):97–112
9. Grassl P, Grégoire D, Solano LR, Pijaudier-Cabot G (2012) Meso-scale modelling of the size effect on the fracture process zone of concrete. Int J Solids Struct 49(13):1818–1827
10. Fakhimi A, Tarokh A (2013) Process zone and size effect in fracture testing of rock. Int J Rock Mech Min Sci 60:95–102
11. Hu X, Li Q, Wu Z, Yang S (2022) Modelling fracture process zone width and length for quasi-brittle fracture of rock, concrete and ceramics. Eng Fract Mech 259:108158

12. Ohno K, Uji K, Ueno A, Ohtsu M (2014) Fracture process zone in notched concrete beam under three-point bending by acoustic emission. Constr Build Mater 67:139–145
13. Saliba J, Loukili A, Regoin JP, Grégoire D, Verdon L, Pijaudier-Cabot G (2015) Experimental analysis of crack evolution in concrete by the acoustic emission technique. Frat Integrità Strutt 9(34)
14. Sutton MA, Hild F (2015) Recent advances and perspectives in digital image correlation. Exp Mech 55(1):1–8
15. Shah S, Chandra Kishen JM (2011) Fracture properties of concrete–concrete interfaces using digital image correlation. Exp Mech 51(3):303–313
16. Wu Z, Rong H, Zheng J, Xu F, Dong W (2011) An experimental investigation on the FPZ properties in concrete using digital image correlation technique. Eng Fract Mech 78(17):2978–2990
17. Alam S, Loukili A, Grondin F (2012) Monitoring size effect on crack opening in concrete by digital image correlation. Eur J Environ Civ Eng 16(7):818–836
18. Alam SY, Saliba J, Loukili A (2014) Fracture examination in concrete through combined digital image correlation and acoustic emission techniques. Constr Build Mater 69:232–242
19. Skarżyński Ł, Tejchman J (2013) Experimental investigations of fracture process using DIC in plain and reinforced concrete beams under bending. Strain 49(6):521–543
20. Le JL, Manning J, Labuz JF (2014) Scaling of fatigue crack growth in rock. Int J Rock Mech Min Sci 72:71–79
21. Erarslan N (2016) Microstructural investigation of subcritical crack propagation and fracture process zone (FPZ) by the reduction of rock fracture toughness under cyclic loading. Eng Geol 208:181–190
22. Dong W, Wu Z, Zhou X, Wang N, Kastiukas G (2017) An experimental study on crack propagation at rock-concrete interface using digital image correlation technique. Eng Fract Mech 171:50–63
23. Bhowmik S, Ray S (2019) An experimental approach for characterization of fracture process zone in concrete. Eng Fract Mech 211:401–419
24. Li D, Huang P, Chen Z, Yao G, Guo X, Zheng X, Yang Y (2020) Experimental study on fracture and fatigue crack propagation processes in concrete based on DIC technology. Eng Fract Mech 235:107166
25. Bhowmik S, Ray S (2016) Effects of fracture process zone on fatigue crack growth in concrete. In: Proceedings of 9th international conference on fracture mechanics of concrete and concrete structures
26. Rajan V, Rossol M, Zok F (2012) Optimization of digital image correlation for high-resolution strain mapping of ceramic composites. Exp Mech 52(9):1407–1421
27. Schreier H, Orteu J-J, Sutton MA (2009) Image correlation for shape, motion and deformation measurements. Springer US, Boston
28. Sutton M, Yan J, Tiwari V, Schreier H, Orteu J-J (2008) The effect of out-of-plane motion on 2D and 3D digital image correlation measurements. Opt Lasers Eng 46(10):746–757
29. Blaber J, Adair B, Antoniou A (2015) Ncorr: open-source 2D digital image correlation MATLAB software. Exp Mech 55(6):1105–1122
30. Skarżyński Ł, Syroka E, Tejchman J (2011) Measurements and calculations of the width of the fracture process zones on the surface of notched concrete beams. Strain 47:e319–e332
31. Scholz CH, Dawers NH, Yu JZ, Anders MH, Cowie PA (1993) Fault growth and fault scaling laws: preliminary results. J Geophys Res Solid Earth 98(B12):21951–21961
32. Vermilye JM, Scholz CH (1998) The process zone: a microstructural view of fault growth. J Geophys Res Solid Earth 103(B6):12223–12237
33. Zhou R, Lu Y, Wang LG, Chen HM (2021) Mesoscale modelling of size effect on the evolution of fracture process zone in concrete. Eng Fract Mech 245:107559
34. Mihashi H, Nomura N (1996) Correlation between characteristics of fracture process zone and tension-softening properties of concrete. Nucl Eng Des 165(3):359–376
35. Hadjab H, Chabaat M, Thimus JF (2007) Use of scanning electron microscope and the non-local isotropic damage model to investigate fracture process zone in notched concrete beams. Exp Mech 47:473–484

36. Hadjab H (2009) Fracture process zone in concrete beams: experimental investigation and numerical modelling. In: Proceedings of the SEM annual conference, June 2009, pp 1–4

Investigation of the Effect of Stiffener Parameters on Post-buckling Strength of Stiffened Composite Panels Using FEA

Srilakshmi Rayasam, Siddarth Patil, and Achchhe Lal

Abstract Composite stiffened panels are widely used in numerous components of aircraft. Stiffeners can recover stability and stiffness efficiently with less gain in the structure's weight when compared to increasing plate thickness. In this paper, finite element analysis of stiffened panels made of composite is carried out using ABAQUS. Damage analysis within the plies is performed by implementing progressive damage modelling based on Hashin's criteria. The initiation and evolution of the adhesive layer damage are modelled using cohesive zone modelling. A convergence study is conducted to find the optimal mesh size. Next, a parametric study is carried out by varying stiffener sections such as *T*, *I*, and *C*. The failure load and the post-buckling stiffness of the structure are taken as performance examiners to evaluate the concert of each stiffener. In this work, optimization of the stiffened panel is carried out by using Rao-1 Algorithm. The optimal dimensions of the stiffener height, the distance between the stiffeners, the thickness of the stiffener, the height of the stiffener, and the thickness of the panel are estimated for the given specified panel size.

Keywords Composite stiffened panel · Finite element analysis · Post-buckling strength

1 Introduction

Stiffened panels made of composites have been extensively used in aerospace applications. These stiffened panels have many advantages such as high specific stiffness and strength. However, these stiffened panels undergo damages such as stiffener debonding, interlaminar damage, and intra-laminar damage during service and manufacturing. These damages affect the performance and reduce the structural integrity. There are studies that report on the failure performance of panels with stiffeners

Srilakshmi R. (✉) · S. Patil · Achchhe Lal
Department of Mechanical Engineering, Sardar Vallabhbhai National Institute of Technology, Surat, Gujarat 395007, India
e-mail: srilakshmi_r@med.svnit.ac.in

S. Sidhardh et al. (eds.), *Advances in Structural Integrity for Mechanical, Civil, and Aerospace Applications*, Lecture Notes in Mechanical Engineering,
https://doi.org/10.1007/978-981-97-6367-2_3

under buckling load [1–3]. One of the most significant challenges facing the aircraft industry is the design of composite panels with stiffeners, that can withstand the compression load and function effectively after buckling. Zhang et al. [4] have studied the damage behaviour of panels with stiffeners using a micromechanics-based material degradation model. The degradation model is validated by the FEM results with experimental data in terms of stiffness, strength, and failure pattern. They have implemented the progressive damage model defined by Hashin criteria accompanied by a material degradation model. Ye et al. [5] have conducted a study on the effect of different bonding methods of stiffener and the panel, on the strength of stiffened panels made of composites after buckling. They estimated the ratio of the interface strength to interlaminar shear strength and concluded that as this ratio increases, the panel fails without debonding. The debonding behaviour of stiffener and panel is an important aspect while designing the stiffened structures [6]. There are different parameters that affect the performance of stiffened panels such as stiffener shape, location, distance between stiffeners, and stiffener size. Out of these, the stiffener shape is the most important one on buckling strength. Naresh Reddy et al. [7] have performed an experimental and numerical analysis of glass fibre-reinforced polymer composite stiffened panels under compression load using T and Hat type of stiffeners. From their study, they found that the T-section-stiffened panel performs better compared to the Hat type of stiffener panel. Wu et al. [8] have investigated the buckling strength of stiffened panels with J type of stiffener. From the literature, it is found that there are few studies on the damage behaviour of different stiffener shapes. They have not considered I-section and C-section as stiffener shapes. Recently, Garcia and Ramos [9] have carried out the influence of fillet radius on buckling analysis of stiffened panels under compression load. They found that the fillet radius inclusion increases the local buckling load. All the above-mentioned literature explains the damage behaviour of stiffened panels made of composites under compression load. The optimisation of stiffened panels plays a major role in weight reduction and improves the strength of the panel. Wang et al. [10] have performed an optimisation analysis of stiffened panels using the ant colony algorithm. They have found the optimal lay-up configuration and number of stiffeners for the given panel size. They have chosen T-stiffeners for their optimisation study. Alhajahmad and Mittelstedt [11] have developed an optimization tool for the minimisation of weight of variably stiffened curvilinear grid-stiffened fuselage panels. They considered height and number of stiffeners as design variables and weight as the objective function. There are few more studies exist on the optimisation of stiffened shells [12–14]. Putra et al. [15] have performed optimisation of the stiffened panel using a hybrid Genetic algorithm (GA) approach. They have considered a number of stiffeners, stiffener shape, thickness of stiffener, and stiffener spacing as design variables and the mass of the stiffener as the objective function. The algorithms mentioned above such as the ant colony algorithm and GA approach are population-dependent algorithms and depend on control parameters. These algorithms follow the method of suitable adjustment of control parameters. Inadequate control parameter adjustment results in premature convergence. In general, the optimisation metaheuristics contain two types of control parameters. The first category is algorithm which depends on specific control parameters, which include mutation and crossover probabilities in genetic algorithms. Another category of values that control the optimisation algorithm has been identified as standard control parameters, and they include generation number and population size. However, choosing an acceptable size of population for different optimisation applications is a time-consuming task. Rao [16] proposed the Rao algorithms, which are algorithm-specific parameter-less algorithms. However, choosing an acceptable population size for various optimisation applications is a time-consuming task.

In this paper, finite element analysis (FEA) of stiffened panels using ABAQUS has been carried out. Both Hashin failure criteria and cohesive zone modelling are implemented numerically. There are different parameters that affect the performance of stiffened panels such as stiffener shape, stiffener thickness, location, height, and distance between the stiffeners. In this work, an effective stiffener shape is identified by carrying out an FEA of the stiffened panel with different stiffener shapes of equal volume. Further in this work, the optimal stiffener thickness, location, height, and distance between the stiffeners for the given panel dimensions are identified by implementing Rao's algorithm. For implementing Rao's algorithm, local buckling and global buckling functions are considered and the weight of the stiffener is considered as the objective function. The optimal dimensions of the stiffener for the given panel size are estimated using the Rao-1 algorithm.

2 Finite Element Analysis of Stiffened Panel

In this section, the damage behaviour of the stiffened panel is discussed. For validation of the FEA model with existing experimental results, the panel and stiffener dimensions are taken as per Ref. [6]. The panel dimensions are 728 × 538 mm with lay-up configuration $[90/0/90/0/\pm 45/90/02/90/\pm 45/\pm 45/90]_s$, and the thickness of each layer is 0.125 mm. In this work, a T-section stiffener is chosen with a height of 45 mm, a base of 55 mm, and a lay-up configuration of $[90/02/90/\pm 45/0/90/\pm 45/0]_s$; each layer thickness is 0.125 mm. The stiffened panel modelling is comprised plate modelling, stiffener modelling, and die modelling which compresses the stiffened panel assembly. Assembly modelling of the stiffened panel is shown in Fig. 1a. The plate and stiffener are modelled using four-noded shell elements, while the die is modelled as a discrete rigid body using four-noded quadrilateral elements. The meshed view of the stiffened panel is shown in Fig. 1b. The material properties of the panel and stiffener are given in Table 1.

2.1 *Damage Analysis of Stiffened Panel*

The interlaminar damage behaviour of the panel and stiffener is predicted using Hashin damage criteria as mentioned in Table 2.

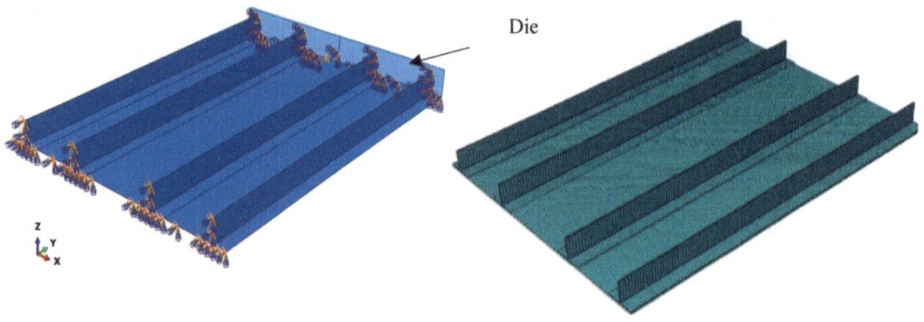

(a) (b)

Fig. 1 Stiffened panel **a** assembly view, **b** meshed view

Table 1 Material properties of T800 924C unidirectional composite [6]

Properties (MPa)		Strength parameters (MPa)		Fracture parameters	
E_1	145,000	X_T	2700	G_{IC}	276 J/m^2
E_2	9500	X_C	1650		
V_{12}	0.3	Y_T	55	$G_{IIC} = G_{IIIC}$	575 J/m^2
G_{12}	5000	Y_C	225		
G_{13}	5000	S_{12}	100		
G_{23}	3700				
ρ	145e−9 tonnes/mm^3				

Table 2 Hashin's failure criteria

Failure modes	Failure criteria	
Matrix tension failure	$\sigma_{22} > 0$	$\left(\frac{\sigma_{22}}{Y_T}\right)^2 + \left(\frac{\tau_{12}}{S_{12}}\right)^2 + \left(\frac{\tau_{23}}{S_{23}}\right)^2 \geq 1$
Matrix compressive cracking	$\sigma_{22} \leq 0$	$\left(\frac{\sigma_{22}}{Y_C}\right)^2 + \left(\frac{\tau_{12}}{S_{12}}\right)^2 + \left(\frac{\tau_{23}}{S_{23}}\right)^2 \geq 1$
Fibre tensile failure	$\sigma_{11} > 0$	$\left(\frac{\sigma_{11}}{X_T}\right)^2 + \left(\frac{\tau_{12}}{S_{12}}\right)^2 + \left(\frac{\tau_{13}}{S_{13}}\right)^2 \geq 1$
Fibre compressive failure	$\sigma_{11} \leq 0$	$\left(\frac{\sigma_{11}}{X_C}\right)^2 \geq 1$
Interlaminar tensile failure	$\sigma_{33} > 0$	$\left(\frac{\sigma_{33}}{Z_T}\right)^2 + \left(\frac{\tau_{23}}{S_{23}}\right)^2 + \left(\frac{\tau_{13}}{S_{13}}\right)^2 \geq 1$
Interlaminar compression failure	$\sigma_{33} \leq 0$	$\left(\frac{\sigma_{33}}{Z_C}\right)^2 + \left(\frac{\tau_{23}}{S_{23}}\right)^2 + \left(\frac{\tau_{13}}{S_{13}}\right)^2 \geq 1$

σ_{ij} $(i, j = 1, 2, 3)$ are stress values, and S_{ij} $(i, j = 1, 2, 3)$ are the strength value

Further, the damage behaviour of adhesive failure is predicted using a cohesive zone (CZ) contact approach. To estimate the debonding behaviour of stiffener from the panel, the CZ modelling with bilinear traction separation response law is chosen which is existing in ABAQUS. Quadratic stress–strain criteria are used to evaluate the damage initiation in cohesive elements as given in Eq. 1. Table 3 represents the material properties of cohesive zone elements.

$$\left(\frac{\sigma_{33}}{Z_T}\right)^2 + \left(\frac{\tau_{23}}{S_{23}}\right)^2 + \left(\frac{\tau_{13}}{S_{13}}\right)^2 \geq 1. \tag{1}$$

Plate, stiffener, and die shell geometries are discretized with four-noded quadrilateral elements. The element size is chosen to be 6 mm initially, which shall be optimized by performing convergence study to have a balance between solution accuracy and solution time. The convergence plot of displacement versus mesh size is shown in Fig. 2. The convergence study is carried out at load step which is mentioned in Table 4. It is observed that the convergence is achieved at a mesh size of 3 mm.

Table 3 Material properties of cohesive zone interface

Stiffness coefficients (kN/m)			Damage initiation stress (MPa)			Evolution (kJ/m^2)
K_{nn}	K_{ss}	K_{tt}	Normal	Shear-1	Shear-2	Fracture energy
10,000	3620	3620	50.5	50.5	50.5	3.47

Fig. 2 Displacement versus mesh size

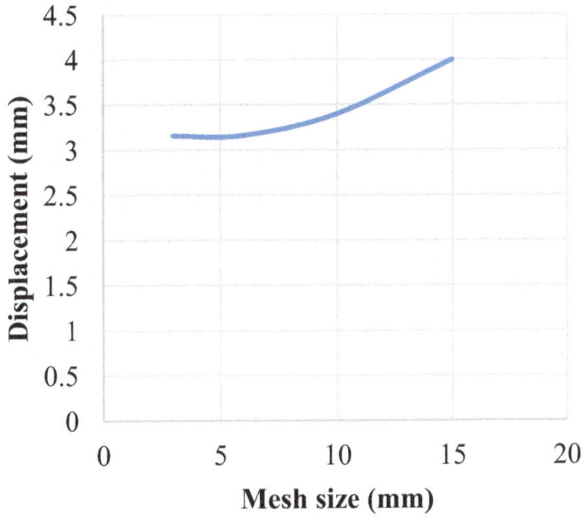

Table 4 Load step for convergence study

Sec	Bottom BC	Top BC	Die BC	Step time	Element size	Mode shape
T	UR1	U2	U2 = − 4 mm	1 s	3 mm	5–4

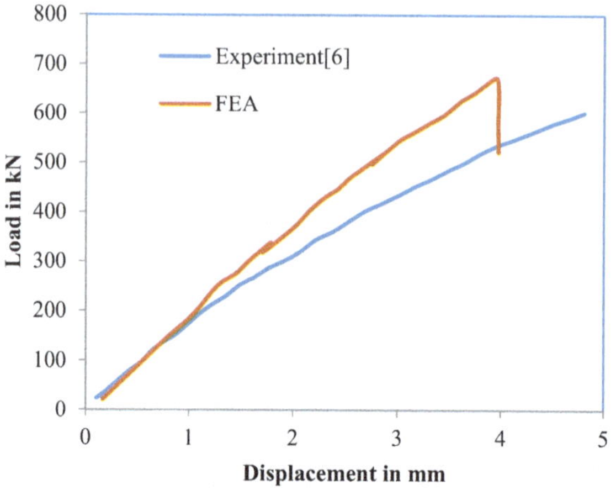

Fig. 3 Load versus displacement plot

2.2 Variation of Load with Respect to Shortening Length

Figure 3 shows the load versus displacement curve. It is observed from the curve that the failure load obtained from FEA is 665 kN. The failure load obtained by experimentally is around 601 kN as mentioned in Ref. [6]. This shows the accuracy of damage modelling with the tested results with 10% error. The failure of stiffened panel is happened by delamination of stiffener and then failure of panel.

Figure 4 shows the delamination behaviour of stiffener from the panel at 350 and 600 kN. Delamination between stiffener and skin panel is more dominant as compared to delamination in skin. Delamination is same for all the stiffener cases.

2.3 Effect of Stiffener Shape

In order to study the effect of stiffener shape, FEA of the stiffened panel is carried out with three different stiffener cross sections such as *T*, *I*, and *C*-section as shown in Fig. 5a. Here, other parameters such as stiffener mass, no. of stiffeners, stiffener location, ply stacking sequence in plate as well as stiffener are maintained constant.

From Fig. 5, it is evident that the stiffness of all three panels is almost the same irrespective of the stiffener shape. The failure load values for '*C* and *T*' stiffeners are 675,834 N and 663,266 N, respectively. Whereas the failure load for '*I*-stiffener'

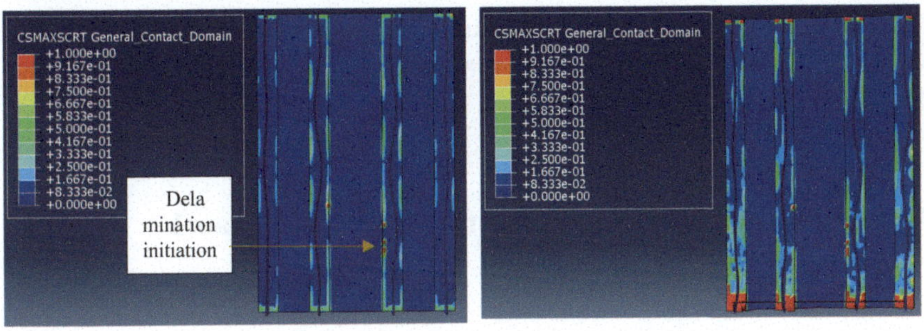

Fig. 4 Delamination between skin panel and stiffener at 350 and 600 kN

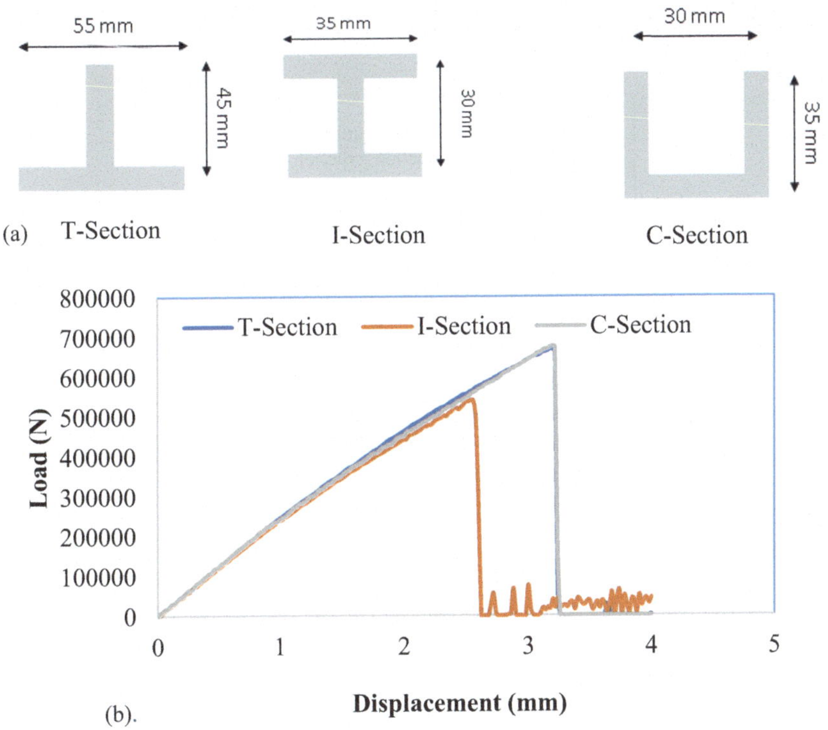

Fig. 5 **a** Different stiffener shapes, **b** load versus displacement plot for plates with different stiffener shapes

panel is 539,647 N which shows that *I* stiffened panel fails first as compared '*T*', '*C*' stiffened panel is due to larger contact area in case of '*T*' and '*C*' stiffeners; hence, more load transfer takes place. Failure load of I-shaped stiffener is less as compared to C-shaped stiffener because C-shaped stiffener has greater second moment of area due to farther than distribution of material from axis of cross section and results in higher radius of gyration. Hence, C-shaped stiffener carries more load than I-shaped stiffener.

3 Optimisation of Stiffener Dimensions Using Rao-1 Algorithm

In this work, the Rao algorithm is implemented in order to find the optimal values of panel thickness, the distance between two stiffeners, stiffener thickness, height, and width. Figure 6 shows the flowchart of Rao-1 algorithm. For this algorithm, no algorithmic-specific parameters are required. The important steps for the optimization of an objective function Z are as follows:

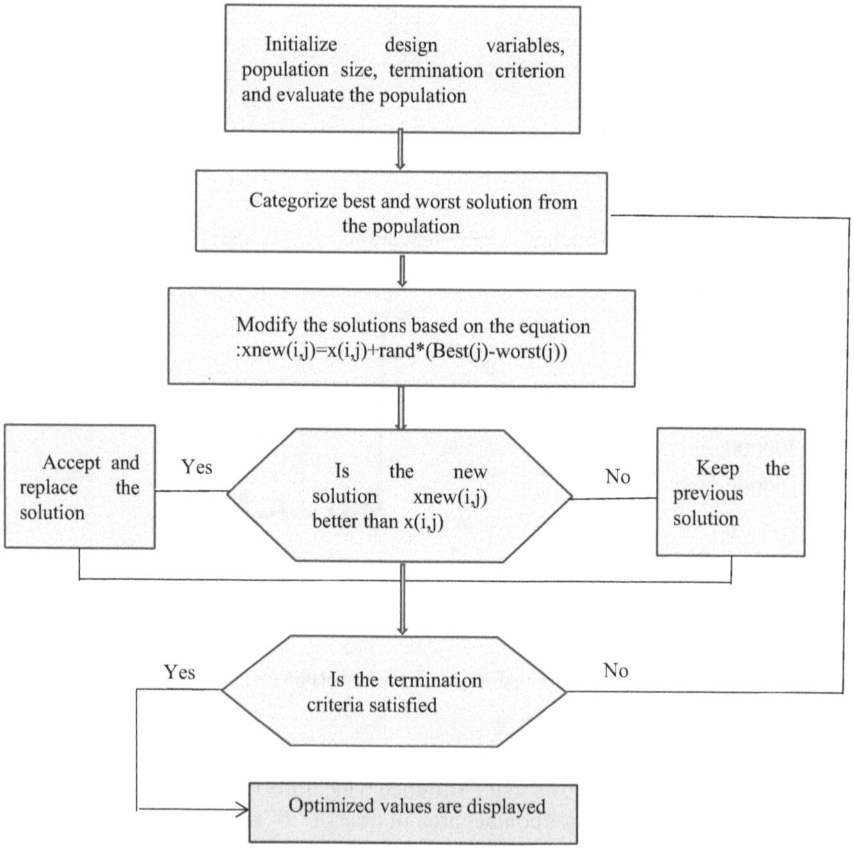

Fig. 6 Rao-1 algorithm [16]

Step-I: Define the number of design variables as six and the population size as 10.
Step-II: Define lower and upper bounds.

Panel thickness (t_{sx}) = 1–4 mm.
Stringer thickness (t_{st}) = 1–4 mm.
Height of stringer h = 30–100 mm.
Width of stringer $b1$ = 30–100 mm.
Number of stringers $n1$ = 4–7.
Distance between stiffeners b = 100–150 mm.

Step-III: Define objective function Z; in this study, weight of the entire panel including stiffeners is the objective function. Z is considered as minimisation function. Evaluation of Z is explained as below.

$$Z = Z_1 + Z_2 + Z_3, \tag{2}$$

$$Z_1 = \rho * A(B * t_{sx} + n * (h * t_{st} + b1 * t_{st})), \tag{3}$$

$$\sigma_{kr} = k_i \sigma_E, \tag{4}$$

$$\sigma_E = \frac{\pi^2 E}{12(1-\mu^2)} \left(\frac{t_{sx}}{B}\right)^2 (N+1)^2, \tag{5}$$

$$\sigma_{vkr} = k_v \sigma_{vE}, \tag{6}$$

$$\sigma_{vE} = \frac{\pi^2 E}{12(1-\mu^2)} \left(\frac{t_{st}}{h}\right)^2, \tag{7}$$

where k_i and k_v are stability coefficients of skin panel and stringer, μ is Poisson's ratio, and σ_{kr} and σ_{vkr} are the critical buckling stress in the skin panel and stiffener. σ_E and σ_{vE} are the Euler's critical stress in the panel and stiffener.

For optimization study, T-shaped stiffener is considered. In this work, the weight of the stiffened panel is minimized by carrying out optimization analysis using Rao-1 algorithm. In this study, single objective optimization analysis is performed. Only buckling constraints are considered in terms of penalty functions that are related to the critical buckling stress of skin panel and stringer panel as mentioned in Eqs. 4, 5, 6, 7,

Where

ρ density of panel.
A panel length = 610 mm.
B panel width = 610 mm.

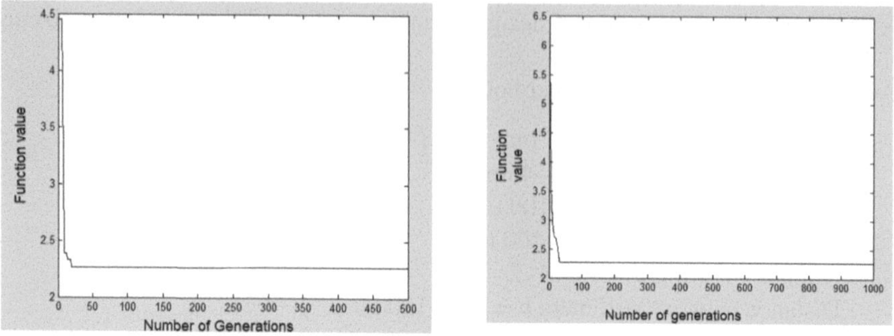

Fig. 7 Convergence plot with number of generations

Table 5 Optimal dimensions arrived from Rao-1 algorithm

	Panel thickness	Stringer thickness	Width of stiffener	Height of stiffener	Number of stiffener	Distance between stiffeners
Optimal value for the given panel size (mm)	1.2	1	33	35	4	110

Local skin buckling and global panel buckling penalty functions Z_2 and Z_3 are estimated from Eqs. 4, 5, 6, 7 as mentioned in Ref. [17].
Step-IV: If the new function value is less than the old one, then the new function value is considered for further steps. If the new function value is more than the old value then it the function value is assigned as the old value.
Step-V: This process will be repeated until the termination criteria reached. Here, it should be emphasized that the interactions between the present solution and the best, worst, and randomly chosen solutions cause the Rao algorithms to modify the population. In Rao-1 algorithm, the best and worst solutions will interact with the current solution. The function versus number of generations are shown in Fig. 7a, b. From the figure, it is observed that the convergence is achieved at around 20 generations for the two cases such as 500 generations and 1000 generations. Table 5 shows the optimal values of design variables.

4 Conclusions

In this work, finite element analysis and optimisation study of stiffened panel has been carried out. From the analysis, the following conclusions are drawn:

- From the progressive damage analysis, it is observed that the stiffened panel fails by debonding of stiffener from the panel, followed by base panel failure. From the load vs displacement plots, it is observed that the failure load obtained by FEA is in good consistency with experimental existing data [6].
- The failure load with different stiffener shapes such as '*T*', '*I*', and '*C*'-section is estimated by finite element simulations, and it is found that '*C*'-section stiffener can withstand highest load as compared to other sections.
- The optimal stiffener dimensions are estimated using Rao-1 algorithm which does not require algorithmic-specific parameters.

Acknowledgements This work is supported by the SVNIT, under the Seed grant project scheme (Project no: SG/DoME/2021/23).

References

1. Zimmermann R, Klein H, Kling A (2006) Buckling and postbuckling of stringer stiffened fibre composite curved panels—tests and computations. Compos Struct 73:150–161
2. Masood N, Vishakh S, Viswamurthy R, Kotresh MG, Sridhar I (2018) Influence of stiffener configuration on post-buckled response of composite panels with impact damages. Compos Struct 194:433–444
3. Ji R, Zhao L, Wang K, Liu F, Gong Y, Zhang J (2021) Effects of debonding defects on the post-buckling and failure behaviours of composite stiffened panel under uniaxial compression. Compos Struct 256:113121
4. Zhang J, Zhou L, Chen Y, Zhao L, Fei B (2015) A micromechanics-based degradation model for composite progressive damage analysis. J Compos Mater 50(16):2271–2287
5. Ye Y, Zhu W, Jiang J, Xu Q, Ke Y (2019) Computational modelling of postbuckling behaviour of composite T-stiffened panels with different bonding methods. Compos B Eng 166:247–256
6. Zhao L, Wang K, Ding F, Qin T, Xu J, Liu F, Zhang J (2019) A post-buckling compressive failure analysis framework for composite stiffened panels considering intra-, inter-laminar damage and stiffener debonding. Results Phys 13:102205
7. Naresh Reddy K, Gangadharan R, Ramji M (2018) Experimental and numerical studies on the buckling and post-buckling behaviour of single blade-stiffened CFRP panels. Compos Struct 196:135–154
8. Wu Q, Hu S, Tang X, Liu X, Chen Z, Xiong J (2023) Compressive buckling and post-buckling behaviours of J-type composite stiffened panel before and after impact load. Compos Struct 304:116339
9. Garcia FG, Ramos R Jr (2019) A methodology to investigate the influence of the fillet radius on the buckling of integrally stiffened panels. Lat Am J Solids Struct 16(6):1679
10. Wang W, Guo S, Chang N, Yang W (2010) Optimum buckling design of composite stiffened panels using ant colony algorithm. Compos Struct 92(3):712–719
11. Alhajahmad A, Mittelstedt C (2021) Buckling performance of curvilinear grid-stiffened tow-placed composite panels considering manufacturing constraints. Compos Struct 260:113271
12. Bisagni C, Lanzi L (2002) Post-buckling optimisation of composite stiffened panels using neural networks. Compos Struct 58(2):237–247
13. Giavotto V, Lanzi L (2006) Post-buckling optimization of composite stiffened panels: computations and experiments. Compos Struct 73(2):208–220

14. Montemurro M, Pagani A, Alberto Fiordilino G, Pailhes J, Carrera E (2018) A general multi-scale two-level optimisation strategy for designing composite stiffened panels. Compos Struct 201:968–979
15. Putra GL, Kitamura M, Takezawa A (2019) Structural optimization of stiffener layout for stiffened plate using hybrid GA. Int J Naval Archit Ocean Eng 11(2):809–818
16. Rao RV, Pawar RB (2020) Self-adaptive multi-population Rao algorithms for engineering design optimization. Appl Artif Intell 34(3):187–250
17. Pistek A, Miroslay I (2008) Optimisation of stiffened panel with the help of mathematical programming experimental verification. In: 26th international congress of the aeronautical sciences

Identifying the Cracks in Beam Structures Using a Simplified Substructure Technique

Eshwar Kuncham, Md Armanul Hoda, and Subhamoy Sen

Abstract Beam definitions are used to simplify complex structures like bridges or skyscrapers. This study highlights the importance of monitoring beams as they are often the critical components and damage in them can result in changes in their dynamic characteristics. This study proposes a substructure-based approach for damage estimation in structural beams, that can later be extended for monitoring high-dimensional structures. The traditional approach of full structure monitoring mandates heavy instrumentation and computational costs. To avoid that, the existing subdomain estimation approaches need to deal with the quasi-static displacement at subdomain boundaries making the estimation a coupled problem. Instead, the proposed approach focuses on only monitoring a subdomain of interest, independently, making it computationally cheaper. The study employs an interacting filtering algorithm with Particle and Ensemble Kalman filters, circumventing the complexities of quasi-static displacement. The proposed approach has been tested numerically and the results are promising for further investigation and improvement.

Keywords Subdomain · Damage estimation · Ensemble Kalman filters · Interacting filtering · Structural responses

1 Introduction

Monitoring a structure and detecting damage at the earliest possible stage is a primary concern in civil, mechanical, and aerospace engineering. The location of the damage must be known beforehand for most of the non-destructive testing (NDT)

E. Kuncham · M. A. Hoda · S. Sen (✉)
i4S Laboratory, Indian Institute of Technology Mandi, Mandi, HP, India
e-mail: subhamoy@iitmandi.ac.in
URL: https://www.i4siitmandi.com/subhamoy-sen

E. Kuncham
e-mail: eshwar.eshwar.research@gmail.com

M. A. Hoda
e-mail: armanhoda22@gmail.com

S. Sidhardh et al. (eds.), *Advances in Structural Integrity for Mechanical, Civil, and Aerospace Applications*, Lecture Notes in Mechanical Engineering,
https://doi.org/10.1007/978-981-97-6367-2_4

approaches such as acoustic or ultrasonic methods, and components of the structure being inspected must be easily accessible. Owing to these constraints, NDT methods can only detect damage at the structure's surface. The need for additional global damage detection methods that can be applied to complex structures has led to the development of a vibration-based method that can be explicitly applied to examine changes in the vibration characteristic of the structure. The change in the structural properties such as stiffness and damping are detected by analyzing changes in the structure's dynamics. Most of the researchers have defined the damage as a localized reduction in structural stiffness or damping [1]. Structural damage is often tragic and may lead to the failure of the structure with huge financial loss. A structural health monitoring (SHM) system can be used in place of expensive, time-consuming, and ineffective structural inspections to avoid such scenarios in engineering structures. In the SHM system, structural responses and mathematical models are used to determine the location, severity, and estimate of the structural health. To minimize discrepancies between predictions and measurement data, Bayesian filters are used.

In the real world, evaluating civil structures like buildings, and bridges often requires mathematical models consisting of many degrees of freedom and unknown parameters. Practically, measuring and identifying the entire structure at once is a challenging process. Thus, a substructuring technique is used in this paper to decompose the large structural system into substructures (with fewer degrees of freedom and unknown parameters) in order to maximize the effectiveness of the identification process. However, substructures interact with the rest of the structure and should therefore be considered in terms of the interaction forces between the substructure of interest and the global structure [2].

The concept of substructural system identification method was presented by Koh et al. [3] by using an Extended Kalman filter to identify unknown structural parameters. As a result of this substructure formulation of system identification, computation time is reduced significantly as well as the convergence of the structural parameters is improved. To use this method, all degrees of freedom at the interface must be measured (i.e., acceleration, velocity, and displacement). According to Koh and Shankar [4], there is no need to measure the interfaces with a substructural identification method. Damage detection for complex structures is made possible with the use of substructural damage identification techniques. However, multiple tests are required to narrow down the damage region or identify the approximate range using other non-destructive tests. Trinh and Koh [5] proposed an alternative method in which only the accelerations on the interfaces were required, relying on the integration of the acceleration time history to compute velocity and displacement. Here, the proposed method uses a similar approach to identify substructural damage using Bayesian filters.

In a recursive Bayesian framework, the identification of damage is primarily viewed as a state-parameter estimation problem, which is naturally a nonlinear one, even for systems whose dynamics fall within the linear range. As a result, it is impossible to evaluate filter distributions analytically, which means methods must be developed that can deal with nonlinearities. This can be accomplished through successive linearizations of the motion equations leading to extended Kalman-type filters. Particle Filters (PF), a sequential Monte Carlo sampling method that operates directly on the nonlinear system equations, provide a numerical evaluation of the filter distributions while also dealing with non-Gaussian distributions. The drawback of PF is that it has a high computational cost when the problem is highly nonlinear. Considering this drawback, IPKF is introduced in which the KF is used for linear state estimation while the PF is used for parameter estimation. Ensemble Kalman Filter (EnKF) has been proven to be efficient in propagating nonlinear system states while simultaneously parallelizing the entire SHM approach [6]. To estimate state parameters, an Interacting Particle Ensemble Kalman Filter (IP-EnKF) has been used. This paper presents an innovative health estimation approach that employs the interacting filtering algorithm IP-EnKF based on the measurement. The component-wise estimation has been made standalone so that the estimation of the complete structural health can be run in parallel.

2 State Space Formulation for the Substructure System

The substructure approach for this study is discussed here. The Governing Differential Equation (GDE) of the linear time-varying beam system can be defined with time-invariant mass $\mathbf{M}$, time-varying stiffness $\mathbf{K}(t)$, damping $\mathbf{C}(t)$ matrices and external force $\mathbf{f}(t)$ as,

$$\mathbf{M}\ddot{\mathbf{q}}(t) + \mathbf{C}(t)\dot{\mathbf{q}}(t) + \mathbf{K}(t)\mathbf{q}(t) = \mathbf{f}(t) \tag{1}$$

where $q\,(t)$, $\dot{q}\,(t)$, and $\ddot{q}(t)$ are the displacement, velocity, and acceleration response at the nodes. The simple substructure technique is demonstrated by considering the beam structure as shown in Fig. 1a and further, it is considered as lumped mass system as represented in Fig. 1b.

Based on the substructure interface degree of freedom (DoF), the dynamic equation of the complete beam structure is expressed as

$$\begin{bmatrix} \mathbf{M}_{rr} & \mathbf{M}_{rb} & 0 \\ \mathbf{M}_{br} & \mathbf{M}_{bb} & \mathbf{M}_{bi} \\ 0 & \mathbf{M}_{ib} & \mathbf{M}_{ii} \end{bmatrix} \begin{Bmatrix} \ddot{\mathbf{q}}_r \\ \ddot{\mathbf{q}}_b \\ \ddot{\mathbf{q}}_i \end{Bmatrix} + \begin{bmatrix} \mathbf{C}_{rr} & \mathbf{C}_{rb} & 0 \\ \mathbf{C}_{br} & \mathbf{C}_{bb} & \mathbf{C}_{bi} \\ 0 & \mathbf{C}_{ib} & \mathbf{C}_{ii} \end{bmatrix} \begin{Bmatrix} \dot{\mathbf{q}}_r \\ \dot{\mathbf{q}}_b \\ \dot{\mathbf{q}}_i \end{Bmatrix} + \begin{bmatrix} \mathbf{K}_{rr} & \mathbf{K}_{rb} & 0 \\ \mathbf{K}_{br} & \mathbf{K}_{bb} & \mathbf{K}_{bi} \\ 0 & \mathbf{K}_{ib} & \mathbf{K}_{ii} \end{bmatrix} \begin{Bmatrix} \mathbf{q}_r \\ \mathbf{q}_b \\ \mathbf{q}_i \end{Bmatrix} = \begin{Bmatrix} \mathbf{f}_r \\ \mathbf{f}_b \\ \mathbf{f}_i \end{Bmatrix} \tag{2}$$

with subscript, r, b, and i signifies remaining, boundary/interface, and internal DoFs respectively. From Eq. (2), the full dynamics system can further be decoupled as:

$$\mathbf{M}_{ii}\ddot{\mathbf{q}}_i + \mathbf{C}_{ii}\dot{q}_i + \mathbf{K}_{ii}\mathbf{q}_i = \mathbf{f}_i - \mathbf{M}_{ib}\ddot{\mathbf{q}}_b - \mathbf{C}_{ib}\dot{\mathbf{q}}_b - \mathbf{K}_{ib}\mathbf{q}_b \tag{3}$$

with the right side of Eq. (3) inclusively considered as an external force acting on the substructure and representing in state space form,

$$\dot{\mathbf{x}}_i\,(t) = \mathbf{F}\,(t)\,\mathbf{x}_i\,(t) + \mathbf{B}\,(t)\,\mathbf{u}(t) + \mathbf{E}\,(t)\,\ddot{\mathbf{q}}_b(t) + \mathbf{G}\,(t)\,\dot{\mathbf{x}}_b\,(t) + \mathbf{v}\,(t) \tag{4}$$

where, $\dot{\mathbf{x}}_i(t) = \begin{Bmatrix} \mathbf{q}_i \\ \dot{\mathbf{q}}_i \end{Bmatrix}$, $\mathbf{F}(t) = \begin{bmatrix} \mathbf{0}_{n_i} & \mathbf{I}_{n_i} \\ -\mathbf{M}_{ii}{}^{-1}\mathbf{K}_{ii} & -\mathbf{M}_{ii}{}^{-1}\mathbf{C}_{ii} \end{bmatrix}$, $\mathbf{B}(t) = \begin{bmatrix} \mathbf{0}_{n_i} \\ \mathbf{M}_{ii}{}^{-1} \end{bmatrix}$, $\mathbf{E}(t) = \begin{bmatrix} \mathbf{0}_{n_i} \\ -(\mathbf{M}_{ii}{}^{-1}\mathbf{M}_{ib}) \end{bmatrix}$, $\mathbf{G}(t) = \begin{bmatrix} \mathbf{0}_{n_b} & \mathbf{0}_{n_b} \\ -\mathbf{M}_{ii}{}^{-1}\mathbf{K}_{ib} & -\mathbf{M}_{ii}{}^{-1}\mathbf{C}_{ib} \end{bmatrix}$, $\mathbf{u}(t) = \mathbf{f}_i$, and $\dot{\mathbf{x}}_b(t) = \begin{Bmatrix} \mathbf{q}_b \\ \dot{\mathbf{q}}_b \end{Bmatrix}$.

The additional term $\mathbf{v}(t)$ represents process uncertainty originating from the model inaccuracies and unmodelled inputs. Similarly, measurement equations as

$$\mathbf{y}_i(t) = \mathbf{H}(t)\mathbf{x}_i(t) + \mathbf{D}(t)\mathbf{u}(t) + \mathbf{L}(t)\ddot{\mathbf{q}}_b(t) + \mathbf{S}(t)\dot{\mathbf{x}_b}(t) + \mathbf{w}(t) \tag{5}$$

here, $\mathbf{H}(t) = \left[-\mathbf{M}_{ii}{}^{-1}\mathbf{K}_{ii} \; -\mathbf{M}_{ii}{}^{-1}\mathbf{C}_{ii}\right]$, $\mathbf{L}(t) = -\mathbf{M}_{ii}{}^{-1}\mathbf{M}_{ib}$, $\mathbf{D}(t) = \mathbf{M}_{ii}{}^{-1}$, $\mathbf{S}(t) = \left[-\mathbf{M}_{ii}{}^{-1}\mathbf{K}_{ib} \; -\mathbf{M}_{ii}{}^{-1}\mathbf{C}_{ib}\right]$, and $\mathbf{w}(t)$ denoting measurement noise. Equations (4) and (5) can also be represented in the discrete time since responses are typically discretely sampled in reality.

$$\begin{aligned} \mathbf{x}_{i,k} &= \mathbf{F}_k\mathbf{x}_{i,k-1} + \mathbf{B}_k\mathbf{u}_k + \mathbf{E}_k\ddot{\mathbf{q}}_{b,k} + \mathbf{G}_k\mathbf{x}_{b,k} + \mathbf{v}_k = \mathcal{F}_k + \mathbf{v}_k \\ \mathbf{y}_{i,k} &= \mathbf{H}_k\mathbf{x}_{i,k} + \mathbf{D}_k\mathbf{u}_k + \mathbf{L}_k\ddot{\mathbf{q}}_{b,k} + \mathbf{S}_k\mathbf{x}_{b,k} + \mathbf{w}_k = \mathcal{Z}_k + \mathbf{w}_k \end{aligned} \tag{6}$$

In order to compute the external force, all accelerations, velocities, and displacements at the interface DoFs must be measured [3]. It is usually preferred to measure accelerations (by accelerometers) instead of velocities and displacements. Hence, Koh and Shankar [4] proposed the concept of quasi-static displacement so as to eliminate the need for velocities and displacements when computing interface forces. Nevertheless, inaccuracy could result from the inversion of the internal stiffness matrix and neglecting the velocity-dependent component. According to Trinh and Koh [5], acceleration measurement can be directly used to compute the interface force vector

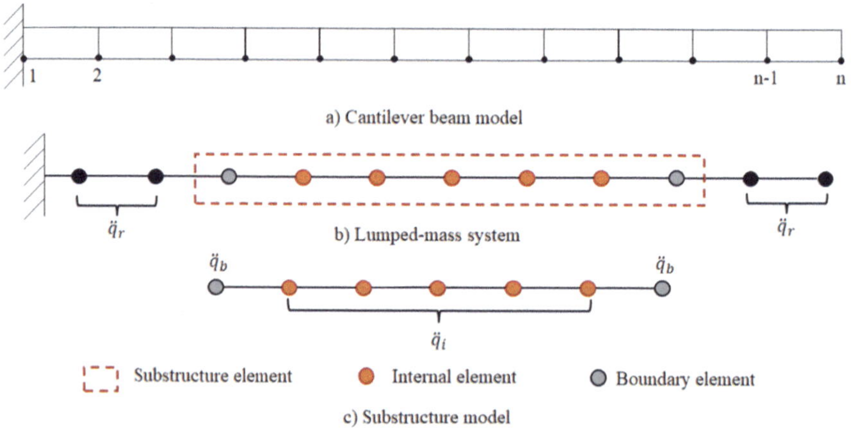

Fig. 1 Cantilever beam model with substructure

directly in order to improve the accuracy of damage identification results. Based on measured interface acceleration, the following equations are used to embed a simple numerical integration scheme to determine interface velocity and displacement:

$$\begin{aligned}\dot{\mathbf{q}}_{b,k+1} &= \dot{\mathbf{q}}_{b,k} + \frac{\Delta t}{2}\left(\ddot{\mathbf{q}}_{b,k} + \ddot{\mathbf{q}}_{b,k+1}\right)\\ \mathbf{q}_{b,k+1} &= \mathbf{q}_{b,k} + \frac{\Delta t}{2}\left(\dot{\mathbf{q}}_{b,k} + \dot{\mathbf{q}}_{b,k+1}\right)\end{aligned} \tag{7}$$

3 State-Parameter Estimation with Interacting Particle Ensemble Kalman Filter

The substructure system and simulation approach to dynamic structures have been demonstrated in Sect. 2. In the IP-EnKF approach, parameter θ is estimated with the PF and the system state $\dot{\mathbf{x}}_k$ is estimated with EnKF. PF propagates at any arbitrary time step k as realizations of a random variable θ k in time as a set of N_p parameter particles $\Xi = [\xi_k^1, \xi_k^1, \ldots, \xi_k^{N_p}]_{m_s \times N_p}$. Every time step k, an arbitrary particle evolves according to ξ_{k-1}^j basically random perturbations around the current position.

$$\xi_k^j = \alpha \xi_{k-1}^j + \mathcal{N}\left(\delta\xi_k, \sigma_k^\xi\right) \tag{8}$$

A Gaussian blurring is applied on ξ_{k-1}^j using a shift $\delta\xi_k = (1-\alpha)\bar{\xi}_{k-1}$ and a spread σ_k^ξ. During estimation, the α parameter controls the turbulence. In order to find the posterior, the evolved particles are weighted based on their likelihood in comparison with the current measurement. The evolution of the parameter particles is eventually automated and independent of the initial distribution for the space of particles; instead, it is based on their likelihood against measurement. Embedded EnKF for state estimation is then applied to the propagated particles in order to estimate the likelihood. EnKF propagates N_e state ensembles through the system (cf. Eq. (10)) with following, prior state ensembles $\mathbf{x}_{k-1|k-1}^{i,j}$ where current parameter ξ_k^j connected with ith ensemble and jth particle. Following are steps involved in EnKF algorithm to calculate mean and covariance

Mean Calculation
Propagated state: $\mathbf{x}_{i,k|k-1}^{i,j} = \mathcal{F}_k + \mathbf{v}_k^{i,j}$ where $\mathbf{v}_k^{i,j} \approx \mathcal{N}(0, \mathbf{Q})$
Estimated/Internal measurement: $\mathbf{y}_{i,k|k-1}^{i,j} = \mathcal{Z}_k + \mathbf{w}_k^{i,j}$ where $\mathbf{w}_k^{i,j} \approx \mathcal{N}(0, \mathbf{R})$
Innovation: $\epsilon_k^j = \frac{1}{N_e}\sum_{i=1}^{N_e} \epsilon_k^{i,j}$ where $\epsilon_k^{i,j} = \mathbf{y}_k - \mathbf{y}_{k|k-1}^{i,j}$
Propagated state estimates: $\mathbf{x}_{k|k-1}^j = \frac{1}{N_e}\sum_{i=1}^{N_e} \mathbf{x}_{k|k-1}^{i,j}$
Estimated internal measurements: $\mathbf{y}_{i,k|k-1}^j = \frac{1}{N_e}\sum_{i=1}^{N_e} \mathbf{y}_{i,k|k-1}^{i,j}$.

Covariance Calculation

State and measurement: $C_k^{j,xy} = \frac{1}{N_e-1}\sum_{i=1}^{N_e}\left(\mathbf{x}_{k|k-1}^{j} - \mathbf{x}_{k|k-1}^{i,j}\right)\left(\mathbf{y}_{k|k-1}^{j} - \mathbf{y}_{k|k-1}^{i,j}\right)^{\mathrm{T}}$

Measurement estimation: $C_k^{j,yy} = \frac{1}{N_e-1}\sum_{i=1}^{N_e}\left(\mathbf{y}_{k|k-1}^{j} - \mathbf{y}_{k|k-1}^{i,j}\right)\left(\mathbf{y}_{k|k-1}^{j} - \mathbf{y}_{k|k-1}^{i,j}\right)^{\mathrm{T}}$

Innovation error: $\mathbf{S}_k^j = C_k^{j,yy} + \mathbf{R}$ and EnKF gain: $\mathbf{G}_k^j = C_k^{j,xy}(\mathbf{S}_k^j)^{-1}$.

Initially, based on the innovation mean, $\in_k^{i,j}$, and covariance, $\mathbf{S}_k^j$ the state ensembles can be updated as follows:

$$\mathbf{x}_{k|k}^{i,j} = \mathbf{x}_{k|k-1}^{i,j} + \mathbf{G}_k^j \in_k^{i,j} \tag{9}$$

Furthermore, again based on the $\in_k^{i,j}$, and $\mathbf{S}_k^j$ the likelihood, $\mathcal{L}(\xi_k^j)$, of each particle is also calculated. For each jth particle, the normalized weight is updated as follows:

$$w(\xi_k^j) = \frac{w(\xi_{k-1}^j)\mathcal{L}\xi_k^j}{\sum_{j=1}^{N_p} w(\xi_{k-1}^j)\mathcal{L}(\xi_k^j)} \tag{10}$$

where $\mathcal{L}(\xi_k^j) = \left((2\pi)^n\sqrt{|\mathbf{S}_k|}\right)^{-1} \mathrm{e}^{-0.5\in_k^{j\,\mathrm{T}}\mathbf{S}_k^{-1}\in_k^j}$. Finally, particle approximations for states and parameters are estimated as follows:

$$\mathbf{x}_{k|k} = \sum_{j=1}^{N_p} w\left(\xi_k^j\right)\mathbf{x}_{k|k}^j \quad \text{and} \quad \theta_{k|k} = \sum_{j=1}^{N_p} w\left(\xi_k^j\right)\xi_k^j \tag{11}$$

4 Numerical Experiment

The numerical analysis is carried out using a finite element model of a prismatic cantilever beam with geometric properties: span = 0.75 m, width (w) = 0.025 m, depth (d) = 0.006 m, and material properties: mass density = 7850 kg/m^3 and elastic modulus (E) = 208 GPa. A beam is divided into 11 equal parts, i.e., elements. The beam is divided into 11 equal parts (cf. Fig. 2), and the length of each element is (l_e) = 0.0682 m. In this study, a lumped mass assumption is employed to replicate the continuous system for the sake of simplicity. The mass and stiffness of each element are calculated as $33\rho w d l_e/140$, and $3EI_e/l_e^3$ following the reference of [7]. Here, $I_e = wd^3/12$ represents the moment of inertia of the beam section.

The proposed algorithm is validated using a damaged beam that contains a crack. In order to simulate the effects of a crack present in the substructure element, an exponential function for stiffness reduction, denoted as $EI(x)$, is employed. Here, x represents the position along the rectangular beam element, and this function accounts for the variations in stiffness as employed in this chapter [8].

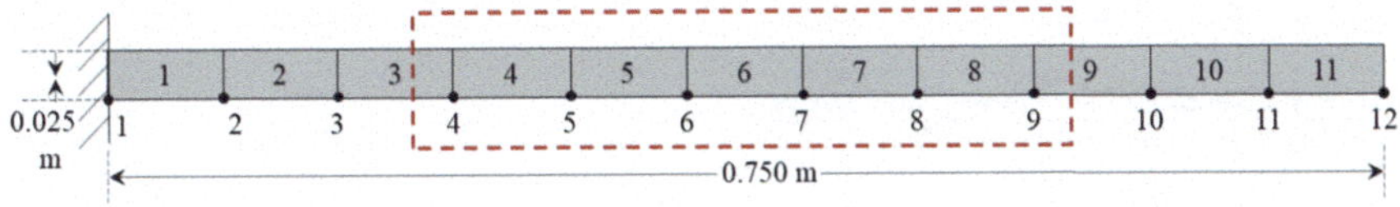

Fig. 2 Schematic diagram of a cantilever beam model with its substructure

$$EI(x) = \frac{EI}{1 + C \exp^{((-2\beta|x - l_c|)/d_c)}} \tag{12}$$

where $C = (I_e - I_c)/I_c$ for $I_c = w(d - d_c)^3/12$. The constant β, estimated from experiments as in [8], is found to be 0.6667. To simulate the crack in the third element of the substructure, specific crack attributes are considered. The crack depth (d_c) is set to 0.0045 m, and its position (l_c) is located at a distance of 0.375 m from the fixed end [9].

The beam is subjected to an SWGN $\mathcal{N}(0, 1e^{-5}\ \text{N})$ force applied to each node of the structure for a time window of 1 s, with a sampling frequency of 1000 Hz. Acceleration responses are recorded at each node of the substructure element. To simulate real-life scenarios, the dynamic response measurements collected at the nodes are contaminated with SWGN of *snr* of 1%. The initial distribution of the parameters followed a normal distribution $\mathcal{N}(1, 0.02)$, with the parameter α set to 0.98 (refer to Eq. (8)). In the numerical experiment, 2000 particles are used for the PF, and 500 ensembles are utilized for EnKF.

In Fig. 3, the convergence of parameters to their true values is shown for both undamaged and damaged elements. Additionally, Fig. 4 compares the estimated and measured acceleration of an internal *dof*. The results indicate that the simple substructure algorithm, assuming the availability of acceleration response, accurately and precisely estimates the measurement response and parameters for damaged cases, even under a noise severity of 1% *snr*. This algorithm offers computational efficiency by monitoring only a specific subdomain of interest while providing prompt estimation. The performance evaluation on a numerical cantilever beam demonstrates that the simple substructure algorithm is accurate, precise, and efficient.

5 Conclusions

The standard process for estimating the health of a structure and detecting damage uses sensors scattered throughout it to collect responses. However, a model-based approach may be difficult to implement because of the associated computational workload with a high number of degrees of freedom. Monitoring only a subdomain of interest would be computationally cheaper while providing prompt estimation.

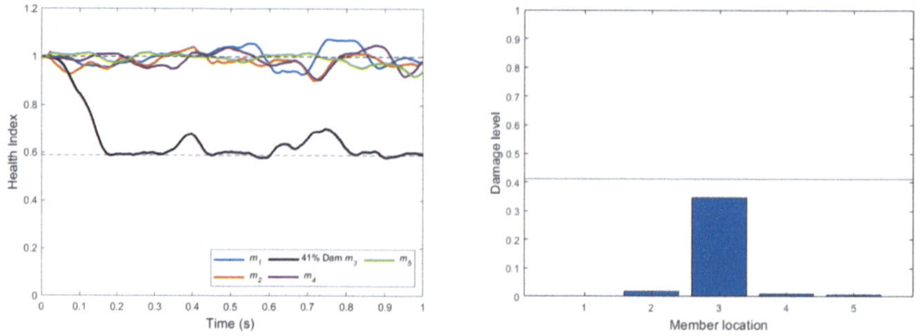

(a) Identifying damage by time domain (b) Bar graph with mean calculation

Fig. 3 Estimated parameter of a substructure beam element for damage case (*dashed lines represent respective actual values*)

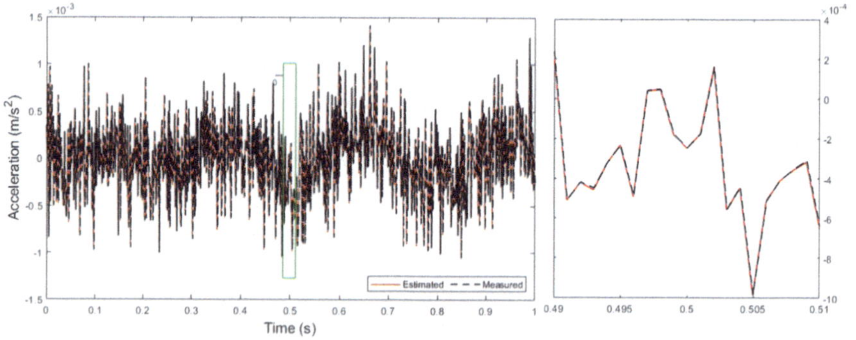

Fig. 4 Recreating the internal response from the estimated states

A novel approach to estimating health is presented that uses interacting filtering algorithms along with particle and ensemble Kalman filters (IP-EnKF). Based on its performance on a numerical cantilever beam, the proposed method is found to be accurate, precise, and prompt.

References

1. Fan W, Qiao P (2011) Vibration-based damage identification methods: a review and comparative study. Struct Health Monit 10(1):83–111
2. Tee KF, Koh CG, Quek ST (2003) System identification and damage estimation via substructural approach. Comput Struct Eng Int J 3(1):1–7
3. Koh CG, See LM, Balendra T (1991) Estimation of structural parameters in time domain: a substructure approach. Earthq Eng Struct Dyn 20(8):787–801
4. Koh CG, Shankar K (2003) Substructural identification method without interface measurement. J Eng Mech 129(7):769–776

5. Trinh TN, Koh CG (2012) An improved substructural identification strategy for large structural systems. Struct Control Health Monit 19(8):686–700
6. Hommels A, Murakami A, Nishimura SI (2009) A comparison of the ensemble Kalman filter with the unscented Kalman filter: application to the construction of a road embankment. Geotechniek 13(1):52
7. Reddy J (1999) On the dynamic behaviour of the Timoshenko beam finite elements. Sadhana 24:175–198
8. Christides S, Barr ADS (1984) One-dimensional theory of cracked Bernoulli–Euler beams. Int J Mech Sci 26(11–12):639–648
9. Khatir S, Dekemele K, Loccufier M, Khatir T, Wahab MA (2018) Crack identification method in beam-like structures using changes in experimentally measured frequencies and particle swarm optimization. C R Mécanique 346(2):110–120

Analysis of the Non-circular Suture Designs on Bio-inspired Materials

C. Sachin Chandran, Pratiksha Rodewad, and S. Anup

Abstract Biological composite materials found in nature like nacre, shells of cephalopods, and woodpecker beaks possess superior mechanical properties compared to their simple constituents having weak mechanical properties. The weak suture interfaces present in these biological composite materials play a significant role in achieving excellent mechanical properties since they govern the deformation and fracture mechanisms. These non-adhesive suture interfaces possess non-linear behaviour depending on friction and the geometrical interlocking. Bio-inspired sutured materials are developed by mimicking the excellent properties of biological materials. The geometry of a curved suture can be developed using the two-dimensional (2D) shape descriptor approach. In the present study, an elliptical geometry of the suture interface is proposed, and the analytical equations and finite element results are developed to capture the non-linear pullout response of the suture.

Keywords Bio-inspired · Suture interface · Shape descriptor approach · Non-linear pullout response

1 Introduction

The biological composites are characterised by comparatively improved mechanical properties than that of their constituents [1]. The weak interfaces and superior architecture seen in biological composite materials such as bones, seashells, cara-

C. Sachin Chandran (✉) · P. Rodewad
Department of Aerospace Engineering, Indian Institute of Space Science and Technology, Valiamala, Thiruvananthapuram, Kerala 695547, India
e-mail: sachinchandran.20@res.iist.ac.in; sachinchandran0@gmail.com

P. Rodewad
e-mail: rodewadpratiksha@gmail.com

S. Anup
Guide, Department of Aerospace Engineering, Indian Institute of Space Science and Technology, Valiamala, Thiruvananthapuram, Kerala 695547, India
e-mail: anup@iist.ac.in

S. Sidhardh et al. (eds.), *Advances in Structural Integrity for Mechanical, Civil, and Aerospace Applications*, Lecture Notes in Mechanical Engineering,
https://doi.org/10.1007/978-981-97-6367-2_5

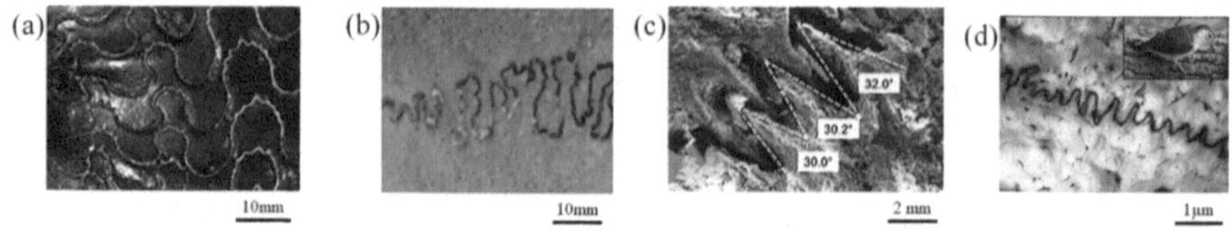

Fig. 1 Natural weak sutured interfaces: some examples **a** ammonite shell, **b** pan troglodytes cranial sutures, **c** osteoderms observed in sea turtle shell, **d** red-bellied wood peckers beak (Reproduced with permission from [4])

pace of turtle, osteoderms of sea turtle, wood peckers beak play a significant role in enhancing the mechanical properties such as fracture toughness, damping compared to there hard constituent materials [2, 5]. The weak interfaces found in some cephalapods like ammonoid are complicated in shapes across the species. The suture geometry having sinusoidal interfaces are more tougher than that of having straight interfaces [9]. Tailored sutured geometries and locking features used in the design of glasses and ceramics, provide significant improvement in deformation behaviour, energy absorption, and toughness [3, 6]. These weak sutured interfaces will channel the propagation of cracks caused even due to impact loads, so that the energy gets dissipated. The non-linear deformation behaviour of the suture is governed by the friction between the interfaces [4]. In this paper set of models having elliptical suture interfaces were compared with those having circular interfaces [5]. The analytical equations for the complete pullout response of both the models are compared with the finite element results based on friction at the contact interfaces (Fig. 1).

2 Analytical Formulation-Elliptical Suture Model

The Elliptical suture geometry was constructed from an ellipse having semi-major axis a, semi-minor axis b, and interlocking angle θ_0. The methodology we employed here is similar to that of circular sutures employed by Malik et al. [4, 5]. The contact area of the elliptical suture tab is higher than circular suture tabs, which will disperse and reduce the contact stresses developed, reducing the fracture and enhancing the efficiency of geometric locking.

The (x, y) coordinates at the point of contact between two ellipses are given by:

$$x = r \cos \theta_0 \tag{1}$$

$$y = r \sin \theta_0 \tag{2}$$

The general equation of an ellipse having major axis a and minor axis b is given by,

$$\frac{x^2}{b^2} + \frac{y^2}{a^2} = 1 \tag{3}$$

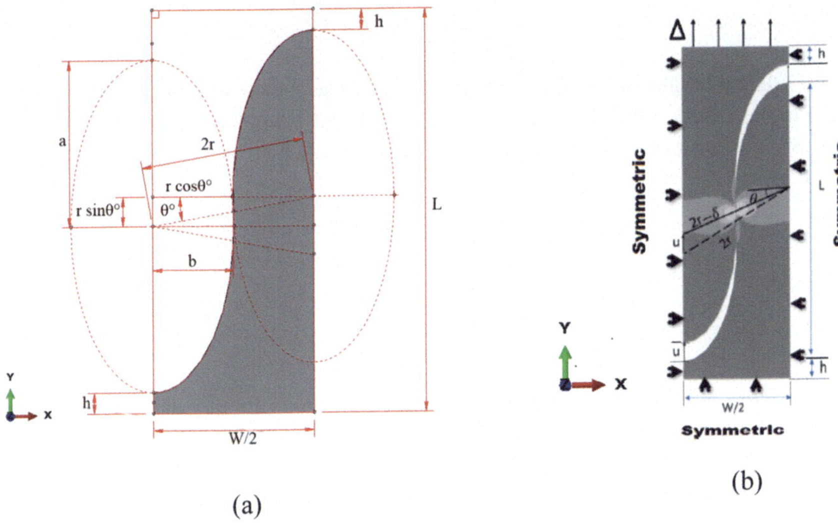

Fig. 2 **a** Modelling of elliptical suture. **b** Model with symmetry and boundary conditions

Substitute Eqs. 1 and 2 in Eq. 3, we get the value of r

$$r = \frac{ab}{\sqrt{a^2(\cos\theta_0)^2 + b^2(\sin\theta_0)^2}} \tag{4}$$

from Fig. 2a, the length and width of the elliptical tab are given by

$$L = 2a + 2r\sin\theta_0 \tag{5}$$

$$w = 4r\cos\theta_0 \tag{6}$$

$$b = r\cos\theta_0 \tag{7}$$

The θ_0 is the only independent geometrical parameter. The extent of interlocking among the tabs is controlled by the interlocking angle θ_0. Geometrical interlocking occurs only when θ_0 is greater than 0°.

The interfaces of the elliptical sutures were assumed to be non-adhesive, so that the mechanics of the contact surfaces, geometrical interlocking and friction will govern the pullout response of the elliptical suture. The suture material is assumed to be linear elastic isotropic and plane stress condition is used for the analysis. The boundary conditions are similar to that employed for circular suture [4, 5]. A displacement controlled boundary condition was used to model the pullout process:

$$u_y(x, -h) = 0 \tag{8}$$

$$u_y(x, L + h) = u \tag{9}$$

where h is the height of the upper and lower portions of the suture. ϵ_x is the average strain in the x direction. Periodic boundary conditions is applied to the left and right sides of a full unit cell having width, W is given by [4, 5]:

$$u_x\left(\frac{w}{2}, y\right) - u_x\left(\frac{-w}{2}, y\right) = w\epsilon_x \tag{10}$$

$$u_y\left(\frac{w}{2}, y\right) = u_y\left(\frac{-w}{2}, y\right) \tag{11}$$

Due to the symmetric loading and geometry conditions about y-axis,the displacements in the x-direction should satisfy

$$u_x(x, y) = -u_x(-x, y) \tag{12}$$

combining Eqs. 8–12, we get the periodic boundary conditions as [4, 5]:

$$u_x(0, y) = 0 \tag{13}$$

$$u_x(w/2, y) = \frac{w}{2}\epsilon_x \tag{14}$$

The $\epsilon_x \neq 0$ would imply elastic deformations in the solid areas above and below the suture, with uniform strains away from the jigsaw tab. However, when h is sufficiently large, the solid regions stiffness in the transverse direction becomes significant, preventing any deformation in the transverse direction and resulting in $\epsilon_x = 0$. So the periodic boundary conditions are simplified to symmetric BCs as mentioned in Fig. 2b then Eqs. 13 and 14 become:

$$u_x(0, y) = 0 \tag{15}$$

$$u_x(w/2, y) = 0 \tag{16}$$

The model with the provided boundary conditions after symmetry is applied. The kinematics and balance of the suture are studied using this reduced geometry. The kinematic relations are obtained by comparing the original and deformed configurations, assuming that elastic deformations are small in compared to the displacement caused by sliding of the suture interfaces and pullout. From Fig. 2b, we get x component [4, 5]:

$$(2r - \delta)\cos\theta = 2r\cos\theta_0 \tag{17}$$

$$2r\cos\theta - \delta\cos\theta = 2r\cos\theta_0 \tag{18}$$

$$\delta = 2r\left[1 - \frac{\cos\theta_0}{\cos\theta}\right] \tag{19}$$

y component [4, 5]

$$u = 2r\sin\theta_0 - (2r - \delta)\sin\theta \tag{20}$$

$$u = 2r\sin\theta_0 - \left(2r - 2r\left(1 - \frac{\cos\theta_0}{\cos\theta}\right)\right)\sin\theta \tag{21}$$

$$u = 2r\sin\theta_0 - 2r\cos\theta_0\tan\theta \tag{22}$$

The non-dimensional pullout distance

$$\frac{u}{L} = \frac{r\sin\theta_0 - r\cos\theta_0\tan\theta}{a + r\sin\theta_0} \tag{23}$$

The locking at the suture geometries produces a pullout force F which is along the y axis when the upper suture is pulled along the y axis. Contact forces at surfaces include a horizontal component as well, resulting in compressive forces along direction the x axis F_c. Because the net horizontal force on the unit cell is zero, the suture region's F_c forces must be balanced by F_T, the tensile forces created on both sides of the suture line [4, 5]. These forces are classified as normal forces P_i and $f P_i$. The equation that convert these contact forces to pullout forces as [4, 5],

$$F = \sum_{i=1}^{N} P_i(\sin\theta + f\cos\theta) \tag{24}$$

The solution for two identical circular disks in contact is approximated for elliptical disks and can be used for finding forces transmitted between the two tabs. The Non-Hertzian contact mechanics problem gives the value of δ [4, 5].

$$\frac{\delta}{r} = \frac{2}{\pi}\frac{P}{rtE}\left(\ln\left(4\pi\left(\frac{rtE}{P} - 1\right)\right)\right) \tag{25}$$

where t is the thickness of the suture tab, numerically solve the above equation to obtain the value of $\frac{P}{rtE}$. Non-Dimensional pullout force is given by

$$\frac{F}{rtE} = \frac{2P(\sin\theta + f\cos\theta)}{rtE} \tag{26}$$

Thus, in this case, we use $\frac{F}{wtE}$ as the direct measure of the average tensile stress applied to the suture, where (wt is cross-sectional area of the suture).

$$\frac{F}{wtE} = \frac{P}{rtE}\frac{(\sin\theta + f\cos\theta)}{2\cos\theta_0} \tag{27}$$

3 Finite Element Analysis of the Suture

Widely used finite element Software, ABAQUS [7] is used for the analysis. The material property of ABS taken (from the literature [4, 5]) with $E = 1.72$ GPa and $\nu = 0.35$ is used. FE Analysis of circular suture with $a = b = 1$, interlocking angle 10° and coefficient of friction, $f = 0.1, 0.2, 0.3$, and 0.4 is carried out. The model is meshed using CPS4R Plane stress element. The interface regions were refined to an element size of $\frac{R}{400}$ to ensure the results to be mesh independent and converged. The elliptical suture were modelled with $a = 2$, $b = 1$ and interlocking angle $\theta = 10°$. Finite element analysis is carried out to find the pullout response developed in the elliptical suture tabs.

4 Result and Discussions

The plot of Non-dimensional pullout force versus non-dimensional distance is done using MATLAB [8] using the values obtained from the ABAQUS software.

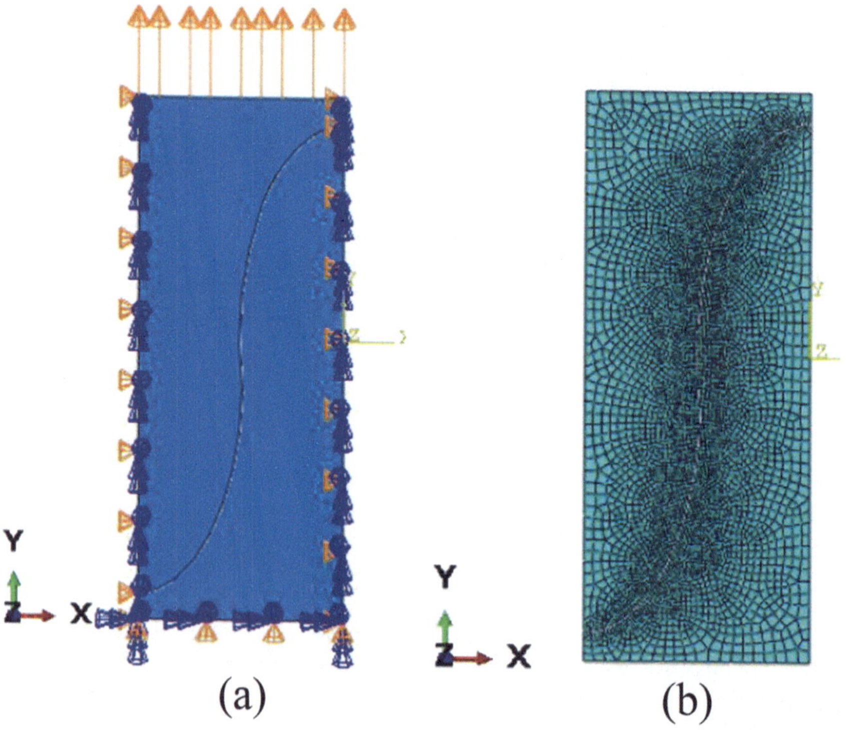

Fig. 3 **a** Displacement controlled boundary conditions (on the left, right and bottom edge symmetric BCs are applied and top edge displacement along y axis is applied). **b** Meshing of elliptical suture model

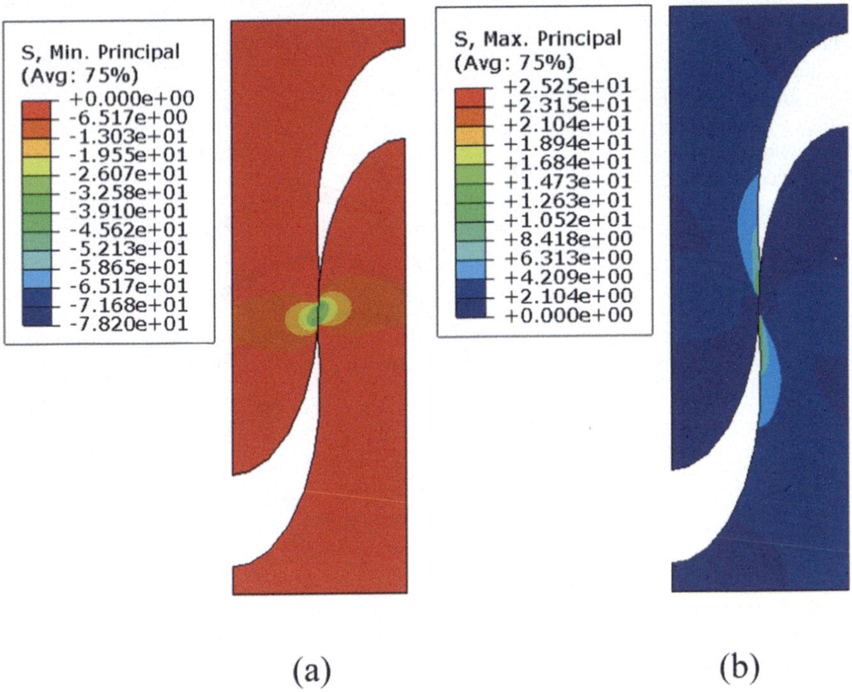

Fig. 4 **a** Contour showing minimum principal stress (maximum pressure). **b** Contour showing maximum principal stress (maximum tensile stress developed in the suture) contour for elliptical suture with $a = 2$, $b = 1$ and $\theta_0 = 10°$

Figure 3 shows the boundary conditions and meshing done on circular suture with radius 1 and for $\theta = 10°$. The displacement controlled BC is given to the suture with bottom edge fixed and top surface displacement. The left and right surfaces are given x-symmetric BC. Figure 4 shows the maximum and minimum principal stress contour plot for the elliptical suture. Figure 5 depicts non-dimensional pullout response for different friction coefficients ($f = 0$, 0.1, 0.2, 0.3, 0.4) for a circular suture with interlocking angle $\theta = 10°$. Reproduced from [4, 5]. The plot shows good agreement between analytical and FE results. The pullout response increases with increase in friction coefficient between the interfaces. The comparison of elliptical and circular suture pullout response is shown in Fig. 6 and their effects on friction coefficient (f). The higher value of frictional coefficients (f) result in comparatively increase in strength due to higher friction at the contact site, but they have no effect on the maximum pullout distance, which depend on geometric values. The Non-dimensional pullout response for the elliptical suture with different frictional coefficients and $\theta_0 = 10°$ is plotted analytically as shown in Fig. 7. A good agreement with the FE results is obtained.

Fig. 5 Non-dimensional pullout force versus non-dimensional displacement for different values of f for circular suture with $\theta_0 = 10°$. Reproduced from [4, 5]

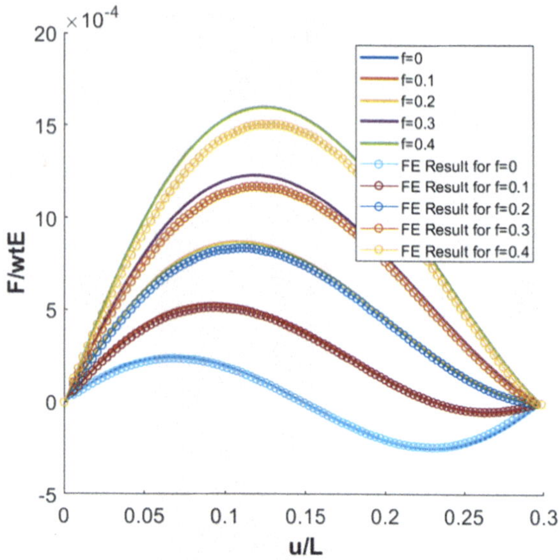

Fig. 6 Comparison of FE results for circular ($R = 1$) and elliptical suture ($a = 2, b = 1$) with $\theta_0 = 10°$ for f = 0.3 and 0.4

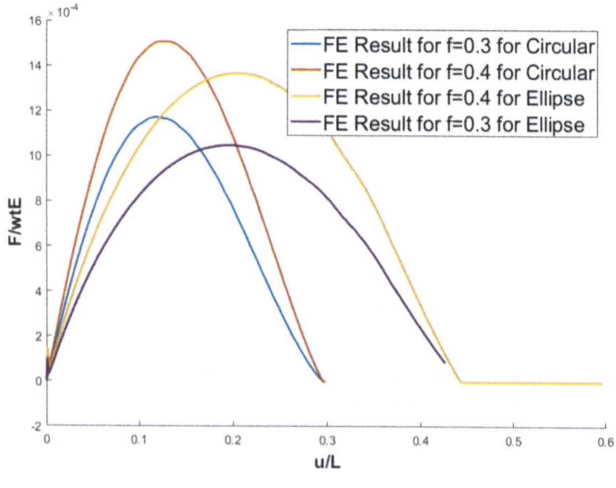

5 Conclusion

Analytical and FE modelling were done to determine the relationship between the suture interface strength, toughness on geometric factors, and the characteristics of the material constituents. The reproduced analytical and FE results of circular sutured interfaces show excellent comparison similar to that shown in literature. Pullout response of elliptical suture is carried out using FE analysis software and compared with the FE solution of circular suture it is found that interlocking angle, a, b and coefficient of friction has important role in pullout analysis. The strain energy absorbed during pullout is more for elliptical suture than circular sutures.

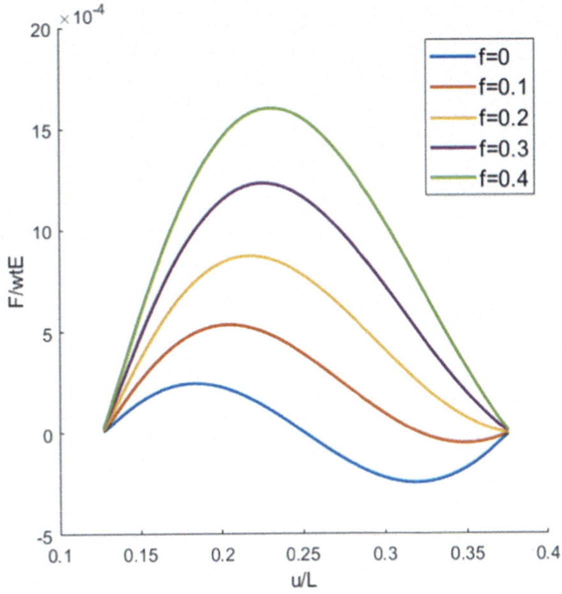

Fig. 7 Analytical plot of the elliptical suture ($a = 2, b = 1$) with $\theta_0 = 10°$ for different values of f

Higher coefficients of friction improve strength due to higher friction at the contact site, but have no effect on the maximum pullout distance, which is determined mainly by the geometry. This study systematically investigates the detailed role of geometry in the mechanical behaviour of bio-inspired suture interfaces. Bio-inspired designs with variation in specific geometric parameters were created in order to understand the direct effects of these parameters.

References

1. Abhirami A, Anup S (2021) Mechanical properties of unidirectional bio-inspired composites with two non-self-similar hierarchical structures. Mech Mater 163:104082
2. Achrai B, Bar-On B, Wagner HD (2014) Bending mechanics of the red-eared slider turtle carapace. J Mech Behav Biomed Mater 30:223–233
3. Anup S (2015) Influence of initial flaws on the mechanical properties of nacre. J Mech Behav Biomed Mater 46:168–175
4. Malik IA, Barthelat F (2018) Bioinspired sutured materials for strength and toughness: pullout mechanisms and geometric enrichments. Int J Solids Struct 138:118–133
5. Malik IA, Mirkhalaf M, Barthelat F (2017) Bio-inspired "jigsaw"-like interlocking sutures: modeling, optimization, 3D printing and testing. J Mech Phys Solids 102:224–238
6. Mirkhalaf M, Dastjerdi AK, Barthelat F (2014) Overcoming the brittleness of glass through bio-inspiration and micro-architecture. Nat Commun 5(1):1–9
7. SIMULIA (2019) ABAQUS/standard user's manual, version 2019. Providence, RI
8. The Mathworks I (2021) MATLAB version 9.9 (R2021b). Natick, MA
9. Zavattieri PD, Hector LG Jr, Bower AF (2008) Cohesive zone simulations of crack growth along a rough interface between two elastic-plastic solids. Eng Fract Mech 75(15):4309–4332

Finite Element Analysis of Drilling Process in Carbon Fiber Composite Material: Mechanical Behavior and Delamination Prevention

Anmol Choudhary and Greegar George

Abstract Carbon fiber-reinforced polymer (CFRP) composites have gained popularity in various fields owing to their strength, durability, and lightweight nature. CFRP is a fiber-reinforced composite material that employs carbon fiber as the primary structural component, making it suitable for aircraft components that require high strength, anisotropy, and thermal conductivity. For the components made from composite laminate, drilling is a crucial machining technique that requires particular attention. During drilling, maintaining processing quality is challenging due to the influence of drilling force, which can lead to delamination, tearing, and other defects. In this paper, we present a finite element analysis of the drilling process in carbon fiber composite and study its mechanical behavior. We developed a three-dimensional model of an M21/T700 carbon fiber-reinforced polymer and simulated the drilling process using Abaqus software. The various force components during drilling are analyzed and compared the stress and plastic equivalent strain incurred during the drilling process with the spindle speed and feed rate. This study can aid in selecting a spindle speed and feed rate combination that reduces the thrust force induced during drilling, preventing interlaminar and intralaminar failure modes at critical thrust forces. This, in turn, can assist in reducing the degradation and development of drilling tools and designing carbon fiber composite materials.

Keywords Finite element analysis · Drilling · Carbon fiber-reinforced polymer

A. Choudhary · G. George (✉)
National Institute of Technology Tiruchirappalli, Tiruchirappalli, India
e-mail: greegar@nitt.edu

A. Choudhary
e-mail: 203220004@nitt.edu

S. Sidhardh et al. (eds.), *Advances in Structural Integrity for Mechanical, Civil, and Aerospace Applications*, Lecture Notes in Mechanical Engineering,
https://doi.org/10.1007/978-981-97-6367-2_6

1 Introduction

Carbon fiber-reinforced polymer (CFRP) is a high-performance composite material that finds applications in various industries like aerospace, defense, automotive, high-speed equipment, and sports equipment. Typically, CFRP composites are fabricated using thermosetting resins such as epoxy, polyester, or vinyl ester [9]. The distinctive properties of carbon fiber composites (CFCs) are low density, high stiffness, and negligible thermal expansion. The weight of carbon fiber mixtures is 1.55 g/cm^3 with epoxy resin accounting for 30% and carbon fiber accounting for 70% [7]. They have high durability. CFCs are produced with standard fabrics, and 0°/90° weave (where 0 and 90° are angles of the lamina), providing a durability of 90.5 GPa (Young's modulus), whereas aluminum provides 69 GPa. Moreover, CFCs are 42% lighter than single aluminum of the same thickness. Carbon fiber composites have a very low coefficient of thermal conductivity having a coefficient of thermal expansion of 2×10^6 °C^{-1} at 20 °C.

Numerous studies have been conducted on carbon fiber composites. Pederson [7] proposed an energy dissipation system that uses carbon fiber composites to protect passengers in passenger and racing cars from direct impact with moving or stationary objects, thereby enhancing car crash safety. Rafi et al. [8] investigated the behavior of reinforced concrete beams fastened with bars of CFRP at high temperatures and proposed a basic model for determining the residual mechanical properties of CFRP bars at high temperatures, which predicted the strength and durability of fiber-reinforced polymer (FRP)-reinforced beams.

Significant work has been done to investigate drilling in CFRP. A review by Hale and Ng [4] emphasized the importance of process monitoring during the drilling of CFRP to ensure the quality of borehole. The paper by Kahwash et al. [5] highlighted the current practice of CFRP cutting process modeling using 2D orthogonal shear. Arola and Ramulu [1] developed a 2D orthogonal model using epoxy-graphite material as the matrix and carbon fiber as the fiber component, finding a positive correlation between predictable and experimental shear resistance, but a negative correlation in the same thrust force.

As composite materials continue to improve, the demand for machining technologies for the material is increasing. The distinctive mechanical and physical properties of composite materials result in substandard cutting performance, causing various plastic changes and processing defects during machining, such as burrs and delamination [2]. This can lead to the wearing of finely tuned tools that affect the accuracy and efficiency of the equipment. The finite element method can be used to simulate the actual drilling process, analyze the parameters that affect defects, and predict defects in the machining of materials before the actual process. These parameters can guide the formation of tools and improve the efficiency of the drilling process by optimizing the factors that affect it.

In this paper, a finite element analysis of the drilling process is performed in an M21/T700 carbon fiber composite, investigating the influence of drilling force on this process. In addition, this study aims to enhance the drilling efficiency in carbon fiber composites, while minimizing delamination effects. The study will also compare the stress and plastic equivalent strain with the spindle speed and feed rate, and establish an optimal combination of spindle speed and feed rate that can prevent interlaminar and intralaminar failure modes.

2 Analysis of Drilling Process

In this study, an analysis of the drilling process is conducted in two stages. Firstly, the drilling force is analyzed to identify the component responsible for causing delamination. Subsequently, the changes in torque and drilling force values during the drilling process are investigated.

2.1 Components of the Drilling Force

In the context of drilling carbon fiber composite materials, the rake angle of the tool refers to the angle between the cutting face of the tool and a line perpendicular to the workpiece. As the cutting edge of the drill bit progresses, the rake angle changes gradually from positive to negative, starting from the chip edge and moving toward the chisel edge. At the chisel edge, the rake angle can reach its maximum negative value. The force on the chisel edge and the main cutting edge of the rake angle acting on the carbon fiber composite material can have two directions: one corresponds to the edge denoted by F_{V_1}, and the other is direct force, denoted by F_{Z_1} (Fig. 1).

Due to the slow linear velocity of the drill bit at the chisel edge, F_{Z_1} cannot play a direct role in the cutting process. However, as the deformation of the carbon fiber composite material exceeds its compressive strength limit, it begins to deform and then break and separate under compression. The chisel edge and the angle components of the fabricated parts are aligned with the direction of the composite fiber and differ from the cut thread type of the fine angle section.

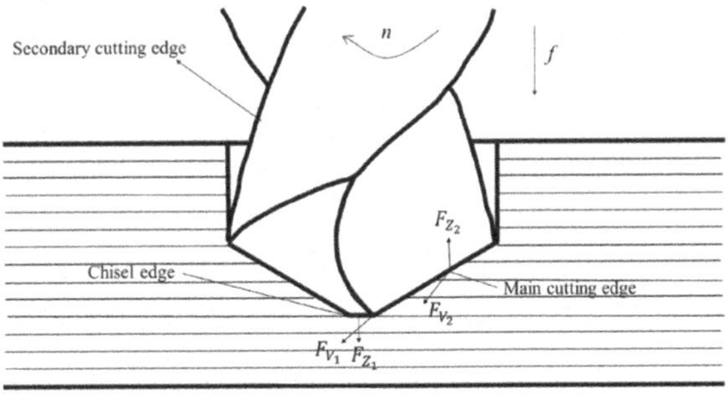

Fig. 1 Schematic representation of various forces acting during drilling process

Although the value of F_{V_1} is small at the chisel edge, the force, F_{Z_1} acts as a major component of the axial force, exerting downward pressure on the workpiece, and is therefore considered a primary cause of delamination and aggressive errors. The rake of the cutting edge is the primary material removal component. On the upper part of the laminated layer, force acting is F_{Z_1} which removes the extracted material and is considered as one of the main causes of delamination and tearing.

2.2 *Variation of Torque and the Drilling Force During Drilling Process*

During the drilling of carbon fiber laminate, the applied force and torque experience fluctuations before the drilling is completed. The drilling process can be delineated into multiple stages, the first of which is the initial contact between the cutting tool and the workpiece. Since the feed rate is high, the drill bit has just initiated contact with the workpiece, and the material has not been processed. As a result, the workpiece experiences a negative impact with the drilling tool at the beginning of the process, leading to a large and rapidly increasing force at this stage.

In the subsequent phase, the drill begins to penetrate the workpiece, resulting in a reduction in the drilling force. The degree of stabilization in the axial force is influenced by the carbon fiber composite chip formation and associated damage. At this stage, the magnitude of the axial force is relatively low, owing to the high fiber content and low volume of the carbon fiber composite structure. The material properties indirectly impact the drilling process, resulting in fluctuations in the cutting force due to the intermittent contact between the cutting tool and the workpiece material. In the final stage, the drilling process enters a steady phase with low penetration depth, accompanied by a continuously changing force value. The stress level undergoes a significant change during the piercing process, and the pressure fluctuates significantly when the cutting tool reaches its maximum cutting strength, ultimately completing the piercing process.

3 Finite Element Modeling of the Composite in Abaqus

This section focuses on the modeling of carbon fiber composite materials and drill bit in Abaqus to simulate the drilling process. The M21/T700 CFRP material is used as the workpiece material, and the material properties are modeled with a focus on key properties. The drilling process parameters are determined, including the cutting speed limit, feed rate, and the use of a polycrystalline diamond drill bit. The crucial role of meshing in the standard feature modeling process is also discussed, and the appropriate mesh size is determined. Finally, the definition of analysis steps and boundary conditions is established for simulating the drilling process.

3.1 Modeling of Carbon Fiber Composite Material

The present study focuses on the M21/T700 CFRP material as the workpiece material. In this study, the material properties of the CFRP are modeled using the Abaqus module, with a focus on key properties such as Young's modulus, Poisson's ratio, coefficient of thermal expansion, thermal conductivity, and radiation resistance. Tables 1 and 2 provide the values of elastic modulus, shear modulus, Poisson's ratio, and other important properties of the M21/T700 CFRP laminate considered in this study [10]. Additionally, the respective elastic modulus in all three directions and damage model parameters of the single lamina are incorporated. Once the composite layer is set up, each layer is oriented at an angle of 0 and 90° to one another, as this configuration has been shown to exhibit a higher shear strength capacity [6].

Composite Layer Setup. Figure 2 presents the model of a CFRP laminate with a total thickness of 8 mm, composed of 8 individual layers. The figure illustrates the orientation, thickness, and size of the laminate. Table 3 provides the properties of each lamina in the composite structure. To model each ply, three integration points are used, and the material properties of CFRP are considered throughout the analysis.

Table 1 Properties of the CFRP lamina

Lamina constant	E_{11}	E_{22}	E_{33}	G_{12}	G_{13}	G_{23}	ν_{12}	ν_{13}	ν_{23}
Value	105.5	7.2	7.2	3.4	3.4	2.52	0.34	0.34	0.378
Unit	GPa	GPa	GPa	GPa	GPa	GPa	–	–	–

Note E_{ij} is elastic modulus in '*i*' and '*j*' direction, G_{ij} is shear modulus in '*i*' and '*j*' direction, and ν_{ij} is Poisson's ratio in '*i*' and '*j*' direction

Table 2 Damage model parameters of the CFRP lamina

S. No.	Damage model parameters of lamina	Value
1	Longitudinal tensile strength	1340 MPa
2	Longitudinal compressive strength	1192 MPa
3	Transverse tensile strength	19.6 MPa
4	Transverse compressive strength	92.3 MPa
5	Longitudinal shear strength	51 MPa
6	Transverse shear strength	23 MPa
7	Longitudinal tensile fracture energy	48.4 Nmm
8	Longitudinal compressive fracture energy	60.3 Nmm
9	Transverse tensile fracture energy	4.5 Nmm
10	Transverse compressive fracture energy	8.5 Nmm

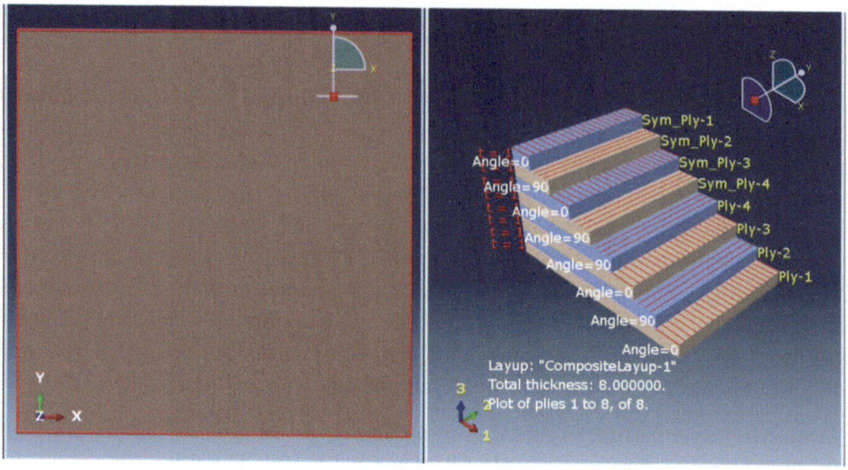

Fig. 2 Composite layer setup of CFRP laminate

Table 3 Details of composite layers in CFRP laminates

S. No.	Ply name	Material	Thickness (mm)	Rotation angle (°)	Integration points
1	Ply-1	CFRP	1	0	3
2	Ply-2	CFRP	1	90	3
3	Ply-3	CFRP	1	0	3
4	Ply-4	CFRP	1	90	3
5	Sym_Ply-1	CFRP	1	0	3
6	Sym_Ply-2	CFRP	1	90	3
7	Sym_Ply-3	CFRP	1	0	3
8	Sym_Ply-4	CFRP	1	90	3

3.2 *Modeling of Drill Bit*

Feito et al. [3] have extensively studied the effects of various drill bit parameters on the drilling process for T300 CFRP laminate, including spindle speed, feed rate, torque, drilling force, and drill bit size. In this study, a cutting speed limit of 150 rad/s and a feed rate of 0.15 m/s are used. The natural temperature during drilling operations is taken as 20 °C, with a shear friction factor of 0.6. A polycrystalline diamond (PCD) drill bit with a diameter of 7.963 mm is employed. Figure 3a depicts the three-dimensional model of the PCD developed in Abaqus.

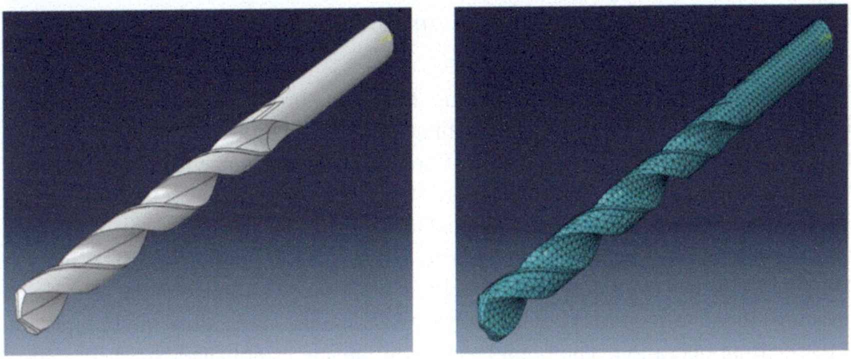

(a) (b)

Fig. 3 **a** 3D model of PCD drill bit in Abaqus, **b** meshed model of drill bit in Abaqus

3.3 Meshing

Meshing plays a crucial role in the standard feature modeling process and significantly influences the accuracy and numerical value of the final results. Reducing mesh size enhances numerical accuracy and durability, albeit at the expense of increased computation time. Determining the appropriate mesh size depends on the study's requirements. In this research, a mesh size of 0.05 mm is selected for the carbon fiber composite. The minimum cell size is determined based on the applied feed rate. Figure 3b displays the meshed model of the PCD cutting tool.

The mesh size selected for the drill bit is 0.05. The meshed workpiece of CFRP composite is shown in Fig. 4.

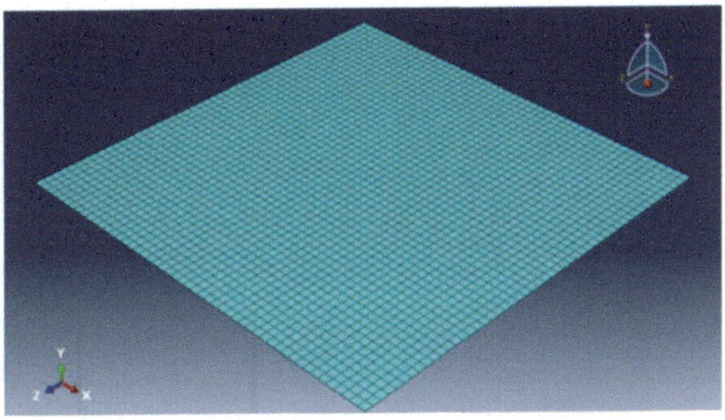

Fig. 4 Finite element mesh of CFRP workpiece modeled in Abaqus

3.4 Definition of Analysis Step and Boundary Condition

The subsequent phase involves simulating the drilling process by specifying the step size for every 1° of drill bit rotation, with an average rotation of approximately 300 steps. The drill bit rotates at a speed of 150 rad/s, and the distance from the bottom of the workpiece to the top of the drill bit is 150 mm. It can be assumed that a complete hole in the workpiece necessitates an axial movement of at least 160 mm. The actual drilling time may be higher due to the grid and automatic timing control. When the drill bit approaches the 5 mm range specified by the stop command in the processing direction, the data is verified, and the process is halted, indicating the completion of the drilling operation. The boundary conditions are established based on a local temperature of 20 °C, while the continuous collision coefficient is considered to be 0.6.

4 Results and Discussion

The assembly of meshed CFRP plate and drill bit modeled in Abaqus is shown in Fig. 5. The reference point is positioned at the end of the drill bit. Figure 6 depicts the variation of stress in the CFRP material after the drilling process, indicating the presence of irregular fragments that remain in the drilled hole, leading to delamination.

Figure 7 illustrates the variation of plastic strain in the CFRP laminate, with plastic strain values serving as a measure of material deformation. The objective of this study is to minimize the plastic equivalent strain to improve the drilling process efficiency and reduce delamination.

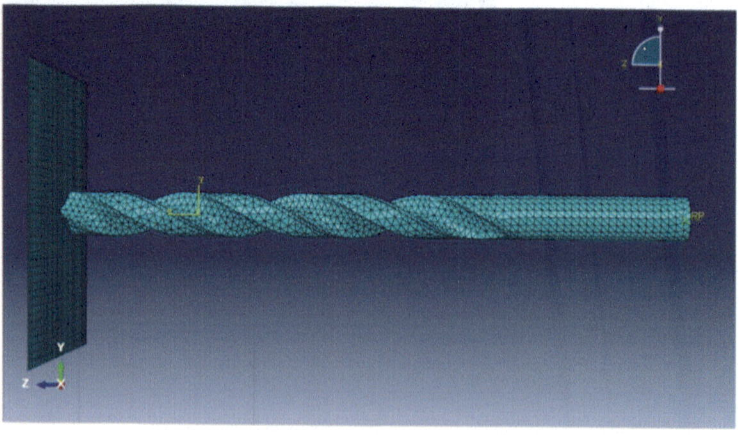

Fig. 5 CFRP and drill bit assembly modeled in Abaqus

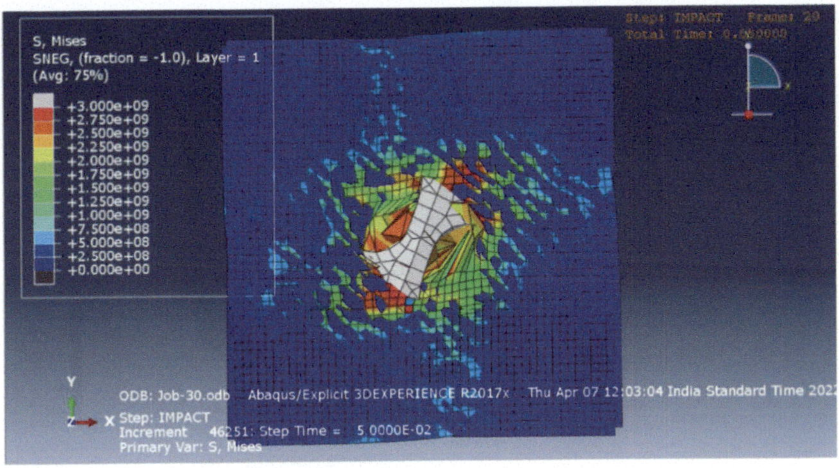

Fig. 6 Stress variation in CFRP composite

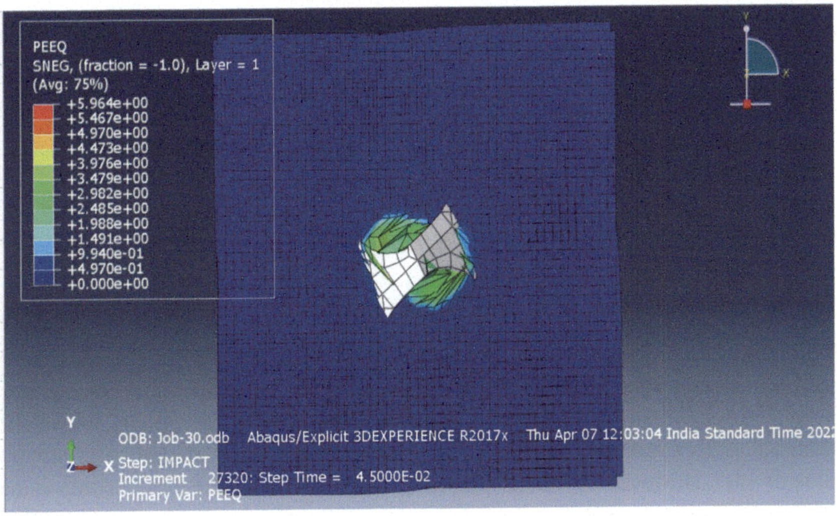

Fig. 7 Plastic equivalent strain variation in CFRP composite

4.1 Plot of von Mises Stress Versus Spindle Speed Versus Feed Rate

A three-dimensional graph is plotted for stress developed in the CFRP plate by varying the spindle speed and feed rate of the drill bit on the CFRP plate. This is illustrated in Fig. 8. The stress value decreases from 3.05 to 2.46 GPa with an increase in spindle speed from 20 to 120 rad/s. However, it starts to increase again,

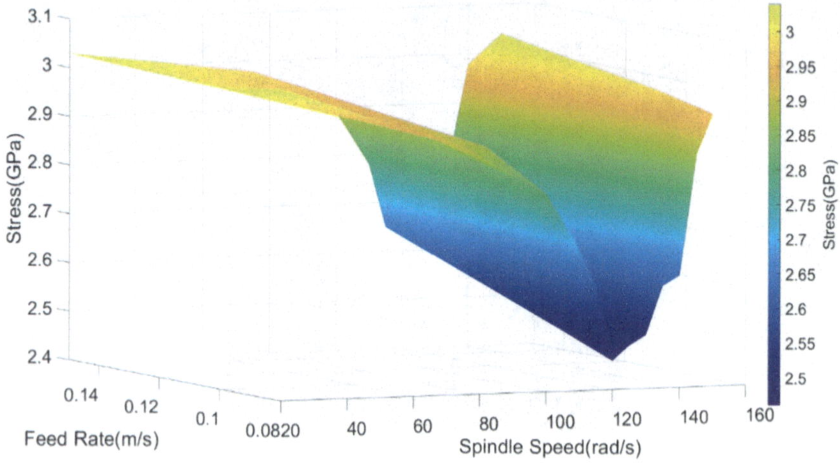

Fig. 8 Plot of von Mises stress versus spindle speed versus feed rate

reaching 3.04 GPa at a spindle speed of 150 rad/s. The stress value is observed to be minimum at a feed rate of 0.08 m/s. Beyond this, the stress value increases, which can be attributed to the occurrence of interlaminar and intralaminar failure modes, ultimately leading to delamination.

4.2 Plot of Plastic Equivalent Strain (PEEQ) Versus Spindle Speed Versus Feed Rate

The extent of delamination in a composite can be determined by analyzing the plastic strain values following the drilling process. A lower plastic strain value corresponds to reduced delamination in the composite. As illustrated in Fig. 9, initially, the plastic strain value decreases from a spindle speed of 20–100 rad/s, attaining a minimum value of 1.05. However, beyond a spindle speed of 100 rad/s, the plastic strain value begins to increase again. After a spindle speed of 120 rad/s, the strain values decrease, reaching a minimum value of 1.06 at a spindle speed of 130 rad/s. Additionally, an increase in feed rate from 0.08 to 0.15 m/s leads to a rise in plastic strain value. This observation can be attributed to the occurrence of interlaminar and intralaminar failure modes, which contribute to an increase in plastic strain and, consequently, delamination.

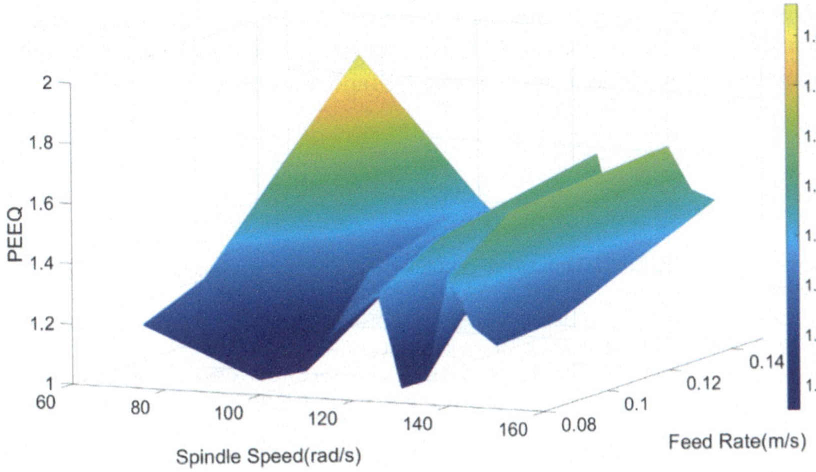

Fig. 9 Plot of plastic equivalent strain (PEEQ) versus spindle speed versus feed rate

5 Conclusions

The versatile nature and strength of carbon fiber-reinforced polymer make it a widely used material in various fields, but the occurrence of defects during the machining process limits its full potential. To investigate this issue, a finite element model of the drilling process is developed using Abaqus, focusing on M21/T700 CFRP composite. The main cause of delamination during drilling is identified as the drilling force exerted by the drill bit, which in turn is affected by the spindle speed and feed rate. The effect of these parameters on plastic equivalent strain and von Mises stress in the CFRP plate is investigated, and a 3D plot is presented to illustrate their variation. The results show that both stress and plastic equivalent strain can be reduced by adjusting the spindle speed and feed rate of the drill bit. Specifically, a feed rate of 0.08 m/s and a spindle speed between 120 and 130 rad/s are found to prevent interlaminar and intralaminar failure modes in the CFRP plate. It should be noted that the recommended values for spindle speed and feed rate are based on the geometry of the components, and the stress and plastic strain plots obtained in this study are dependent on the constant drill bit size used. Future work could involve varying the diameter of the drill bit to investigate its impact on stress and strain, as well as using different CFRP materials in the study.

The present investigation does not incorporate considerations related to temperature effects during the drilling process. In order to assess the influence of temperature on the drilling process within composite materials, it is imperative to conduct numerical simulations at reduced drilling speeds. This approach aims to mitigate the influence of temperature, allowing for a focused examination of delamination phenomena. Subsequently, a comparative analysis between simulations conducted at lower and higher drilling speeds will elucidate the temperature-induced differentials. It is therefore suggested as a potential area for further research.

References

1. Arola D, Ramulu M (1997) Orthogonal cutting of fiber-reinforced composites: a finite element analysis. Int J Mech Sci 39(5):597–613
2. Benammar A, Drai R, Guessoum A (2008) Detection of delamination defects in CFRP materials using ultrasonic signal processing. Ultrasonics 48(8):731–738
3. Feito N, Díaz-Álvarez J, López-Puente J, Miguelez MH (2018) Experimental and numerical analysis of step drill bit performance when drilling woven CFRPs. Compos Struct 184:1147–1155
4. Hale P, Ng EG (2021) 3D finite element model on drilling of CFRP with numerical optimization and experimental validation. Materials 14(5):1161
5. Kahwash F, Shyha I, Maheri A (2015) Modelling of cutting fibrous composite materials: current practice. Procedia CIRP 28:52–57
6. Naresh K, Krishnapillai S, Velmurugan R (2017) Effect of fiber orientation on carbon/epoxy and glass/epoxy composites subjected to shear and bending. Solid State Phenom 267:103–108
7. Pederson J (2008) Finite element analysis of carbon fiber composite ripping using ABAQUS. Doctoral dissertation, Clemson University
8. Rafi MM, Nadjai A, Ali F (2008) Finite element modeling of carbon fiber-reinforced polymer reinforced concrete beams under elevated temperatures. ACI Struct J 105(6):701
9. Sambruno A, Bañon F, Salguero J, Simonet B, Batista M (2019) Kerf taper defect minimization based on abrasive waterjet machining of low thickness thermoplastic carbon fiber composites C/TPU. Materials 12(24):4192
10. Thaiprayoon N, Chalernphon K, Somtua C, Aroonjarattham P (2021) Parametric study of composite wing box for high altitude platform using finite element analysis. In: 11th TSME-international conference on mechanical engineering (TSME-ICoME 2020). IOP Publishing, p 012018

Optimization of Orthogrid-Stiffened Cylinder Under Axial Force and External Pressure

R. Santhosh, P. C. Jain, and Gangadharan Raju

Abstract The optimization problem of the design of orthogrid-stiffened cylinder under combined axial force and external pressure is a complex problem. It involves finding the optimal layout of stiffening rings and stringers (number of rings and stringers which are discrete variables) and sizing of members and skin (continuous variables) while minimizing the mass of the stiffened cylinder and satisfying the buckling constraints. Also, the presence of multiple modes of buckling adds to the complexity of the optimization problem. Further, the design space is nonconvex with multiple local minima, and steps must be taken to avoid sub-optimal solutions. In this paper, a methodology for carrying out systematic optimization for the stiffened cylinder is presented. The optimization problem statement is defined as the minimization of mass with buckling load factor as the constraint, and two methods are presented to evaluate the buckling load. The first method is a simplified analytical method for evaluating individual buckling modes, and the second method is a high-fidelity method where buckling load is calculated using FEM. The optimization problem is solved using both methods. A comparison is made between the optimal solutions by the two fidelity models in terms of accuracy and cost. It is observed that the low-fidelity method is suited for preliminary calculations and screening the design space. The high-fidelity method gives an optimal solution which is having significant mass savings (28.13%) compared to results in the literature.

Keywords Stiffened cylinder · Optimization · Buckling

R. Santhosh (✉) · P. C. Jain
DRDL, DRDO, Hyderabad, India
e-mail: san.drdo@gmail.com

R. Santhosh · G. Raju
IIT Hyderabad, Hyderabad, India

S. Sidhardh et al. (eds.), *Advances in Structural Integrity for Mechanical, Civil, and Aerospace Applications*, Lecture Notes in Mechanical Engineering,
https://doi.org/10.1007/978-981-97-6367-2_7

1 Introduction

The airframe sections of space vehicles like rockets, launch vehicles, and weapon systems like missiles are typically cylindrical in shape. These sections are designed to withstand the flight load without failure by fracture, retain shape without excessive deformation (buckling), and protect the internal components during flight. Cylindrical sections in aerospace structures are often subjected to high axial load and external pressure during their operation. Both these load cases create compressive stresses in the cylinder, and its design is generally found to be critical in buckling. Hence, instead of using a monocoque construction, the cylinders are stiffened to increase the load-carrying capacity with a lower structural mass. Some of the commonly used stiffening schemes in cylindrical shells are isogrid, orthogrid, and anisogrid. Among them, the orthogrid construction offers the advantage of simplicity of the design, ease of manufacturing, and lesser weight and is preferred in the design of metallic airframes. In the orthogrid construction, the cylinder is stiffened in meridional (stringers) and hoop direction (rings). It is important that the size and layout parameters of the orthogrid cylinder are optimized to minimize the mass of the airframe. This will translate to an increase in range or an increase in the payload carrying capacity of the flight vehicle. This paper presents the work carried out on mass optimization of stiffened cylinders considering orthogrid-stiffening pattern.

2 Optimization Problem Statement

The problem of optimization of orthogrid-stiffened cylinder under combined axial force and external pressure is a complex problem. It involves finding the optimal layout of stiffening rings and stringers (number of rings and stringers which are integer valued) and sizing of members and skin (continuous) while minimizing the mass of the stiffened cylinder and satisfying the buckling constraints. A schematic of typical parameters of orthogrid is shown in Fig. 1.

The mathematical problem statement of the optimization problem is provided below.

Objective: Minimize mass.

Constraint

- Buckling load factor (BLF) $\geq$ BLF limit (typically 1.5).
- Practical dimensional limits of skin, stringer, ring.
- Integer number of stringer and ring.

Design Variables (DVs)

1. Number of rings (n_r).
2. Number of stringers (n_s).
3. Thickness of shell (t_{sh}).

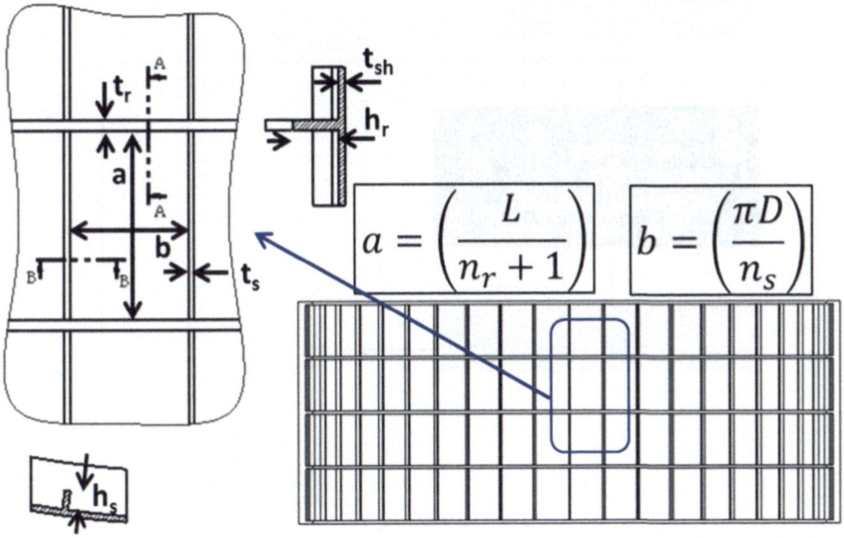

Fig. 1 Scheme of orthogrid construction

4. Thickness of rings (t_r).
5. Thickness of stringers (t_s).
6. Height of stiffeners ($h = h_r = h_s$).

Fixed Parameters

1. Geometry: Length (L), radius.
2. Material properties: Young's modulus, Poisson's ratio.
3. Loads: Axial force (AF), pressure (Pr).
4. Buckling knockdown factor (Kdf).

Here, the following simplifying assumptions are made:

- The cylinder is not stress-critical, and buckling is the dominant failure. Hence, stress constraints are not considered.
- The presence of cut-outs and mounting features are not considered as these features are local in nature.
- Material of construction is given and not a design variable.
- Knockdown factor which is an empirical correction for the difference between experiment and theory is generally taken from NASA SP 8007. For this paper, knockdown of 1 is considered (perfect cylinder) to enable comparison with the literature.

In this problem statement, it is to be noted that the number of rings and stringers must take only integer values, while other design variables can take continuous values. Also, the presence of multiple modes of buckling including the global buckling of the cylinder, local buckling of the ring, stringer, and skin adds to the complexity.

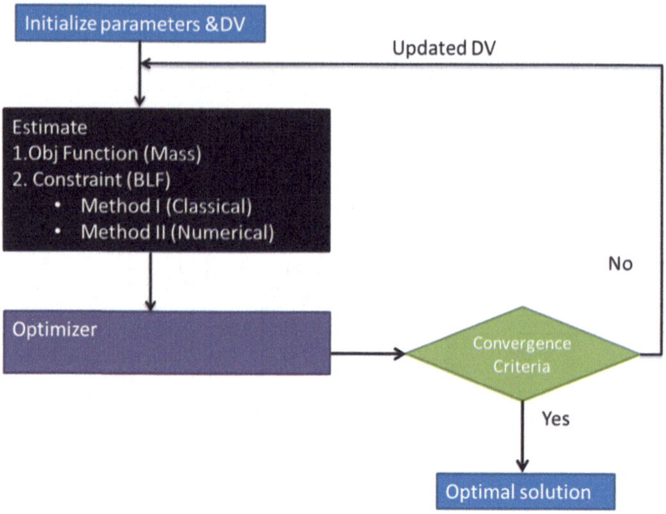

Fig. 2 Optimization workflow

Some of these modes may be closer leading to issues for the optimizer. Also, the design space is having multiple local minima which can lead to choosing sub-optimal solutions.

For carrying out the optimization, the mass of the structure can easily be estimated from geometry. The buckling load of cylinder has to be evaluated for the combination of design variables. In this paper, two approaches are considered for this: I—analytical method and II—numerical (FEM). A schematic of the optimization workflow with either of the methods is shown in Fig. 2. For illustrating the optimization methodology and results, a typical problem from [1] is considered. Here, a stiffened cylinder of $L = 500$ mm, $R = 500$ mm subjected to AF = 1000 kN, and Pr = 0.5 MPa is considered. In [1], the cylinder is constructed with aluminum alloy and it is stiffened with an orthogrid pattern with 7 rings and 45 stringers. The height of the stiffeners is 10 mm with a thickness of 3 mm, and the skin thickness is 2.1 mm. The optimization results are compared against this reference design.

3 Optimization Using Method I

The Method I is a low-fidelity method that enables a quick estimation of buckling loads. The buckling load of the stiffened cylinder is estimated by considering the following different modes of buckling. This method is similar to the approach followed for estimating the buckling load of the isogrid-stiffened cylinder given in the isogrid handbook [2].

Table 1 Range of design variables

Design variable	Lower bound	Upper bound
1. No. of rings (n_r)	No rings	16
2. No. of stringers (n_s)	4	120
3. Thickness of shell (t_{sh}) (mm)	1	8
4. Thickness of rings (t_r) (mm)	1.5	10
5. Thickness of stringers (t_s) (mm)	1.5	10
6. Height of stiffener ($h_r = h_s$) (mm)	5	50

- Global buckling.
- Skin pocket buckling.
- Stringer crippling.
- Rib crippling.

The buckling load is estimated by making the following simplifying assumptions:

- The buckling of each mode is independent (no interaction) and can be considered in isolation.
- The onset of buckling in the lowest mode is assumed critical and is the buckling load of the full structure.

These assumptions are typically considered in the preliminary design of the structure to arrive at the overall configuration of the grid. The buckling load of each mode is estimated using the formulas given in Appendix 1. The lowermost buckling load is considered the buckling mode of the structure.

3.1 Optimization

The optimization problem is solved using buckling load calculated using Method I. The range of design variables considered is given in Table 1. The optimizer used is a gradient-based optimizer. Due to the limitations of the optimizer considered, the number of rings and stiffeners is considered as continuous valued. This assumption is rectified in calculation using Method II.

3.2 Results

The optimization problem is solved with the initial conditions given in literature [1]. The convergence of objective function with iteration is shown in Fig. 3, and the optimal values are given in Table 2. It is observed the buckling load calculated using Method I shows the initial point as an infeasible point. The optimal Point A is

Fig. 3 Objective function convergence for Point A

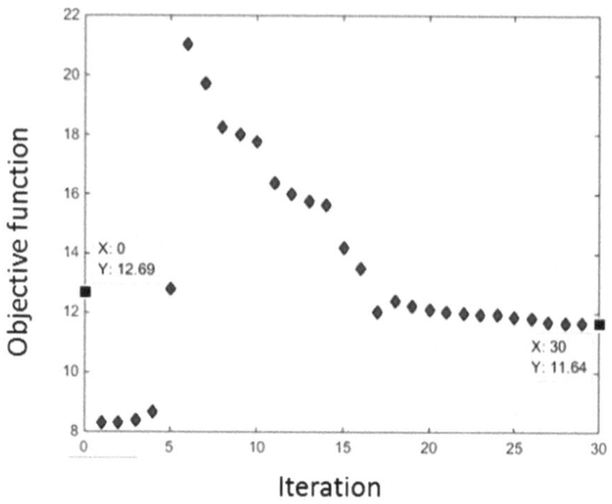

Table 2 Optimization results—Method I

DV/values	A		B		C	
	Initial	Optimal	Initial	Optimal	Initial	Optimal
1. No. of rings (n_r)	6	6	16	14	0	16
2. No. of stringers (n_s)	45	46	120	90	4	4
3. Thickness of shell (t_{sh}) (mm)	2.1	2.07	1	1.21	8	1.6
4. Thickness of rings (t_r) (mm)	3	1.66	1.5	1.52	10	1.5
5. Thickness of stringers (t_s) (mm)	3	1.73	1.5	1.53	10	6.33
6. Height of stiffener ($h_r = h_s$) (mm)	10	12.77	5	11.3	50	12.29
7. Mass (kg)	12.69	11.64	12.85	9.55	35.87	10.03

having a mass of 8.27% reduction. The optimization converged in 30 iterations, and the solution time is 3.5 s. Since the achieved optimum point can be a local minimum, the optimization problem is solved considering the initial point as lower and upper bound points to give optimal Points B and C, respectively. These results are given in Table 2. Of these points, B is giving the lowest mass which is 24.7% lesser than the reference design point.

3.3 *Validation of Optimal Solution*

The results of optimization which are calculated based on low-fidelity Method I are validated against high-fidelity simulation using FEM. The stiffened cylinder is modeled using shell elements. Two adjacent sections are modeled on either side to preclude any local mode due to load application or boundary condition. A brief

Table 3 Buckling load comparison

Case	Optimal A	Optimal B	Optimal C
BLF—Method I	1.005 (global), 1.02 (panel)	1.013 (global), 1.026 (panel)	1.05 (global), 1.05 (panel)
BLF FEM	0.89 (panel + stringer)	0.9 (global)	0.74 (panel + rings)
Buckling mode	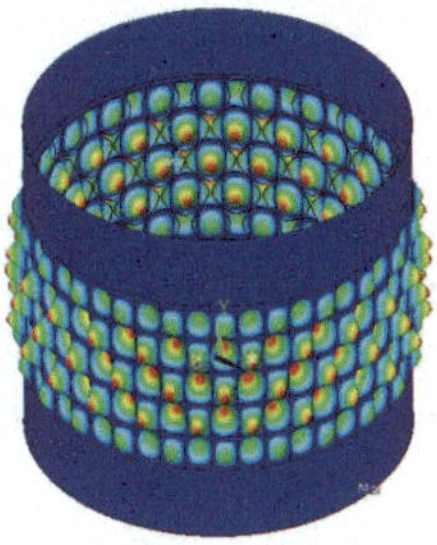	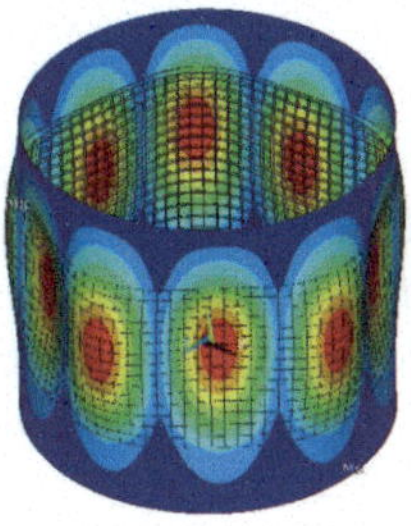	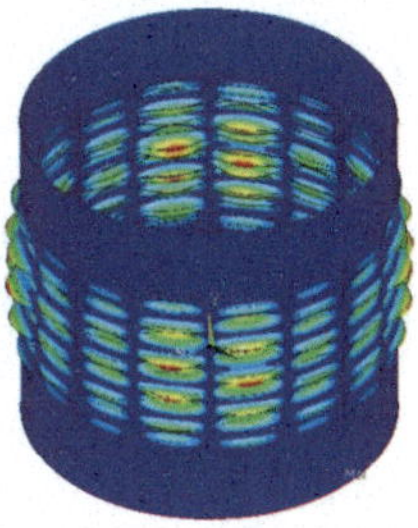

description of FE model is given in Appendix 2. The buckling load factor (BLF) and mode shape of Cases A, B, and C are shown in Table 3.

From this analysis, it is concluded that

- The optimal points calculated using Method I are having lower buckling load (constraint violation).
- This is due to the accuracy of prediction using the low-fidelity Method I.
- The predicted buckling load for problems considered having error as high as 26%.
- Hence, it is required to find the optimal design with high-fidelity simulations.

4 Optimization Using Method II

Method II is a high-fidelity method that uses finite element analysis for the accurate estimation of buckling loads. A parametric finite element model is built using the commercial software, ANSYS. The stiffened cylinder is modeled using four-noded shell elements (6 dof at each node). The ring and stiffener are modeled using shell elements. The buckling load is calculated using the eigen solver, and the lowest mode is considered as the buckling load of the structure. A brief description of the FE model is given in Appendix 2.

The parametric FE model is coupled with an optimizer (hyperstudy) to carry out the optimization studies. The range of design variables considered is the same as Table 1. Optimizer used is Adaptive Response Surface Method (ARSM) as it supports discrete variables.

The Adaptive Response Surface Method works by internally building response surfaces and adaptively updating them as new evaluations are available. The first response surface it builds is a linear regression polynomial, and then it finds the optimum on this surface and validates it with the exact simulation. If the response values from the response surface and the exact simulation are not close, ARSM updates the surface with the new evaluation and finds the optimum in this updated surface. ARSM repeats this loop until it meets the convergence criteria. An exploratory optimizer like genetic algorithm can also be considered.

4.1 Results

For the optimization studies, the identified optimal Points B and C (from Method I) are considered initial points. Point A is not considered as the performance of Points B and C is superior. The convergence of objective function, constraint, and six design variables with iteration for point B is shown in Fig. 4, and the optimal values are given in Table 4.

It is observed the optimal solution using Method II is having lesser mass than Method I while meeting the buckling load factor constraint. Of the points studied, Point B is having minimal mass with a 28.13% reduction over Ref. [1] design which implies that significant mass savings can be obtained by systematic optimization of stiffened cylinders. The solution time of the optimization problem is 1885 s with 28 iterations. This is significantly higher than the Method I optimization. The optimal points calculated are local optima. More initial points can be explored to find whether other lesser mass points are present. The results of optimization at Points B and C are given in Table 4.

5 Conclusion

The optimization of stiffened cylindrical shell under combined axial force and external pressure has been carried out using two methods, and the comparison of the performance of optimal solution by both methods is shown in Table 4. The following conclusions are drawn:

- Optimization by the low-fidelity Method I gives optimal points which are having lower buckling load than desired.
- Optimization by the high-fidelity Method II gives optimal points which meet the performance requirements.
- Both optimal points are having lower mass than the initial design point considered.
- A comparison of the computational time for both methods is shown in Fig. 5.
- The time taken by Method I is 3.5 s with 30 iterations during optimization cycle while Method II is 1885 s for 28 iterations.
- The low-fidelity Method I gives a significant time advantage over high-fidelity Method II.
- The error of Method I in calculating the buckling load is as high as 26% for the problems considered.

Fig. 4 Objective, constraint, and DV convergence for Point B—Method II

Table 4 Optimization results—Method II

DV/values	B		C	
	Initial	Optimal	Initial	Optimal
No. of rings (n_r)	14	13	16	10
No. of stringers (n_s)	90	65	4	4
Thickness of shell (t_{sh}) (mm)	1.21	1.07	1.6	1.8
Thickness of rings (t_r) (mm)	1.52	1.5	1.5	1.56
Thickness of stringers (t_s) (mm)	1.53	1.5	6.33	7.91
Height of stiffener ($h_r = h_s$) (mm)	11.3	13.25	12.29	12.93
Mass (kg)	9.55	9.12	10.03	10.63
BLF (FEM)	0.9	0.998	0.74	0.997

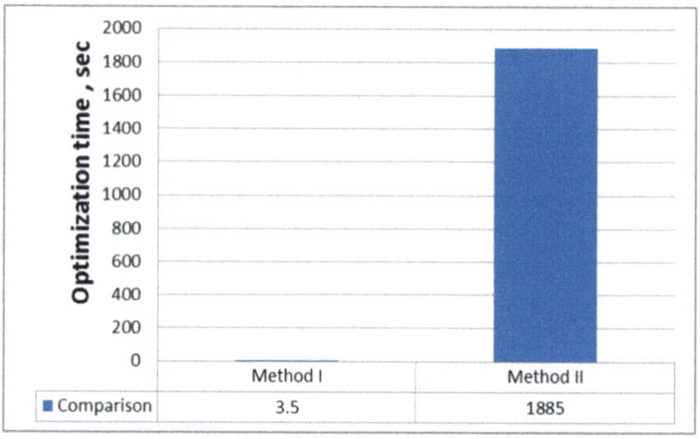

Fig. 5 Optimization time comparison for the two methods

- Hence, Method I can be considered for preliminary design cycles when rough estimates of the mass of structure are required.
- Also, it can be used to evaluate comparative advantage between different configurations.
- Method II is costly and can be used in the detailed design stage where more accuracy is required.

Appendix 1: Buckling Load Calculation—Method I

Mode	Buckling load	Remarks	Scheme of loading
Global buckling	$N_x = \left(\frac{L}{m\pi}\right)^2 \frac{\begin{vmatrix} A_{11} & A_{12} & A_{13} \\ A_{21} & A_{22} & A_{23} \\ A_{31} & A_{31} & A_{33} \end{vmatrix}}{\begin{vmatrix} A_{11} & A_{12} \\ A_{21} & A_{22} \end{vmatrix}},$ $p_{cr} = \frac{r}{n^2} \frac{\begin{vmatrix} A_{11} & A_{12} & A_{13} \\ A_{21} & A_{22} & A_{23} \\ A_{31} & A_{31} & A_{33} \end{vmatrix}}{\begin{vmatrix} A_{11} & A_{12} \\ A_{21} & A_{22} \end{vmatrix}}$	Equivalent orthotropic cylinder Refer [3] for parameters A_{ij}	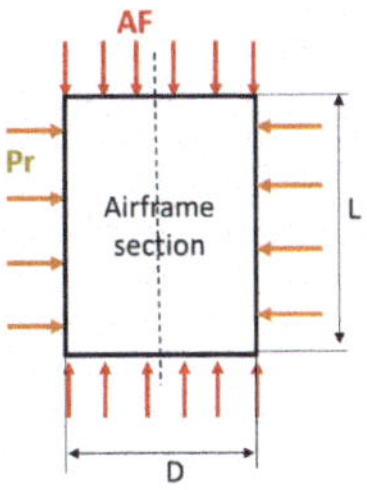
Skin pocket buckling	$\sigma_{xcr}\frac{m^2}{a^2} + \sigma_{ycr}\frac{n^2}{b^2} = D\pi^2\left(\frac{m^2}{a^2} + \frac{n^2}{b^2}\right)^2$	Plate with simply supported edges under biaxial load	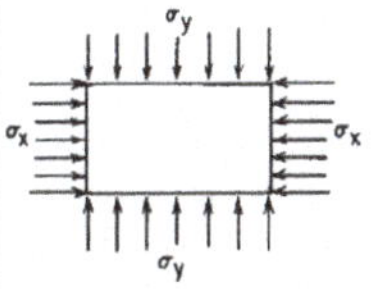
Stringer crippling Rib crippling	$\sigma_{xcr} = k \cdot \frac{D\pi^2}{tb^2}$ $k = \left(0.456 + \frac{b^2}{a^2}\right)$	Plate with three edges simply supported one edge free	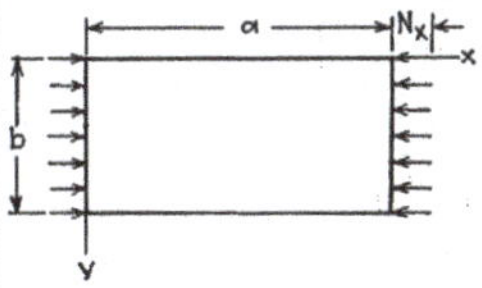

Appendix 2: Description of Parametric FE Model

- Parametric FE model is created in commercial software ANSYS.

- The stiffened cylinder is modeled using four-noded shell elements SHELL181 (6 dof at each node).
- The ring and stringer stiffeners are also modeled using shell elements.
- Dummy sections are modeled to simulate stiffness of adjacent sections.
- The axial force is applied on the top dummy.
- The external pressure is applied on the shell surface.
- The top and bottom ends are arrested in translational degree of freedom.
- The buckling load is calculated using linear buckling analysis, and the lowest mode is considered as the buckling load of structure.
- A code is generated in ANSYS Parametric Development Language (APDL) to generate the geometry and mesh, apply loads, and solve the linear buckling problem in batch mode.
- An image showing the FE model, loads, and boundary condition is shown below.

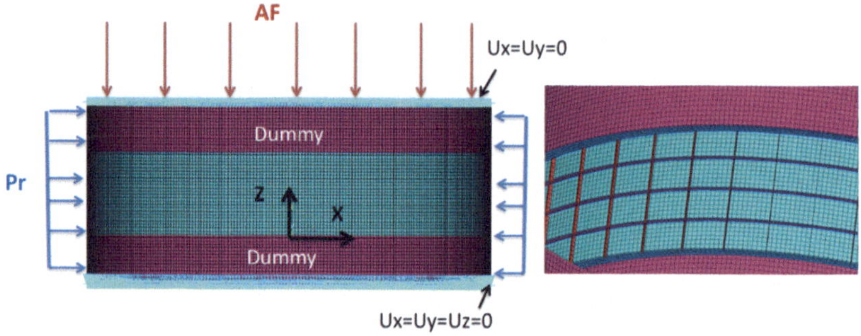

References

1. Wang B et al (2021) Post-buckling behavior of stiffened cylindrical shell and experimental validation under non-uniform external pressure and axial compression. Thin-Walled Struct 161:107481
2. NASA (1973) CR 124075—isogrid design handbook
3. Seide et al (1968) NASA SP 8007—buckling of thin-walled circular cylinders
4. Vescovini R et al (2015) Semi-analytical buckling analysis of omega stiffened panels under multi-axial loads. Compos Struct 120:285–299
5. Young et al. Roark's formulas for stress and strain. McGraw Hill Publication
6. Timoshenko S et al. Theory of elastic stability. Dover Publications
7. Altair. Hyperstudy user manual
8. ANSYS. Mechanical user manual

A Systematic Survey on Dynamic Analysis of Functionally Graded Rotor Systems

Arepalli Sri Rama Murty, Prabhakar Sathujoda, and Neelanchali Asija Bhalla

Abstract "Functionally graded materials" (FGMs) are becoming more popular owing to merits in their thermos-mechanical behaviour and performance. However, designing the systems made of FGM, considering the practical conditions such as thermal conditions, corrosive environments, manufacturing difficulties and manufacturing defects are still evolving. Present work is aimed at reviewing different design and modelling approaches suitable specifically to rotor systems made of "functionally graded" materials. Though research is picking up in this area, limited studies have been reported on FG rotors. A step-by-step concise information has been provided on every aspect of FGM related to rotor systems, which includes brief history of FGM, literature review of fabrication methods, applications of FGM, material modelling due to gradation of materials over the cross section of rotor, temperature variant material properties because of temperature distribution, analytical and finite element methods available to analyse rotor systems subjected dynamic loads, modelling and analysis of defects such as effect of corrosion, porosity and cracks in dynamic behaviour of rotor systems.

Keywords FGM · Functionally graded materials · Rotor system · Rotor-bearing shaft systems · Material gradation · Material modelling · Dynamic analysis

1 Introduction

It is a relentless pursuit for new materials, to improve performance of structures and systems, is a never-ending endeavour of engineering community. Composite materials are being used from ages. Commonly used composites are laminae of constituent materials. Though engineering community used them for advantage over homogenous materials, their performance in long run gets compromised due to the laminae detachment and localized stresses.

A. S. R. Murty (✉) · P. Sathujoda · N. A. Bhalla
Bennett University, Greater Noida, Uttar Pradesh 201310, India
e-mail: E21SOEP0016@bennett.edu.in

S. Sidhardh et al. (eds.), *Advances in Structural Integrity for Mechanical, Civil, and Aerospace Applications*, Lecture Notes in Mechanical Engineering,
https://doi.org/10.1007/978-981-97-6367-2_8

"Functionally graded materials" (FGMs) are class of composite materials made from multiple phases (materials) with a composition varying continuously along single or multiple directions. "Functionally graded materials" were coined in 1986. In 1987, the research plan of space plane in Japan "fundamental studies on the relaxation of Thermal stress by Tailoring Graded Structures" aimed at the capabilities of withstanding a surface temperature of 200 K and an inner surface temperature 100 K across a small thickness 10 mm plate were achieved by FGM [1]. FGMs are being used in several practical and challenging applications to withstand the thermal gradients, corrosive conditions along with the functional mechanical loads. Manufacturing of FGM materials, mechanics of FGM, thermal properties of FGM and different applications of FGM are available in recent international symposium [2]. Majority of the mechanical applications made of FGMs are in general composites generated by particles of constituent materials, where the content of material particles varies along pre-determined directions. Figure 1 pictorially shows the variation particles along radial direction. Particulate composites address the common problem of delamination in the composite materials. Infinitesimally FGMs are homogeneous, but macroscopically, they are heterogeneous because of volume fraction of different material particles is spatially varied.

Majority of the thermos-mechanical applications those adapt FMSs are made from ceramic and metal phases. Works of Reddy et al. assumed the variation of volume fraction of the ceramic particles along the depth/thickness [3]. A similar analogy can be made along radius in case of a circular shaft. A proper dynamic analysis of rotor systems made of FGMs depends mainly on two factors: one how well the spatial and thermal material gradation is modelled, and how well the defects such as corrosion, cracks, etc., are modelled. Current work aimed at providing systematic information to lead the research progression in the domain of rotor systems made of FGM. As

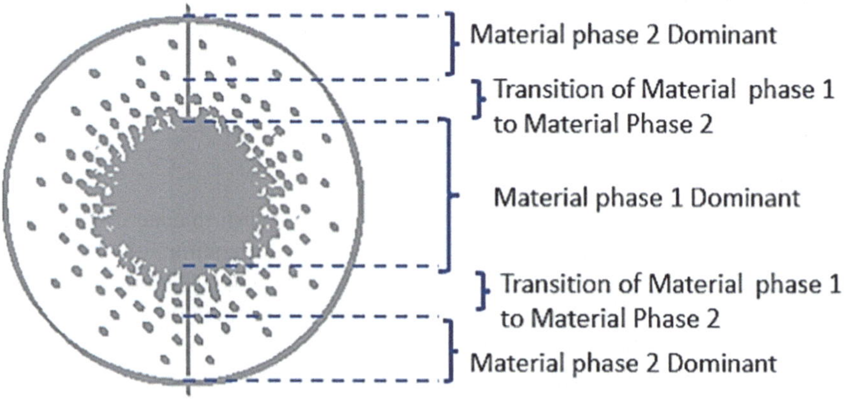

Fig. 1 Spatial varied volume fractions of material particles graded along the radius of a shaft cross section

part of information, current work briefly reviews the fabrication techniques used to manufacture FGM, different modelling techniques used for spatial gradation of materials and various modelling techniques used in thermal gradation and review of few dynamic analysis cases considering the defects like cracks and corrosion.

2 Fabrication Techniques of FGMs

FGM fabrication mainly has two steps:

1. Spatially varying structure gradation (spatially Inhomogeneous) and
2. Deposition of material on substrate.

Naebe et al. [4] reviewed and synthesized the techniques used in preparing FGM. Vastly there are three approaches to attain a material with desired gradation, namely gaseous, liquidous, and solid phase [5]. In gas-based methods, chemical or electrical sources are used to create vapours of required chemical constituents deposited on a substrate. The chemical vapour deposition (CVD) technique, thermal spray technique and surface reaction process are common fabrication methods in the gaseous category. In the liquid-based methods, the molten constituents' phases are deposited on substrate by centrifugal or by chemical reaction or by thermal sources. Both gaseous and liquidous methods are well-suited for coating type of applications. The gradation and constituents are chosen such that they produce specific applications like thermal variant, electrical conductivity variant, thermal conductivity variant, etc. Centrifugal casting, tape casting, combustion, gel casting, electrophoretic deposition method, chemical solution deposition (CSD) method, laser deposition method, directional solidification method and electrochemical gradation are the methods used in liquidous category. In the solid phase processes, constituent materials in the powdered form are disposed and sintered to attain a required geometric shape and the gradation. Spark plasma sintering (SPS) and power metallurgy technique are the common solid phase fabrication methods.

3 Fabrication Techniques of FGMs

The major difference between the regular composites and the FGM is the gradation. Regular composites are modelled by homogenization methods. Usually, the material properties are modelled as an equivalent homogeneous material property. Both thermal and mechanical properties are modelled as constants. Whereas FGM are graded materials, constituents vary along one direction or multiple directions (Fig. 2).

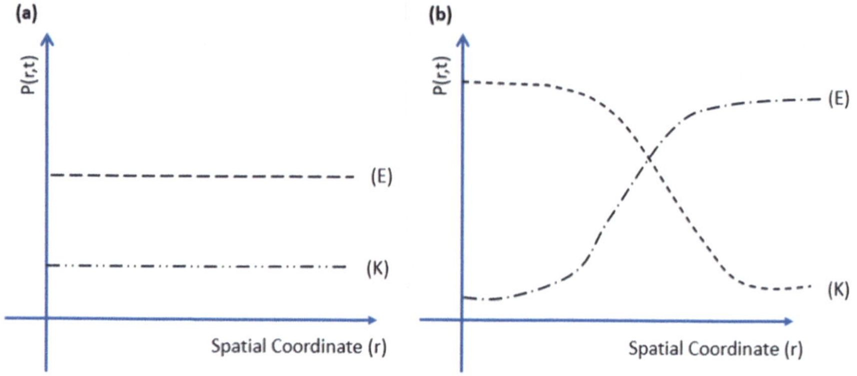

Fig. 2 Schematic representation of varying mechanical property (E) and thermal conductivity (K) along the radius of shaft in case of **a** non-FGM and **b** FGM

4 Temperature-Dependent Properties

Usually for thermo-mechanical applications such as rotary systems, FGMs are made from metals and ceramics. FGMs of such thermo-mechanical applications, are exposed to higher temperatures there by constituents of the FGMs may show up properties, dependent on temperature. Properties of constituents, like modulus of elasticity, density and Poisson's ratio are temperature sensitive. All temperature-dependent properties $P(T)$ are expressed as

$$P(T) = P_o\left(P_{-1}T^{-1} + 1 + P_1T^1 + P_2T^2 + P_3T^3\right), \quad (1)$$

where P_i (where $i = 0, -1, 1, 2, 3$) are coefficients in the curve fitting of the material curve. The material properties are expressed in this way so that the higher-order effects of the temperature on material properties can be readily observed [6]. The material properties are experimentally found and reported by Touloukian [7]. The constants mentioned in the above equation for various metals and ceramics are listed in Table 1 [6, 7].

5 Effective Material Properties

FGMs are not only dependent on temperature but also position-dependent. Gradation in the material is position-dependent. Spatial dependence can be represented by the "volumetric fractions of constituent phases". Properties of the constituents are sensitive to temperature. Effective properties can be expressed as function of both temperature (T) and position (r) as below:

Table 1 Mechanical and thermal properties of different ceramics and metals

		Zirconia	Aluminium oxide	Silicon nitride	Stainless steel
Thermal conductivity (W/mK)	P_0	1.7	– 14.087	13.723	15.379
	P_{-1}	0	– 1123.6	0	0
	P_1	1.276 E−4	0	– 1.032 E−3	– 1.264 E−3
	P_2	6.648 E−8	0	5.466 E−7	2.092 E−6
	P_3	0	0	– 7.876 E−11	– 7.223 E−10
Coeff. of thermal expansion (K)	P_0	12.766 E−6	6.8269 E−6	5.8723 E−6	12.330 E−6
	P_{-1}	0	0	0	0
	P_1	– 1.491 E−3	1.838 E−4	9.095 E−4	8.086 E−4
	P_2	1.006 E−5	0	0	0
	P_3	– 6.778 E−11	0	0	0
Poisson's ratio	P_0	0.2882	0.2600	0.2400	0.3262
	P_{-1}	0	0	0	0
	P_1	1.133 E−4	0	0	– 2.002 E−4
	P_2	0	0	0	3.797 E−7
	P_3	0	0	0	0
Specific heat (J/kg K)	P_0	487.34	1471.97	555.11	496.56
	P_{-1}	0	– 138.87	0	0
	P_1	3.049 E−4	– 2.241 E−5	1.016 E−3	– 1.151 E−3
	P_2	– 6.037 E−8	0	2.920 E−7	1.236 E−6
	P_3	0	0	– 1.670 E−10	– 5.863 E−10
Modulus of elasticity (Pa)	P_0	244.27 E9	349.55 E9	244.27 E9	201.04 E9
	P_{-1}	0	0	0	0
	P_1	– 1.371 E− 3	– 3.853 E−4	– 3.070 E−4	3.079 E−4
	P_2	1.214 E−6	4.027 E−7	2.160 E−7	– 6.534 E−7
	P_3	– 3.681 E−10	– 1.673 E−10	– 8.946 E−11	0

$$P(T, r) = P_{\mathrm{m}}(T) \cdot V_{\mathrm{m}}(r) + P_{\mathrm{c}}(T) \cdot (1 - V_{\mathrm{m}}(r)), \tag{2}$$

where properties P_{m} and P_{c} are temperature-dependent properties of metal (m) and ceramic (c). V_{m} is the volume fraction of the metal. Spatial distribution of the FGM can be modelled mathematically by various distributions that are reported in the literature.

6 Material Properties Based on Material Gradation

Power Law Distribution: Power law distribution is simplest form of distribution and are widely used in computing the material properties of FGM. Properties of plates and circular sections are utilized in the analysis of FG members [6, 8–10]. Material property based on power law, P_{pow} as a function of radius $P_{\mathrm{pow}}\ (r)$ can be simply represented in terms of volume fraction, V_{im} of constituent phases [8–10]

$$P_{\mathrm{pow}}(r) = P_{\mathrm{m}} \cdot V_{\mathrm{m}}(r) + P_{\mathrm{c}} \cdot (1 - V_{\mathrm{m}}(r)), \tag{3}$$

where as

$$V_{\mathrm{m}}(r) = \left(\frac{r - R_i}{R_{\mathrm{o}} - R_{\mathrm{i}}}\right)^k \text{ and } R_{\mathrm{i}} \leq r \leq R_{\mathrm{o}} \text{ and } 0 \leq k \leq \infty,$$

where 'k' is the index of power law. R_{o} is the outer radius of shaft, R_{i} is the inner radius of shaft, $V_{\mathrm{m}}(r)$ is the volume fraction of metal as a function of radius (refer Fig. 3) [6].

Exponential Distribution: Afsar and Go [11] provided way to compute mechanical properties by exponential distribution of constituent phases. As shown in Fig. 3, 'A' represents material near to inner radius of the disc, and 'B' represents the material at outer radius of disc. R_{i} represents the radius of inner wall, and R_{o} represents the radius of extreme wall.

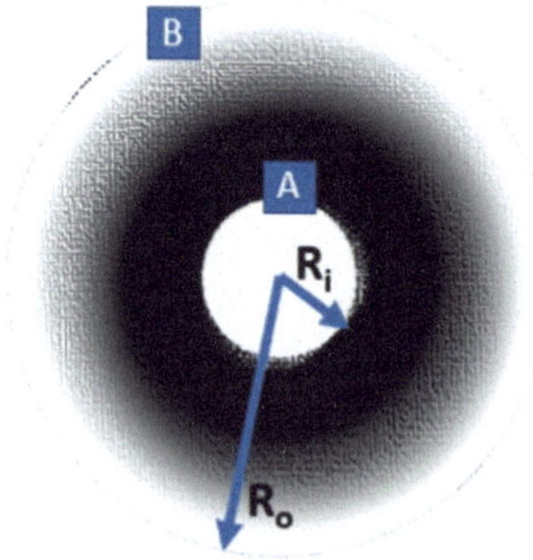

Fig. 3 Showing cross section of cylindrical shaft made of FGM and gradation material A to material B

Mechanical properties are mathematically modelled using an exponential variation. FG material properties mathematically expressed as exponential function. For instance, varying modulus of elasticity (E) along radius $\boldsymbol{r}$ can be written as

$$E(r) = E_0 e^{\beta r}, \tag{4}$$

where $E_0 = E_A e^{-\beta a}$ and $\beta = \frac{1}{R_o - R_i} \ln\left(\frac{E_A}{E_B}\right)$.

We can write the above property, P_{exp} in general, for any property such as modulus of elasticity, density, etc., expressed as follows:

$$P_{exp}(r) = P_m \, e^{\lambda(r - R_i)}, \tag{5}$$

$$\lambda = \frac{1}{R_i - R_o} \ln\left(\frac{P_m}{P_c}\right) \quad R_i \leq r \leq R_o.$$

Sigmoid Function Distribution: Sigmoid function is used in the research work to avoid stress concentration in functionally graded material specially when the material fabricated layer by layer. Mathematically, volume fractions are modelled by two power law functions to provide a smoother solution, which is also called as the sigmoid law gradation [12]. Sigmoid function based, spatial varying properties $P_{sig}(r)$, of a FGM shafts are expressed as

$$P_{sig}(r) = P_c + (P_m - P_c) \cdot \left(1 - \frac{r - R_i}{R_o - R_i}\right)^k, \tag{6}$$

where $\frac{R_i + R_o}{2} \leq r \leq R_o$.

7 Thermo-mechanical Properties of FGM

Nonlinear Temperature-Distribution (NLTD): Temperature distribution is one of the important factors to be considered in analysing the FGM. Nonlinear temperature distribution (NLTD) is a method to find out the temperature dependent materials by using power law. Heat conduction can be represented by Fourier Eq. (7), considered to represent the temperature variation in hallow cylinder [13].

$$\frac{d}{dr}\left(r \cdot K \cdot \frac{dT}{dr}\right) = 0. \tag{7}$$

Considering T_m temperature at outer radius R_o, T_c temperature at inner radius of cylinder R_i, thermal conductivity K at varying radius r, and assuming no heat generation BC. The solution for nonlinear temperature distribution $T(r)$ can be obtained by solving the above Fourier equation considering only first six terms of the polynomial series [14].

$$T(r) = T_m + (T_c - T_m) \cdot \left[\sum_{n=0}^{5} \left\{ \frac{(-1)^n}{n \cdot k + 1} (K_{cm})^n \left(\frac{r - R_i}{R_o - R_i} \right)^{n \cdot k + 1} \right\} \right] / \left[\sum_{n=0}^{5} \frac{(-1)^n}{n \cdot k + 1} (K_{cm})^n \right]. \quad (8)$$

$K_{cm} = (K_c - K_m)/K_m$, where K_c is heat conduction coefficient of outer ceramic and K_m is the heat conduction coefficient of inner metal.

Exponential-Temperature-Distribution (ETD): Based on mathematical exponential law temperature distribution can be derived using the thermal conductivity equation for cylinders with appropriate BC. Gayen et al. [15] proposed an equation for temperature distribution along the cross section of the shaft,

$$T(r) = A + Be^{-\lambda(r - R_i)}, \quad (9)$$

where $A = T_m - \frac{(T_c - T_m)}{e^{-\lambda(R_o - R_i)} - 1}$, $B = \frac{(T_c - T_m)}{e^{-\lambda(R_o - R_i)} - 1}$, $\lambda = \left(\frac{\ln\left(\frac{K_c}{K_m}\right)}{R_o - R_i} \right)$.

8 Dynamic Analysis of FGM Rotor System

Analysis of FGM structures like plates and beams with different loading and boundary conditions is being carried out from a long time academically and commercially. However, analysis of FGM rotors systems is not extensively done. There is a good traction in the last decade. Not many comparison studies are available to determine the dynamic properties of the FG rotor systems. A study by Rao et al. [16] found that FG rotors are more stable compared to conventional steel rotor systems. For the configuration they studied shows the first critical speed occurs at 3891 rpm for steel shaft when for FG shaft first critical speed occurs at 4536 rpm. Also they concluded FG rotors are more suitable for high speeds and high temperatures environments. Another study by Gayen et al. [17] found that for FG rotors that the internal viscous damping is more effective on the system stability than that of hysteretic damping. Also found natural frequencies significantly vary based on the gradation, in this case the index of power law.

Conventionally analysis of composite materials can be performed analytically, numerically, and experimentally. Few of the methods used to analyse the systems made from FGM. More focus is given to dynamic analysis of FGM rotor system. In the literature, there are several methods proposed. This report provides a few of the studies done by classical finite element methods, computational finite element methods, dynamic stiffness method and coupled field displacement method.

9 Classical Finite Element Methods

In this section, some light is provided on the analysis of FGM components with more focus on the rotary shaft systems. Gayen et al. [18] provided a brief history of the finite-element method (FEM) utilized to analyse the FGM rotor systems. Zorzi and Nelson [19] are earlier researchers developed FE simulation of rotor baring systems. They included internal damping also in the analysis. Substantial contributions are made by Nelson [20], in the FEM analysis of rotating shafts considering Gyroscopic effects, rotary-inertia and shear-deformation.

Gayen et al. [18] studied vibration analysis of the FGM rotor-bearing shaft system modelled as linear beam element based on Timoshenko beam theory. Power law-based spatial distribution of materials considered. Analysis is done on rotor-bearing system, flexible bearing modelled as spring and damper boundary conditions.

Derived the stress condition, Strain energy expression of FGM. FE model proposed can be used as a guiding tool for the vibration analysis of the rotor-bearing shaft systems. Major observations of the work are (a) material property distribution for various power law indices, (b) FG shaft systems are more stable than homogeneous steel shafts, (c) Power law gradient index shows the influence of vibration characteristics which were observed through Campbell diagram, (d) vibration responses were better with lower values of power law index and (e) hysteretic damping is less effective when compared to Internal viscous damping.

Prabhakar et al. [21] studied effect of corrosion on vibration behaviour of a rotor system made of FGM (stainless steel and Zirconia, ZrO_2). As corrosion leads to material loss and thereby alters component mass and stiffness. FE analysis based on Timoshenko beam theory is carried out on an FGM rotor-bearing system. Power law-based material gradation is considered and thermal gradient is modelled by nonlinear-temperature-distribution (NLTD). Key findings from the work are (a) if corrosion occurs near to the bearings, effect on natural frequencies is negligible, (b) fundamental frequencies are depended on depth of corrosion, and the distance of corrosion from bearings.

Prabhakar et al. [22] presented FE analysis of FG rotor system and reported the findings due to the presence of porosity, on natural and whirl frequencies. Material gradation and thermal distribution are considered like their previous study [21]. Porosity is expressed as a "volume fraction" of the material volume. Study provided an insight into the influences of porosity and thermal gradient on the whirl and natural frequencies. While calculating thermo-mechanical material properties, reduction of

Table 2 Natural frequencies (Hz) case of simply supported FG (SS-Al_2O_3) shaft and power law index 0.5

Mode	[23]	Gayen [24]	Difference (%)
First	255.99	255.92	0.027
Second	1003.5	1017.71	− 1.4

the property due to porosity is considered, equal to average of material property multiplied by the porosity percentage of volume fraction. Key findings from the work are (a) dimensionless natural frequencies (DNFs) fell with an increase in porosity percentage, (b) up to certain number higher power-law index numbers influenced the DNFs and (c) combined effect of thermal gradient and porosity influenced DNFs notably.

10 Computational Finite Element Methods

Very few studies are carried out on modelling and analysing FG rotor systems by 3D FE models Bose and Sathujoda [23] studied vibration characteristics of rotor system with multiple discs. Commercial software, ANSYS used for modelling and analysing the system. A customized Python code was used to prepare the modelling data such as gradient properties and the thermos-elastic material characteristics. Both power law and exponential-based material gradations are modelled. Two thermal distributions, NLTD and exponential thermal distribution (ETD) were considered. Effects of thermal gradient and material gradation on the natural frequencies are reported. Table 2 shows the good conformance of results between classical and computational finite element method.

11 Analytical Methods

In the literature functionally graded materials are analysed using various classical analytical methods. Wattanasakulpong et al. [25] studied the analysis of free vibration of FG beams by Taylor series based, differential transform method (DTM). Vibration and buckling analysis of FGM deep beams by "differential quadrature" method are studied by Sahree et al. [26]. Zheng et al. [27] performed "fracture analysis of FGM" by hybrid mesh-less displacement-discontinuity method (MDDM). Dynamic stiffness method (DSM) is gaining popularity in analysing FG components. Banerjee et al. [28] derived local and global dynamic stiffness matrix by DSM. Governing differential equation of motion based on FGM properties varies as a function of depth. A relational matrix of force amplitude and displacement at extreme ends of beam forms the dynamic stiffness matrix. Free vibration problem was finally solved by Wittrick–William's algorithm. Bharath and Prabhakar [29] performed vibration analysis rotor-bearing shaft system under the effect of temperature-gradient by DSM. Translational and rotational equilibrium conditions of elastic rotor shaft elements, rigid disc and the boundary conditions of bearing form the global dynamic stiffness matrix. Used Wittrick–William root searching algorithm to find "all" the natural and whirl frequencies.

12 Porosity and Cracks

In addition to the gradation, porosity also has influence on the dynamic characteristics of the FG rotor shaft systems. Usually, porosity is modelled as volume fraction, like that of material gradation. A few recent studies by Vaka et al. [30, 31] show that the natural whirl frequencies fall with an increase in the porosity volume. Steady State dynamic analysis reveals critical speeds decreases with increased porosity volume and temperature. Studies by Prabhakar et al. [32] also reveal that the natural frequencies and dynamic responses of rotors can be used to detect cracks efficiently. These frequencies responses are influenced by factors such as depth and location of cracks and temperature gradients. The same paper [32] has the way to model the slant cracks using Local Flexibility Matrix.

13 Conclusion

Literature survey done on evolution of FGM and applications provided insight into the research happening in this field. In the literature survey, more focus is given on the FG materials used as rotor-bearing systems. Although there is a lot of research happened on the FG materials, still a lot of research may have to extend the advantages FG to different fields. Future directions for applications related to the dynamic analysis of FG systems are

- There are limited experimental results available with respect to functionally graded rotor systems. More experimental results could benefit future analytical and numerical methods to analyse complex shapes, environmental conditions and configurations of rotor systems.
- Little work is available on implementing analytical techniques such as Dynamic Stiffness Method and Coupled Field Displacement, etc., for dealing with rotor shaft systems.
- There is ample scope to improve and enrich the model and analyse the FG rotor systems for complex failure mechanisms.

References

1. Niino M, Maeda S (1990) Recent development status of functionally gradient materials. ISIJ Int 30(9):699–703. https://doi.org/10.2355/isijinternational.30.699
2. Van der Biest O, Gasik M, Vleugels J (2004) Functionally graded materials VIII. In: Proceedings of the eighth international symposium on multifunctional and functionally graded materials. Mater Sci Forum 492–493. Trans Tech Publications Ltd., Uetikon-Zurich, Switzerland
3. Reddy JN (2000) Analysis of functionally graded plates. Int J Numer Meth Eng 47(1–3):663–684. https://doi.org/10.1002/(sici)1097-0207(20000110/30)47:1/3%3c663::aid-nme787%3e3.0.co;2-8
4. Naebe M, Shirvanimoghaddam K (2016) Functionally graded materials: a review of fabrication and properties. Appl Mater Today 5:223–245. https://doi.org/10.1016/j.apmt.2016.10.001
5. Sasaki M, Hirai T (1991) Fabrication and properties of functionally gradient materials. J Ceram Soc Jpn 99(1154):1002–1013. https://doi.org/10.2109/jcersj.99.1002
6. Reddy JN, Chin CD (1998) Thermomechanical analysis of functionally graded cylinders and plates. J Therm Stresses 21(6):593–626. https://doi.org/10.1080/01495739808956165
7. Touloukian YS (ed) (1967) Thermophysical properties of high temperature solid materials, vol 3. Macmillan, New York
8. Zuiker JR (1995) Functionally graded materials: choice of micromechanics model and limitations in property variation. Compos Eng 5(7):807–819. https://doi.org/10.1016/0961-9526(95)00031-H
9. Reddy JN (1998) Thermomechanical behavior of functionally graded materials. Texas A and M University, College Station, Department of Mechanical Engineering
10. Reddy JN, Chin CD (1998) Thermomechanical analysis of functionally graded cylinders and plates. J Therm Stresses 21(6):593–626. https://doi.org/10.1080/01495739808956165
11. Afsar AM, Go J (2010) Finite element analysis of thermoelastic field in a rotating FGM circular disk. Appl Math Model 34(11):3309–3320. https://doi.org/10.1016/j.apm.2010.02.022
12. Chung YL, Chi SH (2001) The residual stress of functionally graded materials. J Chin Inst Civ Hydraul Eng 13(2001):1–9
13. Sathujoda P, Batchu A, Obalareddy B, Canale G, Maligno A, Citarella R (2020) Free vibration analysis of a thermally loaded porous functionally graded rotor-bearing system. Appl Sci 10(22):8197. https://doi.org/10.3390/app10228197
14. Lanhe W (2004) Thermal buckling of a simply supported moderately thick rectangular FGM plate. Compos Struct 64(2):211–218. https://doi.org/10.1016/j.compstruct.2003.08.004
15. Gayen D, Tiwari R, Chakraborty D (2019) Static and dynamic analyses of cracked functionally graded structural components: a review. Compos B Eng 173:106982. https://doi.org/10.1016/j.compositesb.2019.106982
16. Rao DK; Roy T (2016) Vibration analysis of functionally graded rotating shaft system. Procedia Eng 144:775–780. https://doi.org/10.1016/j.proeng.2016.05.084
17. Gayen D, Roy T (2014) Finite element based vibration analysis of functionally graded spinning shaft system. Proc Inst Mech Eng C J Mech Eng Sci 228(18):3306–3321
18. Gayen D, Roy T (2014) Finite element based vibration analysis of functionally graded spinning shaft system. Proc Inst Mech Eng C J Mech Eng Sci 228(18):3306–3321. https://doi.org/10.1177/0954406214527923
19. Zorzi ES, Nelson HD (1977) Finite element simulation of rotor-bearing systems with internal damping. https://doi.org/10.1115/1.3446254
20. Nelson HD (1980) A finite rotating shaft element using Timoshenko beam theory. ASME J Mech Des 102(4):793–803. https://doi.org/10.1115/1.3254824
21. Sathujoda P, Obalareddy B, Batchu A, Canale G, Maligno A, Citarella R (2020) Effect of corrosion on the natural and whirl frequencies of a functionally graded rotor-bearing system subjected to thermal gradients. Materials 13(20):4546. https://doi.org/10.3390/ma13204546
22. Sathujoda P, Batchu A, Obalareddy B, Canale G, Maligno A, Citarella R (2020) Free vibration analysis of a thermally loaded porous functionally graded rotor-bearing system. Appl Sci 10(22):8197. https://doi.org/10.3390/app10228197

23. Bose A, Sathujoda P (2020) Effect of thermal gradient on vibration characteristics of a functionally graded shaft system. Math Model Eng Probl 7(2):212–222. https://doi.org/10.18280/mmep.070207
24. Gayen D, Chakraborty D, Tiwari R (2017) Whirl frequencies and critical speeds of a rotor-bearing system with a cracked functionally graded shaft—finite element analysis. Eur J Mech A/Solids 61:47–58. https://doi.org/10.1016/j.euromechsol.2016.09.003
25. Wattanasakulpong N, Ungbhakorn V (2012) Free vibration analysis of functionally graded beams with general elastically end constraints by DTM. World J Mech 2(06):297. https://doi.org/10.4236/wjm.2012.26036
26. Sahraee S, Saidi AR (2009) Free vibration and buckling analysis of functionally graded deep beam-columns on two-parameter elastic foundations using the differential quadrature method. Proc Inst Mech Eng C J Mech Eng Sci 223(6):1273–1284. https://doi.org/10.1243/09544062JMES1349
27. Zheng H, Sladek J, Sladek V, Wang SK, Wen PH (2021) Fracture analysis of functionally graded material by hybrid meshless displacement discontinuity method. Eng Fract Mech 247:107591. https://doi.org/10.1016/j.engfracmech.2021.107591
28. Banerjee JR, Ananthapuvirajah A (2018) Free vibration of functionally graded beams and frameworks using the dynamic stiffness method. J Sound Vib 422:34–47. https://doi.org/10.1016/j.jsv.2018.02.010
29. Obalareddy B, Sathujoda P, Citarella R (2022) Dynamic stiffness matrix approach to free vibration analysis of functionally graded rotor bearing system subjected to thermal gradients. Materials 15(4):1540. https://doi.org/10.3390/ma15041540
30. Vaka V, Sathujoda P, Bhalla NA, Kumar YS (2022) Steady-state and transient vibration analysis of a thermally loaded power law-based functionally graded rotor system with induced porosities. Forces Mech 9:100123
31. Vaka V, Sathujoda P, Bhalla NA, Satish Kumar YV, Citarella R (2022) A study on dynamic behaviour of thermally distributed exponentially graded rotor system with induced porosities. Math Probl Eng 2022
32. Sathujoda P, Batchu A, Canale G, Citarella R (2022) Steady state and transient vibration analysis of an exponentially graded rotor bearing system having a slant crack. Appl Sci (Switzerland) 12(14). https://doi.org/10.3390/app12146900
33. Khan T, Zhang N, Akram A (2019) State of the art review of functionally graded materials. In: 2019 2nd international conference on computing, mathematics and engineering technologies (iCoMET), Jan 2019. IEEE, pp 1–9. https://doi.org/10.1109/ICOMET.2019.8673489
34. Dai HL, Rao YN (2014) Nonlinear dynamic behavior of a long temperature dependent FGM hollow cylinder subjected to thermal shocking. Sci Eng Compos Mater 21(2):267–280. https://doi.org/10.1515/secm-2012-0072
35. Shen HS, Wang ZX (2012) Assessment of Voigt and Mori–Tanaka models for vibration analysis of functionally graded plates. Compos Struct 94(7):2197–2208. https://doi.org/10.1016/j.compstruct.2012.02.018

Mass Optimization of Bracket

Kavadi Ravi Teja, R. Santhosh, and P. C. Jain

Abstract Brackets, crucial components in aerospace structures, serve the dual purpose of load transfer and mounting packages onto airframe shells. Achieving the delicate balance between minimizing mass and meeting stringent strength and stiffness requirements is a perennial design challenge. This paper details the optimization of brackets, employing solid modeling software for intricate design and ANSYS packages for rigorous analysis and optimization. The process integrates finite element analysis to assess structural integrity, topology optimization to refine material distribution, and parametric optimization to fine-tune geometric parameters. The resulting design achieves optimal efficiency within the constraints of confined space and attachment requirements dictated by various subsystems. Beyond optimization, the design undergoes thorough manufacturability scrutiny, considering material properties and fabrication processes. The successful realization of a prototype hardware validates the proposed methodology, showcasing a seamless transition from conceptualization to a technically robust and functional bracket design.

Keywords Optimum design · Topology optimization · Sizing optimization · Aerospace bracket

1 Introduction

Attachment brackets, integral components within aerospace structures, play a pivotal role in maintaining structural integrity and supporting various aerospace elements. Achieving a meticulous balance between strength, weight efficiency, and cost-effectiveness is imperative in their design. The potential consequences of bracket failure underscore the critical importance of precise design, rigorous testing, and meticulous manufacturing processes to ensure the success of aerospace missions and the safety of stakeholders. Traditionally guided by designer intuition, the aerospace

K. R. Teja (✉) · R. Santhosh · P. C. Jain
Directorate of Flight Structures, DRDL, DRDO, Kanchanbagh, Hyderabad 500058, India
e-mail: krteja.drdl@gov.in

S. Sidhardh et al. (eds.), *Advances in Structural Integrity for Mechanical, Civil, and Aerospace Applications*, Lecture Notes in Mechanical Engineering,
https://doi.org/10.1007/978-981-97-6367-2_9

component design, especially attachment brackets, is undergoing a transformative shift to meet the evolving demands of aerospace engineering. This paper delves into the nuanced considerations of attachment bracket design, emphasizing key factors such as strength, stiffness, material selection, manufacturing tolerances, and proper installation. These considerations are pivotal for ensuring the overall structural resilience and dependability of aerospace vehicles.

Departing from conventional practices, this study focuses on designing a bracket housing a Reaction Control System (RCS) for aerospace maneuvering. The outlined methodology, driven by commercially available tools, achieves a notable 49.7% reduction in mass compared to conventional design practices, signaling a paradigm shift in aerospace bracket design. This underscores the irreplaceable role of optimized parameters and technical ingenuity in ensuring the success and safety of aerospace missions.

In the pursuit of advancing aerospace bracket design, the study embraces a comprehensive optimization approach, moving away from reliance on intuition in design. The systematic methodology incorporates cutting-edge techniques, commencing with topology optimization—a transformative process aimed at determining an optimal material layout and maximizing strength and stiffness while minimizing weight. Building upon insights gained from topology optimization, the cross section of bracket members is fixed, and a further optimization strategy is employed for sizing bracket elements. This phase considers objectives such as minimum mass with constraints, including strength and stiffness, ensuring a holistic approach to bracket design. The refinement process culminates in a thorough finite element analysis (FEA) to assess the structural integrity and performance of the optimized design. FEA, a powerful computational tool, simulates and analyzes the bracket's behavior under diverse loading conditions, providing invaluable insights for further design enhancement. Theoretical advancements achieved through optimization are translated into tangible results with the fabrication and testing of hardware, validating theoretical models against real-world performance. This pragmatic approach affirms the efficacy of the optimized design in practical aerospace applications.

In essence, this study represents a departure from conventional design practices, introducing a paradigm shift with the incorporation of topology optimization, multi-objective optimization, and finite element analysis. This ushers in a new era of precision and efficiency in aerospace bracket design, marking a significant milestone in the evolution of aerospace engineering methodologies.

2 Literature Review

The exploration of topology optimization by researchers, as discussed by Bendsoe and Sigmund [1], has been pivotal in achieving optimal layouts for structural components across diverse applications. Their work introduces various topology optimization theories, with a notable focus on the Solid Isotropic Material with Penalization

(SIMP) approach. The application of SIMP, emphasizing compliance minimization, is particularly well-suited for the initial stiffness-based design of aerospace brackets.

In the domain of additive manufacturing, the study by Barbieri et al. [2] showcases the utilization of topology optimization techniques to elevate the performance of an engine piston. The objective here is to replace conventional aluminum pistons, known for their low mechanical properties and production through casting or forging, with steel pistons manufactured through additive processes. The primary optimization goal is mass reduction, with stiffness constraints applied to critical components.

Topology optimization extends its reach to the design of pipeline brackets, as evidenced by Samson Raj [3]. Raj's work on mass reduction through topology optimization presents three optimized designs, comparing results with unmodified and modified bracket designs. The findings reveal a substantial 65% reduction in bracket mass without compromising structural stability, underscoring the effectiveness of the optimization process.

Bodog et al. [4] utilizing ANSYS, contribute to the literature by optimizing plate structures without sacrificing functionality. Their approach integrates topology optimization with parametric validation and response surface techniques, ensuring the structural integrity of the optimized plate structures.

Exploring dimensional sensitivity and optimizing natural frequency, Xiangjie et al. [5] employ ANSYS Workbench. Their study, centered on columns, identifies parameters with a significant impact on natural frequency. Through optimization, they achieve a noteworthy increase in natural frequency, signaling advancements in structural performance.

In a study by Abhinav et al. [6], a bracket undergoes optimization for a 40% reduction in mass, utilizing structural steel. The resulting weight reduction leads to a substantial decrease in von-Mises stress and deformation, reported to be nearly 33%. This study underscores the practical implications of mass optimization in enhancing the mechanical performance of aerospace brackets.

This literature review provides insights into the diverse applications and methodologies employed in topology optimization across various engineering contexts. It underscores the significance of topology optimization in achieving efficient and high-performance structural designs, aligning with the transformative shift outlined in the introduction, and marking a significant milestone in the evolution of aerospace engineering methodologies.

3 Problem Statement

The design imperatives for the bracket are outlined with precision to establish clear design requirements:

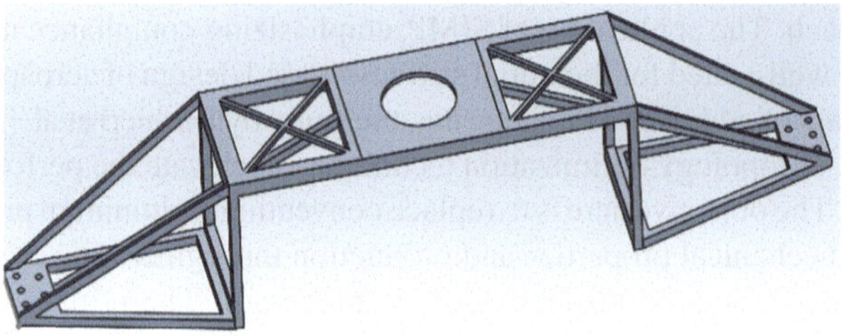

Fig. 1 Existing design of thruster bracket

(a) Structural performance: The bracket is mandated to exhibit adequate stiffness, with deformation not exceeding 5 mm under operational conditions. (b) Material safety margin: The Factor of Safety (FOS) concerning the yield strength of the material must surpass a value of 2, ensuring a robust safety margin. (c) Buckling load factor: The buckling load factor (BLF) should exceed 1.5, incorporating a knockdown factor of 0.6 to guarantee stability against buckling phenomena. (d) Spatial constraints: The bracket configuration must conform to the available space envelope, avoiding interference with adjacent subsystems. The design should ensure compatibility and coexistence within the designated spatial constraints. (e) Fabrication considerations: Facilitating ease of fabrication is paramount, emphasizing a design that is not only structurally optimized but also practical and efficient in manufacturing processes.

The current configuration of the existing thruster bracket, adhering to these design requirements and possessing a mass of 2.45 kg, is depicted in Fig. 1. The challenge at hand involves the redesign of this thruster bracket to achieve further optimization in mass. Notably, the spatial envelope is delimited by attachment locations and neighboring packages, resulting in a design characterized by legs and a central bottom that is hollow. To address this, a numerically advanced methodology is proposed for the design of the new bracket. This approach aims to leverage numerical simulations and optimization techniques to iteratively refine the bracket design, ultimately achieving an optimum mass while upholding the specified design requirements within the constrained spatial envelope.

4 Proposed Methodology

The redesign of the bracket is outlined through a structured methodology, encompassing three pivotal stages to achieve minimum mass:

Step 1. Topology Optimization:

- The initial step involves employing topology optimization with the Solid Isotropic Material with Penalization (SIMP) approach. This methodology is geared toward determining the material layout and design configuration of the bracket in alignment with the available space and attachment requirements.

- The material layout configuration, derived from topology optimization, represents the initial stiff design configuration within the feasible design space.

Step 2. Sizing Optimization:

- The stiffer design configuration obtained undergoes sizing optimization to meet specific design requirements, including allowable deflection and stress. Parametric techniques are applied to bracket dimensions as design variables.
- Constrained sizing optimization is executed with the primary objective of achieving minimum mass while adhering to the predefined design specifications.

Step 3. Detailed Design and Validation:

- Following sizing optimization, the stiffer design configuration is further optimized for mass. The resulting design variables represent the optimum dimensions for the bracket, ensuring minimum mass and satisfying specified design requirements.
- The optimized bracket design undergoes thorough scrutiny for manufacturability and integration requirements. Detailed design is conducted, incorporating minor modifications to enhance manufacturability and integration considerations.
- The final phase involves Finite Element Analysis (FEA) to rigorously validate the redesigned bracket's structural integrity and performance. This comprehensive analysis ensures alignment with the specified design requirements.

This proposed methodology adopts a systematic and iterative approach, initiating with topology optimization to establish an efficient material layout and design configuration. Subsequent steps progressively refine and optimize the bracket's dimensions, culminating in a detailed design that is thoroughly validated through FEA. This methodological framework aims to deliver a final bracket design that not only meets stringent performance criteria but also aligns with manufacturability and integration considerations.

5 Topology Optimization

It is the method to find out optimal material distribution within predefined design domain. There are many approaches to solve topology optimization problem; among them, ANSYS uses Solid Isotropic Material with Penalization (SIMP) where material properties are constant and fictitious element density for each element is a design variable. These artificial material densities of element (ρ_e) have continuous values in the range of $0 < \rho_{min} \leq \rho_e \leq 1$. To avoid singularity of the solution, ρ_{min} has to be specified by the user. This means that the volume of an element is multiplied by artificial density to produce its actual volume; if there are n number of nodes, then the volume of design domain is given by

$$V = \sum_{e=1}^{n} \rho_e \times v_e.$$

From the resultant volume, mass of the object is calculated using actual material density.

In SIMP, modulus of elasticity is modeled as the relative densities power P multiplied by original material modulus of elasticity.

$$E_e = \left(\rho_e^p\right)E.$$

P is the penalty factor, which converts the problem to discrete value problem. User has to specify value of P in ANSYS.

The method of minimizing compliance has been shown to be the most effective for increasing stiffness. It is not only easy to implement but also computationally efficient compared to other approaches. Minimizing compliance is maximizing stiffness, as we also have stiffness-based design requirement for bracket, we have chosen minimization of compliance as objective function for topology optimization. Formulation used by ANSYS is follows:

$$\text{Objective: Min } C(\rho_e) = U^T K U = \sum_{e=1}^{n} \left(\rho_e^P\right)(u_e)^T k(u_e)$$

$$\text{Subjected to: } \sum_{e=1}^{n} \rho_e \times v_e \le V_s$$

$$: KU = F$$

$$: 0 < \rho_{\min} \le \rho_e \le 1,$$

where

U Global displacement.
K Global stiffness.
F Force vector.
u_e Elemental displacement.
v_e Elemental volume.
V_s Input volume constraint.
n Total number of FE elements used to represent the design domain.

Steps involved in topology optimization are (a) defining design domain, (b) finite element analysis (FEA) at every iteration, (c) sensitivity analysis at every iteration, (d) design update with optimization routine, and (e) convergence, which results in optimized layout of design space.

Fig. 2 Geometric design domain

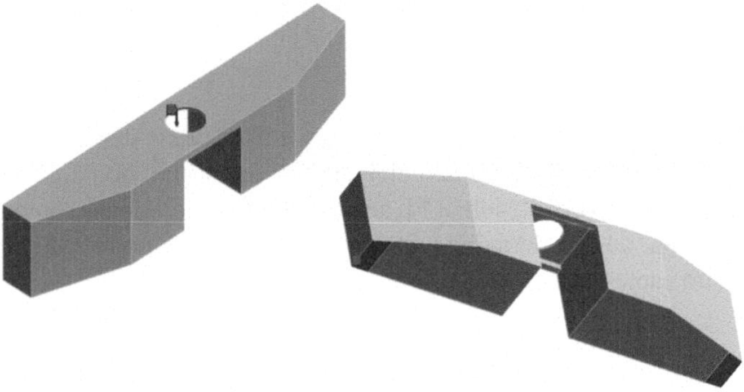

Fig. 3 Showing constraints and load face in design domain

5.1 Defining Design Domain

Solid block is created as design domain as shown in Fig. 2. This block is created keeping the space constraint. Tapering on this block is due to engineering mounting constraint and space constraint. Rectangular slot below design space is because space for other component and circular hole is provided to fit the thruster in bracket.

The faces attaching to other components with bolts are constrained in design domain and forces acting due to thrusters are applied downwards on circular hole as shown in Fig. 3.

Initially, aluminum with $E = 71{,}000$ MPa is considered.

Input specified to ANSYS is response constraint of allowable volume, $V_s = 5\%$.

Minimum element density, $\rho_{\min} = 0.001$, and penalization $P = 3$.

Initial mass and volume of the design domain are 44.484 kg and 1.6059e + 007 mm^3, respectively.

Fig. 4 Result of FE analysis on initial design domain

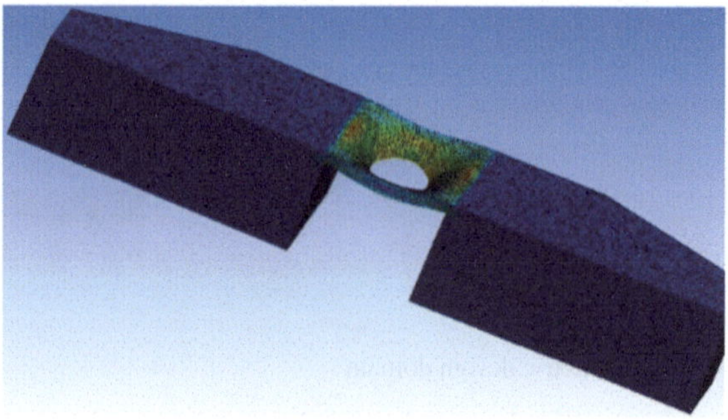

5.2 FEA

FE analysis is carried out on this design space for every iteration using solid elements. Initial number of nodes and elements of the design domain is 1,530,495 and 1,109,664, respectively. Result of FE analysis for first iteration is shown in Fig. 4. For every iteration, design domain is updated, and for every update, FE analysis is carried out with same boundary conditions and forces acting.

5.3 Sensitivity Analysis

Sensitivity analysis is to check which elemental densities are more sensitive to compliance. These element densities are updated accordingly to minimize compliance. Gradient-based method is used to know which element should be increased in material density and which element densities have to be decreased. Optimality criteria method is specified to ANSYS for updating design domain by changing the design variables, i.e., element densities (ρ_e). Update move size, and other coefficients for updating elemental densities are already predefined in ANSYS.

5.4 Convergence

Initially, FE analysis on design space is carried out. Followed by sensitivity analysis by updating design variable (ρ_e) and material properties results in new design domain. Again, FE analysis on new design happens and the loop continues. Convergence of this loop results in optimized FE elements' layout of design space. Convergence accuracy specified to ANSYS is 0.1%. Convergence plot with respect to every iteration is shown in Fig. 5.

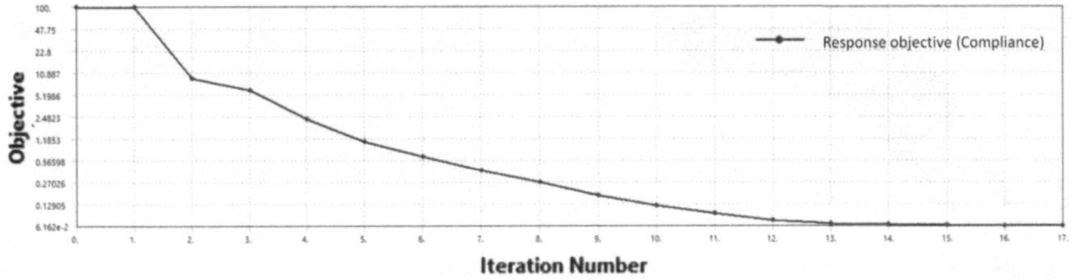

Fig. 5 Showing convergence with respect to number of iterations in topology optimization

Fig. 6 Topology-optimized structure

5.5 *Results of Topology Optimization*

With topology optimization, 5% of volume is retained from the block, with the layout shown in Fig. 6. The structure obtained from ANSYS topology has to be modified further for sizing optimization.

6 Sizing Optimization

Optimum structures are usually stress critical or deformation critical designs. In this paper, it is considered addressing deformation criticality first followed by stress criticality. Topology optimization using compliance minimization gives stiffest design layout possible in the given design domain. By this method, minimum possible deformation layout is achieved, which can be possible in the given design domain. Next, the design layout must be sized appropriately to meet all the design requirements. Steps followed in sizing optimization follows.

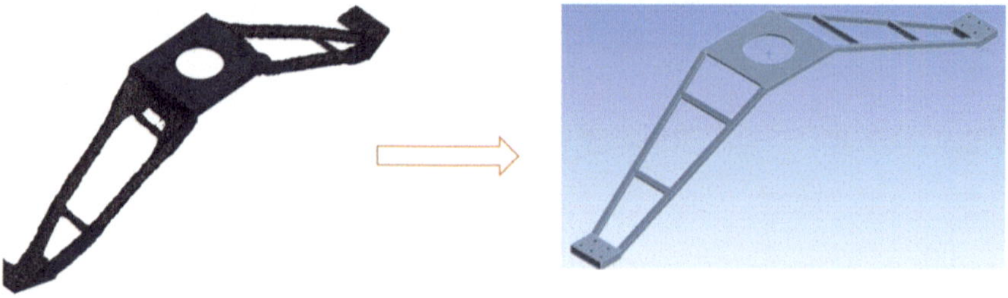

Fig. 7 Showing structure synthesizing

6.1 Layout Synthesizing

The structure obtained from ANSYS topology optimization is coerced surfaced, and directly assigning parameters to this structure is difficult for parametric sizing optimization. Output layout from topology optimization is synthesized in solid modeling software to make input to parametric optimization by assigning variable dimensions to the structure and optimized for mass with design-specified constraint. The structure synthesized is shown in Fig. 7.

6.2 Assigning Design Variable

Structure is idealized like two beams with width and height running continuously between the bolted joints. Three design variables one height (H) and two widths ($V1$ and $V2$) are assigned to the sizing optimization model as shown in Fig. 8. Variable assignment is done during modeling in SolidWorks, and parameters are defined in ANSYS (Fig. 9).

6.3 Mathematical Formulation of Sizing Optimization

Bracket is optimized for minimum mass with constraints as stress and deflection. Formulation of the problem is shown below.

$$\begin{aligned}
\text{Objective: } & \text{Min(Mass)} = f(H, V1, V2). \\
\text{Subject to: } & \sigma(\max) \leq \sigma(\text{Allowable}) \\
: & \ \delta(\max) \leq \text{Allowable deformation.} \\
\text{Design Variables: } & \text{Dimensions}(H, V1, V2).
\end{aligned}$$

Response surface is created with generating data points with multiple combinations of parameters (H, $V1$, and $V2$) and optimized for minimum mass. Output of sizing optimization is optimum dimensions of design variables H, $V1$, $V2$. Mixed Integer Sequential Quadratic Programming (MISQP) is used for sizing optimization. MISQP solves nonlinear programming problems by a modified sequential quadratic programming (SQP) method. This code is also applicable to non-convex optimization problem; thus, it is selected for this requirement.

For stress constraint, three global stress concentration zones are probed and defined as constrained variables shown in Fig. 10. This avoids ending of optimization solution prematurely due to stress concentration caused by FEM.

Allowable stress in structure is given as 180 MPa (UTS of aluminum). Allowable deflection is 5 mm; this value is based on design requirement. Bounds for H, $V1$, and $V2$ are considered as 7–15 mm each. These bounds are based on intuitive feeling. Convergence percentage is specified as 0.0001%, and constraint tolerance is 0.001.

6.4 Results of Sixing Optimization

Optimum values of H, $V1$, and $V2$ are obtained for allowable stress and deflection. Mass obtained after this step is 1.0029 kg. Solution points are shown in Table 1. Mass variation with respect to number of iterations is shown in Fig. 11. Deformation variation and stress variations with respect to number of iterations are shown in Fig. 12.

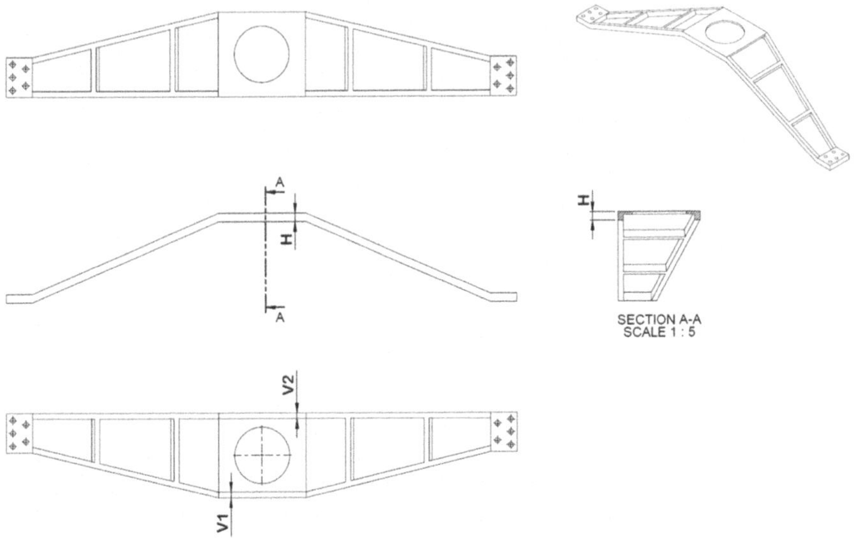

Fig. 8 Drawing showing assigning design variables

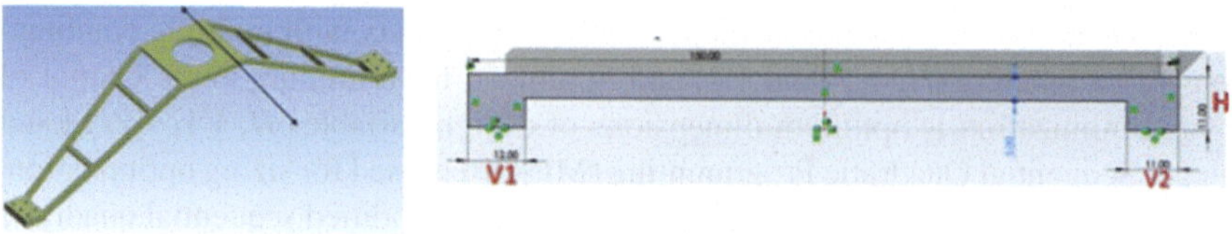

Fig. 9 Showing design variables in cross section view with respect to line shown in left figure

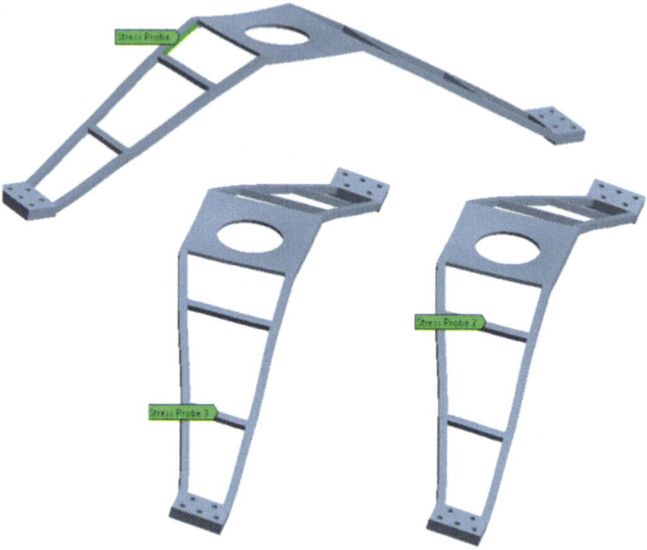

Fig. 10 Stress points for getting stress for solving constraint

Table 1 Optimal points obtained from sizing optimization

	Resultant points
H (mm)	9.0673
$V1$ (mm)	10.758
$V2$ (mm)	10.953

7 Detail Design

With optimum layout from topology optimization and optimum dimensions from sizing optimization, detail design of thruster bracket is carried out taking manufacturing constrain and integration requirements. The structure is made by welding, after dividing it into segments. Bolt-holes are made to attach the bracket with next components. Material considered is Aluminum 2219 because of weldability. Material properties are, Young's modulus—73.8 GPa, UTS—455 MPa, YS—350 MPa,

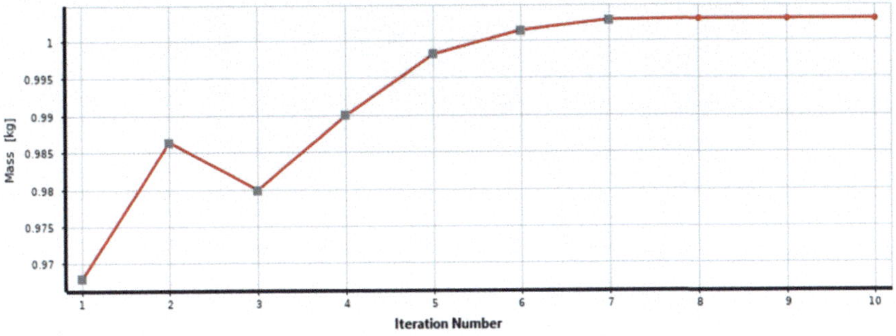

Fig. 11 Mass behavior with respect to number of iterations in sizing optimization

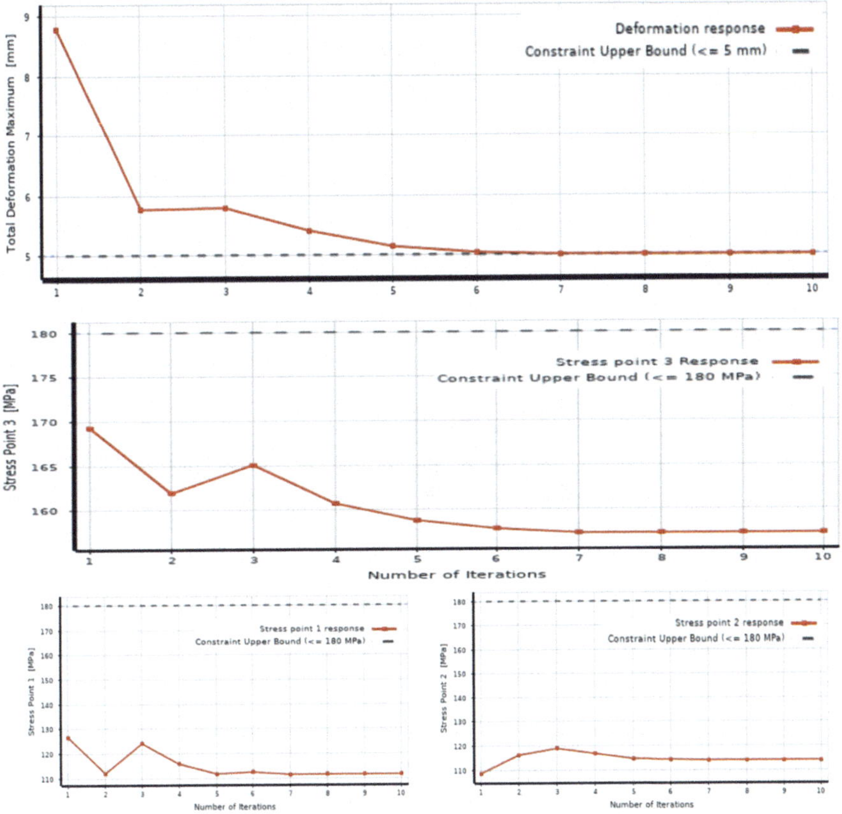

Fig. 12 Deformation and stress behavior with respect to number of iterations in sizing optimization

weld UTS—180 MPa, weld YS—140 MPa, and Poisson's ratio—0.33. FE analysis on final structure is carried out for FOS and BLF (Figs. 13 and 14).

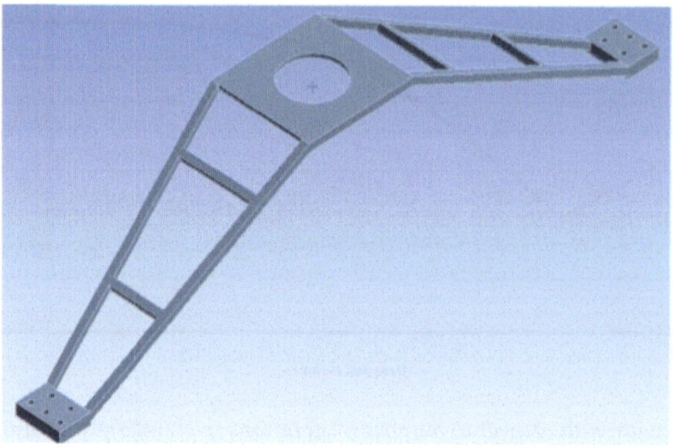

Fig. 13 Detailed design

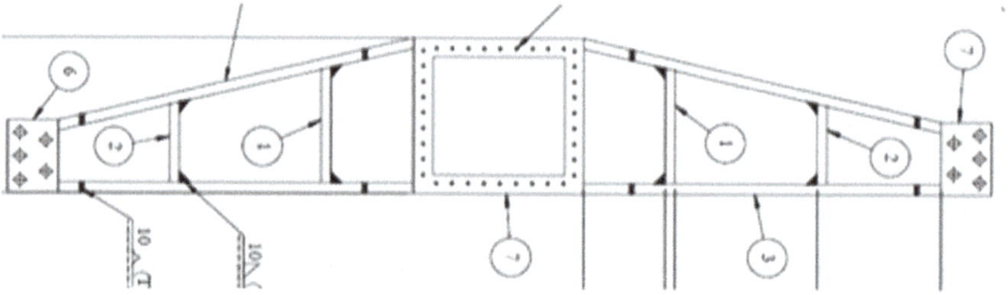

Fig. 14 Welding zones are highlighted in this figure

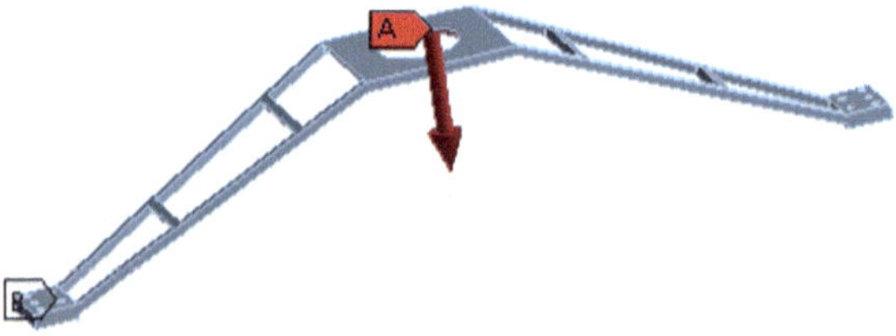

Fig. 15 Loading on bracket

FE analysis is carried out for the bracket. All the components are modeled with solid elements in ANSYS workbench. The bolt-holes of the bracket are fixed in all directions, and axial load is applied on the circular slot (Fig. 15).

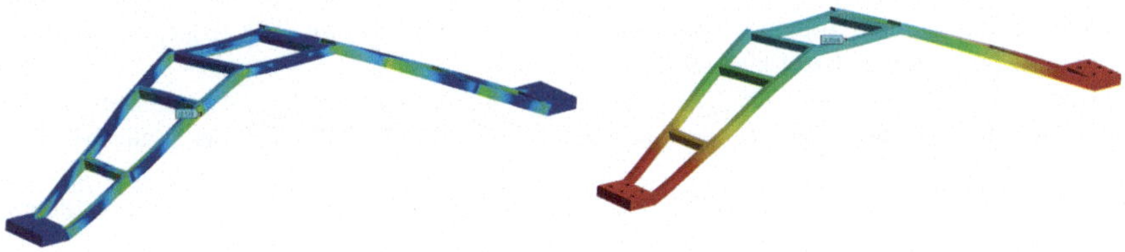

Fig. 16 Stress and deformation plots

8 Results

The mass of the bracket achieved is 1.23kg, taking all engineering aspects into consideration. The stress in the bracket is 70 MPa in the parent material and 82 MPa in the welded zone. The factor of safety (FOS) in the parent material is 6.5 based on ultimate tensile strength (UTS) and 5 based on yield strength (YS). Similarly, FOS in weld zone is 2.1 and 1.7 on YS and UTS, respectively. Deformation is 2.7 mm as shown in Fig. 16. Buckling load factor with knockdown factor of 0.6 is 3.012 which is greater than required 1.5.

9 Conclusion

Highly optimized bracket design is achieved with mass reduction of 49.7% compared to traditionally designed bracket. It is observed that the process flow explained in this paper is efficient to optimizing structures using currently available commercial software. It is also observed that designing structures fully automatic is not possible with this setup for achieving all required objectives; manual intervention is required in between to analyze the process flow and correct them. Automating this process, involving all types of optimizations, autonomous classification, and intelligent selection, will result in revolutionary method of generative design of structure.

Future Work
Design for additive manufacturing; generative design.

References

1. Bendsoe MP, Sigmund O (2004) Topology optimization: theory, methods, and applications, 2nd edn. Springer
2. Barbieri SG et al (2017) A design strategy based on topology optimization techniques for an additive manufactured high performance engine piston. Procedia Manuf 11:641–649. https://doi.org/10.1016/j.promfg.2017.07.162

3. Samson Raj J (2020) Topology optimization of pipeline bracket for mass reduction. Int Res J Eng Technol (IRJET) 07(09):948–953
4. Bodog AB, Grebenişan G (2022) An integrated approach of topological optimization with parametric validation and response surfaces, using ANSYS. IOP Conf Ser Mater Sci Eng 1256:012014
5. Xiangjie J, Cai Z, Chong W (2020) Response surface optimization of machine tool column based on ANSYS workbench. Acad J Manuf Eng 18(2)
6. Abhinav, Arun Kumar B, Hegde AA (2020) Topology optimization of mechanical bracket using finite element analysis. Int J Innov Res Multidiscip Field 6(6). ISSN: 2455-0620

Stress Transfer in Two-Hierarchical Non-self-similar Bio-inspired Composites

A. J. Abhirami and S. Anup

Abstract The innovations and advancements in composite materials have significantly been backed by the design of biological composites such as spider silk, nacre, bone, antler, and bamboo, owing to their magnificent mechanical properties. Even though the constituents at their basic level are weak, biological composites are characterized by superior mechanical properties like high toughness and strength which makes them unique. Bio-inspired composites, inspired by biological composites, are engineered by mimicking the essential features of the biological composites, which are responsible for their superior mechanical properties. These include geometric and material properties such as the aspect ratio of platelet, Young's moduli ratio of platelet to the matrix, and the staggering and hierarchical arrangement of platelets. This study deals with the comparison of stress distribution in a multi-scale finite element model of a two hierarchical (2H) non-self-similar bio-inspired composite. The results of non-self similar composite which is a 2H Stairwise staggered composite with Regular staggered composite as platelets (SR) are compared with that of the models without hierarchy.

Keywords Bio-inspired · Hierarchical · Non-self-similar · Stress-transfer

1 Introduction

The stunning and sublime self-defense mechanisms observed in nature greatly inspired the invention of new materials. For example, the horns of a sheep can withstand an impact force as high as 3400 N, in the time of collisions for competing territory and hunting for food [12]. Not only animals but various plants and fruits also exhibit excellent strength and energy absorption characteristics. For example, the energy dissipation of the mesocarp spongy layer of a pomelo fruit is around 80 J, which is obtained from free-fall tests for no visible outer damage on the peel

A. J. Abhirami (✉) · S. Anup
Indian Institute of Space Science and Technology, Valiamala, Thiruvananthapuram, Kerala 695547, India
e-mail: abhiramiaj.17@res.iist.ac.in

S. Sidhardh et al. (eds.), *Advances in Structural Integrity for Mechanical, Civil, and Aerospace Applications*, Lecture Notes in Mechanical Engineering,
https://doi.org/10.1007/978-981-97-6367-2_10

[7]. Biological composites refer to a wide range of natural materials such as nacre, antler, bone, spider silk, bamboo, and fish skin which possess superior mechanical properties [8, 17, 20]. Their excellent mechanical properties include high strength, stiffness, and toughness [5, 8, 15].

Various pieces of research have reported the possible reasons behind the outstanding mechanical properties of biological composites. The pattern of arrangement of constituents at the elemental level of the composite is one of the major factor responsible for the excellent mechanical properties of biological composites [10, 13, 14, 21]. For example, the toughness of nacre in terms of energy is three orders of magnitude higher than that of the platelets of which it is composed of, which is chiefly attributed to its peculiar staggering structure [6]. Figure 1 shows the brick and mortar arrangement observed in biological composites. In the brick and mortar structure of nacre or mother of pearl, the bricks are aragonite platelets and mortar is a protein matrix, with a platelet volume fraction of 95–99% [4, 11]. For bones, the basic constituents are mineral platelets and collagen matrix, with a platelet volume fraction of around 40–45% [11]. The other factors which impart the superior mechanical properties to biological composites include the high Young's moduli ratio between platelet and matrix, the optimum length to width ratio (aspect ratio) of platelet, the interface properties, and the nanoscale size and hierarchical arrangement of the constituents [9, 13].

Artificial composites derived from the inspiration of biological composites are collectively called bio-inspired composites. Even though the studies have found various properties that can be mimicked to obtain a bio-inspired composite, it has still been impossible to get an artificial material as perfect as its natural inspiration. For this, researchers are still exploring the optimum parameters of the different factors affecting the final properties of a bio-inspired composite.

Out of the different types of arrangements of platelets, so-called staggering, the most observed types in nature are the regular staggering, and the stairwise staggering [22]. In a regular staggered model (RSM), the platelet pattern repeats after two rows, as shown in Figs. 1 and 2b, whereas in a stairwise staggered model (SSM) (see Fig. 2a), the pattern repeats after 'n' number of rows of platelets. The non-self-similarity, i.e., different kinds of staggering at different levels of hierarchy, also has a significant role in achieving superior mechanical properties at the final level of the composite [3].

Only a few studies have been conducted on hierarchical structures in the open literature [2, 16, 18, 23]. Also, the FE analysis of multi-scale models of non-self-similar hierarchical structures is an area that is still to be looked into. Here, we conduct

Fig. 1 Schematic illustration of a brick and mortar structure in a biological composite with regular staggering. Reproduced with permission from [1]

a Finite Element (FE) analysis using the commercially available software ABAQUS [19] to compare the variation of stresses in a nacre-inspired two hierarchical (2H) non-self-similar composite with that of the staggered composites without hierarchy viz. regular staggered and stairwise staggered composites.

2 Methodology

Figure 2 shows the schematic diagram of a 2H stairwise staggered composite with regular staggered composite as platelet (SR). For the present study we choose an optimized model for maximum toughness for a platelet volume fraction of 80% and the material properties as shown in Table 1 using the studies by [2]. The geometry details of the 2H SR model is shown in Table 2. The number of platelets in a period n is chosen as 5 for the present study.

The material properties for the platelet and matrix is assigned the same as that of the widely used materials for experimental works on bio-inspired composites, Verowhite and Agillus 30, respectively. The material properties adopted for the FE analysis including the Young's modulus, Poisson's ratio, normal strength of platelet, and shear strength of matrix are shown in Table 1.

The details of geometry and meshing adopted for the models is given in Table 2. The 3D models of the composites for the three configurations (2H SR, SSM, and RSM) with an out of plane thickness of 5 mm are developed in ABAQUS. The meshing is done using the 3D element type C3D8R which is an eight noded linear brick element.

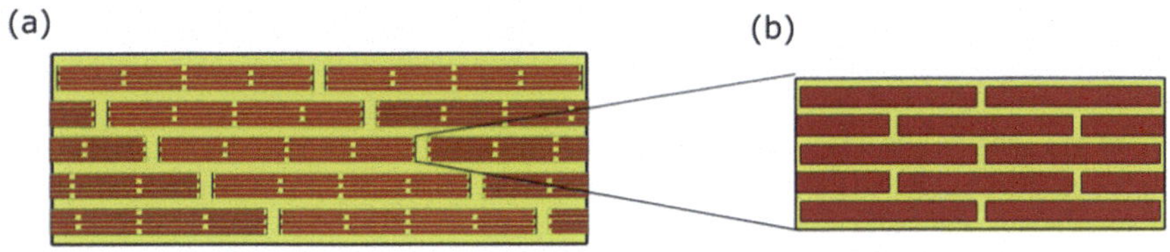

Fig. 2 Schematic representation of **a** two hierarchical (2H) stairwise staggered composite ($n = 5$) composed of regular staggered composite platelets (SR), **b** zoomed view of the platelet which is a regular staggered model (RSM)

Table 1 Material Properties adopted for the FEA

Part	Material	Young's modulus (MPa)	Poisson's ratio	Strength (MPa)
Platelet	Verowhite	2600	0.33	47.5 (normal)
Matrix	Agillus 30	0.56	0.49	0.67 (shear)

Table 2 Geometry and mesh details of the FE model

Model	Volume fraction	Platelet aspect ratio	Meshing element	
			Type	No. of elements
2H SR	0.8	20 (first level of hierarchy), 32 (second level of hierarchy)	C3D8R	375,555
SSM		32		397,790
RSM		20		104,670

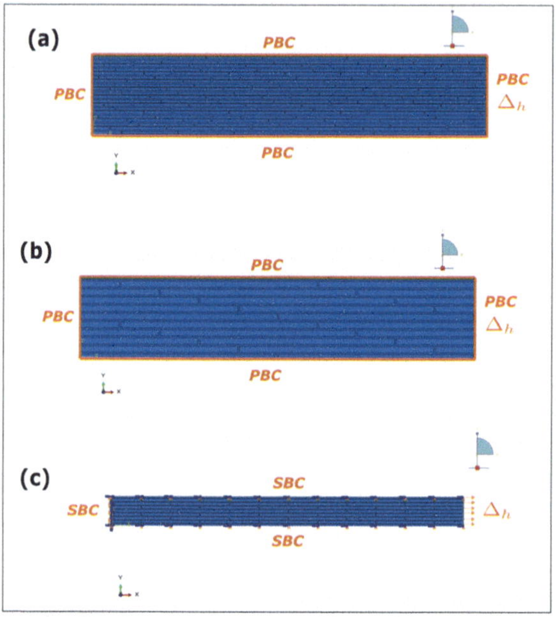

Fig. 3 Boundary conditions applied in different models: **a** 2H SR model with Periodic Boundary Conditions (PBC) at the top, bottom, left, and right edges with a displacement along x-direction $\Delta_h = 0.01$ at the right edge; **b** stairwise staggered model with PBCs at top, bottom, left, and right edges with a displacement along x-direction $\Delta_h = 0.01$ at the right edge; **c** regular staggered model with Symmetric Boundary Conditions (SBC) at top, bottom and left edges, with a displacement along x-direction $\Delta_h = 0.01$ at the right edge

Figure 3 shows the FE models of the 2H SR, stairwise staggered, and regular staggered composites with the details of the applied boundary conditions on the four edges. We apply a displacement $\Delta_h = 0.01$ with periodic boundary conditions (PBC) at all edges, owing to the periodicity of the stairwise staggering in the 2H SR, and stairwise staggered model (SSM) with no hierarchy as shown in Fig. 2a, and b respectively. For the simple regular staggered model (RSM), symmetric boundary

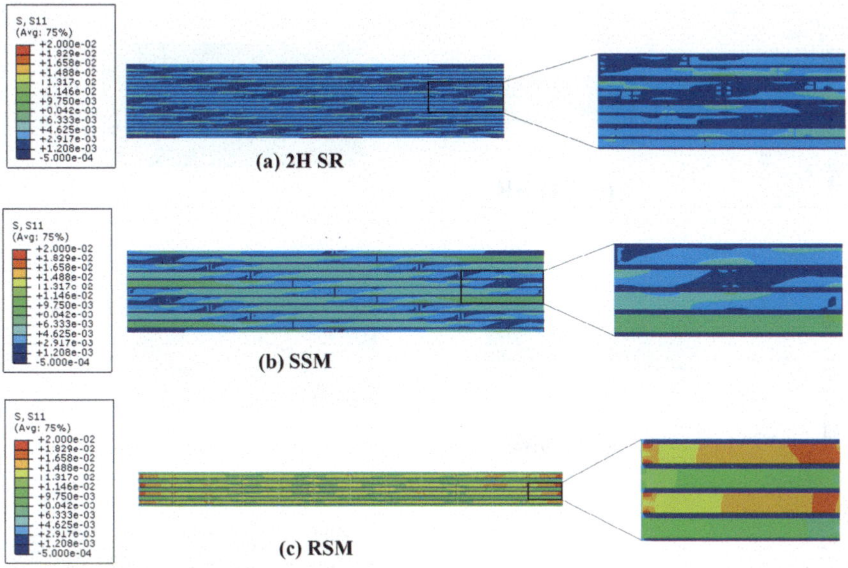

Fig. 4 σ_{11} (normal stress along x-direction) contours obtained from the Output Data Base (ODB) for **a** 2H SR model, **b** SSM, and **c** RSM

conditions are applied at the top, bottom, and left edges, with a displacement of $\Delta_h = 0.01$ at the right edge, as shown in Fig. 3c.

3 Result and Discussions

The stress contours obtained for σ_{11} (stress along x-direction) from the Output Data Base (ODB) of the three models are shown in Fig. 4. It can be observed from Fig. 4 that the normal stress distribution in a 2H SR composite (Fig. 4a) is uniform when compared with that of the simple stairwise staggered (Fig. 4b) or regular staggered model (Fig. 4c).

Also, stress concentrations are observed toward the end of the models in the simple staggered models with stairwise and regular staggering configurations (Fig. 4b and c), whereas it is reduced to a great extent in the 2H SR model (Fig. 4a).

The contours obtained for the shear stress in the three models are shown in Fig. 5. The shear stress distribution in the two hierarchical model (Fig. 5a) is observed to be a combination of the shear stress distributions observed in the stairwise staggered model (Fig. 5b) and the regular staggered model (Fig. 5c). However, in the 2H SR model, more uniform shear stress distribution and lesser regions with high stress concentrations are observed than in the other models with no hierarchy.

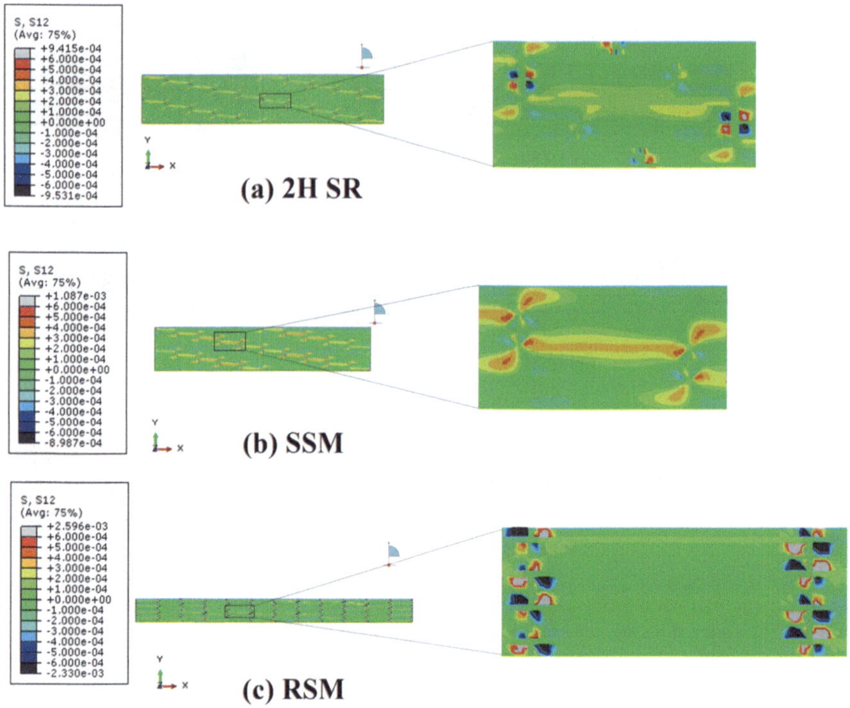

Fig. 5 τ_{12} (shear stress) contours obtained from the Output Data Base (ODB) for **a** 2H SR model, **b** SSM, and **c** RSM

The shear stress (τ_{12}) contour of the stairwise staggered model (Fig. 5b) shows that the regions with more shear stresses are at the platelet matrix horizontal interface and at the corner regions. But, in a regular staggered model (Fig. 5c), maximum shear stress is observed to be at the platelet corner regions.

4 Conclusion

From the results and discussion of the present study, it can be concluded that the hierarchical arrangement of platelets could reduce the stress concentrations significantly and assure a more uniform stress distribution inside the structure, compared to the structures without hierarchy. This can help the structure to withstand impact loads, and improve the defect tolerance.

The current study will be extremely useful in the multi-scale modeling and simulation of bio-inspired hierarchical staggered structures subjected to various loads. Also, this could provide a reasonably good approximation of results which can be used as a preliminary estimation before conducting experimental works.

The influence of stress transfer on the overall macroscopic mechanical properties like stiffness, strength, and toughness is a broad area left for future studies.

References

1. Abhirami AJ, Anup S (2019) Elastic properties of non-self-similar two hierarchical bio-inspired unidirectional composites. In: 64th congress of the Indian Society of Theoretical and Applied Mechanics (ISTAM). School of Mechanical Sciences and School of Basic Sciences (Mathematics), Indian Institute of Technology Bhubaneswar
2. Abhirami A, Anup S (2021) Mechanical properties of unidirectional bio-inspired composites with two non-self-similar hierarchical structures. Mech Materi, 104082
3. An B, Zhao X, Zhang D (2014) On the mechanical behavior of bio-inspired materials with non-self-similar hierarchy. J Mech Behav Biomed Mater 34:8–17
4. Barthelat F, Espinosa H (2007) An experimental investigation of deformation and fracture of nacre-mother of pearl. Exp Mech 47(3):311–324
5. Currey J (1977) Mechanical properties of mother of pearl in tension. Proc R Soc Lond B: Biol Sci 196(1125):443–463
6. Espinosa HD, Rim JE, Barthelat F, Buehler MJ (2009) Merger of structure and material in nacre and bone-perspectives on de novo biomimetic materials. Progr Mater Sci 54(8):1059–1100
7. Fischer SF, Thielen M, Loprang RR, Seidel R, Fleck C, Speck T, Bührig-Polaczek A (2010) Pummelos as concept generators for biomimetically inspired low weight structures with excellent damping properties. Adv Eng Mater 12(12):B658–B663
8. Fratzl P, Gupta H, Paschalis E, Roschger P (2004) Structure and mechanical quality of the collagen-mineral nano-composite in bone. J Mater Chem 14(14):2115–2123
9. Gao H, Ji B, Jäger IL, Arzt E, Fratzl P (2003) Materials become insensitive to flaws at nanoscale: lessons from nature. Proc Natl Acad Sci 100(10):5597–5600
10. Ji B (2008) A study of the interface strength between protein and mineral in biological materials. J Biomech 41(2):259–266
11. Ji B, Gao H (2004) Mechanical properties of nanostructure of biological materials. J Mech Phys Solids 52(9):1963–1990
12. Kitchener A (1988) An analysis of the forces of fighting of the blackbuck (*Antilope cervicapra*) and the bighorn sheep (*Ovis canadensis*) and the mechanical design of the horn of bovids. Jo Zool 214(1):1–20
13. Mathiazhagan S, Anup S (2016) Influence of platelet aspect ratio on the mechanical behaviour of bio-inspired nanocomposites using molecular dynamics. J Mech Behav Biomed Mater 59:21–40
14. Melaibari A, Wagih A, Basha M, Kabeel A, Lubineau G, Eltaher M (2021) Bio-inspired composite laminate design with improved out-of-plane strength and ductility. Compos A Appl Sci Manuf 144:106362
15. Meyers MA, Chen PY, Lin AYM, Seki Y (2008) Biological materials: structure and mechanical properties. Progr Mater Sci 53(1):1–206
16. Mirzaeifar R, Dimas LS, Qin Z, Buehler MJ (2015) Defect-tolerant bioinspired hierarchical composites: simulation and experiment. ACS Biomater Sci Eng 1(5):295–304
17. San Ha N, Lu G (2020) A review of recent research on bio-inspired structures and materials for energy absorption applications. Compos B Eng 181:107496
18. Sen D, Buehler MJ (2011) Structural hierarchies define toughness and defect-tolerance despite simple and mechanically inferior brittle building blocks. Sci Rep 1:35
19. SIMULIA (2019) ABAQUS/standard user's manual, version 2019. Providence, RI
20. Wegst UG, Bai H, Saiz E, Tomsia AP, Ritchie RO (2015) Bioinspired structural materials. Nat Mater 14(1):23
21. Wu K, Zheng Z, Zhang S, He L, Yao H, Gong X, Ni Y (2019) Interfacial strength-controlled energy dissipation mechanism and optimization in impact-resistant nacreous structure. Mater Des 163:107532

22. Zhang Z, Liu B, Huang Y, Hwang K, Gao H (2010) Mechanical properties of unidirectional nanocomposites with non-uniformly or randomly staggered platelet distribution. J Mech Phys Solids 58(10):1646–1660
23. Zhang Z, Zhang YW, Gao H (2011) On optimal hierarchy of load-bearing biological materials. Proc R Soc Lond B: Biol Sci 278(1705):519–525

Detection and Prediction of Bond Degradation for Piezo Impedance-Based Structural Health Monitoring (PISHM) Using Hybrid Deep Learning Model

Lukesh Parida, Sumedha Moharana, and Sourav Kumar Giri

Abstract In past years, sensor-based health diagnostics have shown significant potential for electromechanical impedance (EMI) techniques with many practical implementations. In order to enhance the effectiveness of health monitoring systems, it is crucial to examine the bonding interface between the piezoelectric sensor (PZT) and the host structure. A promising approach to achieve this goal is the utilization of Sn–Ag bonding, which offers improved resistance to moisture, chemicals, and temperature compared to traditional epoxy adhesive bonding. The Sn–Ag bond layer presents a novel attachment method to the structural design, which may ultimately enhance the sensitivity of the health monitoring system. This study used EMI techniques to instrument a unique Sn–Ag alloy-based bond and compared it against an epoxy adhesive bond for better sensing and actuation. Later, both the bonds were cured, and the bond performances were analyzed through conductance and susceptance signatures. Moreover, three different bond layer damage conditions, i.e., incipient, moderate, and severe, were chosen to check the performance of the bond layer through admittance signatures. The EMI techniques rely on pristine data, which presents significant constraints when applied to pre-existing structures. Furthermore, the experimental results were used to predict baseline data through deep learning classifications, i.e., convolutional neural network (CNN), long short-term memory (LSTM), and CNN-LSTM hybrid model for all the bond layer damage conditions. The experimental results were compared against the chosen deep learning model for

L. Parida (✉)
Research Scholar, Department of Civil Engineering, Shiv Nadar Institute of Eminence, Dadri 201314, India
e-mail: lp617@snu.edu.in

Present Address:
Postdoctoral Researcher, Indian Institute of Technology Bombay, Mumbai, India

S. Moharana
Associate Professor, Department of Civil Engineering, Shiv Nadar Institute of Eminence, Dadri 201314, India

S. K. Giri
Assistant Professor, School of Computer Engineering, KIIT Deemed to be University, Patia, Bhubaneswar, Odisha 756001, India

S. Sidhardh et al. (eds.), *Advances in Structural Integrity for Mechanical, Civil, and Aerospace Applications*, Lecture Notes in Mechanical Engineering,
https://doi.org/10.1007/978-981-97-6367-2_11

futuristic prediction. The efficiency of all the deep learning models is quantified using performance metrics like accuracy and R^2 score. The present study demonstrates that the hybrid model yields superior performance compared to the other two models for predicting Sn–Ag and adhesive bond layer damage. The proposed framework aims to replace traditional optimization algorithms to achieve a more simplified effective prototype for the futuristic prediction of EMI signals.

Keywords Bond · Deep learning · EMI · Sn–Ag bond · CNN · LSTM · Piezo sensor

1 Introduction

Sensor-based health diagnostics have significantly advanced in recent years, giving potential structural health monitoring (SHM) solutions in various engineering applications [1–4]. Among the many sensor technologies, electromechanical impedance (EMI) approaches have emerged as a potent tool for analyzing structural integrity and health [5–8]. EMI uses the electromechanical coupling phenomena in piezoelectric materials, such as piezoelectric transducers (PZTs), to monitor structural problems by monitoring changes in impedance spectra [9–12]. This non-destructive testing approach has practical applications in identifying damage, strength, and fatigue in civil, aeronautical, and mechanical structures. By examining alterations in the admittance signatures under different conditions, the overall condition of the structure can be evaluated [13–15].

The bonding interface between the PZT sensor and the host structure is critical in improving the effectiveness of health monitoring systems. The bond layer connecting this two must be carefully considered to ensure the transmission of signals between the host structure and the PZT. Therefore, establishing a solid and stable bond between the sensor and the structure is paramount for the reliable in situ installation of PZT patches. The interface bond layer among a PZT and the host structure should be perfect and the patch's rigidity must be greater than that of the structure, allowing the sensor to cause severe strains in the structural system [16]. The quality of this bond significantly affects the sensitivity and reliability of EMI-based health monitoring systems [17, 18]. The effect of shear lag on surface-attached piezoelectric patches becomes evident, in line with the one-dimensional theory of bonded connections, when subjected to shear stresses across the adhesive layer. This adhesive layer is critical for load transfer [19]. Several analytical and experimental models have been conducted to check the adhesive bonding conditions of piezo transducers [20]. The thickness and modulus of the adhesive layer have an apparent impact on the operation of the piezoelectric disks when using the electromechanical impedance approach [21]. Peak frequencies may be changed by adjusting the stiffness and mass of the sensor by manipulating steel plate and adhesive layer thicknesses [22, 23]. The performance of PZT sensors was studied experimentally and numerically for assessing the interfacial delamination behavior [24, 25]. Epoxy adhesive bonding has traditionally been used to attach PZT sensors to structural surfaces. While epoxy adhesives are commonly used, they have several constraints, especially in terms of resistance to environmental conditions such as moisture, chemicals, and temperature changes. These restrictions could affect the sensors' long-term performance and perseverance in real-world applications.

To address these difficulties, this work investigates an innovative strategy to improve capabilities of health monitoring systems. The emphasis is on Sn–Ag alloy-based bonding as an alternative to typical epoxy adhesives. The Sn–Ag bond layer is more resistant to external stresses, improving its lifespan and robustness. This innovative bonding process can transform health monitoring devices' structural design, eventually improving sensitivity and consistency. The introduction of Ag into the PZT ceramic material resulted in the deterioration of PZT sensors with Silver-Palladium (Ag–Pd) conductive components [26]. An analogous investigation was carried out to understand the dynamic characteristics of the adhesive bond layer [27]. Furthermore, Zhili et al. [28] investigated the relation between bonding temperatures and adhesion strength in the context of gold wire thermosonic attachment to bond layers. Additionally, previous research has shown the significant influence of bond degradation, bond layer thickness, and modulus on impedance monitoring systems. Identifying and assessing the progressive reduction of sensor performance can be extremely difficult, especially when dealing with a degree of bond area degradation. The primary objective of this study is to evaluate the performance of the Sn–Ag alloy-based bond layer to that of the standard epoxy adhesive bond layer when combined with EMI methods.

One of the challenges encountered in EMI-based health monitoring systems is their reliance on pristine signatures, which restricts their application to existing structures. In reality, pristine data representing the structure's undamaged condition can be challenging to acquire, especially for pre-existing structures. As a result, this research looked at the potential of employing deep learning algorithms to forecast baseline data without pristine data. Researchers are increasingly interested in utilizing deep learning as an improved intelligence-driven approach to address pattern recognition challenges, such as the detection of decay in civil infrastructure. In recent times, deep learning approaches have significantly advanced, demonstrating great benefits in various sectors, including crack detection [29], medical science [30], and so on. With the advantages of artificial intelligence, deep learning-based models are utilized in spectral imaging [31], smart cities [32], education systems [33], agricultural systems [34], and energy-efficient buildings [35]. Deep learning is integrated more into civil infrastructure planning, construction, and management [36]. The utilization of convolutional neural network (CNN) in structural health monitoring (SHM) has gained immense attention as a prominent deep learning technique. Its effectiveness has been extensively researched and is now being applied in real-life scenarios [37]. The CNN has only been employed for simple deterioration circumstances and impedance-based structural fault diagnosis of a plate [38]. A data-driven model machine learns the data and predicts the model based on the number of data and input parameters. The deep learning model has been utilized to solve challenges in civil engineering, such as calculating the strength of concrete [39], geotechnical engineering [40], transport systems [41], and estimating and predicting the selling price of new buildings [42]. Although deep learning has shown potential in other domains and has been successfully used for simple deterioration circumstances and impedance-based fault diagnosis of structural plates, its utilization in civil infrastructure for predicting baseline data without pristine data still needs to be explored. The potential of deep learning algorithms to forecast baseline data without pristine signatures should considerably increase the application of such systems to a more extensive range of structures. As a result, there is a need for additional research to investigate and develop hybrid deep learning-based approaches specifically suited to this problem, potentially leveraging improvements in deep learning algorithms observed in other fields and adapting them to the unique challenges of infrastructure construction and SHM. The paper uses experimental results to predict baseline data using deep learning techniques such as CNN, LSTM, and CNN-LSTM hybrid model to solve this challenge.

In this study, a novel Sn–Ag bonding layer was implemented for the first time to check the performance of the PZT patch attached to the host structure. The signatures of the adhesive bond and Sn–Ag bond were compared. Three different bond layer damage conditions, i.e., incipient, moderate, and severe, were used to monitor the performance of the Sn–Ag alloy-based bond layer to that of the standard epoxy adhesive bond layer. The admittance signatures were acquired to visualize the performance of different bond layer degradations. Moreover, a novel deep learning-based hybrid model was implemented to predict the EMI data for different bond layer damage conditions. In this study, 80% of the datasets were used to train the model, and the rest 20% of the data were used for prediction. The efficacy of these deep learning models was assessed using performance indicators such as R^2 score and MAPE values. The outcomes of the deep learning models were compared to determine their applicability for forecasting bond layer degradation in both Sn–Ag and epoxy adhesive bonds. This study aimed to find the most effective model for future predictive applications in EMI-based health monitoring systems.

2 Materials and Methods Used

This section highlights the material specifications and different methods used to build the predictive models using deep learning techniques to predict frequency-based bond layer damages. This study utilized different deep learning models, including LSTM, CNN, and CNN-LSTM, to predict the EMI signatures for different bond layer damage conditions. The CNN extracts features while the LSTM learns the pattern of the data sequence.

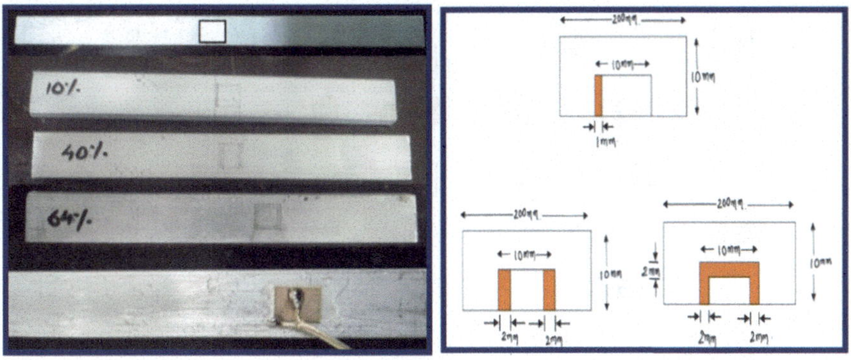

Fig. 1 Schematic diagram for different bond layer damage conditions

2.1 *Material Specifications*

An aluminum block (48 × 48 × 10 mm) was used in the experiments to monitor the performance of an adhesive and Sn–Ag-based bonding using a piezo impedance-based EMI techniques. The PZT patch has been instrumented at mid-places on the upper side of the aluminum block. The researcher considered a fully bonded and bond layer damage conditions of Sn–Ag and adhesive bond for this investigation. The schematic diagram of different bond layer damage in terms of bond layer degradation is shown in Fig. 1. The Soft PZT transducers of grade PIC, 300 (developed by PI ceramic), with dimensions of 15 × 5 × 0.3 mm have been chosen for all sensor setups. In some instances, a thin epoxy adhesive coating was used to attach PZT transducers, while Sn–Ag alloy layers were used in other cases. The authors have examined a new piezo configuration using Sn–Ag-based alloy bonding as bond layer materials to observe the overall structural condition using the EMI technique. The suggested design started with an Sn–Ag alloy-based bond, which also featured the usage of glue to lessen binding rigidity and create a special bond that offers a middle ground between rigid and flexible bonds. With a thermo-activated technique, an alloy bond based on Sn–Ag has been made. The bond damage was done by reducing the areas of the bond layer for both bonding cases.

2.2 *LSTM Architecture*

The LSTM model [43] utilizes both short-term and long-term memories to perform efficient computations via multiple gates. The LSTM cell's design is illustrated in Fig. 2. In this diagram, x_t denotes the input for the current frequency step, h_{t-1} represents the output of the previous LSTM unit, C_{t-1} stands for the memory of

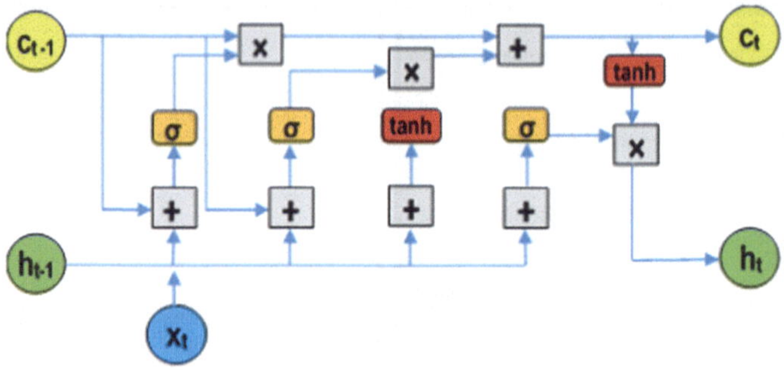

Fig. 2 LSTM architecture

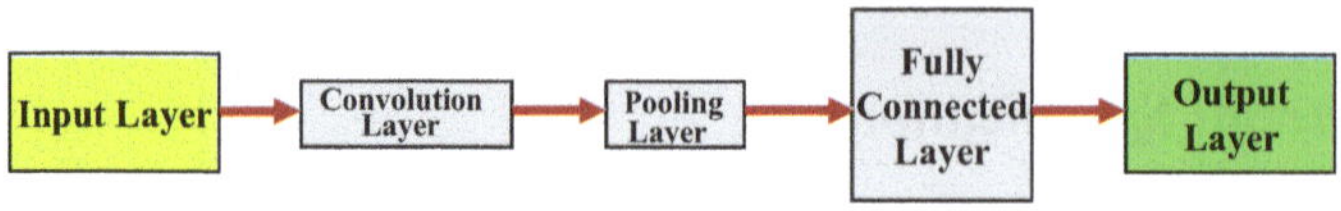

Fig. 3 CNN architecture

the previous unit, h_t indicates the output of the current LSTM unit, and C_t represents the memory of the current unit. Element-wise multiplication and element-wise summation are denoted by $\times$ and $+$, respectively.

2.3 CNN Architecture

The CNN architecture (as illustrated in Fig. 3) consists of three layers: the convolutional layer (CONV), pooling layer (POOL), and fully connected layer (FC). The convolutional layer uses kernels or filters to perform computations. Meanwhile, the pooling layer reduces the spatial size, effectively minimizing the number of computations required. Finally, the fully connected layer processes the outputs from the pooling layer to produce the final output.

2.4 CNN-LSTM Architecture

In this research, a combined model was developed to predict the frequency-based EMI signals using a conductance dataset. The architecture of CNN-LSTM consists of a series of LSTM layers placed on top of a CNN layer. The feature extraction process is carried out by the CNN layer, while the LSTM unit is responsible for predicting sequence data. Within an LSTM unit, dropout layers control the model overfitting

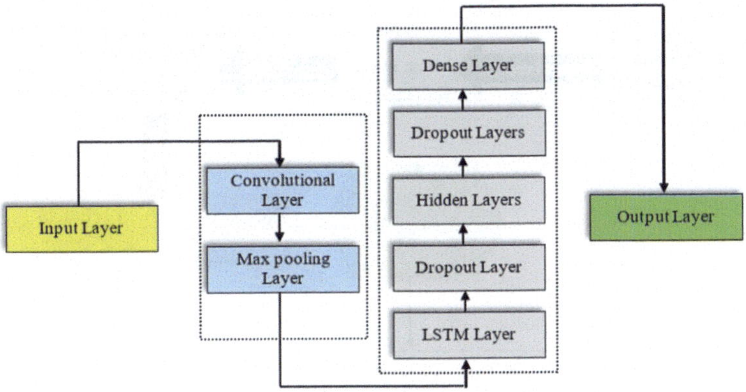

Fig. 4 CNN-LSTM architecture

issues. Figure 4 illustrates the CNN-LSTM hybrid model architecture consisting of various phases.

2.5 *Proposed Framework*

The proposed framework consists of all steps involved in developing predictive models starting from acquisition of data to validation of models. The data recorded from the experiment are exported in .csv format per the schema shown in Table 1.

Data of adhesive and Sn–Ag bond for four bond layer damage cases, viz. baseline, incipient, moderate, and severe, are considered and exported in .csv file for developing the predictive models. Each .csv file has two features, i.e., frequency and conductance. Exploratory Data Analysis (EDA) was performed on each dataset to get valuable insights into the dataset. Min–max normalization techniques were applied to normalize the data as per Eq. (1).

$$x_{\text{scaled}} = \frac{x - x_{\min}}{x_{\max} - x_{\min}}. \tag{1}$$

Using normalized data produces better predictive models in terms of accuracy and efficiency compared to raw data. The datasets were divided into training and test units using data splitting with a ratio of 8:2. The model was trained using the optimized hyperparameters with the training dataset to create predictive models. The test dataset is used to validate the models by comparing the results of predictive

Table 1 Schema of the datasets

Frequency (kHz)	Conductance (*G*)
30000	0.0000483867

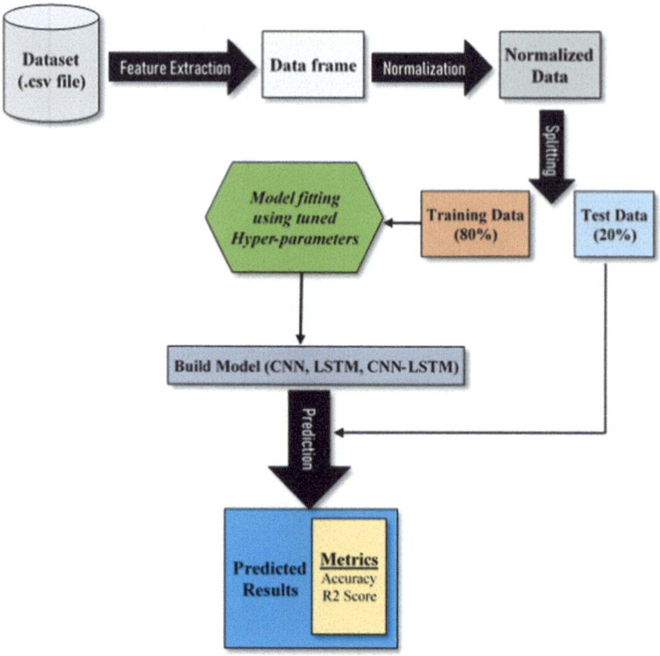

Fig. 5 Proposed framework of predictive models

models applied to the test data and the original test data. Figure 5 illustrates the structure of the proposed study.

2.6 Software Requirements

Python 3.8 version and Jupyter Notebook installed in Windows 11 machine are used to execute all the above predictive models implemented in Python programming. Major Python packages used in this work are NumPy, pandas, sklearn, matplotlib, TensorFlow, etc.

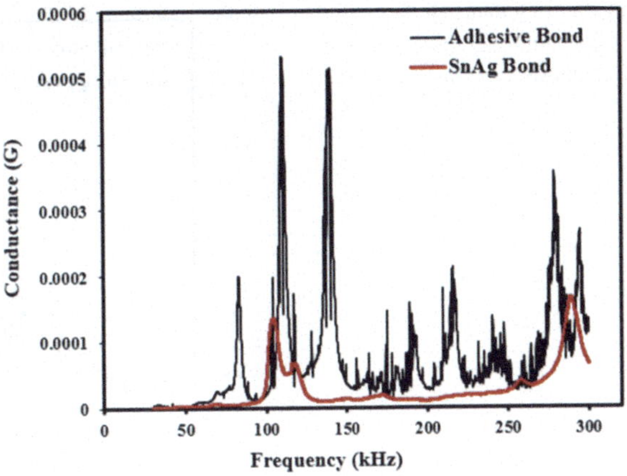

Fig. 6 Plot between conductance and frequency for adhesive and Sn–Ag bond

3 Result and Discussions

3.1 Sensor Stabilization of Both the Bonding Conditions

Primarily impedance-based monitoring is suitable for the health monitoring of any structural system. In EMI techniques, effect of the bond layer plays a vital role in the transfer of signals from PZT to structure. Here, the author focuses on different bond damage conditions for adhesive and Sn–Ag bond conditions. Figure 6 shows the plot between conductance and frequency for both cases. The overall pattern follows the same trend for adhesive and Sn–Ag bond cases. The conductance values are less in the case of a Sn–Ag bond than in an adhesive bond, but structural peaks are more prominent for both cases. Looking for the efficiency of both the bonding conditions, the Sn–Ag bond is quite feasible for high temperature and moisture conditions where placing PZT through an adhesive bond is not feasible. The susceptance versus frequency signatures are shown in Fig. 7 and follow the same pattern as the conductance signature. The peaks are less prominent than conductance signatures. However, the signatures are utilized to calculate phase angle spectrum and inverse-based structural parameter calculations.

3.2 Effect of Bond Damage for Different Bonding Conditions

Sn–Ag bond has never been employed as a bonding agent for the impedance-based monitoring system. The effect is relatively noticeable from the perspective of conductance peaks and susceptance graph slopes. The effect of bond damage for Sn–Ag and adhesive bonding cases considerably impacts conductance and susceptance signatures. Figures 8 and 9 show conductance and susceptance signatures for adhesive bond cases. For adhesive bonding, the frequency shifts rightward with a

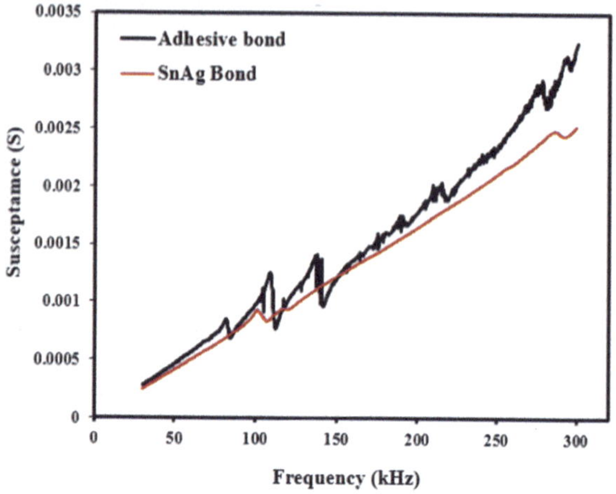

Fig. 7 Plot between susceptance and frequency for adhesive and Sn–Ag bond

higher conductance value for incipient, moderate, and severe damage conditions, respectively. Although the influence on susceptance is negligible, the movement of peaks and frequency for the adhesive bonding cases are highly noticeable. The concern signatures of conductance and susceptance for the Sn–Ag bond are shown in Figs. 10 and 11. The graph indicates that the frequency is shifting rightward with bond damages and attains a higher conductance value for severe damage conditions. Instead, the Sn–Ag bond shows a lower conductance value, but the variations of signatures indicate the behavior of bond damages for incipient, moderate, and severe damage conditions. Hence, the effect of bond damage plays a vital role in impedance-based monitoring.

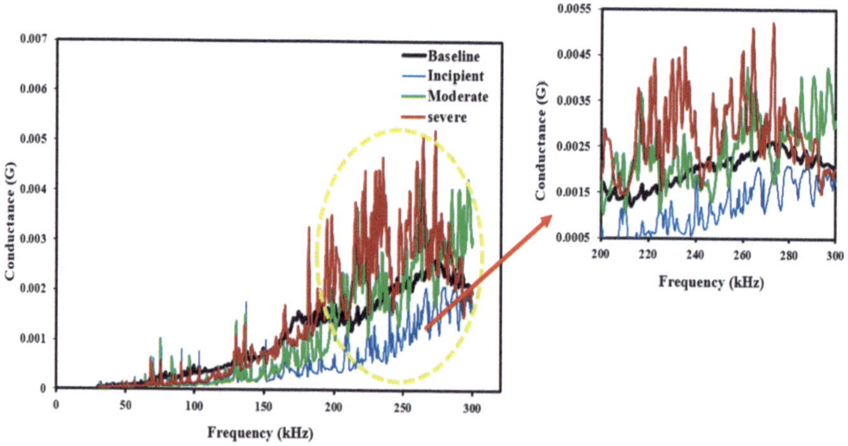

Fig. 8 Plot between conductance (G) and frequency for different adhesive bond damage conditions

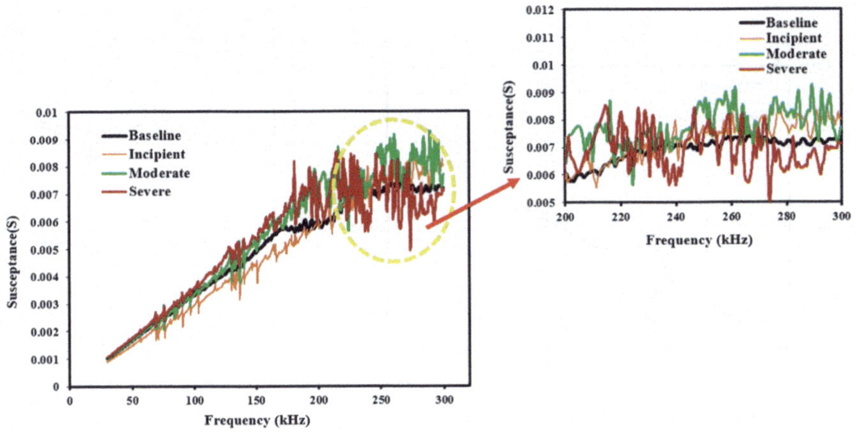

Fig. 9 Plot between susceptance (S) and frequency for different adhesive bond damage conditions

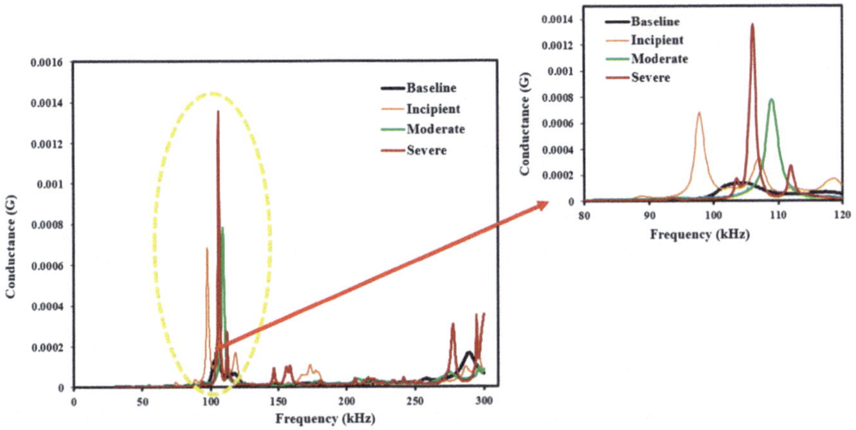

Fig. 10 Plot between conductance (G) and frequency for different Sn–Ag bond damage conditions

3.3 Deep Learning Predictive Model for the Effect of Bond Damage

As shown above, the effect of bond degradation reflects through variation of EMI signals. The effect of bond damage directly affects the transfer of waves from PZT to host structure. The deep learning-based architecture provides a better optimization approach for piezo impedance signal prediction. The author utilized improved deep learning algorithms for the prediction of conductance data at different bond layer damage scenarios, considering the shear lag effect between the structure and

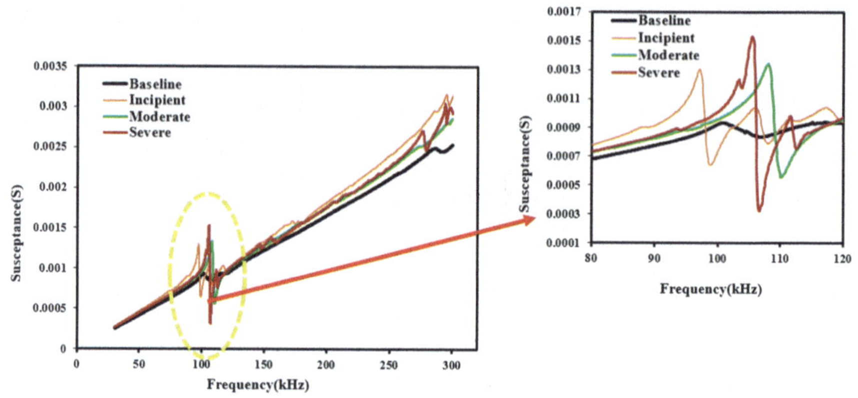

Fig. 11 Plot between susceptance (S) and frequency for different Sn–Ag bond damage conditions

Table 2 Optimized hyperparameter values for deep learning models

Hyperparameters	LSTM/CNN model	
	Adhesive bond	Sn–Ag bond
Learning rate	0.0003/0.0003	0.0060/0.0987/
Dropout rate	0.0694/0.0904	0.3301/0.0875
No hidden layers	1/2	1/1
Neurons per layer	500/500	355/355
Batch size	64/64	128/128
Activation function	ReLU/ReLU	ReLU/ReLU
Optimizer	ADAM/ADAM	ADAM/SGD

PZT patch. Predictive models are constructed using 80% of the original data as training data, while the remaining 20% is utilized as test data for validation. The hyperparameters used by the predictive models are depicted in Table 2.

In this work, the performance of the predictive models was compared considering two metrics, i.e., MAPE and R^2 score. Performance metrics were used to determine the accuracy of the predictive models in terms of prediction, which can be calculated by the formula shown in Eqs. (2) and (3).

$$\text{Accuracy } (\%) = 100 - \text{MAPE}.$$

$$\text{MAPE} = \frac{100}{n} \sum_{i=1}^{n} \left| \frac{y_i - \hat{y}_i}{y_i} \right|. \tag{2}$$

Here, y_i and $\hat{y}_i$ represent the actual and predicted values, respectively. Equation (3) depicts the formula to find R^2 score.

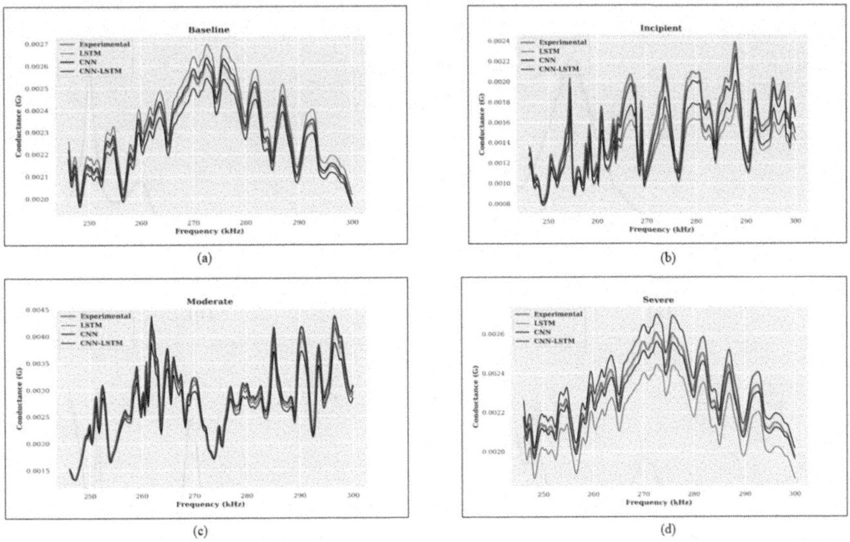

Fig. 12 Prediction results: experimental, LSTM, CNN, CNN-LSTM for adhesive bond **a** baseline, **b** incipient, **c** moderate, **d** severe

$$R^2 = 1 - \frac{\sum_{i=1}^{n} (y_i - \hat{y}_i)^2}{\sum_{i=1}^{n} (y_i - \overline{y_i})^2}, \tag{3}$$

where y_i, $\overline{y_i}$, and $\hat{y}_i$ represent the actual, mean, and predicted values, respectively.

It is evident from Figs. 12a–d and 13a–d that the curve of actual versus predicted data seems very close for the hybridized CNN-LSTM model for adhesive bond and Sn–Ag bond datasets, respectively. The results in Tables 3 and 4 demonstrate that the CNN-LSTM predictive models have higher accuracy than the other two models for all types of damage in both adhesive bond and Sn–Ag bond. In CNN-LSTM model, CNN layer is responsible for extracting precise features from the dataset and LSTM unit is responsible for sequence data prediction taking past values into consideration.

4 Conclusions

In this paper, the effects of different bond layer degradation and deep learning model-based prediction have been reported. The experimental results show that the admittance signature is capable enough to detect the bond layer damages. The purpose of the current work is to predict Sn–Ag and adhesive bond degradation that replace

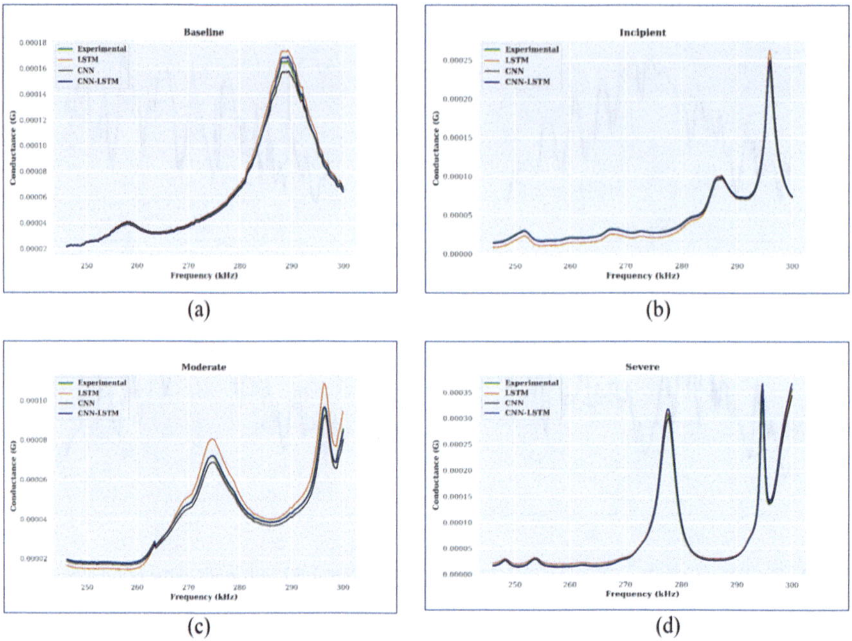

Fig. 13 Prediction results: experimental, LSTM, CNN, CNN-LSTM for Sn–Ag bond **a** baseline, **b** incipient, **c** moderate, **d** severe

Table 3 Accuracy of deep learning models

Model	Adhesive bond damage				Sn–Ag bond damage			
	Baseline	Incipient	Moderate	Severe	Baseline	Incipient	Moderate	Severe
LSTM	97.34	84.72	97.14	94.55	97.76	89.98	95.18	95.13
CNN	98.28	88.40	94.56	96.50	97.88	90.68	82.28	86.23
CNN-LSTM	99.18	97.00	97.19	98.67	98.82	99.14	99.20	97.02

Table 4 R^2 score of deep learning models

Model	Adhesive bond damage				Sn–Ag bond damage			
	Baseline	Incipient	Moderate	Severe	Baseline	Incipient	Moderate	Severe
LSTM	0.85	0.78	0.97	0.68	0.96	0.98	0.98	0.994
CNN	0.93	0.65	0.90	0.87	0.95	0.97	0.96	0.990
CNN-LSTM	0.98	0.96	0.98	0.96	0.99	0.99	0.99	0.994

experimental and field observations. The reported experimental findings show satisfactory and validated through conductance and susceptance signatures. From the results, it is clear that CNN-LSTM, a hybrid deep learning model performs better than the standalone LSTM and CNN models. For adhesive bond damage, the accuracy of the CNN-LSTM model is 99.18%, 97.00%, 97.19%, and 98.67% for baseline, incipient, moderate, and severe damage conditions, respectively. Similarly, for Sn–Ag Bond damage, the model outperforms others with the accuracy of 98.82%, 99.14%, 99.20%, and 97.02% for baseline, incipient, moderate, and severe damage, respectively.

Different hybrid models with suitable hyperparameter optimization techniques may be used in future to develop even better predictive models. Also, a more robust PISHM system is needed to improve the sensitivity and reliability of bond degradation detection and prediction. This can be achieved by developing new transducer designs, optimizing the sensor placement, and improving the signal processing algorithms. Advanced data preprocessing techniques are needed to improve the accuracy and efficiency of PISHM. One way to accomplish this is through the utilization of sophisticated signal processing methods like wavelet analysis and time–frequency analysis to extract significant characteristics from the original data. In addition, the creation of comprehensible deep learning models can resolve the problem of interpretability. Techniques such as attention mechanisms can be used to highlight the important features in the input data that led to the model's prediction. This can provide valuable insights into the underlying mechanisms of bond degradation in PISHM.

Acknowledgements The Department of Science and Technology (DST), India, is acknowledged with gratitude by the authors for their support in this research. Furthermore, the Science and Engineering Research Board (ECR/2018/001662) from India also provided support for this study.

References

1. Mishra M, Lourenço PB, Ramana GV (2022) Structural health monitoring of civil engineering structures by using the internet of things: a review. J Build Eng 48:103954
2. Song G, Wang C, Wang B (2017) Structural health monitoring (SHM) of civil structures. Appl Sci 7(8):789
3. Feng D, Feng MQ (2018) Computer vision for SHM of civil infrastructure: from dynamic response measurement to damage detection—a review. Eng Struct 156:105–117
4. Li HN, Li DS, Song GB (2004) Recent applications of fiber optic sensors to health monitoring in civil engineering. Eng Struct 26(11):1647–1657
5. Parida L, Moharana S (2022) Monitoring loading effects of construction steel structure using piezo transducer for electromechanical impedance techniques. In: Recent advances in materials, mechanics and structures: select proceedings of ICMMS 2022. Springer Nature Singapore, Singapore, pp 371–380
6. Parida L, Moharana S (2023) A comprehensive review on piezo impedance based multi sensing technique. Results Eng 101093
7. Parida L, Moharana S, Ferreira VM, Giri SK, Ascensão G (2022) A novel CNN-LSTM hybrid model for prediction of electro-mechanical impedance signal based bond strength monitoring. Sensors 22(24):9920

8. Parida L, Moharana S (2023) Comparative assessment of a multitudinal piezo arrangement for non-destructive evaluation of construction steel: an experimental study. Measurement 113592
9. Parida L, Moharana S, Giri SK (2023) Machine learning approach for predicting impedance signatures of construction steel structures in various tensile pull actions. Mater Today Proc
10. Moharana S, Bhalla S (2019) Development and evaluation of an external reusable piezo-based concrete hydration-monitoring sensor. J Intell Mater Syst Struct 30(18–19):2770–2788
11. Banerjee T, Moharana S, Parida L (2023) Early detection of thermal instability in railway tracks using piezo-coupled structural signatures. J Infrastruct Intell Resil 2(4):100063
12. Li G, Luo M, Huang J, Li W (2023) Early-age concrete strength monitoring using smart aggregate based on electromechanical impedance and machine learning. Mech Syst Signal Process 186:109865
13. Lim YY, Soh CK (2014) Electromechanical impedance (EMI)-based incipient crack monitoring and critical crack identification of beam structures. Res Nondestr Eval 25(2):82–98
14. Tawie R, Park HB, Baek J, Na WS (2019) Damage detection performance of the electromechanical impedance (EMI) technique with various attachment methods on glass fibre composite plates. Sensors 19(5):1000
15. Rocha H, Semprimoschnig C, Nunes JP (2021) Sensors for process and structural health monitoring of aerospace composites: a review. Eng Struct 237:112231
16. Gandhi MV, Thompson BD (1992) Smart materials and structures. Springer Science & Business Media
17. Sbarufatti C, Manes A, Giglio M (2014) Application of sensor technologies for local and distributed structural health monitoring. Struct Control Health Monit 21(7):1057–1083
18. Schubert KJ, Herrmann AS (2012) On the influence of moisture absorption on Lamb wave propagation and measurements in viscoelastic CFRP using surface applied piezoelectric sensors. Compos Struct 94(12):3635–3643
19. Blackshire JL, Cooney A (2006) Evaluation and improvement in sensor performance and durability for structural health monitoring systems. In: Advanced sensor technologies for nondestructive evaluation and structural health monitoring II, Mar 2006, vol 6179. SPIE, pp 137–146
20. Parpe A, Saravanan TJ (2021) New refined analytical models for various bonding conditions of an adhesively bonded smart PZT transducer using the EMI technique. Smart Mater Struct 30(12):125015
21. Qing XP, Chan HL, Beard SJ, Ooi TK, Marotta SA (2006) Effect of adhesive on the performance of piezoelectric elements used to monitor structural health. Int J Adhes Adhes 26(8):622–628
22. Priya CB, Saravanan TJ, Balamonica K, Gopalakrishnan N, Rao ARM (2018) EMI based monitoring of early-age characteristics of concrete and comparison of serial/parallel multi-sensing technique. Constr Build Mater 191:1268–1284
23. Gopalakrishnan N (2019) Piezoelectric sensor–based damage progression in concrete through serial/parallel multi-sensing technique. Struct Health Monit 1:18
24. Blackshire JL, Martin SA, Na JK (2007) Disbonding effects on elastic wave generation and reception by bonded piezoelectric sensor systems. In: Sensor systems and networks: phenomena, technology, and applications for NDE and health monitoring 2007, Apr 2007, vol 6530. SPIE, pp 196–203
25. Lanzara G, Yoon Y, Kim Y, Chang FK (2009) Influence of interface degradation on the performance of piezoelectric actuators. J Intell Mater Syst Struct 20(14):1699–1710
26. van den Ende DA, Bos B, Groen WA, Dortmans LMJG (2009) Lifetime of piezoceramic multilayer actuators: interplay of material properties and actuator design. J Electroceram 22(1):163–170
27. Han L, Wang XD, Sun Y (2008) The effect of bonding layer properties on the dynamic behaviour of surface-bonded piezoelectric sensors. Int J Solids Struct 45(21):5599–5612
28. Zhili L, Lei H, Yunxin W, Jue Z (2005) Experiment study of temperature parameter effect on bonding process and quality in thermosonic wire bonding. In: 2005 6th international conference on electronic packaging technology. IEEE, pp 408–413

29. Swaminathan B, Mishra SK, Jana S, Parida L, Kalyani N, Sultanuddin SJ (2022) A deep learning-based approach for surface crack detection using convolutional neural network. In: 2022 international conference on edge computing and applications (ICECAA), Oct 2022. IEEE, pp 1229–1234
30. Liu T, Siegel E, Shen D (2022) Deep learning and medical image analysis for COVID-19 diagnosis and prediction. Annu Rev Biomed Eng 24
31. Huang L, Luo R, Liu X, Hao X (2022) Spectral imaging with deep learning. Light Sci Appl 11(1):1–19
32. Bhattacharya S, Somayaji SRK, Gadekallu TR, Alazab M, Maddikunta PKR (2022) A review on deep learning for future smart cities. Internet Technol Lett 5(1):e187
33. Zhong L, Qi C, Gao Y (2022) Deep learning-assisted performance evaluation system for teaching SCM in the higher education system: performance evaluation of teaching management. Inf Resour Manag J (IRMJ) 35(3):1–22
34. Kumari KS, Haleem SA, Shivaprakash G, Saravanan M, Arunsundar B, Pandraju TKS (2022) Agriculture monitoring system based on internet of things by deep learning feature fusion with classification. Comput Electr Eng 102:108197
35. Tien PW, Wei S, Darkwa J, Wood C, Calautit JK (2022) Machine learning and deep learning methods for enhancing building energy efficiency and indoor environmental quality—a review. Energy AI 10:100198
36. Agyemang IO, Zhang X, Acheampong D, Adjei-Mensah I, Kusi GA, Mawuli BC, Agbley BLY (2022) Autonomous health assessment of civil infrastructure using deep learning and smart devices. Autom Constr 141:104396
37. De Oliveira MA, Monteiro AV, Vieira Filho J (2018) A new structural health monitoring strategy based on PZT sensors and convolutional neural network. Sensors 18(9):2955
38. Mitiche I, Nesbitt A, Conner S, Boreham P, Morison G (2020) 1D-CNN based real-time fault detection system for power asset diagnostics. IET Gener Transm Distrib 14(24):5766–5773
39. Zeng Z, Zhu Z, Yao W, Wang Z, Wang C, Wei Y et al (2022) Accurate prediction of concrete compressive strength based on explainable features using deep learning. Constr Build Mater 329:127082
40. Hong C, Luo G, Chen W (2022) Safety analysis of a deep foundation ditch using deep learning methods. Gondwana Res
41. Chen MY, Chiang HS, Yang KJ (2022) Constructing cooperative intelligent transport systems for travel time prediction with deep learning approaches. IEEE Trans Intell Transp Syst
42. Chiu SM, Chen YC, Lee C (2022) Estate price prediction system based on temporal and spatial features and lightweight deep learning model. Appl Intell 52(1):808–834
43. Greff K, Srivastava RK, Koutník J, Steunebrink BR, Schmidhuber J (2016) LSTM: a search space odyssey. IEEE Trans Neural Netw Learn Syst 28(10):2222–2232
44. Rawat W, Wang Z (2017) Deep convolutional neural networks for image classification: a comprehensive review. Neural Comput 29(9):2352–2449

Modelling and Diagnosis of Faults in Deep Groove Ball Bearing

Naveen Kumar, Samrat Mandal, Chintamani Mishra, and Nirmal Baran Hui

Abstract Deep groove bearing is widely used in rotating machines as it can operate at a wide range of load and speed combinations. This article presents two simulation models for deep groove ball bearings. The first model of bearing is a 5-DOF system block diagram model developed using MATLAB Simulink environment and the second model of bearing is a detailed three-dimensional multi-body dynamic model of bearing where contact between races, slip and effects due to friction are taken into consideration and it is developed using multi-body simulation (MBS) software ADAMS. Different geometrical faults are given to the different parts of the bearing. Envelope analysis is done to the simulated signal and the result are than interpreted to find out the condition of bearing. These models of deep groove ball bearing elements can be integrated to different machine components to find the condition of industrial rotating machines.

Keywords Bearing dynamics · Modelling and simulation · Signal processing · Condition monitoring

1 Introduction

Bearings (ball, roller, journal, etc.) are used widely in different rotating machine, such as wheels, fans, motors, turbines, engines, due to their ability to sustain at different ranges of load, speed, different temperature, operating conditions, and different environmental conditions. Moreover, they have a low friction property which makes rolling easier. The main parts of rolling element bearings (Fig. 1) consists of rolling element (rollers or balls), cage (to hold the rolling element in position), inner race, and outer race (they both are grooves or flat paths in which rolling element rolls). Any type of fault in any of these parts can lead to complete failure or even seizure of the machine. So, monitoring and maintenance of rolling element bearing health are one of the most important factors to increase the machine's life and workability.

N. Kumar · S. Mandal · C. Mishra (✉) · N. B. Hui
NIT Durgapur, Durgapur, India
e-mail: chintamani.mishra@me.nitdgp.ac.in

S. Sidhardh et al. (eds.), *Advances in Structural Integrity for Mechanical, Civil, and Aerospace Applications*, Lecture Notes in Mechanical Engineering,
https://doi.org/10.1007/978-981-97-6367-2_12

Fault symptoms of rolling element bearing can be found out by analysing the vibration signal or acoustics signal and/or motor current signal. In the present study, vibration analysis is used for detecting the different characteristics of the faults in rolling element bearing. The acceleration signal of outer race is collected and further simplified in MATLAB.

Different signal processing steps are applied to successfully extract the fault features of the signal. An impulse of vibration signal arises when a rolling bearing element strikes any discontinuities in its path. The bearing characteristic frequency (BCF) provides a hypothetical approximation of frequency to be projected in the spectra when there is any fault in any parts of rolling element bearing. Different vibration generation mechanism are proposed by different researchers depending on the design and operational parameters.

Vibration analysis is successfully used to detect faults in rolling element bearing [1, 2]. Vibration signatures for either simulation or experimental results are analysed to get the condition of any machines. Some important contributions in fault diagnosis schemes of rolling element are higher-order spectrum technique [3], empirical mode decomposition [4], averaging of time-synchronous [5], wavelet analysis [6], envelope analysis [7], spectral kurtosis [8], etc. Earlier, some work on bearing has also been reported using multi-body dynamics model MBS software ADAMS [9, 10]. In modelling, Mishra et al. [11] developed a multi-body dynamics model of ball bearing using bond graph multi-body simulation tool. Smith et al. [12] successfully diagnosed the faults in rolling element bearing at slow speed operation.

In the present work, the first model, a 5-DOF model is based on the earlier reported paper [13], which is done in MATLAB Simulink environment and the second model is an extension of earlier reported work [14], which is done in MBS-ADAMS environment.

2 Objectives

A brief literature review of various works, reported by different researchers is presented in previous section. The main focus of different research activities can be either on modelling or on signal processing techniques to extract faults in different operating and loading conditions. Based on the above literature survey, the focus of this work is given to the following areas.

- Modelling of ball bearing: Develop model of a ball bearing that takes care of ball and cage dynamics, inner race, and outer race dynamics, ball dynamics, contact force, non-linear contact stiffness, compression, slip, and traction force. Different types of faults are introduced in the bearing so that the synthetically generated signal can be used to find out the fault symptoms of the bearing.
- Diagnosis scheme: Propose a suitable diagnosis scheme that can diagnose faults in rolling element bearing at constant speed operation.

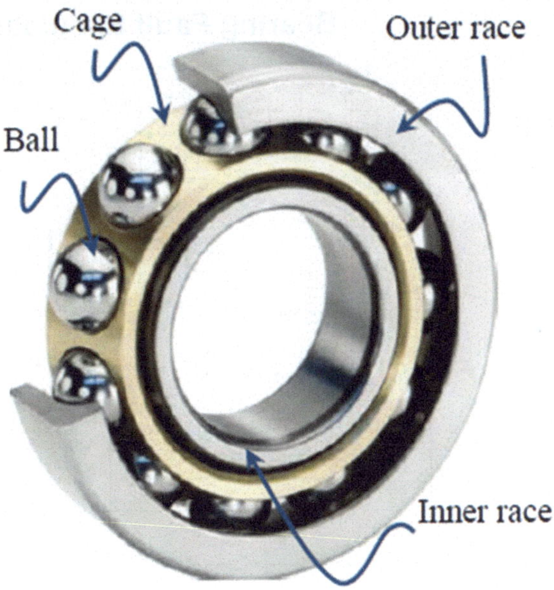

Fig. 1 Important parts of rolling element bearing

The above-mentioned objectives are achieved by (a) a 5-DOF MATLAB Simulink model and (b) a CAD-based simulation software ADAMS. The model is further extended by introducing different types of faults such as ball, inner, outer races faults.

3 5-DOF Matlab Simulink Model of Ball Bearing

The method proposed in [13] is used to develop the 5-DOF MATLAB Simulink model (Fig. 2) of bearing. The outer and inner races have 2-DOF of each in x and y direction, and the inner race has one extra DOF due to rotation about the z-axis. Inertia of inner and outer races, contact force and localized fault are considered while developing this model. It is assumed in [8, 9], and most of the bearing models available in the paper that the balls are spaced at equal angular spacing with a consideration that the cage as well as balls rotate at constant angular speed. In modelling, the outer race is constraint to move by connecting it to the pedestal/foundation by spring damper systems and this spring damper system represents structural resistance and damping.The inner race rotates at a constant angular speed whereas the outer race is fixed. Hertzian contact theory is used to find the normal force between mating surfaces assuming contact between the different mating surfaces as surface contact. Again in modelling the slip (tangential forces) between the contacting surfaces is neglected.

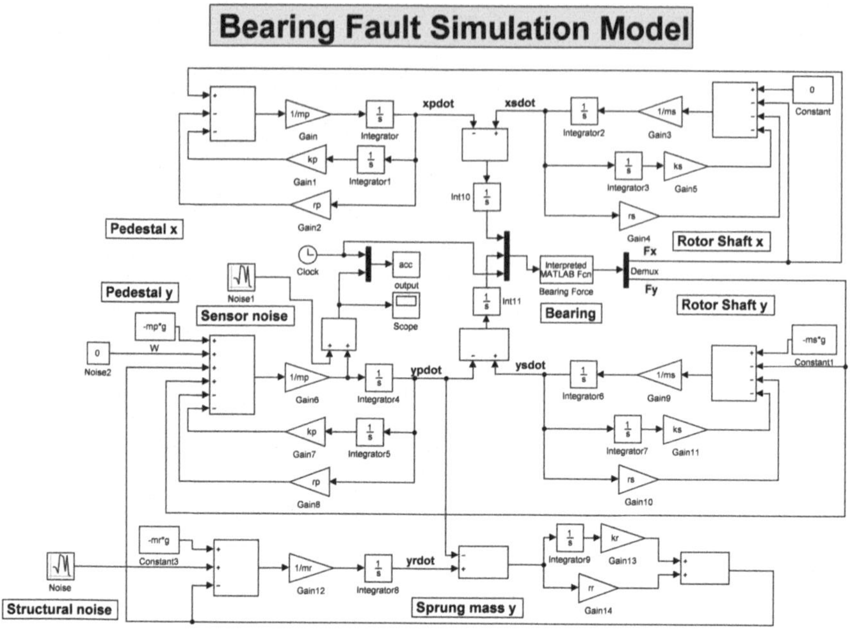

Fig. 2 MATLAB Simulink model of the rolling element bearing

4 Multi-body Dynamics CAD Model of Ball Bearing

The deep grooved ball bearings and test rig, which is used for bearing fault detection in MSC Adams Simulation Software is shown in Fig. 3. In developing the model, it is assumed that the outer race, inner race, balls, and cage are rigid. A flexible coupling is used to connect the rotor shaft of the motor to the connecting rod. As the rotor shaft is rigid, it is assumed to be rigid under torsion as well as bending. A small parallel misalignment of 0.5 mm is introduced in the model by shifting the axis of the rotor and the axis of the connecting shaft of the motor. The connecting rod is supported by two deep groove ball bearings.

Each of the two bearing is mounted on a pedestal. A good bearing is installed at the near end whereas the faulty bearing is mounted at the far end of the motor. The model assumes inner race and the connecting rod moves at a constant speed and it can be achieved by introducing a fixed joint between the two.

All the parameters and conditions, used in designing of deep groove ball bearing in [14], is used to design the ball bearing in present research work. The simulation data was collected for 10 s. with a step size of 0.0001 s. The rotor was introduced with a constant angular speed of 10 Hz, and data are collected in time domain. The signal is then post processed by using envelope analysis (passed through a band pass filter and then FFT) to check the response of the system with and without faults.

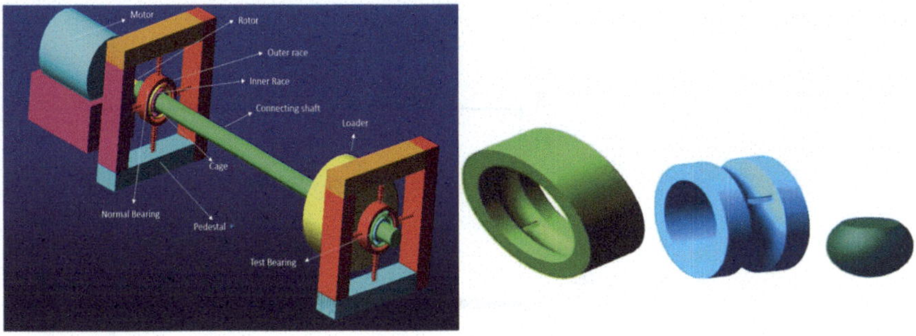

Fig. 3 MSC ADAMS model of ball bearing with the test rig

Table 1 Contact parameters used in ADAMS model [14]

Parameters	Values
Contact stiffness (K)	1.0×10^{-8} N/m
Contact damping (R)	1.0×10^{4} Ns/m
Dynamic coefficient of friction (μ_k)	0.3
Static coefficient of friction (μ_s)	0.1
Pedestal bushing stiffness (K_x, K_y, K_z)	1.0×10^{8} N/m
Pedestal bushing damping (R_x, R_y, R_z)	1.0×10^{4}Ns/m
Coupling stiffness (K_x, K_y, K_z)	1.0×10^{5} N/m
Coupling damping (R_x, R_y, R_z)	1.0 Ns/m
Pedestal stiffness	1.0×10^{8} N/m
Pedestal damping	1000 Ns/m
Force exponent (e)	2.2
Boundary penetration	1.0×10^{-4}

5 Validation of Model and Simulation Setup

The parameters of the bearing are given in the Table 1 [14]. In design, the stiffness and damping of the pedestal are taken as 1.0×10^{8} N/m and 1000 Ns/m, respectively. The contact constraints, used in the contact model are given in Table 1 [14].

The frequency of shaft speed (inner race), ball speed and cage speed are represented by $\dot{\theta}_i$, $\dot{\theta}_b$, $\dot{\theta}_c$ in Fig. 4. Using the simple kinematic analysis we can easily find out the speed of the rotating elements in the rolling element bearing. The results obtain in MSC ADAMS model nearly matches the value of the corresponding speed in case of simple kinematics-based model.

In Fig. 4, it can be observed that inner race and cage speed is positive whereas the ball speed is negative. Thus, the balls rotate in the opposite direction to that of the inner race and the cage. This validates the basic principles of rotation of ball bearing parts.

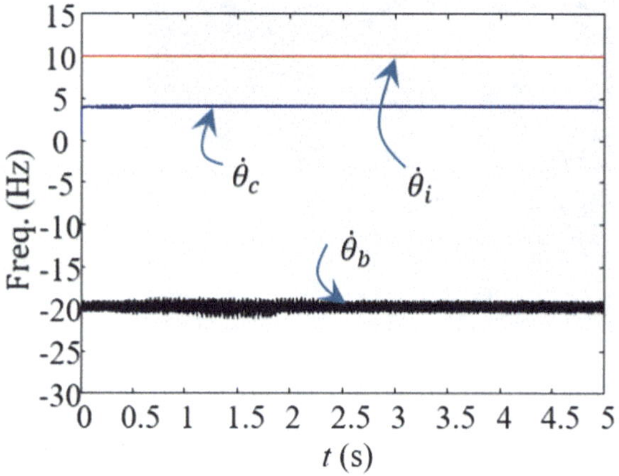

Fig. 4 Simulated result of constant speed rotor supported by bearing ADAMS model

6 Results and Discussion

Two models ((a) 5-DOF MATLAB Simulink model and (b) ADAMS model) of bearing is used for simulation of different faults. The simulation of the bearing is done for 10 s. with a step size of 0.00005 s. and the rotor shaft rotates at a constant speed of 10 Hz.

The bearing characteristics frequencies [14] can be calculated as:

Inner-race defect frequency:

$$f_{\mathrm{bpfi}} = \frac{z(f_{\mathrm{o}} - f_{\mathrm{i}})\left(1 + \frac{d_{\mathrm{b}}}{D_{\mathrm{p}}}\cos(\theta)\right)}{2} \tag{1}$$

Outer-race defect frequency:

$$f_{\mathrm{bpfo}} = \frac{z(f_{\mathrm{o}} - f_{\mathrm{i}})\left(1 - \frac{d_{\mathrm{b}}}{D_{\mathrm{p}}}\cos(\theta)\right)}{2} \tag{2}$$

Cage rotational frequency:

$$f_{\mathrm{c}} = \frac{f_{\mathrm{i}}\left(1 - \frac{d_{\mathrm{b}}}{D_{\mathrm{p}}}\cos(\theta)\right)}{2} + \frac{f_{\mathrm{o}}\left(1 + \frac{d_{\mathrm{b}}}{D_{\mathrm{p}}}\cos(\theta)\right)}{2} \tag{3}$$

Ball spin frequency:

$$f_{\mathrm{bsf}} = \frac{f_{\mathrm{O}} - f_{\mathrm{i}}}{2}\frac{D_{\mathrm{p}}}{d_{\mathrm{b}}}\left(1 - \frac{d_{\mathrm{b}}}{D_{\mathrm{p}}}\cos(\theta)\right)^{2} \tag{4}$$

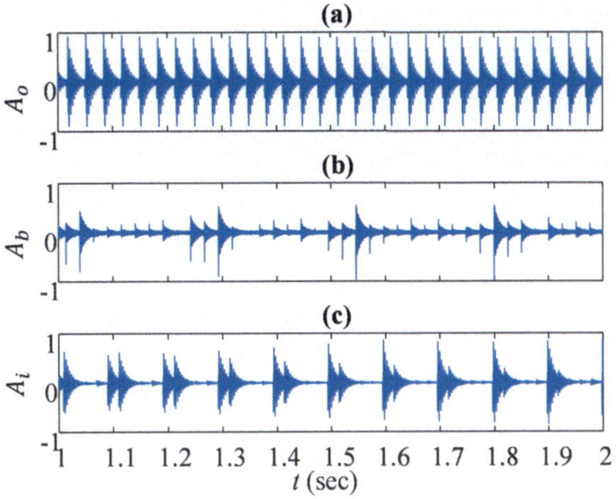

Fig. 5 A zoomed figure of normalized vibration signal from MATLAB Simulink model: **a** vibration from outer race fault, **b** vibration from ball fault and **c** vibration from inner race fault

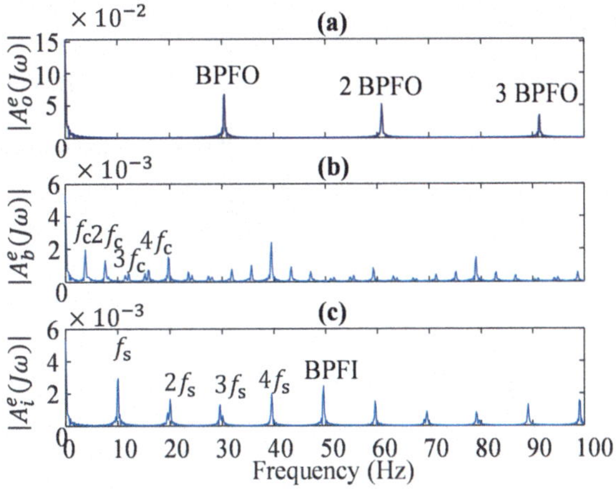

Fig. 6 Enveloped spectra of vibration signal from MATLAB Simulink model: **a** vibration from outer race fault, **b** vibration from ball fault and **c** vibration from inner race fault

Since outer race is fixed, so $f_0 = 0$, $f_{\mathrm{bpfi}} = 4.955\ f_{\mathrm{i}} = 49.55\,\mathrm{Hz}$, $f_{\mathrm{bpfo}} = 3.044$, $f_i = 30.44\,\mathrm{Hz}$, $f_c = 0.38\ f_{\mathrm{i}} = 3.80\,\mathrm{Hz}$ and $f_{\mathrm{bsf}} = 1.594$, $f_{\mathrm{i}} = 15.94\,\mathrm{Hz}$.

Among the bearing characteristics frequencies the BPFO and BPFI depends on the number of balls whereas the ball spin frequency (BSF) and cage frequency (CF) does not depends on the number of balls.

A zoomed portion of simulated signal for 5-DOF system with outer race, ball fault and inner race fault are shown in Fig. 5. The envelope spectra of the corresponding signal are shown in Fig. 6.

In case of outer race fault BPFO (ball pass frequency of outer race) and its harmonics are exposed in the envelope spectra. The BPFI (ball pass frequency of inner race), frequency of shaft, and its harmonics and sidebands are exposed in the enveloped spectra. In case of ball fault, cage frequency and its harmonics are exposed in the envelope spectra.

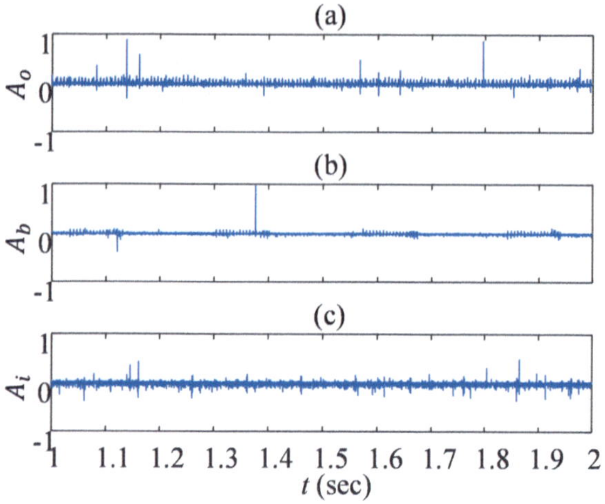

Fig. 7 A zoomed figure of normalized vibration signal from MSc ADAMS model: **a** vibration from outer race fault, **b** vibration from ball fault and **c** vibration from inner race fault

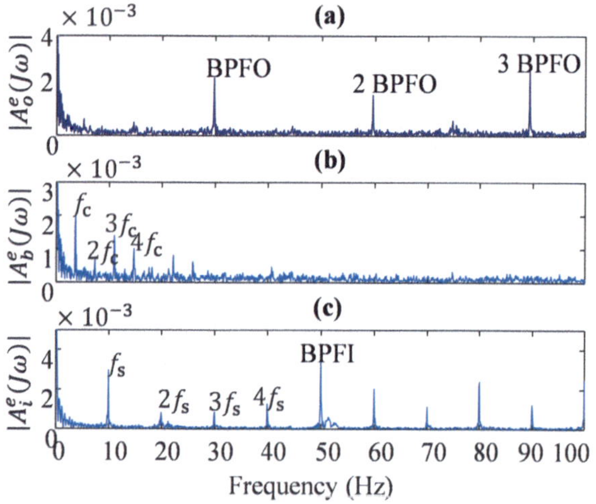

Fig. 8 Enveloped spectra of vibration signal from MSc ADAMS model: **a** vibration from outer race fault, **b** vibration from ball fault and **c** vibration from inner race fault

The primary difference between all the envelope spectra is that the shaft frequency and its harmonics can be observed in the case of inner race fault. Zoomed portion of normalized vibration signal from MBS-ADAMS model with outer race fault, ball fault, and inner race fault, are shown in Fig. 7 with respective enveloped spectra of vibration signal in Fig. 8.

A noisier signal can be observed in Fig. 7 as compared to Fig. 5 due to the fact that the ADAMS model takes care all the dynamics of the ball bearing model compared to a Simulink model. If we compare envelope spectra results (Figs. 6 and 8), frequencies of ball pass frequencies of outer race, cage and inner race, it can be observed in the corresponding signals in case of outer race, inner race and ball faults and these results are also matched with the numerical values of these frequencies calculated using kinematic analysis. For example, the numerical value of BPFO is 3.044

$f_i = 30.44$ Hz if we take the frequency of inner race is 10 Hz, in both Figs. 6 and 8, BPFO comes almost 30.44 Hz and these results are also the same for BPFI, BSF, and cage rotational frequencies [14].

7 Conclusions

Two models of the deep groove ball bearing (Rexnord MB-ER-12K 3/4") are designed and simulated. Between the two models, one is a 5-DOF system deep groove ball bearing model is developed in MATLAB Simulink environment and the other is a multi-body dynamics CAD model of deep groove ball bearing developed using multi-body simulation software MSC ADAMS. Different types of faults such as outer race fault, ball fault, and inner race fault are introduced in the developed models. Comparing the results of both the models in frequency domain and time domain, it can be observed that both results are in agreement with each other and also these results exactly matches with the theoretical bearing characteristics frequencies of bearing. The contribution of this work can be summarized as:

- The model developed in MSC ADAMS can be useful for simulating very complex faults, which is difficult to do practically or experimentally.
- Development of model using MSC ADAMS is too easy, and it does not require a knowledge of bearing kinematics/dynamics. The model can be used for a variety of problems with rotors and motors.
- The frequency domain characteristics of bearings under different types of defects are studied using the developed models, and the diagnosis schemes are tested using the simulated signals.
- The MSC ADAMS model of deep groove ball bearing is the one which can give vibrational response of the good as well as defected bearing under non-lubrication or greased contact condition.
- The model can be further used to test the effect of different lubrication and the vibrational response of the defected bearing.
- Between the two models the 5-DOF Simulink is the simple one but it is successful to model fault and deliver response at high-speed rotor operation.

References

1. Tandon N, Choudhury A (1999) A review of vibration and acoustic measurement methods for the detection of defects in rolling element bearings. Tribol Int 32(8):469–480
2. Jardine AK, Lin D, Banjevic D (2006) A review on machinery diagnostics and prognostics implementing condition-based maintenance. Mech Syst Signal Process 20(7):1483–1510
3. Howard IM (1997) Higher-order spectral techniques for machine vibration condition monitoring. Proc Inst Mech Eng G J Aerosp Eng 211(4):211–219

4. Bin G, Gao J, Li X, Dhillon B (2012) Early fault diagnosis of rotating machinery based on wavelet packets-empirical mode decomposition feature extraction and neural network. Mech Syst Signal Process 27:696–711
5. McFadden P, Toozhy M (2000) Application of synchronous averaging to vibration monitoring of rolling element bearings. Mech Syst Signal Process 14(6):891–906
6. Mishra C, Samantaray A, Chakraborty G (2016) Rolling element bearing defect diagnosis under variable speed operation through angle synchronous averaging of wavelet de-noised estimate. Mech Syst Signal Process 72:206–222
7. Su YT, Lin SJ (1992) On initial fault detection of a tapered roller bearing: frequency domain analysis. J Sound Vib 155(1):75–84
8. Antoni J, Randall RB (2006) The spectral kurtosis: application to the vibratory surveillance and diagnostics of rotating machines. Mech Syst Signal Process 20(2):308–331
9. Cakmak O, Sanliturk K (2011) A dynamic model of an overhung rotor with ball bearings. Proc Inst Mech Eng K J Multi-body Dyn 225(4):310–321
10. Sakaguchi T, Harada K (2006) Dynamic analysis of cage behavior in a tapered roller bearing. J Tribol-Trans ASME 128:604–611
11. Mishra C, Chakraborty G, Samantaray AK (2013) Rolling element bearing fault modelling to develop a diagnosis scheme for oscillating and non-uniform shaft rotation. In: The first international and sixteenth national conference on machines and mechanisms, Roorkee, pp 12–18
12. Smith J (1982) Vibration monitoring of bearings at low speeds. Tribol Int 15(3):139–144
13. Sawalhi N, Randall R (2008) Simulating gear and bearing interactions in the presence of faults: Part I. The combined gear bearing dynamic model and the simulation of localised bearing faults. Mech Syst Signal Process 22(8):1924–1951
14. Mishra C, Samantaray A, Chakraborty G (2017) Ball bearing defect models: a study of simulated and experimental fault signatures. J Sound Vib 400:86–112

Numerical Study of Bulging Instability in a Porous Tube Under Internal Pressure

Bobby Dudhe, Arun Kumar Singh, Pawan Kumar Soni, and Vadapalli Surya Prasanth

Abstract Bulging instability is a mechanical failure that happens when a thin tube undergoes excessive localized deformation before ultimately failing. Understanding this process is crucial in ensuring the stability of various mechanical and biological systems, such as fluid-carrying rubber tubes and aneurysms in the human brains. In this numerical study, we used the finite element method and a hyperelastic material to simulate bulging instability in a closed-end tube. Our findings reveal that inflation only occurs when the internal pressure in the tube reaches a critical value, which depends on the tube's thickness but not its length. Further analysis of the poro-hyperelastic tube demonstrates that the presence of porosity lowers the critical pressure required for inflation by reducing the stresses in the tube material.

Keywords Bulging/inflation instability · Poro-hyperelastic materials · Large deformation analysis

1 Introduction

Soft rubbers and elastomers are known for their large deformation and low strength and are generally classified as hyperelastic materials [1]. These materials are renowned for their flexibility, tear resistance, durability, etc. [1]. When a hyperelastic tube is pressurized internally, it expands axisymmetrically, and further increase in internal pressure leads to bulging instability [2]. An interesting phenomenon related to such a solid is brain aneurysm, which arises due to increased pressure in blood vessels [3]. Blood vessels and veins are also examples of hyperelastic materials; however, these materials are porous in nature. A porous material can be visualized as a solid with microvoids [4, 5]. Furthermore, fluid in a pipe/tube causes stresses

B. Dudhe · A. K. Singh (✉) · P. K. Soni · V. S. Prasanth
Visvesvaraya National Institute of Technology, Nagpur, India
e-mail: aksinghb@gmail.com

S. Sidhardh et al. (eds.), *Advances in Structural Integrity for Mechanical, Civil, and Aerospace Applications*, Lecture Notes in Mechanical Engineering,
https://doi.org/10.1007/978-981-97-6367-2_13

in the wall of the pipes/tubes. Two types of stresses, namely hoop/circumferential stress and longitudinal stress, develop in the wall of a circular tube [2]. However, in the case of a thick cylinder, radial stress also becomes significant [2].

A literature survey reveals a widespread study on the bulging behavior of rubber tubes and spheres [6–8]. Mallock (1891) was perhaps the first to report experiments on the bulging of tubes and balloons made of India rubber using fluid. Numerical simulations of the inflation of rubber tubes due to internal pressure have also been reported in the literature [7, 9]. The critical pressure, the pressure at which bulging begins, is computed numerically and found to be of lower magnitude than the experimental value [7, 9]. The nonlinear behavior of an axisymmetric membrane results in the generation and spread of bulges in cylindrical rubber balloons [10]. Sang et al. (2016) have proposed a theoretical model for the bulging of a rubber tube with a nonlinear theory of elasticity [2]. They have also derived an expression correlating internal pressure and circumferential principal stretch. However, very limited study on the bulging of a poro-hyperelastic tube has been reported in the literature. Selvadurai and Suvorov (2016, 2018) have studied a column of the porous material, and fluid was used in the experiment to observe that internal fluid pressure dissipates slowly with time [4, 5]. The aim of the present study is to understand the bulging behavior of an internally pressurized poro-hyperelastic tube under different conditions.

2 Methodology

A thin tube made of hyperelastic materials is simulated numerically using finite element analysis with Abaqus software [11]. Outer diameter and thickness of porous-hyperelastic cylindrical tube are $d = 20.0$ mm and $h = 1.0$ mm respectively, and length l of the tube $l = 0.70$ m is considered in the model. This matches the condition of thin cylinder which is generally compared with diameter-to-thickness ratio of the tube ($d/h \leq 20$). The study incorporates the two-parameter Mooney–Rivlin material model to describe the behavior of hyperelastic solids. According to this model, the strain energy density of the material is represented by a polynomial function of the principal stresses or strains [1]. The magnitudes of the coefficients of this model are taken to be $C_{10} = 0.69$, $C_{01} = 0.173$, and $D_1 = 0.0124$ [12]. Effect of porosity in the material model is also considered in terms of void fraction Φ of the tube material in the numerical simulations. In addition, quadrilateral mesh was used to model the tube. Both ends of porous hyperelastic tube are closed and fixed as a boundary condition. Static pressure was applied inside the tube internally until bulging occurs.

Figure 1 presents the convergence analysis of the numerical simulations. It is observed that hoop stress σ_h initially depends on number of elements n_0, but that becomes constant at $n_0 = 3000$ elements. As a result, at least 3000 mesh elements were chosen for the analysis of the tube at different internal pressures, p_0.

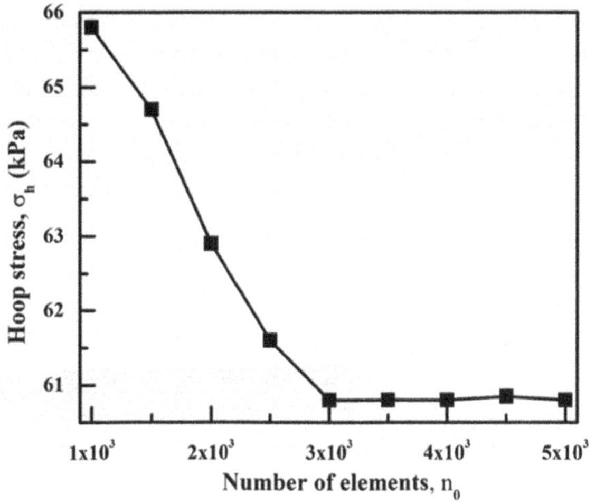

Fig. 1 Hoop stress versus number of elements for convergence analysis

3 Results and Discussion

3.1 *Effect of Tube Length on Hoop Stress of Mooney–Rivlin Material*

Figure 2 presents the plot concerning hoop stress σ_h versus tube length l. It is observed that σ_h initially depends on length l of the tube, but the same becomes independent of l. Similar observation was seen in the case of longitudinal stress σ_l versus tube length l. A possible reason for the observation is attributed to end effect of the tube in case of tubes having smaller length, in that case.

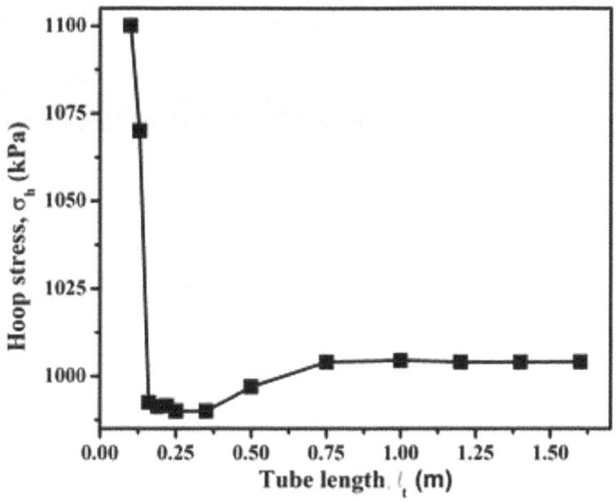

Fig. 2 Presents the effect of tube length l on hoop stress σ_h of a hyperelastic tube at a constant internal pressure, p_0

Figure 3 illustrates that lateral deformation of the tube remains cylindrical as long as $p_0 < p_{cr}$. At $p_0 = p_{cr}$, deformation in the central region of the tube tends to be localized in the form of bulging. Further increase of pressure that is, $p_0 > p_{cr}$, size of the bulge grows symmetrically, while the region closer to the boundaries is almost free of the internal pressure in Fig. 3. Although it is not reported here, it has been observed that critical pressure p_{cr} remains independent of tube length l.

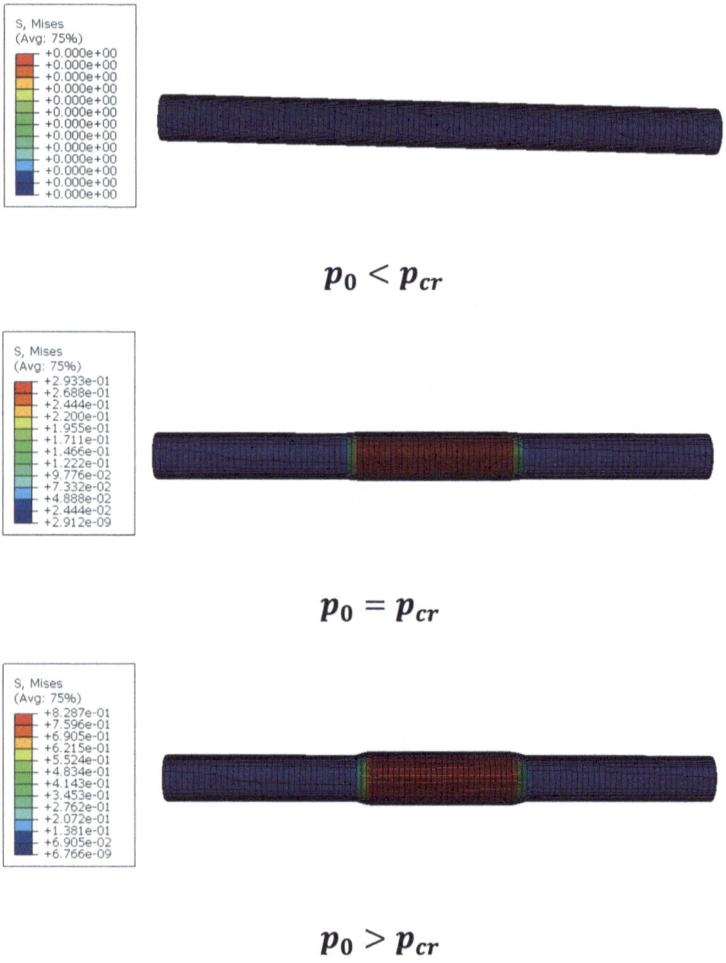

Fig. 3 Shows the bulging process of a cylindrical tube as function of internal pressure p_0

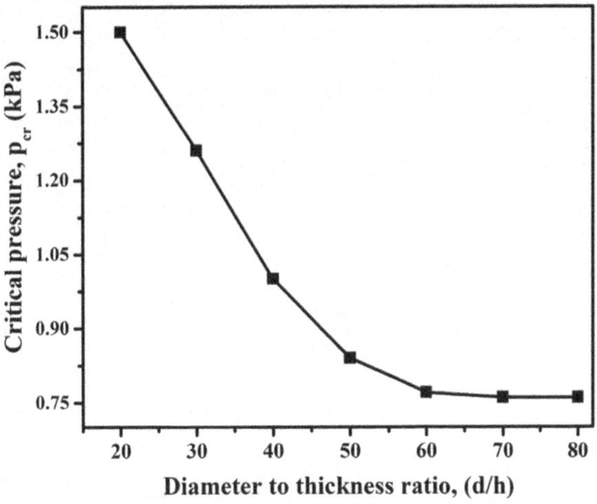

Fig. 4 Presents the relationship between (d/h) ratio and critical pressure p_{cr}

Figure 4 presents the parametric study concerning the effect of diameter-to-thickness ratio (d/h) on critical pressure p_{cr}. It is observed that p_{cr} decreases with increasing d/h ratio, before becoming constant at $\frac{d}{h} = 60$. It is also concluded from Fig. 4 that wall thickness h of the tube affects the bulging process more that diameter d of the tube. This is owing to increased thickness results in existence of radial stress, thus resulting in higher p_{cr} [2].

3.2 *Effect of Void Fraction on Critical Pressure of a Poro-Hyperelastic Tube*

Figure 5 illustrates the effect of void fraction Φ (%) on p_{cr} and σ_h of the porous tube. It is concluded that p_{cr} decreases upon increasing Φ in the material of the tube. As a result, the porous tube bulges at lower p_0 than the tube with $\Phi = 0\%$. The corresponding hoop stress σ_h also decreases with Φ. A possible reason for the observation is attributed to diffusion of internal pressure due to presence of voids in the tube material.

3.3 *Effect of Internal Pressure on Hoop and Longitudinal Stresses in Poro-Hyperelastic Tube*

Figure 6 presents the variation of hoop stress σ_h as well as longitudinal stress σ_l as function of p_0. At lower p_0, magnitudes of both match closely with stresses in the

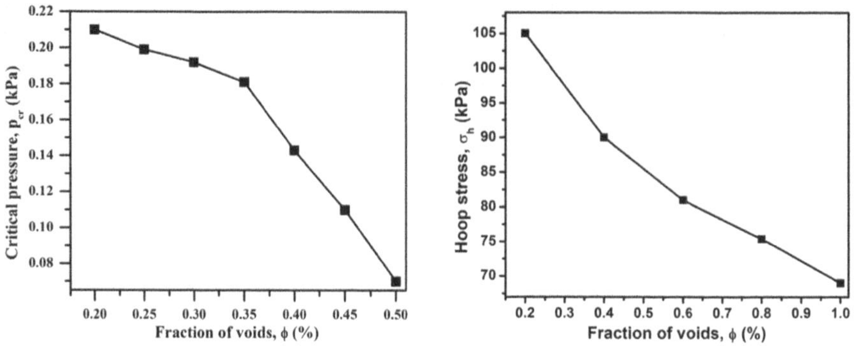

Fig. 5 Effect of void fraction Φ (%) on critical pressure p_{cr} and hoop stress σ_h

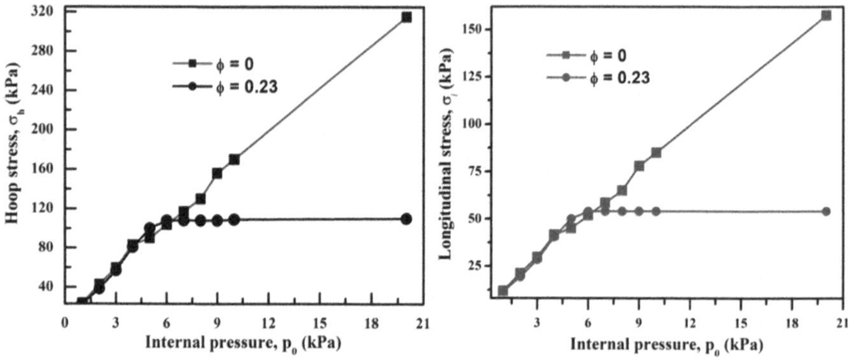

Fig. 6 Illustrates the effect of internal pressure p_0 on hoop stress σ_h and longitudinal stress σ_l for void fraction $\Phi = 0\%$ and 0.23%

tube without porosity at lower. However, both stresses begin to deviate each other at higher. Also, both become constant at ~ 7.0 kPa in the porous tube. This observation may be attributed to increased size of voids due to pressure, thus failing to sustain the internal pressure.

4 Conclusions

- Critical pressure, at which bulging begins, is independent of tube length. Moreover, critical pressure of poro-hyperelastic tube (8.0 kPa) is less than the hyperelastic tube (100.0 kPa).
- As d/h ratio of the tube increases, critical pressure decreases. Diameter of the tube does not affect the critical pressure.
- Critical pressure of poro-hyperelastic tube decreases with increasing void fraction.

- Magnitudes of both hoop and longitudinal stresses are less in the porous tube than the tube without porosity. These stresses do not increase after a certain pressure ~ 7.0 kPa.

References

1. Treloar LR (1976) The mechanics of rubber elasticity. Proc R Soc Lond A Math Phys Sci 351:301–330
2. Sang J, Xing S, Liu H, Li X, Wang J, Lv Y (2016) Large deformation analysis and stability analysis of a cylindrical rubber tube under internal pressure. J Theo Appl Mech 55:177–188
3. Fu YB, Xie YX (2012) Effects of imperfections on localized bulging in inflated membrane tubes. Philos Trans R Soc A Math Phys Engg Sci 370:1896–1911
4. Selvadurai APS, Suvorov AP (2016) Coupled hydro-mechanical effects in a poro-hyperelastic material. J Mech Phys Solids 91:311–333
5. Selvadurai APS, Suvorov AP (2018) On the development of instabilities in an annulus and a shell composed of a poro-hyperelastic material. Proc R Soc A Math Phys Eng Sci 474:20180239
6. Mallock A (1891) Note on the instability of India-rubber tubes and balloons when distended by fluid pressure. Proc R Soc Lond 49:458–463
7. Pamplona DC, Goncalves PB, Lopes SRX (2006) Finite deformations of cylindrical membrane under internal pressure. Int J Mech Sci 48:683–696
8. Wang S, Guo Z, Zhou L, Li L, Fu Y (2019) An experimental study of localized bulging in inflated cylindrical tubes guided by newly emerged analytical results. J Mech Phys Solids 124:536–554
9. Lin Z, Li L, Ye Y (2020) Numerical simulation of localized bulging in an inflated hyperelastic tube with fixed ends. Int J Appl Mech 12:2050118
10. Lestringant C, Audoly B (2018) A diffuse interface model for the analysis of propagating bulges in cylindrical balloons. Proc R Soc A Math Phys Eng Sci 474:20180333
11. Dudhe B (2022) Bulging instability of hyper-elastic tubes under internal pressure. M. Thesis, VNIT Nagpur
12. Abaqus user's guide, http://130.149.89.49:2080/v6.14/books/usb/default.htm

Damage Assessment in Coated Femoral Stem Using Numerical Analysis

Samiksha Moharana, R. Gnanamoorthy, and Yuichi Otsuka

Abstract The deterioration of bone cartilage in hip joints caused by osteoarthritis and rheumatoid arthritis is the most prevalent cause of total hip replacement. The femoral stem in the implant is fixed to the bone at the time of surgery. It is coated with biocompatible ceramics such as hydroxyapatite (HAp) to enable faster bone growth. The coating experiences relative motion during the initial period, before sufficient bone growth occurs, and loosens in many cases. Suspension plasma-sprayed hydroxyapatite (HAp)-based coatings such as HAp/Titania on titanium substrates have been proven suitable for both mechanical properties and biocompatibility. Slip occurs primarily at the coating and metallic implant interface (the bonded interface), accelerating the coating damage due to fretting. The current investigation is directed at understanding the influence of assembly load, coating property, and delamination length on the interface strength between coating and implant. A two-dimensional finite element model of a 170 mm-long femoral stem coated with HAp/Titania coating is analyzed using commercial software. The geometrical model is like the original joint condition of the human hip implant. The numerical analysis showed that the first contact edge is where the most stress and contact slip happen because that is where the stress is concentrated during an assembly loading condition. The variation in load and contact slip conditions will result in coating failure in the implant.

Keywords Hip implant · HAp/Titania coatings · Reliability · Contact slip · FEA

S. Moharana (✉) · R. Gnanamoorthy
Department of Mechanical Engineering, Indian Institute of Technology Madras, Chennai 600036, India
e-mail: me18d002@smail.iitm.ac.in

R. Gnanamoorthy
e-mail: gmoorthy@smail.iitm.ac.in

Y. Otsuka
Department of Mechanical Engineering, Nagaoka University of Technology, Nagaoka 940 2188, Japan
e-mail: otsuka@vos.nagaokaut.ac.jp

S. Sidhardh et al. (eds.), *Advances in Structural Integrity for Mechanical, Civil, and Aerospace Applications*, Lecture Notes in Mechanical Engineering,
https://doi.org/10.1007/978-981-97-6367-2_14

1 Introduction

Total hip replacements are a widely used surgical procedure in modern times aimed at addressing the consequences of accidents or the degenerative effects associated with arthrosis. Titanium and its alloys are often thought to be the best materials for prosthetic joints because they are strong, lightweight, and compatible with biological systems [1]. However, some issues must be addressed despite the answer appearing to be the best option. As the lifespan of prosthetics progresses, the release of metal ions and debris from their surface gives rise to an intricate interaction with the bloodstream and body fluids [2]. The mentioned interaction, although not readily observable at a more significant level, has the potential to initiate adverse effects on the patient's health. The outcomes are frequently evident as osteolysis, which refers to the breakdown of bone tissue, and metallosis, which is the widespread distribution of metallic debris throughout the body [3]. Although titanium is known for its ability to improve the reliability of joint implants, the presence of metal debris remains a notable concern. Achieving a delicate equilibrium between mechanical durability and mitigating potential health hazards associated with metal release is of utmost importance in the advancement of prosthetic technologies, hence guaranteeing the efficacy and safety of total hip replacements [4].

The artificial implant is given different treatments and surface modifications to increase biocompatibility and early bone growth. Different implant fixation methods are used since the implant cannot be linked to human bone directly, cementless and cement [5]. Bone cement is used for fixing implants. However, because of the brittle nature of the materials, if the bone cement fails, the femoral stem will disintegrate [6]. Surgeons typically choose a fixation of the implant to the bone that is not cemented. Many failures are reported due to fretting wear and corrosion of the coatings.

Hydroxyapatite (HAp)-based calcium phosphate coatings are usually applied as implant coating materials [7] as they contain the organic component of the human bone [8]. The suspension plasma spray (SPS) process is a recent development, as the coatings produced have reduced porosity, high interfacial strength, and smooth surfaces [9]. The bioactive coating must not deteriorate for some time in order to maximize the efficacy of bone growth [10]. Poor fixation, implant design, or high daily activity levels contribute to failure between two mating surfaces under high tangential loads. Metallosis, osteolysis, and loosening of the bone-implant interface have been reported due to the wear debris produced after prolonged use [11]. This fixation loss during the initial implantation period causes wear and degradation of the coating material. Failure of the coating material can increase the risk of implant alteration. There has been very little research on understanding the nature of contact slip at the bone coating interface caused by adequate bone growth. Contact slip is primarily estimated at the coating and implant interface, which is a significant cause of femoral implant delamination (Fig. 1). The cyclic load on the coating surface can cause slip conditions on the coated implant.

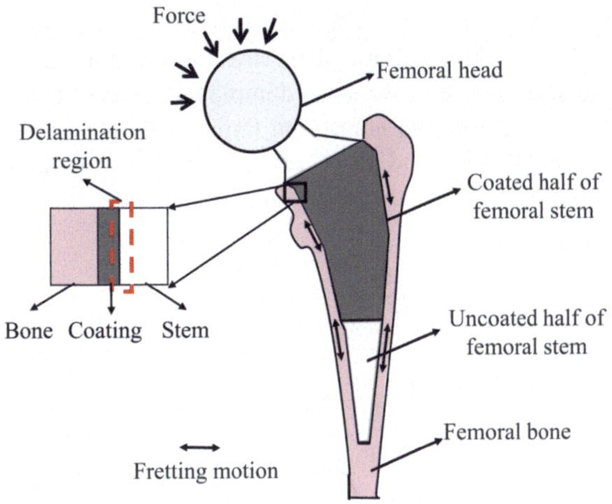

Fig. 1 Femoral stem and femur bone assembly and expected delamination region

The present work implements finite element analysis to investigate the influence of assembly loading, coating quality, and delamination length on the interfacial region between coatings and implants.

2 Material and Finite Element Model

Numerical simulations were performed to assess the contact slip at the delamination locations of the stem-bone contact. The material used for the numerical analysis is HAp and HAp-based composite coatings on titanium substrates for the cementless implant. The implant is coated on the top half portion with a coating thickness of 100 microns. The material properties considered for current studies are given in Table 1.

Using commercial finite element software ABACUS®, a 2D finite element model of the implant with a 165 mm-long femoral stem was made, put together, and analyzed (Fig. 2). The femoral head has a diameter of 28 mm, a femoral neck that is 135° bent, and a 12–14 mm taper. The femur bone is assumed to be immobile (all degrees of freedom are set to zero) in the current numerical analysis. A commercial finite element tool checks how much the femoral stem and the femur bone move relative

Table 1 Material properties used in the numerical simulation [12]

Material	Elastic modulus (GPa)	Poisson's ratio
Titanium (femoral stem)	110	0.36
Bone (femoral bone)	17	0.31
HAp coating	75	0.24
HAp/Ti_2O_3 coating	90	0.24

to each other when different loads are put on them (Fig. 3). There were 26,525 eight-noded biquadratic plane stress-reduced integration elements in the model. The femur and the metallic-coated implant are meant to create solid contact as the former is inserted into the latter. The femur bone is assumed to be fixed with DOF = 0. Loads of 784 N and 981 N are equivalent to human body weights of 80 and 100 kg and are applied to the taper femoral neck region.

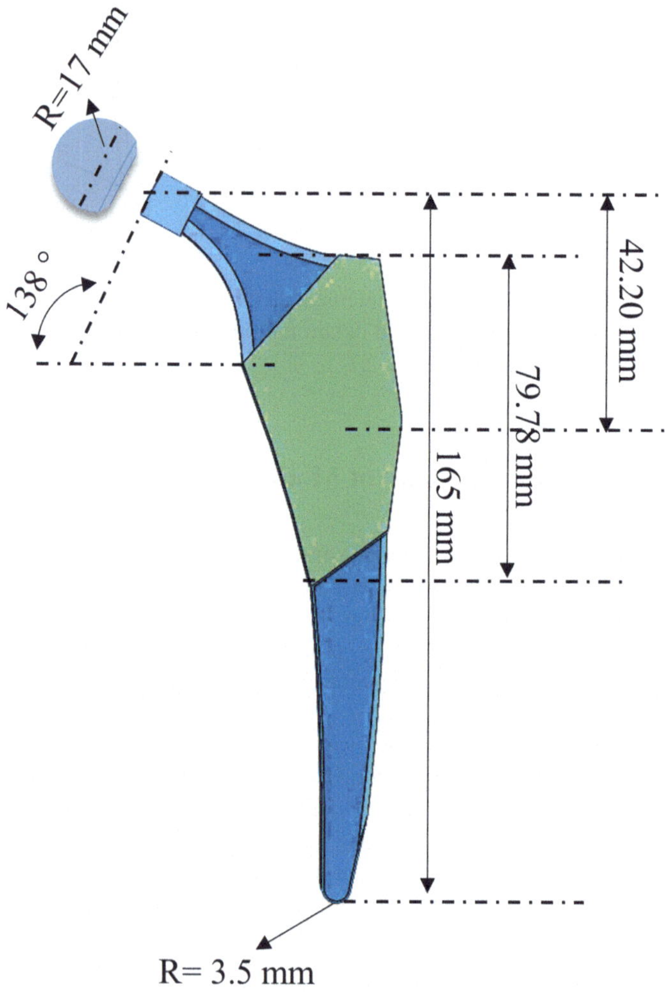

Fig. 2 Two-dimensional representation of the femoral neck and head

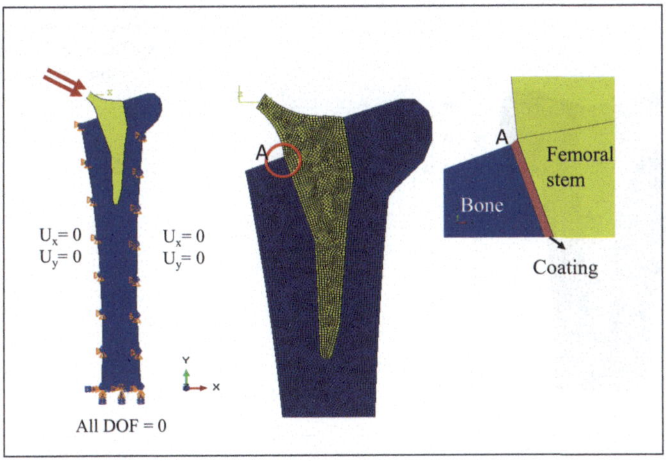

Fig. 3 Loading and boundary conditions applied to femoral stem and femur bone assembly

3 Results and Discussion

3.1 Effect of Load on the Shear Stress Distribution

When two different body weights are put on the area between A and B, Fig. 4 shows how the shear stress is distributed on the surface of the HAp/Titania coating that is bonded to the titanium substrate. Shear stress is induced by the compression of the femoral implant, which may cause the coating to separate from the substrate. The shear stress exceeds the interfacial stress value at the initial contact region of the coating and substrate in the case of a coating thickness of 100 microns considered in the present study [13]. The stress concentration in the coated area causes this increase in shear stress. This concentration zone is prone to delamination and fails under body weight. The failure initiation region is defined as the point at which the shear stress at the interface exceeds the interfacial stress of the coating.

3.2 Contact Slip Estimation with Varying Delamination Lengths

In order to mitigate inadequate bonding caused by coating problems, the research utilizes the concept of 'delamination lengths'. The lengths mentioned in this context function as a modeling parameter, serving to depict potential areas of local failure. The study focuses on the analysis of three discrete contact delamination lengths, which are strategically positioned at locations where the shear stress surpasses the

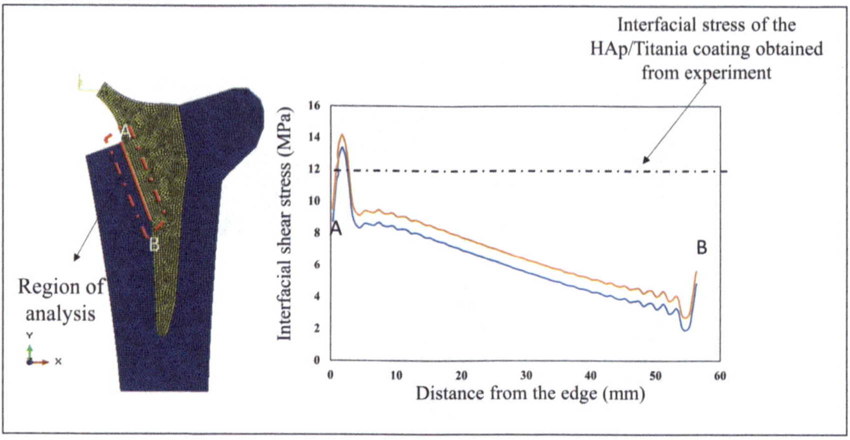

Fig. 4 Shear stress distribution in the bonded region between coating and metallic stem (**A**, **B**)

maximum interfacial stress. The utilization of this methodology allows for the identification of relative motion occurring at the interface, providing significant insights into possible vulnerabilities and contributing to the comprehension of problems stemming from insufficient bonding of coatings. The 2D finite element model and the delamination zone are mentioned in Fig. 5a.

Three potential contact delamination lengths were considered for the numerical simulation: 0.2, 0.5, and 1 mm. The contact slip estimation on the bonded surface for 784 N and 981 N is shown in Fig. 5b and c. The contact slip indicates a maximum value for the contact delamination length of 1 mm and a minimum value for the contact delamination length of 0.2 mm for both loading conditions. At the contact edge, a noticeable contact slip can be seen, which gradually decreases as it approaches the region where the tie contact first occurs. Prediction of delamination and slide at the interface of the contact pad and substrate under fatigue loading conditions has been described by Nagentrau et al. [14]. It is stated that the initial edge of the contact was where the delamination and contact slide were highest.

The presence of a sharp contacting edge serves as a focus point that intensifies stress concentration, resulting in increased contact slip and making the area more prone to failure. The concentration at the edge has a significant effect on the stress distribution along the bonded surface. This causes the load to rise and the coating to start stretching. The present strain induces increased contact slip values, exhibiting a significant peak at a delamination length of 1 mm. The higher contact slip is because the coating and substrate do not interact with each other as much, and their interfacial strength is decreasing. The increased slippage and delamination are two risks that need to be considered: They damage the interface and speed up the wear in the coating. The complicated relationship between stress concentration, contact slip, and delamination shows how important it is to control these weaknesses to make sure that the bonded materials are solid and last a long time.

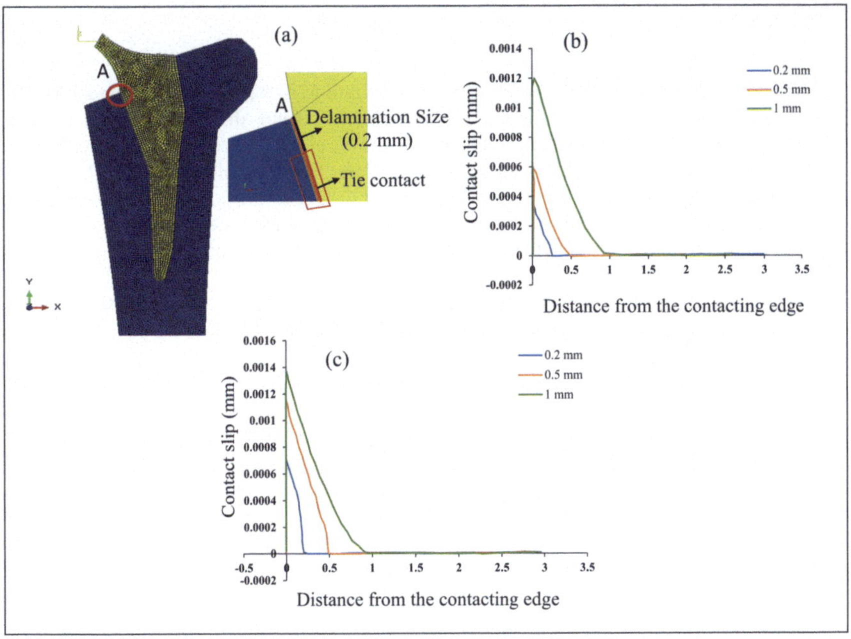

Fig. 5 **a** Model showing delamination region and bonded region, and elemental slip distribution at contact zone under different assembly loads, **b** 784 N, **c** 981 N

4 Conclusion

The impact of assembly load and contact delamination length on the contact slip estimation at the coating/substrate interface in a HAp/Titania coating with a thickness of 100 microns is explored based on numerical simulation studies.

- The role of having sharp corners in the implant is crucial, as it leads to stress concentration that results in noticeable contact slip occurring in close proximity to these edges. These corners are identified as probable locations where failure may initiate.
- As the length of delamination increases, there is a corresponding rise in contact slippage, indicating a progression toward coating breakdown. Moreover, an increased load on the implant results in the amplification of contact slip at the interface, hence amplifying the relative motion.
- The interaction of assembly load, delamination length, and contact slip yields significant insights into the susceptibilities of the interface between the coating and substrate, hence revealing significant factors that impact the durability and performance of the material.

References

1. Moghadasi K, Mohd Isa MS, Ariffin MA, Mohd Jamil MZ, Raja S, Wu B, Yamani M, Bin Muhamad MR, Yusof F, Jamaludin MF, Bin Ab Karim MS, Binti Abdul Razak B, Bin Yusoff N (2022) A review on biomedical implant materials and the effect of friction stir-based techniques on their mechanical and tribological properties. J Mater Res Technol 17:1054–1121
2. Family R, Solati-Hashjin M, Nik SN, Nemati A (2012) Surface modification for titanium implants by hydroxyapatite nanocomposite. Caspian J Intern Med 3:460–465
3. Kim KT, Eo MY, Nguyen TTH, Kim SM (2019) General review of titanium toxicity. Int J Implant Dent 5:1–12
4. Quinn J, McFadden R, Chan CW, Carson L (2020) Titanium for orthopedic applications: an overview of surface modification to improve biocompatibility and prevent bacterial biofilm formation. IScience 23:101745
5. Hsieh MF, Perng LH, Chin TS (2002) Hydroxyapatite coating on Ti6Al4V alloy using a sol-gel derived precursor. Mater Chem Phys 74:245–250
6. Herrera A, Mateo J, Gil-Albarova J, Lobo-Escolar A, Ibarz E, Gabarre S, Más Y, Gracia L (2015) Cementless hydroxyapatite coated hip prostheses. BioMed Res Int 2015(1):386461
7. Søballe K (1993) Hydroxyapatite ceramic coating for bone implant fixation: mechanical and histological studies in dogs. Acta Orthop 64:1–58
8. Sun L, Berndt CC, Gross KA, Kucuk A (2001) Material fundamentals and clinical performance of plasma-sprayed hydroxyapatite coatings: a review. J Biomed Mater Res 58:570–592
9. Pawlowski L (2009) Suspension and solution thermal spray coatings. Surf Coat Technol 203:2807–2829
10. Chung YY, Ki SC, So KY, Kim DH, Park KH, Lee YS (2009) High revision rate of hydroxyapatite-coated ABG-I prosthesis. J Orthop Sci 14:543–547
11. Langton DJ, Sidaginamale R, Lord JK, Nargol AVF, Joyce TJ (2012) Taper junction failure in large-diameter metal-on-metal bearings. Bone Joint Res 1:56–63
12. Samiksha M, Gnanamoorthy R, Otsuka Y (2022) Fretting wear characteristics of suspension plasma-sprayed hydroxyapatite coating on titanium substrates for orthopedic applications. J Mater Eng Perform 31:7290–7901
13. Ahmed F (2012) Deformation and damaging mechanisms in diamond thin films bonded to ductile substrates DOKTOR-INGENIEUR
14. Nagentrau M, Tobi ALM, Jamian S, Otsuka Y (2019) Contact slip prediction in HAp coated artificial hip implant using finite element analysis. Mech Eng J 6(3):18–00562

Analysis of Prestressed Shape Memory Alloy Actuator for Compliant Mechanism

G. Jayabharath Reddy and S. Maniprakash

Abstract Comparing other possible actuators for compliant mechanisms, shape memory alloy has an advantage in terms of compactness and the specific actuation energy. In this work, actuation capacity of an SMA wire under different prestressed condition has been studied. To perform this study, due to the elastic nature of the compliant mechanism, the mechanism has been modeled as a spring and considered the entire system to be a spring and SMA wire assembled in series. With a suitable constitutive model for SMA, the displacement of spring has been simulated for cyclic thermal loading. This analysis has been repeated for different prestressed condition. The simulation results show the strong dependence of prestressed value with the actuator displacement.

Keywords Shape memory alloy · Smart material · SMA actuator · Compliant mechanism

1 Introduction

Compliant mechanisms offer a novel alternative to traditional industrial actuators by replacing moving parts with flexible components, reducing complexity. They can withstand large deflection, which helps in accomplishing the required actuation [1]. These mechanisms can perform multiple functions with relatively fewer moving links, leading to less wear, tear and noise [2]. Therefore they can be made as a single piece with reduced cost [3]. Because of their flexibility, they are used for smart applications with multiple stable positions possible, providing a range of displacement [4]. With the advancement in manufacturing, they can be manufactured on a microscale for application in micro-electro-mechanical systems [5].

The links of compliant mechanisms are generally passive members where the electric or pneumatic actuators provide the motion. Instead of these, smart materials like piezoelectric, shape memory alloys, or magnetostrictive can be used as actuators.

G. J. Reddy · S. Maniprakash (✉)
Indian Institute of Information Technology, Design and Manufacturing, Kurnool, India
e-mail: smp@iiitk.ac.in

S. Sidhardh et al. (eds.), *Advances in Structural Integrity for Mechanical, Civil, and Aerospace Applications*, Lecture Notes in Mechanical Engineering,
https://doi.org/10.1007/978-981-97-6367-2_15

Smart materials have coupling between different physical domains where a change in the state of one domain, like temperature, can change the state in another domain like strain [6]. Because of this, they provide the advantage of having both the actuation and control. Piezoelectric actuators are used if high accuracy and fast response speed are required [7]. For the present work, shape memory alloys have been considered due to its large stroke for actuation [8].

Shape memory alloys (SMA) can recover large strains of around 6–7%, and because of this, they are exploited in many robotic and medical applications [9, 10]. They possess high specific actuation energy density [11] which makes them an ideal choice for compact actuators. At high temperature, SMA exhibit a stable phase called austenite. At low temperatures, they have another phase structure called martensite. When deformed at low temperatures, they undergo large inelastic strain. When the temperature increases, the alloy transforms into its parent phase austenite. During this process, the inelastic strain present in the alloy is recovered. It is known as the shape memory effect. This recoverable strain is utilized for actuation. Because of the thermomechanical coupling, they can be controlled by the temperature. At high temperatures, the material also exhibits another unique behavior called pseudoelasticity.

Because of their ability to perform as both sensing and work producing device, they can be used as actuators in compliant mechanisms [12]. It is utilized for designing a sun tracking mechanism to position a solar receptor without any external power source [13]. A finger-like mechanism is designed where bending SMA are used as compliant links providing the actuation [14]. The shape memory effect is utilized in designing the variable stiffness compliant mechanism [15]. Compliant tendon cables along with SMA wires are used in rehabilitation robotics to mimic the hand applications [16]. The compliant SMA actuators are used in rehabilitation and surgical robotics, to increase response speed and output range outperforming traditional SMA actuators [17]. The design of a compliant mechanism using SMA-based hinge actuators allows for reduced weight, size, and complexity of a gripper which demonstrates good adaptability for grasping objects of different sizes [18]. The performance is analyzed for changing the lengths of flexible matrix and length of rigid segments. Results show that a longer rigid link increases the angle of actuation whereas a longer flexible matrix decreases the actuation. A self-sensing gripper prototype made of 3D-printed parts using SMA actuator technology can provide improved performance compared to a standard pneumatic gripper [19]. In the above applications, the SMA is embedded in the structure. It can be used as a separate member for providing the actuation at discrete points. Multiple SMA wires are utilized to obtain different configurations [20]. By training the alloy, maximum actuation of the compliant mechanism can be achieved by exploiting the shape memory effect and pseudoelasticity [21].

In this work, we have discussed the analysis of the response of shape memory alloy actuator for compliant mechanisms. The actuation of the mechanism is controlled by varying the temperature of shape memory alloys. In this article, we propose that by prestressing the SMA wire, the performance of the compliant mechanism can be enhanced. This hypothesis has been established by performing various numerical simulations. To simulate the behavior of shape memory alloy, a phenomenon independent constitutive model developed and briefly discussed in Sect. 2. A simplified model is provided for the compliant mechanism to describe the kinetics in Sect. 3. Finally, the results for the influence of prestressing are discussed in Sect. 4.

2 Constitutive Model for SMA

In this section, the modeling of shape memory alloys is discussed. To effectively utilize the properties of shape memory alloys, a mathematical model to describe their behavior is essential. The underlying phenomenon called martensite transformation is diffusion less solid phase transformation [22]. The characterization of the phase transformation is the key for modeling their behavior [23, 24]. The hysteresis behavior of phase transformation adds to their complexity. This can be modeled at different length scales from single crystal to polycrystal, or bulk material [25]. Though single crystal models [26, 27] provide an accurate understanding of the behavior but are computationally intensive. The phenomenological approach is generally considered for engineering practice where constitutive relations are used for describing the behavior [28]. Various attempts have been made by using phase diagrams and a free energy approach for simulating the complex behavior of Shape memory alloys [29]. In this regard, some initial attempts were made by Tanaka [30] by utilizing the $\sigma - T$ (Stress–Temperature) phase diagram to identify different zones of transformation. In their work, an exponential relation was used to describe the kinetics of phase transformation with the help of an internal variable known as martensite volume fraction. Later this model was improvised by Liang and Rogers [31] by proposing a cosine relation for the evolution of the internal variable. But these have the drawback of not being able to differentiate between twinned and detwinned martensite. To overcome this, a model was proposed by Brinson [32] with two internal variables differentiating the evolution of twinned and detwinned martensite. A modified kinetic relation of internal variables was proposed. A thermodynamically consistent model was proposed by Byod and Lagoudas [33], where a thermodynamic force conjugate based on Gibbs free energy was used for the evolution of internal variables. In a model proposed by Nallathambi et al. [34], where three variants of the alloy are used to describe the state of the material, the evolution kinetics between these phases is used to describe all the possible phase transformations. In the present work, the authors have considered a model proposed by them for its thermodynamical consistency and its computational efficiency.

For this study, a one-dimensional thermodynamically consistent model was used to describe the constitutive behavior of the alloy. For modeling the transformation behavior two internal variable martensite strain (ε^{mt}) and austenite volume fraction (ν^{a}) are used. This is a stress-based model with temperature as another independent state variable. The total strain is additively decomposed into elastic and inelastic part, where elastic strain is defined by the state of stress as follows:

$$\varepsilon = \varepsilon^e + \varepsilon^{ie} = \varepsilon^e + \varepsilon^{mt}(1 - \nu^a). \tag{1}$$

$$\mathcal{D} \equiv -\rho\dot{G} - \dot{\sigma}\varepsilon - \rho s\dot{T} \geq 0 \tag{2}$$

The model proposed is thermodynamically consistent, which satisfies the Clausius-Duhem inequality given by Eq. (2) where σ, ε is stress and strain, s, T is entropy and temperature and ρ is density. Two kinds of behavior occur in the alloy. One is a transformation between two solid phases, Austenite and Martensite, called twinning [35] and other way round called reverse transformation. Another is deformation in the alloy called detwinning between the different variants of martensite [36]. To accommodate these two behavior, two internal variables are chosen. The stress and temperature determine the state of the material in the transformation. In terms of these variable, the Gibbs free energy is proposed which is additively decomposed into a reversible and irreversible part. The reversible free energy is given in terms of elastic compliance S, thermal expansion coefficient α, specific heat capacity c and s_0, u_0 are specific entropy and internal energy at reference temperature T_0 as given below.

$$\begin{aligned}
G(\sigma, \varepsilon^{mt}, \nu^a, T) &= G^r(\sigma, \varepsilon^{mt}, \nu^a, T) + G^{ir}(\varepsilon^{mt}, \nu^a, T) \\
&= \frac{-1}{2\rho}\sigma^2 S(\nu^a) - \frac{1}{\rho}\sigma\left[\alpha(\nu^a)(T - T_0) + \varepsilon^{mt}(1 - \nu^a)\right] \\
&\quad + c(\nu^a)\left[(T - T_0) - \ln\left(\frac{T}{T_0}\right)\right] \\
&\quad + \frac{a}{\rho}\left[\frac{\varepsilon^{mt}}{\varepsilon_s^{mt}}.\tanh^{-1}\left(\frac{\varepsilon^{mt}}{\varepsilon_s^{mt}}\right) + \frac{1}{2}\ln\left|1 - \left(\frac{\varepsilon^{mt}}{\varepsilon_s^{mt}}\right)^2\right|\right] \\
&\quad + \frac{b}{2\rho}\left[(2\nu^a - 1)\tanh^{-1}(2\nu^a - 1) + \frac{1}{2}\ln\left|1 - (2\nu^a - 1)^2\right|\right] \\
&\quad + \frac{c_1\ln(c_2 - \nu^a)}{\rho}(T - T^\circ)
\end{aligned} \tag{3}$$

The reversible energy is due to the elastic deformation of the material and irreversible energy is due to the transformation strain of the alloy. An appropriate G^{ir} is chosen such that it effectively captures the behavior of the SMA. By following the standard thermodynamic arguments [37] the relation for strain, entropy and the dissipation rate is given below.

$$\begin{aligned}
\varepsilon = -\rho\frac{\partial G}{\partial \sigma} &= \sigma S(\nu^a) + \alpha(\nu^a)(T - T_0) + \varepsilon^{mt}(1 - \nu^a) \\
&\approx \sigma S(\nu^a) + \varepsilon^{mt}(1 - \nu^a)
\end{aligned} \tag{4}$$

Table 1 Model parameters used in simulations for NiTi Alloy (50.1% Ni)

Parameter	Value	Unit
$S_A = 1/E_A$	1/62.701	GPa^{-1}
$S_M = 1/E_M$	1/30.591	GPa^{-1}
T_c	6	K
σ_c	$(1-\nu)\times 135+\nu\times 435$	MPa
T^0	299.5	K
c_1	0.7	$MPa\ K^{-1}$
c_2	3.4	–
ε_s^{mt}	0.05	–
a	80	MPa
b	6	K

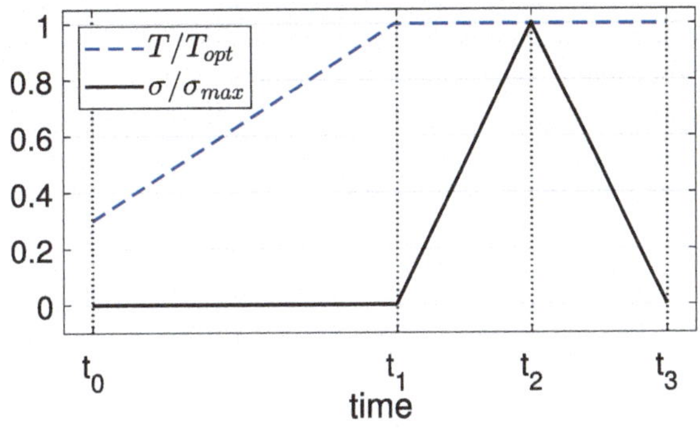

Fig. 1 The isothermal tensile loading applied for simulating stress strain behavior at different operating temperatures

$$\begin{aligned}\mathcal{D} &= -\rho\frac{\partial G}{\partial\varepsilon^{mt}}\dot{\varepsilon}^{mt} - \rho\frac{\partial G}{\partial\nu^{a}}\dot{\nu}^{a} \geq 0\\ &= \left[\sigma(1-\nu^{a}) - \rho\frac{\partial G^{ir}}{\partial\varepsilon^{mt}}\right]\dot{\varepsilon}^{mt} + \left[\frac{\sigma^{2}}{2}\Delta S - \sigma\varepsilon^{mt} - \rho\frac{\partial G^{ir}}{\partial\nu^{a}}\right]\dot{\nu}^{a} \geq 0 \quad (5)\end{aligned}$$

The evolution equations are chosen such that they satisfy the dissipation inequality given by Eq. (5). To demonstrate the efficacy of the model, the simulation results of uniaxial tensile tests are performed. The material parameters chosen are given in Table 1. The thermomechanical loading applied for the tests is shown in Fig. 1. As the deformation of the SMA alloy is dependent on operating temperature therefore its behavior is simulated at different operating temperatures.

At low temperatures the strain present in the alloy is not recovered completely on unloading as shown in Fig. 2a. But at high temperatures, the strain occured due to detwinning is recovered as observed at 2b and c. The results are compared with experiments done by Shaw and Kyridas [38]. The material is in the martensite phase at low temperatures, upon loading it is transformed into stress-induced martensite, forming the stress plateau called detwinning. During this phase transformation, there

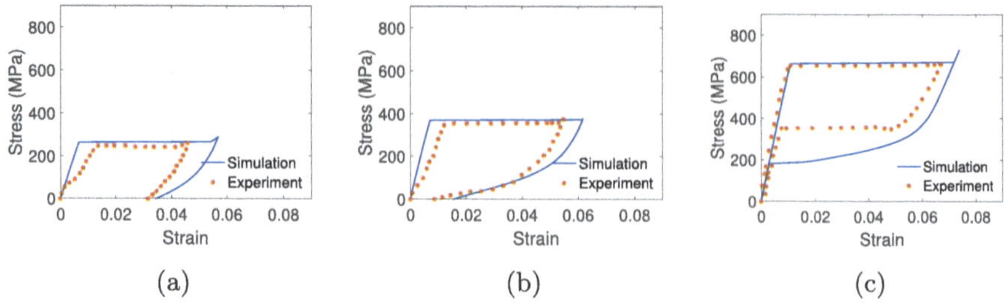

Fig. 2 The isothermal stress strain response from t_1 to t_3 as shown in Fig. 1 compared with the experimental results [38] at different temperatures. **a** Response at 293 K where the detwinned strain present is not recovered. **b** Response at 303 K showing partial recovery when unloaded. **c** Response at 353 K showing pseudoelasticity where complete strain is recovered upon unloading

is also an intermediate phase known as R-Phase which is not accounted for in this model. This causes the variation in slope during the initial loading stage. Also, the experiments performed are strain controlled where the maximum applied strain is given as a function of temperature. As the proposed model is stress controlled the end of detwinning (stress plateau) is not explicitly defined as a model parameter causing deviations during unloading.

3 Design and Analysis of Mechanism-Actuator System

In this section, the description of the compliant mechanism along with the SMA actuator is given. Following that, the formulation for the response of the compliant mechanism under cyclic thermal loading has been presented. Finally, an illustration is given to establish a fact that the prestress condition will enhance the response of the actuator.

3.1 Description of the Mechanism-Actuator System

For this analysis, the compliant mechanism along with the SMA actuator has been simplified into a spring and SMA wire arrangement as shown in Fig. 3. During the initial stage of the actuation, the SMA wire will be positioned with a prestrained condition. At this stage, the material will be in a detwinned martensite state. Then, upon heating, the material undergoes a phase transformation from martensite to austenite. Due to this, the inelastic strain will be recovered, and that results in the displacement of point P from position P_0 to P_1 as shown in Fig. 3b. This strain recovery process also exerts a force on the spring. Upon cooling, the material once again goes back to martensite state. At this stage, the spring will provide the bias force to retain its initial state by bringing back the point P toward its initial position

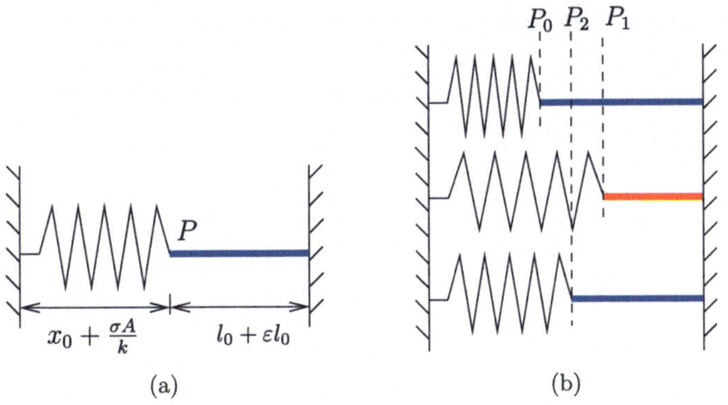

Fig. 3 **a** The schematic of the simplified representation of the compliant mechanism along with the SMA actuator. **b** The various configuration of the system under different thermal condition

P_0. However, the complete recovery is not possible for the subsequent cycles and therefore the point P reaches the position P_2, see Fig. 3b. This is due to the fact that the system will choose another equilibrium configuration at a lower temperature. Further heating and cooling cycle takes the point P between P_1 and P_2.

3.2 Formulation

In this mechanism-actuator arrangement, l_0, A represents the unstrained length and area of the SMA wire respectively, x_0, k is the free length and stiffness of the spring respectively. By ignoring all the dynamical effects, under any condition of actuation process, the force exerted on the spring can be assumed to be as same as the force on the SMA wire at position P. For this analysis, the response of the spring is assumed to be linear and the force balance condition leads to the following equation.

$$k\varepsilon_{\mathrm{w}} l_0 = \sigma_{\mathrm{w}} \, . \, A \tag{6}$$

where σ_{w}, ε_{w} represents stress and strain in the SMA wire. The strain value for the current load step can be calculated for a given stress and the temperature along with the internal values of the previous load step. Therefore, the constitutive relation can be compactly represented as $\varepsilon = \mathcal{E}(\sigma, T, \varepsilon^{mt}, \nu^a)$. The total length of the mechanism is given by the sum of free length of spring and extension in the spring, free length and extension in the SMA wire as given below.

$$L = l_0 + (\varepsilon_{\mathrm{w}})l_0 + x_0 + \frac{\sigma_{\mathrm{w}} A}{k} \tag{7}$$

Table 2 Algorithmic sketch for shape memory alloy actuator

$Given : T_{n+1}, \varepsilon_n^{mt}, \nu_n^a, \varepsilon_n, \sigma_n, m, c$
$Compute :$
$\quad \sigma_{n+1}, \varepsilon_{n+1}^{mt}, \nu_{n+1}^a, \varepsilon_{n+1}$
$\quad$ set $k = 1, error = 1, tol = 1e^{-9}, \sigma_{n+1}^k = \sigma_{res}$
$While : error \geq tol \;\&\; k \leq 100$
$\quad \varepsilon = \mathcal{E}(\sigma_n^k, T_{n+1}, \varepsilon_n^{mt}, \nu_n^a)$
$\quad \sigma_n^{k+1} = m\varepsilon + c \qquad error = \lvert\sigma_n^k - \sigma_n^{k+1}\rvert$
$\quad k = k + 1$
End
$\sigma_{n+1} = \sigma_n^{k+1}$

As the total length of this arrangement is constant in all instances of Fig. 3. From the position in Fig. 3a the kinematic constraint can be represented using the equation as follows.

$$\mathcal{L} = l_0 + (\varepsilon_{\text{res}})l_0 + x_0 + \frac{\sigma_{\text{res}} A}{k} \tag{8}$$

By combining the Eqs. (6), (7) and (8), the following equation can be obtained.

$$f(T, \sigma, \varepsilon) = l_0 + [\varepsilon(T, \sigma)]\, l_0 + x_0 + \frac{\sigma A}{k} - \mathcal{L} = 0 \tag{9}$$

By solving the above equation, the displacement of point A in the mechanism-actuator arrangement as a function of temperature can be calculated. The parameters given in Eq. (9) can be reduced into two constants as shown below.

$$\sigma = \underbrace{\frac{-l_0 k}{A}}_{m} \varepsilon(T, \sigma) + \underbrace{\frac{k(\mathcal{L} - x_0 - l_0)}{A}}_{c} \tag{10}$$

Since Eq. (10) is a nonlinear equation, a fixed point iterative procedure is used to solve the equation. The algorithmic procedure is given in Table 2.

3.3 *Influence of Prestressed Condition*

In Fig. 4, the schematic of stress and strain graph of SMA wire at different operating temperature conditions. The graph also contains the schematic of force displacement relation of the spring normalized by the cross sectional area and the length of the SMA wire.

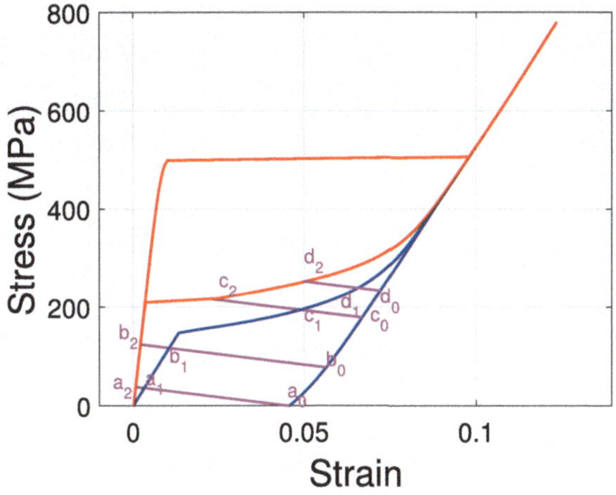

Fig. 4 Schematic representation of the effect of prestressing the SMA wire on the recoverable strain demonstrated at no prestress and at arbitrary prestress

The intersection points (Points a and b of Fig. 4) of the stress strain curve of the SMA and the normalized force displacement curve of the spring represent the equilibrium point at the corresponding operating temperature. It can be observed that by increasing the prestress of SMA the recoverable strain can be increased from a_1-a_2 to b_1-b_2. Upon further increasing the prestress the recoverable strain decreases as observed from c_1-c_2 to d_1-d_2.

4 Results

The actuation of the compliant mechanism is achieved by increasing and decreasing the temperature of the SMA. The temperature is increased until the material is completely transformed into austenite. The loading profile for the SMA actuator has two parts, one is before connecting to the mechanism, and the other is after the connection. In the first part, SMA is subjected to a thermal cycle for the shape setting process. After that, the alloy will undergo a stress cycle to convert twinned martensite to detwinned martensite. The material in this state is connected to the mechanism, and the repeated thermal cycle will be applied. While rising the temperature from the detwinned state, the alloy will recover the inelastic strain which causes the actuation. To simulate thermomechanical behavior, stress and temperature inputs are chosen as shown in Fig. 5. Here, upto T_A represents the first part of loading. The remaining, which is upto T_B represents the second part of the loading. During the second part of loading, the stress is not known explicitly. For this analysis, the physical properties of the mechanism are given by two values $m = -860$ MPa and $c = 64.3$ MPa at no prestress as given in Eq. (10).

As the temperature is gradually increased from 250 to 340 K (from T_A), the reverse transformation from detwinned martensite to austenite occurs. During this process, strain recovery occurs as shown in Fig. 6a. At the same time, due to extension, the bias

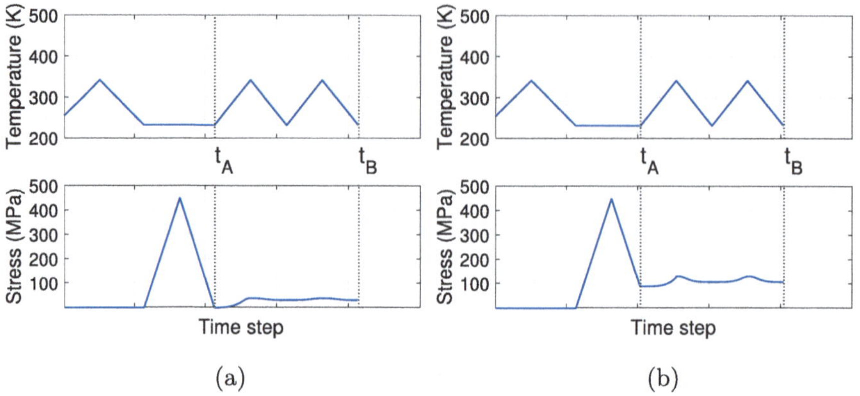

Fig. 5 Thermal and mechanical loading on shape memory alloy actuator where t_0 to t_A represents before connecting to compliant mechanism and t_A to t_B after connecting to mechanism. **a** With complete unloading when there is no prestress applied at t_A. **b** With a prestress of 90 MPa at t_A

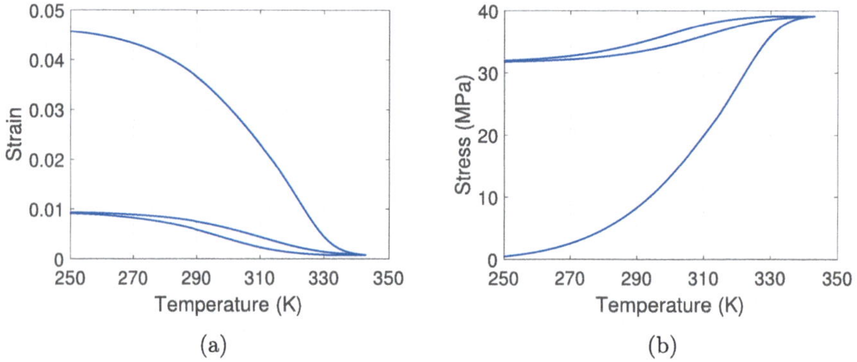

Fig. 6 The response of the SMA during thermal actuation for no prestress: **a** The change in strain with increase and decrease in temperature. **b** The change in stress with increase and decrease in temperature

force in the spring increases the stress in the SMA, as observed in Fig. 6b. For the other half of the cycle, martensite is formed in the SMA when the temperature is decreased. Because of the presence of bias force, it is again strained or transformed to detwinned martensite. However, during this transformation, the SMA is only partially strained. This causes the actuator to reach only upto point a_1 instead of point a_0 (Fig. 4). After that, When the temperature increases and decreases, it oscillates between points a_1 and a_2 as shown in Fig. 6. The strain between a_1 and a_2 will be the actual recoverable strain during the application.

The recoverable strain can be improved by increasing the prestress of the SMA. This is because, at the start of actuation, the spring is expanded due to the presence of stress in the SMA. When the temperature is increased, it further elongates the spring increasing the bias force. When the prestress of 90 MPa is applied, the maximum

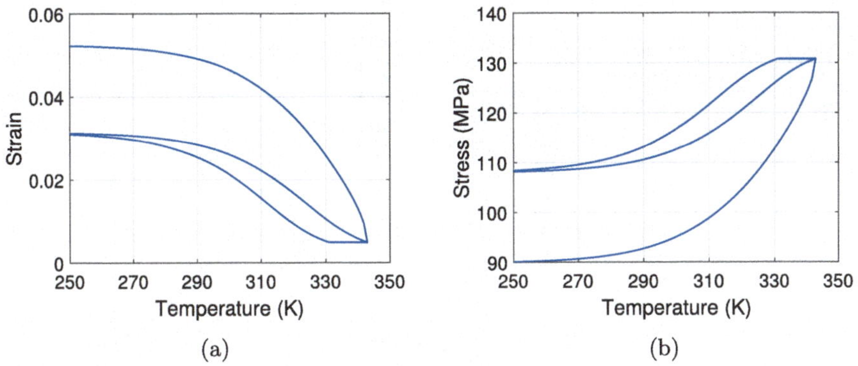

Fig. 7 The response of the SMA during thermal actuation at $\sigma_{res} = 90$ MPa. **a** The change in strain with increase and decrease in temperature. **b** The change in stress with increase and decrease in temperature

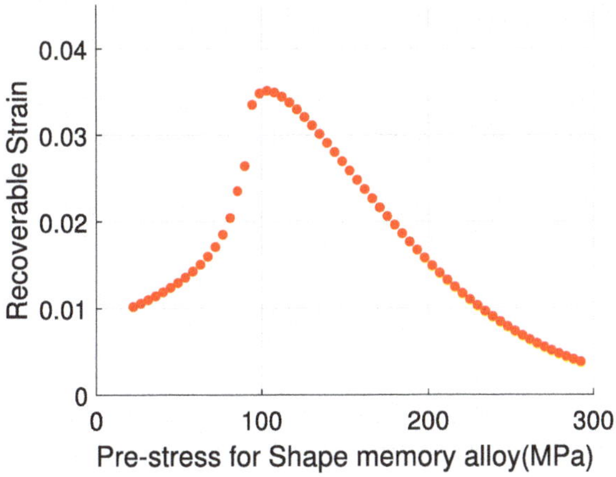

Fig. 8 The available recoverable strain for actuation with the increase in prestress of SMA

stress of 131 MPa is reached as shown in Fig. 7b. In the previous case of no prestress, the maximum stress is only 40 MPa. Therefore, due to prestressing, the SMA gains more than 2% of recoverable strain compared to no prestressed condition, as shown in Fig. 7.

As we increase the prestress, the available recoverable strain is shown in Fig. 8. It can be observed that as we increase the prestress, the recoverable strain increases upto a certain limit. The maximum recoverable strain of 3.5% is achieved at prestress of 103 MPa. This limitation comes from the maximum possible inelastic strain during the twinned to detwinned process. Upon further increasing the recoverable strain decreases as shown in Fig. 8. This is because of decrease in width between $c_1 - c_2$ and $d_1 - d_2$ as seen in Fig. 4. The maximum temperature applied can be utilized for modifying the recoverable strain required for the actuator. The variation is given in Fig. 9 which can be used appropriately. And the level of prestress can also be chosen according to the requirement.

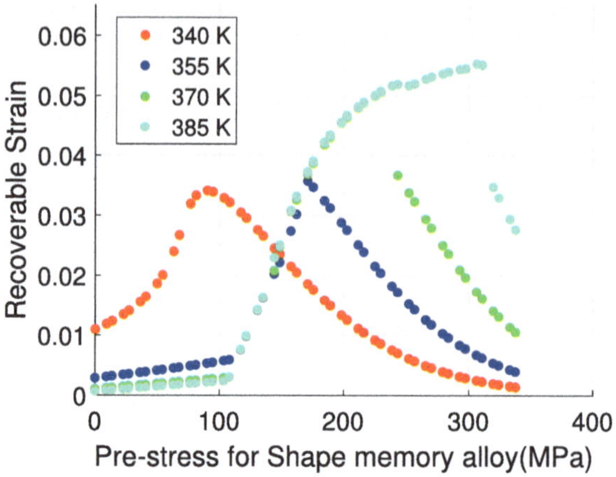

Fig. 9 The effect of maximum temperature on recoverable strain at different prestress values

5 Conclusions

With the recent advancements in product technology the requirement of compact and light weight actuators are needed. In this regard shape memory alloys are one of the important actuators for compliant mechanisms. The state of the alloy used in actuator determines its behavior. Prestressing the alloy while assembling it along with the mechanism can enhance the actuation stroke. In this work, the effect of the prestressed condition is analyzed. To perform this analysis, first, a suitable constitutive model is chosen. Then, the compliant mechanism is simplified into a bias spring. After that, the assemblage of bias spring and SMA wire in series is analyzed under cyclic thermal loading. Also By changing the maximum temperature the extent of strain recovery used for actuation can be altered. The simulation results show that increasing prestress increases the recovering strain significantly upto a saturation limit. This limit comes from the material's maximum inelastic strain during the twinned to detwinned transformation.

Acknowledgements This work was supported by a grant from Science and Engineering Research Board (SERB) under the SRG program, Project no.: SRG/2021/001672.

References

1. Howell LL (2013) Compliant mechanisms. In: Michael MJ (ed) 21st Century kinematics. Springer, London, pp 189–216 (2013)
2. Midha A, Norton TW, Howell LL (1994) On the nomenclature, classification, and abstractions of compliant mechanisms. J Mech Des 116(1):270–279
3. Howell LL (2001) Compliant mechanisms, 1st edn. Wiley-Interscience, New York
4. Zhu B, Zhang X, Zhang H, Liang J, Zang H, Li H, Rixin Wang (2020) Design of compliant mechanisms using continuum topology optimization: a review. Mech Mach Theory 143:103622

5. Kota S, Joo J, Li Z, Rodgers SM, Sniegowski J (2001) Design of compliant mechanisms: applications to MEMS. Analog Integr Circuits Signal Process 29(1):7–15
6. Leo DJ (2007) Engineering analysis of smart material systems. Wiley
7. Lyu Z, Xu Q (2021) Recent design and development of piezoelectric-actuated compliant microgrippers: a review. Sens Actuators A Phys 331:113002
8. Lu Y, Xie Z, Wang J, Yue H, Wu M, Liu Y (2019) A novel design of a parallel gripper actuated by a large-stroke shape memory alloy actuator. Int J Mech Sci 159:74–80
9. Sreekumar M, Nagarajan T, Singaperumal M, Zoppi M, Molfino R (2007) Critical review of current trends in shape memory alloy actuators for intelligent robots. Ind Robot Int J 34(4):285–294
10. Mohd Jani J, Leary M, Subic A, Gibson MA (2014) A review of shape memory alloy research, applications and opportunities. Mater Des (1980–2015) 56:1078–1113
11. Lagoudas DC (2008) Shape memory alloys: modeling and engineering applications. Springer
12. Garner LJ, Wilson LN, Lagoudas DC, Rediniotis OK (2000) Development of a shape memory alloy actuated biomimetic vehicle. Smart Mater Struct 9(5):673–683
13. Ganesh NJ, Maniprakash S, Chandrasekaran L, Srinivasan SM, Srinivasa AR (2011) Design and development of a sun tracking mechanism using the direct SMA actuation. J Mech Des 133(7)
14. Ali HFM, Khan AM, Baek H, Shin B, Kim Y (2021) Modeling and control of a finger-like mechanism using bending shape memory alloys. Microsyst Technol 27(6):2481–2492
15. Mekaouche A, Chapelle F, Balandraud X (2018) A compliant mechanism with variable stiffness achieved by rotary actuators and shape-memory alloy. Meccanica 53(10):2555–2571
16. Bundhoo V, Haslam E, Birch B, Park EJ (2009) A shape memory alloy-based tendon-driven actuation system for biomimetic artificial fingers, part I: design and evaluation. Robotica 27(1):131–146
17. Guo Z, Pan Y, Wee LB, Yu H (2015) Design and control of a novel compliant differential shape memory alloy actuator. Sens Actuators A Phys 225:71–80
18. Wang W, Rodrigue H, Kim H-I, Han M-W, Ahn S-H (2016) Soft composite hinge actuator and application to compliant robotic gripper. Compos B Eng 98:397–405
19. Scholtes D, Seelecke S, Rizzello G, Motzki P (2020) Design of a compliant industrial gripper driven by a bistable shape memory alloy actuator. In: Smart materials, adaptive structures and intelligent systems. American Society of Mechanical Engineers Digital Collection
20. Chaudhry Z, Rogers CA (1991) Bending and shape control of beams using SMA actuators. J Intell Mater Syst Struct 2(4):581–602
21. Sreekumar M, Nagarajan T, Singaperumal M (2009) Application of trained NiTi SMA actuators in a spatial compliant mechanism: experimental investigations. Mater Des 30(8):3020–3029
22. Otsuka K, Wayman CM (1999) Shape memory materials. Cambridge University Press
23. Lu ZK, Weng GJ (1997) Martensitic transformation and stress-strain relations of shape-memory alloys. J Mech Phys Solids 45(11):1905–1928
24. Šittner P, Heller L, Sedlák P, Chen Y, Tyc O, Molnárová O, Kadeřávek L, Seiner H (2019) B2 $\Rightarrow$ B19$'$ $\Rightarrow$ B2^T martensitic transformation as a mechanism of plastic deformation of NiTi. Shape Memory and Superelasticity 5:383–396
25. Cheikh Cisse, Wael Zaki, Tarak Ben Zineb (2016) A review of constitutive models and modeling techniques for shape memory alloys. Int J Plasticity 76:244–284
26. Etienne Patoor, Lagoudas Dimitris C, Entchev Pavlin B, Catherine Brinson L, Xiujie Gao (2006) Shape memory alloys, Part I: General properties and modeling of single crystals. Mech Mater 38(5–6):391–429
27. Anand L, Gurtin ME (2003) Thermal effects in the superelasticity of crystalline shape-memory materials. J Mech Phys Solids 51(6):1015–1058
28. Lagoudas DC, Entchev PB, Popov P, Patoor E, Catherine BL, Gao X (2006) Shape memory alloys, Part II: Modeling of polycrystals. Mech Mater 38(5–6):430–462
29. Ashish Khandelwal, Vidyashankar Buravalla (2009) Models for shape memory alloy behavior: an overview of modeling approaches. Int J Struct Changes Solids 1(1):111–148

30. Tanaka K, Nagaki S (1982) A thermomechanical description of materials with internal variables in the process of phase transitions. Ingenieur-Archiv 51(5):287–299
31. Liang C, Rogers CA (1990) One-dimensional thermomechanical constitutive relations for shape memory materials. J Intell Mater Syst Struct 8(4):285–302
32. Brinson LC (1993) One-dimensional constitutive behavior of shape memory alloys: thermomechanical derivation with non-constant material functions and redefined martensite internal variable. J Intell Mater Syst Struct
33. Boyd JG, Lagoudas DC (1996) A thermodynamical constitutive model for shape memory materials. Part II. The SMA composite material. Int J Plasticity 12(7):843–873
34. Nallathambi AK, Doraiswamy S, Chandrasekar AS, Srinivasan SM (2009) A 3-species model for shape memory alloys. Int J Struct Changes Solids 1(1):149–170
35. Bhattacharya K (1992) Self-accommodation in martensite. Arch Rational Mech Anal 120(3):201–244
36. Duerig TW, Melton KN, Stöckel D (2013) Engineering aspects of shape memory alloys. Butterworth-Heinemann
37. Malvern LE (1969) Introduction to the mechanics of a continuous medium. Prentice-Hall
38. Shaw JA, Kyriakides S (1995) Thermomechanical aspects of NiTi. J Mech Phys Solids 43(8):1243–1281

Free Vibration Analysis of Hybrid Fibre Metal Laminated Panels

Darshan Singh Bisht, Nikesh Chelimilla, Naresh Kali, and Srikanth Korla

Abstract Fibre metal laminates (FMLs) are advanced multilayered composite materials. In recent days, the demand of using these materials increased rapidly in aerospace, structural and automotive industries due to their high strength and lightweight attributes. These FMLs may be subjected to various vibratory loads and prone to damages during their working condition; in the present work, a numerical comparative study between Glass-Aluminium-reinforced Laminates (GLARE) and Carbon-Aluminium-reinforced Laminates (CARALL) with 5 mm, 10 mm, 15 mm, 20 mm length cracks and without cracks is conducted under free vibration analysis; the effect of boundary condition on natural frequency is also studied. From the results, it is clearly noticed that CARALL exhibited higher natural frequency values when compared to GLARE. It may be due to high strength and stiffness nature of the carbon fibre. It is also depicted that, as the crack length increases at the fixed position the natural frequency for different modes decreases, which may be due to decrease in stiffness of the FML.

Keywords Fibre metal laminates · GLARE · CARALL · Free vibration analysis

1 Introduction

Many aerospace and automotive industries have been searching for substitute alternatives to traditional materials over the last ten years in order to boost the strength and lifespan of the component due to the desire for lightweight and more durable materials. Utilizing lightweight materials in the automotive and aerospace industry may allow for a certain reduction in fuel consumption, and increased fuel efficiency will assist reduce carbon dioxide emissions. Thin metal and composite fibre-reinforced

D. S. Bisht (✉) · N. Chelimilla · N. Kali · S. Korla
Department of Mechanical Engineering, NIT Warangal, Warangal, Telangana, India
e-mail: dsme21507@student.nitw.ac.in

S. Sidhardh et al. (eds.), *Advances in Structural Integrity for Mechanical, Civil, and Aerospace Applications*, Lecture Notes in Mechanical Engineering,
https://doi.org/10.1007/978-981-97-6367-2_16

polymer plies make up FMLs [1]. They are widely employed in various industrial applications, such as the automotive, medical and aerospace industries, due to their outstanding fracture and fatigue resistance, impact qualities and strength-to-weight ratio [2, 3]. FMLs are widely employed in the production of numerous structural elements for the aerospace industry, including the engine and fuselage.

The aramid/aluminium-reinforced FML (ARALL) was introduced by the Delft University of Technology (DUT) in 1978 [4, 5]. The first metal sheet reinforcement in FMLs is aluminium because of its advantages such as better mechanical properties and its low weight [6, 7]. The utility of GLARE in the Airbus A380 owing to its low weight, good damage resistance and enhanced fatigue resistance made it one of the well-known fibre metal laminates [8, 9]. However, the existence of multiple layers (carbon/glass fibres, metal sheets) in the FMLs may lead to weak interface formation resulting in delamination of FMLs [10, 11]. In addition, FMLs are susceptible to damage during manufacturing and loading conditions. It is observed that delamination is the most common failure among all damages associated with FMLs [12]. Delamination hampers the load-bearing capacity of the material and plays part in reduction of its stiffness [13, 14]. The utilization of FMLs in structures associated with rivets, bolts, fuel pipes, passage cable inspection, etc., requires holes/cut-outs [15, 16]. Such requirements affect the stiffness of FMLs especially during loading conditions (compression, tension, shear, etc.). Also, the presence of notches results in reduction of static strength of the FMLs [17]. The existence of open holes leads to damage evolution and decrease in the peak impact force owing to high stress agglomerations in the open hole region. Hence, FMLs require sophisticated numerical and experimental analysis to effectively model their dynamic modal behaviour. In this context, free vibration analysis is one of the vibration techniques used to estimate modal parameters such as natural frequencies and mode shapes [18, 19]. Changes in modal parameters distinguish structures without and with damages such as cracks. A decrease in natural frequency for damages such as small crack or notch can be noticed due to release of stresses carried by the material [20].

It can be inferred from the above literature survey that dynamic behaviour of FRPs is extensively studied. Also, the numerical and analytical studies concerning to the impact and tensile behaviour of FMLs are broadly researched. However, the dynamic behaviour of FMLs is scarcely reported. The present study aims to fill this gap by focusing on the numerical vibration analysis of GLARE and CARALL for different cases, i.e. with and without cracks. Since the fabrication and testing of fibre metal laminates are laborious, finite element analysis can be a potential alternative to capture the dynamic behaviour of the FMLs.

Table 1 Mechanical properties of GFRP and CFRP composite materials [20] and Aluminium 6061

Materials	E_1 (GPa)	E_2 (GPa)	υ	G_{12} (GPa)	G_{13} (GPa)	ρ (kg/m^3)
Aluminium 6061	72.4	72.4	0.33	28.00	28.00	2780
GFRP	7.00	7.00	0.22	1.00	2.50	1720
CFRP	17.50	17.50	0.24	2.00	10.50	1430

Table 2 Specimens dimensions used for vibration analysis

Materials	Dimensions (mm^3)
Aluminium 6061	250 × 25 × 0.80
GFRP	250 × 25 × 0.54
CFRP	250 × 25 × 0.54
Al/GFRP/Al/GFRP/Al	250 × 25 × 3.48
Al/CFRP/Al/CFRP/Al	250 × 25 × 3.48

2 Materials and Methodology

2.1 Materials

A GLARE (Al/GFRP/Al/GFRP/Al) and CARALL (Al/CFRP/Al/CFRP/Al) FMLs are modelled in the present study. The mechanical properties of materials and their dimensions were showed in Tables 1 and 2, respectively.

2.2 Geometrical Modelling of GLARE and CARALL

In the present study, fixed–free and fixed–fixed specimens are modelled with the dimensions as mentioned in Table 2 using Abaqus CAE. During this analysis, three aluminium metal sheets and two glass fibre layers are used to model the GLARE FML and two carbon fibres are used with three aluminium layers to model CARALL FML as shown in Fig. 1. The crack is modelled 50 mm from the fixed end with 1 mm of width and varying crack lengths (5 mm, 10 mm, 15 mm and 20 mm crack lengths) as shown in Fig. 2a. In order to avoid any relative motion between the layers of aluminium and composite material in GLARE and CARALL, all the layers in both the specimens are connected together using the tie constraint. Linear perturbation and Lanczos method were used to extract eigenvalues for the first five natural frequencies. Aluminium sheet has meshed with 6295 linear hexahedral elements of type C3D8R and the composite layer with 6295 linear hexahedral elements of type SC8R as shown in Fig. 2. Partition has been done around the crack, and fine mesh is created to get better results. The plate dimensions of 250 mm × 25 mm × thickness are chosen from

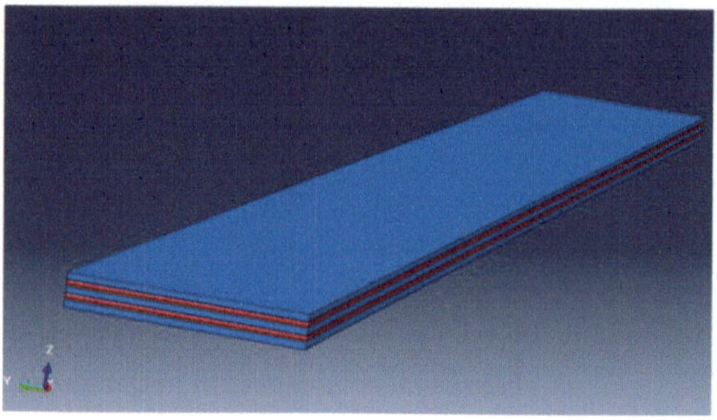

Fig. 1 FML plate (Al/GFRP/Al/GFRP/Al)

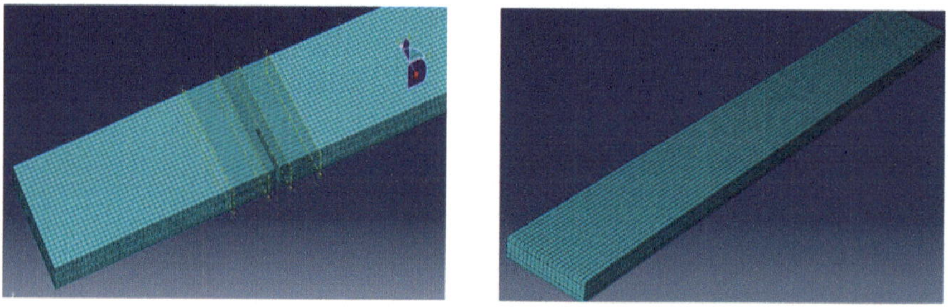

(a) (b)

Fig. 2 Meshing for **a** damaged beam, **b** healthy beam

reference standards as shown in Table 2 for modelling the GLARE and CARALL FMLs, but the effective length has been evaluated as 230 mm and considering a fixed length of 20 mm. The Encastre (U1 = U2 = U3 = UR1 = UR2 = UR3 = 0) boundary condition is used to fix the one side in fixed–free and both sides for fixed–fixed specimens as shown in Fig. 3.

3 Results and Discussion

GLARE and CARALL FMLs are taken into consideration in this analysis to understand their free vibration response with variable crack lengths of 5 mm, 10 mm, 15 mm and 20 mm created at a length of 50 mm from the fixed end, and the boundary condition effect is also studied. For each fracture length, the change in natural frequency and mode shapes is determined. It is observed from Figs. 4a and 5a that CARALL healthy specimens exhibit a higher natural frequency value from mode 1 to mode 5 in

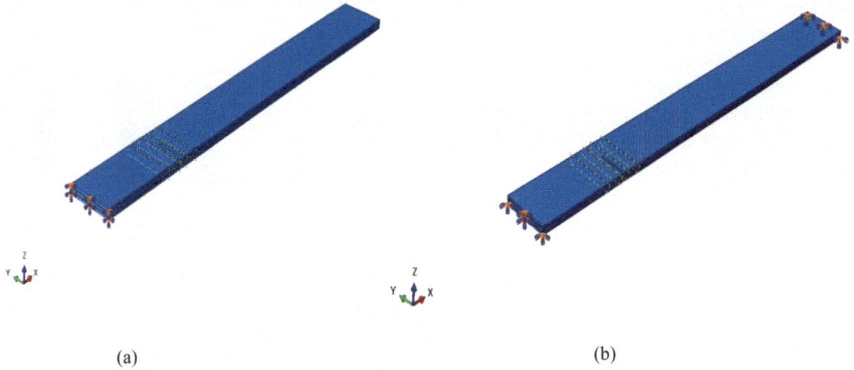

Fig. 3 **a** Fixed–free beam, **b** fixed–fixed beam with crack

comparison with GLARE in both fixed–free and fixed–fixed conditions, respectively. It may be attributed to high strength and stiffness nature of the carbon fibre, and it is also noticed that the natural frequency values are increasing as mode number is increasing in both the boundary conditions.

Figure 4b–d represents the free vibration behaviour of GLARE and CARALL from mode 1 to mode 3 under fixed–free condition. It is noticed from these results that with increase in crack length the natural frequency is decreasing for almost each mode. However, in case of mode 3 at 15 mm crack length, it is noticed that there is slight enhancement in natural frequency when compared to 10 mm crack length. In the similar way, it is observed that as crack length increases, the natural frequency values are reducing under fixed–fixed condition but again a slight increment in natural frequencies at 15 mm crack length is also observed. From Tables 3 and 4, it is depicted that the reduction in natural frequencies in cracked specimens from mode 1 to mode 3 may be due to loss of stiffness of the FML specimens. However, the reduction in natural frequencies not only depends on stiffness but also may rely on the mode shape of the FML as shown in Figs. 6 and 7. In addition, when boundary conditions are taken into consideration, fixed–fixed condition is exhibiting higher natural frequency values for all the conditions carried out in the present study.

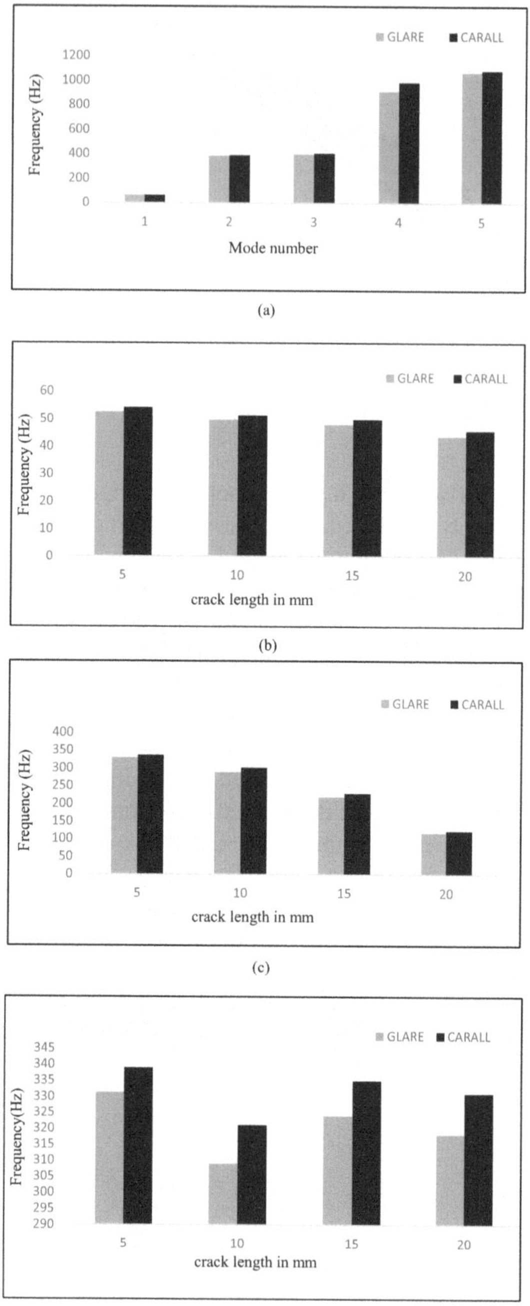

Fig. 4 **a** First five natural frequencies for healthy GLARE and CARALL under fixed–free condition, **b**, **c**, **d** variation in frequency with varying crack length for mode 1, mode 2 and mode 3

Fig. 5 **a** First five natural frequencies for healthy GLARE and CARALL under fixed–fixed condition, **b**, **c**, **d** variation in frequency with varying crack length for mode 1, mode 2 and mode 3

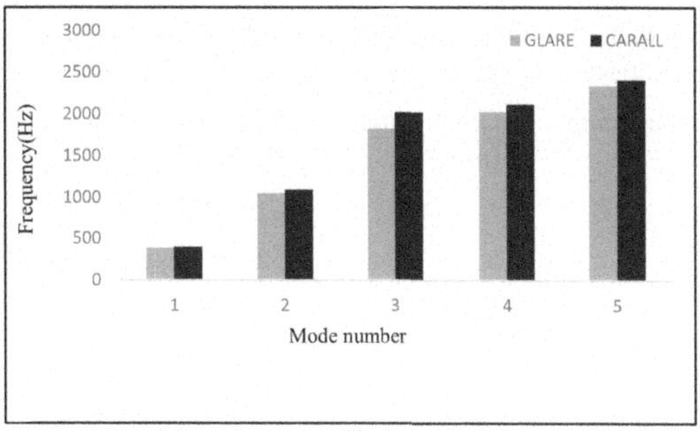

(a)

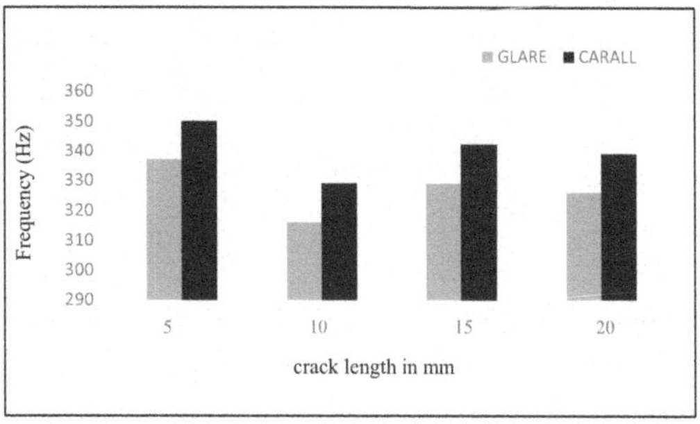

(b)

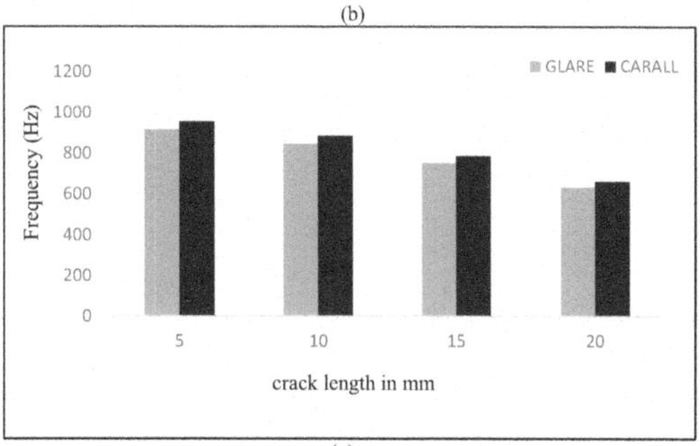

(c)

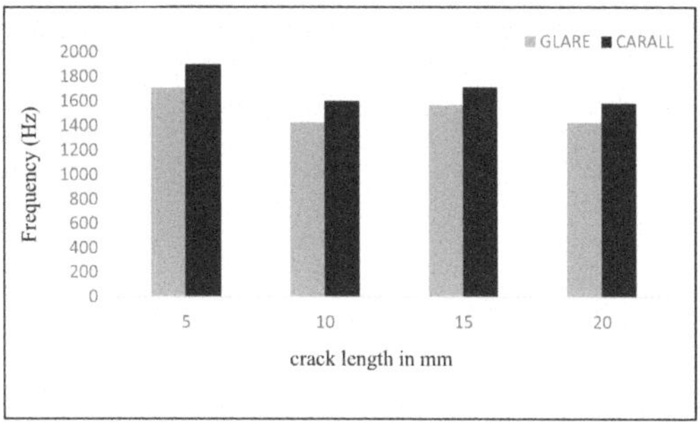

(d)

Table 3 Natural frequencies of GLARE and CARALL with varying crack length for fixed–free condition

Specimens	Crack length (mm)	Mode 1 (Hz)	Mode 2 (Hz)	Mode 3 (Hz)	Mode 4 (Hz)	Mode 5 (Hz)
GLARE	No crack	61.534	383.65	394.81	912.9	1066.3
	5	52.516	329.28	331.27	678.54	834.69
	10	49.643	287.90	308.97	668.45	847.60
	15	47.921	218.55	323.97	740.39	874.55
	20	43.631	116.81	318.08	631.46	846.42
CARALL	No crack	61.94	387.44	404.96	983.96	1082.6
	5	54.21	335.84	339.45	691.54	849.75
	10	51.234	300.84	320.94	718.19	883.37
	15	49.78	228.18	335.26	807.78	911.93
	20	45.74	121.9	331.22	709.52	885.74

Table 4 Natural frequencies of GLARE and CARALL with varying crack length for fixed–fixed condition

Specimens	Crack length (mm)	Mode 1 (Hz)	Mode 2 (Hz)	Mode 3 (Hz)	Mode 4 (Hz)	Mode 5 (Hz)
GLARE	No crack	383.32	1046.02	1826.78	2025.05	2339.47
	5	337.5	916.25	1710.74	1771.53	2056.22
	10	316.45	845.24	1428.42	1685.24	2052.01
	15	329.75	750.42	1567.43	1659.64	2044.00
	20	326.23	633.42	1426.65	1592.57	2021.24
CARALL	No crack	395.45	1085.42	2022.77	2120.69	2412.53
	5	350.75	955.40	1901.34	1869.35	2119.42
	10	329.72	884.97	1603.02	1786.77	2144.14
	15	342.04	785.55	1713.71	1748.41	2135.72
	20	339.54	660.41	1582.21	1678.67	2111.46

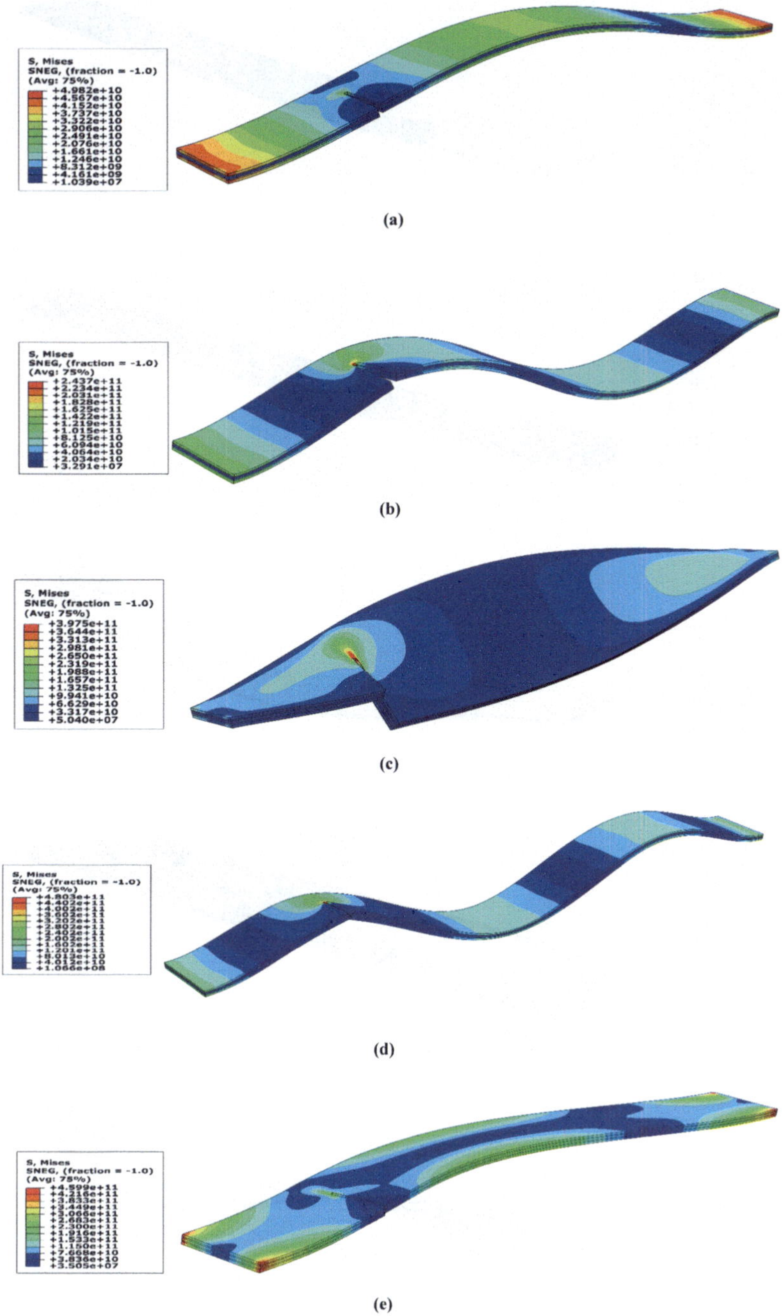

Fig. 6 **a** First mode shape, **b** second mode shape, **c** third mode shapes, **d** fourth mode shape, **e** fifth mode shape for GLARE with 10 mm crack length under fixed–fixed boundary condition

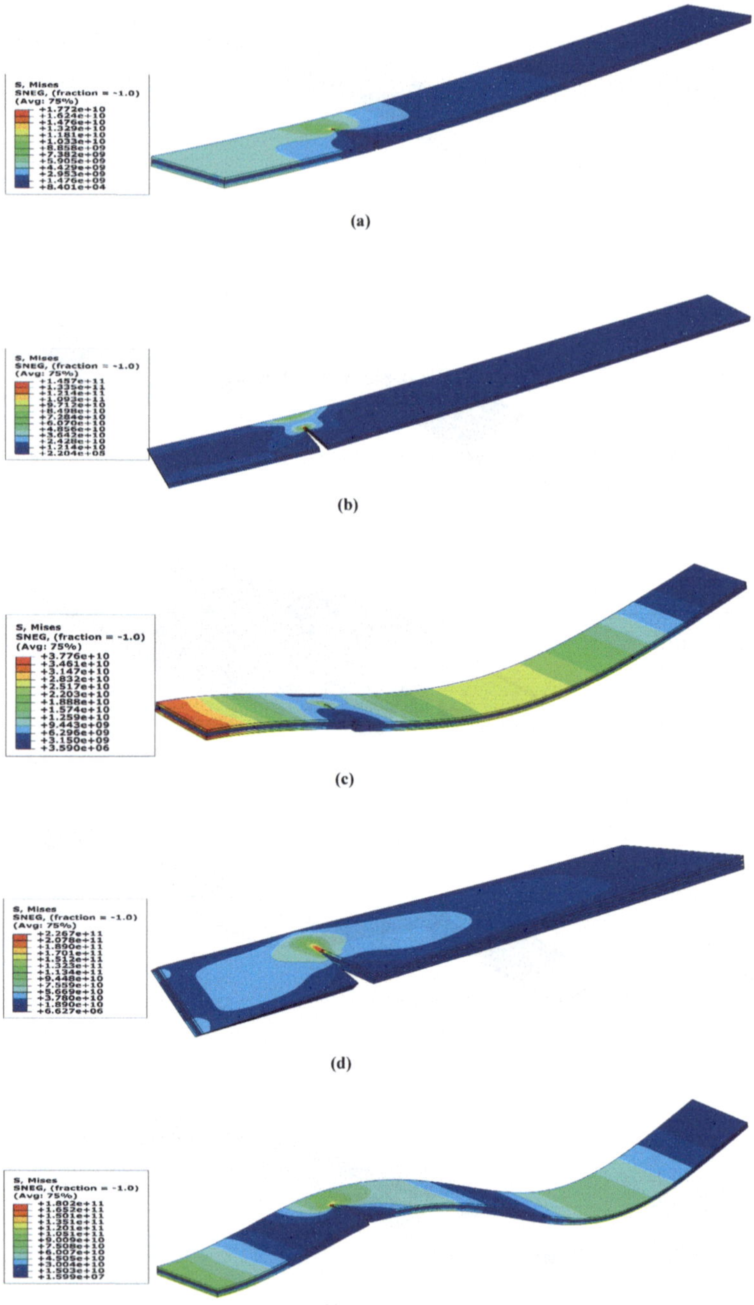

Fig. 7 **a** First mode shape, **b** second mode shape, **c** third mode shapes, **d** fourth mode shape, **e** fifth mode for GLARE with 10 mm crack length under fixed–free boundary condition

4 Conclusion

In the present study, a comparative numerical free vibration analysis between GLARE and CARALL was carried out with different crack lengths along the width of the specimen and influence of boundary conditions (fixed–free and fixed–fixed) on natural frequency is studied.

In the present study, the following results are observed.

1. It is depicted from the results that, CARALL exhibited higher natural frequency values when compared to GLARE. It may be due to higher stiffness and strength nature of the carbon fibre.
2. A reduction in natural frequency values is observed in both GLARE and CARALL as crack length increases at the fixed position for both the fixed–free and fixed–fixed boundary condition. This may be caused by stiffness reduction in FMLs.
3. It is also noticed that the reduction in natural frequency values is influenced by the crack lengths besides the mode shapes.
4. Fixed–fixed boundary condition showed higher natural frequency values when compared to that of fixed–free condition.

References

1. Sharma AP, Velmurugan R, Shankar S, Ha SK (2021) High-velocity impact response of titanium-based fiber metal laminates. Part I : experimental investigations. Int J Impact Eng 152(February). https://doi.org/10.1016/j.ijimpeng.2021.103845
2. Kali N, Korla R, Korla S (2022) Influence of nanoparticles on mechanical properties of hybrid polymer-layered fibre metal laminates (FMLs). Trans Indian Inst Met. https://doi.org/10.1007/s12666-022-02584-8
3. Kali N, Korla R, Korla S (2022) Impact behaviour of nano-hybrid (carbon/glass) fibre metal laminates: an experimental study. Arab J Sci Eng. https://doi.org/10.1007/s13369-022-07317-z
4. Sinmazçelik T, Avcu E, Bora MÖ, Çoban O (2011) A review: fibre metal laminates, background, bonding types and applied test methods. Mater Des 32(7):3671–3685. https://doi.org/10.1016/j.matdes.2011.03.011
5. Kali N, Pathak S, Korla S (2019) Effect on vibration characteristics of fiber metal laminates sandwiched with natural fibers. Mater Today Proc 28(January):1092–1096. https://doi.org/10.1016/j.matpr.2020.01.088
6. Chandrasekar M, Ishak MR, Jawaid M, Leman Z, Sapuan SM (2017) An experimental review on the mechanical properties and hygrothermal behaviour of fibre metal laminates. J Reinf Plast Compos 36(1):72–82. https://doi.org/10.1177/0731684416668260
7. Mansor MA, Ahmad Z, Abdullah MR (2020) Experimental studies on the impact characteristics of seamless fibre metal laminate (FML) tubes. Mater Today Proc 39:1077–1081. https://doi.org/10.1016/j.matpr.2020.06.360
8. Ding Z, Wang H, Luo J, Li N (2021) A review on forming technologies of fibre metal laminates. Int J Light Mater Manuf 4(1):110–126. https://doi.org/10.1016/j.ijlmm.2020.06.006
9. Zhang X et al (2017) Effect of surface treatment on the corrosion properties of magnesium-based fibre metal laminate. Appl Surf Sci 396:1264–1272. https://doi.org/10.1016/j.apsusc.2016.11.131

10. Megahed M, Abd El-baky MA, Alsaeedy AM, Alshorbagy AE (2019) An experimental investigation on the effect of incorporation of different nanofillers on the mechanical characterization of fiber metal laminate. Compos Part B Eng 176(February):107277. https://doi.org/10.1016/j.compositesb.2019.107277
11. Pan Y, Wu G, Cheng X, Zhang Z, Li M, Ji S (2016) Mode I and Mode II interlaminar fracture toughness of CFRP/magnesium alloys hybrid laminates. Compos Interfaces 6440. https://doi.org/10.1080/09276440.2016.1144911
12. Moslemi M, Khoshravan M (2015) Cohesive zone parameters selection for mode-i prediction of interfacial delamination. Stroj Vestnik/J Mech Eng 61(9):507–516. https://doi.org/10.5545/sv-jme.2015.2521
13. Kaveh A (2014) Determining the crack propagation in DCB composite specimens by fiber bridging modeling. Rev Tech Fac Ing Univ Zulia 37(2):1–9
14. Moheimani R, Sarayloo R, Dalir H (2020) Failure study of fiber/epoxy composite laminate interface using cohesive multiscale model. Adv Compos Lett 29:2633366X2091015, 2020, https://doi.org/10.1177/2633366x20910157.
15. Taha R (2016) The effect of a hole on natural frequencies of laminated composite plate. Introduction, no April 2017. https://doi.org/10.13140/RG.2.2.12192.79366
16. Kali N, Chelimilla N, Korla S (2022) Influence of damages (hole/slits) on tensile behavior of Al 7075–T6/glass fibre-based hybrid laminates. Mater Today Proc 56:631–636. https://doi.org/10.1016/j.matpr.2021.12.399
17. Kötter B, Karsten J, Körbelin J, Fiedler B (2020) CFRP thin-ply fibre metal laminates: influences of ply thickness and metal layers on open hole tension and compression properties. Materials (Basel) 13(4). https://doi.org/10.3390/ma13040910
18. Kali N, Korla S (2022) Numerical studies on mode I delamination and its effect on the vibrational characteristics in fibre metal laminates. IOP Conf Ser Mater Sci Eng 1225(1):012039. https://doi.org/10.1088/1757-899x/1225/1/012039
19. Saini NK, Ahmad F (2019) Vibration analysis of fiber metal laminate beam in ANSYS parametric design language. Int J Appl Eng Res 14(9):149–155
20. Merzuki MNM, Rejab MRM, Sani MSM, Zhang B, Quanjin M (2019) Experimental investigation of free vibration analysis on fibre metal composite laminates. J Mech Eng Sci 13(4):5753–5763. https://doi.org/10.15282/jmes.13.4.2019.03.0459

Finite Element Model Updating Using Modal Data

Rajpurohit Kiran and Sahil Bansal

Abstract Finite element (FE) model updating is a framework for identification of FE model and its parameters consistent with the system measurements. For updating FE model with uncertain model parameters, the Bayesian model updating approach is widely adopted which uses probability logic for uncertainty quantification. This approach using modal data is conducted by mode matching or without mode matching approaches. Mode matching approach requires solving the eigenvalue problem and relative weighting factors for modal frequencies and mode shapes. Whereas, without mode matching approach adopts the eigenvalue equation instead of solving it and does not require weighting factors. Due to limitations in gathering data from the structure in a single setup, model updating can be performed using data from multiple setups. Sampling techniques are commonly adopted for generating samples from posterior distribution of model parameters and its statistics. Among various sampling techniques, Gibbs sampling is appropriate for multi-dimensional problems when conditional distributions are possible for the parameters. The present work aims to perform FE model updating using modal data from multiple setups. The effectiveness and efficiency of a modified Gibbs sampling method in Bayesian model updating are illustrated by 10 degrees of freedom (DOF) numerically simulated freedom (DOF) numerically simulated example.

Keywords Bayesian model updating · Uncertainty quantification · Gibbs sampling · Multiple setups

R. Kiran (✉) · S. Bansal
Indian Institute of Technology Delhi, Hauz Khas, Delhi 110016, India
e-mail: Rajpurohit.Kiran@civil.iitd.ac.in

S. Bansal
e-mail: sahil@civil.iitd.ac.in

S. Sidhardh et al. (eds.), *Advances in Structural Integrity for Mechanical, Civil, and Aerospace Applications*, Lecture Notes in Mechanical Engineering,
https://doi.org/10.1007/978-981-97-6367-2_17

1 Introduction

Mathematical models are widely assumed to derive the numerically simulated response of real systems of science and engineering. These models, termed as FE models, are used for analysis, design, and response predictions of structural systems like buildings, bridges, dams, etc. These FE models have numerous model parameters, which are usually unknown due to incomplete knowledge and limitations in accurate modeling of its real system. The need for FE model updating [1] arises for identifying these model parameters and its associated FE model. FE model updating of a structure is not only concerned with identifying model parameters and system identification but also with robust prediction of the structural system and damage assessment.

There are broadly two approaches for updating a structural model, namely deterministic and probabilistic. The deterministic approach is basically an optimization problem where the optimal parameters of the model are obtained by optimizing the objective function. In contrast, the probabilistic approach has the ability to quantify model uncertainties due to modeling error and measurement noise, handling incomplete modal data. Uncertain model parameters in the probabilistic approach are considered as probability distribution functions (PDFs) [2]. According to this approach, the updated or posterior PDF of model parameters is given based on system data (likelihood function) and prior PDF of model parameters using Bayes theorem,

$$p(\theta|D) = \frac{p(D|\theta)p(\theta)}{p(D)}, \tag{1}$$

Bayesian model updating can be performed with time history response, frequency domain, or modal measurements, namely modal frequencies, mode shapes, and damping ratios of the structure. For the data in the modal domain, model updating is implemented in one stage, whereas for time-domain or frequency-domain data, the approach is implemented in one stage [3] or in two stages depending on whether the data is explicitly represented with model parameters or data is represented with modal parameters which are in turn represented with model parameters [4].

Mode matching approaches of model updating involve minimizing the measure of difference between predicted and measured modal parameters. Christrodoulou and Papadimitrou [5] and Goller [6] presented the likelihood function considering mode matching with weighting factors for different modal parameters. Behmanesh et al. [7] utilized mode matching in hierarchical model updating, which not only accounts for modeling and measurement uncertainty but also considers structural parameters' variability due to environmental, temperature, or other external effects. Modal measurements are difficult to identify as to which mode it corresponds to after the damage, so it is difficult to adopt mode matching for this case. Without mode matching approach [8] overcomes this difficulty as it involves Eigen equation in the likelihood function. However, there is an increase in parameter space using this approach by considering modal parameters also as updating parameters. Without mode matching approaches can be further carried out assuming or neglecting damping in the formulation [9–12]. However, consideration of damping increases the computational cost of the methodology but represents the real system. Since modal measurements are partial or incomplete and gathering data from the single setup is insufficient for modal updating, multiple setups are needed to gather modal data from the whole structure where each of the setups gathers data from different parts of the structure. Some of the key works in Bayesian model updating using multiple setups are in Refs. [13–17]. Further, in an attempt to reduce computational effort by considering eigen problem corresponding to measured degrees of freedom, model updating based on the dynamic condensation technique was proposed in Refs. [18–20].

Sampling techniques under the class of Markov Chain Monte Carlo (MCMC) simulation are adopted to overcome the analytical and numerical difficulties in evaluating the posterior PDF of single or multiple peaks in very small neighborhoods. Techniques such as Adaptive Metropolis Hastings (AMH) [21] and Transitional Markov Chain Monte Carlo [20] were proposed by considering number of sampling levels based on judgment to map the distance between prior PDF and posterior PDF. Gibbs sampling [8] is an alternate technique suitable for high dimensional PDFs where samples are extracted for the different model parameters iteratively. The main drawback of Gibbs sampling is that it requires conditional PDF for different parameters, and MH in Gibbs [10] sampling overcomes this drawback. Hybrid Monte Carlo [22] is another technique that is based on Hamiltonian dynamics suitable for sampling correlated model parameters. This paper proposes a FE model updating using modal data using dynamic condensation and data from multiple setups, and posterior samples are extracted by a modified Gibbs sampling method.

2 Proposed Approach

Dynamic condensation reduces the structural model to a reduced model corresponding to measured degrees of freedom. The proposed technique directly adopts the partial mode shapes and eliminates the need to select the measured degrees of freedom from the full mode shapes. A Gibbs-based sampling technique is adopted to extract samples from posterior PDF. In the proposed approach, the uncertain parameters considered are structural parameters, prediction error variances, modal frequencies, and partial mode shapes, which are updated using modal data. The modal data is considered as posterior statistics of modal parameters (most probable value and covariance matrix) given by $D = \{\tilde{\omega}_{r,i}, \tilde{\phi}_{r,i}, \tilde{C}_{\omega_{r,i}}, \tilde{C}_{\phi_{r,i}}; r = 1, ...N_m; i = 1, ...R\}$, where N_m is the number of measured modes of vibration, and $\tilde{\omega}_{r,i} \in R^+$, $\tilde{\phi}_{r,i} \in R^{n_i}$, $\tilde{C}_{\omega_{r,i}} \in R^+$, and $\hat{C}_{\phi_{r,i}} \in R^{n_i \times n_i}$ are the most probable modal frequency, most probable normalized mode shape, covariance matrix of modal frequency, and covariance matrix of mode shape, respectively, for r-th mode and i-th setup. Although the modal data are observed from multiple setups, combined statistics of mode shape $\tilde{\phi}_r \in R^{n_0}$, $\hat{C}_{\phi_r} \in R^{n_0 \times n_0}$, where n_0 is the number of measured DOF from all setups are considered, and similarly, the combined statistics of modal frequency $\tilde{\omega}_r$, $\tilde{C}_{\omega_r}$ are considered.

2.1 Model Reduction Based on Dynamic Condensation

For a structural system with N_d-DOF, mass matrix $M(\theta_s) \in R^{n_d \times n_d}$, and stiffness matrices $K(\theta_s) \in R^{n_d \times n_d}$ parameterized by structural model parameter set $\theta_s \in R^{n_{\theta s}}$, the eigenvalue equation for the reduced system with nominal parameters of structural model parameters θ_s^n modal frequency λ_r, and mode shape ψ_r as in Bansal [20] is given by

$$(K_r^R(\varpi_r, \theta_s^n) - \lambda_r^2 M_r^R(\varpi_r, \theta_s^n))\psi_r^m = 0, \tag{2}$$

where $K_r^R(\varpi_r, \theta_s^n)$ and $M_r^R(\varpi_r, \theta_s^n)$ are the reduced system matrices of r-th mode, and superscripts m, s are master and slave DOFs. These master DOFs are generally the measured DOFs in the FE model of structural systems.

2.2 Model Prediction Error

Equation (2) considered is for the nominal system, but for a system with uncertain parameters, the eigenvalue equation is given by replacing $[\theta_s^n, \lambda_r, \psi_r]$ with $[\theta_s, \omega_r, \phi_r]$ as

$$(K_r^R(\theta_s) - \omega_r^2 M_r^R(\theta_s))\phi_r = \varepsilon_r, \tag{3}$$

where the prediction error ε_r is assumed to zero mean Gaussian distribution with covariance matrix $\Sigma_r = \sigma_r^2 I_{n_0}$.

The PDF for ε_r is given by

$$p(\varepsilon_r) = \frac{1}{\sqrt{|2\pi\sigma_r^2 I_{n_o}|}} \exp(-\frac{1}{2\sigma_r^2}\phi_r^T (K_r^R(\theta_s) - \omega_r^2 M_r^R(\theta_s))^T (K_r^R(\theta_s) - \omega_r^2 M_r^R(\theta_s))\phi_r), \tag{4}$$

where $p(\varepsilon_r)$ can also be denoted as $p(\omega_r, \phi_r|\theta_s, \sigma_r^2)$ since ω_r and ϕ_r are dependent on θ_s and σ_r^2.

2.3 Probability Model for Modal Parameters

The prior PDF for modal frequency and mode shape is given by Gaussian PDF with the posterior statistics of modal parameters (modal data) given by

$$p_0(\omega_r|D) = \frac{1}{\sqrt{|2\pi \tilde{C}_{\omega_r}|}} \exp(-0.5(\omega_r - \tilde{\omega}_r)^T \tilde{C}_{\omega_r}^{-1} (\omega_r - \tilde{\omega}_r)), \tag{5}$$

$$p_0(\phi_r|D) = \frac{1}{\sqrt{|2\pi \tilde{C}_{\phi_r}|}} \exp(-0.5(\phi_r - \tilde{\phi}_r)^T \tilde{C}_{\phi_r}^{-1} (\phi_r - \tilde{\phi}_r)), \tag{6}$$

where $p_o(\omega_r|D)$ and $p_o(\phi_r|D)$ are hypothetical posterior PDF for modal frequency and mode shapes, since ω_r and ϕ_r are the additional uncertain parameters that are dependent on θ_s and σ_r^2. So according to the double counting principle in Au and Zhang [4], posterior PDFs for ω_r and ϕ_r are considered with a uniform prior PDF. Assuming the modal frequencies and mode shapes are independent for different modes, then $p_0(\omega|D) = \prod_{r=1}^{N_m} p_0(\omega_r|D)$ and $p_0(\phi|D) = \prod_{r=1}^{N_m} p_0(\phi_r|D)$, where $\omega = [\omega_1, \omega_2, ...\omega_{N_m}]$ and $\phi = [\phi_1, \phi_2, ...\phi_{N_m}]$.

2.4 Prior PDFs for Structural Parameters and Prediction Error Variances

The prior PDF for structural parameters θ_s is assumed as Gaussian PDF with mean, μ_{θ_s}, and covariance matrix, P_{θ_s}.

$$p(\theta_s) = \frac{1}{\sqrt{|2\pi P_{\theta_s}|}} \exp\left(-0.5\left(\theta_s - \mu_{\theta_s}\right)^T P_{\theta_s}^{-1} \left(\theta_s - \mu_{\theta_s}\right)\right). \tag{7}$$

The prior PDF for prediction error variance is assumed as inverse gamma distribution with shape and scale parameters ρ_0, k_0 taken according to Bayesian conjugate priors is given by

$$p(\sigma_r^2) = \frac{k_0^{\rho_0}}{\Gamma(\rho_0)} \left(\sigma_r^2\right)^{-\rho_0 - 1} \exp\left(-\frac{k_0}{\sigma_r^2}\right). \tag{8}$$

Since prediction error variance is assumed independent for different modes, then $p(\sigma^2) = \prod_r p\left(\sigma_r^2\right)$, where $\sigma^2 = \left[\sigma_1^2, \sigma_2^2, ...\sigma_{N_m}^2\right]$.

2.5 Posterior PDF

The posterior PDF for uncertain parameters is given by

$$p\big(\theta_s, \omega, \phi, \sigma^2|D\big) \propto p\big(\theta_s, \sigma^2\big)p\big(\omega, \phi|\theta_s, \sigma^2\big)p(D|\omega, \phi). \tag{9}$$

Since uniform PDF is assumed for probability model of modal parameters

$$p\big(\theta_s, \omega, \phi, \sigma^2|D\big) = cp(\theta_s)p\big(\sigma^2\big)p_0(\omega|D)p_0(\phi|D)p\big(\omega, \phi|\theta_s, \sigma^2\big), \tag{10}$$

where c is normalizing constant, assuming statistical independence for different modes gives

$$p(\theta_s, \omega, \phi|D) = cp(\theta_s)\prod_r p\big(\sigma_r^2\big)p_0(\omega_r|D)p_o(\phi_r|D)p\big(\omega_r, \phi_r|\theta_s, \sigma_r^2\big). \tag{11}$$

3 Iterative Sampling

The proposed algorithm is according to Gibbs-based sampling to extract samples iteratively, i.e., the current sample for each parameter is sampled while taking previously updated samples for other parameters.

Since the considered four uncertain parameters sets are $\theta_s, \sigma^2, \omega, \phi$ is as given below

a.θ_s at j-th step is sampled from σ^2, ω, ϕ at (j-1)-th step
b.σ^2 at j-th step is sampled from θ_s at j-th step and ω, ϕ at (j-1)-th step
c.ω at j-th step is sampled from θ_s, σ^2 at j-th step and ϕ at (j-1)-th step
d.ϕ at j-th step is sampled from θ_s, σ^2, ϕ at j-th step.

These iterative steps (a–d) are performed for the required number of samples. The samples for uncertain parameters for which well conditional PDF (e.g., Normal distribution, Gamma distribution, Lognormal distribution, etc.) are possible which are extracted from their conditional PDF. For the examples considered with dynamic condensation, conditional PDF is possible for mode shape and error variance. Conditional PDF for an uncertain parameter is obtained by assuming other uncertain param-

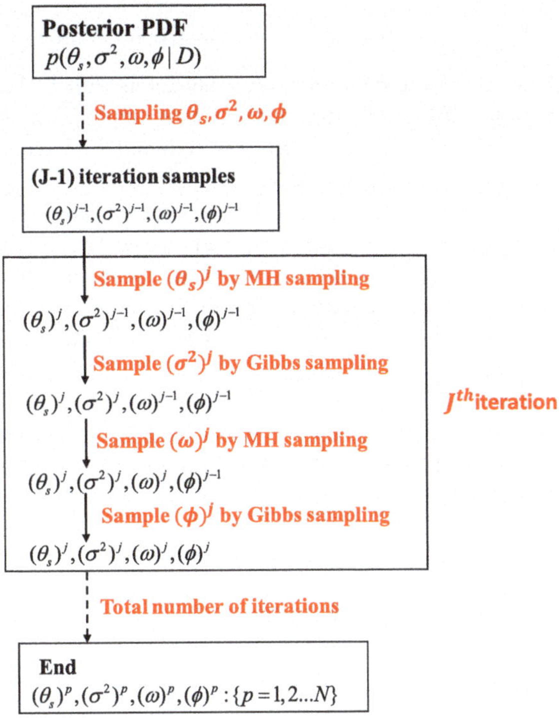

Fig. 1 Flowchart for the proposed iterative sampling

eters known as their previously updated samples. MH sampling is implemented for the uncertain parameters for which conditional PDF is not possible. Structural parameters,θ_s, are sampled using MH sampling with Gaussian proposal PDF with mean as previous θ_s sample and covariance matrix as the covariance matrix of prior PDF of θ_s. ω is sampled using MH sampling with prior PDF for modal frequencies as proposal PDF. The flowchart of the proposed iterative sampling is shown in Fig. 1.

4 Illustrative Example

The example of ten DOF shear building with story masses $m_0 = 1000$ kg and inter-story stiffness at each floor $k_0 = 8000$ N/m. Assuming three modes are measured, and modal data are considered from two setups with a maximum of four sensors available. DOFs measured in 1st setup are 1, 2, 3, and 4 and in 2nd setup are 4, 7, 10, i.e., and six DOFs are measured in total.

For identification, the stiffness matrix is parameterized as $K(\theta_s) = \sum_{i=1}^{3} \theta_i K_i$, where three substructure matrices are considered with K_1 having DOF 1, 2, 3, 4 from the nominal stiffness matrix and rest zero similarly, K_2 having DOF 5, 6, 7 from the nominal stiffness matrix and rest zero and K_3 having DOF 8, 9, 10 from nominal stiffness matrix and rest zero.

The four groups of uncertain parameters considered are θ_s, ω, ϕ, and σ^2 with $\theta_s = [\theta_1, \theta_2, \theta_3]^T$, $\omega = [\omega_1, \omega_2, \omega_3]^T$, $\sigma^2 = [\sigma_1^2, \sigma_2^2, \sigma_3^2]^T$, and $\varphi = [\varphi_1^T, \varphi_2^T, \varphi_3^T]^T$, where $\varphi_r \in R^{n_o \times 1}$ is there mode shape parameters for r-th mode and $n_o = 6$ in this example. Posterior statistics of modal data are obtained by finding the statistics of modal parameters for 100 datasets by perturbing the actual stiffness parameters in each dataset.

The prior PDF of structural parameters is assumed as Gaussian with each parameter having mean one and coefficient of variation (COV) 15%. Prior PDF of error variance is assumed as non-informative inverse gamma distribution with $\rho_o = 0$, $k_o = 0$.

4.1 Results and Conclusion

Table 1 shows the statistical properties of the posterior PDF of the stiffness scaling parameters and modal frequencies. It shows that the posterior mean of structural, modal frequencies, and mode shapes are close to actual values. Posterior COV of structural parameters is between 1.1 and 1.5%, which is quite less than the assumed initial COV of structural parameters, and posterior COV of modal frequencies is approximately 0.25%. It is observed that the COV of mode shape parameters is approximately 1.5%. The COV for the 5-th DOF of φ_2 is obtained 972% as it is a nodal point, and the actual value is close to zero. Figure 2 shows the plot matrix of structural parameters and modal frequencies; it is clearly observed that the posterior distribution is Gaussian with the posterior statistics. Figure 3 shows the plot of error variance in the log scale. It is observed that error variance is also following a stationary distribution after the burn in samples. Figure 4 shows the kernel density plot of stiffness scaling parameters. It is observed that during different runs, the same distribution function is obtained; this shows that the obtained structural parameters are globally identifiable.

Table 1 Statistical properties of posterior stiffness scaling parameters, modal frequencies, and mode shape parameters

Parameters	Actual values	Posterior mean	Posterior COV (%)
θ_s	[1,1,1]	[0.9987,1.0014, 1.0002]	[1.3181, 1.1893, 1.4060]
ω	[0.4227, 1.258, 2.0667]	[0.4228, 1.2589, 2.0669	[0.2655,0.2588, 0.2702]
φ_1	[0.0970, 0.1919, 0.2825, 0.3668, 0.5638, 0.6492]	[0.0971, 0.1924, 0.2827, 0.3671, 0.5636, 0.6490]	[0.973, 0.669, 0.478, 0.347, 0.144, 0.182]
φ_2	[− 0.2271, − 0.4092, − 0.5102, − 0.5102, 0.0000, 0.5102]	[− 0.2267, − 0.4093, − 0.5098, −0.5098, 0.0003, 0.5111]	[0.81, 0.44, 0.22, 0.27, 972, 0.54]
φ_3	[− 0.3533, − 0.5179, − 0.4061, − 0.0774, 0.4498, − 0.4835]	[− 0.3517, − 0.5161, − 0.4047, − 0.0761, 0.4493, − 0.4840]	[1.53, 1.17, 1.15, 5.76, 1.41, 1.55]

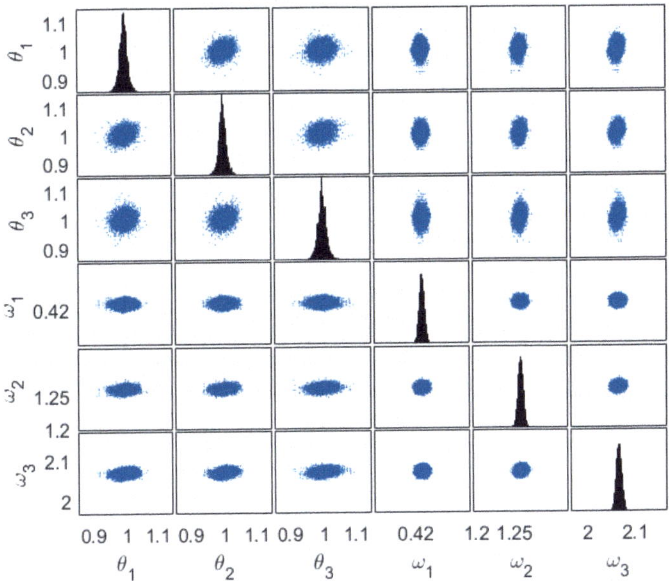

Fig. 2 Plot matrix of stiffness scaling parameters and modal frequencies

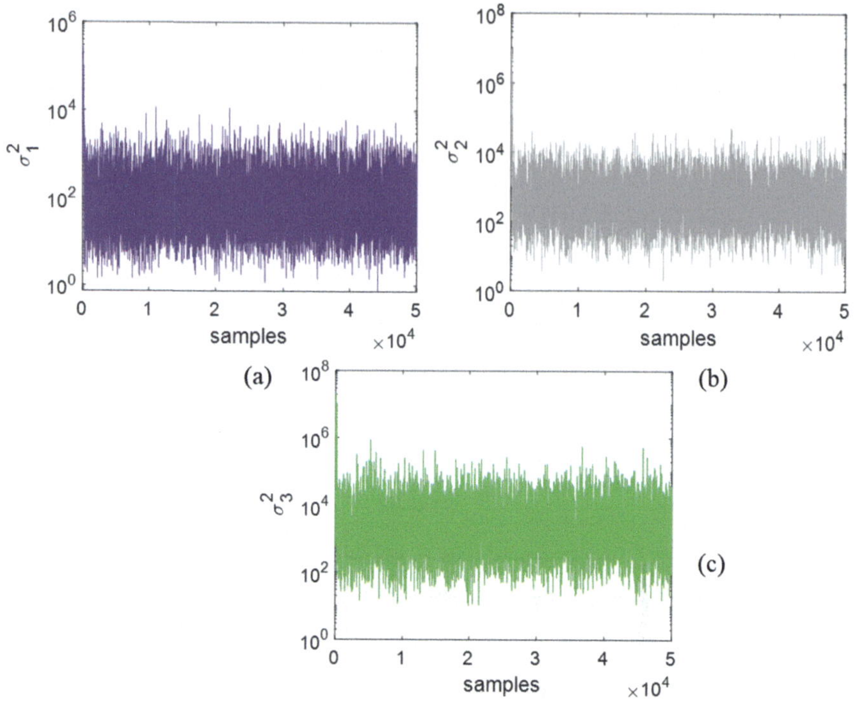

Fig. 3 Plot for Markov Chain samples of error variance for different modes in log-scale: **a** 1st mode, **b** 2nd mode, **c** 3rd mode

The present work focused on Bayesian model updating using modal data for multiple setups. Model reduction technique is employed, and the updated parameters of the structural parameters, modal frequencies, and partial mode shapes are obtained. A numerically simulated example is presented with the formulation.

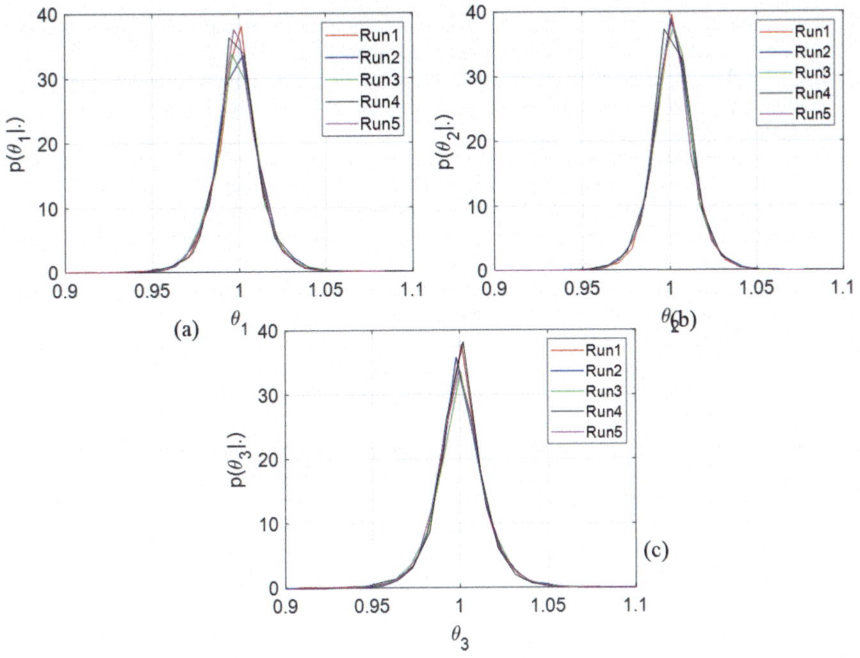

Fig. 4 Kernel density plot for stiffness scaling parameters for five runs

Acknowledgements The authors would like to gratefully acknowledge the Grant-in-aid scheme received from the Aeronautics R&D Board, Government of India.

Data Availability Statement The experimental data will be made available on request.

Declaration of Competing Interest The authors declare that they have no competing financial interests or personal relationships that could have appeared to influence the work reported in this paper.

References

1. Friswell M, Mottershead JE (1995) Finite element model updating in structural dynamics. Springer Science & Business Media
2. Beck JL, Katafygiotis LS (1998) Updating models and their uncertainties. I: Bayesian statistical framework. J Eng Mech 124:455–461
3. Vanik MW, Beck JL, Au S-K (2000) Bayesian probabilistic approach to structural health monitoring. J Eng Mech 126:738–745
4. Christodoulou K, Papadimitriou C (2007) Structural identification based on optimally weighted modal residuals. Mech Syst Signal Process 21:4–23
5. Goller B, Beck JL, Schueller GI (2012) Evidence-based identification of weighting factors in Bayesian model updating using modal data. J Eng Mech 138:430–440

6. Behmanesh I, Moaveni B, Lombaert G, Papadimitriou C (2015) Hierarchical Bayesian model updating for structural identification. Mech Syst Signal Process 64:360–376
7. Song M, Renson L, Noël JP, Moaveni B, Kerschen G (2018) Bayesian model updating of nonlinear systems using nonlinear normal modes. Struct Control Heal Monit. https://doi.org/10.1002/stc.2258
8. Ching J, Muto M, Beck JL (2006) Structural model updating and health monitoring with incomplete modal data using Gibbs sampler. Comput Civ Infrastruct Eng 21:242–257
9. Yuen KV, Beck JL, Katafygiotis LS (2006) Efficient model updating and health monitoring methodology using incomplete modal data without mode matching. Struct Control Heal Monit Off J Int Assoc Struct Control Monit Eur Assoc Control Struct 13(1):91–107
10. Bansal S, Cheung SH (2016) Stochastic sampling based Bayesian model updating with incomplete modal data. Int J Uncertain Quantif 6:229
11. Cheung SH, Bansal S (2017) A new Gibbs sampling based algorithm for Bayesian model updating with incomplete complex modal data. Mech Syst Signal Process 92:156–172
12. Das A, Debnath N (2020) A Bayesian model updating with incomplete complex modal data. Mech Syst Signal Process 136:106524
13. Hızal Ç, Turan G (2020) A two-stage Bayesian algorithm for finite element model updating by using ambient response data from multiple measurement setups. J Sound Vib 469:115139
14. Das A, Debnath N (2021) Gibbs sampling for damage detection using complex modal data from multiple setups. ASCE-ASME J Risk Uncertain Eng Syst Part A Civ Eng 7(2):04021018
15. Zhang F, Ni Y, Lam H (2017) Bayesian structural model updating using ambient vibration data collected by multiple setups. Struct Control Heal Monit 24:e2023
16. Yan W-J, Katafygiotis LS (2015) A novel Bayesian approach for structural model updating utilizing statistical modal information from multiple setups. Struct Saf 52:260–271
17. Bansal S (2015) A new Gibbs sampling based Bayesian model updating approach using modal data from multiple setups. Int J Uncertain Quantif 5:361
18. Sun H, Büyüköztürk O (2016) Probabilistic updating of building models using incomplete modal data. Mech Syst Signal Process 75:27–40
19. Yin T, Jiang Q-H, Yuen K-V (2017) Vibration-based damage detection for structural connections using incomplete modal data by Bayesian approach and model reduction technique. Eng Struct 132:260–277
20. Bansal S (2020) Bayesian model updating using modal data based on dynamic condensation. J Eng Mech 146:4019123
21. Beck JL, Au S-K (2002) Robust reliability of stochastic structural systems under stochastic. In: Structural dynamics: EURODYN 2002: proceedings of the 4th [ie 5th] international conference on structural dynamics, Munich, Germany, 2–5 September 2002, p 331. CRC Press
22. Cheung SH, Beck JL (2009) Bayesian model updating using hybrid Monte Carlo simulation with application to structural dynamic models with many uncertain parameters. J Eng Mech 135:243–255
23. Au S-K, Zhang F-L (2016) Fundamental two-stage formulation for Bayesian system identification, Part I: general theory. Mech Syst Signal Process 66:31–42

An Approximate Analytical Solution for Metal-FRP Circular Toroidal Pressure Vessel

Mohan Krishna Paleti, S. Suriya Prakash, and Vijayabaskar Narayanamurthy

Abstract The internal fluid pressure holding capacity of the isotropic toroidal pressure vessel (TPV) with a circular cross-section can be enhanced by the external wrapping of fiber-reinforced polymer (FRP) composites. An analytical solution for this metal-FRP hybrid TPV is not readily available. This study presents an approximate analytical solution to predict strains and stresses in the base metal and FRP layer in the metal-FRP TPV. The solution is derived based on modified-linear membrane theory (MLMT). This energy-based approach overcomes the deficiency of singular displacements commonly observed in a pure linear membrane theory. The solution considers an equivalent metal thickness for the FRP layer. The proposed solution is cross-verified with results of finite element analysis. Three-dimensional FE models were created and analyzed. Comparison of FE results with analytical predictions shows that the assumption of FRP as an equivalent isotropic material results in a more straightforward solution besides providing the better prediction of strains and stresses in the metal-FRP TPV.

Keywords FEA · Internal pressure · Modified-linear membrane solution · Toroidal pressure vessel

M. K. Paleti (✉) · S. Suriya Prakash
Department of Civil Engineering, Indian Institute of Technology Hyderabad, Sangareddy 502285, India
e-mail: ce20resch13002@iith.ac.in

S. Suriya Prakash
e-mail: suriyap@ce.iith.ac.in

V. Narayanamurthy
Research Centre Imrat, Hyderabad 500069, India
e-mail: v.nmurthy@gov.in

S. Sidhardh et al. (eds.), *Advances in Structural Integrity for Mechanical, Civil, and Aerospace Applications*, Lecture Notes in Mechanical Engineering,
https://doi.org/10.1007/978-981-97-6367-2_18

List of Symbols

N_ϕ *and* N_θ	Meridional and circumferential stress resultants
R	Toroidal radius
r	Meridional radius
r_0	Circumferential radius
ϕ	Meridional angle
θ	Circumferential angle
t_a	Thickness of aluminum liner
t_f	Actual thickness of CFRP layer
t_{ef}	Equivalent thickness of CFRP in terms of aluminum
t_{eq}	Equivalent thickness of the pressure vessel
E	Modulus of elasticity of aluminum
ν	Poisson's ratio of aluminum
E_{11}	Longitudinal elastic modulus
E_{22}, E_{33}	Transverse elastic modulus
ν_{12}, ν_{13} and ν_{23}	Poisson's ratio of CFRP
G_{12}, G_{13} and G_{23}	Shear modulus
u	Horizontal displacements of the meridian
w	Vertical displacements of the meridian
Δ_n	Normal deflections
Δ_t	Tangential deflections
Δ	Total membrane displacement
ψ	Rotation of the membrane element
e_ϕ and e_θ	Second-order meridional and circumferential strains
$e_{\phi a}$ and $e_{\theta a}$	Second-order meridional and circumferential strains in aluminum
$e_{\phi f}$ and $e_{\theta f}$	Second-order meridional and circumferential strains in CFRP
ϵ_ϕ and ϵ_θ	Meridional and circumferential strains
$N_{\phi a}$ and $N_{\theta a}$	Meridional and circumferential stress resultants in aluminum
$N_{\phi f}$ and $N_{\theta f}$	Meridional and circumferential stress resultants in CFRP
F	Potential energy function
U	Strain energy stored in the membrane
W	Work done by the internal pressure
p	Internal pressure
a_i and b_i	Constants of displacement function

1 Introduction

A pressure vessel is a container used to store and transport gases and liquids under high pressures. Generally, they are available in various shapes like cylindrical, spherical, and toroidal. The composite cylindrical and spherical pressure vessels are being studied from the past few decades [1–4]. The behavior of metallic toroidal pressure vessels (TPV) is studied and understood well [5–8]. However, information on composite (i.e., fiber reinforced polymer over wrapped) TPV are scarce, and its behavior has been explored in detail. A toroid is a closed cylinder bent into a complete circle obtained by the surface of revolution of closed curve/meridian about axis of revolution. Toroidal-shaped pressure vessels have many advantages over the traditional cylindrical pressure vessels [9, 10].

Fiber-reinforced polymers (FRPs) are known to have excellent mechanical properties and high strength-to-weight ratio. FRP laminates have been successfully utilized in various fields like civil, mechanical, and aerospace structures for performance enhancement of structural components [11–14]. The use of FRP laminate strips on the bottom of reinforced concrete beams will improve the load-carrying capacity and ductility [12]. Hadi [11] studied the behavior of normal strength concrete columns with circumferential warping of carbon fiber reinforced polymers. The results show that there is an increase in axial load-carrying capacity and ductility of the columns. Wakayama et al. [15] experimentally investigated the residual burst pressure of filament wound cylindrical pipe after impact and concluded that low modulus pitch-based carbon fibers are effective in resisting impact. It is clear from the review of above works that wrapping of FRP on the structural components can enhance their capacity.

Researchers have studied the behavior of TPVs made of metals [6–8, 16]. The analytical solutions are available only for pure isotropic and orthotropic materials. Flugge [16] developed the membrane solution of TPV by using the linear membrane theory (LMT). However, the limitation of LMT is that the deformations are discontinuous at $\phi=0^\circ$ of the toroid. Sanders and Liepins [17] derived a set of nonlinear equations using nonlinear membrane theory for circular TPV under internal pressure. The equations are then solved by asymptotic integration to arrive at the solution. Tielking [5] developed modified-linear membrane theory (MLMT) for the TPV and concluded that circumferential stress is not constant and changes with the meridional angle. MLMT overcomes the limitations of LMT. Geuskens et al. [18] explored the application of MLMT for the orthotropic toroidal and multi-toroidal pressure vessels. Paleti et al. [19] developed a membrane solution for metal-FRP hybrid TPVs under internal pressure.

In this paper, an approximate membrane solution for a metal-FRP TPV is derived to find the stresses and strains in metal liner and FRP. The proposed analytical solution is based on the assumption of treating the orthotropic material as an equivalent isotropic material. The accuracy of the proposed analytical solution is cross-verified with results from finite element analysis (FEA).

2 Proposed Analytical Solution

The proposed analytical solution for metal-FRP TPV is based on the formulation of MLMT for metallic TPV. The equivalent thickness (t_{eq}) for metal-FRP TPV is introduced here. Thereafter, the MLMT for metal-FRP TPV is introduced by minimizing the potential energy function. After that, displacements, stresses, and strains in the FRP and metal are found by using strain–displacement and stress–strain relations. The schematic of metal-FRP TPV with membrane stress resultants on an element is shown in Fig. 1a, and the actual cross-section and the assumed cross-section in the current formulation are shown in Fig. 1b and c.

2.1 Formulation

FRP is treated as quasi-isotropic material, i.e., the thickness of FRP is converted into equivalent thickness of metal using the longitudinal modulus of FRP. This assumption results in having a good approximation for finding meridional and von-Mises stresses. To have maximum utilization of the FRP layer, the fiber direction is orientated in meridional direction (since, $N_\phi > N_\theta$). The meridional stress (σ_ϕ) and circumferential stress (σ_θ) are the stresses acting in the meridional and circumferential directions as shown in Fig. 1. It is assumed that the toroidal shell is thin, and to calculate the stresses in FRP, the bond between aluminum liner and FRP layer is considered as rigid. In the current study, the equivalent thickness of CFRP in terms of aluminum is calculated as

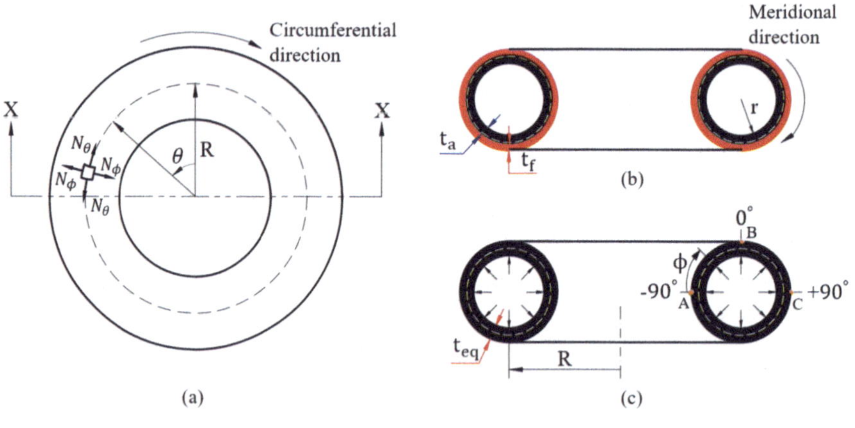

Fig. 1 Schematic of metal-FRP TPV: **a** TPV with stresses in an element, **b** actual cross-section, and **c** assumed cross-section

$$t_{\mathrm{eq}} = t_a + t_{\mathrm{ef}}, \tag{1}$$

$$t_{\mathrm{ef}} = \frac{E_{11}}{E} t_f. \tag{2}$$

Strain–Displacement Relations. The deformation of the meridian is shown in Fig. 2. Let w and u be respectively the vertical and horizontal displacements of the meridional element; Δ_{n} and Δ_{t} be the normal and tangential deflections, respectively; ψ and Δ are the rotation and total displacement perpendicular to the mid surface of the membrane element [5]. These deformations can be expressed as

$$\Delta_{\mathrm{n}} = u \sin\phi + w \cos\phi, \tag{3}$$

$$\Delta_{\mathrm{t}} = u \cos\phi - w \sin\phi, \tag{4}$$

$$\Delta = \Delta_{\mathrm{n}} + \psi \Delta_{\mathrm{t}}, \tag{5}$$

$$-\psi = \frac{u' \sin\phi + w' \cos\phi}{r}, \tag{6}$$

where (′) indicates the differentiation of the variables with respect to ϕ. The second-order strain–displacement relation, from the moderate rotation theory [5], is given as

$$e_\theta = \epsilon_\theta \text{ and } e_\phi = \epsilon_\phi + \frac{1}{2}\psi^2, \tag{7}$$

where

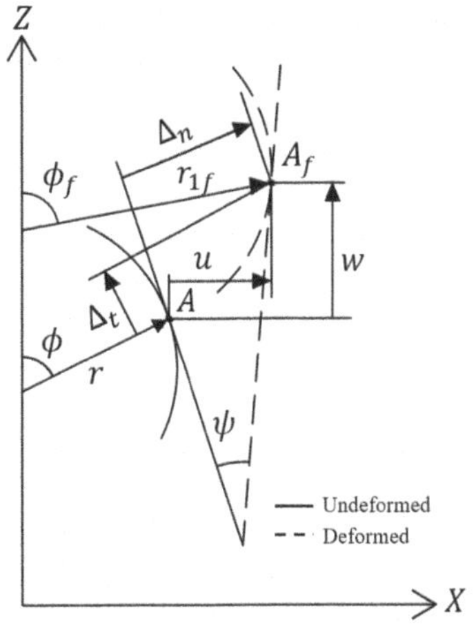

Fig. 2 Deformation of the meridian

$$\epsilon_\theta = \frac{u}{r_0} \text{ and } \epsilon_\phi = \frac{u' \cos\phi - w' \sin\phi}{r}. \tag{8}$$

Stress–Strain Relations. For a linear isotropic material, the membrane stress resultant–strain relation is given as

$$N_{\phi a} = \frac{Et_a}{1-\nu^2}\left(e_{\phi a} + \nu e_{\theta a}\right);\ N_{\theta a} = \frac{Et_a}{1-\nu^2}\left(e_{\theta a} + \nu e_{\phi a}\right), \tag{9}$$

where E is the modulus of elasticity, ν is the Poisson's ratio, and t_a is the thickness of the base metal of the TPV.

For the FRP, the membrane stress resultant–strain relation is given in Eqs. (10 and 11). However, these relations are not used in writing the potential energy functional F, but used in calculating meridional and circumferential stresses in the FRP based on the assumption that the strains in the FRP and metal are same.

$$N_{\phi f} = t_f\left(Q_{\phi\phi} e_{\phi f} + Q_{\phi\theta} e_{\theta f}\right), \tag{10}$$

$$N_{\theta f} = t_f\left(Q_{\theta\theta} e_{\theta f} + Q_{\phi\theta} e_{\phi f}\right), \tag{11}$$

$$Q_{\phi\phi} = \frac{E_{11}}{1-\nu_{12}\nu_{21}};\ Q_{\theta\theta} = \frac{E_{22}}{1-\nu_{12}\nu_{21}} \text{ and } Q_{\phi\theta} = \frac{\nu_{12}E_{22}}{1-\nu_{12}\nu_{21}}.$$

For the equivalent section, the stress resultant–strain relation is given as

$$N_\phi = \frac{Et_{\text{eq}}}{1-\nu^2}\left(e_\phi + \nu e_\theta\right);\ N_\theta = \frac{Et_{\text{eq}}}{1-\nu^2}\left(e_\theta + \nu e_\phi\right). \tag{12}$$

The circumferential and meridional stresses in the FRP and metal (i.e., $\sigma_{\theta f}, \sigma_{\phi f}, \sigma_{\theta a}, \sigma_{\phi a}$ and) of a metal-FRP TPV are obtained by dividing the corresponding stresses resultant with their respective thickness.

The area of the deformed mid surface of the membrane element is given as

$$\mathrm{d}A_d = \left(1 + e_\phi\right)(1 + e_\theta)\mathrm{d}A \approx \mathrm{d}A = r_0 r \mathrm{d}\phi \mathrm{d}\theta, \tag{13}$$

where dA is the undeformed area of the shell element.

Potential Energy Function. The potential energy function applicable for the metal-FRP TPV may be expressed as

$$F = U - W. \tag{14}$$

The work done W by the internal pressure is

$$W = p \int_A \Delta \mathrm{d}A. \tag{15}$$

Using Eqs. (3–6, 13, and 15) and integrating in circumferential direction, the work done by the internal pressure can be expressed as

$$W = 2\pi p \int_{\phi_a}^{\phi_b} \left\{ (u \sin\phi + w \cos\phi) - \left(\frac{u' \sin\phi + w' \cos\phi}{r} \right) (u \cos\phi - w \sin\phi) \right\} r_0 r \mathrm{d}\phi. \tag{16}$$

The strain energy of the uniformly load membrane is expressed as

$$U = \frac{1}{2} \int_A (N_\phi e_\phi + N_\theta e_\theta) \mathrm{d}A. \tag{17}$$

The strains and rotations are small compared to unity for the stiff shell. Therefore, the terms containing ψ^4 can be ignored. Using Eqs. (7, 8, 12, and 13) and integrating in circumferential direction, the strain energy Eq. (17) becomes

$$U = \pi \int_{\phi_a}^{\phi_b} \left\{ \begin{array}{l} \frac{E\, t_{eq}}{(1-\nu^2)} \left(\left(\frac{u' \cos\phi - w' \sin\phi}{r} \right)^2 + 2\,\nu\, \frac{u}{r_0} \left(\frac{u' \cos\phi - w' \sin\phi}{r} \right) \right) \\ + \left(\frac{u}{r_0} \right)^2 + N_\phi \left(\frac{u' \sin\phi + w' \cos\phi}{r} \right)^2 \end{array} \right\} r_0 r\, \mathrm{d}\phi. \tag{18}$$

Therefore, the potential energy functional using Eqs. (14), (16), and (18) becomes

$$F = \int_{\phi_a}^{\phi_b} \left[\begin{array}{l} \left(\frac{E\, t_{eq}}{1-\nu^2} \right) \left\{ \left(\frac{u' \cos\phi - w' \sin\phi}{r} \right)^2 + 2\,\nu\, \frac{u_1}{r_0} \left(\frac{u' \cos\phi - w' \sin\phi}{r} \right) + \left(\frac{u}{r_0} \right)^2 \right\} \\ + N_\phi \left(\frac{u' \sin\phi + w' \cos\phi}{r} \right)^2 \\ -\, 2p \left\{ (u \sin\phi + w \cos\phi) - \left(\frac{u' \sin\phi + w' \cos\phi}{r} \right) (u \cos\phi - w \sin\phi) \right\} \end{array} \right] r_0 r\, \mathrm{d}\phi, \tag{19a}$$

where

$$r_0 = R + r \sin\phi \tag{19b}$$

N_ϕ represents the meridional stress based on the LMT and is considered as $(p.r)$ according to Tielking [5]. The integration boundaries for TPVs are set as $[\phi_a, \phi_b] = [-\pi/2, \pi/2]$ due to the single axis of symmetry of the geometry of TPV.

The Rayleigh–Ritz method is employed to minimize the Eq. (19a), utilizing finite series approximation for both u and w. The displacement functions u and w are represented as Fourier series, ensuring that they conform to the pressure vessel's boundary conditions.

For the TPV, the boundary conditions are as follows:

$$w(\pm\pi/2) = u'(\pm\pi/2) = w'(\pm\pi/2) = 0. \tag{20}$$

The vertical and horizontal displacement functions that meet the specified boundary conditions, expressed in terms of Fourier series, are as follows:

$$w_n(\phi) = \sum_{i=0}^{m} b_i \sin\left[i\left(\phi + \frac{\pi}{2}\right)\right], \tag{21}$$

$$u_n(\phi) = \sum_{i=0}^{m} a_i \cos\left[i\left(\phi + \frac{\pi}{2}\right)\right]. \tag{22}$$

When Eqs. (21) and (22) are substituted into Eq. (19a), the solution is obtained with the product of coefficients (a_i or b_i). These coefficients in Eq. (19a) can be determined by minimizing the Eq. (19a), utilizing the Ritz method, as follows:

$$\frac{\partial F}{\partial a_i} = 0 \quad i = 0, 1, 2, \ldots, n, \tag{23}$$

$$\frac{\partial F}{\partial b_i} = 0 \quad i = 0, 1, 2, \ldots, n. \tag{24}$$

Equations (23) and (24) bring down Eq. (19a) into 2n or 2n + 1 number of linear equations based on the selection of u and w. By solving Eqs. (23) and (24), the a_i and b_i can be found. Upon getting the coefficients, then solution for the displacements is obtained by using Eqs. (21) and (22), strains from Eq. (7), and membrane stresses from Eqs. (9–11).

3 Numerical Modeling

A three-dimensional finite element model has been developed in ABAQUS [20] to cross-verify the proposed solution. A full geometry of the toroid is modeled for the analysis, and its FE model is shown in Fig. 3. The membrane elements (M3D4R) are used for having a consistent comparison of results, since the proposed solution is based on membrane formulation. The toroidal shell is discretized with number of elements in meridional and circumferential direction as 50 and 158 elements, respectively. The tie interaction property is used as contact condition between metal

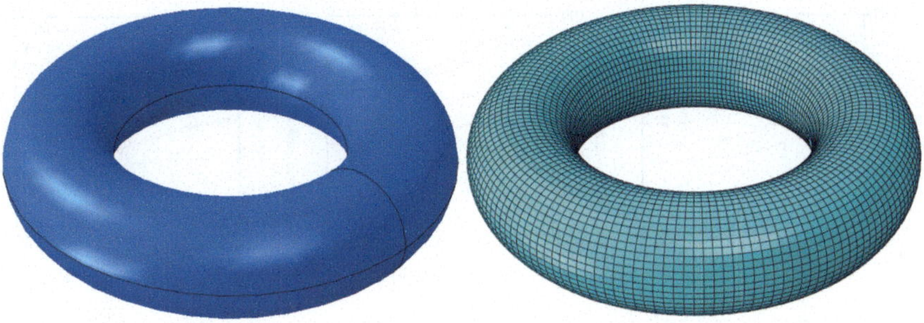

Fig. 3 FEA model of metal-FRP TPV

and FRP for metal-FRP TPV. The chosen contact condition acts as rigid contact. A static-linear analysis is carried out by using general static solver.

4 Illustrative Example

To illustrate the proposed solution, a metal-FRP TPV with following geometric properties is used. The toroidal radius $R = 99.5$ mm, meridional radius $r = 31.5$ mm, thickness of base metal $t_a = 4$ mm, and thickness of CFRP layer $t_f = 0.2$, 1, and 2 mm (corresponding to 1, 5, and 10 number of layers (n), respectively) are considered. The base metal is made up of aluminum, and its modulus of elasticity and Poisson's ratio is 69204 MPa and 0.33, respectively. The properties of CFRP are given in Table 1 and are used for the analysis of metal-FRP TPV.

A 13-term displacement approximation is chosen in the proposed solution based on the convergence study. The results, i.e., circumferential and meridional stresses and strain in the aluminum and FRP and the von-Mises stress in the aluminum along the meridian, A to C (-90° to $+90^\circ$) shown in Fig. 1c, are plotted as shown in Figs. 4 and 5.

The meridional strains in aluminum and CFRP are matching closely with the FEA results as shown in Fig. 4(a, i). The circumferential strains from the proposed solution do not align well with the findings from the FEA results, as shown in Fig. 4(b, i). This is because of the assumption in calculating the equivalent thickness of the CFRP. The equivalent thickness is calculated by using the longitudinal modulus E_{11} of CFRP (i.e., CFRP is treated as isotropic material with modulus of elasticity E) where the transverse modulus E_{22} of CFRP is almost 1/18 of E_{11}. Hence, the circumferential

Table 1 Material properties of CFRP lamina [21]

E_{11} (GPa)	$E_{22} = E_{33}$ (GPa)	$\nu_{12} = \nu_{13}$	ν_{23}	$G_{12} = G_{13}$ (GPa)	G_{23} (GPa)
90.49	4.98	0.234	0.3	1.91	1.56

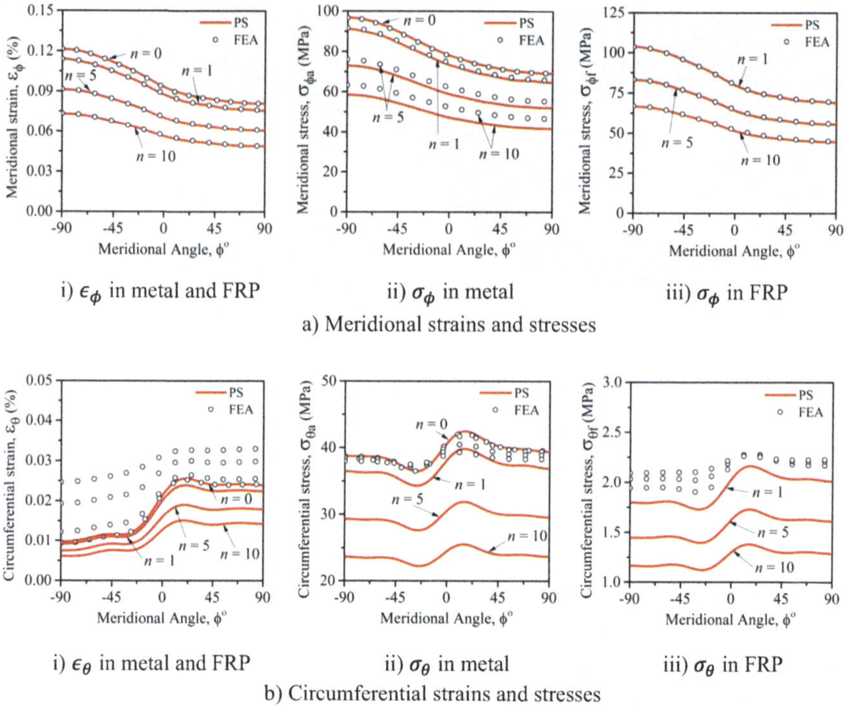

i) ϵ_ϕ in metal and FRP ii) σ_ϕ in metal iii) σ_ϕ in FRP

a) Meridional strains and stresses

i) ϵ_θ in metal and FRP ii) σ_θ in metal iii) σ_θ in FRP

b) Circumferential strains and stresses

Fig. 4 Stress and strain response of the metal-FRP TPV ($n = 1$, 5, and 10) at a pressure of 100 Bar

Fig. 5 von-Mises stress in the base metal of metal-FRP TPV ($n = 1$, 5, and 10) at a pressure of 100 Bar

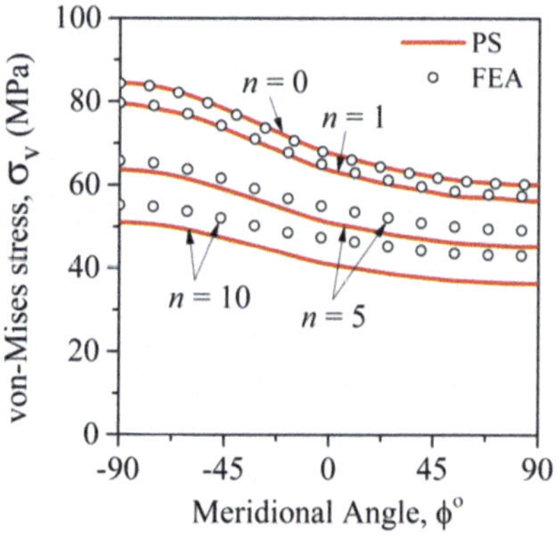

stresses and strains obtained from the proposed solution are lower and not matching with the FEA results as shown in Fig. 4b. However, the meridional stresses in the aluminum are matching with FEA results with a reasonable accuracy of maximum percentage error of 7.5% for metal-FRP TPV with $n = 10$ and the deviation from the FEA results is because of the Poisson's effect, Eq. (9). The deviation observed in

Table 2 Comparison of yield pressure from proposed solution and FEA

Number of layers (n)	Yield pressure P_y (MPa)		
	Proposed solution	FEA	% Error
1	38.17	38.09	0.21
5	47.65	46.12	3.32
10	59.39	54.96	8.06

meridional stresses in aluminum is not seen in meridional stresses in CFRP (Fig. 4(a, iii)). This is because both $Q_{\phi\theta}$ and $e_{\theta f}$ are very small and have negligible effect in calculation of meridional stresses in CFRP. It is observed that von-Mises stresses are also matching with the FE results with a reasonable accuracy as shown in Fig. 5. As the number of layer of CFRP is increasing, the deviation of meridional and von-Mises stresses is also increasing because of under prediction of circumferential strains in the aluminum. The yield pressure P_y obtained from the proposed solution and its percentage error from FEA results, shown in Table 2. The maximum error observed in calculating yield pressure is 8% for metal-FRP TPV with $n = 10$. Therefore, it is observed that proposed solution is agreeing well with the FEA results with a reasonable accuracy for lesser layers of CFRP.

5 Conclusions

A simplified approximate membrane solution for metal-FRP TPV by considering CFRP as an equivalent isotropic material is presented. The outcomes from the proposed solution are compared with the FE results. The proposed solution is helpful in approximating the yield pressure of metal-FRP TPV. The analytical and FEA outcomes lead to the following conclusions.

- The proposed model predicts the meridional strains with an acceptable error of less than 1% for the case of TPV strengthened with 10 layers of CFRP wrapping.
- The under prediction of circumferential strains is because of the assumption in calculating the equivalent thickness and treating the CFRP as an equivalent isotropic material.
- The Poisson's effect from circumferential strain led to the variation of meridional stresses predictions by 7.5% compared to the FE results.
- When the number of layers of CFRP is more than 10, the error in finding the yield pressure increases beyond 10%.
- The developed analytical solution has to be validated with experimental results for accessing the effectiveness of the proposed solution. The future scope of the work involves design and development of metal-FRP TPV and conduct an internal pressure test on metal-FRP TPV to validate the proposed analytical solution.

References

1. Zu L, Xu H, Zhang Q, Jia X, Zhang B, Li D (2019) Design of filament-wound spherical pressure vessels based on non-geodesic trajectories. Compos Struct 218:71–78
2. Harada S, Arai Y, Araki W, Iijima T, Kurosawa A, Ohbuchi T et al (2018) A simplified method for predicting burst pressure of type III filament-wound CFRP composite vessels considering the inhomogeneity of fiber packing. Compos Struct 190:79–90
3. Berro Ramirez JP, Halm D, Grandidier JC, Villalonga S, Nony F (2015) 700 bar type IV high pressure hydrogen storage vessel burst—simulation and experimental validation. Int J Hydrog Energy 40(38):13183–13192
4. Wang L, Zheng C, Luo H, Wei S, Wei Z (2015) Continuum damage modeling and progressive failure analysis of carbon fiber/epoxy composite pressure vessel. Compos Struct 134:475–482
5. Tielking JT, McIvor IK, Clark SK (1971) A modified linear membrane theory for the pressurized toroid. J Appl Mech ASME 38(2):418–422
6. Jordan PF (1962) Stresses and deformations of the thin-walled pressurized torus. J Aerosp Sci 29(2):213–225
7. Vu VT, Blachut J (2009) Plastic instability pressure of toroidal shells. J Press Vessel Technol Trans ASME 131(5):051203
8. Błachut J (2005) Plastic loads for internally pressurized toroidal shells. J Press Vessel Technol Trans ASME 127(2):151–156
9. Zu L, Zhang D, Xu Y, Xiao D (2012) Integral design and simulation of composite toroidal hydrogen storage tanks. Int J Hydrog Energy 37(1):1027–1036
10. Zu L (2012) Stability of fiber trajectories for winding toroidal pressure vessels. Compos Struct 94(5):1855–1860
11. Hadi MNS (2006) Behaviour of FRP wrapped normal strength concrete columns under eccentric loading. Compos Struct 72(4):503–511
12. Askar MK, Hassan AF, Al-Kamaki YSS (2022) Flexural and shear strengthening of reinforced concrete beams using FRP composites: a state of the art. Case Stud Constr Mater 17:e01189
13. Teng JG, Hu YM (2007) Behaviour of FRP-jacketed circular steel tubes and cylindrical shells under axial compression. Constr Build Mater 21(4):827–838
14. Date S, Abe Y, Okabe T (2022) Effects of fiber properties on aerodynamic performance and structural sizing of composite aircraft wings. Aerosp Sci Technol 124:107565
15. Wakayama S, Kobayashi S, Imai T, Matsumoto T (2006) Evaluation of burst strength of FW-FRP composite pipes after impact using pitch-based low-modulus carbon fiber. Compos Part A Appl Sci Manuf 37(11):2002–2010
16. Flügge W (1973) Stress in shells, 2nd edn. Springer, Heidelberg
17. Sanders JL, Liepins AA (1963) Toroidal membrane under internal pressure. AIAA J 1(9):2105–2110
18. Geuskens FJJMM, Tielking JT, Koussios S, Beukers A (2013) Modified linear-membrane formulation for the pressurized torus and multitorus. AIAA J 51(9):2114–2125
19. Paleti MK, Prakash SS, Narayanamurthy V (2023) A Theoretical solution for metal-FRP hybrid toroidal pressure vessel based on membrane approach. Thin-Walled Struct 188:110866
20. ABAQUS (2014) Abaqus v. 6.14–2 Analysis user's guide—online documentation
21. Matta S, Chinthapenta V, Ramji M (2017) A novel approach to analyse adhesive layer strain field in a stepped lap repaired carbon fiber reinforced polymer panel using digital image correlation. J Adhes Sci Technol 31(19–20):2180–2201

Effect of Focused Ultrasound on Shear Wave Propagation in Nonlinear Media

Aniket Sabale, Mohd Suhail Rizvi, Viswanath Chinthapenta, and Avinash Eranki

Abstract Focused ultrasound (FUS) is an exciting and emerging technology that can ablate or fractionate tissues and is used to treat solid tumors alone or in combination with other therapies. FUS waves at the focus can produce shear waves that can be used to understand the stiffness of the tissue or track treatment progression or completion. Herein, using finite element model (FEM)-based simulations, we test multiple ultrasound frequencies at different focal pressures, producing shear waves in a nonlinear medium. We quantified the displacements and shear stress produced by the shear wave propagation. We observed that for the 0.75 MPa FUS transducer surface pressure, the peak shear stress production was substantially different between 250 and 500 kHz. Furthermore, the time to peak for shear stress waves was lower for the 250 kHz and 0.75 MPa conditions than the 1 MPa condition at 250 kHz and the 1 MPa condition at 500 kHz. In summary, we have developed a framework to evaluate the effect of different FUS transducers in producing shear waves in nonlinear media.

Keywords Focused ultrasound · High-intensity focused ultrasound · Shear wave · Neo-Hookean

1 Introduction

Focused ultrasound (FUS) is a non-invasive, non-ionizing radiation, ultrasound-based therapeutic technology that uses acoustic wave focused at a region of interest. The focused waves are used in multiple flavors of therapeutic ultrasound, such as thermal ablation [1, 2], mild hyperthermia [3, 4], boiling histotripsy [5–8], cavitation histotripsy [9], and low-intensity focused ultrasound [10], among others. These effects are classified either into thermal or non-thermal. They tend to have a wide

A. Sabale · M. S. Rizvi (✉) · A. Eranki
Department of Biomedical Engineering, IIT Hyderabad, Kandi 502285, India
e-mail: suhailr@bme.iith.ac.in

V. Chinthapenta
Department of Mechanical and Aerospace Engineering, IIT Hyderabad, Kandi 502285, India

S. Sidhardh et al. (eds.), *Advances in Structural Integrity for Mechanical, Civil, and Aerospace Applications*, Lecture Notes in Mechanical Engineering,
https://doi.org/10.1007/978-981-97-6367-2_19

variety of outcomes in targeted tissues, such as increased blood flow or shutting vasculature [3, 11], breaking blood clots [12], upregulation of immune and genetic markers [8], increasing cellular permeability to drugs [13], as well as drug delivery. These flavors of focused ultrasound are used to treat solid tumors [2, 14], deep vein thrombosis [15], essential tremors [16], Parkinson's disease [17], kidney stones [18], and benign prostate hyperplasia [19, 20]. All of these indications were treated by FUS alone or in combination with encapsulated drugs or adjuvant radiation, chemotherapy, or immunotherapy.

Acoustic pressures at the focus of FUS have been seen to range between 2 $MPa^{\pm}$ (superscripts indicate peak positive and/or peak negative values) in the linear wave propagation regime to 100 MPa^{+} and 20 MPa^{-} in the nonlinear acoustic regime [9, 21]. This could lead to heating at the focal region and cavitation bubble formation. The extent of these effects can vary with acoustic pressure and its nonlinearity. In addition to these effects, transverse wave (shear wave) propagation is expected along the transverse axis originating from the focal zone [22]. Shear wave production and propagation vary with the type of media and require the generation of mechanical stimulation. Ultrasound [23], magnetic resonance imaging (MRI) [24], or optical tracking [25] has been used to image and track shear waves in soft tissues. While it has shown promise, MRI elastography tends to be expensive and may not have the temporal resolution ultrasound-based elastography offers. Ultrasound-based methods such as shear wave elasticity imaging (SWEI) [26], acoustic radiation force imaging (ARFI) [27], supersonic shear imaging (SSI) [28], harmonic motion imaging (HME), and comb-push ultrasound shear elastography (CUSE) [29] have been used to quantify mechanical properties of the tissues. Some of these techniques have been approved by the U.S. Food and Drug Administration (FDA) for quantifying liver stiffness (e.g., liver cirrhosis, tumors, etc.), breast tumors, and thyroid lesions, among others. These methods assume an isotropic, incompressible, and linearly elastic media (disregarding shear wave dispersion). However, assuming soft tissues to be linear may not be sufficient to understand the FUS at the tissue level and its applications, such as drug delivery, especially in heterogeneous tissues such as tumors. Ultrasound-based methods rely on low acoustic pressure (often less than 1 MPa). However, in therapeutic ultrasound using FUS, the acoustic pressure exceeds 1 MPa, and the subsequent shear wave propagation and displacement in nonlinear media mimicking soft tissues are poorly characterized. Therefore, understanding the effect of shear waves in vivo for potential cellular effects and drug delivery applications is very important. Recently, lower frequencies in the range of 250–500 kHz have been used to open the blood–brain barrier, ablate bone tumors, and help penetrate deeper tissues. In addition, acoustic pressures at the focus between 2 and 4 MPa have been generated to produce a low-intensity focused ultrasound effect demonstrating a therapeutic outcome.

This work presents a computational framework to simulate the effects of frequency and mechanical pressure applied by the ultrasound transducer on the shear wave generation in the soft tissue using ABAQUS Explicit. The finite element-based framework tested longitudinal (acoustic) 250 and 500 kHz frequencies. The displacement and shear stress at different locations along the axis transverse to the longitudinal wave propagation were quantified. The outcomes of the present study provide insights into FUS therapy planning and understanding the role of shear wave propagation in the surrounding tissues.

2 Methods and Materials

A finite element model (FEM) was created to simulate wave propagation from a focused ultrasound transducer into a neo-Hookean material (mimicking soft tissue) in ABAQUS Explicit (version 6.9, Dassault Systemes, France), as shown in Fig. 1a.

A single-element axisymmetric ultrasound transducer was considered, with an outer diameter (OD) of 64 mm and a focal length of 63.2 mm. Direct contact between the active surface of the FUS transducer and the neo-Hookean material was assumed. A simulation was carried out for the entire model using axisymmetric analysis due to the axisymmetric shape of the geometry and loading. The simulated neo-Hookean material was divided into two domains: the first one with a width of 32 mm and a length of 73.2 mm, and a second one with a width of 32 mm and a length of 15 mm was added toward the end of the first material to dampen the waves. The smallest mesh element size for the first neo-Hookean domain considered was 50 μm to spatially resolve the wave propagation in our simulations. In contrast, the mesh element size in the second domain (damping material) was kept at 200 μm. This difference in mesh size between both domains is to reduce overall computational time. A four-node, bilinear axisymmetric quadrilateral mesh element with reduced integration (CAX4R) was used in both domains with hourglassing control to prevent

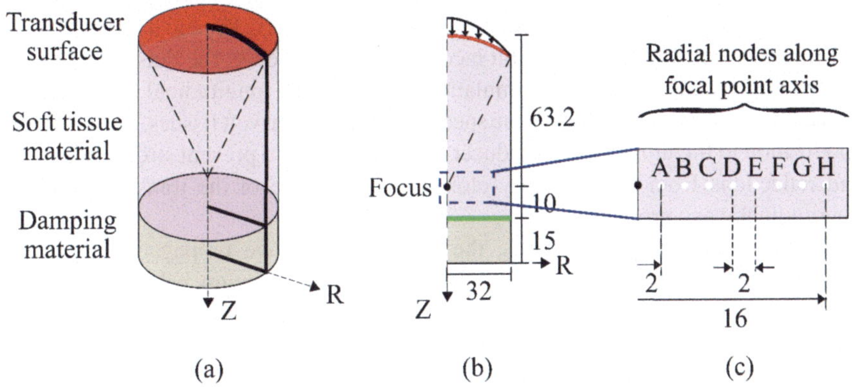

Fig. 1 **a** Schematic of focused ultrasound (axisymmetric model). The top curved boundary represents the transducer surface. **b** Axisymmetric cross-section. **c** Eight locations were selected at a distance of 2 mm from the center of the focal zone normal to the acoustic wave propagation direction. (All dimensions are in mm)

zero-energy deformation modes. A surface-based tie constraint at the interface of the two domains, represented as the green line in Fig. 1b, fused the two domains of dissimilar mesh sizes. The top neo-Hookean material was used as a surrogate for soft tissue media. The strain energy density function W, for Neo-Hookean material, is given as:

$$W = \frac{\mu}{2}\left(\overline{I_1} - 3\right) + \frac{K}{2}(J - 1)^2, \tag{1}$$

where $\overline{I_1}$ is the first strain invariant of the left Cauchy–Green deformation tensor, and J is the determinant of the deformation gradient. The term μ is the shear modulus, and K is the bulk modulus of the material.

The damping material domain (second domain) was extended by introducing Rayleigh damping in addition to soft tissue material properties. The Rayleigh damping matrix $[C]$ is defined as:

$$[C] = \alpha[M] + \beta[S], \tag{2}$$

where $[M]$ is the mass matrix and $[S]$ is the stiffness matrix. Mass proportional damping coefficient α, having a value of 2.6×10^6, is considered for the simulations, as it does not affect the stable time increment and reduces the computational time. The stiffness proportional damping coefficient, β, leads to a significant increase in computational time and is therefore avoided in the present simulations. Also, the automatic stable time increment option was used to automatically adjust the temporal resolution in these simulations.

To simulate the effect of longitudinal waves and resulting shear waves in human soft tissues, the mechanical properties of the neo-Hookean domain are considered to have a density of 1050 kg/m^3, shear modulus (μ) of 3000 Pa, and a bulk modulus (K) of 2.15 GPa. Soft tissues exhibit incompressibility; hence, the Poisson's ratio of 0.499999 was considered in this simulation. The resulting longitudinal wave speed is expected to be 1540 m/s, which is the speed of ultrasound in soft tissues. Furthermore, apodization to the ultrasound transducer was considered to prevent strong side lobes and reflections from the boundary closer to the surface of the transducer during longitudinal wave propagation.

In the absence of body forces, the equation for wave propagation in a 2D-axisymmetric can be obtained from the following momentum balance equations in terms of material density (ρ_0), , displacement vector field (u), and Cauchy stress tensor (σ).

$$\rho_0 \frac{\partial^2 u}{\partial t^2} = \nabla \cdot \sigma \tag{3}$$

The Cauchy stress tensor, σ, for the neo-Hookean material

Table 1 Table below presents FUS transducer frequencies, surface pressure used, and corresponding focal pressure

Transmit frequency (kHz)	Surface pressure P_0 (MPa)	Focal pressure (MPa)
250	0.75	2
250	1	3
500	0.75	4

$$\sigma = \frac{\mu}{J^{5/3}}\left[\lambda^2 - \frac{I_1}{3}\right] + K(J-1) \tag{4}$$

is substituted in Eq. (3), and the resulting equations are solved in ABAQUS Explicit solver.

The ultrasound transducer was excited with a five-cycle sinusoidal pulse, with a transmit frequency of 250 and 500 kHz. The pressure at the surface of the transducer was apodized along the transverse direction using Eq. 5.

$$P = P_0 \cos\left[\pi\left(\frac{r}{\mathrm{OD}}\right)\right]. \tag{5}$$

Focal pressures of 2 and 3 MPa were generated for the 250 kHz transducer and 4 MPa for the 500 kHz transducer. The surface pressure of 0.75 and 1 MPa was used to produce focal pressure of relevance, as shown in Table 1. Shear wave displacement and shear stress were computed for each of the above conditions at eight-node points located in the vicinity of the focal region. These points were spaced at 2 mm intervals along the transverse axis, starting from the center of the focal region, as illustrated in Fig. 1c.

3 Results

One of the primary goals of this study is to understand the effect of acoustic pressures at the focal region produced by FUS transducers operated at 250 kHz (at 0.75 and 1 MPa) and 500 kHz (at 0.75 MPa) on shear wave generation and propagation. The displacement at eight locations (see Fig. 1c) along the transverse direction varied in their behavior across different parameters (Fig. 2). At 250 kHz and 1 MPa, the displacements observed were the highest and took longer to reach the peak value. On the other hand, the lowest displacements were observed at 250 kHz and 0.75 MPa conditions at all node points, with a reduction of almost 30% relative to those at 250 kHz and 1 MPa.

The peak displacement due to the shear wave propagation at each node is plotted in Fig. 3a, b for 250 kHz at 0.75/1 MPa and 500 kHz at 0.75 MPa, respectively. Peak displacement is the greatest for the node closest to the center of the focus at 0.15 microns magnitude for 250 kHz frequency and 1 MPa surface pressure.

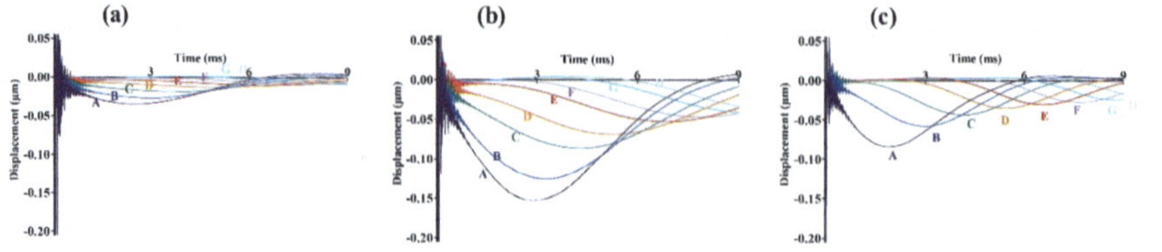

Fig. 2 Displacement curves measured at eight locations (labeled A through H) along the transverse axis originating at the center of the focal zone for **a** 250 kHz at 0.75 MPa, **b** 250 kHz at 1 MPa, and **c** 500 kHz at 0.75 MPa. In **a** peak displacement is up to 0.04 microns in magnitude. In **b** peak displacement is about 0.15 microns, higher than the 0.75 MPa case due to a higher focal pressure (2 vs. 3 MPa). And in **c** displacement curves at all eight locations have a lower displacement compared to the curves in (**b**)

The maximum magnitude of the peak displacement at 250 kHz and 0.75 MPa is 0.04 microns. Peak displacement appeared to decay exponentially with the distance between the first node at 2 mm and the last node at 16 mm for 250 kHz frequency and 1 MPa surface pressure with a threefold reduction in peak displacement (Fig. 3a). Similar measurements made for 500 kHz and 0.75 MPa showed a maximum magnitude of peak displacement of 0.09 microns (Fig. 3b). The exponential decrease in the magnitude of peak displacement over all the eight-node points was similar to the 250 kHz and 1 MPa case.

The peak shear stress values at 250 kHz at both 0.75 MPa (red circles in Fig. 4) and 1 MPa (blue squares in Fig. 4) and 500 kHz at 0.75 MPa (green triangle in Fig. 4) at all eight locations along the transverse direction are plotted in Fig. 4. At 250 kHz transmit frequency and 0.75 MPa surface pressure, the peak shear stress is, on average, about 0.03 Pa. However, at 1 MPa, the peak shear stress is distributed between 0.035 and 0.06 Pa. Peak shear stress measured at all these eight locations for 500 kHz and 0.75 MPa surface pressure is distributed between 0.025 and 0.07 Pa, with most of the locations presenting peak shear stress around 0.03 Pa.

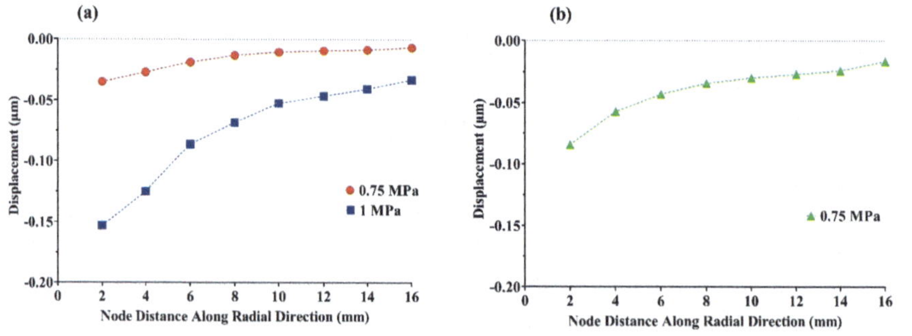

Fig. 3 Peak displacement curves are measured at all eight locations (labeled A through H in Fig. 2) along the transverse axis originating at the center of the focal zone for 250 kHz at **a** 0.75 MPa and 250 kHz at 1 MPa and **b** 500 kHz at 0.75 MPa

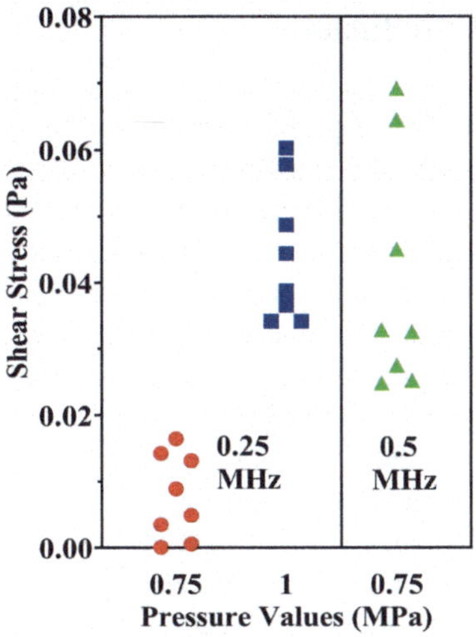

Fig. 4 Peak shear stress was measured at all eight locations along the transverse axis originating at the center of the focal zone for 250 kHz at 0.75 MPa (red circles), 250 kHz at 1 MPa (blue squares), and 500 kHz at 0.75 MPa (green triangles). The peak shear stress is lower at 0.75 MPa compared to 1 MPa for a frequency of 250 kHz

To understand the effect of wave dispersion, the time-to-peak for displacement and shear stress curves at 250 kHz and 500 kHz were also measured at different transverse locations (Fig. 5). The time to peak showed a linear increase with a distance from about 2–9 ms for the displacement of each node from the center of the focal zone along the transverse direction as shown in Fig. 5a. There was no noticeable difference between 250 kHz and 0.75 MPa, 250 kHz and 1 MPa, and 500 kHz and 0.75 kPa. The time to peak for shear stress was calculated from the shear stress waveform, as shown in Fig. 5b.

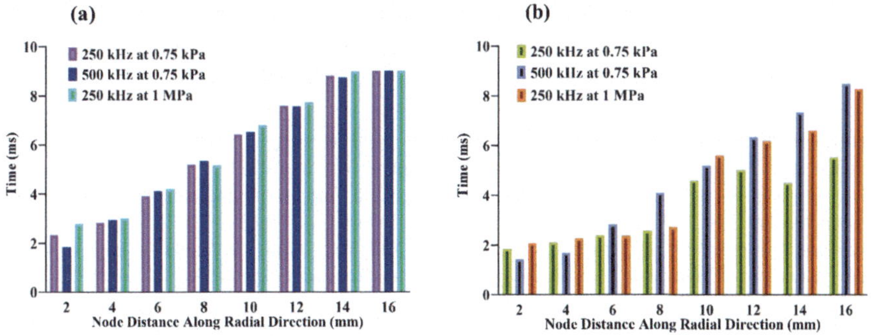

Fig. 5 Time to peak for: **a** displacement and **b** shear stress, at eight locations for different transmit frequencies and transducer surface pressures

4 Discussion and Conclusion

In this work, FUS-derived wave propagation in soft tissues is investigated. The focused nature of the incident ultrasound wave shows an increase in the pressure at the focus. From the results presented here, an approximate three to fourfold increase in the pressure at the center of the focus was observed. This increase in the pressure at the focal point is dependent on the operational parameters of the transducer and the mechanical properties of the biological tissues involved.

The results of this investigation demonstrate the generation of a shear wave from the center of the focus propagating along the transverse direction. The generation of this shear wave can be attributed to the elastic nature of the tissue, where the application of pressure from the transducer results in localized oscillatory axial displacements at the focus, producing the shear wave. Expectedly, the displacement and shear stress magnitudes decrease as the shear wave propagates along the transverse direction. This can be due to the axisymmetric geometry of the domain resulting in a decrease in energy density as the shear wave propagates. One interesting observation from this study is the difference between the peak shear stress values for the same pressure (750 kPa) at two different transmit frequencies. A remarkably smaller peak shear stress was seen for 250 kHz than 500 kHz. This frequency dependence on the shear stress values in FUS can have therapeutic applications where the transmit frequency can be used as an operational parameter to modulate applied shear stress. Currently, the frequency of the transducer is used as an operational parameter to control the focus size and not the pressure. The decrease in the peak shear stress with distance can also be utilized to estimate the effect of the shear stress due to FUS therapy in the surrounding tissue.

In this work, the material model considered is a homogeneous neo-Hookean elastic material, which is a simplification compared to actual tissue behavior. For more realistic representation, one must perform a similar analysis with viscoelastic-layered media, representing different tissue layers between the transducer (usually outside the patient's body) and the tumor where FUS is applied. The future scope of this work is to understand the role of shear stress on cellular and genetic expression. This may be quite useful as it has been demonstrated that selective mutations in breast cancer or tumor types may lead to worse outcomes than others [30, 31].

Acknowledgements Avinash Eranki acknowledges funding support from the Department of Science and Technology—Technology Development Program (BDTD) [TDP/BDTD/24/2021 (G)]—Department of Science and Technology, Science and Engineering Research Board—Startup Research Grant (SRG/2021/001061), and Indian Council for Medical Research (ICMR)—Center of Excellence Grant 5/3/8/5/2021/MDMS/ICMR-DHR COE. Md. Suhail Rizvi acknowledges support from Science and Engineering Research Board—Startup Research Grant (SRG/2021/001020).

References

1. Chapman A, Ter Haar G (2007) Thermal ablation of uterine fibroids using MR-guided focused ultrasound-a truly non-invasive treatment modality. Eur Radiol 17(10):2505–2511
2. Zhou YF (2011) High intensity focused ultrasound in clinical tumor ablation. World J Clin Oncol 2(1):8–27 PubMed PMID: 21603311
3. Li L, ten Hagen TL, Hossann M et al (2013) Mild hyperthermia-triggered doxorubicin release from optimized stealth thermosensitive liposomes improves intratumoral drug delivery and efficacy. J Control Release 168(2):142–150
4. Yarmolenko PS, Moon EJ, Landon C et al (2011) Thresholds for thermal damage to normal tissues: an update. Int J Hyperth 27(4):320–343
5. Eranki A, Farr N, Partanen A et al (2017) Boiling histotripsy lesion characterization on a clinical magnetic resonance imaging-guided high-intensity focused ultrasound system. PLoS ONE 12(3):e0173867
6. Khokhlova TD, Canney MS, Khokhlova VA et al (2011) Controlled tissue emulsification produced by high-intensity focused ultrasound shock waves and millisecond boiling. J Acoust Soc Am 130(5):3498–3510
7. Eranki A, Farr N, Partanen A et al (2018) Mechanical fractionation of tissues using microsecond-long HIFU pulses on a clinical MR-HIFU system. Int J Hyperth 34(8):1213–1224
8. Eranki A, Srinivasan P, Ries M et al (2020) High-intensity focused ultrasound (HIFU) triggers immune sensitization of refractory murine neuroblastoma to checkpoint inhibitor therapy. Clin Cancer Res 26(5):1152–1161
9. Vlaisavljevich E, Maxwell A, Warnez M et al (2014) Histotripsy-induced cavitation cloud initiation thresholds in tissues of different mechanical properties. Ultrason, Ferroelectr Freq Control IEEE Trans 61(2):341–352
10. Saha S, Bhanja P, Partanen A et al (2014) Low-intensity focused ultrasound (LOFU) modulates unfolded protein response and sensitizes prostate cancer to 17AAG. Oncoscience 1(6):434
11. Partanen A, Yarmolenko PS, Viitala A et al (2012) Mild hyperthermia with magnetic resonance-guided high-intensity focused ultrasound for applications in drug delivery. Int J Hyperth 28(4):320–336
12. Burgess A, Huang Y, Waspe AC et al (2012) High-intensity focused ultrasound (hifu) for dissolution of clots in a rabbit model of embolic stroke. PLoS ONE 2012:e42311
13. Kim A, Sharma K, Yarmolenko P, et al (2017) Phase 1 trial of lyso-thermosensitive liposomal doxorubicin (LTLD) and magnetic resonance-guided high-intensity focused ultrasound (MR-HIFU) for pediatric refractory solid tumors. Am Soc Clin Oncol
14. van den Bijgaart RJ, Eikelenboom DC, Hoogenboom M et al (2017) Thermal and mechanical high-intensity focused ultrasound: perspectives on tumor ablation, immune effects and combination strategies. Cancer Immunol Immunother 66(2):247–258
15. Barnat N, Grisey A, Lecuelle B et al (2019) Non-invasive vascular occlusion with HIFU for venous insufficiency treatment: preclinical feasibility experience in rabbits. Phys Med Biol 64(2):025003
16. Elias WJ, Huss D, Voss T et al (2013) A pilot study of focused ultrasound thalamotomy for essential tremor. N Engl J Med 369(7):640–648
17. Bond AE, Shah BB, Huss DS et al (2017) Safety and efficacy of focused ultrasound thalamotomy for patients with medication-refractory, tremor-dominant Parkinson disease: a randomized clinical trial. JAMA Neurol 74(12):1412–1418
18. Ikeda T, Yoshizawa S, Koizumi N, et al (2016) Focused ultrasound and lithotripsy. In: Therapeutic ultrasound, pp 113–129
19. Chaussy C, Thüroff S (2001) Results and side effects of high-intensity focused ultrasound in localized prostate cancer. J Endourol 15(4):437–440
20. Poissonnier L, Chapelon J-Y, Rouviere O et al (2007) Control of prostate cancer by transrectal HIFU in 227 patients. Eur Urol 51(2):381–387

21. Canney MS, Khokhlova VA, Bessonova OV et al (2010) Shock-induced heating and millisecond boiling in gels and tissue due to high-intensity focused ultrasound. Ultrasound Med Biol 36(2):250–267
22. Iwasaki R, Takagi R, Nagaoka R et al (2016) Monitoring of high-intensity focused ultrasound treatment by shear wave elastography induced by two-dimensional-array therapeutic transducer. Jpn J Appl Phys 55(7S1):07KF05
23. Taljanovic MS, Gimber LH, Becker GW et al (2017) Shear-wave elastography: basic physics and musculoskeletal applications. Radiographics 37(3):855–870
24. Bishop J, Poole G, Leitch M et al (1998) Magnetic resonance imaging of shear wave propagation in excised tissue. J Magn Reson Imaging 8(6):1257–1265
25. Wang S, Larin KV (2014) Shear wave imaging optical coherence tomography (SWI-OCT) for ocular tissue biomechanics. Opt Lett 39(1):41–44
26. Sarvazyan AP, Rudenko OV, Swanson SD et al (1998) Shear wave elasticity imaging: a new ultrasonic technology of medical diagnostics. Ultrasound Med Biol 24(9):1419–1435
27. Gallotti A, D'onofrio M (2010) Acoustic radiation force impulse (ARFI) technique in ultrasound with virtual touch tissue quantification of the upper abdomen. Radiol Med (Torino) 115(6):889–897
28. Bercoff J, Tanter M, Fink M (2004) Supersonic shear imaging: a new technique for soft tissue elasticity mapping. IEEE Trans Ultrason Ferroelectr Freq Control 51(4):396–409
29. Song P, Zhao H, Manduca A et al (2012) Comb-push ultrasound shear elastography (CUSE): a novel method for two-dimensional shear elasticity imaging of soft tissues. IEEE Trans Med Imaging 31(9):1821–1832
30. Vigneshwaran G, Hasan QA, Kumar R, Eranki A (2022) Analysis of single-nucleotide polymorphisms in genes associated with triple-negative breast cancer. Front Genet 13:1071352
31. Sotiriou C, Pusztai L (2009) Gene-expression signatures in breast cancer. N Engl J Med 360(8):790–800

Operational Modal Identification Using Bayesian Approach

Shakir Rather and Sahil Bansal

Abstract This work addresses the problem of identification of modal properties, namely partial modeshapes, damping ratios, and natural frequencies, of a non-classically damped system under ambient excitation using a Bayesian probabilistic method. Ambient vibration tests eliminate the need for expensive dynamic experiments which take a lot of time, require permission, and cause problems to people who avail the benefits of the structure. Under operating conditions, naturally occurring vibrations can be measured and subsequent modal identification can be performed. For a Multiple Degree of Freedom (MDOF) system, the first few modes make the greatest contribution to the structural response. The operational excitation is assumed to be broadband enough to excite the most important modes of the structure so that the response can be effectively used for modal identification. In this work, a probabilistic approach based on Bayesian inference is presented to identify the complex modal parameters of a linear MDOF structure using output-only response data in the time domain. Computational obstacles in the estimation of probabilities are also overcome by using an approximate expansion for the likelihood function. The effectiveness of the approach is tested on simulated response data from a 2-DOF system.

Keywords Time-domain data · Bayesian · Complex modes · Modal identification

1 Introduction

Mechanical structures are prone to premature performance loss and unforeseen changes in expected behavior under operating conditions. This happens because the design of the structure is based on a model that fails to account for the uncertainties encountered in ambient excitation and modal properties. Identification of modal

S. Rather (✉) · S. Bansal
Indian Institute of Technology Delhi, Delhi, India
e-mail: cey217539@civil.iitd.ac.in

S. Bansal
e-mail: sahil@civil.iitd.ac.in

S. Sidhardh et al. (eds.), *Advances in Structural Integrity for Mechanical, Civil, and Aerospace Applications*, Lecture Notes in Mechanical Engineering,
https://doi.org/10.1007/978-981-97-6367-2_20

parameters helps in structural health monitoring, updating of structural models, and control of structural damage. Numerous methodologies, both output-only and input–output, have been developed to predict structural modal parameters. An extensive review of these methods is presented in [1].

Output-only methods provide the advantage of not requiring any elaborate dynamic testing setup and cause very little inconvenience to the people benefitting from the structure in addition to being cheap. In such cases, response data from structures is analyzed to arrive at estimates of their modal properties. Output-only methods are classified into two broad classes: (1) frequency-domain methods and (2) time-domain methods. In frequency-domain methods, response analysis is carried out in the frequency domain by making use of response power spectral densities (PSDs) or Fast-Fourier Transform (FFT) of time histories while in time-domain methods, response autocorrelation functions are analyzed through different techniques. Some examples of frequency-domain methods are given in [2, 3] while those of time-domain methods are given in [4, 5]. Modal identification of systems under ambient excitations has also attracted the attention of researchers (see [6]).

Most of these studies focus on systems with classical damping. However, in real structures, the conditions of classical damping are almost never satisfied. Thus, the need for frameworks dealing with non-classical damping is justified. Reference [7] presented an algorithm that used the concept of modal sweep to predict modal properties of non-classically damped systems from recorded responses. A second-order blind identification (SOBI) technique was used in [8] for identification of non-classically damped systems using ambient vibration tests.

In this study, a Bayesian probabilistic technique is used to identify complex modal parameters of a non-classically damped linear dynamic system subjected to ambient vibration. The ambient vibration is assumed to be an uncorrelated stationary stochastic process best modeled as a zero-mean broadband Gaussian white noise process. First, the formulation of response autocorrelation for a multiple degree-of-freedom (MDOF) system with non-classical viscous damping subjected to random excitation is presented. Subsequently, a formulation for likelihood function, its computational requirements and a way of circumventing computational intensiveness is presented. Lastly, the developed formulations are used with simulated response data from a 2-DOF system and results of modal identification are examined.

2 Non-classically Damped MDOF System

The equation of motion of a non-classically damped linear N-degree-of-freedom (N-DOF) subjected to random excitation is given by

$$\mathbf{M}\ddot{\mathbf{x}}(t) + \mathbf{C}\dot{\mathbf{x}}(t) + \mathbf{K}\mathbf{x}(t) = \mathbf{p}(t), \tag{1}$$

where $\mathbf{M}, \mathbf{C}, \mathbf{K} \in \mathbb{R}^{N\times N}$ are the symmetric mass, damping and stiffness matrices, respectively, and $\mathbf{x}(t)$, $\dot{\mathbf{x}}(t)$, $\ddot{\mathbf{x}}(t)$ and $\mathbf{p}(t) \in \mathbb{R}^N$ are the displacement, velocity, acceleration and excitation vectors, respectively. The above system is said to possess non-classical damping if the matrices $\mathbf{M}^{-1}\mathbf{C}$ and $\mathbf{M}^{-1}\mathbf{K}$ do not commute and thereby cannot be diagonalized by the same eigenvector matrix [1]. Therefore, to diagonalize this equation, it needs to be written in the state space which is of twice the dimensions. The state-space representation of (1) is given by

$$\mathbf{A}\dot{\mathbf{y}}(t) + \mathbf{B}\mathbf{y}(t) = \mathbf{f}(t), \tag{2}$$

where

$$\mathbf{A} = \begin{bmatrix} \mathbf{0} & \mathbf{M} \\ \mathbf{M} & \mathbf{C} \end{bmatrix} \in \mathbb{R}^{2N\times 2N}, \tag{3}$$

and

$$\mathbf{B} = \begin{bmatrix} -\mathbf{M} & \mathbf{0} \\ \mathbf{0} & \mathbf{K} \end{bmatrix} \in \mathbb{R}^{2N\times 2N} \tag{4}$$

are the symmetric state space matrices,

$$\mathbf{y}(t) = \begin{bmatrix} \dot{\mathbf{x}}(t) \\ \mathbf{x}(t) \end{bmatrix} \in \mathbb{R}^{2N} \tag{5}$$

is the state vector, and

$$\mathbf{f}(t) = \begin{bmatrix} \mathbf{0}_{N\times N} \\ \mathbf{p}(t) \end{bmatrix} \in \mathbb{R}^{2N} \tag{6}$$

is the state space excitation vector. The characteristic equation corresponding to Eq. (2) is

$$(\lambda_j\mathbf{A} + \mathbf{B})\boldsymbol{\psi}_j = 0, \quad j = 1, 2, \cdots, 2N, \tag{7}$$

where $\lambda_j \in \mathbb{C}$ are the eigenvalues and $\boldsymbol{\psi}_j \in \mathbb{C}^{2N}$ are the eigenvectors of the equation. For an underdamped system, the eigenvalues and eigenvectors can be shown to exist as complex-conjugate pairs [2]. The eigenvectors can be combined into a matrix $\boldsymbol{\Psi} \in \mathbb{C}^{2N\times 2N}$ which is given by

$$\boldsymbol{\Psi} = \begin{bmatrix} \boldsymbol{\psi}_1 & \boldsymbol{\psi}_2 & \cdots & \boldsymbol{\psi}_N & \boldsymbol{\psi}_1^* & \boldsymbol{\psi}_2^* & \cdots & \boldsymbol{\psi}_N^* \end{bmatrix} = \begin{bmatrix} \boldsymbol{\Phi}\Lambda & \boldsymbol{\Phi}^*\boldsymbol{\Lambda}^* \\ \boldsymbol{\Phi} & \boldsymbol{\Phi}^* \end{bmatrix}, \tag{8}$$

where $(\cdot)^*$ denotes the complex conjugate operation. This form of $\boldsymbol{\Psi}$ follows from the definition of $\mathbf{y}(t)$. $\boldsymbol{\Phi} \in \mathbb{C}^{N\times N}$ is the modeshape matrix of the system and $\boldsymbol{\Lambda} \in \mathbb{C}^{N\times N}$ is a diagonal matrix containing the N distinct complex eigenvalues of the system. The eigenvalues in $\boldsymbol{\Lambda}$ correspond to the modeshapes in $\boldsymbol{\Phi}$.

Substituting $\mathbf{y}(t) = \Psi\mathbf{q}(t)$ in Eq. (2), where $\mathbf{q}(t) \in \mathbb{C}^{2N}$ is the complex modal coordinate vector, gives

$$\mathbf{A}\Psi\dot{\mathbf{q}}(t) + \mathbf{B}\Psi\mathbf{q}(t) = \mathbf{f}(t). \tag{9}$$

The above equation, in combination with the characteristic equation of the system gives (after math)

$$\mathbf{X}(i\omega) = \begin{bmatrix} \boldsymbol{\Phi} & \boldsymbol{\Phi}^* \end{bmatrix} \begin{bmatrix} (i\omega\mathbf{I} - \boldsymbol{\Lambda})^{-1} & \\ & (i\omega\mathbf{I} - \boldsymbol{\Lambda}^*)^{-1} \end{bmatrix} \tilde{\mathbf{F}}(i\omega) = \overline{\mathbf{H}}(i\omega)\tilde{\mathbf{F}}(i\omega), \tag{10}$$

where $\mathbf{X}(i\omega)$, $\tilde{\mathbf{F}}(i\omega)$ are the Fourier transforms of $\mathbf{x}(t)$ and $\tilde{\mathbf{f}}(t)$, respectively, and

$$\tilde{\mathbf{f}}(t) = (\mathbf{A}\Psi)^{-1}\mathbf{f}(t). \tag{11}$$

$$\overline{\mathbf{H}}(i\omega) = \begin{bmatrix} \boldsymbol{\Phi} & \boldsymbol{\Phi}^* \end{bmatrix} \begin{bmatrix} (i\omega\mathbf{I} - \boldsymbol{\Lambda})^{-1} & \\ & (i\omega\mathbf{I} - \boldsymbol{\Lambda}^*)^{-1} \end{bmatrix} \tag{12}$$

is the transfer matrix between $\mathbf{x}(t)$ and $\tilde{\mathbf{f}}(t)$.

2.1 Response Autocorrelation

It is assumed that the random excitation force vector $\mathbf{p}(t)$ consists of zero-mean Gaussian white noise signals acting independently along each DOF. So, PSD matrix of $\mathbf{p}(t)$ is given as follows.

$$\mathbf{S}_{\mathbf{p}}(\omega) = \mathbf{S}_{\mathbf{p0}} = \begin{bmatrix} S_1 & & & \\ & S_2 & & \\ & & \ddots & \\ & & & S_N \end{bmatrix} \in \mathbb{R}^{N\times N}. \tag{13}$$

The response $\mathbf{x}(t)$ will be a zero-mean Gaussian process with PSD matrix at any angular frequency ω_k given by

$$\mathbf{S}_{\mathbf{x}}(\omega_k) = \mathbf{X}(i\omega_k)\mathbf{X}^H(i\omega_k) = \overline{\mathbf{H}}(i\omega_k)\mathbf{S}_{\tilde{\mathbf{f}}}(\omega_k)\overline{\mathbf{H}}^H(i\omega_k), \tag{14}$$

where $(\cdot)^H$ denotes the conjugate transpose operation,

$$\mathbf{S}_{\tilde{\mathbf{f}}}(\omega_k) = (\mathbf{A}\Psi)^{-1}\mathbf{S}_{\mathbf{f}}(\omega_k)((\mathbf{A}\Psi)^{-1})^H, \text{ and}$$
$$\mathbf{S}_{\mathbf{f}}(\omega_k) = \mathbf{F}(i\omega_k)\mathbf{F}^{\#}(i\omega_k) = \begin{bmatrix} \mathbf{0} & \mathbf{0} \\ \mathbf{0} & \mathbf{S}_{\mathbf{p0}} \end{bmatrix}. \tag{15}$$

3 Probabilistic Formulation-Likelihood Function

Assume that discrete response data is available at $N_o(\leq N)$ DOFs and that a prediction error exists on account of measurement noise and modeling error that can be represented mathematically as a discrete Gaussian white noise signal with zero mean. We can express the measured response as

$$\tilde{\mathbf{y}}_n = \mathbf{L}_0\mathbf{x}_n + \mathbf{n}_n, \tag{16}$$

where $\tilde{\mathbf{y}}_n$, $\mathbf{n}_n \in \mathbb{R}^{N_0}$, $\mathbf{x}_n \in \mathbb{R}^N$ are the measured response, prediction error and model prediction, respectively, at time $t = n\Delta t$, Δt is the sampling time step, and .$\mathbf{L}_0$. is an $N_o \times N$ observation matrix consisting of zeros and ones. The prediction error has the covariance matrix $\mathbf{\Sigma}_\in$ which can be found from

$$\mathrm{E}[\mathbf{n}(m)\mathbf{n}^T(p)] = \mathbf{\Sigma}_\in\delta_{m,p} \tag{17}$$

where
$\delta_{m,p} = \begin{cases} 1 \; if \;\; m = p \\ 0 \; if \;\; m \neq p \end{cases}$ is the Kronecker delta.

Let $\tilde{\mathbf{Y}}_{n,n'}$ be a vector that contains the discrete response data from $t = n\Delta t$ to $t = n'\Delta t$ in reverse order, i.e.,

$$\tilde{\mathbf{Y}}_{n,n'} = [\tilde{\mathbf{y}}_{n'}^T, \tilde{\mathbf{y}}_{n'-1}^T, \cdots, \tilde{\mathbf{y}}_n^T]^T, \quad n \leq n'. \tag{18}$$

$\tilde{\mathbf{Y}}_{n,n'}$ is a random vector since both $\mathbf{x}$ and $\mathbf{n}$ are random processes.

Let $D = \{\tilde{\mathbf{y}}_n : n = 1, \cdots, N_T\}$ denote the complete response data. The probability density function (PDF) of D given $\boldsymbol{\theta}$ will follow an N_oN_T—variate Gaussian distribution (GD) with zero mean and covariance matrix $\mathbf{\Sigma}_{\tilde{\mathbf{Y}}_{1,N_T}}$.

$$p(D|\boldsymbol{\theta}) = p(\tilde{\mathbf{Y}}_{1,N_T}|\boldsymbol{\theta}) = \frac{1}{(2\pi)^{\frac{N_0N_T}{2}}\left|\mathbf{\Sigma}_{\tilde{\mathbf{Y}}_{1,N_T}}\right|^{\frac{1}{2}}}\exp(-\frac{1}{2}\tilde{\mathbf{Y}}_{1,N_T}^T\mathbf{\Sigma}_{\tilde{\mathbf{Y}}_{1,N_T}}^{-1}\tilde{\mathbf{Y}}_{1,N_T}), \tag{19}$$

With

$$\mathbf{\Sigma}_{\tilde{\mathbf{Y}}_{1,N_T}} = \mathrm{E}(\tilde{\mathbf{Y}}_{1,N_T}\tilde{\mathbf{Y}}_{1,N_T}^T) = \begin{bmatrix} \mathbf{\Gamma}_{N_T,N_T} & & sym \\ \vdots & \ddots & \\ \mathbf{\Gamma}_{1,N_T} & \cdots & \mathbf{\Gamma}_{1,1} \end{bmatrix}. \tag{20}$$

where

$$\mathbf{\Gamma}_{j,k} = \mathrm{E}\left[\tilde{\mathbf{y}}_j\tilde{\mathbf{y}}_k^T\right] = \mathbf{L}_0\mathrm{E}\left[\mathbf{x}_j\mathbf{x}_k^T\right]\mathbf{L}_0^T + \mathbf{\Sigma}_\epsilon\delta_{j,k} = \mathbf{L}_0\mathbf{R}_\mathbf{x}^{j,k}\mathbf{L}_0^T + \mathbf{\Sigma}_\epsilon\delta_{j,k}, \tag{21}$$

$$\mathbf{R}_\mathbf{x}^{j,k} = \mathbf{R}_\mathbf{x}((k-j)\Delta t). \tag{22}$$

Using Bayes' theorem, the updated PDF of $\boldsymbol{\theta}$ given some response data $\tilde{\mathbf{Y}}_{1,N_T}$ can be written as

$$p(\boldsymbol{\theta}|\tilde{\mathbf{Y}}_{1,N_T}) = k.p(\boldsymbol{\theta})p(\tilde{\mathbf{Y}}_{1,N_T}|\boldsymbol{\theta}), \tag{23}$$

where $p(\boldsymbol{\theta})$ is the prior PDF of the parameters to be determined. Absent any prior information, $p(\boldsymbol{\theta})$ can be assumed to be constant. k is a normalizing constant that ensures that the area under the updated PDF is unity. The optimal parameters $\hat{\theta}$ are found by maximizing $p(\boldsymbol{\theta})p(\tilde{\mathbf{Y}}_{1,N_T}|\boldsymbol{\theta})$. Since $p(\boldsymbol{\theta})$ is treated as a constant, our objective is to maximize $p(\tilde{\mathbf{Y}}_{1,N_T}|\boldsymbol{\theta})$.

The right-hand side of Eq. (19) can be expanded as a product of probabilities using the product rule as follows.

$$p(\tilde{\mathbf{Y}}_{1,N_T}|\boldsymbol{\theta}) = p(\tilde{\mathbf{Y}}_{1,N_p}|\boldsymbol{\theta}) \prod_{i=N_p+1}^{N_T} p(\tilde{\mathbf{Y}}_{i,i}|\tilde{\mathbf{Y}}_{1,i-1}, \boldsymbol{\theta}). \tag{24}$$

In order to reduce computational complexity, we introduce the below approximation

$$p(\tilde{\mathbf{Y}}_{i,i}|\tilde{\mathbf{Y}}_{1,i-1}, \boldsymbol{\theta}) \approx p(\tilde{\mathbf{Y}}_{i,i}|\tilde{\mathbf{Y}}_{i-N_P,i-1}, \boldsymbol{\theta}). \tag{25}$$

With this approximation, the data points further back in time than N_p time steps are not counted as having any effect on the statistics of the present point. Hence, Eq. (24) becomes

$$p(\tilde{\mathbf{Y}}_{1,N_T}|\boldsymbol{\theta}) = p(\tilde{\mathbf{Y}}_{1,N_p}|\boldsymbol{\theta}) \prod_{i=N_p+1}^{N_T} p(\tilde{\mathbf{Y}}_{i,i}|\tilde{\mathbf{Y}}_{i-N_P,i-1}, \boldsymbol{\theta}). \tag{26}$$

The joint PDF $p(\tilde{\mathbf{Y}}_{1,N_P}|\boldsymbol{\theta})$ follows an $N_o \times N_p$ variate zero-mean GD with the following covariance matrix.

$$\boldsymbol{\Sigma}_{00} = \boldsymbol{\Sigma}_{\tilde{\mathbf{Y}}_{1,N_P}} = \mathrm{E}\left[\tilde{\mathbf{Y}}_{1,N_P}\tilde{\mathbf{Y}}_{1,N_P}^T\right] = \begin{bmatrix} \boldsymbol{\Gamma}_{N_P,N_P} & & sym \\ \vdots & \ddots & \\ \boldsymbol{\Gamma}_{1,N_P} & \cdots & \boldsymbol{\Gamma}_{1,1} \end{bmatrix}, \tag{27}$$

and

$$p(\tilde{\mathbf{Y}}_{1,N_p}|\boldsymbol{\theta}) = \frac{1}{(2\pi)^{\frac{N_O N_P}{2}}\left|\boldsymbol{\Sigma}_{\tilde{\mathbf{Y}}_{1,N_p}}\right|^{\frac{1}{2}}}\exp(-\frac{1}{2}\tilde{\mathbf{Y}}_{1,N_p}^T\boldsymbol{\Sigma}_{\tilde{\mathbf{Y}}_{1,N_p}}^{-1}\tilde{\mathbf{Y}}_{1,N_p}). \tag{28}$$

We can express the conditional probabilities in the product in the same manner. To start with, the random vector $\tilde{\mathbf{Y}}_{i-N_p,i}$ is a zero-mean vector with covariance matrix

$$\boldsymbol{\Sigma}_{\tilde{\mathbf{Y}}_{i-N_P,i}} = \mathrm{E}\left[\tilde{\mathbf{Y}}_{i-N_P,i}\tilde{\mathbf{Y}}_{i-N_P,i}^T\right] = \begin{bmatrix} \boldsymbol{\Gamma}_{i,i} & \cdots & \boldsymbol{\Gamma}_{i,i-N_P} \\ & \ddots & \vdots \\ sym & & \boldsymbol{\Gamma}_{i-N_P,i-N_P} \end{bmatrix}. \tag{29}$$

This matrix can be partitioned in the following manner

$$\boldsymbol{\Sigma}_{\tilde{\mathbf{Y}}_{i-N_P,i}} = \begin{bmatrix} \boldsymbol{\Sigma}_{11,i} & \boldsymbol{\Sigma}_{12,i} \\ sym & \boldsymbol{\Sigma}_{22,i} \end{bmatrix}, \tag{30}$$

where $\boldsymbol{\Sigma}_{11,i}$, $\boldsymbol{\Sigma}_{12,i}$, and $\boldsymbol{\Sigma}_{22,i}$ have sizes $N_o \times N_o$, $N_o \times N_oN_p$, and $N_oN_p \times N_oN_p$, respectively.

If the response is stationary, $\boldsymbol{\Sigma}_{\tilde{\mathbf{Y}}_{i-N_P,i}}$ will be independent of i. So, it can be written as

$$\boldsymbol{\Sigma}_{\tilde{\mathbf{Y}}_{i-N_P,i}} = \begin{bmatrix} \boldsymbol{\Sigma}_{11} & \boldsymbol{\Sigma}_{12} \\ sym & \boldsymbol{\Sigma}_{22} \end{bmatrix}. \tag{31}$$

Since the response is assumed to be a zero-mean process, the mean of $\tilde{\mathbf{Y}}_{i,i}$ given $\tilde{\mathbf{Y}}_{i-N_P,i-1}$ can be written as ([3])

$$\bar{\mathbf{y}}_i = \mathrm{E}(\tilde{\mathbf{Y}}_{i,i}|\tilde{\mathbf{Y}}_{i-N_P,i-1},\boldsymbol{\theta}) = \boldsymbol{\Sigma}_{12}\boldsymbol{\Sigma}_{22}^{-1}\tilde{\mathbf{Y}}_{i-N_P,i-1}, \tag{32}$$

and the covariance matrix $\boldsymbol{\Sigma}_{e,i}$ of the prediction error $\mathbf{e}_i = \tilde{\mathbf{y}}_i - \bar{\mathbf{y}}_i$ can be written as

$$\boldsymbol{\Sigma}_{e,i} = \boldsymbol{\Sigma}_e = \mathrm{E}(\mathbf{e}_i\mathbf{e}_i^T|\tilde{\mathbf{Y}}_{i-N_P,i-1},\boldsymbol{\theta}) = \boldsymbol{\Sigma}_{11} - \boldsymbol{\Sigma}_{12}\boldsymbol{\Sigma}_{22}^{-1}\boldsymbol{\Sigma}_{12}^T. \tag{33}$$

Hence, each of the conditional probabilities $p(\tilde{\mathbf{Y}}_{i,i}|\tilde{\mathbf{Y}}_{i-N_P,i-1},\boldsymbol{\theta})$ follows an N_o-variate GD whose mean is $\bar{\mathbf{y}}_i$ and covariance matrix is $\boldsymbol{\Sigma}_e$ So,

$$p(\tilde{\mathbf{Y}}_{i,i}|\tilde{\mathbf{Y}}_{i-N_P,i-1},\boldsymbol{\theta}) = \frac{1}{(2\pi)^{\frac{N_o}{2}}|\boldsymbol{\Sigma}_e|^{\frac{1}{2}}}\exp\left[-\frac{1}{2}(\tilde{\mathbf{y}}_i - \bar{\mathbf{y}}_i)^T\boldsymbol{\Sigma}_e^{-1}(\tilde{\mathbf{y}}_i - \bar{\mathbf{y}}_i)\right]. \quad (34)$$

The optimal (or most probable) parameters $\hat{\theta}$ are found by minimizing the negative log-likelihood function (NLLF), which is preferred over maximizing $p(\tilde{\mathbf{Y}}_{1,N_T}|\boldsymbol{\theta})$ for computational reasons. The NLLF can be written as

$$\begin{aligned} L(\boldsymbol{\theta}) &= \frac{1}{2}\ln\left|\boldsymbol{\Sigma}_{\tilde{\mathbf{Y}}_{1,N_P}}\right| + \frac{1}{2}\tilde{\mathbf{Y}}_{1,N_p}^T\boldsymbol{\Sigma}_{\tilde{\mathbf{Y}}_{1,N_p}}^{-1}\tilde{\mathbf{Y}}_{1,N_p} \\ &+ \frac{1}{2}\sum_{i=N_p+1}^{N_T}\left[\ln|\boldsymbol{\Sigma}_e| + (\tilde{\mathbf{y}}_i - \bar{\mathbf{y}}_i)^T\boldsymbol{\Sigma}_e^{-1}(\tilde{\mathbf{y}}_i - \bar{\mathbf{y}}_i)\right] \\ &= \frac{1}{2}\ln\left|\boldsymbol{\Sigma}_{\tilde{\mathbf{Y}}_{1,N_P}}\right| + \frac{1}{2}\tilde{\mathbf{Y}}_{1,N_p}^T\boldsymbol{\Sigma}_{\tilde{\mathbf{Y}}_{1,N_p}}^{-1}\tilde{\mathbf{Y}}_{1,N_p} + \frac{N_T - N_p}{2}\ln|\boldsymbol{\Sigma}_e| \\ &+ \frac{1}{2}\sum_{i=N_p+1}^{N_T}(\tilde{\mathbf{y}}_i - \bar{\mathbf{y}}_i)^T\boldsymbol{\Sigma}_e^{-1}(\tilde{\mathbf{y}}_i - \bar{\mathbf{y}}_i). \end{aligned} \quad (35)$$

4 Results and Discussion

Simulated displacement data from both DOFs of a 2-DOF non-classically damped spring-mass system is used to identify parameters corresponding to both the modes of the system. The modal parameter vector to be identified is denoted by $\boldsymbol{\theta} = [|\phi_2^1|, \alpha_1^\phi, |\phi_2^2|, \alpha_2^\phi, |\lambda_1|, \alpha_1^\lambda, |\lambda_2|, \alpha_2^\lambda]^T$, where $|\phi_2^i|$, α_i^ϕ are the absolute value and angle of the (2, *i*)-*th* element of $\boldsymbol{\Phi}$, respectively, and $|\lambda_i|$, α_i^λ are the absolute value and angle of the *i-th* diagonal entry of $\boldsymbol{\Lambda}$. The elements of the first row of the mode-shape matrix are made unity for ease of identification and reduction of unknowns. Response data are generated for a system with natural frequencies 438.465 rad/s and 917.956 rad/s, and damping ratios 0.0128 and 0.0319 for the two modes (Fig. 1).

The sampling time interval is chosen as 1/2000s and the response is obtained for 15s. The spectral densities of forces at the two DOFs are $8 \times 10^5 N^2 s$ and $9 \times 10^5 N^2 s$, respectively. The value of N_p should be such that the statistical dependency of the current point on a point more than N_p time steps further back is almost negligible. This is to say that the response autocorrelation has decayed to a small value at time $N_p \Delta t$.

However, a value of N_p that covers more than one fundamental period would suffice for reliable modal identification. Additionally, the mean and covariance of conditional probability terms in the likelihood function stabilize for a certain value of N_p beyond which the approximation introduced in Sect. 3 doesn't give improved accuracy.

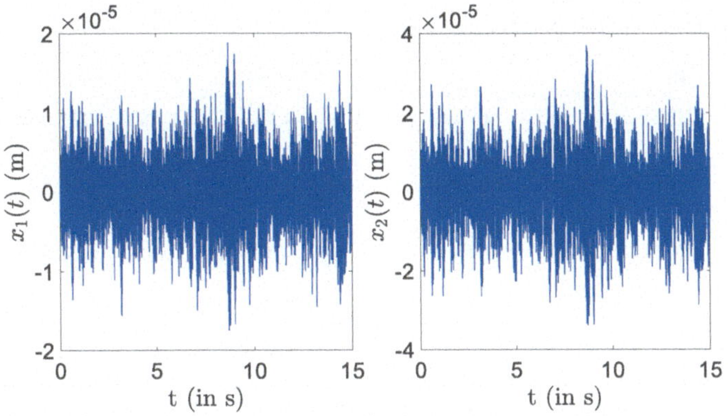

Fig. 1 Displacement responses at DOF 1 and DOF 2

As data from both DOFs is used to identify the vibration modes of the system, the accuracy of the identification process will depend on how strongly the modes appear in the simulated response. One can use forces with different PSDs to excite different modes. In this example, it is seen that the random force described above is sufficient to help identify the parameters of concern of the system. Figure 2 shows how the modes appear as peaks in the response PSD.

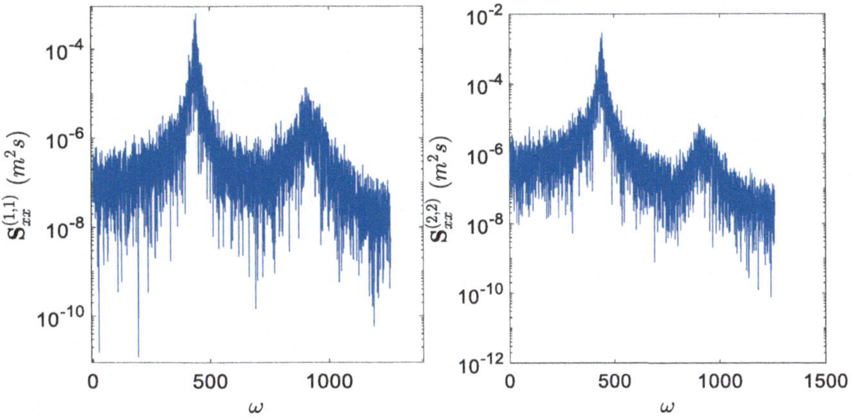

Fig. 2 PSD of response at 1st and 2nd DOFs

Table 1 Results of optimization for $N_p = 30$

Parameter	$\|\phi_2^1\|$	α_1^ϕ	$\|\phi_2^2\|$	α_2^ϕ	$\|\lambda_1\|$	α_1^λ	$\|\lambda_2\|$	α_2^λ
Actual value	2.144	−0.015	0.746	−3.108	438.465	1.583	917.956	1.602
Optimal value	2.140	−0.0013	0.760	−3.100	438.793	1.583	916.598	1.601
Standard deviation	0.074	0.041	0.021	0.037	0.982	0.008	1.235	0.008

The complex parameters to be identified are represented in polar form since in the optimization routine, setting the bounds for the unknowns is easier in polar coordinates. Natural frequencies can be roughly guessed from the PSD figures and bounds may be set accordingly for the absolute values of $\lambda_j s$. Optimization is carried out in MATLAB using *fmincon*. The results are tabulated in Table 1.

The damping ratios are found from

$$\xi_j = -\frac{\mathrm{Re}(\lambda_j)}{|\lambda_j|} \tag{36}$$

It is clear from Table 1 that all the parameters are identified well except for α_1^ϕ. Since the phase lag between the oscillations at the two DOFs is very small in the first mode, α_1^ϕ is identified poorly. Increasing this lag would entail a corresponding increase in damping, and excessive damping would render the time-domain response unsuitable for identification. The standard deviations of the parameters give information about the certainty of identification. Natural frequencies and modal damping ratios are observed to be identified with small uncertainty values and hence the proposed approach is best at picking them.

5 Conclusion

A Bayesian time-domain approach for the identification of modal parameters of non-classically damped linear systems subjected to ambient vibration is presented. Noise-contaminated response data is used in the likelihood function formulation and minimization of negative log-likelihood function is carried out to determine the optimal system modal parameters. For computational efficiency, an approximate expansion of the likelihood function is used that relies on the statistical independence of widely-separated data points in time. Excitation of all important modes to be identified is ensured by using appropriate Gaussian white noise signals as nodal forces. Optimal parameters are found to be a close approximation of the actual parameters. The posterior PDF can be updated and the uncertainties in identified modal parameters can be used to revise and better the finite-element models of the system.

References

1. Maia NMM, Montalvão Silva e Silva JM (1997) Theoretical and experimental modal analysis. Research Studies Press
2. Crawford R, Ward HS (1964) Determination of the natural period of buildings. Bull Seismol Soc Am 54:1743–1756
3. Brincker R, Zhang L, Andersen P (2001) Modal identification from ambient responses using frequency domain decomposition. 18th Int Modal Anal Conf
4. Young PC (1970) An instrumental variable method for real-time identification of a noisy process. Automatica 6:271–287. https://doi.org/10.1016/0005-1098(70)90098-1
5. Asmussen JC (1997) Modal analysis based on the random decrement technique. Application to civil engineering structures. Aalborg University
6. Safak E, Cakti E, Kaya Y (2010) Recent developments on structural health monitoring and data analyses. Geotech Geol Earthq Eng 17:331–355. https://doi.org/10.1007/978-90-481-9544-2
7. Tan RY, Cheng WM (1993) System identification of a non-classically damped linear system. Conputers Struct 46:67–75. https://doi.org/10.1016/0045-7949(93)90168-D
8. Abazarsa F, Nateghi F, Ghahari SF, Taciroglu E (2013) Blind modal identification of non-classically damped systems from free or ambient vibration records. Earthq Spectra 29:1137–1157. https://doi.org/10.1193/031712EQS093M
9. Caughey TK, O'Kelly MEJ (1965) Classical normal modes in damped linear dynamic systems. J Appl Mech 32:583–588. https://doi.org/10.2514/2.1409
10. Foss KA (1958) Coordinates which uncouple the equations of motion of damped linear dynamic systems. J Appl Mech 25
11. Brockwell P, Davis R (2009) Time series theory and methods. https://doi.org/10.2307/1270380

Influence of Geometric Variation of Internal Block on Dynamic Characteristics of Base-Isolated Rectangular Liquid Tank

Jyoti Ranjan Barik and Kishore Chandra Biswal

Abstract The existence of submerged components can significantly influence the seismic response of a liquid storage tank. Consequently, due to the modification of dynamic properties, the seismic behavior of liquid tank systems with internal components can be significantly benefited. In this research, a potential-based finite element technique has been adopted to study the slosh dynamics of a partially filled rigid rectangular container with a bottom-mounted submerged block under harmonic excitation. The liquid domain is governed by the Laplace equation where a velocity potential-based Galerkin's approach is adopted for the liquid domain formulation. The obtained results of the developed finite element model are validated with the published results with a decent agreement. The phenomenal sloshing behavior of the base-isolated tank for different widths and depths of the block is emphasized in the present analysis by the application of the laminated rubber bearing. The parametric research of tank-liquid-block-isolation systems demonstrates the significance of varying dimensions of the submerged structural component.

Keywords Submerged block · Laminated rubber bearing · Parametric study

1 Introduction

The liquid sloshing in a tank has been extensively examined for a very long time due to its fundamental significance in flow physics and its practical importance in a wide range of applications, including in ships, satellites, rockets, trucks, and even stationary petroleum containers. The liquid structures, sometimes appears with internal obstacles forming a complicated geometry. Due to these submerged components, the liquid dynamic properties may significantly differ and an anti-slosh effect is created due to the presence by effectively altering the tank geometry [1].

J. R. Barik (✉) · K. C. Biswal
Department of Civil Engineering, National Institute of Technology Rourkela, Rourkela, Odisha, India
e-mail: jyotiranjanbarik0@gmail.com

S. Sidhardh et al. (eds.), *Advances in Structural Integrity for Mechanical, Civil, and Aerospace Applications*, Lecture Notes in Mechanical Engineering,
https://doi.org/10.1007/978-981-97-6367-2_21

Numerous researchers have performed both analytical and experimental investigations on the sloshing behavior of liquid in a rectangular tank [2–5]. The physical significance of internal submerged component is briefly emphasized in the majority of investigations. In summary, research has shown that presence of internal block in liquid tank is quite effective at reducing vibration and improving dynamic response of a system. However, for the past few decades, base isolation technique has been adopted as a proven earthquake mitigation standard to obtain a seismically safe liquid container [6–9]. Moreover, the study on base-isolated tanks with the presence of internal block is insignificant. In addition, the numerical methodology on liquid sloshing has not been emphasized exhaustively earlier for the assessment of impulsive and convective parameters of the base-isolated tank [10].

The current study aims to examine the effects of changing the block geometry with respect to the width and depth to comprehend the seismic nature of both non-isolated and isolated liquid containers under harmonic excitation. The study further highlights the effectiveness of laminated rubber bearing with regard to the seismic quantification of the predominant responses of the tank as manifested in terms of impulsive and convective components.

2 Geometry of Tank-Block-Isolation System

A liquid-filled rectangular tank with a block is shown in Fig. 1. The liquid boundaries are manifested in terms of tank walls (Γ_w), block walls (Γ_{Bw}), tank base (Γ_b), block surface (Γ_{Bt}), and liquid free surface (Γ_s), respectively. The base of the tank is decoupled with laminated rubber bearings. The liquid in the tank is assumed to be incompressible, inviscid, and irrotational. The governing differential equation for the liquid in terms of velocity potential variable is as follows:

$$\nabla^2 \Phi = 0 \tag{1}$$

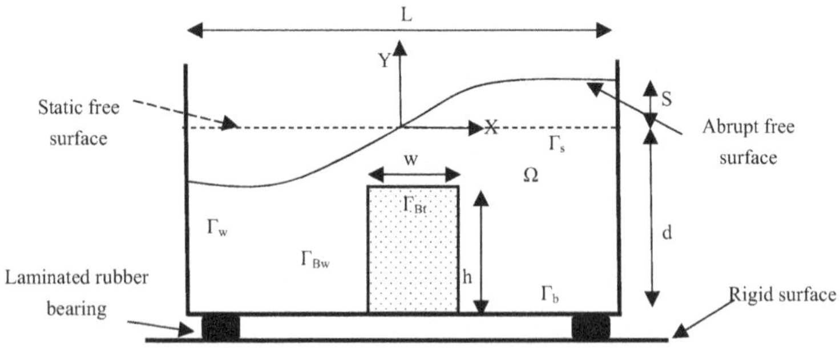

Fig. 1 Tank-block-isolation model

Following time-dependent boundary conditions are adopted to simplify Eq. (1),

(a) On liquid-wall interfaces (W_s),

$$\frac{\partial \Phi}{\partial n} = V_i \quad \text{on } W_\text{s} = \Gamma_\text{w} \cup \Gamma_\text{Bw} \tag{2}$$

V_i represents the instant horizontal velocity on wall interface produced by the longitudinal component of earthquake whereas, n is the perpendicular direction to the boundary wall (W_s). The tank and block wall boundaries are implicitly represented by W_s.

(b) On tank base (S_s),

$$\frac{\partial \Phi}{\partial n} = 0 \quad \text{on } S_\text{s} = \Gamma_\text{b} \cup \Gamma_\text{Bt} \tag{3}$$

where the bottom surface boundaries are simultaneously represented by S_s.

(c) At free surface (Γ_s),

$$\frac{\partial^2 \Phi}{\partial t^2} + \mu \frac{\partial \Phi}{\partial t} + g \frac{\partial \Phi}{\partial y} = 0 \quad \text{on } \Gamma_s \tag{4}$$

Here, μ denotes the kinematic viscosity of the liquid and g is the acceleration due to gravity.

3 Numerical Model Formulation

The tank peripheries are illustrated in Fig. 1, where the meshing of liquid region (Ω) is done by quadrilateral elements (four node). The time varying function Φ can be represented as,

$$\Phi \approx \Phi(x, y, t) = \sum_{j=1}^{n} \Phi_j(t) N_j(x, y) \tag{5}$$

Every node is associated with a potential function and shape functions. The implementation of Galerkin's-weighted-residual method gives rise to the governing equation as follows,

$$\int_{\Omega} N_i \left(\frac{\partial^2 \Phi}{\partial x^2} + \frac{\partial^2 \Phi}{\partial y^2} \right) \mathrm{d}\Omega = 0 \tag{6}$$

The integral is expanded by using Stoke's and Green's theorems and hence, it becomes,

$$\frac{1}{g}\int_{\Gamma_s} N_i \sum_{j=1}^{n} N_j \ddot{\Phi} \mathrm{d}\Gamma_s + \frac{\mu}{g}\int_{\Gamma_s} N_i \sum_{j=1}^{n} N_j \dot{\Phi} \mathrm{d}\Gamma_s$$
$$+ \int_{\Omega} \left[\frac{\partial N_i}{\partial x} \sum_{j=1}^{n} \frac{\partial N_j}{\partial x} + \frac{\partial N_i}{\partial y} \sum_{j=1}^{n} \frac{\partial N_j}{\partial y} \right] \Phi_j \mathrm{d}\Omega$$
$$= \int_{W_s} N_i V_n \mathrm{d}W_s \tag{7}$$

In terms of the matrix representation,

$$[E]\ddot{\Phi} + [C]\dot{\Phi} + [G]\Phi = \{F\}V_n, \tag{8}$$

where $E_{ij} = \sum E^e_{ij}$, $C_{ij} = \sum C^e_{ij}$, $G_{ij} = \sum G^e_{ij}$ and $F_i = \sum F^e_i$ are the global matrices for different elements.

$$E^e_{ij} = \frac{1}{g}\int_{\Gamma_s} N_i N_j \mathrm{d}\Gamma_s \tag{9}$$

$$C^e_{ij} = \frac{\mu}{g}\int_{\Gamma_s} N_i N_j \mathrm{d}\Gamma_s \tag{10}$$

$$G^e_{ij} = \int_{\Omega} \left[\frac{\partial N_i}{\partial x}\frac{\partial N_j}{\partial x} + \frac{\partial N_i}{\partial y}\frac{\partial N_j}{\partial y} \right] \mathrm{d}\Omega \tag{11}$$

$$F^e_i = -\int_{W_s} N_i \, \mathrm{d}W_s \tag{12}$$

The magnitudes of wave amplitude (S), total pressure (p) and total base shear (B_s) are obtained from the following expressions.

$$S = -\frac{1}{g}\left(\frac{\partial \Phi}{\partial t} + \mu\Phi\right) \tag{13}$$

$$p = -\rho\left(\frac{\partial \Phi}{\partial t} + \mu\Phi\right) \tag{14}$$

$$B_s = \int_{W_s} p n_x \mathrm{d}W_s \tag{15}$$

The impulsive component of hydrodynamic pressure can be obtained independently by neglecting the liquid sloshing.

$$\frac{\partial \Phi}{\partial t}(x, 0, t) = 0 \tag{16}$$

Thus, using the boundary condition (Eq. (16)), the impulsive pressure (p_{imp}) and base shear ($B_{\text{s\_imp}}$) are calculated using Eqs. (14) and (15), respectively, in a similar manner. Moreover, the magnitude of convective pressure (p_{con}) and base shear ($B_{\text{s\_con}}$) are evaluated using Eqs. (17) and (18), respectively.

$$p_{\text{con}} = p - p_{\text{imp}} \tag{17}$$

$$B_{\text{s\_ con}} = B_{\text{s}} - B_{\text{s\_ imp}} \tag{18}$$

4 Attributes of the Isolator

Laminated rubber bearing is adopted for the present study due to its wide applicability. However, the manufacturing is such that the rubber layers generate flexibility and the steel plates maintain stability. The ideal relationship between the displacement and applied force is expressed by,

$$F_b = c_b \dot{x}_b + k_b x_b \tag{19}$$

The displacement and velocity of the isolator are designated by x_b and $\dot{x}_b$, respectively. The time period (T_{b}) and fundamental frequency of the isolator (ω_{b}) are obtained through Eq. (20) and the damping of the isolator is calculated through Eq. (21). The liquid mass (m_{b}) is assumed to be exerted on the isolator.

$$\omega_{\text{b}} = \sqrt{\frac{2\pi}{T_{\text{b}}}} = \sqrt{\frac{k_{\text{b}}}{m_{\text{b}}}} \tag{20}$$

$$c_{\text{b}} = 2 m_{\text{b}} \omega_{\text{b}} \xi_{\text{b}} \tag{21}$$

5 Dynamic Equations

A set of characteristics equations are established (Eqs. (22) and (23)) for the coupled system of isolated tank with varying dimension of the block.

Table 1 Period of sloshing

h/*d* ratio	Antisymmetric mode	T_{slosh} (seconds)	
		Choun and Yun [2]	Present FEM
0	1	3.738	3.740
	2	2.067	2.055
0.5	1	4.223	4.219
	2	2.070	2.058
0.8	1	5.675	5.650
	2	2.119	2.106

$$m_b\ddot{x}_b + F_b + B_s = -m_b\ddot{x}_g \tag{22}$$

$$[E]\ddot{\Phi} + [C]\dot{\Phi} + [G]\Phi = \{F\}\left(\dot{x}_g + \dot{x}_b\right) \tag{23}$$

F_i and B_s denote the force subjected by the isolator and the base shear force. The excitation velocity and ground acceleration are represented by $\dot{x}_g$ and $\ddot{x}_g$, respectively.

6 Competency of the Developed Model

The outputs of the established non-isolated tank model are verified with the results of available literature for different tank-block conditions [2]. Choun and Yun examined the sloshing problem analytically whereas the present approach emphasizes the finite element modeling. The tank dimension is 10mx5m with a shallow rectangular shape where a 2 m wide rigid block is mounted centrally ($a_1 = a_2$) at the tank floor. The time period of sloshing are obtained for different antisymmetric modes as mentioned in Table 1 for different block geometries based on a free vibration analysis. A decent correlation is contrasted in between the results of two different approaches, which interprets that the present model can be suitably implemented to develop a coupled tank-block-isolation system to investigate the liquid sloshing behavior.

7 Numerical Results and Discussion

The dimension of the tank is kept same to that of the model established for validation. The tank confines water ($\rho = 1000$ kg/m^3) as a reference liquid. For assessing efficient results, an ideal mesh size is determined prior to the investigation through a convergence study on the natural slosh frequency and presented in Table 2. A typical discretization with 20×20 mesh dimension is observed to be efficient for the present study. At present, a finite element model is developed in MATLAB 2020b platform

Table 2 Convergence test of rectangular tank

Mesh dimension	Total No. of elements	Slosh frequency (Hz)
12 × 12	144	0.2682
16 × 16	256	0.2679
20 × 20	400	0.2674
24 × 24	576	0.2674

where the effect of varying width and depth of block is investigated under a sinusoidal harmonic excitation ($-A_{max}\sin\omega t$) provided at the tank base where the maximum amplitude (A_{max}) is considered as 0.2 g.

7.1 Parametric Analysis on Isolator Characteristics

A parametric analysis is scrutinized on the dynamic response parameters of different tank-block configurations. Unlike base-isolated building structures, no certain guidelines are available for obtaining the isolator properties for base-isolated tanks. Therefore, to obtain appropriate time period (T_b) and damping of laminated rubber bearing (ξ_b), the parametric study is emphasized where majority of the parameters are considered. A series of investigations are made on the dynamic response parameters primarily by changing ξ_b (0.1–0.4) with a constant value of T_b (say 1 s) and subsequently the study is advanced explicitly for different values of T_b (2–4 s). The results showed that the variation in the damping of isolator results in an insignificant variation in the key parameters at a constant T_b. Therefore, an average value of ξ_b (0.2) is prescribed initially to predict the optimum value of T_b through the analysis.

It has been observed that the optimum value of time period lies between 2 and 3 s where majority of the parameters are attaining almost identical values. Among all of the parametric results conducted, the influence of varying isolation period with the key responses at $b/L = 0.4$ for different block depths is illustrated in Fig. 2 and presently, the effective value of time period is suitably prescribed as 3 s. After obtaining the converged value of T_b, the parametric study for different values of ξ_b is shown. However, the damping effect of the isolator is almost negligible for the majority of parameters as obtained from the study. Figure 3 indicates the parametric responses corresponding to the isolation damping and an average damping ratio of 0.2 is suitably prescribed for the present study. Figure 4 indicates the variation of bearing displacements for different isolator characteristics for typical configurations of tank-block-isolator system.

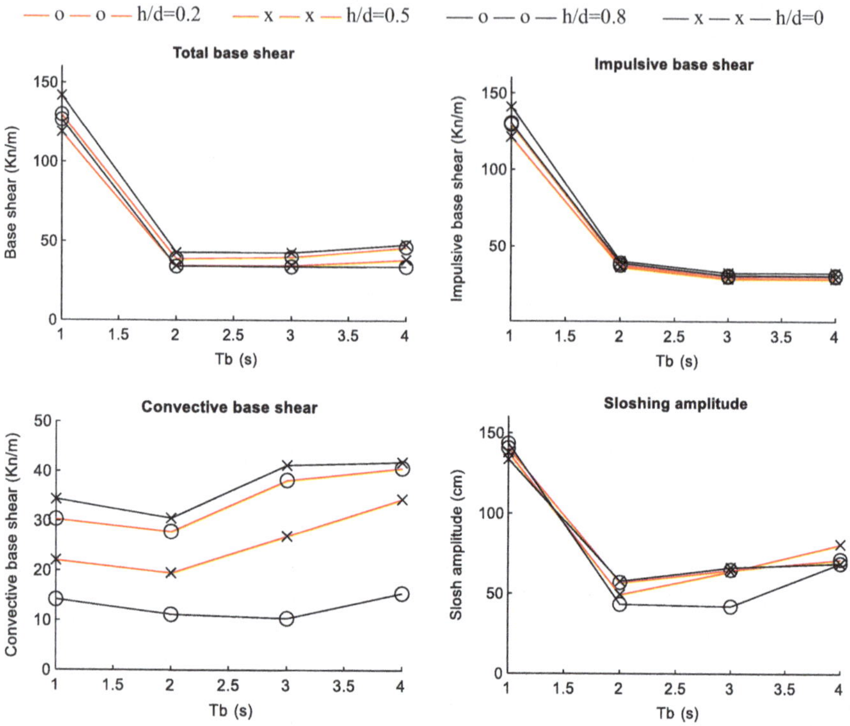

Fig. 2 Isolation period versus key response components for $b/L = 0.4$ at different block depths

7.2 Sloshing Amplitude

Table 3 presents the maximum sloshing amplitudes of different tank-liquid-block configurations. It can be observed that the amplitude is decreasing with the increasing h/d ratio for a specified width of block. The reason may be attributed to the anti-slosh effect due to the modified geometry in the presence of block as well as it causes some additional means of damping. However, the effect of varying width on sloshing amplitude is minimal for a certain depth of the block. The amplitude for isolated tank is significantly less than the non-isolated tank and the decrease is ranging over 50–60% for different geometry of the block which emphasizes the efficacy of laminated rubber bearing. The tendency is inevitable as the ground acceleration transferred to the tank is reduced due to base isolation. In general, maximum reduction of slosh amplitude for isolated tank is shown when $h/d = 0.8$ irrespective of the width of the block. Figures 5 and 6 illustrate the time history of sloshing amplitudes of isolated and non-isolated tanks for different block geometries. It is evident that the amplitude is more sensitive toward the depth of the block rather than the width. Consequently, a significant phase lag is observed in the responses as the block height increases.

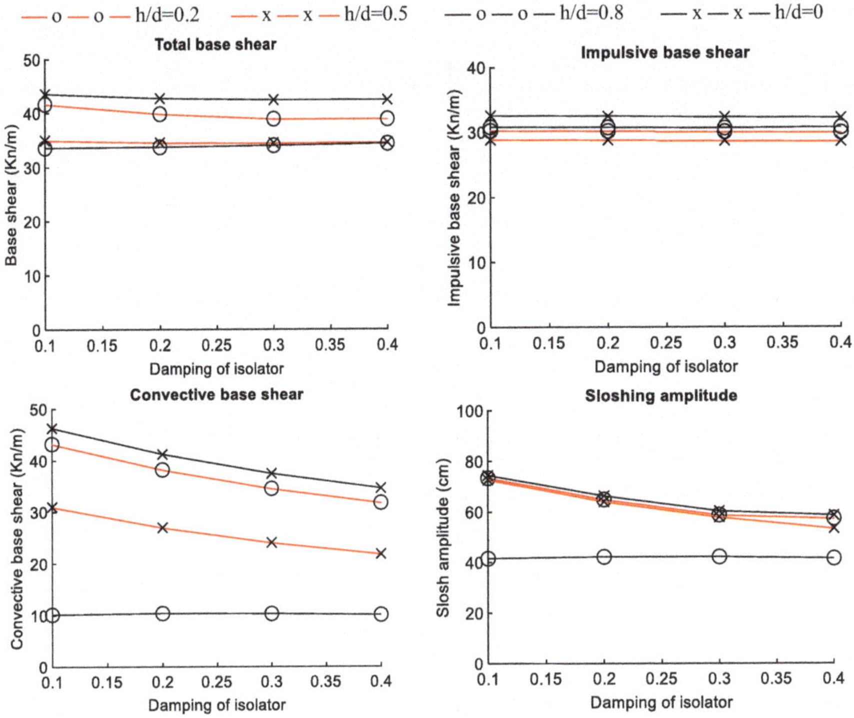

Fig. 3 Isolation damping versus key response components for $b/L = 0.4$ at different block depths

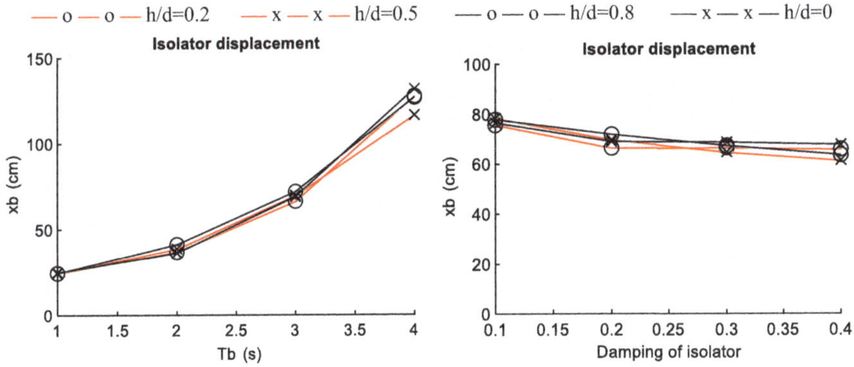

Fig. 4 Isolation period and damping versus isolator displacement at $b/L = 0.4$ at different block depths

Table 3 Absolute peak sloshing amplitude (cm)

h/*d* ratio	*b*/*L* = 0.2		*b*/*L* = 0.4		*b*/*L* = 0.6		*b*/*L* = 0.8	
	Non-isolated	Isolated	Non-isolated	Isolated	Non-isolated	Isolated	Non-isolated	Isolated
0.2	61.79	31.50	64.80	34.47	65.34	31.61	65.39	31.89
0.5	62.49	33.96	56.75	33.72	56.89	32.92	56.46	31.92
0.8	57.19	24.52	57.62	22.25	59.88	23.42	63.87	23.76

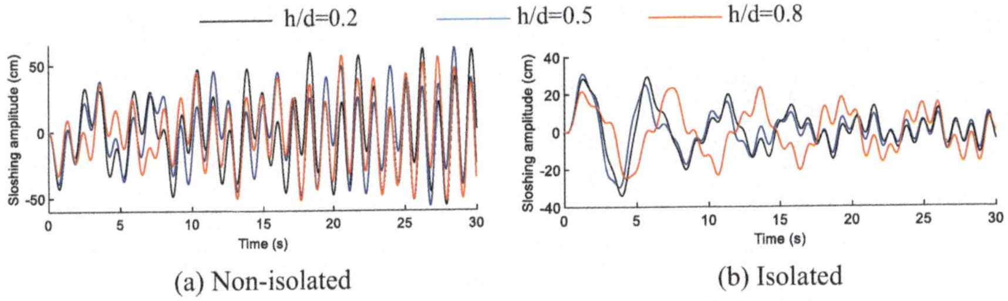

Fig. 5 Temporal variation of slosh amplitude for different *h*/*d* ratios at $b/L = 0.2$

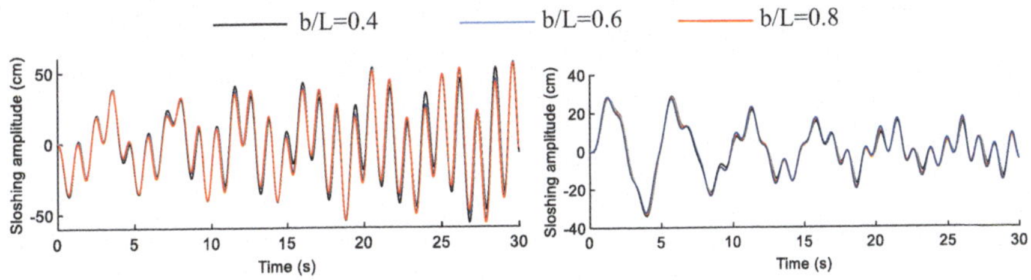

Fig. 6 Temporal variation of slosh amplitude for different *b*/*L* ratios at $h/d = 0.5$

7.3 *Hydrodynamic Pressure*

Figure 7 highlights the variation of pressure distribution across the left wall of the tank for different width and depth of the block. The effect of depth of the block is significant when a particular width of block ($b/L = 0.2$) considered as obtained from Fig. 7a whereas, the effect of width is comparatively small for non-isolated tank response and negligible for isolated tank response as shown in Fig. 7b. The reason may be attributed to the sharing of hydrodynamic pressure along the tank and block wall. The pressure distribution along the block walls act in the opposite direction with a relatively lower magnitude to that of the distribution along tank walls for which a net reduction in the component is attained. The application of base isolation is found to be very effective in reducing the pressure acting along the wall irrespective of the varying block geometries that is directly associated to the phenomenon of reduction in ground acceleration.

7.4 *Base Shear*

The absolute base shear values are presented in Tables 4, 5 and 6 for different tank-block-isolation systems where, it is evident that both the width and depth of block significantly affecting the base shear components. The variation of total, impulsive and convective components of base shear due to the varying depth of block is identical

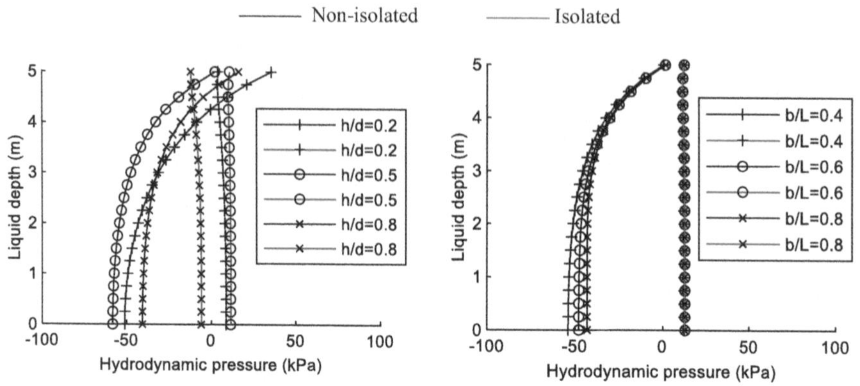

(a) Varying height atconstant b/L=0.2 (b) Varying width at constant h/d=0.5

Fig. 7 Pressure distribution for different tank-block-isolation system

for non-isolated tank results. On the contrary, the total and impulsive base shear for isolated tank is significantly smaller than non-isolated tank response and the decreasing percentage ranges from 40 to 70 for different geometries of the block. Furthermore, the convective base shear component attains minimal value for isolated tank in all of the configurations except at $h/d = 0.5$.

Figures 8 and 9 illustrate the base shear response for different configurations of tank-block-isolation system. It can be observed that the time history response has shown a significant phase lag when the geometry of the block is changed with respect to depth rather than the width for both non-isolated and isolated tanks.

7.5 *Isolator Displacement*

Table 7 presents the absolute peak isolator displacement for different configurations of tank-block-isolation system. The difference in global peak values for different tank-block configurations is almost negligible. Figures 10 and 11 illustrate the time variation of isolator displacement for typical tank-block-isolation system. It is evident that the geometrical change due to the block depth significantly influences the time variation of displacement whereas the effect of block width is negligible.

Table 4 Absolute dynamic peak total base shear (kN/m)

h/*d* ratio	*b*/*L* = 0.2		*b*/*L* = 0.4		*b*/*L* = 0.6		*b*/*L* = 0.8	
	Non-isolated	Isolated	Non-isolated	Isolated	Non-isolated	Isolated	Non-isolated	Isolated
0.2	56.04	24.16	52.74	23.82	49.10	22.61	45.25	21.94
0.5	54.36	23.65	47.18	20.93	38.33	17.38	28.82	14.10
0.8	60.00	25.34	47.76	21.38	34.48	16.30	20.14	9.92

Table 5 Absolute dynamic peak impulsive base shear (kN/m)

h/*d* ratio	*b*/*L* = 0.2		*b*/*L* = 0.4		*b*/*L* = 0.6		*b*/*L* = 0.8	
	Non-isolated	Isolated	Non-isolated	Isolated	Non-isolated	Isolated	Non-isolated	Isolated
0.2	47.32	19.99	44.16	18.96	40.62	17.74	36.80	16.37
0.5	48.91	20.75	41.32	18.22	32.56	15.01	23.04	11.15
0.8	56.51	23.43	45.36	19.90	32.33	15.20	17.68	9.00

Table 6 Absolute dynamic peak convective base shear (kN/m)

h/*d* ratio	*b*/*L* = 0.2		*b*/*L* = 0.4		*b*/*L* = 0.6		*b*/*L* = 0.8	
	Non-isolated	Isolated	Non-isolated	Isolated	Non-isolated	Isolated	Non-isolated	Isolated
0.2	19.53	16.33	18.94	16.02	18.63	15.93	18.54	15.99
0.5	14.46	14.89	12.98	13.89	12.21	13.07	11.97	12.90
0.8	10.19	7.24	7.35	4.79	6.10	4.43	5.52	4.05

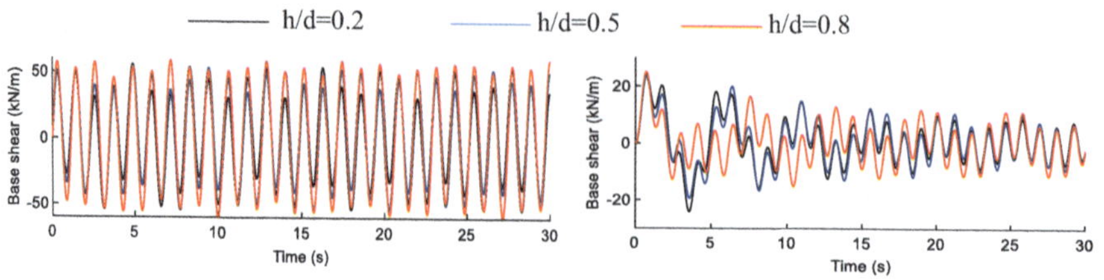

(a) Non-isolated (b) Isolated

Fig. 8 Total base shear behavior for different *h*/*d* ratios at $b/L = 0.2$

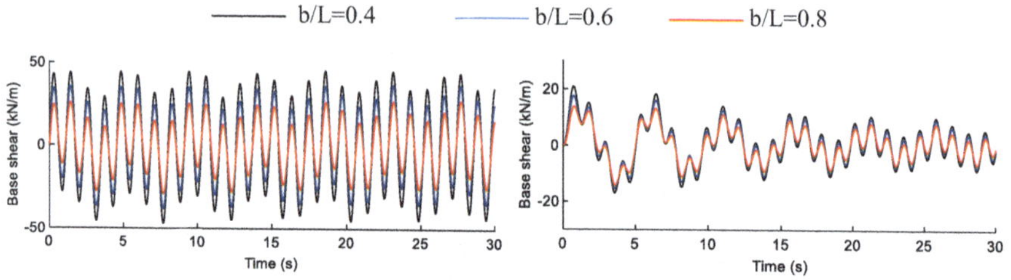

(a) Non-isolated (b) Isolated

Fig. 9 Total base shear behavior for different *b*/*L* ratios at $h/d = 0.5$

Table 7 Absolute peak isolator displacement

h/*d* ratio	Isolator displacement (cm)			
	$b/L = 0.2$	$b/L = 0.4$	$b/L = 0.6$	$b/L = 0.8$
0.2	25.53	25.30	25.06	24.82
0.5	25.54	25.04	24.46	23.76
0.8	25.92	25.22	24.34	23.22

Fig. 10 Isolator displacement profile for different *h*/*d* ratios at $b/L = 0.2$

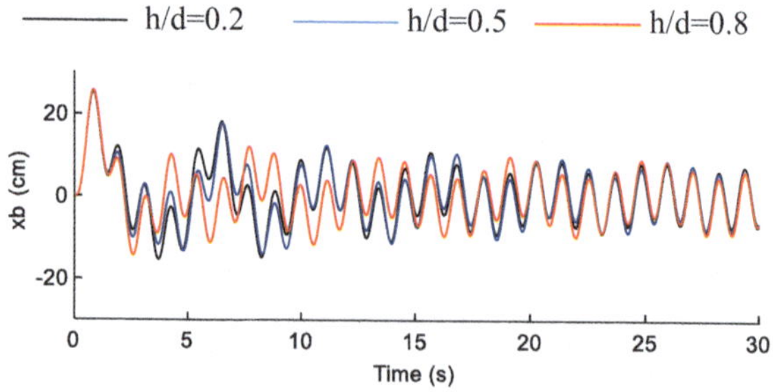

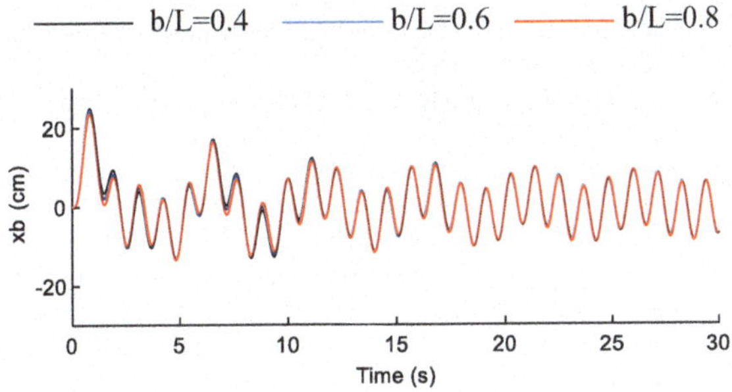

Fig. 11 Isolator displacement profile for different *b*/*L* ratios at *h*/*d* = 0.5

8 Summary of the Work

The influence of geometrical change in the block on the dynamic responses for both type of tanks is examined for different configurations. It is found that the height of the submerged structure has a very strong influence on the slosh dynamics as compared to the depth. However, the following conclusions are made,

- The effect of block height has a significant impact on the variation in sloshing amplitude and base shear components.
- The application of laminated rubber bearing reduces the sloshing amplitudes by 50–60% and the base shear components by 40–70% depending upon different block geometries.
- The effect of depth of the block is more significant for both types of tanks whereas, the effect of width is comparatively small for non-isolated tank response and negligible for isolated tank response.
- The application of base isolation is found to be very effective in reducing the magnitude of pressure irrespective of the varying block geometries.
- The geometrical change corresponding to block depth has a substantial influence on the isolator movement whereas the effect of block width is negligible.

References

1. George A, Cho IH (2021) Anti-slosh effect of a horizontal porous baffle in a swaying/rolling rectangular tank: analytical and experimental approaches. Int J Naval Architect Ocean Eng 13:833–847
2. Choun YS, Yun CB (1999) Sloshing analysis of rectangular tanks with a submerged structure by using small-amplitude water wave theory. Earthquake Eng Struct Dynam 28(7):763–783
3. Arafa M (2007) Finite element analysis of sloshing in rectangular liquid-filled tanks. JVC/J Vibr Control 13(7):883–903
4. Nayak SK, Biswal KC (2013) Quantification of seismic response of partially filled rectangular liquid tank with submerged block. J Earthquake Eng 17(7)
5. Cheng X, Jing W, Gong L (2019) Liquid sloshing problem in a concrete rectangular LSS with a vertical baffle. Arab J Sci Eng 44(5):4245–4256

6. Malhotra PK (1997) New method for seismic isolation of liquid-storage tanks. Earthquake Eng Struct Dynam 26(8):839–847
7. Shrimali MK, Jangid RS (2004) Seismic analysis of base-isolated liquid storage tanks. J Sound Vib 275(1–2):59–75
8. Seleemah AA, El-Sharkawy M (2011) Seismic response of base isolated liquid storage ground tanks. Ain Shams Engineering Journal 2(1):33–42
9. Kalantari A, Nikoomanesh MR, Goudarzi MA (2019) Applicability of mass-spring models for seismically isolated liquid storage tanks. J Earthquake Tsunami 13(1)
10. Shekari MR, Hekmatzadeh AA, Amiri SM (2019) On the nonlinear dynamic analysis of base-isolated three-dimensional rectangular thin-walled steel tanks equipped with vertical baffle. Thin-Walled Struct 138:79–94

All Mean Field Homogenization Methods Are Approximate: Some Might Be Useful

Atul Jain, Rahul Agarwal, Dhruvil Changani, Deepjyoti Dhar, Triparna Mahata, Hitesh Patil, and Diwakar Swaroop

Abstract A fundamental problem in mechanics of materials is homogenization which involves the calculation of the effective response of heterogenous materials. Mean field homogenization (MFH) methods are easy to setup, computationally cheap, and reasonably accurate making them quite attractive. However, there are persistent questions about their accuracy, physically admissibility, and versatility. In this paper, the experience of the co-authors over past 3 years is presented. MFH methods are studied from a wide range of perspectives from predictive abilities to confirming physically admissibility for a range of composites with varied fiber architectures. Algorithms are developed for microstructure generation with complex fiber architectures and finite element (FE) models are created by automation using python scripts. The MFH methods are benchmarked against the full FE solutions with comparisons at several length scales—effective modulus, stresses in individual inclusions and stresses in the interphase. Using this modus operandi, a wide range of MFH schemes is studied from full Mori–Tanaka (MT) formulation to pseudo-grain discretized version of MT to various multi-step MT formulations. It is concluded that each of the MFH variants is approximate with reasonable predictions for some microstructure and large errors for some microstructures. The accuracy of these methods decrease as the length scale of comparison is reduced. MFH methods are extended to model composites with discontinuous curved fibers which involve transformation of fibers to equivalent assembly of inclusions. An unbiased comparison formulation methodology is developed to validate the MFH methods involving transformation of fibers to equivalent assembly of inclusions.

Keywords Mean field homogenization · Finite element analysis · Microstructure · Interphase

A. Jain (✉) · R. Agarwal · D. Changani · D. Dhar · T. Mahata · H. Patil · D. Swaroop
Mechanical Engineering Department, Indian Institute of Technology Kharagpur, Kharagpur 721302, India
e-mail: atuljain@mech.iitkgp.ac.in

S. Sidhardh et al. (eds.), *Advances in Structural Integrity for Mechanical, Civil, and Aerospace Applications*, Lecture Notes in Mechanical Engineering,
https://doi.org/10.1007/978-981-97-6367-2_22

1 Introduction

Mean field homogenization (MFH) are a family of homogenization methods where starting from the Eshelby's conjecture [1], one estimates the effective properties based on the mean field assumption. This method was first proposed by Mori and Tanaka in the year 1973 [2] and remains popular to this day. MFH methods have now been adapted and multiple variations of the same have been developed over the years. Key driver of the new approaches includes mathematical problems with the classical MT formulation, desire for higher accuracy and accommodation of newer material types. The most common reported mathematical problems with the MT are the yielding of non-symmetric stiffness tensor [3], violation of the Hashin–Shtrikman bounds [4, 5].

The mathematical problems of the MT formulation are circumvented by the two step homogenization method proposed by Camacho et al. [6], in this method the volume element (VE) is discretized to a number of bi-phase composites. Following a two-step homogenization procedure, the bi-phase composites are first homogenized and subsequently the composite consisting of several bi-phase composites is homogenized. Usually one uses MT formulation for the first step while the Voigt assumption-based volume weighted average is deployed in the second step of homogenization. This method is commonly known as the MT-Voigt homogenization. Most commercial software such as DIGIMAT, Moldflow, etc. have used the MT-Voigt homogenization approach. The key advantage of this approach is that it is circumvents the mathematical problems associated with the MT formulation and is perhaps computationally cheaper compared to the traditional MT approach.

A second popular method is the incremental multistep Mori–Tanaka (MT) formulation [7–9] in which one starts with isotropic matrix and adds an inclusion phase and homogenization is performed. In the next iteration next inclusion phase is added and subsequently homogenized. Other variants of the MT formulation include Self-consistent schemes [10], double inclusion (DI) method [11, 12] regularization approach [13].

The family of MFH methods have been used for almost all types of composites including short fiber composites (both randomly orientated and aligned), unidirectional continuous fiber composites, nano-tubes reinforced composites, textile composites etc. A detailed review of the application of MFH methods for composites has been presented by Jain [14]. Besides effective modulus, MT based methods have been used for damage modeling [15], matrix nonlinearity [16] even fatigue modeling [17]. In these applications, the stress localization capabilities of the MFH formulations are of paramount importance.

A brief but complete description of the mathematical formulation of the MT method is provided below. First, the dilute strain concentrations are calculated based on the Eshelby's tensor.

$$A_m^i = \left[I + S^i(C_m)^{-1}(C_i - C_m)\right]^{-1} \tag{1}$$

here I is the identity matrix, S^i is the Eshelby's tensor for the ith inclusion, C_m and C_i are the stiffness matrices of the matrix phase and the ith inclusion phase, respectively.

Next, based on the mean field assumption the strain concentration tensors (A^i) for inclusions is calculated.

$$A^i = A^i_m \left(V_m I + \sum_{i=1}^{N} v_i A^i_m \right)^{-1} \tag{2}$$

here V_m is the volume fraction of the matrix, I is the identity matrix, V_i is the volume fraction of the individual inclusion phase. The image strain, i.e., the strain in an inclusion due to the presence of other inclusions is represented by the summation terms. Finally, the effective stiffness matrix for the composite, C_{eff}, is calculated.

$$C_{\text{eff}} = C_m + \sum_{i=1}^{N} v_i (C_i - C_m) A^i \tag{3}$$

In this paper, the experience of the co-authors in the domain of MFH is presented. Each section presents a new question and the insights gathered by the co-authors. Overall conclusions and future outlook are presented in the final section. For all the analysis presented in the paper, the Young's modulus of the inclusion and matrix is 72,000 MPa and 3000 MPa, respectively, while the Poisson's ratio is 0.22 and 0.37, respectively, unless otherwise specified.

2 MT-Voigt Versus MT

As described in sect. 1, there are certain mathematical and physical admissibility problems while employing the MT formulation for homogenization of multi-phase composites. It is to be noted that the MT formulation has no such shortcomings when employed for bi-phase composites. To overcome this challenge, one often resorts to a two-step homogenization where the representative volume elements (RVE) are discretized to bi-phase composites (sometimes also referred to as pseudo-grains). The first step of homogenization involves MT-based homogenization of the bi-phase composites followed by homogenization of the bi-phase composites together using Voigt assumption which leads to volume weighted average. A schematic representation of the two schemes is shown in Fig. 1. In this section, we compare the predictive abilities of the two schemes viz. MT and MT-Voigt for effective modulus and stresses in individual inclusions.

Using the Random Sequential Algorithm (RSA) [18], RVE comprising of inclusions with Aspect Ratio (AR = 3) and AR = 15 are generated. The inclusions with AR = 3 are randomly oriented while those with AR = 15 are aligned. The total volume fraction of the inclusions is 10%. A schematic representation of the RVE is

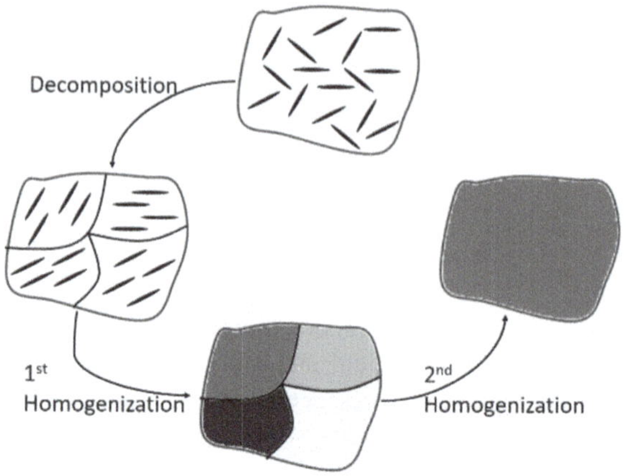

Fig. 1 Schematic representation of the two step MT-Voigt scheme

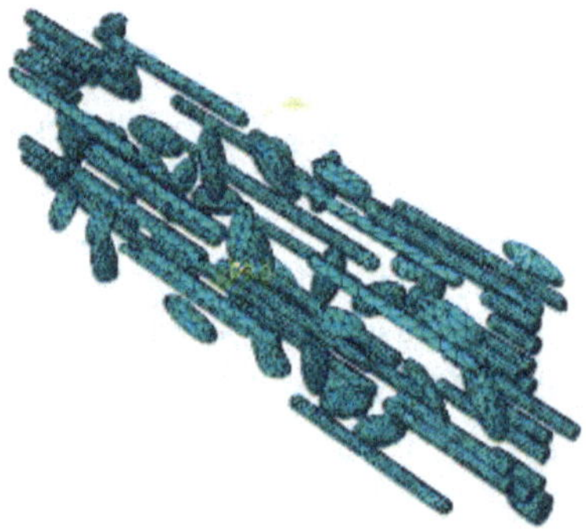

Fig. 2 RVE chosen for the comparison between MT and MT-Voigt homogenization approaches

shown in Fig. 2. Periodic boundary conditions are used, uniaxial tension loading is simulated. The elements used are C3D10M, which is 10 noded quadratic tetrahedron elements.

Figure 3 shows that the predictions by the two schemes are very similar and that the FE results have significant scatter which is not captured by either of the two approaches. The effective modulus predicted by the two schemes is 6.85 and 6.79 GPa, respectively, while the FE predictions are 5.7 GPa. It is concluded that the predictions of the two schemes are very similar to each other with significant mismatch with the FE results.

For more complex cases involving different elastic properties for the two inclusion phases, the predictions of the two schemes become more deviant with no one scheme out-performing the other. Detailed study regarding this comparison has been presented in [19].

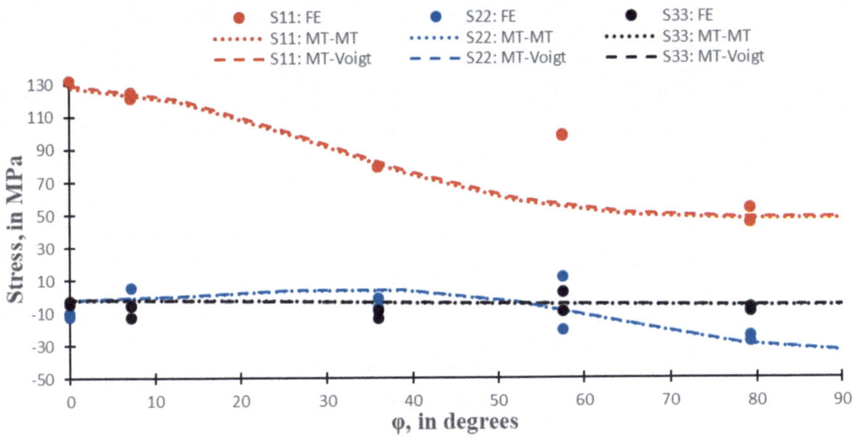

Fig. 3 Comparison of stresses in individual inclusions by the MT, MT-Voigt, and FE-based calculations

3 Pseudo-Grain Assumption

In Sect. 2, the predictive abilities of the MT and MT-Voigt homogenization schemes were studied. In this section, the key assumption involving the two-step homogenization is critically examined. Before the results of the analysis is reported, it is useful to recall the "pseudo-grain" assumption.

In the two step MT-Voigt homogenization approach, the RVE is discretized into a number of bi-phase composites and the bi-phase composites are homogenized individually (see Fig. 1). During the homogenization, the strain concentration tensors for the individual inclusions are calculated. The strain concentration tensors for a particular inclusion phase is independent of the presence of other inclusions. In other words, the strains/stresses in an inclusion is the same if it were part of an RVE with random orientation distribution of inclusions or if it were part of some preferred orientation distribution as long as the far-field applied strain is the same. The orientation distribution of the inclusions is usually characterized by second order orientation tensor, a_{ij} [20]. If the diagonal terms of the tensor a_{ij} are equal to 0.33 each, then it is said that the inclusions are uniformly random in 3 planes. Similarly, the diagonal terms of the tensor $a_{ij} = 0.5$ represents uniform random distribution in 2 planes. Other values of the orientation tensor denote some preferred randomness. The orientation distribution plots for $a11 = 0.35$, 0.5 and 0.65 are given in Fig. 4.

In this paper, for inclusion volume fraction of 25%, two planar orientation distributions corresponding to $a11 = 0.35$ and 0.65, respectively, are considered. The microstructure is generated using an in-house developed algorithm which ensures faithful capturing of the orientation distribution. The far-field strain in both the RVEs is the same and is equal to $\varepsilon_x = 0.01$, $\varepsilon_y = 0$, $\varepsilon_z = 0$. The stresses in individual inclusions is calculated and plotted as a function of the orientation (Fig. 5).

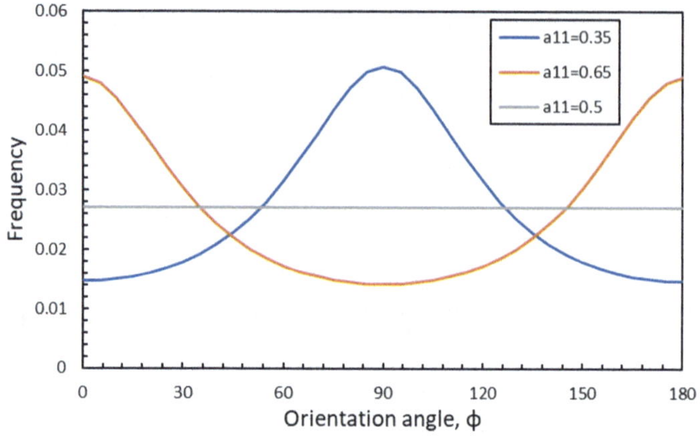

Fig. 4 Planar orientation distribution of inclusions for different values of orientation tensor term a11

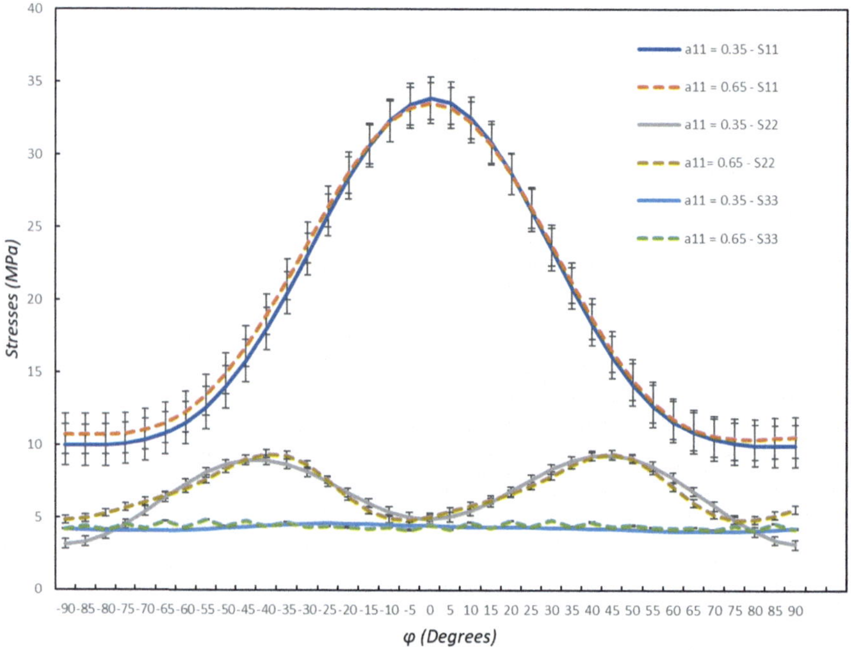

Fig. 5 Variation of the average stresses in individual inclusions as a function of the orientation, AR = 3

Fig. 6 Microstructure consisting of spheres and cylinders

The variation of the stresses in the inclusions as a function of the orientation is shown in Fig. 5, three realizations of each RVE is considered to ensure that there were at least 5 inclusions per orientation for reliable estimate of the mean and standard deviation. The results of the two preferred random orientation distributions are almost co-incident. From the results it can be concluded that the pseudo-grain assumption is reasonable for random media with low to moderate volume fractions and modulus mismatch (Young's modulus of inclusion/Young's modulus of matrix) of about 25 (typical for glass fiber–epoxy composites). However, for inclusions with higher AR, or if the modulus mismatch is higher or if the inclusions are softer than the matrix, it was seen that the pseudo-grain assumption is no longer reasonable as the stress distribution for different orientation distributions was higher and sometimes off the standard deviation limits.

4 Multi-step MT

Apart from the two-step homogenization discussed in sect. 2 and 3, there is another variant of the MT formulation involving multiple phases. In the so-called incremental multi-step MT [7, 8]. In this variant of the MT formulation, one starts with isotropic matrix and adds an inclusion phase and homogenization is performed. In the next iteration next inclusion phase is added and subsequently homogenized. This is repeated till all the inclusion phases have been accounted for. This multi-step MT scheme was validated using a special RVE comprising on spheres and infinite cylinders. Spheres and cylinders have extreme values of the aspect ratio (1 and infinity). While there have been several studies involving inclusions having finite aspect [21–23]; such comparisons usually did not yield conclusive conclusions about suitability of one scheme or the other. Comparison of the multi-step MT for extreme AR configurations, one can derive comprehensive evaluation of the homogenization schemes [24].

Periodic RVE comprising of spheres and cylinder was created using the RSA algorithm (see Fig. 6). The proportion of the sphere is varied and the effective modulus, stresses in individual phases is derived and compared with 4 implementations of the multi-step MT, namely MT-Voigt, MT-Reuss and MSMT 1 and MSMT 2. In MSMT-1, the sphere phase is homogenized first and the cylinders are added later, while the reverse is done for MSMT-2.

It is seen that the different multistep MT schemes predict with reasonable accuracy the axial Young's modulus. However, the same cannot be said for the transverse modulus and the stresses in individual phases. Also, the order of homogenization is seen to have a significant effect on the final predictions. The effective modulus predictions of two step MT do not necessarily converge to single step MT when the volume fraction of one of the phases tends to zero, this is particularly true for the transverse modulus predictions (Fig. 7).

5 Damage Modeling

Apart from the effective modulus and the stresses in individual inclusions, the MFH schemes can also be used for modeling damage in the inclusions. The most common damage mode in composites with short fibers is fiber matrix debonding which is implemented by first extracting the interphase stresses from the phase average stresses by the use of traction and displacement continuity equations. The interphase stress at every point can be calculated based on the local outward normal (see Fig. 8). For any angle θ, the outward normal are calculated using equation

$$\begin{aligned} n_1 &= \frac{\cos\theta}{\sqrt{1+\left(p^4-1\right)\sin^2\theta}} \\ n_2 &= \frac{p^2\sin\theta}{\sqrt{1+\left(p^4-1\right)\sin^2\theta}} \\ n_3 &= 0 \end{aligned} \tag{4}$$

Once the normal vector and the stress tensor at any point is known, one can evaluate the normal and shear stress components from them as proposed by Huysmans et al. [25].

$$\begin{aligned} \sigma_n &= \sigma_{ij} n_j n_i \\ \tau_n &= \sqrt{|\sigma_{ij}|^2 - \sigma_n^2} \end{aligned} \tag{5}$$

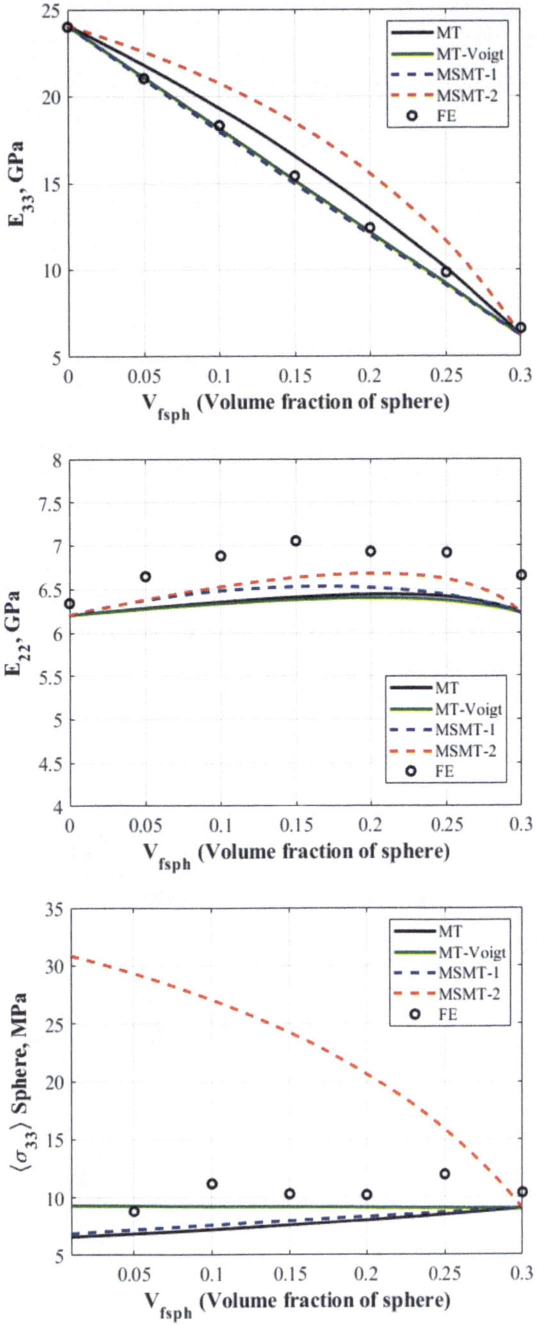

Fig. 7 Predictions of the multi-step MT formulations involving varying proportion of spherical phase, **a** effective axial modulus, **b** effective transverse modulus, **c** stresses in the spherical phase, loading is *z* (Axial)-direction, and **d** stresses in the spherical phase, loading is *y* (transverse)-direction

Fig. 8 Outward normal for an ellipse

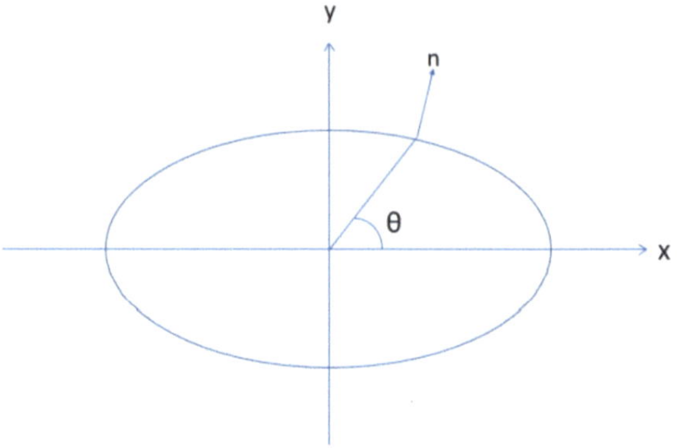

In this section, the stresses in the interphase as predicted by FE and using MT formulations are compared. First, RVEs with regular packing of ellipsoids are generated, appropriate meshing is undertaken and the stresses in the matrix region adjacent to the inclusions as a function of angle θ is estimated. The local FE results are compared with the stresses in the interphase as derived using the MT average stresses in the inclusions and FE-average stresses in the inclusions (Figs. 9 and 10).

It is seen that the MT model does a reasonable job at estimating the stresses in the interphase and captures the overall trend very well. The local interphase stresses in the FE and those calculated using the average stresses are very similar, thus validating

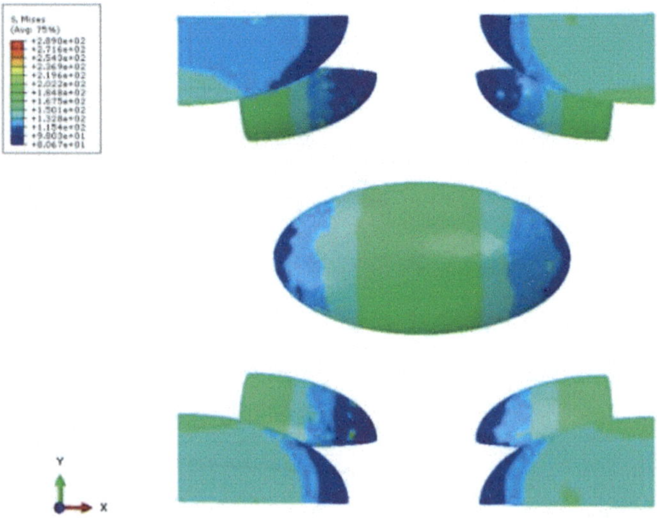

Fig. 9 FE results of stresses in the inclusions, notice the periodic arrangement of the inclusions and the variation of stresses within an inclusion

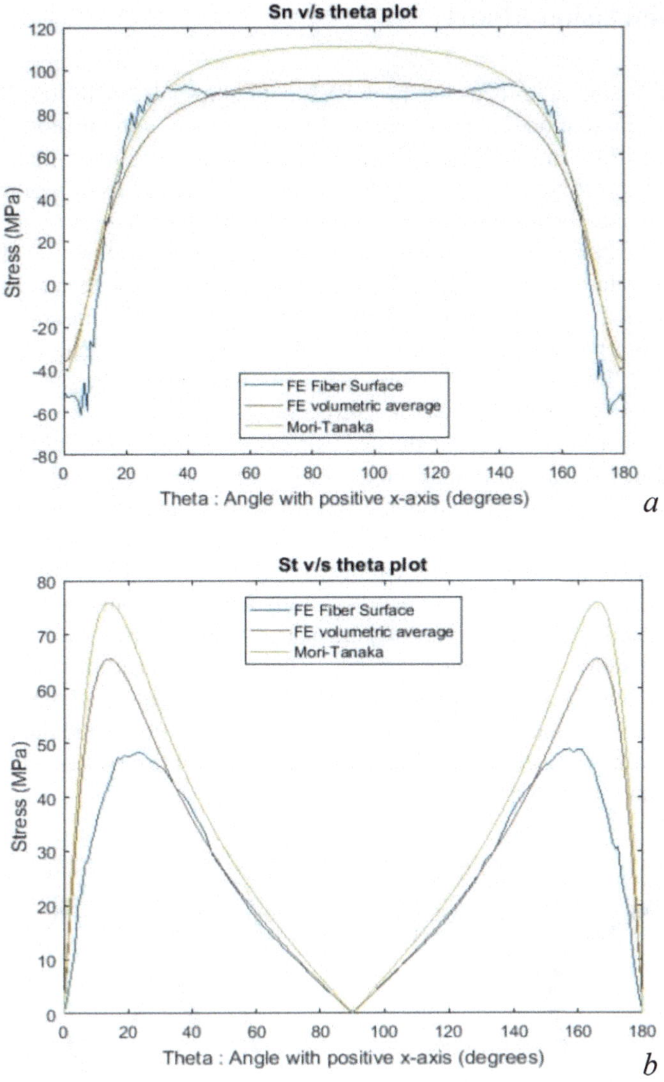

Fig. 10 Comparison of stresses in the interphase, **a** normal stresses, **b** shear stresses

the approach proposed in literature [15, 26, 27]. Overall, it can be said that the stresses in the interphase are over-predicted by the MT schemes, which in turn implies that the onset of damage in the interphase will be over-predicted by MT based damage modeling.

6 Poly-inclusion Model

So far in this paper, the focus has been RVE with ellipsoidal inclusions. However, the utility of these models can be extended to composites with curved fibers using the "poly-inclusion model [26, 28]". In this approach, the curved fibers are first transformed into an equivalent assembly of ellipsoid of inclusions by segmenting the curved fibers into a number of segments and replacing the segment by an ellipsoid with same orientation as the segment and the AR calculated as a function of the local radius of curvature. The RVE with the assembly of inclusions is then homogenized using the single or multi-step MT formulations. A schematic representation of the poly-inclusions model is given in Fig. 11. In this paper, two segmentation strategies are employed. First involves the poly-inclusion model, while the other is a crude segmentation of the inclusion with no corrections for aspect ratio.

The poly-inclusion model is validated using curved fibers with single curvature. Using modified RSA algorithm, periodic RVE with random distribution is created and subjected to uni-axial normal and shear loading to calculate the equivalent elastic constants. An example of the microstructure generated, including meshing is shown in Figure 12. For this validation, the properties of the curved fibers are assumed to be 72,000 MPa, while that of the matrix is 1308 MPa. And value of the Poisson's ratio is 0.22 and 0.37 for the fibers and the matrix, respectively.

Figure 13 shows the comparison between the FE results and the different poly-inclusion variations for the axial and transverse Young's Modulus and shear Modulus. It is seen that while the predictions for the transverse and shear modulus is in good agreement with the FE results, the same cannot be said for the axial modulus, where the error is about 20% or so.

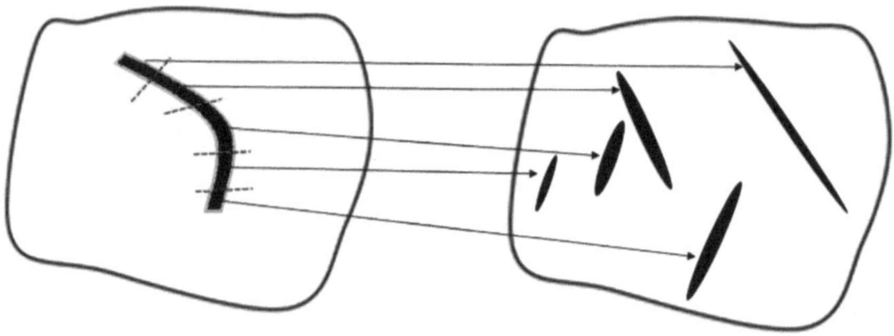

Fig. 11 Schematic representation of the poly-inclusion model, a single curved fiber is replaced by an assembly of inclusions

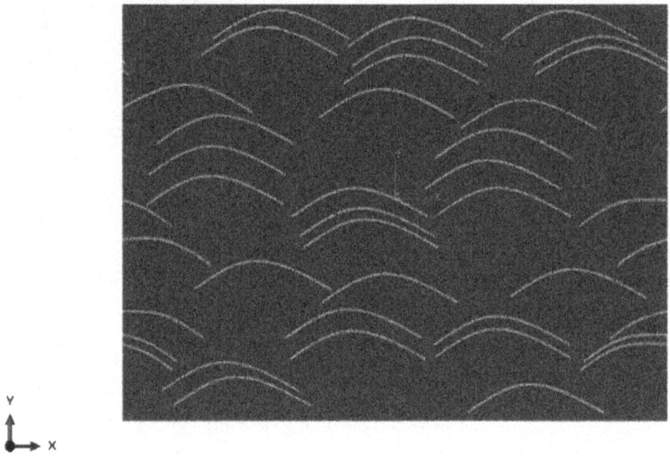

Fig. 12 Microstructure with curved fibers, notice that the microstructure is periodic, with one fiber intersecting one edge is mirrored at the other end

7 Conclusions

An overview of the activities in the field of mean field homogenization has been presented in this paper. Different topics including examination of the predictive abilities of the single step and multi-step MT, the scrutiny of the underlying assumptions, and extension of the MT methods to curved inclusions has been introduced. For each problem, different strategy for microstructure generation has been adopted and FE models have been built to compare and study the different MT methods.

The overall conclusion is that the different mean field homogenization methods are approximate in some cases and quite accurate in other cases. The methods are versatile and can be sued for a wide range of heterogenous materials. Even though somewhat approximate, the utility of the methods and possibilities are immense.

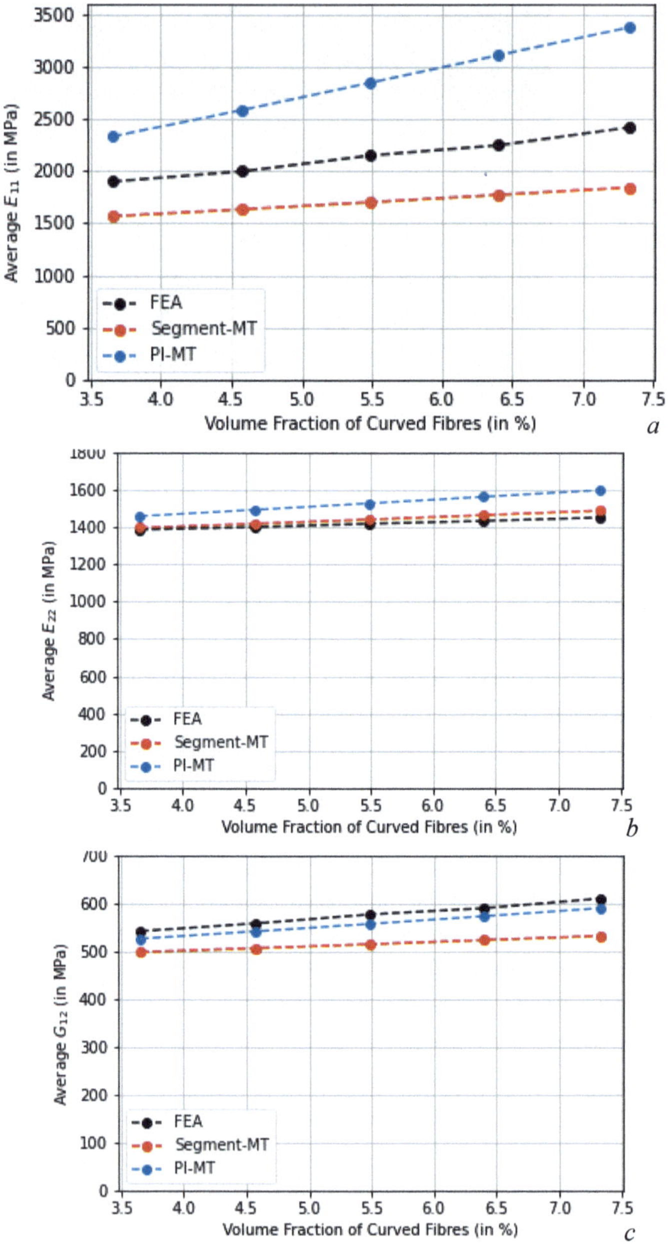

Fig. 13 Variation of the effective elastic constants as a function of volume fraction, predictions by various homogenization approaches, **a** axial Young's modulus, **b** transverse Young's modulus, and **c** shear Modulus

Acknowledgements IIT Kharagpur is thanked for funding this research as part of project "Multi-scale Modeling and Characterization of 3D Printed Short Fiber Composites" funded under the aegis of Institute Scheme for Innovate Research and Development.

References

1. Eshelby JD (1957) The determination of the elastic field of an ellipsoidal inclusion, and related problems. In: Proceedings of the royal society of London series A mathematical and physical sciences. 241(1226): 376–396
2. Mori T, Tanaka K (1973) Average stress in matrix and average elastic energy of materials with misfitting inclusions. Acta Metall 21:571–574
3. Schjodt-Thomsen J, Pyrz R (2001) The Mori-Tanaka stiffness tensor: diagonal symmetry, complex fibre orientations and non-dilute volume fractions. Mech Mater 33:531–544. https://doi.org/10.1016/S0167-6636(01)00072
4. Pierard O, Friebel C, Doghri I (2004) Mean-field homogenization of multi-phase thermo-elastic composites: a general framework and its validation. Compos Sci Technol 64:1587–1603. https://doi.org/10.1016/j.compscitech.2003.11.009
5. Benveniste Y (1987) A new approach to the application of Mori-Tanaka theory in composite-materials. Mech Mater 6:147–157. https://doi.org/10.1016/0167-6636(87)90005-6
6. Camacho CW, Tucker CL, Yalvaç S, McGee RL (1990) Stiffness and thermal expansion predictions for hybrid short fiber composites. Polym Compos 11:229–239. https://doi.org/10.1002/pc.750110406
7. Yang Q-S, Tao X, Yang H (2007) A stepping scheme for predicting effective properties of the multi-inclusion composites. Int J Eng Sci 45:997–1006. https://doi.org/10.1016/j.ijengsci.2007.07.005
8. Müller V, Kabel M, Andrä H, Böhlke T (2015) Homogenization of linear elastic properties of short-fiber reinforced composites—a comparison of mean field and voxel-based methods. Int J Solids Struct 67–68:56–70. https://doi.org/10.1016/j.ijsolstr.2015.02.030
9. Dhar D, Jain A (2023) Improved micromechanical prediction of short fibre reinforced composites using differential Mori-Tanaka homogenization. Mech Mater 185:104768. https://doi.org/10.1016/j.mechmat.2023.104768
10. Laws N, McLaughlin R (1979) Effect of fiber length on the overall moduli of composite-materials. J Mech Phys Solids 27:1–13. https://doi.org/10.1016/0022-5096(79)90007-3
11. Nemat-Nasser S, Hori M (1999) Micromechanics: overall properties of heterogeneous materials. 2nd edn. Elsevier
12. Lielens G (1999) Micro-macro modeling of structured materials. Université Catholique de Louvain, Belgium
13. Li JY (1999) On micromechanics approximation for the effective thermoelastic moduli of multi-phase composite materials. Mech Mater 31:149–159. https://doi.org/10.1016/s0167-6636(98)00071-4
14. Jain A (2019) Micro and mesomechanics of fibre reinforced composites using mean field homogenization formulations: a review. Mater Today Commun 21:100552. https://doi.org/10.1016/j.mtcomm.2019.100552
15. Jain A, Abdin Y, Van Paepegem W, Verpoest I, Lomov SVSV (2015) Effective anisotropic stiffness of inclusions with debonded interface for Eshelby-based models. Compos Struct 131:692–706. https://doi.org/10.1016/j.compstruct.2015.06.007
16. Sekkate Z, Aboutajeddine A, Seddouki A (2020) Elastoplastic mean-field homogenization: recent advances review. Mech Adv Mater Struct. https://doi.org/10.1080/15376494.2020.1776431

17. Jain A, Veas JMJM, Straesser S, Van Paepegem W, Verpoest I, Lomov SVSV (2016) The Master SN curve approach—a hybrid multi-scale fatigue simulation of short fiber reinforced composites. Compos Part A Appl Sci Manuf 91:510–518
18. Feder J (1980) Random sequential adsorption. J Theor Biol 87:237–254. https://doi.org/10.1016/0022-5193(80)90358-6
19. Dhar D, Lomov SV, Jain A (2024) Predictive abilities of pseudodiscretization and pseudograin discretization schemes of the Mori-Tanaka homogenization, benchmarked against real and virtual RVEs. Mech Adv Mater Struct 31(13):2743–2758. https://doi.org/10.1080/15376494.2022.2163438
20. Advani SG, Tucker CL III (1987) The use of tensors to describe and predict fiber orientation in short fiber composites. J Rheol (N Y N Y) 31:751–784
21. Naili C, Doghri I, Kanit T, Sukiman MS, Aissa-Berraies A, Imad A (2020) Short fiber reinforced composites: unbiased full-field evaluation of various homogenization methods in elasticity. Compos Sci Technol 187:107942. https://doi.org/10.1016/j.compscitech.2019.107942
22. Jain A, Abdin Y, Van Paepegem W, Lomov SV (2020) Direct Mori-Tanaka calculations of strains in ellipsoidal inclusions with multiple orientations—Comments on the papers: Naili G, et al. Comp Sci Tech. 187: 107942. (https://doi.org/10.1016/j.compscitech.2019.107942) and Jain A, et al. Comp Sci T Compos Sci Technol, p 190. https://doi.org/10.1016/j.compscitech.2020.108068
23. Klusemann B, Svendsen B (2010) Homogenization methods for multi-phase elastic composites: comparisons and benchmarks. Tech Mech J Eng Mech 30:374–386
24. Swaroop D, Dhar D, Suriyan A, Jain A (2022) Of spheres and infinite cylinders: a critical relook at multi-step mean-field homogenization formulations. Mech Mater 174:104447. https://doi.org/10.1016/j.mechmat.2022.104447
25. Huysmans G, Verpoest I, Van Houtte P (2001) A damage model for knitted fabric composites. Compos Part A Appl Sci Manuf 32:1465–1475. https://doi.org/10.1016/s1359-835x(01)00045-8
26. Huysmans G, Verpoest I, Van Houtte P (1998) A poly-inclusion approach for the elastic modelling of knitted fabric composites. Acta Mater 46:3003–3013. https://doi.org/10.1016/s1359-6454(98)00021-4
27. Meraghni F, Blakeman CJ, Benzeggagh ML (1996) Effect of interfacial decohesion on stiffness reduction in a random discontinuous-fibre composite containing matrix microcracks. Compos Sci Technol 56:541–555
28. Abdin Y, Jain A, Verpoest I, Lomov SV (2016) Mean-field based micro-mechanical modelling of short wavy fiber reinforced composites. Compos Part A Appl Sci Manuf 91:472. https://doi.org/10.1016/j.compositesa.2016.03.022

Damage Initiation and Strength Predictions of Randomly Oriented Strands of Prepreg-Based Discontinuous Composites

Akshat Bagla, P. R. Krishna Mohan, and P. M. Mohite

Abstract Composite materials have been heavily used in major industrial applications like defence, automobiles and wind turbine blades for the past few decades due to their tailorability to meet specific design needs. The high-cost left-over prepreg (3–5% waste) during composite manufacturing is usually land-filled. These left-overs can be chopped into rectangular chips and, through random spatial distribution in the mould, can be compression moulded into flat, curved plates or any other complex structural part. These randomly oriented strands (ROS) reinforced transversely isotropic composite can replace conventional metals and unidirectional composite for aircraft's secondary and tertiary loading components. In the present work, the stochastic spatial distribution of chips in compression moulded ROS reinforced composite is modelled through Representative Volume Element (RVE). The behaviour of ROS composite is analysed by performing finite element analysis. A continuum mechanics-based linear-elastic damage model is developed involving damage parameters to degrade the effective property of individual constituents of composite for constitutive material modelling through user material, UMAT in commercial finite element solver Abaqus/Standard. This work mainly focuses on failure in the constituents of the ROS composite. The four failure modes in the chip are longitudinal breakage, transverse splitting, in-plane and out-plane shear failure, and the failure mode in the matrix is isotropic. The uni-axial tensile loaded ROS reinforced composite shows transverse chip splitting and matrix failure as dominant modes of failure, which were in good agreement with the in-house experimental results.

Keywords Damage mechanics · Finite element analysis · Randomly oriented strands

---

A. Bagla · P. R. Krishna Mohan (✉) · P. M. Mohite
Department of Aerospace Engineering, Indian Institute of Technology Kanpur, Kanpur, Uttar Pradesh 208016, India
e-mail: prkmohan@iitk.ac.in

A. Bagla
e-mail: akshatbagla2021@iitkalumni.org

P. M. Mohite
e-mail: mohite@iitk.ac.in

S. Sidhardh et al. (eds.), *Advances in Structural Integrity for Mechanical, Civil, and Aerospace Applications*, Lecture Notes in Mechanical Engineering,
https://doi.org/10.1007/978-981-97-6367-2_23

1 Introduction

The randomly oriented strands (ROS) reinforced composites are found as promising alternatives to conventional metals for aircraft, space and automotive fields in secondary and some primary loading applications [1–3]. Many researchers have recently focused on the characterization, and failure analysis of compression moulded chopped ROS composites. However, the complex arrangement of the constituents in the randomly oriented strands composite obtained by compression moulding makes it challenging to estimate the composite's elastic properties and local failure behaviour.

In the literature, one can find different approaches to packing the chips in the RVE and predicting the elastic properties of the chopped chip-reinforced composites. Mohan et al. have used Random Sequential Adsorption (RSA) algorithm to model the packing of randomly oriented fibre-reinforced composites computationally [4]. RSA method packs chips sequentially into space with random locations, avoiding overlapping the new chip with existing chips. Other packing methods, like the Monte Carlo method, work on packing fibres with random orientation first and then use perturbation for their location and angles to avoid overlapping [5]. Li et al. used the data of the fibre bundles obtained from optical microscopic images to generate the desired mesostructure with Voronoi tessellation [6].

2 Stochastic Mesoscale RVE Construction

In this section, the micromechanical modelling of chopped chip reinforced composites is done through Representative Volume Element (RVE) generation. The randomly oriented chips are packed inside the RVE using Random Sequential Adsorption (RSA) algorithm.

RVE represents a domain in the material that contains a sufficient number of micro-features so that the microstructure characteristics and properties of different RVE realizations become statistically equivalent. In the present work, the dimension of RVE is chosen as $3L$ (where L is the in-plane dimension of the chip) based on the convergence study performed on RVE through elastic properties estimation using the Standard Mechanics approach [4]. The constituent material properties of the composite used in the analysis, i.e. chip (AS4/3501-6) and matrix (3501-6), are shown in Tables 1 and 2, respectively which is taken from the work of Soden et al. The actual unidirectional laminate chip (extracted from prepreg) is idealized as homogenized transversely isotropic material through two-phase idealizations whose details can be found in Mohan et al. [4]. These set of unidirectional laminate, constituting *chip* phase, are randomly oriented inside RVE to constitute the reinforcement of matrix material to form the discontinuous composite. We will study the failure behaviour mechanism of such generated RVE as shown in Fig. 1a.

Table 1 Material properties of chip (AS4/3501-6)

Property	Values
E_1 (GPa)	90.04
E_2 (GPa)	7.52
υ_{12}	0.28
G_{12} (GPa)	6.94
G_{23} (GPa)	2.92

Table 2 Material properties of epoxy

Property	Values
E (GPa)	4.20
υ	0.35

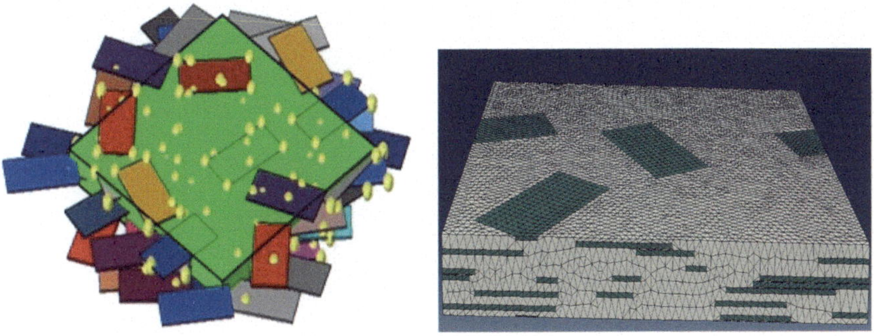

(a) (b)

Fig. 1 **a** Assembled RVE, **b** meshed RVE

The flowchart for the RSA algorithm for RVE generation can be found in Fig. 2. The stochastic structural arrangement is approximated by the sequential placement of chips through the generation of center points and angles using a random generator function in MATLAB. This in-house developed code ensures the periodicity of RVE based on their interaction with the face, edge and corner of RVE. One can find more details in the work of [4]. The assembled chip-matrix model of the RVE and its tetrahedral mesh model (after the removal of excess chip from RVE boundary) are shown in Fig. 1a, b, respectively.

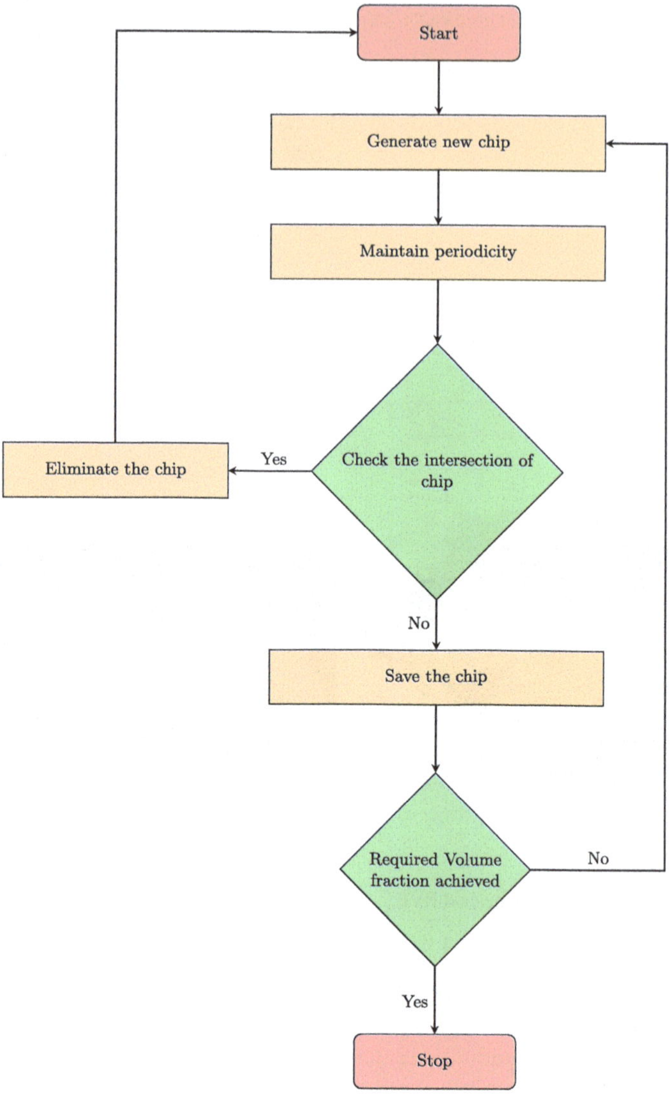

Fig. 2 Flowchart for RSA algorithm

3 Critical Region Determination

The understanding of the failure mechanism of ROS reinforced composites is quite complex due to the random distribution arrangement of strands/chips inside the RVE. This section attempts to determine the critical location inside RVE based on the strain concentration factor study over the mid-plane of RVE as shown in Fig. 3, under uniaxial loading.

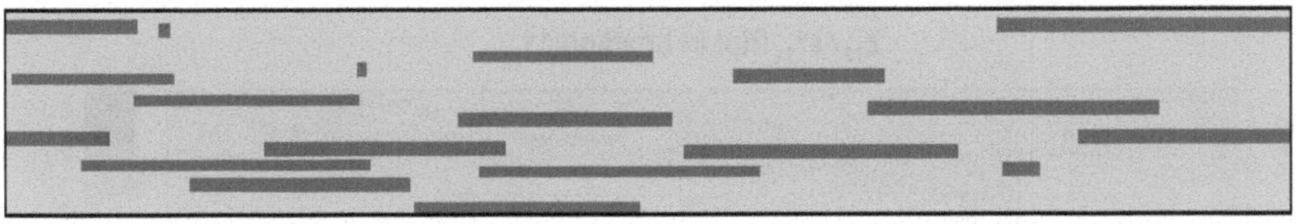

Fig. 3 Mid-plane of RVE under analysis

The strain data is generated using displacement boundary conditions similar to the *Standard Mechanics* approach for RVE effective properties estimation [4]. The six displacement load cases, with each corresponding to the respective strain component of the strain column vector, form the Strain Concentration Factor (SCF) column vector. Numerically, SCF = $\epsilon_{\text{micro}}/\epsilon_{\text{macro}}$, where ϵ_{micro} is the elemental strain and ϵ_{macro} is the applied strain. The generated SCF vector for each element is then, component-wise, linearly interpolated over the midplane ($X = 0$) of RVE to generate the corresponding component-wise SCF plots. The mid-plane is chosen for the sake of achieving unbiased randomness.

The key points obtained from this analysis are given below:

1. The critical region is found near the chip–matrix interface region due to the high load-carrying capacity of chip compared to its surrounding matrix.
2. The disturbance of high-strain zones in the matrix near boundary is due to rigid boundary conditions at edges and random placement of chips.
3. The Poisson's effect during normal tensile load cases is also responsible for lateral displacement, causing high strains near the interface due to these two effects:
   (a) The matrix zone between chips is restricted to compress at the chip interface, causing tensile hotspot locations.
   (b) Matrix-rich zones tend to compress laterally, causing compressive hotspot zones.

The generated SCF plots for applied uniaxial loading in x-direction (load case-1) generates SCF column vector with six components. The in-plane SCF components are shown in Fig. 4. In a similar manner, the remaining five displacement load cases generate corresponding five SCF column vectors. The surface interpolation of these five SCF vectors component-wise gives similar plots of critical location. To be clear, these six displacement load scenarios create six SCF column vectors, which is equal to a 6×6 SCF matrix.

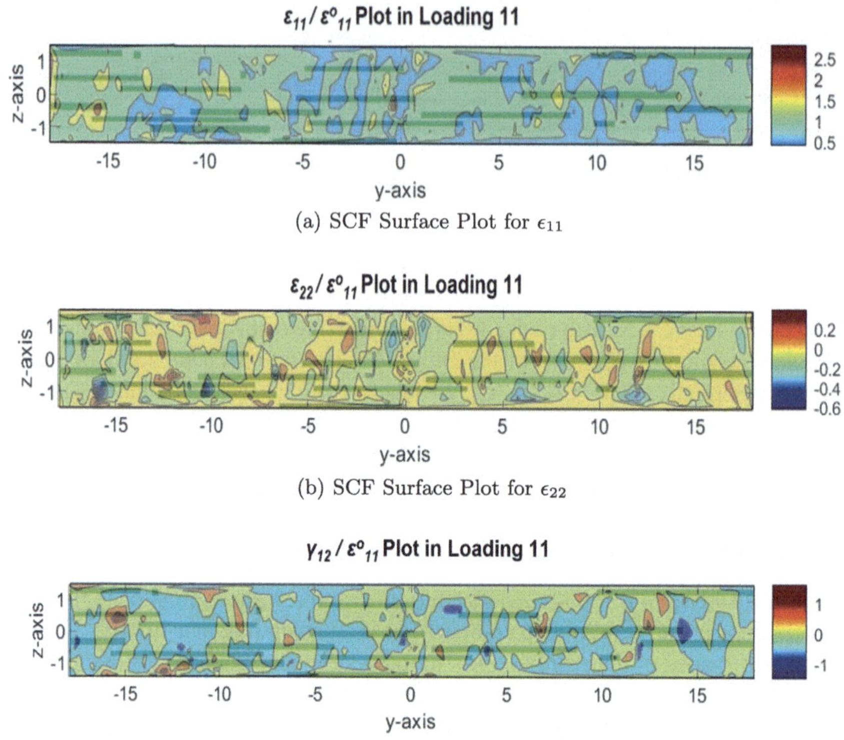

(a) SCF Surface Plot for ϵ_{11}

(b) SCF Surface Plot for ϵ_{22}

(c) SCF Surface Plot for ϵ_{12}

Fig. 4 Surface-plots obtained from loading in longitudinal direction

4 Continuum Damage Model Framework

The proposed framework for continuum damage modelling is established on the mesostructural level. The continuum-damage mechanics describes material failure as the reduction of corresponding material stiffness through incorporating the damage parameter. This section introduces two constitutive material models to implement the finite-element analysis through user subroutine, *UMAT*, implemented in commercial finite element solver $ABAQUS$.

4.1 Chip (ROS): Elastic Damage Model

The chip phase failure model is proposed using the Tsai–Wu criterion. The carbon fiber chip is modelled as a transversely isotropic material. The chip is assumed to be extracted from a unidirectional prepreg having a fiber volume fraction of 40%. The strength properties of the AS4/3501-6 chip incorporated for the chip damage model can be found in Table 1. The RVE can be visualized as bundles of unidirectional fibers (called a chip) with random orientation immersed inside matrix material. The effective properties of the chip are extracted using a two-step homogenization process. The details can be found in Mohan et al. [4]. The constitutive damage model of the chip is developed using state-dependent damage parameters to degrade the corresponding elastic modulus based on the Tsai–Wu failure criterion. The Failure Index, FI variable, decides the chip failure and its corresponding components, which helps us determine the chip failure mode. The chip strength parameter values are taken from Soden et al. [7].

The Tsai–Wu criterion [8]-based Failure Index, FI can be found below:

$$\mathrm{FI} = F_1\sigma_{11} + F_2\sigma_{22} + F_{11}\sigma_{11}^2 + F_{22}\sigma_{22}^2 + 2 \times F_{12}\sigma_{11}\sigma_{22} + F_{66}\sigma_{12}^2 \tag{1}$$

$F_1 = \frac{1}{X_t} + \frac{1}{X_c}$; $F_{11} = -\frac{1}{X_t X_c}$

$F_2 = \frac{1}{Y_t} + \frac{1}{Y_c}$; $F_{22} = -\frac{1}{Y_t Y_c}$

$F_{66} = \frac{1}{S^2}$; $F_{12} = -0.5 \times (F_{11}F_{22})^{1/2}$

where

X_t and X_c are ultimate tensile and compressive strengths of chip in longitudinal direction;

Y_t and Y_c are ultimate tensile and compressive strength of chip along transverse direction;

S depicts the in-plane and out-plane shear strengths of chip.

We split the components of FI based on its contribution in longitudinal, transverse and shear failure as:

$$f_1 = F_1\sigma_{11} + F_{11}\sigma_{11}^2 + F_{12}\sigma_{11}\sigma_{22} \tag{2}$$

$$f_2 = F_2\sigma_{22} + F_{22}\sigma_{22}^2 + F_{12}\sigma_{11}\sigma_{22} \tag{3}$$

$$f_3 = F_{66}\sigma_{12}^2 \tag{4}$$

For Failure Index > 1, any of the three modes could be the dominant mode of failure.

- *Longitudinal Failure*

if $f_1 > f_2$ and $f_1 > f_3$; $d_{11} = 1$

Otherwise, $d_{11} = 0$.

We hereby propose the degraded Young's modulus of chip along longitudinal direction.

$$E_{11} = E_{11}^{o}(1 - d_{11})$$

- *Transverse Chip Splitting*

$$\text{if } f_2 > f_1 \quad \text{and} \quad f_2 > f_3; \quad d_{22} = 1$$

Otherwise, $d_{22} = 0$.

We propose the degraded Young's modulus of chip along transverse direction.

$$E_{22} = E_{22}^{o}(1 - d_{22})$$

- *Shear Failure*

$$\text{if } f_3 > f_1 \quad \text{and} \quad f_3 > f_2; \quad d_{12} = 1$$

Otherwise, $d_{12} = 0$.

We propose the degraded in-plane shear modulus of chip in case of failure.

$$G_{12} = G_{12}^{o}(1 - d_{12})$$

where

E_{11}^{o}, E_{22}^{o} and G_{12}^{o} are original material properties;

E_{11}, E_{22} and G_{12} are degraded material properties;

d_{11}, d_{22} and d_{12} are chip damage parameters.

4.2 Epoxy Matrix: Elastic Damage Model

The matrix phase failure model is proposed using the von Mises criterion. The matrix damage is modelled as isotropic damage behaviour. The failure criterion for the matrix considers degradation of the elastic modulus of the matrix, E, the stiffness representative. Generally, matrix cracks are of two types transverse and longitudinal, based on the load-applied direction. The damage parameter d_m taken here considers both of these effects in locations of higher strain concentration region, mainly chip corners and interface region.

The epoxy matrix is known to take better shear loads than the tensile loads. Therefore, the maximum distortion energy-based von Mises criterion is used to predict the matrix failure based on distortion energy density reaching the critical value. This matrix damage model provides instant failure to the chip once it reaches its yield strength.

The von Mises stress is depicted below:

$$V_{\mathrm{m}}^{\mathrm{stress}} = [0.5 \times \{(\sigma_{xx} - \sigma_{yy})^2 + (\sigma_{yy} - \sigma_{zz})^2 + (\sigma_{zz} - \sigma_{xx})^2\} + 3 \times (\tau_{xy} + \tau_{yz} + \tau_{zx})]^{(1/2)} \tag{5}$$

where the stress state components have its usual meaning.

5 Results and Discussion

This section consists of the implementation of the continuum damage model over its constituents, chips and matrix through the degradation of effective properties using state-dependent damage parameters. It discusses the failure mechanism for the elastic matrix phase and chip phase, followed by the strength prediction of the composite laminate.

5.1 Damage Behaviour Estimation

The dominant failure modes were found to be matrix failure originating at different critical locations of the higher strain concentration region and the transverse splitting of the chip.

The failure pattern for uniaxial tensile loading (load case: 1) is captured in three stages, i.e. damage initiation, propagation and ultimate failure of the composite, which is described below:

1. *Damage Initiation Phase*—The damage can be visualized to be initiated in the matrix phase at the chip corner and regions between consecutive chips. These regions are explained in detail in the critical region section. Simultaneously, the chip starts failing through the transverse failure which even enhances the matrix damage. This is mainly because the transverse strength of chip 48 MPa is even lower than the matrix yield strength 69 MPa. This can be compared in material properties table described earlier.
2. *Damage Propagation Phase*—Subsequently, interface matrix cracks between chips and near chip boundaries started to pile up and form crack fronts at random places in RVE.
3. *Ultimate Failure Phase*—Several such crack-fronts start accumulating due to the discontinuous spatial distribution of chips. The extensive merging of such

(a) (b)

Fig. 5 **a** Computational RVE failure, **b** experimental RVE failure [1, 2]

different crack-fronts creates a thorough RVE crack-based damage, causing the ultimate failure of the ROS-reinforced composite laminate.

The damage process can be visualized from Fig. 5a where matrix crack can be seen as propagating near chip interfaces. This RVE failure can be compared with the experimental testing of RVE with similar geometric dimensions.

5.2 Uniaxial Tensile Strength Prediction

This section discusses the strength prediction of RVE using the proposed damage model. This stress–strain curve is plotted by taking average nodal force behaviour at displacement loading faces. For the strength estimation, we took RVE with chip AR: 1 and a volume fraction of 32%. The material properties used for the constituents AS4/3501-6 are obtained from Tables 1 and 2. The volume fraction is limited due to the computational inefficiency of the RSA algorithm with constant aspect ratio chips.

Three RVE models with similar configurations are analyzed using tensile-loaded displacement boundary conditions. The remaining four faces were constrained along transverse direction displacements to maintain the pure displacement boundary condition. The chip's transverse tensile strength and the matrix's yield strength are quite similar and relatively low. Therefore, damage initiation was simultaneously caused by matrix failure and transverse chip splitting at the value of an average 70% of total deformation. The subsequent damage propagation in the matrix region and extensive accumulation of damage fronts causes a sudden drop in the stress–strain curve, causing the ultimate failure of composite as shown in Fig. 6. The strength prediction through computational tensile testing of composite is shown in Table 3 where 1 and 2 are in-plane directions. The estimated average strength is comparable to the in-house

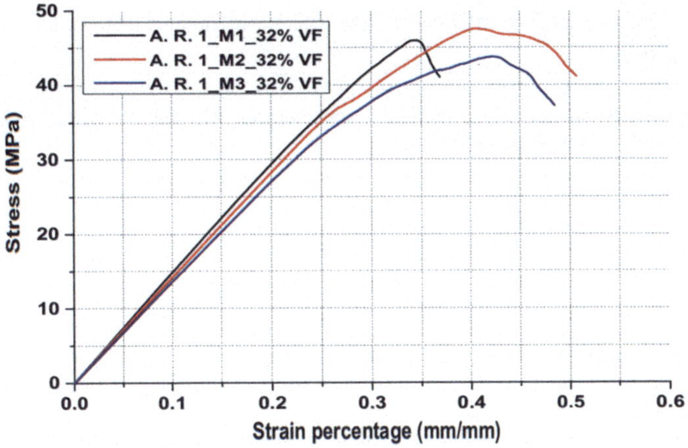

Fig. 6 Stress–strain plot of RVE with AR: 1 and V_f: 32%

Table 3 Strength properties and chip spatial distribution from different RVE models

Model	Stress (MPa)	Strain (%)	# chips1	# chips2	% chips1	% chips2
Model 1	36.29	0.51	17	13	56.67	43.33
Model 2	45.63	0.41	15	15	50	50
Model 3	40.1	0.43	20	10	66.67	33.33
Mean	**40.67**	**0.45**	17.33	12.67	57.78	42.22
SD	4.7	0.05	2.52	2.52	8.39	8.39
COV	11.54	11.16	14.54	19.89	14.52	19.87

tensile test of compression moulded chopped chips reinforced discontinuous composites. A detailed description of manufacturing is found in Mohan et al. [1, 2]. The proposed continuum damage model can be extended to predict the compressive and pure shear failure and strength parameters by incorporating additional constraints.

6 Conclusion

The main objective of this micromechanical analysis is to predict the strength and failure behaviour of randomly oriented strands (ROS) reinforced prepreg-based discontinuous composite through the development of a continuum damage model. Matrix failure and transverse chip splitting were considered the dominant failure modes. The damage initiation hotpots were studied through mid-plane strain concentration plots showing hotspot regions at chip corners, chip–matrix interfaces and boundary regions. The stress–strain behaviour through averaging of internal forces depicts the brittle behaviour of ROS-reinforced composite with linear nature before achieving the ultimate stress.

Acknowledgements We thank and gratefully acknowledge AR&DB for partial support under research project ARDB/01/1051932/M/I.

References

1. Mohan PRK, Kumar MA, Kumar SG, Mohite PM (2019) Development of in-house unidirectional carbon/epoxy prepregs and its characterization for aerospace applications. Procedia Struct Integr 14:176–183
2. Krishna Mohan PR, Piyush, Mohite PM, Modi K, Sharma D (2022) Characterisation of composites made by prepreg waste. In: Kumar R, Chauhan VS, Talha M, Pathak H (eds) Machines, mechanism and robotics. Lecture notes in mechanical engineering, LNME. Springer, Singapore, pp 635–642
3. Sute P, Krishna Mohan PR, Anil Kumar M, Mohite PM (2022) Characterization of composites made with in-house prepregs at different curing cycle. In: Kumar R, Chauhan VS, Talha M, Pathak H (eds) Machines, mechanism and robotics. Lecture notes in mechanical engineering, LNME. Springer, Singapore, pp 733–742
4. Mohan PRK, Kumar MA, Mohite PM (2020) Representative volume element generation and its size determination for discontinuous composites made from chopped prepregs. Compos Struct 252:112633
5. Mori T, Tanaka K (1973) Average stress in matrix and average elastic energy of materials with misfitting inclusions. Acta Metall 21(5):571–574
6. Li Y, Chen Z, Su L, Chen W, Jin X, Xu H (2018) Stochastic reconstruction and microstructure modeling of SMC chopped fiber composites. Compos Struct 200:153–164
7. Soden PD, Hinton MJ, Kaddour AS (1998) Lamina properties, lay-up configurations and loading conditions for a range of fibre-reinforced composite laminates. Compos Sci Technol 58:1011–1022
8. Tsai SW, Wu EM (1971) A general theory of strength for anisotropic materials. J Compos Mater 5(1):58–80

Analysis of Bistable Arches Connected at the Centre with Pinned Boundary Conditions

Rajat Goswami, Unnikrishnan, and Safvan Palathingal

Abstract Interconnected arches are conceived by fusing two or more planar arches at their centres; like the ribs of an umbrella. They offer three-dimensional structural robustness of shells and analytical simplicity of planar arches. Furthermore, just as an umbrella flips upside down in strong winds, interconnected arches are observed to be bistable. In this paper, a shallow interconnected bistable arch with pinned boundary conditions is subjected to a transverse load at the centre. We analytically model the resulting post-buckling behaviour and describe it by obtaining the force–displacement characteristics. The analytical model accounts for bending, compression, and torsional energies developed in the interconnected arch as it switches between its stable states. The bistable behaviour and dependence of the geometric parameters of the connected arch predicted by the analytical model are validated with the results from nonlinear finite element analysis and table-top experiments.

Keywords Bistability · Snap-through · Buckling · Compliant · Flexible

1 Introduction

Bistable structures have two stable force-free equilibrium states. They find applications wherein two distinct configurations need to be maintained without constantly applying power, such as switches and grippers [1, 2, 5].

Arches and shells are the two known compliant monolithic structures that can be designed to be bistable solely due to their as-fabricated shape. Bistable arches are well studied in the literature [3, 4, 7–10]. Their analytical models and closed-form solutions that facilitate their design and analysis are also known. On the other hand, analytical modelling of bistable shells is mathematically involving [12] and is not as widely studied. However, bistable shells offer better practical utility in terms of their robustness and versatility when used in engineering applications.

R. Goswami · Unnikrishnan · S. Palathingal (✉)
Department of Mechanical and Aerospace Engineering, Indian Institute of Technology Hyderabad, Kandi, Telangana, India
e-mail: safvan@mae.iith.ac.in

S. Sidhardh et al. (eds.), *Advances in Structural Integrity for Mechanical, Civil, and Aerospace Applications*, Lecture Notes in Mechanical Engineering,
https://doi.org/10.1007/978-981-97-6367-2_24

In this work, we present a third form of bistable structure by interconnecting bistable arches. Interconnected bistable arches are formed by combining two or more planar arches just like the ribs of an umbrella at their centre. Thus, they are truly three-dimensional (3D) unlike bistable arches. They offer three-dimensional structural robustness of shells and analytical simplicity of arches. Prototypes based on interconnected arches are already conceived to be used as soft grippers for application in space programs [6]. We develop an analytical model of the interconnected bistable arch with 4 arms and hinged boundary conditions. We do this by modifying an analytical model used for capturing the mechanics of shallow spatial arches [9]. The post-buckling deformation pathways obtained from the analysis of connected arches is compared to the results from nonlinear finite element analysis (FEA) and table-top experiments. The interconnected arches used for the experiments are 3D-printed, and its force–deformation characteristics is obtained by using a load cell.

The rest of the paper is organized as follows: Sect. 2 presents the analysis of a spatial bistable arch with a concentrated load acting at the centre. A brief summary of the analysis of planar bistable arches is given in Sect. 2.1. The additional energy terms required to model interconnected bistable arches as compared to planar arches is given in Sect. 2.2. The post-buckling analysis for connected arches is given in Sect. 2.3. Here, we discuss the necessary conditions for bistability and the analytical relation between the derived mode weights used for approximating the deformed profile. The results from the model is verified using FEA in Sect. 3.1 and by using experiments in Sect. 3.2. The summary of the work is given in the last section.

2 Analysis of Bistable Arches Connected at the Centre

We consider a 4-arm arch connected at the centre, which deforms spatially in the 3D space. The 4-arm arch maybe viewed as two planar arches in two orthogonal planes fused at their midspan. We first summarize analysis of planar arches [7] by using energy method before extending the planar model to 3D space.

2.1 *Summary of Analysis for Planar Arches*

Let the as-fabricated profile of arch with thickness t, span L, and width b, is given by $w_0(x)$ as shown in Fig. 1.

The height at the mid-span is h_{mid}, and the deformed arch profile is denoted by $w(x)$. By taking a_i as the mode weight, we can approximate $w_0(x)$ as

$$w_0(x) = \sum_{i=1}^{n} a_i \sin\left(i\pi \frac{x}{L}\right), \tag{1}$$

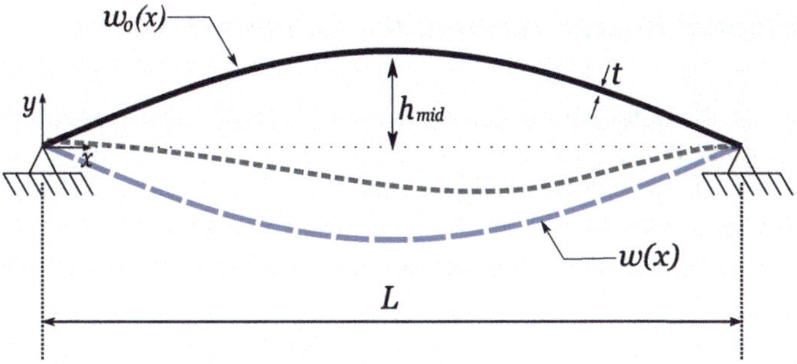

Fig. 1 Schematic of a shallow planar bistable arch pinned at its ends

where n is the number of modes considered to approximate $w_0(x)$ and x varies between 0 and L. Similarly, the deformed shape of the arch is given by

$$w(x) = \sum_{i=1}^{m} A_i \sin\left(i\pi\frac{x}{L}\right), \tag{2}$$

where A_i is the mode weight for the deformed arch profile and m is the number of modes shapes used for approximating the deformed profile.

For a concentrated transverse load acting at the mid-point of the arch, the total potential energy of the planar arch is [7]:

$$\begin{aligned} \mathrm{PE} = {} & \frac{EI}{2}\int_0^L \left(\frac{\mathrm{d}^2 w_0}{\mathrm{d}x^2} - \frac{\mathrm{d}^2 w}{\mathrm{d}x^2}\right)^2 \mathrm{d}x \\ & + \frac{Ebt}{8L}\left\{\int_0^L \left[\left(\frac{\mathrm{d}w_0}{\mathrm{d}x}\right)^2 - \left(\frac{\mathrm{d}w}{\mathrm{d}x}\right)^2\right]\mathrm{d}x\right\}^2 - F\delta_y, \end{aligned} \tag{3}$$

where δ_y is the displacement of the centre point of the arch and is given by

$$\delta_y = w_0\left(\frac{L}{2}\right) - w\left(\frac{L}{2}\right). \tag{4}$$

Equation (3) comprises strain energy due to bending in the arch, compression along the length of the arch, and work potential due to the external transverse force F applied at the mid-point. The equilibrium equations can be obtained as

$$\frac{\partial \mathrm{PE}}{\partial A_i} = 0, \quad \text{where } i = 1, 2, 3, \ldots m \tag{5}$$

2.2 Additional Energy Terms in the Connected Arches

The validity of the potential energy expression given by Eq. (3) is limited to planar deformations. While writing the potential energy in the case of interconnected bistable arches, along with the energy terms (Eq. 3) in their respective planes, we also consider strain energy terms arising from out-of-plane bending and torsion. This is because the interconnected arch deforms spatially, i.e., exhibits out-of-the-plane deformation.

We consider the shallow interconnected arch shown in Fig. 2. Let us denote the projections of the central axis of the as-fabricated profile of the arches (the arch in the xy plane) in the xy and xz planes as $w_0(x)$ and $u_0(x)$, respectively. The projections of the arch's deformed centre line in the xy and xz planes are $w(x)$ and $u(x)$, respectively. The cross section's initial twist is zero, and $\phi(x)$ represents its rotational displacement as the arch deforms. By denoting the curvatures of $w_0(x)$, and $u_0(x)$ as κ_{xy0}, and κ_{xz0}, respectively, using the extended St. Venant and Michell's relationship, the curvatures and rotational displacement are related by [9]

$$\Delta\kappa_{xz} = \frac{\mathrm{d}^2 u}{\mathrm{d}x^2} - \left(\kappa_{xz0} + \phi\kappa_{xy0}\right), \tag{6}$$

and

$$\Delta\kappa_{xy} = \frac{\mathrm{d}^2 w}{\mathrm{d}x^2} - \left(\kappa_{xy0} - \phi\kappa_{xz0}\right). \tag{7}$$

Additionally, Eqs. (6) and (7) are expanded by substituting $\kappa_{xz0} = \mathrm{d}^2 u_0/\mathrm{d}x^2$, $\kappa_{xy0} = \mathrm{d}^2 w_0/\mathrm{d}x^2$, assuming ϕ to be small (i.e. sin $(\phi) \approx \phi$) and simplified by disregarding higher-order terms as

$$\Delta\kappa_{xz} = \frac{\mathrm{d}^2 u}{\mathrm{d}x^2} - \left(\kappa_{xz0} + \phi\kappa_{xy0}\right) = \frac{\mathrm{d}^2 u}{\mathrm{d}x^2} - \frac{\mathrm{d}^2 u_0}{\mathrm{d}x^2} - \phi\frac{\mathrm{d}^2 w_0}{\mathrm{d}x^2}, \tag{8}$$

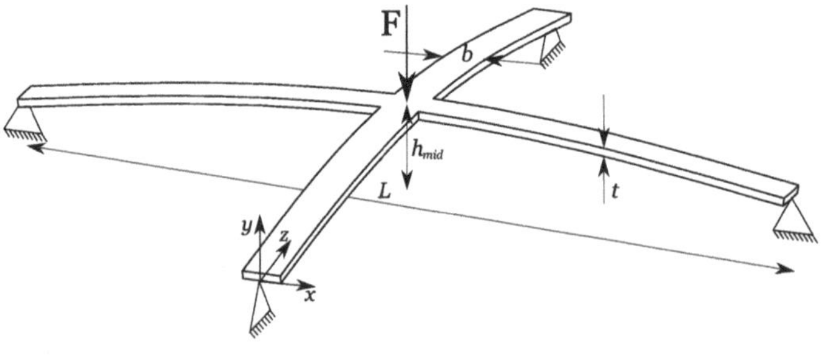

Fig. 2 Schematic of an interconnected arch formed by joining two arches at the centre with hinged boundary conditions

and

$$\Delta\kappa_{xy} = \frac{\mathrm{d}^2 w}{\mathrm{d}x^2} - \left(\kappa_{xy0} - \phi\kappa_{xz0}\right) = \frac{\mathrm{d}^2 w}{\mathrm{d}x^2} - \frac{\mathrm{d}^2 w_0}{\mathrm{d}x^2} + \phi\frac{\mathrm{d}^2 u_0}{\mathrm{d}x^2}. \tag{9}$$

For a rectangular cross-sectional arch with the second moment of area about z-axis and y-axis, I_z and I_y and having Young's modulus of E, the strain energy due to bending is given by

$$\mathrm{SE_b} = \frac{EI_z}{2}\int_0^L \left(\Delta\kappa_{xy}\right)^2 \,\mathrm{d}x + \frac{EI_y}{2}\int_0^L (\Delta\kappa_{xz})^2 \,\mathrm{d}x. \tag{10}$$

By substituting Eqs. (8) and (9) in Eq. 10, the bending strain energy becomes

$$\begin{aligned}\mathrm{SE_b} = {} & \frac{EI_z}{2}\int_0^L \left(\frac{\mathrm{d}^2 w}{\mathrm{d}x^2} - \frac{\mathrm{d}^2 w_0}{\mathrm{d}x^2} + \phi\frac{\mathrm{d}^2 u_0}{\mathrm{d}x^2}\right)^2 \mathrm{d}x \\ & + \frac{EI_y}{2}\int_0^L \left(\frac{\mathrm{d}^2 u}{\mathrm{d}x^2} - \frac{\mathrm{d}^2 u_0}{\mathrm{d}x^2} - \phi\frac{\mathrm{d}^2 w_0}{\mathrm{d}x^2}\right)^2 \mathrm{d}x,\end{aligned} \tag{11}$$

which, along with in-plane bending, also accounts for strain energy due to out-of-plane bending. The rotation of the cross sections, ϕ, results in torsional strain energy given by

$$\mathrm{SE_t} = \frac{GJ}{2}\int_0^L \left(\frac{\mathrm{d}\phi}{\mathrm{d}x}\right)^2 \mathrm{d}x, \tag{12}$$

where G is the shear modulus and J is the polar moment of inertia.

2.3 *Post-buckling Analysis of Bistable Arches Connected at the Centre*

In order to write the potential energy of a deforming spatial arch, we approximate the expressions for the interconnected arch profile as

$$w_0(x) = h_{\mathrm{mid}}\sum_{i=1}^{\infty} a_i w_i(x), \tag{13}$$

$$u_0(x) = h_{\mathrm{mid}}\sum_{i=1}^{\infty} b_i u_i(x), \tag{14}$$

$$w(x) = h_{\text{mid}} \sum_{i=1}^{\infty} A_i w_i(x), \tag{15}$$

$$u(x) = h_{\text{mid}} \sum_{i=1}^{\infty} B_i u_i(x), \tag{16}$$

$$\phi(x) = \sum_{i=1}^{\infty} C_i \phi_i(x), \tag{17}$$

where w_i, u_i, and ϕ_i are the basis set formed to satisfy the boundary constraints of w, u, and ϕ, respectively. For given set of a_i and b_i, the mode weights A_i, B_i, and C_i of the deformed profile are determined by minimizing the potential energy. In this work, we take an arch with an in-plane initial shape of a first mode sine-curve and out-of-plane initial profile to be zero. We use three modes to approximate deformations in the plane and one mode for capturing deformations out of the plane. Thus, we have

$$w_0(x) = h_{\text{mid}} a_1 \sin\left(\frac{\pi x}{L}\right), \tag{18}$$

$$w(x) = h_{\text{mid}} \left[A_1 \sin\left(\frac{\pi x}{L}\right) + A_2 \sin\left(\frac{2\pi x}{L}\right) + A_3 \sin\left(\frac{3\pi x}{L}\right) \right], \tag{19}$$

$$u_0(x) = 0, \tag{20}$$

$$u(x) = h_{\text{mid}} B_1 \left[\frac{1}{2} - \frac{1}{2} \cos\left(\frac{2\pi x}{L}\right) \right], \tag{21}$$

$$\phi(x) = C_1 \left[\frac{1}{2} - \frac{1}{2} \cos\left(\frac{2\pi x}{L}\right) \right]. \tag{22}$$

As the shallow arch deforms, compression energy is built up in the arch due to the axial force, p, denoted by SE_c and is given by

$$\text{SE}_\text{c} = \frac{1}{2} p \left(s_{\text{initial}} - s \right), \tag{23}$$

where s_{initial} is the arc length of the arch in the initial profile, and s is the arch's length as it deforms. For linear elastic material

$$p = Ebt \left(\frac{s_{\text{initial}} - s}{L} \right) \tag{24}$$

$$s = \int_0^L \sqrt{1 + \left(\frac{dw}{dx}\right)^2 + \left(\frac{du}{dx}\right)^2}\, dx \tag{25}$$

As arch is considered to be shallow, i.e. $(dw_0/dx)^2 \ll 1$ and $(du_0/dx)^2 \ll 1$, s and s_{initial} is approximated as

$$s \approx \int_0^L \left[1 + \frac{1}{2}\left(\frac{dw}{dx}\right)^2 + \frac{1}{2}\left(\frac{du}{dx}\right)^2\right] dx \tag{26}$$

$$s_{\text{initial}} \approx \int_0^L \left[1 + \frac{1}{2}\left(\frac{dw_0}{dx}\right)^2 + \frac{1}{2}\left(\frac{du_0}{dx}\right)^2\right] dx \tag{27}$$

The centre of the arch is subjected to a point force F. The midpoint's displacements along the y-axis, as well as the work potential caused by those displacements, are given by

$$\delta_y = w_0\left(\frac{L}{2}\right) - w\left(\frac{L}{2}\right) \tag{28}$$

$$= \frac{A_3}{200} - \frac{A_1}{200} + \frac{1}{200}, \tag{29}$$

and

$$\text{WP} = -F\delta_y. \tag{30}$$

Equations (10), (12), and (23) describe strain energies due to the deformation of the arch that is in the xy plane, which is only one of the connected arch. We capture the strain energy in the other arch by observing the kinematic relation at the centre. At any instant, at the centre of the arch, the slope of the deformed profile of one arch in its plane is equal to the rotational displacement of the other arch and vice versa. Therefore, at the mid-span of the arch, we have the condition

$$\phi\left(\frac{L}{2}\right) = \left.\frac{d\omega}{dx}\right|_{\frac{L}{2}}. \tag{31}$$

The total potential energy, PE, includes bending, compression, and torsional energies, of both the arches and the work potential. Thus, the total potential energy is given by

$$\text{PE} = 2\text{SE}_b + 2\text{SE}_c + 2\text{SE}_t + \text{WP}. \tag{32}$$

Substituting the respective components of strain energies and work potential, Eqs. (18)–(22) in Eq. (32), potential energy expression yields as

$$\begin{aligned}\mathrm{PE} = {} & \frac{2\pi^4 A_2^2 G J h_{\mathrm{mid}}^2}{L^3} - F h_{\mathrm{mid}} (1 - A_1 + A_3) \\ & + \frac{E\pi^4 \left(A_1^2 - 2A_1 + 16A_2^2 + 81A_3^2 + 1\right) I_x h_{\mathrm{mid}}^2}{2L^3} \\ & + \frac{\pi^4 b E t \left(A_1^2 + 4A_2^2 + 9A_3^2 + B_1^2 - 1\right)^2 h_{\mathrm{mid}}^4}{16L^3} \\ & + \frac{\pi^4 E I_y h_{\mathrm{mid}}^2 \left(128 A_2 B_1 L h_{\mathrm{mid}} + 25\pi^2 A_2^2 h_{\mathrm{mid}}^2 + 40 B_1^2 L^2\right)}{20L^5}.\end{aligned} \tag{33}$$

Note that Eq. (31) is used in eliminating C_1 to write the potential energy expression given by Eq. (33). The equilibrium equations are obtained by minimizing the potential energy with respect to the unknown mode weights, A_1, A_2, A_3, and B_1 as

$$\begin{aligned}\frac{\mathrm{dPE}}{\mathrm{d}A_i} &= 0, \qquad i = 1, 2, 3, \\ \frac{\mathrm{dPE}}{\mathrm{d}B_1} &= 0.\end{aligned} \tag{34}$$

Thus, we have,

$$\begin{aligned}\frac{\mathrm{dPE}}{\mathrm{d}A_1} = {} & h_{\mathrm{mid}} \left(\frac{\pi^4 (A_1 - 1) E I_x h_{\mathrm{mid}}}{L^3} + F\right) \\ & + \frac{\pi^4 A_1 b E t \left(A_1^2 + 4A_2^2 + 9A_3^2 + B_1^2 - 1\right) h_{\mathrm{mid}}^4}{4L^3} = 0,\end{aligned} \tag{35}$$

$$\begin{aligned}\frac{\mathrm{dPE}}{\mathrm{d}A_2} = {} & \frac{4\pi^4 A_2 G J h_{\mathrm{mid}}^2}{L^3} \\ & + \frac{\pi^4 E h_{\mathrm{mid}}^2 \left(5A_2 \left(5\pi^2 I_y h_{\mathrm{mid}}^2 + 32 I_x L^2\right) + 64 B_1 I_y L h_{\mathrm{mid}}\right)}{10L^5} \\ & + \frac{\pi^4 A_2 b E t \left(A_1^2 + 4A_2^2 + 9A_3^2 + B_1^2 - 1\right) h_{\mathrm{mid}}^4}{L^3} = 0,\end{aligned} \tag{36}$$

$$\begin{aligned}\frac{\mathrm{dPE}}{\mathrm{d}A_3} = {} & \frac{81\pi^4 A_3 E I_x h_{\mathrm{mid}}^2}{L^3} - F h_{\mathrm{mid}} \\ & + \frac{9\pi^4 A_3 b E t \left(A_1^2 + 4A_2^2 + 9A_3^2 + B_1^2 - 1\right) h_{\mathrm{mid}}^4}{4L^3} = 0,\end{aligned} \tag{37}$$

$$\begin{aligned}\frac{\mathrm{dPE}}{\mathrm{d}B_1} = {} & \frac{4\pi^4 E I_y h_{\mathrm{mid}}^2 (8A_2 h_{\mathrm{mid}} + 5B_1 L)}{5L^4} \\ & + \frac{\pi^4 b B_1 E t \left(A_1^2 + 4A_2^2 + 9A_3^2 + B_1^2 - 1\right) h_{\mathrm{mid}}^4}{4L^3} = 0.\end{aligned} \tag{38}$$

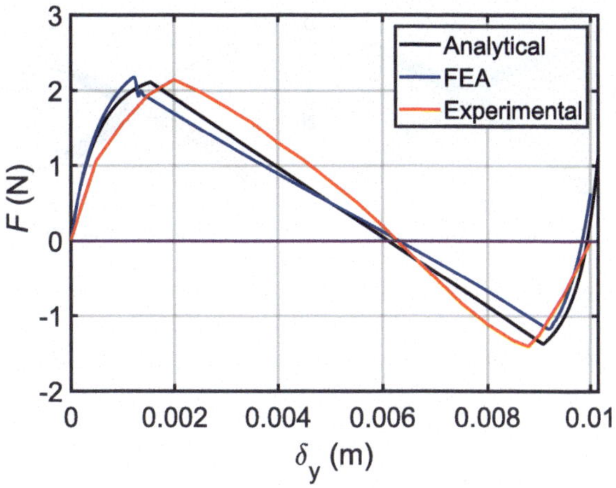

Fig. 3 Comparison of the force–displacement curves from analytical modelling (black), FEA (blue), and experiments (red)

The unknown mode weights A_2, A_3, and B_1, and force F are determined in terms of A_1 by simultaneously solving Eqs. (35)–(38). Thereafter, variation of F versus δ_y (Eq. 29) is plotted (black curve in Fig. 3) by varying A_1 from 1 to -1.

3 Validation of Analytical Results

Here, the analytical modelling described in the last section is compared to the results from FEA and preliminary tabletop experiments.

3.1 Finite Element Analysis

By using the ABAQUS/CAE 2018 software [11], FEA is carried out for the 4-arms interconnected bistable arch, its deformation is studied, and the force–displacement curve is obtained. As mentioned before, the initial arch profile is taken to be a sine curve, and the other geometrical parameters of the arch are given in Table 1.

Table 1 Geometrical parameters of the sine profile interconnected arch

Parameter	Value (mm)
Length (L)	100
Width (b)	5
Thickness (t)	1
Midrise (h)	5

(a) initial profile (b)intermediate profile (c)toggled profile

Fig. 4 Deformation of interconnected bistable arch from the FEA simulation in ABAQUS

Table 2 Properties of the black resin material used for 3D-printing

Property	Value
Tensile modulus	2.16 GPa
Poisson's ratio	0.35

The geometrically nonlinear FEA is implemented by using the dynamic solver available in ABAQUS implicitly by using a quasi-static displacement control. The model is meshed with the 8-noded linear brick element (C3D8R) that has reduced order integration and hourglass control. The deformation of the interconnected bistable arch from the FEA simulation is shown in Fig. 4. The force–displacement curve obtained from the simulations is included in Fig. 3.

3.2 Experiments

In order to obtain force–displacement characteristics experimentally, an interconnected arch is 3D-printed with an elastic black resin material. The material properties of the material are listed in Table 2.

The Instron 5966 Universal Testing System with an Instron (50 N) load-cell mounted is used for conducting the force–displacement experiment. The pin-point load cell fixture is used to impart the force on the 3D-printed specimen at the rate of 1 mm/min, and the corresponding displacement values were recorded. Snapshots of the deformations of the interconnected arch are given in Fig. 5. The initial, intermediate, and toggled profiles are consistent with the ones observed during simulations (see Fig. 4). The force–displacement curve measured from the experiment is included in Fig. 3.

As mentioned before, the superimposed curve in blue is obtained from FEA and the one in red is measured from experiments. We observe that these two curves are in good agreement with the analytical force–displacement curve.

(a) initial profile (b)intermediate profile (c)toggled profile

Fig. 5 Experimental setup used for measuring the force–displacement curve of an interconnected 4-arms bistable arch

4 Summary and Discussion

An interconnected bistable arch is modelled by using analytical model for shallow spatial arches [9]. The initial in-plane profile of the arch is taken as a sine curve, and as it toggles to the second stable state, it deforms spatially, by taking an out-of-plane deformation that can be approximated by using cosine mode shapes. An appropriate number of modes is considered to capture the deformation of the arch. After accounting for the energy terms causing the in-plane and out-of-plane deformation and applying the principle of minimum potential energy, the force–displacement curve is obtained analytically.

The simulation of interconnected arch is done with geometrically nonlinear finite element analysis using a commercially available FEA package, ABAQUS/CAE 2018 software. The table-top experiment was performed on 3D-printed interconnected arch using black resin material with an Instron load cell setup. The results acquired analytically, from FEA simulations, and from table-top experiments are in good agreement with each other.

Acknowledgements We are grateful to the funding received for this work from the Department of Science and Technology, Government of India (SRG/2021/001463). We thank Prof Viswanath, Dr Sasidhar, Mr Raju, Ms Smita, and Mr Manu Sebastian for their help in the fabrication, setup, and conduct of the experiments.

References

1. Banik D, Palathingal S, Ananthasuresh G, Ghosh A (2021) A mechanical or gate using pinned-pinned bistable arches. In: Mechanism and machine science. Springer, pp 861–874
2. Chi Y, Li Y, Zhao Y, Hong Y, Tang Y, Yin J (2022) Bistable and multistable actuators for soft robots: structures, materials, and functionalities. Adv Mater 34(19):2110384
3. Fung YC, Kaplan A (1952) Buckling of low arches or curved beams of small curvature. National Advisory Committee for Aeronautics, Washington, DC
4. Hoff NJ, Bruce VG (1953) Dynamic analysis of the buckling of laterally loaded flat arches. Stud Appl Math 32(1–4):276–288

5. Hu N, Burgueño R (2015) Buckling-induced smart applications: recent advances and trends. Smart Mater Struct 24(6):063001
6. Liu Y, Luo K, Wang S, Song X, Zhang Z, Tian Q, Hu H (2022) A soft and bistable gripper with adjustable energy barrier for fast capture in space. Soft Robot. https://doi.org/10.1089/soro.2021.0147
7. Palathingal S, Ananthasuresh G (2017) Design of bistable arches by determining critical points in the force-displacement characteristic. Mech Mach Theory 117:175–188
8. Palathingal S, Ananthasuresh G (2018) A bilateral relationship between stable profiles of pinned-pinned bistable shallow arches. Int J Solids Struct 143:183–193
9. Palathingal S, Ananthasuresh G (2019) Analytical modelling of spatial deformation pathways in planar and spatial shallow bistable arches. Proc R Soc A 475(2227):20190164
10. Qiu J, Lang JH, Slocum AH (2004) A curved-beam bistable mechanism. J Microelectromech Syst 13(2):137–146
11. SIMULIA (2014) Abaqus 6.14 ABAQUS/CAE user's manual. Dassault Systèmes
12. Sobota P, Seffen K (2017) Effects of boundary conditions on bistable behaviour in axisymmetrical shallow shells. Proc R Soc A Math Phys Eng Sci 473(2203):20170230

A New Reduced Order Model of Soil-Structure Interaction Problem Using Deep Learning

Chandan Bharti and Debraj Ghosh

Abstract The computational cost in analyzing a dynamical system increases with the dimension of the model. This cost further increases where repeated solutions for different parameter values of the system are required, such as uncertainty quantification and optimization. In recent years, a proper orthogonal decomposition (POD)-based reduced order model (ROM) has been developed that reduces the cost significantly. This ROM is intrusive in nature, that is, it requires access to the computer codes used for solving the high-dimensional model. However, for many complex problems such as soil-structure interaction (SSI), such access is not available, which poses a limitation to the ROM. To address this issue, a non-intrusive ROM is developed in this work that does not require access to the source codes. The objective of this paper is to develop a non-intrusive ROM for SSI problems considering uncertainty in the excitations that can be further used for the uncertainty quantification of SSI under earthquake load or random excitations. This ROM is developed using deep neural networks. The application of neural networks makes the model learn features of an earthquake excitation efficiently. Finally, the accuracy of the proposed ROM is tested numerically on a beam on Winkler foundation.

Keywords Neural network · Reduced order model · Regression · Soil-structure interaction

1 Introduction

The computational cost of solving a large dynamical system is often very high and increases with the dimensions of the model. This cost further increases where repeated solutions of the system are required, such as uncertainty quantification and

C. Bharti (✉) · D. Ghosh
Department of Civil Engineering, Indian Institute of Science Bengaluru,
Bengaluru, Karnataka 560012, India
e-mail: cbharti@iisc.ac.in

D. Ghosh
e-mail: dghosh@iisc.ac.in

S. Sidhardh et al. (eds.), *Advances in Structural Integrity for Mechanical, Civil, and Aerospace Applications*, Lecture Notes in Mechanical Engineering,
https://doi.org/10.1007/978-981-97-6367-2_25

optimization [1, 2]. To address this issue, reduced order models (ROMs) are developed that reduce the computational cost greatly. Among many ROMs [3–9], the proper orthogonal decomposition (POD) based ROM has become very popular in recent years and is successfully used in solving many structural and fluid mechanics problems [3–5, 10–13]. In this ROM, first, a high dimension model (HDM) is solved for different parameter values, and a set of snapshots is selected. Then, a singular value decomposition is performed on this set, and a set of orthogonal bases—known as POD—is found. The ROM is finally found by projecting the HDM on these bases. This POD-based ROM is intrusive in nature, that is, it requires access to the original computer codes used for solving the HDM. However, for many complex problems such as soil-structure interaction, such access is not available. This inaccessibility poses a limitation to the intrusive ROM.

In this work, the aforementioned problem is addressed, and a non-intrusive ROM is developed that does not require access to the computer codes. In non-intrusive ROM [14–19], first, solutions of HDM are found for a set of parameter values. These solutions are then projected on a lower dimensional subspace, and the coefficients of projection are found. These coefficients are interpolated or regressed for the new parameter value in the lower dimensional subspace, and the new coefficients of projection are found. These new coefficients are again projected back to the original dimensions to find the solutions of HDM. Audouze et al. [14] proposed a non-intrusive ROM using a radial basis function (RBF) interpolation method for the static case. Xiao et al. [15] further extended this to time-dependent nonlinear partial differential equations using two-step RBF interpolation. The efficiency of these ROMs depends on the kernel of RBF, which depends on the problems at hand. Besides, RBF is applicable to problems where variation in the parameters is low. Guo et al. [16] presented a regression model using the Gaussian process. However, this is limited to static problems. This regression model is further extended to time-dependent problems by Guo et al. [18] using POD twice. This regression model, although good for uncertainty in system parameters, is not suitable where the excitations are random. In recent years, the application of artificial neural networks (ANN) as a regression model has caught significant attention. Hesthaven et al. [20] and Jacquier et al. [21] developed non-intrusive ROMs using ANN. These ROMs are developed for static problems and remain to be explored for dynamic cases.

All the above discussed non-intrusive ROMs are developed interpolating the system parameters, that is, uncertainty is considered in the material or geometric parameters of the system. However, for a large class of problems, the major source of uncertainty is random excitations or earthquake load. The existing ROMs' application in random excitations is nearly impractical. This issue is addressed in this paper, and a non-intrusive ROM is developed for random excitations using ANN. The application of ANN helps in efficiently predicting the features of an earthquake or random excitation. The accuracy of the proposed ROM is tested numerically on a beam on Winkler foundation. This foundation model has been successfully used to study the soil-structure interaction behavior in the literature.

The organization of the paper is as follows: Sect. 1 describes the introduction of POD-based non-intrusive ROMs and the current challenges with the existing ROMs.

Section 2 gives a brief overview of POD and the construction of ROMs. The proposed ROM is described in Sect. 3. The efficiency of the proposed ROM is presented in Sect. 4. Finally, conclusions are made in Sect. 5.

2 POD Based ROM

In this section, a brief overview and general formulation of POD-based non-intrusive ROM are presented. In this ROM, first, a set of snapshots is collected by solving the HDM for different values of the parameters $\boldsymbol{\mu} \in \mathbb{R}^p$, where p is the dimension of the parameter space. These snapshots are then projected on a lower dimensional subspace using a set of orthogonal bases. Let this set be $\boldsymbol{\Phi}$, and $\mathbf{X}$ contains the snapshots. The projection of $\mathbf{X}$ on $\boldsymbol{\Phi}$ leads to

$$\mathbf{X}(\boldsymbol{\mu}, t) = \boldsymbol{\Phi}\mathbf{X}_r(\boldsymbol{\mu}, t) \tag{1}$$

where $\mathbf{X}_r(\boldsymbol{\mu}, t)$ is the coefficient of projection. $\mathbf{X}_r(\boldsymbol{\mu}, t)$ is also called reduced order solution. For a new parameter value $\boldsymbol{\mu}^*$, the $\mathbf{X}_r(\boldsymbol{\mu}^*, t)$ is found by the interpolation or regression. After finding $\mathbf{X}_r(\boldsymbol{\mu}^*, t)$, the HDM solution for the new parameter $\mathbf{X}(\boldsymbol{\mu}^*, t)$ is computed using Eq. (1).

The success of the ROM depends on two choices: (i) a set of bases and (ii) an interpolation or regression method. The choice of bases governs how efficiently the solution space can be captured in lower dimensional subspace, whereas the choice of interpolation or regression method governs the accuracy of the ROM. As discussed earlier, POD bases are the most suitable choice as these minimize the projection error of the snapshots in the L_2 norm among all other bases. In this paper, POD bases are used. Computation of POD bases is briefly described next.

2.1 *Computation of POD Bases*

Let the snapshots of solutions be collected in a matrix $\mathbf{X} \in \mathbb{R}^{N\times n}$.

$$\mathbf{X} = [\mathbf{x}_1 \mid \mathbf{x}_2 \mid \ldots \mid \mathbf{x}_n] \tag{2}$$

The POD bases $\boldsymbol{\Phi}$ then minimizes the following quantity

$$\sum_{i=1}^{n} \left\| \mathbf{x}_i - \boldsymbol{\Phi}\boldsymbol{\Phi}^T \mathbf{x}_i \right\|_2^2 \tag{3}$$

It is shown in [22] that the bases that minimize the expression given in Eq. (3) is the left singular vectors of $\mathbf{X}$. That is, POD bases can be found by performing singular value decomposition (SVD) on $\mathbf{X}$. The SVD of $\mathbf{X}$ leads to

$$\mathbf{X} = \mathbf{\Phi}\mathbf{\Sigma}\mathbf{\Psi}^T \tag{4}$$

where $\mathbf{\Phi}$ and $\mathbf{\Psi}$ contain the left and right singular vectors of $\mathbf{X}$, respectively and $\mathbf{\Sigma}$ is a diagonal matrix containing the singular values $\sigma_1 \geq \sigma_2 \geq \cdots \geq \sigma_n \geq 0$. This $\mathbf{\Phi}$ serves as a set of bases for the projection in Eq. (1). However, to construct an efficient ROM, singular values are truncated to the desired accuracy, and the corresponding left singular vectors are taken as POD bases. If m is the number of singular values that are retained after the truncation, then the error introduced due to truncation is given by

$$\epsilon_{\text{trunc}} = \frac{\sum_{i=m+1}^{n} \sigma_i^2}{\sum_{i=1}^{n} \sigma_i^2} \tag{5}$$

It was mentioned earlier that along with the choice of bases, the success of non-intrusive ROM depends on the interpolation or regression method. In this paper, a regression approach is taken to approximate the solution in the lower dimensional subspace. An artificial neural network is used as a tool for regression. The application of ANN makes the model learn the features of the parameters of the system more efficiently. The formulation of the proposed non-intrusive ROM with the application of ANN is described in detail in the next section.

3 The Proposed Non-intrusive ROM

As discussed earlier, existing non-intrusive ROMs are not suitable for excitation uncertainty. The existing nature of ROMs is practically very challenging as it requires interpolation or regression in the high-dimensional space. In order to understand why random excitations require interpolation in high-dimensional space, let us consider the generation of a random excitation from a given power spectral density function G. The random excitation can be generated using the following equation [23–25].

$$f(t) = \sum_{i=1}^{N_\omega} \sqrt{\frac{\Delta\omega}{\pi} G(\omega_i)\eta_i} \sin(\omega_i t + \phi_i) = \sum_{i=1}^{N_\omega} A_i \sin(\omega_i t + \phi_i) \tag{6}$$

where $A_i = \sqrt{\frac{\Delta\omega}{\pi} G(\omega_i)\eta_i}$, ω_i and ϕ_i, $i = 1, 2, \ldots, N_\omega$ are the angular frequency and phase of the corresponding harmonic excitation, $\Delta\omega$ is the discretization interval between two consecutive sampling frequencies, and $f(t)$ is the generated random excitation. η_i is a standard Gaussian random variable. The ϕ_i is generated randomly from a known probability distribution function. N_ω is the number of sample points, which is chosen very high—a few thousand—to generate the random excitation accurately. These A_i and ϕ_i can be considered as parameters of the excitation. It is evident from Eq. (6) that a large number of parameters are required to represent such excitation, and interpolation or regression among these many parameters using traditional methods is practically not feasible. To address this issue, an artificial neural network is used for the regression in the current work. The ANN is found to be very effective in handling the high parametric dimensions.

It is necessary to emphasize here that the computational gain in ROM is achieved by the *offline–online* framework. In the offline stage, a ROM is developed using different training points, which is then used to evaluate the HDM solutions for new parameter values in the online stage. The online stage is independent of the dimension of the HDM and hence computationally very efficient. The proposed method is formulated in the same framework and described next.

3.1 Formulation of the Proposed Method

Let $\mathbf{X}^{[j]} = [\mathbf{x}_1^j \mid \mathbf{x}_2^j \mid \ldots \mid \mathbf{x}_n^j]$ denote the HDM solution for the parameter value $\boldsymbol{\mu}_j$, where $\mathbf{x}_t^j$, $t = 1, 2, \ldots, n$ is the snapshot of the solution at time instant t for the parameter $\boldsymbol{\mu}_j$. Let us further assume that the HDM solutions are available for different training points $\boldsymbol{\mu}_1, \boldsymbol{\mu}_2, \ldots, \boldsymbol{\mu}_{N_{\text{tp}}}$. The snapshot matrix $\mathbf{X}$ is then given by

$$\mathbf{X} = [\mathbf{X}^{[1]} \mid \mathbf{X}^{[2]} \mid \ldots \mid \mathbf{X}^{[N_{\text{tp}}]}] \tag{7}$$

where N_{tp} is the number of training points. The SVD of $\mathbf{X}$ leads to

$$\mathbf{X} = \boldsymbol{\Phi}\boldsymbol{\Sigma}\boldsymbol{\Psi}^T \tag{8}$$

The projection of HDM solution for each training parameter on $\boldsymbol{\Phi}$ yields

$$\mathbf{X}^{[j]} = \boldsymbol{\Phi}\mathbf{X}_r^{[j]}, \qquad j = 1, 2, \ldots, N_{\text{tp}} \tag{9}$$

where $\mathbf{X}_r \in \mathbb{R}^{m \times n}$ is the coefficient of projection and $m \ll N$ is the dimension of the lower dimensional subspace. The dimension m can be found using Eq. (5) based on truncation error.

In the regression-based approach, we need to find an approximation $\hat{\mathbf{X}}_r(\boldsymbol{\mu})$ for the projection coefficients $\mathbf{X}_r(\boldsymbol{\mu})$ that will be evaluated for the new value of parameters in the online stage. In this paper, $\mathbf{X}_r$ is approximated by approximating each component of $\mathbf{X}_r$. It should be noted here that each component of $\mathbf{X}_r$ is a time history. Let the component of $\mathbf{X}_r$ be $\mathbf{X}_r^{(i)}$, $i = 1, 2, \ldots, m$. Since the same procedure is adopted for the approximation of each component, the approximation of only one component using regression is described. The approximation of other components follows the same procedure. In the following section, the ith component of $\mathbf{X}_r$ is denoted by $\mathbf{y} \in \mathbb{R}^n$, that is, $\mathbf{y} = \mathbf{X}_r^{(i)}$, $i = 1, 2, \ldots, m$.

Regression of $\mathbf{y}$. The purpose of regression is to find a function $\hat{\mathbf{y}}(\boldsymbol{\mu})$ that will approximate a given function $\mathbf{y}(\boldsymbol{\mu})$. In regression, $\hat{\mathbf{y}}(\boldsymbol{\mu})$ is found using the known data set of (*arg, value*), where *arg* is the parameter where the true value of $\mathbf{y}(\boldsymbol{\mu})$ is known, and *value* is the true value of $\mathbf{y}(\boldsymbol{\mu})$. This data set is also known as the training data set. In order to approximate $\mathbf{y}(\boldsymbol{\mu})$, we first need to prepare the data set. The procedure for preparing the data set from the HDM solution is described below.

Preparation of Data Set. Let us assume that HDM solutions are given for different values of the parameter of the system. Let $\mathbf{X}^{[j]}$ be the HDM solution corresponding to parameter $\boldsymbol{\mu}_j$. Using Eq. (9), $\mathbf{X}_r^{[j]}$ can be found as

$$\mathbf{X}_r^{[j]} = \boldsymbol{\Phi}^T \mathbf{X}^{[j]} \tag{10}$$

From Eq. (10), the data set $(\boldsymbol{\mu}_j, \mathbf{X}_r^{[j](i)}) = (\boldsymbol{\mu}_j, \mathbf{y}^{[j]})$, $j = 1, 2, \ldots, N_{\text{tp}}$ is prepared for each component of $\mathbf{X}_r$.

ANN as Regression. After preparing the data set, the approximation of $\mathbf{y}$ is found using the regression approach. As discussed earlier, ANN is used as a tool for the regression. In this paper, a feed-forward fully connected neural network is used. The input to the ANN is the random excitation, and output is $\mathbf{y}$ found using Eq. (10). Although the implementation of ANN is simple; it is computationally very expensive if input and output data are used in their raw form. Since input and output are the time histories, ANN will require a very large number of parameters to be trained, and it will be computationally very expensive. To alleviate these issues, the training data (input and output) are first compressed using principal component analysis before feeding them to the network. The purpose of compressing the data is to extract the important features of the data and leave the less important features. Besides, the compression of data reduces the dimension of the data and accordingly reduces the number of parameters to be learned during the training of the neural network. The architecture of the network is shown in Fig. 1.

After finding the approximation of each component of $\mathbf{X}_r$, the HDM solution for a new parameter $\boldsymbol{\mu}^*$ is found as

$$\mathbf{X}(\boldsymbol{\mu}^*) = \boldsymbol{\Phi} \hat{\mathbf{X}}_r(\boldsymbol{\mu}^*) \tag{11}$$

Details of ANN. A feed-forward, fully connected neural network is adopted, consisting of an input, hidden, and output layer. This kind of architecture is widely used in establishing an accurate mapping between high-dimensional spaces within the tractable computational resources. The network considered here uses the Sigmoid activation function with no regularization. The output layer is linear. The loss function is chosen as the mean square error. The objective of an ANN is to minimize the loss function with respect to weights $\mathbf{W}$ and biases $\mathbf{b}$ associated with each layer. The number of hidden layers is decided using the trial and error method. In this work, four layers are chosen for the first example, and three layers are chosen for the second example. If we decrease the number of layers or neurons in the layers, the network may become overfit, which will increase the generalization error.

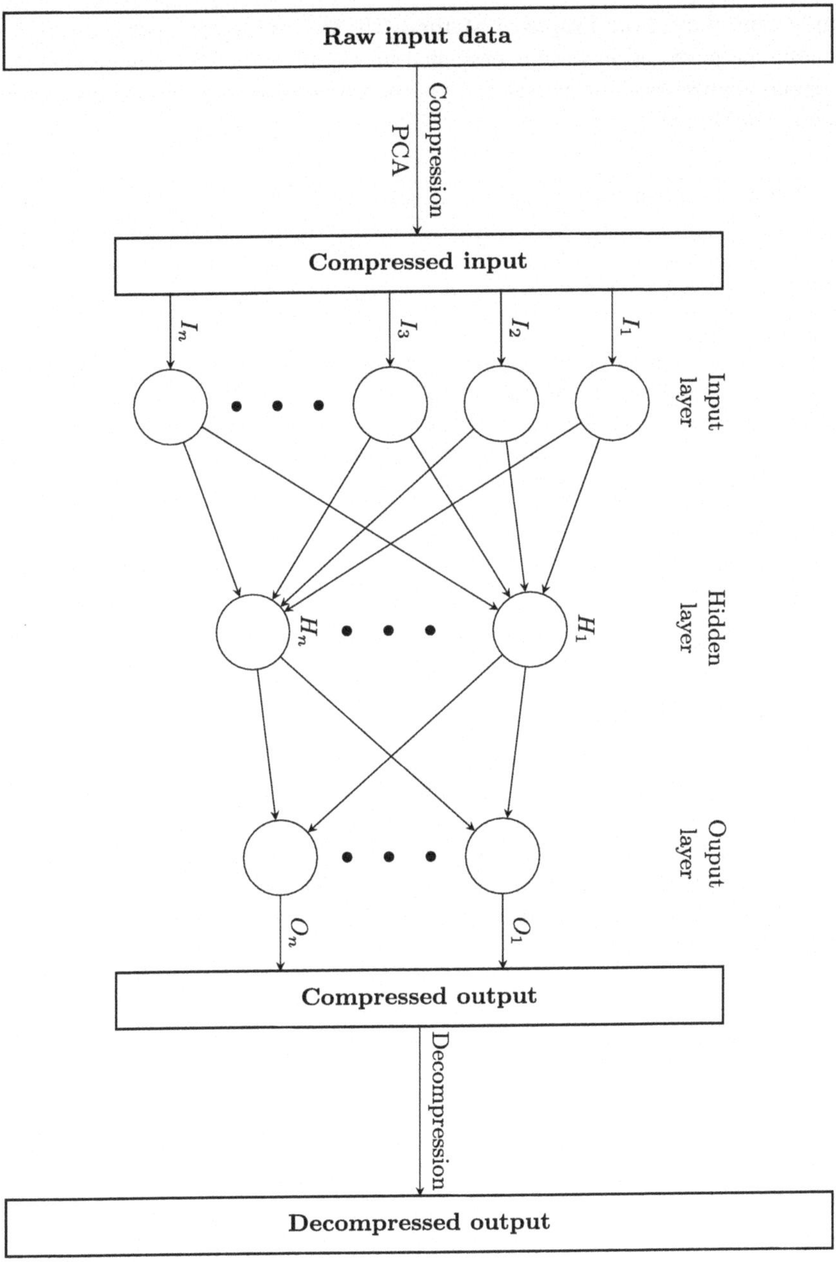

Fig. 1 Architecture of the artificial neural network

Implementation of the Proposed Method. The step by step procedure for implementing the proposed method is presented in Algorithm 1. All the numerical calculations are performed in Matlab 2021a. The neural network is trained using *train* function in Matlab.

Algorithm 1 Implementation of the proposed method

Input: A set of training parameters $\boldsymbol{\mu} = \{\mu_1, \mu_2 \ldots, \mu_{N_{tp}}\}$, HDM solution $\mathbf{X}^{[j]}$ for each training parameter $\boldsymbol{\mu}_j$.
Output: Regression based non-intrusive ROM

Offline stage:
1: Construct the matrix $\mathbf{X} = [\mathbf{X}^{[1]} \mid \mathbf{X}^{[2]} \mid \ldots \mid \mathbf{X}^{[N_{tp}]}]$.
2: Perform SVD on $\mathbf{X}$ and find the POD bases $\boldsymbol{\Phi}$.
3: Project each HDM solution $\mathbf{X}^{[j]}$ on $\boldsymbol{\Phi}$ and find the reduced order solution $\mathbf{X}_r^{[j]}$.
4: Prepare the data set $(\boldsymbol{\mu}_j, \mathbf{X}_r^{[j](i)})$ for each component of $\mathbf{X}_r$.
5: Find approximation for each $\mathbf{X}_r^{(i)}$ and finally find $\hat{\mathbf{X}}_r(\boldsymbol{\mu})$.

Online stage:
1: Evaluate $\hat{\mathbf{X}}_r(\boldsymbol{\mu})$ for the new parameter value $\boldsymbol{\mu}^*$.
2: Compute HDM solution for the new parameter value using Eq. (11).

4 Numerical Studies

In this section, the accuracy of the proposed non-intrusive ROM is numerically tested using two different examples. First, the accuracy of the ANN is tested for a single degree of freedom (SDOF) dynamical system, and then the efficiency of the proposed ROM is established for a beam on Winkler foundation. In both cases, the excitation is considered as the parameter. In this study, white noise band-limited random excitations are considered.

4.1 ANN for SDOF

The accuracy of ANN in predicting the response of SDOF is tested here. The governing equation is given as

$$m\ddot{x} + c\dot{x} + kx = f(t) \tag{12}$$

where m, c, and k are mass, damping, and stiffness of the system, respectively, $f(t)$ is the input force, and x is the response. A dot over x denotes the derivative with respect to time. In Eq. (12), m, c, and k are considered constants, and $f(t)$ is considered as random white noise excitation. The numerical values of the system parameters are as follows: $m = 5$ kg, $c = 3.54$ Ns/m and $k = 250$ N/m. The ANN consists of one

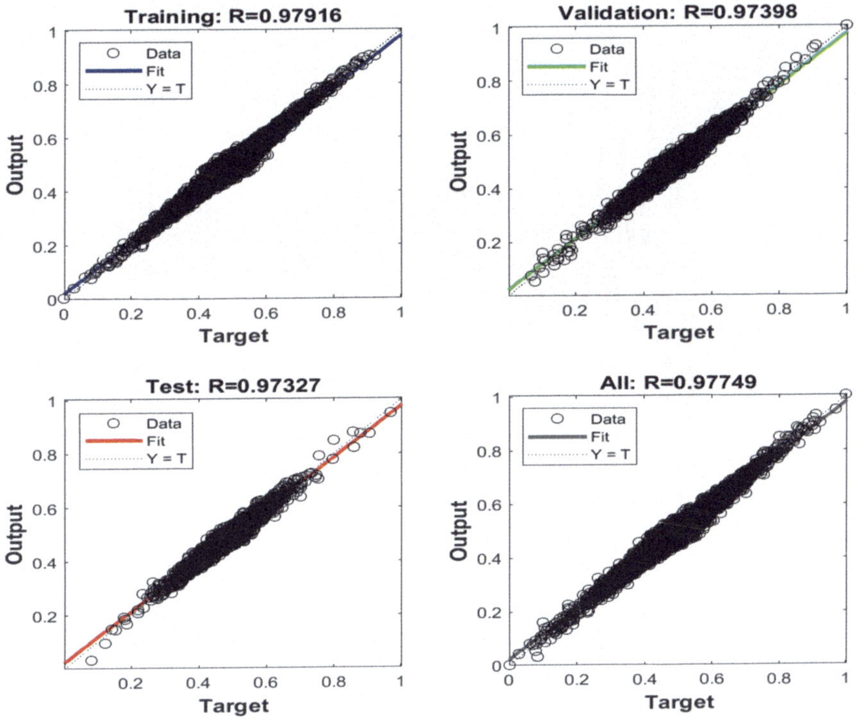

Fig. 2 Performance of the trained network for the SDOF. This network is trained using 2000 training points

input, four hidden, and one output layer. The number of neurons in hidden layers is 200, 200, 100 and 60. The number of neurons in the input layer and output layer is the same as the dimension of the input and output vector. The input to the ANN is random excitation $f(t)$ generated from band-limited white noise of band (15–20) rad/s, and output $x(t)$ is the response obtained by solving Eq. (12). The ANN is trained with 2000 training data points. The performance of the trained network is shown in Fig. 2. It can be observed in this figure that the R value of this network is around 97% for both training and testing points, and this network can be taken as a good model for regression. Further, the accuracy of this regression model is tested for a new excitation, and the comparison of the responses is presented in Fig. 3. It can be shown from this figure that ANN captures the true solution accurately.

4.2 *Beam on Winkler Foundation*

It is established in the previous example that the ANN captures the true solution accurately. In this section, the proposed ROM is tested for a beam on Winkler foundation (BWF). The BWF model is broadly used to study the soil-structure interaction

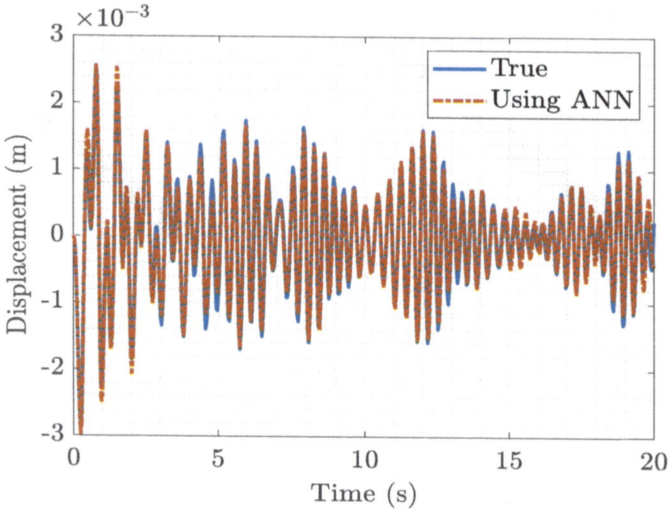

Fig. 3 Comparison of ANN response with true solution for the SDOF

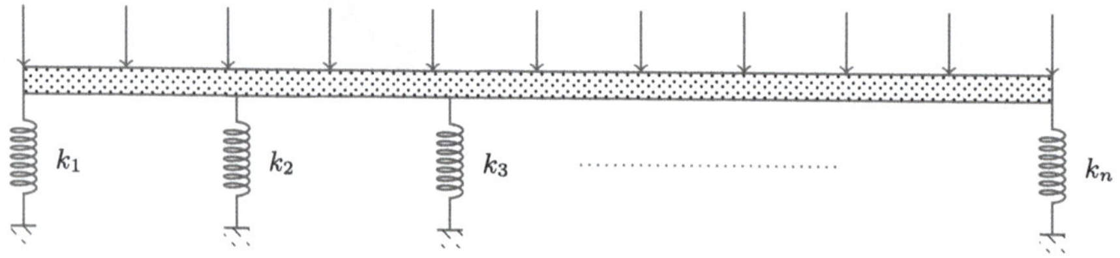

Fig. 4 Diagram of a beam on Winkler foundation model

behavior [26, 27]. This model consists of a beam resting on a set of springs, as shown in Fig. 4. The foundation is modeled by a beam, and the soil is replaced by a set of springs. The properties of the springs are governed by the properties of the soil. In this numerical study, an Euler-Bernoulli beam is considered, and the springs are assumed to be linear. The specifications of the beam are as follows: $L = 10$ m, $A_r = 0.09$ m^2, $\rho = 7800$ kg/m^3, $E = 210 \times 10^9$ Pa and $I = 1.2 \times 10^-6$ m^4, where L, A_r, ρ, E, and I are the length, cross-section area, density, modulus of elasticity, and area moment of inertia of the beam, respectively. The total stiffness of the springs is $3 \times 10^2 EI/L^3$ N/m.

The beam is excited by random signals generated from band-limited (15−20) rad/s white noise. A non-intrusive ROM is formed for this model by following the procedure given in Algorithm 1. In the offline stage, 4000 training samples were used to find the POD bases and to train the neural network. The network consists of 3 hidden layers with 140, 140, and 70 neurons, respectively. The ROM is tested for a new random excitation generated from the same band-limited white noise. The accuracy of the ROM in capturing the response is shown in Fig. 5. This figure

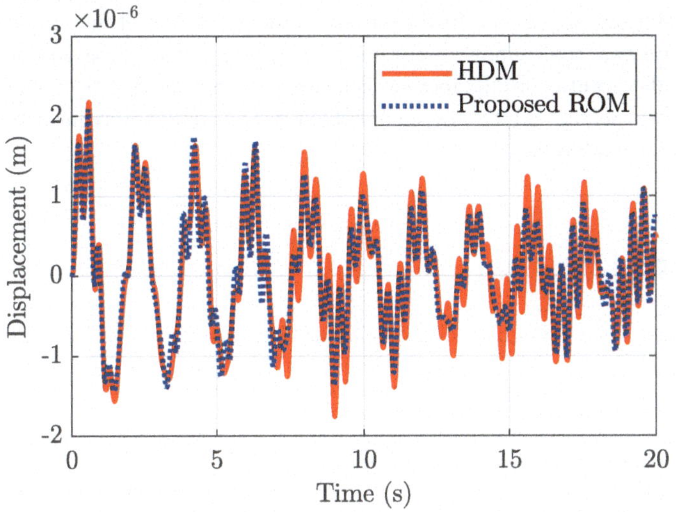

Fig. 5 Comparison of the proposed ROM with the true solution for the beam on Winkler foundation problem. Here true solution means the HDM solution

compares displacements at the left end of the beam obtained using the ROM and the HDM. It can be observed that the proposed ROM is able to capture the HDM solution accurately.

4.3 Discussion on the Accuracy of the Proposed Model

The accuracy of the proposed model depends on how accurately the ANN is trained. A well-trained network yields a low generalization error. If the error of the network is less, the proposed model will give accurate results. The generalization error depends on the number of training samples, its architecture, the type of activation functions, and its hyperparameters. Thus, selecting a suitable set of parameters is essential for reasonable accuracy. The numerical results show that for the chosen problems, the ANN architecture is sufficiently accurate. It is necessary to discuss here that the accuracy of the model is obtained for the narrow band of the excitation. However, its accuracy for the broad band excitation needs to be explored.

5 Conclusions

In this paper, a new non-intrusive ROM is proposed to address the uncertainty in the excitations. This novel ROM uses ANN for the regression of the reduced space solutions. It is observed through numerical studies that the proposed ROM captures the true solutions accurately. Besides it, the implementation of the proposed ROM is straightforward and requires only minor modifications to the existing codes. In this work, the accuracy of the method is tested for linear problems; however, many engineering problems exhibit nonlinear behavior. The applicability in such problems remains to be explored.

References

1. Quarteroni A, Rozza G (2014) Reduced order methods for modeling and computational reduction. Springer
2. Hesthaven J, Rozza G, Benjamin S (2016) Certified reduced basis methods for parametrized partial differential equations. Springer
3. Kunisch K, Volkwein S (2002) Galerkin proper orthogonal decomposition methods for a general equation in fluid dynamics. SIAM J Numer Anal 40(2):492–515
4. Pettit C, Beran P (2002) Application of proper orthogonal decomposition to discrete Euler equations. Int J Numer Methods Eng 55(4):479–497
5. Benner P, Gugercin S, Willcox K (2015) A survey of projection-based model reduction methods for parametric dynamical systems. SIAM Rev 57(4):483–531. https://doi.org/10.1137/130932715
6. Rowley C (2005) Model reduction for fluids, using balanced proper orthogonal decomposition. Int J Bifurcation Chaos 15(3):997–1013. https://doi.org/10.1142/S0218127405012429
7. Haasdonk B, Ohlberger M (2011) Efficient reduced models and a posteriori error estimation for parametrized dynamical systems by offline/online decomposition. Math Comput Model Dyn Syst 17(2):145–161
8. Hossain MN, Sarkar S, Ghosh D (2015) Identification of dominant modes in random dynamical and aeroelastic systems. J Sound Vib 357:128–144. https://doi.org/10.1016/j.jsv.2015.07.021
9. Billaud-Friess M, Nouy A (2017) Dynamical model reduction method for solving parameter-dependent dynamical systems. SIAM J Sci Comput 39(4):A1766–A1792
10. Taira K, Brunton S, Dawson S, Rowley C, Colonius T, McKeon B, Schmidt O, Gordeyev Sand Theofilis V, Ukeiley L (2017) Modal analysis of fluid flows: an overview. AIAA J, 4013–4041. https://doi.org/10.2514/1.J056060
11. Amsallem D, Farhat C (2008) Interpolation methods for adapting reduced order models and application to aeroelasticity. AIAA J 46(7):1803–1813. https://doi.org/10.2514/1.35374
12. Amsallem D, Zahr MJ, Farhat C (2012) Nonlinear model order reduction based on local reduced-order bases. Int J Numer Methods Eng 92(10):891–916. https://doi.org/10.1002/nme.4371
13. Hossain MN, Ghosh D (2020) Adaptive reduced order modeling for nonlinear dynamical systems through a new a posteriori error estimator: application to uncertainty quantification. Int J Numer Methods Eng 121(15):3417–3441. https://doi.org/10.1002/nme.6365
14. Audouze C, De Vuyst F, Nair PB (2009) Reduced-order modeling of parameterized PDEs using time-space-parameter principal component analysis. Int J Numer Methods Eng 80(8):1025–1057. https://doi.org/10.1002/nme.2540. https://onlinelibrary.wiley.com/doi/abs/10.1002/nme.2540
15. Xiao D, Fang F, Pain C, Navon I (2017) A parameterized non-intrusive reduced order model and error analysis for general time-dependent nonlinear partial differential equations and its applications. Comput Methods Appl Mech Eng 317:868–889. https://doi.org/10.1016/j.cma.2016.12.033. https://www.sciencedirect.com/science/article/pii/S0045782516309471
16. Guo M, Hesthaven JS (2018) Reduced order modeling for nonlinear structural analysis using gaussian process regression. Comput Methods Appl Mech Eng 341:807–826. https://doi.org/10.1016/j.cma.2018.07.017. https://www.sciencedirect.com/science/article/pii/S0045782518303487

17. Jacquelin E, Baldanzini N, Bhattacharyya B, Brizard D, Pierini M (2019) Random dynamical system in time domain: a POD-PC model. Mech Syst Signal Process 133:106,251. https://doi.org/10.1016/j.ymssp.2019.106251. https://www.sciencedirect.com/science/article/pii/S0888327019304662
18. Guo M, Hesthaven JS (2019) Data-driven reduced order modeling for time-dependent problems. Comput Methods Appl Mech Eng 345:75–99. https://doi.org/10.1016/j.cma.2018.10.029. https://www.sciencedirect.com/science/article/pii/S0045782518305334
19. Sun X, Choi J (2021) Non-intrusive reduced-order modeling for uncertainty quantification of space-time-dependent parameterized problems. Comput Math Appl 87:50–64. https://doi.org/10.1016/j.camwa.2021.01.015
20. Hesthaven J, Ubbiali S (2018) Non-intrusive reduced order modeling of nonlinear problems using neural networks. J Comput Phys 363:55–78. https://doi.org/10.1016/j.jcp.2018.02.037. https://www.sciencedirect.com/science/article/pii/S0021999118301190
21. Jacquier P, Abdedou A, Delmas V, Soulaïmani A (2021) Non-intrusive reduced-order modeling using uncertainty-aware deep neural networks and proper orthogonal decomposition: application to flood modeling. J Computational Phys 424:109,854. https://doi.org/10.1016/j.jcp.2020.109854. https://www.sciencedirect.com/science/article/pii/S0021999120306288
22. Liang Y, Lee H, Lim S, Lin W, Lee K, Wu C (2002) Proper orthogonal decomposition and its applications-part I: theory. J Sound Vib 252(3):527–544. https://doi.org/10.1006/jsvi.2001.4041. https://www.sciencedirect.com/science/article/pii/S0022460X01940416
23. Preumont A (1980) A method for the generation of artificial earthquake accelerograms. Nucl Eng Des 59(2):357–368. https://doi.org/10.1016/0029-5493(80)90205-8. https://www.sciencedirect.com/science/article/pii/0029549380902058
24. Zentner I (2014) A procedure for simulating synthetic accelerograms compatible with correlated and conditional probabilistic response spectra. Soil Dyn Earthq Eng 63:226–233. https://doi.org/10.1016/j.soildyn.2014.03.012. https://www.sciencedirect.com/science/article/pii/S0267726114000748
25. Boudina A, Hammoutene M (2020) Generation of seismic excitations compatible with target spectrum: application to Eurocode 8. World J Eng
26. Beskou ND, Theodorakopoulos DD (2011) Dynamic effects of moving loads on road pavements: a review. Soil Dyn Earthq Eng 31(4):547–567. https://doi.org/10.1016/j.soildyn.2010.11.002
27. Bhattacharya S, Demirci HE, Nikitas G, Prakhya GKV, Lombardi D, Alexander NA, Aleem M, Amani S, Mylonakis G (2021) Chapter 11—Physical modeling of interaction problems in geotechnical engineering. In: Samui P, Kumari S, Makarov V, Kurup P (eds) Modeling in geotechnical engineering. Academic Press, pp 205–256. https://doi.org/10.1016/B978-0-12-821205-9.00017-4

Seismic Fragility Analysis of Dam Reservoir System by Including Spatial Variability of Ground Motion

P. Varsha and C. S. Manohar

Abstract The problem of random vibration analysis of gravity dam reservoir system subjected to spatially varying multi-component ground motion is considered. The work is done in two parts. First part deals with seismic wave propagation modelling in a three-dimensional elastic medium caused due to a seismic disturbance at a specified location in the interior. The disturbance is modelled as an effective point source of slip characterised by an impulsive double couple varying in time as a spatially localised infinite duration ramp function on the fault plane. A frequency domain wave propagation model that incorporates damping behaviour via a complex modulus representation is employed. The time domain representation is obtained further using Fourier integral. An ensemble of ground motions is simulated by treating various source parameters and properties of the intervening medium as uncertain. In the second part of work, governing equations for the dam reservoir system are formulated in which the ground motions derived in the previous step appear as boundary conditions. This set of equations is subsequently analysed using the finite element method to characterise the dynamic response of the system considering fluid structure interaction. This, in turn, forms the basis for investigating reliability of the dam structure with respect to a set of limit states. Furthermore, the analysis also leads to the determination of seismic fragility curves. While the problem of seismic response of dam reservoir system has been widely studied in the existing literature, present study permits us to examine the role played by spatial variability of support motion and their non-stationary and non-Gaussian features.

Keywords Fluid structure interaction · Fragility analysis · Spatial variability · Wave propagation

P. Varsha (✉) · C. S. Manohar
Indian Institute of Science, Bengaluru 560012, India
e-mail: pvarsha@iisc.ac.in

C. S. Manohar
e-mail: manohar@iisc.ac.in

S. Sidhardh et al. (eds.), *Advances in Structural Integrity for Mechanical, Civil, and Aerospace Applications*, Lecture Notes in Mechanical Engineering,
https://doi.org/10.1007/978-981-97-6367-2_26

1 Introduction

Investigation of the dynamic behaviour of dams under seismic loads is essential because concrete dams can fail catastrophically under strong ground motion. The movement of the dam causes a movement of the reservoir, which is in contact with the face of the dam. As a result, the fluid structure assembly constitutes a coupled system, and the interaction needs to be taken into account in the analysis. Westergaard [1] introduced the concept that the hydrodynamic pressure acting on the upstream face of a rigid dam due to horizontal ground motion can be interpreted as the inertia forces associated with an added mass of water moving with the dam. Chopra [2], Chopra and Chakrabarti [3] have discussed the significance of taking into account the coupled vibration of the dam reservoir system (fluid compressibility) while evaluating the hydrodynamic pressure. Fenves and Chopra [4] have studied the effects of dam reservoir foundation rock interaction and sediments accumulated at the reservoir bottom. The results of the system demonstrated that the absorptive reservoir bottom materials have a major effect on the earthquake response.

Another important factor to be considered in the seismic analysis is the role of spatial variability in ground motion. A dam is a massive structure extending over large distances in both directions and supported on extended foundations [5]. The fluid and foundation domain also extends over a long distance in the upstream direction. Hence the dam reservoir system would be subjected to spatially varying ground motion, i.e. differences in the amplitude and phase of seismic motions at various locations on the ground surface due to the general transmission of waves from the source through the different earth strata to the ground surface [6]. As the supports of these structures experience differential motion during an earthquake, it is crucial to take this into account when designing elongated structures (lifelines) like pipelines, tunnels, dams, and bridges [7–9]. However, there aren't many earthquake accelerograms recorded at closely spaced locations, making it difficult to describe the spatially changing ground motion for the design of significant structures. Hence, they are usually modelled as stochastic processes based on assumptions such as stationarity, homogeneity, Gaussianity, and isotropy. This study uses seismic wave propagation modelling for ground motion simulation to overcome these assumptions. Aki [10] theoretically calculated the ground displacement brought on by an earthquake fault in an infinite media. Subsequently, extensive studies have been carried out on the creation of ground motion due to earthquake faults in infinite medium [11], semi-infinite space [5], and layered space [12–14]. Closed-form analytical expressions for Frequency-Wave number spectra have been derived for the three displacement components at the ground surface when the seismic ground motion is produced by a source located in a single elastic layer of soil and characterised by a double couple [5]. Raghukanth [15] developed an analytical source mechanism model to compute the ground motion in a multi-layered semi- infinite space due to a buried source represented by a series of double couples varying in time as ramp function on the fault plane.

The problem of seismic response analysis of dam reservoir system and the simulation of ground motion from wave propagation modelling have been widely studied in the literature but have yet to be used in analysing the effect of spatially varying ground motion on a dam reservoir system. The present study permits us to examine the role played by spatial variability of support motion along the reservoir length and their non-stationary and non-Gaussian features. Two-dimensional analysis is performed on a concrete gravity dam subjected to ground motion simulated by source mechanism model using a scenario-based approach.

2 Simulation of Spatially Varying Ground Motion

Seismic ground motion produced by a source located deep inside an strata elastic half space is generated using the concept of wave propagation modelling. The source is modelled as a sequence of double couples. The model considers the propagation of three-dimensional linear waves and their reflection at the ground surface. Nonlinear efects are neglected as only a single strata of uniform soil is considered and for ease of calculation. Analytical expressions for the frequency wavenumber spectra have been derived for the three displacement components at the ground surface which are used for digital simulation of seismic ground motion [5, 16].

$$P\text{-wave equation: } \nabla^2 \emptyset + k_\alpha^2 \emptyset = 0 \quad k_\alpha = \frac{\omega}{\alpha} \quad \alpha = V_p\left(1 - \mathrm{i}\frac{1}{Q_p}\right) \tag{1}$$

$$S\text{-wave equation: } \nabla^2 \Psi + k_\beta^2 \Psi = 0 \quad k_\beta = \frac{\omega}{\beta} \quad \beta = V_s\left(1 - \mathrm{i}\frac{1}{Q_s}\right) \tag{2}$$

where ω is the frequency, α and β are the complex P-wave and S-wave velocities. V_p and V_s are elastic P and S wave velocities. Q_p and Q_s are the corresponding attenuation factors and i is the imaginary unit. The time dependence of the seismic moment is taken as a ramp function (Fig. 1) with rise time of 2 s. The source and intervening medium parameters are given in Table 1.

Fig. 1 Ramp function

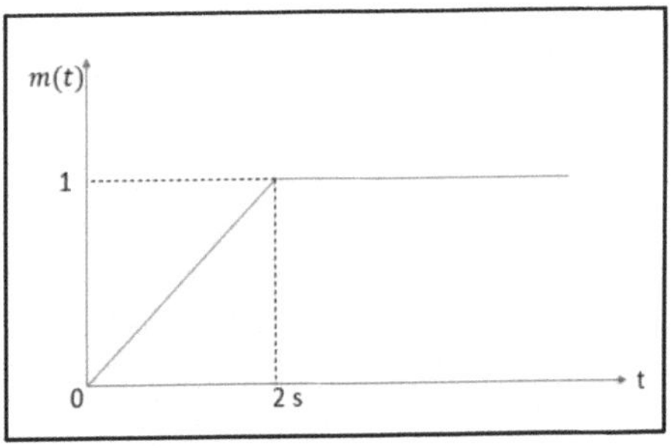

Table 1 Model parameters

Source parameters	
Seismic moment, Mo	10^{11} N m
Rise time, tr	2 s
Depth of source, z_o	10 km
Cut off frequency, ω_c	1 rad/s
Intervening medium parameters	
Density, ρ	3333 kg/m^3
Modulus of rigidity, G	3×10^{10} N/m^2
Elastic P-wave velocity, c_p	5196 m/s
Elastic S-wave velocity, c_s	3000 m/s
Attenuation factor	100

The displacement field at the free surface ($z = 0$) of the half space due to an effective point source of slip located at a point at a depth z_0 and characterised by double couple is written as:

$$u_x(x, y, z, t) = \iiint_{-\infty}^{\infty} \widetilde{u_x}\left(k_x, k_y, z, \omega\right) \exp\left[-\mathrm{i}k_x x - \mathrm{i}k_y y + \mathrm{i}\omega t\right] \mathrm{d}k_x \mathrm{d}k_y \mathrm{d}\omega \quad (3)$$

$$u_y(x, y, z, t) = \iiint_{-\infty}^{\infty} \widetilde{u_y}\left(k_x, k_y, z, \omega\right) \exp\left[-\mathrm{i}k_x x - \mathrm{i}k_y y + \mathrm{i}\omega t\right] \mathrm{d}k_x \mathrm{d}k_y \mathrm{d}\omega \quad (4)$$

$$u_z(x, y, z, t) = \iiint_{-\infty}^{\infty} \widetilde{u_z}\left(k_x, k_y, z, \omega\right) \exp\left[-\mathrm{i}k_x x - \mathrm{i}k_y y + \mathrm{i}\omega t\right] \mathrm{d}k_x \mathrm{d}k_y \mathrm{d}\omega \quad (5)$$

where,

$$\widetilde{u_x}\left(k_x, k_y, \omega\right) = M_0 m(\omega) \widetilde{G_x}\left(k_x, k_y, \omega\right) \quad (6)$$

$$\widetilde{u_y}\left(k_x, k_y, \omega\right) = M_0 m(\omega) \widetilde{G_y}\left(k_x, k_y, \omega\right) \quad (7)$$

$$\widetilde{u_z}\left(k_x, k_y, \omega\right) = M_0 m(\omega) \widetilde{G_z}\left(k_x, k_y, \omega\right) \quad (8)$$

$$\widetilde{G_x}\left(k_x, k_y, \omega\right) = -\frac{1}{8G\pi^3} \times \frac{\left[4k_x^2 \gamma \upsilon \mathrm{e}^{-\mathrm{i}\upsilon z_0} + 4k_x^2 \gamma \upsilon \mathrm{e}^{-\mathrm{i}\gamma z_0} - 4k^2 \gamma^2 \mathrm{e}^{-\mathrm{i}\gamma z_0} + k_\beta^4 \mathrm{e}^{-\mathrm{i}\gamma z_0}\right]}{\left[4k^2 \gamma \upsilon + \left(k_\beta^2 - 2k^2\right)^2\right]} \quad (9)$$

$$\widetilde{G}_y\left(k_x, k_y, \omega\right) = \frac{1}{8G\pi^3} \frac{\left(-4k_x k_y \gamma v e^{-\mathrm{i} v z_0} + 4k_x k_y \gamma v e^{-\mathrm{i}\gamma z_0}\right)}{\left[4k^2\gamma v + \left(k_\beta^2 - 2k^2\right)^2\right]} \tag{10}$$

$$\widetilde{G}_z\left(k_x, k_y, \omega\right) = \frac{1}{8G\pi^3} \frac{\left(2vk_x\left(k_\beta^2 - 2k^2\right) e^{-\mathrm{i} v z_0} + 2vk_x\left(k_\beta^2 - 2\gamma^2\right) e^{-\mathrm{i}\gamma z_0}\right)}{\left[4k^2\gamma v + \left(k_\beta^2 - 2k^2\right)^2\right]} \tag{11}$$

$$v = \sqrt{k_\alpha^2 - k^2}, \quad \gamma = \sqrt{k_\beta^2 - k^2}, \quad k = \sqrt{k_x^2 + k_y^2} \tag{12}$$

k_x and k_y are real wave numbers. Fast Fourier Transform is used to determine the displacement field at free surface. The seismic ground motion generated over an area of 80 km due to a source 10 km below the earth's surface in Fig. 2. The time histories of ground acceleration acting at a point 1 km from the source is as shown in Fig. 4 and the peak ground acceleration at different points located 150 m apart from each other is depicted in Fig. 3 which indicates the spatial variation in ground motion. The present study considers the earth strata as a single layer which can be extended to multi layer differing in their properties. Also, a point source of slip was considered in the present study which could be extended to finding ground motion due to dip slip or strike slip.

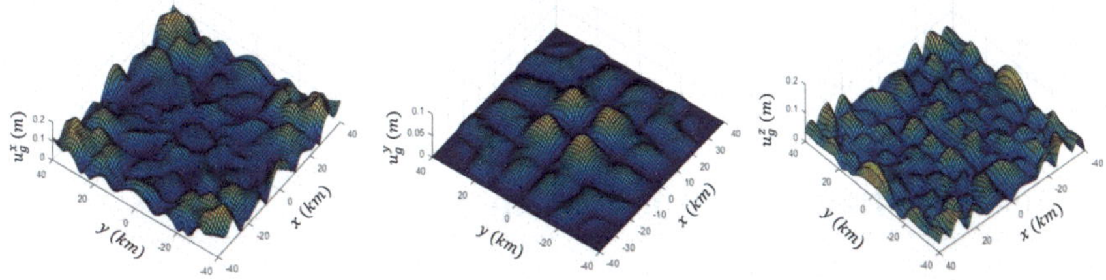

(a) x **(b) y** **(c) z**

Fig. 2 Snapshot of x, y and z component of simulated ground wave at $t = 2$ s over an area of 80 km in both direction

Fig. 3 PGA at different locations

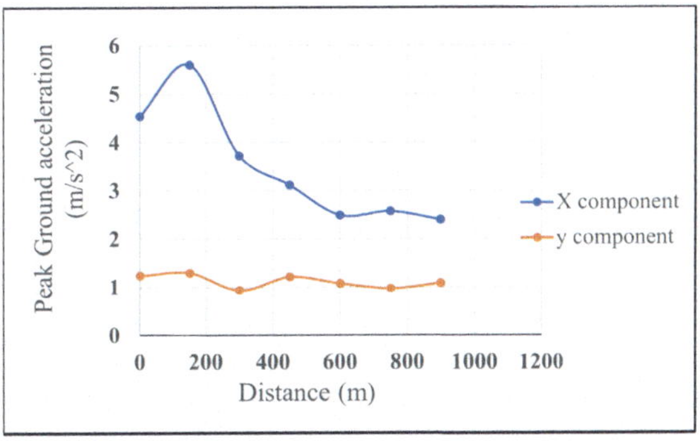

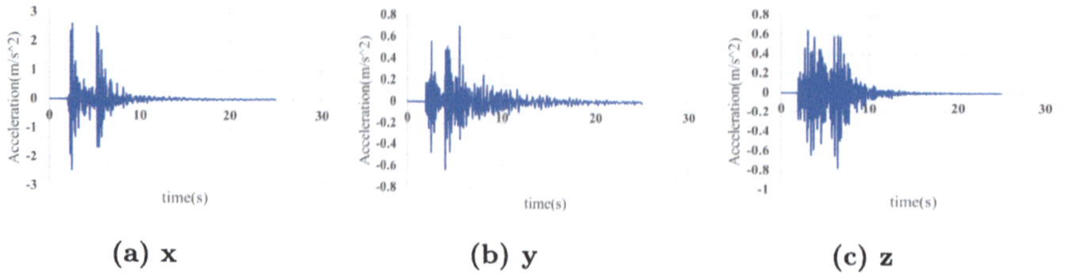

Fig. 4 x, y and z component of ground acceleration time histories

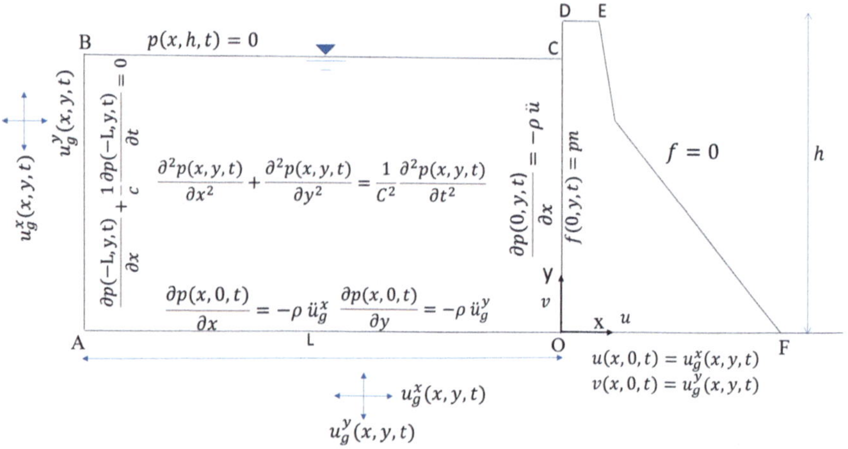

Fig. 5 Problem description

3 Modelling Dam Reservoir System

The governing equations and the boundary conditions for the solid and fluid domains are defined and this set of equations (Fig. 5) are subsequently analysed using the finite element method using Abaqus software to characterise the dynamic response of the dam reservoir system considering fluid structure interaction. Linear elastic 2-D plain strain model of the dam subjected to various loads is analysed. Concrete is assumed to be within the elastic limit. Effects due to surface waves and earthquake ground motion which acts when the upstream end of the dam meets ground are ignored.

3.1 *Free Vibration Analysis*

The dam section considered has dimensions similar to the Koyna dam, as shown in Fig. 6 with a height of 103 m and base dimension of 70 m, and has concrete properties as given in Table 2. Natural frequencies of the dam section computed by assuming

Fig. 6 Koyna dam section

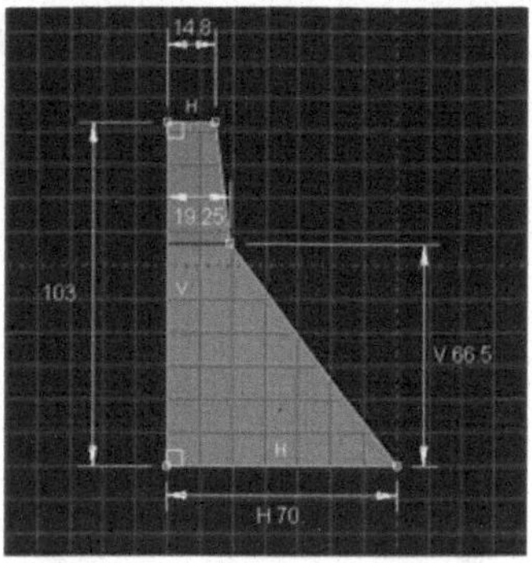

Table 2 Material properties of dam

Density	2643 kg/m^3
Poisson's ratio	0.2
Damping ratio	3%
Elastic modulus	31.027×10^9 N/m^2

Table 3 Natural frequencies of the dam

Mode	Frequency (Hz)	Chopra (Hz) [3]
1	3.0681	3.067
2	8.1052	8.196
3	11.074	10.752
4	15.927	15.872

Table 4 Coupled frequencies of the dam reservoir system

Mode	Frequency (Hz)	Acoustic fraction
1	2.5149	0.39082
2	3.5381	0.97565
3	3.6867	0.95724
4	3.9474	0.96364
5	4.3087	0.97317

plain strain conditions neglecting the fluid structure interaction are shown in Table 3 and compared with the values given in Chopra and Chakravarthi [3]. The reservoir considered is rectangular with a height of 103 m and length 1000 m, resulting in a total 1070 m base dimension. Free vibration analysis of the dam reservoir system is done to obtain the natural frequencies and vibration modes to study the influence of the reservoir on the dynamic behaviour of concrete gravity dams.

Each mode obtained has contributions from both the fluid(reservoir) and the solid(dam) domain. The coupled frequency and mode shapes of vibration as shown in Table 4 and Fig. 7 respectively. Acoustic fraction indicates the contribution or influence from the fluid part in the solid-fluid coupled system. From Table 4 it is clear

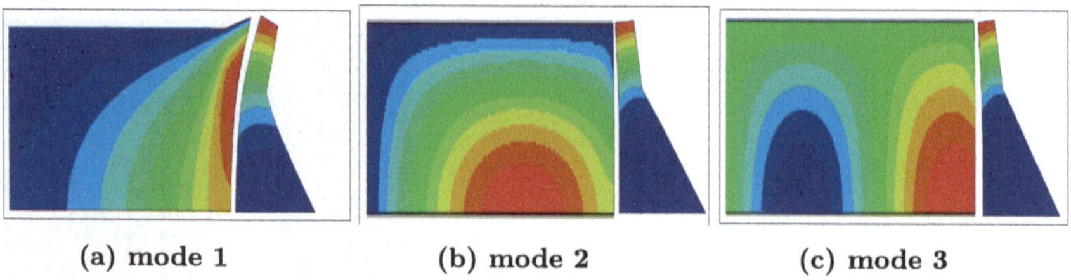

Fig. 7 Mode shapes of the coupled dam reservoir model

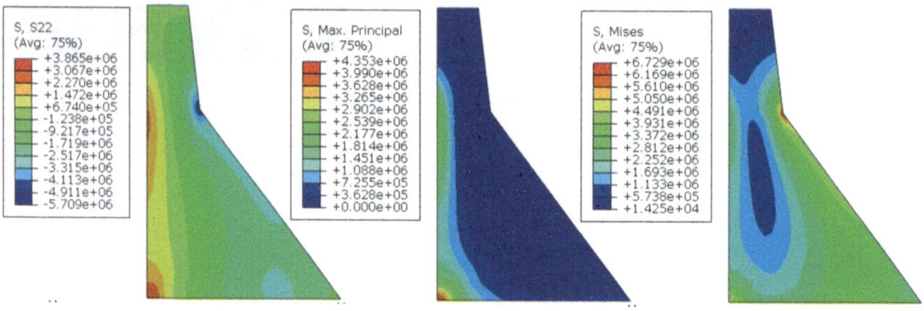

Fig. 8 Dam section **a** normal stress contour; **b** maximum principal stress contour; and **c** Von-Mises stress contour

that mode 1 is dominated by the dam and thus have very less influence due to the reservoir. As the values get closer to one, contributions from the acoustic component becomes more significant.

3.2 Dam Subjected to Uniform Simulated Ground Motion (UGM)

The dam reservoir system is subjected to a uniform ground motion in the x and y direction simulated from the wave propagation model. The ground motion applied throughout the dam reservoir base is due to a fault located at 10 km below the ground and acting at the heel of the dam. The effect of fluid structure interaction is considered. Figure 8 shows various stress contours of the dam section subjected to uniform ground motion.

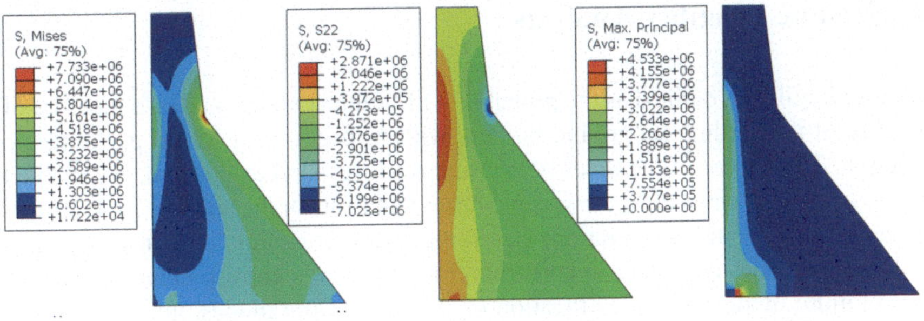

(a) Von-Mises stress contour; (b) Normal stress contour; and (c) Maximum principal stress contour

Fig. 9 Dam section **a** Von-Mises stress contour; **b** normal stress contour; and **c** maximum principal stress contour

Table 5 Comparison of stresses and displacements due to SVGM and UGM

	SVGM	UGM
Relative horizontal displacement (mm)	40.02	32.075
Relative vertical displacement (mm)	7	6.054
Maximum Von-Mises stress at the kink (MPa)	7.734	6.767
Maximum Von-Mises stress at the toe (MPa)	2.958	2.697
Maximum principal stress at kink (MPa)	2.421	1.4938
Maximum principal stress at toe (MPa)	0.623	0.4468
Maximum normal stress at kink (MPa)	− 7.062	− 5.9785
Maximum normal stress at toe (MPa)	− 1.673	− 0.55915

3.3 *Dam Reservoir System Subjected to Spatially Varying Simulated Ground Motion (SVGM)*

To study the effect of fluid structure interaction and the stresses and displacements on a dam reservoir system subjected to spatially varying ground motion (simulated from wave propagation model) in x and y direction. Figure 9 shows various stress contours of the dam section subjected to spatially varying ground motion.

From Table 5 it is clear that, when spatially varying ground motion is considered Vonmises, principal and normal stresses at the kink and toe shows higher value than uniform ground motion. The horizontal displacement at the crest relative to the ground increases by 20% when SVGM is considered. The vertical displacement at crest is comparatively less than horizontal displacement.

4 Seismic Fragility Analysis

Seismic fragility is defined as the probability of failure, conditioned on a measure of severity of the earthquake ground motion, with respect to a specified performance function. The choice of the severity measure of the ground excitations, is governed by the parameter chosen for the seismic hazard analysis of the site like peak ground acceleration, spectral acceleration, etc. Typically, a structural component has several failure modes and one can construct fragility curves corresponding to each of these failure modes. The identification of relevant failure modes, thus, constitutes an important aspect of fragility analysis. Uncertainties in structure specification and loads play a significant role in the safety of structures and need to be considered in the fragility analysis. Random vibration theories are used for the structural analysis.

Hariri-Ardebili and Saouma [17] have presented seismic fragility curves for gravity dams with or without ground motion vertical component within the context of performance based earthquake engineering where a smeared crack model was used for simulation of concrete cracking. Hariri-Ardebili [18] has studied the spatial distribution of random variables in the context of random field theory by spatially varying the concrete material properties.

4.1 *Failure Modes*

A dam failure can be any collapse or movement of part of the dam or foundation so that it fails to fulfil its intended purpose of retaining water. There are various modes in which a concrete dam can fail.Hariri-Ardebili et al. [19] have identified few potential failure modes of the dam like the concrete cracking at neck, subjected to seismic excitation and formulated limit state functions for them in the framework of both linear and nonlinear analyses. Sliding or shear failure is the most common failure for dams constructed on rock. Concrete cannot withstand sustained tensile stress, and if the tension that develops in the concrete exceeds its tensile strength, it could lead to ultimate failure. Also, excessive displacement at the crest leads to the failure of dam.

Limit states considered in the linear finite element analyses considered in this study are tensile stresses at heel, toe and kink exceeding limit (> 2 MPa = LSi) and excessive deflection of the top of the dam relative to heel. Maximum Von-Mises stress at the kink and maximum horizontal displacement at the crest for different intensity levels of ground motion (deterministic analysis) is plotted for 4 cases (Figs. 10 and 11):

(1) Spatially varying ground motion and fluid structure interaction.
(2) Uniform ground motion with fluid structure interaction.
(3) Spatially varying ground motion with no fluid structure interaction.
(4) Uniform ground motion with no fluid structure interaction.

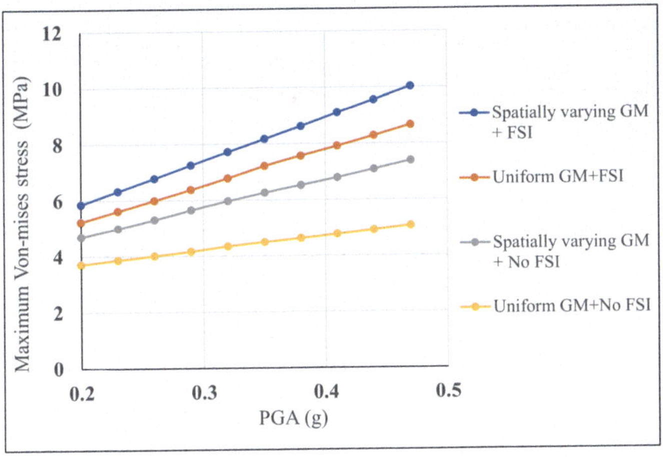

Fig. 10 Maximum Von-Mises stress for different intensity of ground motion

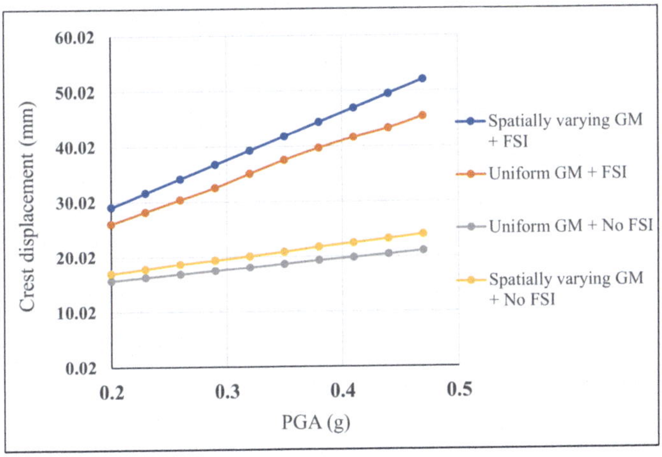

Fig. 11 Maximum crest displacement for different intensity of ground motion

4.2 *Uncertainity Modelling*

Probabilistic framework is necessary to manage the various sources of uncertainty that may impact dam performance and decisions related. Uncertainties in the ground motion simulation parameters (both source and intervening medium) is considered and 500 ground motions are simulated and applied in to the dam reservoir model to study the response. Random variables chosen are shown in Table 6. Different response parameters such as stresses and displacements are extracted and compared with the limit state to plot the fragility curve as shown in Fig. 12.

Table 6 Random parameters

Density (kg/m^3)	$N(3000, 500)$
Shear modulus (Pa)	$N(3 \times 10^{10}, 1 \times 10^{10})$
Attenuation factor	$N(100, 50)$
Rise time (s)	$N(2, 1)$

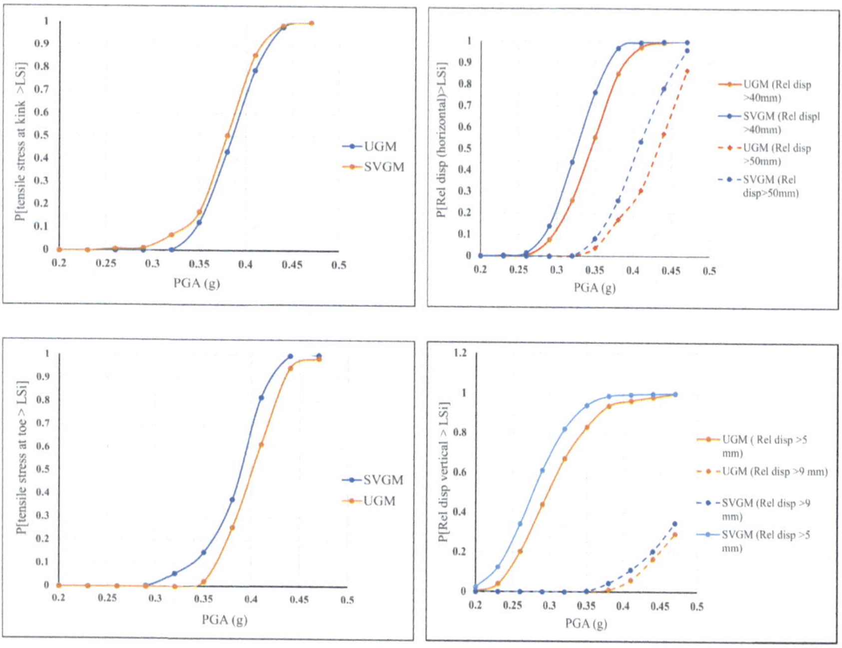

Fig. 12 Fragility curves

5 Concluding Remarks

(a) Dam reservoir system subjected to spatially varying and uniform ground motion is analysed separately, to study the impact of SVGM. According to deterministic analysis (Figs. 10 and 11) taking into account SVGM resulted in approximately 15% increase in stress and relative displacement.

(b) To study the effect of fluid structure interaction, dam reservoir system were analysed with and without fluid structure interaction. According to deterministic analysis (Figs. 10 and 11) taking into account FSI resulted in approximately 40% increase in stress and relative displacement.

(c) A probabilistic assessment of the dam reservoir system was done considering ground motion as random and seismic fragility curves were plotted. Peak ground acceleration is taken as the intensity measure in all cases. The conditional probability of failure at a given intensity level for all the limit states are found to be higher when spatially varying ground motion is considered. For, e.g. at an intensity level of 0.35 g, the probability that horizontal displacement at crest relative to ground exceeds a maximum value of 50 mm is 76% when SVGM is considered and 55% when uniform ground motion is considered (Fig. 12). This implies the significance of taking SVGM and FSI into account when analysing the dam.

References

1. Westergaard HM (1933) Water pressures on dams during earthquakes. Trans Am Soc Civ Eng 98(2):418–433
2. Chopra AK (1967) Hydrodynamic pressures on dams during earthquakes. J Eng Mech Div 93(6):205–223
3. Chopra AK, Chakrabarti P (1972) The earthquake experience at Koyna dam and stresses in concrete gravity dams. Earthq Eng Struct Dyn 1(2):151–164
4. Fenves G, Chopra AK (1987) Simplified earthquake analysis of concrete gravity dams. J Struct Eng 113(8):1688–1708
5. Deodatis G, Shinozuka M, Papageorgiou A (1990) Stochastic wave representation of seismic ground motion. I: F-K spectra. J Eng Mech 116(11):2363–2379
6. Harichandran RS (1987) Correlation analysis in space-time modeling of strong ground motion. J Eng Mech 113(4):629–634
7. Zerva A, Ang AH-S, Wen Y (1988) Lifeline response to spatially variable ground motions. Earthq Eng Struct Dyn 16(3):361–379
8. Zerva A (1991) Effect of spatial variability and propagation of seismic ground motions on the response of multiply supported structures. In: Stochastic structural dynamics, vol 2. Springer, pp 307–336
9. Harichandran RS, Hawwari A, Sweidan BN (1996) Response of long-span bridges to spatially varying ground motion. J Struct Eng 122(5):476–484
10. Aki K (1968) Seismic displacements near a fault. J Geophys Res 73(16):5359–5376
11. Haskell NA (1969) Elastic displacements in the near-field of a propagating fault. Bull Seismol Soc Am 59(2):865–908
12. Haskell NA (1953) The dispersion of surface waves on multilayered media. Bull Seismol Soc Am 43(1):17–34
13. Bouchon M (1979) Predictability of ground displacement and velocity near an earthquake fault: an example: the Parkfield earthquake of 1966. J Geophys Res Solid Earth 84(B11):6149–6156
14. Zhang R, Yong Y, Lin Y (1991) Earthquake ground motion modeling. I: deterministic point source. J Eng Mech 117(9):2114–2132
15. Kanth SR (2008) Source mechanism model for ground motion simulation. Appl Math Model 32(7):1417–1435
16. Deodatis G, Shinozuka M, Papageorgiou A (1990) Stochastic wave representation of seismic ground motion. II: simulation. J Eng Mech 116(11):2381–2399
17. Hariri-Ardebili M, Saouma V (2016) Collapse fragility curves for concrete dams: comprehensive study. J Struct Eng 142(10)
18. Hariri-Ardebili M, Seyed-Kolbadi S, Saouma V, Salamon J, Rajagopalan B (2018) Random finite element method for the seismic analysis of gravity dams. Eng Struct 171:405–420
19. Hariri-Ardebili M, Saouma V (2015) Quantitative failure metric for gravity dams. Earthq Eng Struct Dyn 44(3):461–480

Elastic Response of Hydrogels Under Finite Deformations

Vivek Kumar Singh and Krishnendu Haldar

Abstract Hydrogels are three-dimensional network structures consisting of crosslinked polymer chains. When immersed in a solvent, hydrogels swell such that they retain the structural integrity due to presence of crosslinks. The elastic properties of hydrogels depend on the density of crosslinks: the higher the crosslink density, the stiffer the gel. They derive elastic properties from the polymer network and swelling properties from the migration of solvent molecules through the network. Due to high swellability and deformability, hydrogels have found applications in various engineering fields, such as drug delivery systems, tissue engineering, wound dressing, and soft actuators. We present here a continuum framework to describe the elastic responses of hydrogels undergoing finite deformations. An expression for free energy density of hydrogels has been utilised. Constitutive equations of hydrogels subjected to swelling are derived in a finite deformation framework. The model is calibrated by fitting the stress-strain curve of Polyrotaxane (PR) hydrogels in uniaxial tension. The results are verified with the available experimental results.

Keywords Hydrogels · Crosslinks · Swellability · Deformability · Polyrotaxane (PR)

1 Introduction

A gel is a semi-solid substance with properties ranging from soft and weak to hard and tough. Phenomenologically, it is characterised as a soft, semi-solid "jelly-like" material with two or more components, one of which is liquid, present in substantial quantity. It has properties of both a fluid and a solid. It has an elastic behaviour because it deforms when stressed and returns to its original shape when stress is removed.

V. K. Singh · K. Haldar (✉)
Department of Aerospace Engineering, Indian Institute of Technology Bombay, Mumbai 400076, India
e-mail: krishnendu@aero.iitb.ac.in

V. K. Singh
e-mail: vivekkrsingh.mail@gmail.com

S. Sidhardh et al. (eds.), *Advances in Structural Integrity for Mechanical, Civil, and Aerospace Applications*, Lecture Notes in Mechanical Engineering,
https://doi.org/10.1007/978-981-97-6367-2_27

It also exhibits viscous properties derived from the dissolved liquid. Hydrogels are three-dimensional network of polymer chains that are linked together by chemical and physical crosslinks. When immersed in a solvent, they swell such that they retain the structural integrity due to presence of crosslinks (Fig. 1). Hydrogels have found applications in various engineering fields, such as drug delivery systems, tissue engineering, making contact lenses, wound dressing, and soft actuators [1, 2]. Swelling of hydrogels is a coupled process. The coupling is between mechanical and chemical fields [3], stimulated by various stimuli such as temperature, electric field, magnetic field, and UV radiations. Until now, many works have been elaborated on modelling of swelling of hydrogels. This can be classified into two groups: the mixing-theory and the continuum-based framework. Flory and Rehner [4] studied the interaction of solvent molecules with the network structure. They investigated hydrogels undergoing large deformation with the solvent molecules growing uniformly, and the free energy function of swelling of gels. Hong et al. [5] proposed a theoretical framework that coupled mass transfer and large deformation. They showed that the free energy of the gel results from two processes: one, when the polymer network gets stretched and two, when the long networks are mixed with the small solvent molecules. Following this, Duda et al. [6] presented a theory for the behaviour of hydrogels undergoing two independent processes: a macroscopic deformation due to stretching of polymer networks and a process at microscopic level due to the migration of a chemical species through the network. Lucantonio et al. [7] developed a theoretical model that dealt with large deformation in polymer gels induced by swelling: free as well as forced swelling. Duan et al. [8] investigated transient behaviours of gel swelling and deswelling based on the coupling between mechanical deformation and solvent diffusion. Several attempts have been made to study the mechanical response of hydrogels under multiaxial deformation [9–11]. Many of them successfully described elastic responses of hydrogels coupled with solvent diffusion, however, their applicability is limited to uniaxial deformation. Moreover, very few studies examined the accuracy of theoretical model by comparing them with experimental results. This work presents a theoretical continuum framework to study the elastic response of hydrogels undergoing finite deformations. The objective of this paper is two fold: (i) To develop an energy function that takes into account the strain energy of polymer chains, the mixing energy and the interaction energy between chains and solvent molecules, and (ii) To implement this formulation to the uniaxial and equi-biaxial loading of hydrogels.

The paper is organised as follows: In Sect. 2, the theory of swelling of gel has been described briefly based on the continuum framework. The macroscopic deformation gradient tensor is multiplicatively decomposed into two deformation gradient tensors, the elastic and swelling part. In Sect. 3, we propose an energy function following the continuum modelling of hydrogels by Hong et al. [5]. A hyperelastic model is taken for describing the strain energy density of polymer networks. In Sect. 4, model parameters are computed by fitting the stress-strain plot under uniaxial tension. Subsequently, the analysis of deformation under equi-biaxial tension and pure shear is presented in Sect. 5. Conclusions are made in Sect. 6.

2 Continuum Theory of Swelling of Gel

Hydrogels constitute two immiscible phases: a solid phase and a liquid phase. The mass transfer between these two phases is neglected. Further, these two phases are considered as interpenetrating continuum. The solid phase consists of long polymer network chains constrained by junctions. When a hydrogel immersed in the solvent is loaded, it gets deformed and the molecules of solvent diffuse into the network. Consequently swelling is induced and the load gets redistributed.

2.1 *Kinematics*

We start with an undeformed dry gel, which is the initial configuration at the beginning of deformation. The initial and current configurations are denoted by $\mathcal{B}_d$ and $\mathcal{B}_t$, respectively. The material point is denoted by $\boldsymbol{X} \in \mathcal{B}_d$, which on deformation takes the spatial position $\boldsymbol{x} = \boldsymbol{\varphi}(\boldsymbol{X}, t)$, where t is the time, and φ is the deformation map. The macroscopic deformation gradient is defined as $\boldsymbol{F} = \nabla_X \boldsymbol{\varphi}$. Further, $\boldsymbol{F}$ is multiplicatively decomposed as

$$\boldsymbol{F} = \boldsymbol{F}^e \boldsymbol{F}^s, \tag{1}$$

where $\boldsymbol{F}^e$ is the elastic deformation gradient that is defined with respect to the stress free configuration $\mathcal{B}_s$, which serves as the reference configuration and $\boldsymbol{F}^s$ is the swelling deformation gradient which is defined with respect to the initial configuration $\mathcal{B}_d$, that characterises the deformation induced by chemical swelling [6] (Fig. 2). Further, we consider isotropic stretches λ^s due to swelling [12]. Thus,

$$\boldsymbol{F}^s = \lambda^s \boldsymbol{I}, \quad \lambda^s > 0. \tag{2}$$

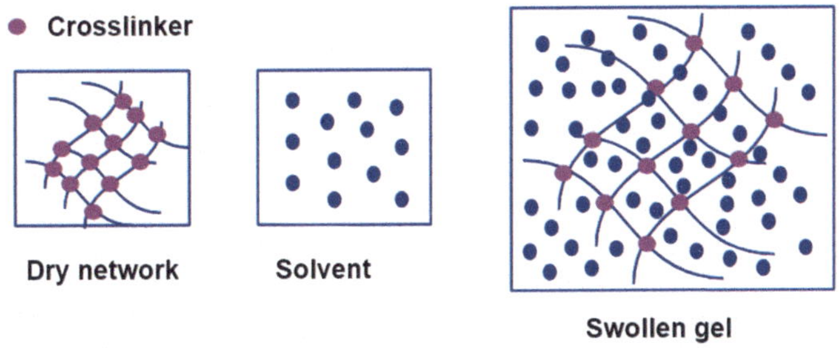

Fig. 1 Schematic diagram of a hydrogel

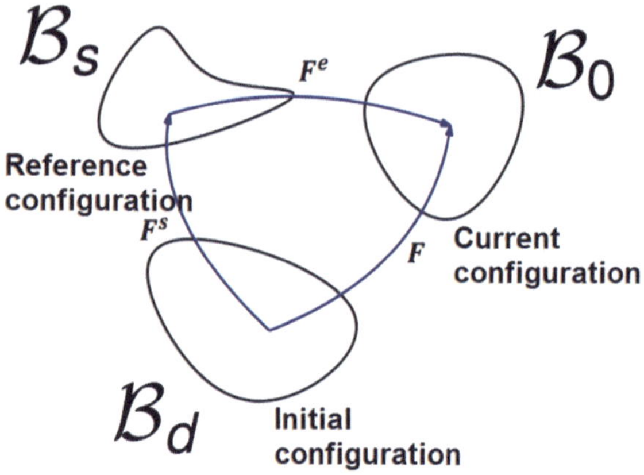

Fig. 2 Multiplicative decomposition of the deformation gradient into the elastic and swelling part

Furthermore, we assume that the polymer network is incompressible [13], thus

$$\det \boldsymbol{F}^e = 1. \tag{3}$$

Consequently, $\det \boldsymbol{F} = \det \boldsymbol{F}^s = (\lambda^s)^3$.

2.2 Molecular Incompressibility Condition

Let C_R be the concentration of solvent molecules (number of molecules per unit volume) in the initial configuration and v is the characteristic volume of water. The increase in volume of the hydrogel is only due to migration of water molecules into the gel. Subsequently, it is assumed that all molecules in a gel are incompressible and the volume of gel is the sum of volume of dry network and the volume of pure liquid solvent. This is called the molecular incompressibility condition [13]. This can be stated as

$$\det \boldsymbol{F} = \det \boldsymbol{F}^s = 1 + vC_R. \tag{4}$$

2.3 Balance Laws

In the current configuration, the concentration of solvent molecules per unit volume is given by [14]

$$c = \frac{C_R}{J} = \frac{C_R}{J^s}. \tag{5}$$

The following equation governs the diffusion of solvent molecules across the polymer network,

$$\boldsymbol{j} = -\tfrac{Dc}{k_B T}\nabla\mu \tag{6}$$

where $\boldsymbol{j}$ is the flux-vector in current configuration and μ is chemical potential of the solvent. The constant D is diffusivity (m^2/s) and $k_B T$ is energy (J) in terms of temperature. We pull back j in the initial configuration $\mathcal{B}_d$

$$\boldsymbol{j}_0 = -\frac{DC_R}{k_B T}\boldsymbol{F}^{-1}\cdot\nabla_0\mu\cdot\boldsymbol{F}^{-1}, \tag{7}$$

where $\boldsymbol{j}_s$ is the flux-vector in reference configuration. We consider the deformation of hydrogel to be quasi-static. Thus, the first Piola-Kirchhoff stress tensor $\boldsymbol{P}$ fulfils the equilibrium condition as

$$\nabla_0\cdot\boldsymbol{P} + \boldsymbol{b}_0 = 0 \tag{8}$$

inside $\mathcal{B}_d$. Further, the force balance equation at the boundary $\partial\mathcal{B}_d$ is given by

$$\boldsymbol{P}\boldsymbol{n}_0 + \boldsymbol{t}_0 = 0. \tag{9}$$

Here, $\boldsymbol{b}$ is the body force, $\boldsymbol{t}$ is surface traction and $\boldsymbol{n}_0$ is the outward normal vector at $\partial\mathcal{B}_d$. Further, the rate change of number of solvent molecules per unit volume is given by

$$\frac{\partial C_R}{\partial t} + \nabla_0\cdot\boldsymbol{j}_0 = R \tag{10}$$

inside $\mathcal{B}_d$ and

$$\boldsymbol{n}_0\cdot\boldsymbol{j}_0 + r = 0 \tag{11}$$

at $\partial\mathcal{B}_d$. R and r are rate of injection of solvent molecules per unit volume and per unit area (at the boundary) respectively.

3 Constitutive Modelling

We take the free energy density as $\Psi_R(\boldsymbol{F}, C_R, t)$, where an explicit dependence on time has been shown that accounts for rearrangement of polymer networks because of solvent diffusion into the network. The local form of second law of thermodynamics is given as

$$\dot{\Psi}_R \leq \boldsymbol{P} : \dot{\boldsymbol{F}} + \mu\dot{C}_R - \boldsymbol{j}_0\cdot\nabla_0\mu. \tag{12}$$

Further, we enforce the molecular incompressibility condition by adding a term $\Pi\,(1 + vC_R - \det \boldsymbol{F})$ to the free energy function, where $\Pi = \Pi(\boldsymbol{X}, t)$ is the Lagrange multiplier. Subsequently, the inequality (12) takes the form

$$\frac{\partial \Psi_R}{\partial t} + \left(\frac{\partial \Psi_R}{\partial \boldsymbol{F}} - \Pi(\det \boldsymbol{F})\boldsymbol{F}^{-T} - \boldsymbol{P}\right) : \dot{\boldsymbol{F}} + \left(\frac{\partial \Psi_R}{\partial C_R} + \Pi v - \mu\right) C_R + j_0 \cdot \nabla_0 \mu \leq 0 \tag{13}$$

Applying the Coleman-Noll procedure, for arbitrary $\dot{\boldsymbol{F}}$ and $\dot{C}_R$, we have the first Piola-Kirchhoff stress

$$\boldsymbol{P} = \frac{\partial \Psi_R}{\partial \boldsymbol{F}} - \Pi(\det \boldsymbol{F})\boldsymbol{F}^{-T} \tag{14}$$

and the chemical potential is given by

$$\mu = \frac{\partial \Psi_R}{\partial C_R} + \Pi v, \tag{15}$$

provided the inequality

$$\frac{\partial \Psi_R}{\partial t} + \boldsymbol{j}_0 \cdot \nabla_0 \mu \leq 0 \tag{16}$$

holds true.

3.1 Free Energy Function

The free energy of hydrogel takes the form

$$\Psi_R(\boldsymbol{F}^e, C_R, t) = \Psi_{\text{elastic}}(\boldsymbol{F}^e, t) + \Psi_{\text{mixing}}(C_R), \tag{17}$$

where $\Psi_{\text{elastic}}(\boldsymbol{F}^e, t)$ denotes the elastic strain energy of the network and $\Psi_{\text{mixing}}(C)$ denotes the mixing energy. We start with the neoHookean model for the strain energy density of the polymer network:

$$\Psi_{\text{elastic}} = \frac{G}{2}\left[(I_1^e - 3) - \frac{1}{2}\ln I_3^e\right], \tag{18}$$

where G is the shear modulus of the polymer network and I_1^e and I_3^e are the first and third invariants of elastic left Cauchy-Green tensor $\boldsymbol{B}^e$ respectively. The specific mixing energy is given by

$$\Psi_{\text{mixing}} = -\frac{k_B T}{v}\left[\frac{1-\phi}{\phi}\ln(1-\phi) + \chi(1-\phi)\right], \tag{19}$$

where χ is a dimensionless constant called the Flory interaction parameter [15], k_B is the Boltzmann constant, T is the absolute temperature, v is the characteristic volume of the solvent molecule, and ϕ is the volume fraction of solid phase (polymer network), which is given by

$$\phi = \frac{1}{1 + vC_R}. \tag{20}$$

Thus, the volume fraction of liquid phase (solvent) is given by

$$\phi_f = 1 - \phi = \frac{vC_R}{1 + vC_R} \tag{21}$$

Now, we multiplicatively decompose $\boldsymbol{F}$ as

$$\boldsymbol{F} = (1 + vC_R)^{\frac{1}{3}} \boldsymbol{F}^e, \tag{22}$$

such that $\det F = 1 + vC_R = (\lambda^s)^3$. Thus, $\boldsymbol{B} = (1 + vC_R)^{\frac{2}{3}} \boldsymbol{B}^e$. From Eq. (14), we have the Cauchy stress tensor

$$\boldsymbol{\sigma} = -\Pi \boldsymbol{I} + \frac{1}{\det \boldsymbol{F}} \frac{\partial \Psi}{\partial \boldsymbol{F}} \boldsymbol{F}^T. \tag{23}$$

We consider elastic deformation of hydrogel from the reference state. Thus, using Eq. (22), we have

$$\boldsymbol{\sigma} = -\Pi \boldsymbol{I} + \frac{1}{1 + vC_R}\left(\beta \frac{\partial \Psi_{\text{elastic}}}{\partial \boldsymbol{F}^e} \boldsymbol{F}^{eT}\right), \tag{24}$$

where

$$\beta = \frac{(1 + vC_R)^{\frac{1}{3}}}{\lambda^s}. \tag{25}$$

4 Curve Fitting Under Uniaxial Tension

We start with Polyrotaxane (PR) hydrogels for uniaxial loading taken from Ref. [13]. The PR gels were prepared by intermolecular crosslinking of PRs using 1,4-butanediol diglycidyl ether as a cross-linker. The cross-linker and PR were dissolved in an aqueous solution of 1.5 N NaOH to get a PR concentration of 0.15 g/mL. The crosslink concentration was 5 or 10%. The gelation was conducted at room temperature for 20 h. The resultant PR gel was allowed to swell fully in distilled water. Then, the swollen PR gel was gradually dried. Further, the fully dried PR network was swollen by adding a controlled amount of dimethylsulfoxide (DMSO) to get a PR concentraton of 34 wt%. This is the asprepared state of PR hydrogel. The uniaxial and equi-biaxial tests were conducted in this state, when the volume fraction of solid phase was fixed, $\phi = 0.34$. The uniaxial tension is given by the deformation gradient

$$\boldsymbol{F}^e = \lambda^e \boldsymbol{e}_1 \otimes e_1 + \frac{1}{\sqrt{\lambda^e}}(\boldsymbol{e}_2 \otimes e_2 + \boldsymbol{e}_3 \otimes \boldsymbol{e}_3) \tag{26}$$

Thus, for the neoHooken model (Eq. 18), the stress tensor becomes

$$\boldsymbol{\sigma} = -\Pi \boldsymbol{I} + \frac{G\beta}{1 + \upsilon C_R} \left[\boldsymbol{B}^e - \boldsymbol{I}\right]. \tag{27}$$

For uniaxial tension, we determine the value of Π from the condition $\sigma_{22} = \sigma_{33} = 0$. Thus, the engineering stress (PK1) is $P = \frac{\sigma_{11}}{\lambda^e}$, given by

$$P = \frac{G\beta}{1 + \upsilon C_R} \left(\lambda^e - \frac{1}{\lambda^{e2}}\right). \tag{28}$$

We fix $\beta = 1$, then we use the least square technique to determine the material constant G. The best fit stress versus stretch curve under uniaxial tension is shown in Fig. 3a. The calibrated material parameter is: $G = 6.65$ kPa. We now validate our model under equi-biaxial tension of PR hydrogel.

5 Model Prediction

5.1 Equi-biaxial Tension

When the hydrogel specimen is loaded under equi-biaxial tension, the elastic deformation gradient is given by

$$\boldsymbol{F}^e = \lambda^e(\boldsymbol{e}_1 \otimes \boldsymbol{e}_1 + \boldsymbol{e}_2 \otimes \boldsymbol{e}_2) + \lambda^{e^{-2}} \boldsymbol{e}_3 \otimes \boldsymbol{e}_3. \tag{29}$$

The equi-biaxial stress P is given by

$$P = \frac{G\beta}{1 + \upsilon C_R} \left(\lambda^e - \frac{1}{\lambda^{e5}}\right). \tag{30}$$

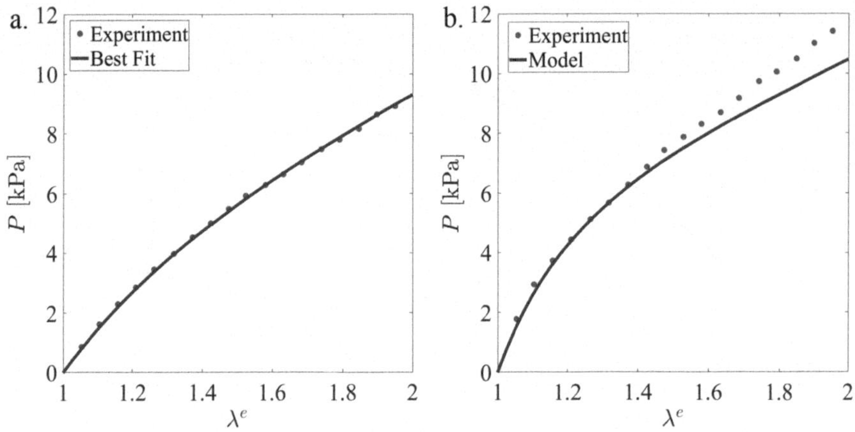

Fig. 3 Stress versus stretch curve of PR hydrogel in the asprepared state, $\phi = 0.34$. **a** Best fit curve under uniaxial tension, **b** model validation under equi-biaxial tension. The experimental data has been taken from Ref. [13]. In the equi-biaxial stretching, the gel sheet was equally stretched along two orthogonal directions, $\lambda_x^e = \lambda_y^e$

The stress versus stretch plot for equi-biaxial loading of PR hydrogel in the asprepared state ($\phi = 0.34$) is plotted in Fig. 3b. We observe that for equi-biaxial tension, the result of current model is in agreement with the experiment till $\lambda^e = 1.5$, then there is a slight deviation.

5.2 *Pure Shear*

Under pure shear loading of the hydrogel, the deformation gradient is given by

$$\boldsymbol{F}^e = \lambda^e \boldsymbol{e}_1 \otimes \boldsymbol{e}_1 + \boldsymbol{e}_2 \otimes \boldsymbol{e}_2 + \lambda^{e^{-1}} \boldsymbol{e}_3 \otimes \boldsymbol{e}_3. \tag{31}$$

The stresses in the longitudinal (PS1) and transverse (PS2) directions are given by

$$P^{\text{PS1}} = \frac{G\beta}{1 + vC_R}\left(\lambda^e - \frac{1}{\lambda^{e3}}\right), \quad \text{and} \tag{32}$$
$$P^{\text{PS2}} = \frac{G\beta}{1 + vC_R}\left(1 - \frac{1}{\lambda^{e2}}\right)$$

respectively. The stress versus stretch response is shown in Fig. 4. We observe a good agreement between the theoretical and experimental results in case of PS1. However, in case of PS2, till $\lambda^e = 1.6$, the elastic response is close to the experimental results. On further stretching, a slight deviation is observed.

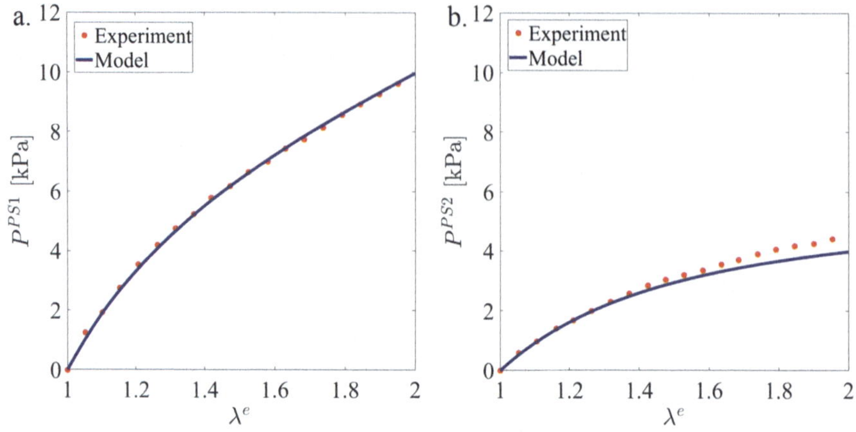

Fig. 4 Stress versus stretch curve under pure shear of PR hydrogel in the asprepared state, $\phi = 0.34$. **a** Longitudinal direction (PS1), **b** transverse direction (PS2). The experimental data has been taken from Ref. [13]. In the pure shear deformation, the gel sheet was stretched in $x-$direction while prohibiting lateral shrinking in y-direction, $\lambda_y^e = 1$

6 Conclusion

A continuum framework is presented to investigate the elastic responses of hydrogels under finite strain. The stress-strain relations are derived from the free energy density function involving elastic strain energy, mixing energy, and interaction energy. In this work, only isotropic swelling is considered having the swelling stretch (λ_s) equal in each direction. The material parameter is calibrated by fitting the observations on PR hydrogel (in the asprepared state) under uniaxial tension. Finally, the model is validated with the experimental results in equi-biaxial and pure shear loading conditions.

References

1. Deligkaris K, Tadele TS, Olthuis W, van den Berg A (2010) Hydrogel-based devices for biomedical applications. Sens Actuators B Chem 147(2):765–774. https://doi.org/10.1016/j.snb.2010.03.083
2. Buenger D, Topuz F, Groll J (2012) Hydrogels in sensing applications. Prog Polym Sci 37(12):1678–1719. https://doi.org/10.1016/j.progpolymsci.2012.09.001
3. Hong W, Liu Z, Suo Z (2009) Inhomogeneous swelling of a gel in equilibrium with a solvent and mechanical load. Int J Solids Struct 46(17):3282–3289. https://doi.org/10.1016/j.ijsolstr.2009.04.022
4. Flory PJ, Rehner J Jr (2004) Statistical mechanics of cross-linked polymer networks I. Rubberlike elasticity. J Chem Phys 11(11):512–520. https://doi.org/10.1063/1.1723791. https://pubs.aip.org/aip/jcp/article-pdf/11/11/512/8106485/512_1_online.pdf

5. Hong W, Zhao X, Zhou J, Suo Z (2008) A theory of coupled diffusion and large deformation in polymeric gels. J Mech Phys Solids 56(5):1779–1793. https://doi.org/10.1016/j.jmps.2007.11.010
6. Duda FP, Souza AC, Fried E (2010) A theory for species migration in a finitely strained solid with application to polymer network swelling. J Mech Phys Solids 58(4):515–529. https://doi.org/10.1016/j.jmps.2010.01.009
7. Lucantonio A, Nardinocchi P, Teresi L (2013) Transient analysis of swelling-induced large deformations in polymer gels. J Mech Phys Solids 61(1):205–218. https://doi.org/10.1016/j.jmps.2012.07.010
8. Duan Z, Zhang J, An Y, Jiang H (2013) Simulation of the transient behavior of gels based on an analogy between diffusion and heat transfer. J Appl Mech 80(4):041017 (2013). https://doi.org/10.1115/1.4007789. https://asmedigitalcollection.asme.org/appliedmechanics/article-pdf/80/4/041017/6077334/jam_80_4_041017.pdf
9. Dolbow J, Fried E, Ji H (2004) Chemically induced swelling of hydrogels. J Mech Phys Solids 52(1):51–84. https://doi.org/10.1016/S0022-5096(03)00091-7
10. Chester SA (2015) Gel mechanics: a thermo-mechanically coupled theory for fluid permeation in elastomeric materials. In: IUTAM symposium on mechanics of soft active materials. Proc IUTAM 12:10–19. https://doi.org/10.1016/j.piutam.2014.12.003
11. Srivastava V, Chester SA, Ames NM, Anand L (2010) A thermo-mechanically-coupled large-deformation theory for amorphous polymers in a temperature range which spans their glass transition. In: Special issue in honor of Lallit Anand. Int J Plast 26(8):1138–1182. https://doi.org/10.1016/j.ijplas.2010.01.004
12. Lubarda VA (2004) Constitutive theories based on the multiplicative decomposition of deformation gradient: thermoelasticity, elastoplasticity, and biomechanics. Appl Mech Rev 57(2):95–108. https://doi.org/10.1115/1.1591000. https://asmedigitalcollection.asme.org/appliedmechanicsreviews/article-pdf/57/2/95/5440572/95_1.pdf
13. Bitoh Y, Akuzawa N, Urayama K, Takigawa T, Kidowaki M, Ito K (2011) Peculiar nonlinear elasticity of polyrotaxane gels with movable cross-links revealed by multiaxial stretching. Macromolecules 44(21):8661–8667. https://doi.org/10.1021/ma201530z
14. Drozdov AD, Christiansen JD (2013) Constitutive equations in finite elasticity of swollen elastomers. Int J Solids Struct 50(9):1494–1504. https://doi.org/10.1016/j.ijsolstr.2013.01.031
15. Flory PJ (1953) Principles of polymer chemistry. Cornell University Press

Topology Optimization of Pressure-Loaded Multi-material Structures

Prabhat Kumar

Abstract Permitting multiple materials within a topology optimization setting increases the search space of the technique, which facilitates obtaining relatively high-performing and efficient optimized designs. Structures with multiple materials involving fluidic pressure loads find various applications. However, dealing with the design-dependent pressure loads is challenging in topology optimization, which gets even more pronounced with a multi-material framework. This paper provides a density-based topology optimization method to design fluidic pressure loadbearing multi-material structures. The design domain is parameterized using hexagonal elements as they ensure nonsingular connectivity. Pressure modeling is performed using the Darcy law with a conceptualized drainage term. The flow coefficient of each element is determined using a smooth Heaviside function considering its solid and void states. The consistent nodal loads are determined using the standard finite element methods. Multiple materials are modeled using the extended SIMP scheme. Compliance minimization with volume constraints is performed to achieve optimized loadbearing structures. A few examples are presented to demonstrate the efficacy and versatility of the proposed approach. The optimized results contain the prescribed volume of different materials.

Keywords Topology optimization · Multi-material · Extended SIMP · Design-dependent load · Honeycomb tessellation

1 Introduction

Topology optimization (TO) is a systematic design technique that determines the optimized material distribution within a given design domain by extremizing an objective with physical/geometrical constraints. The method has proven its relevance and utility in various applications [1, 2]. TO can be used for problems that

P. Kumar (✉)
Department of Mechanical and Aerospace Engineering, Indian Institute of Technology Hyderabad, Hyderabad, Telangana 502285, India
e-mail: pkumar@mae.iith.ac.in

S. Sidhardh et al. (eds.), *Advances in Structural Integrity for Mechanical, Civil, and Aerospace Applications*, Lecture Notes in Mechanical Engineering,
https://doi.org/10.1007/978-981-97-6367-2_28

involve partial differential equations (PDEs) to represent their physics. One solves the associated PDE in each TO iteration; thus, a ubiquitous option is to use finite element analysis. Herein, `HoneyMesher` code provided within `HoneyTop90` [3] is employed to parameterize the design domains by hexagonal elements. Such elements provide nonsingular connectivity between two juxtaposed elements; thus, they offer checkerboard-free and point-connection-free optimized designs [3–6]. Wachspress shape functions are used for the finite element analysis. These shape functions are pretty rich compared to bilinear shape functions [3, 7, 8]. Each element is associated with a design variable that decides different material phases for the element [3].

Lightweight and high-performing structures are typically constructed using different materials and find applications in aircraft and the military. When designing such structures using TO, a multi-material formulation within the TO setting is deemed to be included. On a parameterized level, the optimizer can choose to provide different materials to an element. That generally increases the optimization search space, which facilitates a better-performing optimized structure, perhaps with lightweight. In addition, with the unprecedented advancement of additive manufacturing techniques, optimized structures with multi-material can be readily fabricated [9]. The loads acting on such structures can be either design-independent or design-dependent (e.g., fluidic pressure load [10–12], self-weight [13], etc.). In various applications, structures experience design-dependent pressure loads. Such loads are challenging to model in a TO setting [10, 12, 14]. In addition, to the best of the author's knowledge, to date, only the method presented in [15] optimizes design-dependent pressure loadbearing structures with many candidate materials using the integer programming method and quad-elements. This paper leverages the benefits provided by honeycomb tessellation in the widely acclaimed density-based TO method. It integrates the tessellation with design-dependent pressure loads to design structures capable of bearing such loads.

Sigmund and Torquato [16] proposed a multi-material-based TO approach for designing materials with extreme thermal expansion. They extended the Solid Isotropic Material with Penalization (SIMP) scheme for modeling multiple materials. The procedure requires additional design variables per the number of candidate materials. The Recursive Multiphase Material Interpolation (RMMI) with mass constraints is proposed for many candidate materials TO in [17]. To design multi-material microstructures with different bulk moduli, Gibiansky and Sigmund [18] used the extended SIMP scheme, whereas Fujii et al. [19] used a homogenization-based method. An approach using a peak function was presented by Yin and Ananthasuresh [20]. The method does not require additional design variables as the number of candidate materials increases; however, a gradual adjustment of the parameters is needed to achieve multiple peaks. Zuo and Saitou [21] presented an ordered SIMP scheme that uses one design variable per element even for various materials; however, the interpolation scheme has a non-differentiable nature. An approach using a neural network for multi-material TO was presented in [22]. Sivapuram et al. [15] offered a TO process for multiple materials structures using integer programming and the extended SIMP approach. They also optimized multi-material pressure loadbearing structures in their paper. Cherrière et al. [23] presented a multi-material TO approach using the Wachspress interpolation. Besides the SIMP method, multi-material TO using the level-set method [24, 25], the phase field approach [26] and ESO method [27] also exist. We employ the extended SIMP scheme for multi-material modeling for designing pressure loadbearing structures herein.

Pressure loads alter their direction, magnitude, and/or location as TO advances; thus, they are challenging to model in a TO setting [10, 14]. Hammer and Olhoff [14] were the first to present a TO method to design structures involving pressure loads. The technique uses iso-density contour curves/surfaces to detect the loading surface. Fuchs and Shemesh [28] presented a method wherein they used an additional variable for pressure boundary evolution. Zhang et al. [29] proposed an element-based approach to locate the pressure boundary. Lee and Martins [30] presented a method that is not dependent upon the initial and final points for the pressure curves. Ibhadode et al. [31] provided an approach with boundary identification and load evolution. The aforementioned method evaluated load sensitivities using the finite difference method or completely neglected them from overall sensitivities. However, it is demonstrated in [10, 12] that the load sensitivities affect the final topologies. A fictitious thermal loading approach was presented by Chen and Kikuchi [32]. Sigmund and Clausen [33] presented an approach using the mixed finite element method that needs to satisfy the Babuska-Brezzi condition [34]. Zheng et al. [35] proposed a pseudo electrical potential scheme. Picelli et al. [36] used the level-set method in their approach. Kumar et al. [10] presented a density-based TO approach. They solved pressure loadbearing structures and pressure-actuated compliant mechanisms and considered load sensitivities within the approach. The approach [10] is successfully extended for 3D problems in [11], with the robust TO formulation in [37], with the material masks overlay approach for 0–1 optimized designs in [38] and recently, `TOPress` [12] and `SoRoTop` [39] MATLAB codes are presented using the method. For a comprehensive list of TO approaches involving design-dependent pressure loads, one may refer to [10, 11, 36] and references therein. We adopt the method presented in [10, 12, 38] to model the pressure loads with honeycomb tessellation [3].

The remainder of the manuscript is structured as follows. Section 2 provides the multi-material modeling formulation employed in this paper. Pressure load modeling is presented in Sect. 3; one can refer to [10] for a detailed description. Section 4 presents the topology optimization formulation for the multi-material. The success and efficacy of the proposed approach are demonstrated in Sect. 5 by showing different numerical results and discussions. Lastly, Sect. 6 draws the concluding remarks.

2 Multi-material Modeling

The different methods to model multiple materials with various TO settings are introduced in the previous section. We use a density-based TO method and the extended SIMP interpolation scheme to formulate the multiple materials. The number of design variables assigned to each element is the same as that of the candidate materials. The modified SIMP scheme for a two-phase (single material) TO is

$$E_i = E_{\min} + \tilde{\rho}_i^p (E_1 - E_{\min}) = (1 - \tilde{\rho}_i^p) E_{\min} + \tilde{\rho}_i^p E_1 \tag{1}$$

where $E_{\min}$ and E_1 are Young's moduli of the void and candidate material, respectively. $\tilde{\rho}_i$ is the filtered density of the ρ_i, and p is the SIMP parameter. Candidate materials with E_1 and E_2 Young's moduli (three-phase), Eq. 1 can be written as

$$E_i = (1 - \tilde{\rho}_{i1}^p) E_{\min} + \tilde{\rho}_{i1}^p \left(\left(1 - \tilde{\rho}_{i2}^p\right) E_1 + \tilde{\rho}_{i2}^p E_2 \right) \tag{2}$$

where $E_{\min} = 10^{-6} \times \min(E_1, E_2)$. $\tilde{\rho}_{i1}$ decides the solid and void phases of the material layout. $\tilde{\rho}_{i1} = 0$ ensures the void phase, whereas $\tilde{\rho}_{i1} = 1$ gives the solid phase. $\tilde{\rho}_{i1} = 1$ and $\tilde{\rho}_{i2} = 1$ imply the second material, whereas $\tilde{\rho}_{i1} = 1$ and $\tilde{\rho}_{i2} = 0$ indicate the first material. $\tilde{\rho}_{i1}$ can be considered as the topology variable and $\tilde{\rho}_{i2}$ is the material selection variable. Typically, the extended SIMP approach can be used for a higher number of different materials; however, the nonlinearity of the material model increases [15]. The total volume of the solid phase is determined as

$$V_s = \sum_{i=1}^{\mathrm{Nel}} \frac{v_i \tilde{\rho}_{i1}}{\mathrm{Nel}} \tag{3}$$

where v_i is the volume of element i, and Nel is the total number of elements used to parameterize the design domain. From Eq. (2), the density of material 1 is $\tilde{\rho}_{i1}\,(1 - \tilde{\rho}_{i2})$, and that of material 2 is $\tilde{\rho}_{i1}\tilde{\rho}_{i2}$. The number of design variables is doubled in the three-phase material model (Eq. 2) compared to the two-phase material model (Eq. 1). Likewise, one can determine the interpolated Young's modulus for the four-phase conditions [15].

To circumvent the checkerboard patterns and apply the minimum length scale control on the members of the optimized designs, we employ the classical density filter [40] herein. The filtered design variable $\tilde{\rho}_i$ of ρ_i is defined as

$$\tilde{\rho}_i = \frac{\sum_{j=1}^{\mathrm{Nel}} v_j \rho_j w(\mathbf{x}_j)}{\sum_{j=1}^{\mathrm{Nel}} v_j w(\mathbf{x}_j)}, \tag{4}$$

where v_j is the volume of neighboring element j. $w(\mathbf{x}_j) = \max\left(0,\ 1 - \frac{||\mathbf{x}_i - \mathbf{x}_j||}{r_{\mathrm{fill}}}\right)$, is the weight function, wherein $||(.)||$ is a Euclidean distance between centroids $\mathbf{x}_i$ and $\mathbf{x}_j$ of elements i and j, respectively. r_{fill} is the filter radius. In matrix form, Eq. (4) is written as

$$\tilde{\boldsymbol{\rho}} = \mathbf{H}\boldsymbol{\rho} \tag{5}$$

The derivative of $\tilde{\rho}_i$ (Eq. 4) with respect to ρ_j can be evaluated as

$$\frac{\partial \tilde{\rho}_i}{\partial \rho_j} = \frac{v_j w(\mathbf{x}_j)}{\sum_{k=1}^{\mathrm{Nel}} v_k w(\mathbf{x}_k)}, \quad \text{i.e.,} \quad \frac{\partial \tilde{\boldsymbol{\rho}}}{\partial \boldsymbol{\rho}} = \mathbf{H}^\top. \tag{6}$$

Note that at the beginning of the optimization, we evaluate **H** (Eq. 5) and store it as it does not change with the optimization iterations.

3 Pressure Load Modeling

This section presents the pressure modeling technique–the Darcy model with the conceptualized drainage term. Interested readers can refer to [10, 11] for a complete read with material. The approach is extended herein for multi-material modeling. We know the pressure boundary conditions, i.e., a pressure difference across the domain. Also, elements can be treated as a porous medium in the initial optimization stage. Therefore, a ubiquitous choice to model pressure field is the Darcy law [10], wherein the flux $\mathbf{q}$ is determined in terms of the permeability of medium κ, the fluid viscosity μ, and the pressure difference ∇p as:

$$\mathbf{q} = -\frac{\kappa}{\mu}\nabla p = -K(\tilde{\rho})\nabla p, \tag{7}$$

where $\tilde{\rho}$ is the filtered design variable (Eq. 5). $K(\tilde{\rho})$, the elemental flow coefficient, is determined for element i as

$$K(\tilde{\rho}_i) = K_v\left(1 - (1-\epsilon)\mathcal{H}(\tilde{\rho}_{i1},\ \beta_\kappa,\ \eta_\kappa)\right), \tag{8}$$

where $\mathcal{H}(\tilde{\rho_{i1}},\ \beta_\kappa,\ \eta_\kappa) = \frac{\tanh(\beta_\kappa\eta_\kappa)+\tanh(\beta_\kappa(\tilde{\rho}_{i1}-\eta_\kappa))}{\tanh(\beta_\kappa\eta_\kappa)+\tanh(\beta_\kappa(1-\eta_\kappa))}$. K_s is the flow coefficient of the solid state, and that of the void phase is K_v. $\epsilon = \frac{K_s}{K_v}$ is called flow contrast [11]. $\{\eta_\kappa,\ \beta_\kappa\}$ are called the flow parameters, which define the step position and slope of $K(\tilde{\rho}_i)$ respectively. Remark that in the case of multi-material modeling, we note that $\tilde{\rho}_{i1}$ decides the solid and void state of element i; therefore, the flow coefficient of element i (Eq. 8) is defined only in terms of $\tilde{\rho}_{i1}$ and is independent of the number of candidate materials are taken for the optimization.

It is noted in [10–12] that the Darcy law alone may not provide the meaningful pressure drop for a TO setting; therefore, we introduce a drainage term, Q_{drain}, that helps achieve the suitable pressure field variation within the design domain as TO advances. Q_{drain} is mathematically defined as

$$Q_{\text{drain}} = -D(\tilde{\rho}_e)(p - p_{\text{ext}}) \tag{9}$$

The final balanced equation of the Darcy formulation with the drainage term can be written as:

$$\nabla\cdot\mathbf{q} - Q_{\text{drain}} = 0. \tag{10}$$

When Eq. (10) is solved using the fundamentals of the finite element analysis and by applying the boundary conditions provided on the pressure load, we have

$$\mathbf{Ap} = \mathbf{0}, \tag{11}$$

where **A** and **p** are the global flow matrix and pressure vector respectively. By solving Eq. (11), one gets the pressure field within the design domain as TO progresses. Now, this obtained field is converted to the consistent nodal loads as

$$\mathbf{F} = -\mathbf{Tp}, \tag{12}$$

where **F** is the global force vector, and **T** is the transformation matrix [10]. Using the modified Darcy law, we get the pressure field, i.e., an implicit description of the loading surface and consistent loads for the TO formulation, which is discussed next.

4 Topology Optimization Formulation

As mentioned before, we use the density-based formulation [1] for TO, wherein the design domain is described by hexagonal elements [3]. We consider three-phase pressure load problems. The number of design variables for each element is the same as that of the candidate materials employed, i.e., two variables for each element. The following optimization problem is solved with the volume constraints:

$$\left.\begin{array}{ll}
\min\limits_{\rho} & \mathbf{u}^\top \mathbf{K}\mathbf{u} \\
\text{such that:} & \mathbf{Ap} = \mathbf{0} \\
 & \mathbf{Ku} = \mathbf{F} = -\mathbf{Tp} \\
\text{volume constraint:} & \sum\limits_{i=1}^{\mathtt{Nel}} v_i \tilde{\rho}_{i1} \le \left(v_{f_1} + v_{f_2}\right) \sum\limits_{i=1}^{\mathtt{Nel}} v_i \\
 & \sum\limits_{i=1}^{\mathtt{Nel}} v_i \tilde{\rho}_{i2} \le v_{f_2} \sum\limits_{i=1}^{\mathtt{Nel}} v_i \\
 & \mathbf{0} \le \tilde{\boldsymbol{\rho}} \le \mathbf{1} \\
\text{Data:} & v_{f_1},\ v_{f_2},\ E_1,\ E_2,\ p
\end{array}\right\}, \tag{13}$$

where **u** is the global displacement vector, and **K** is the global stiffness matrix, which is evaluated by assembling the elemental stiffness matrix. As mentioned before, $\tilde{\rho}_{i1}$ decides the topology of the optimized design, whereas $\tilde{\rho}_{i2}$ determines material selection. Two linear volume constraints are formulated using $\tilde{\rho}_{i1}$ and $\tilde{\rho}_{i2}$. The first volume constraint controls the total amount of the solid phase in the final topology, whereas the second fixes only solid phase 2. Alternatively, one may also choose to use separate volume fractions for each phase; however, those constraints will be nonlinear in $\tilde{\rho}_{i1}$ and $\tilde{\rho}_{i2}$.

The optimization problem noted in Eq. (13) is solved using a gradient-based optimizer, the Method of Moving Asymptotes (MMA, cf. [41]). Therefore, sensitivities of the objective and constraints are needed with respect to the design variables. The sensitivities are determined using the adjoint-variable method in view of Sects. 2 and 3. One can find a complete description of the sensitivities in [10–12].

5 Numerical Examples and Discussions

In this section, we present pressure-loaded design problems with two and three materials to demonstrate the success and efficacy of the presented method. We solve externally pressurized arch and pressure-loaded piston designs (Fig. 1). One may also consider other benchmark problems from Ref. [12] to solve. The flow contrast $\epsilon = 1 \times 10^-7$ is used. The out-of-plane thickness is fixed to 0.001 m, and the plane-stress conditions are assumed. Poisson's ratio $\nu = 0.40$ is used. The flow parameters $\{\eta_\kappa, \beta_\kappa\} = \{0.20, 10\}$ and the drainage parameters $\{\eta_d, \beta_d\} = \{0.20, 10\}$ are taken.

Figure 1 displays the design domains of the externally pressurized arch and pressure-loaded piston structures. Γ_p and Γ_{p_0} indicate the full and zero pressure load boundaries, respectively. The displacement boundary conditions are also depicted in the figure. $L_x = 0.2$ m and $L_y = 0.1$ m are taken for the arch design (Fig. 1a), whereas for the piston design, $L_x = 0.12$ m and $L_y = 0.04$ m are set (Fig. 1b). The arch design is solved considering the full model; this is performed to note any deviation from the symmetric nature of the result. However, we exploit the symmetric nature of the piston design; thus, we analyze and optimize only its symmetric half part. $N_\mathrm{ex} \times N_\mathrm{ey} = 200 \times 100$ FEs and $N_\mathrm{ex} \times N_\mathrm{ey} = 180 \times 120$ FEs are used to describe the arch and piston domains, respectively. N_ex and N_ey are respectively the numbers of hexagonal elements in x- and y-directions. $r_\mathrm{fill} = 3 \times \left(\frac{L_x}{N_\mathrm{ex}}, \frac{L_y}{N_\mathrm{ey}}\right)$ and $r_\mathrm{fill} = 3.6\sqrt{3} \times \left(\frac{L_x}{N_\mathrm{ex}}, \frac{L_y}{N_\mathrm{ey}}\right)$ are employed for the arch and piston structures. Different

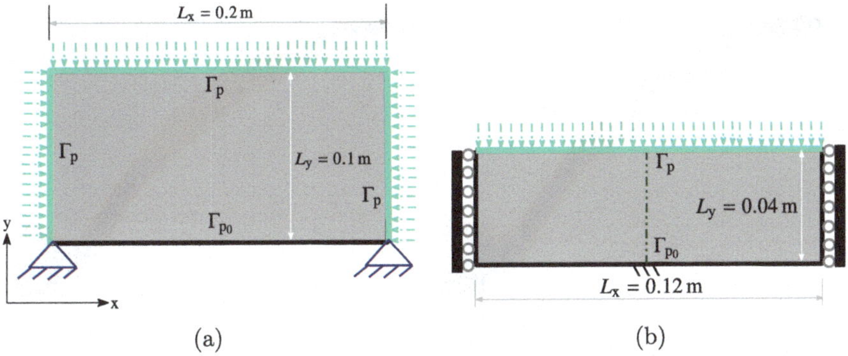

Fig. 1 Design domains **a** externally pressurized arch structure and **b** Piston structure. Displacement and pressure boundary conditions are also depicted. Γ_p and Γ_{p_0} indicate edges with 1 bar and 0 bar pressure loads, respectively

r_{fill} are employed to achieve different minimum length scales for the problems. The maximum number of the MMA iterations is fixed to 100. The external move limit for the MMA optimizer is set to 0.1 [41]. Kumar et al. [42] extend the proposed method herein for optimizing pressure-actuate multi-material compliant mechanism.

5.1 Two-Material Cases

Here, we consider two candidate materials with $E_1 = 40$ MPa and $E_2 = 100$ MPa for designing the arch and piston structures. $v_{f_1} = 0.1$ and $v_{f_2} = 0.1$ are considered.

Figures 2 and 3 depict optimized loadbearing arch and piston results. The optimized designs contain both materials (three-phase: void, material 1, and material 2). In addition, the symmetric nature of the arch structure is maintained in its optimized design (Fig. 2). The optimizer places material with high elastic stiffness at the displacement boundary to reduce the deformation, and hence, it minimizes the objec-

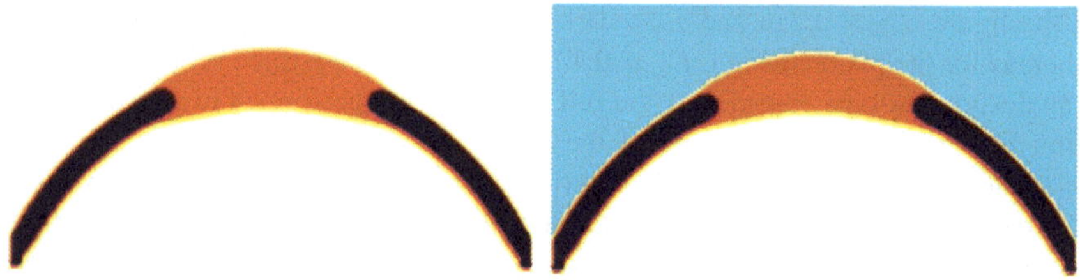

(a) (b)

Fig. 2 Optimized results of the externally pressurized arch structure with two candidate materials **a** optimized material layout **b** optimized material layout with the final pressure field. Black → Material 2 ($E_2 = 100$ MPa), Orange → Material 1 ($E_1 = 40$ MPa)

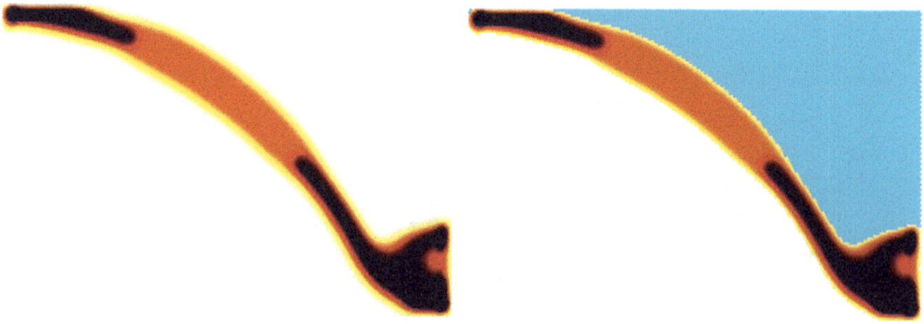

(a) (b)

Fig. 3 Symmetric half optimized pressure loadbearing piston design with two candidate materials **a** optimized material layout and **b** optimized material layout with the final pressure field. Black → Material 2 ($E_2 = 100$ MPa), Orange → Material 1 ($E_1 = 40$ MPa)

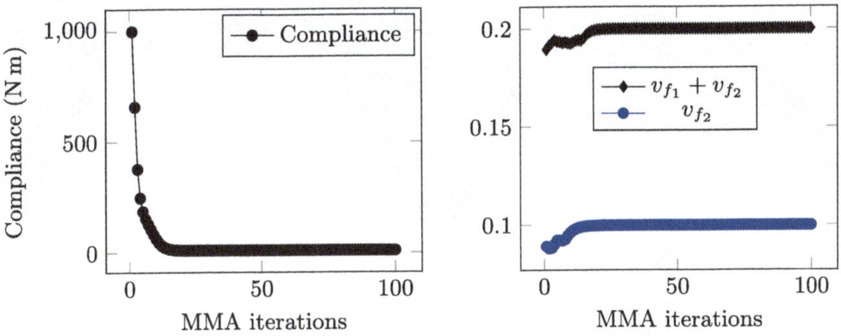

(a) Objective convergence history

(b) Convergence plots of volume constraints

Fig. 4 Convergence plots of the objective and volume constraints of the arch problem

tive. Figures 2b and 3c depict the optimized material layouts of the arch and piston structures with the final pressure fields. The optimizer provides the final material layouts to contain the pressure loads and form the stiffest designs. As in TO approaches, the boundaries of the optimized designs are constituted via the edges of the FEs employed to describe the domains. Therefore, typically, optimized results contain undulated boundaries. To achieve relatively smooth boundaries of the optimized designs, one can use the boundary smoothing scheme per [43] within the proposed approach. The two-material arch design's objective and constraints history plots are shown in Fig. 4. The objective convergence for the problem is rapid and smooth. One can note that both volume constraints remain active at the end of optimization.

5.2 *Three-Material Cases*

For three candidate materials, we consider $E_1 = 10$ MPa, $E_2 = 40$ MPa and $E_3 = 100$ MPa for designing pressure loadbearing arch and piston structures. $v_{f_1} = 0.1$, $v_{f_2} = 0.1$ and $v_{f_3} = 0.05$ are set.

The optimized loadbearing arch and piston designs with three materials are depicted in Figs. 5a and 6a, respectively. The final pressure fields with the optimized material layout for these designs are also shown in Figs. 5b and 6b, respectively. The optimized shape of the arch design maintains symmetry. Material with $E_3 = 100$ MPa (black) forms the fixed boundaries in the optimized designs. The convergences for objective and volume constraints for the piston problem are depicted in Fig. 7. The objective convergence is smooth and rapid, as observed for the two-material arch problem (Fig. 4). The volume constraints remain active at the end of the optimization. The solved TO problems are non-convex, i.e., their solutions depend upon the initial guesses; thus, the optimized designs with three-material (Figs. 5 and 6) are different from those with two-material (Figs. 2 and 3).

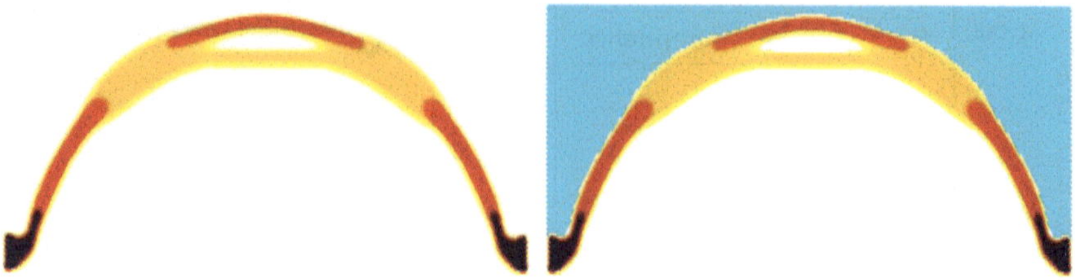

(a) (b)

Fig. 5 Optimized results of the externally pressurized arch structure with three candidate materials **a** optimized material layout **b** optimized material layout with the final pressure field. Black → Material 3 ($E_3 = 100\,\text{MPa}$), Orange → Material 2 ($E_2 = 40\,\text{MPa}$), Gold → Material 1 ($E_1 = 10\,\text{MPa}$)

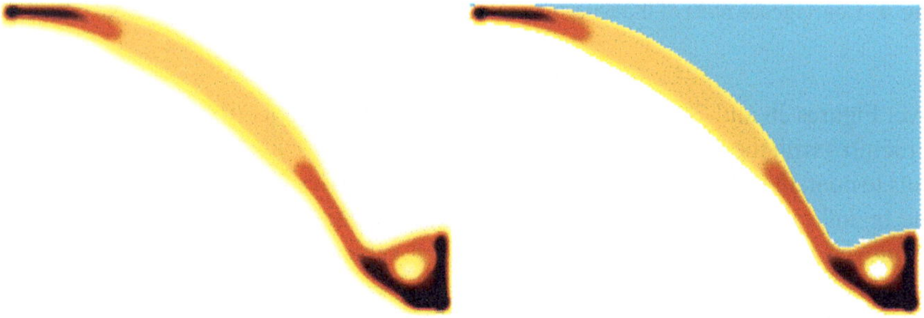

(a) (b)

Fig. 6 Symmetric half optimized pressure loadbearing piston design with three candidate materials **a** optimized material layout and **b** optimized material layout with the final pressure field. Black → Material 3 ($E_3 = 100\,\text{MPa}$), Orange → Material 2 ($E_2 = 40\,\text{MPa}$), Gold → Material 1 ($E_1 = 10\,\text{MPa}$)

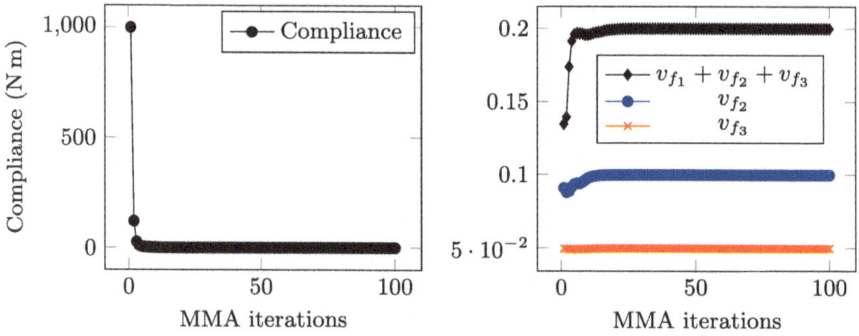

(a) Objective convergence (b) Convergence plots of volume constraints

Fig. 7 Convergence plots of the objective and volume constraints of the three-material piston problem

6 Concluding Remarks

This paper presents a topology optimization approach to designing fluidic pressure-loaded structures with multi-material. The versatility of the proposed method is demonstrated by solving two structural problems subjected to fluidic pressure loads with two and three candidate designs. The extended SIMP approach is employed for the modeling of the multi-material. Compliance is minimized with two and three linear volume constraints for two and three candidate materials. Honeycomb tessellation describes the design domain that inherently provides nonsingular connectivity between two juxtaposed elements. In the optimized designs, the symmetric nature of the problems remains intact. In the optimized designs, the material with the highest elastic stiffness forms the fixed boundary conditions, thus reducing the deformation and compliance of the structures. The objective convergence is smooth and rapid. Volume constraints remain active at the end of the optimization.

Pressure load is modeled using the Darcy law with the conceptualized drainage term as per [10], wherein the flow coefficient of an element is determined using its solid and void phases irrespective of the number of candidate materials. Likewise, the drainage term is defined. The presented scheme works well, as noted by solving two problems. We envision presenting a topology optimization approach for pressure-driven multi-material 3D loadbearing structures and compliant mechanisms as one of the future directions.

References

1. Sigmund O, Maute K (2013) Topology optimization approaches. Struct Multidiscip Optim 48(6):1031–1055
2. Van Dijk NP, Maute K, Langelaar M, Van Keulen F (2013) Level-set methods for structural topology optimization: a review. Struct Multidiscip Optim 48:437–472
3. Kumar P (2023) HoneyTop90: a 90-line MATLAB code for topology optimization using honeycomb tessellation. Optim Eng 24(2):1433–1460
4. Langelaar M (2007) The use of convex uniform honeycomb tessellations in structural topology optimization. In: 7th world congress on structural and multidisciplinary optimization. Seoul, South Korea, May, pp 21–25
5. Saxena A (2011) Topology design with negative masks using gradient search. Struct Multidiscip Optim 44(5):629–649
6. Talischi C, Paulino GH, Le CH (2009) Honeycomb wachspress finite elements for structural topology optimization. Struct Multidiscip Optim 37(6):569–583
7. Wachspress EL (1975) A rational finite element basis
8. Sukumar N, Tabarraei A (2004) Conforming polygonal finite elements. Int J Numer Methods Eng 61(12):2045–2066
9. Bandyopadhyay A, Heer B (2018) Additive manufacturing of multi-material structures. Mater Sci Eng R Rep 129:1–16
10. Kumar P, Frouws J, Langelaar M (2020) Topology optimization of fluidic pressure-loaded structures and compliant mechanisms using the Darcy method. Struct Multidiscip Optim 61:1637–1655

11. Kumar P, Langelaar M (2021) On topology optimization of design-dependent pressure-loaded three-dimensional structures and compliant mechanisms. Int J Numer Methods Eng 122(9):2205–2220
12. Kumar P (2023) TOPress: a MATLAB implementation for topology optimization of structures subjected to design-dependent pressure loads. Struct Multidiscip Optimiz 66(4):97
13. Kumar P (2022) Topology optimization of stiff structures under self-weight for given volume using a smooth Heaviside function. Struct Multidiscip Optimiz 65(4):1–17
14. Hammer VB, Olhoff N (2000) Topology optimization of continuum structures subjected to pressure loading. Struct Multidiscip Optimiz 19(2):85–92
15. Sivapuram R, Picelli R, Yoon GH, Yi B (2021) On the design of multimaterial structural topologies using integer programming. Comput Methods Appl Mech Eng 384:114000
16. Sigmund O, Torquato S (1997) Design of materials with extreme thermal expansion using a three-phase topology optimization method. J Mech Phys Solids 45(6):1037–1067
17. Gao T, Zhang W (2011) A mass constraint formulation for structural topology optimization with multiphase materials. Int J Numer Methods Eng 88(8):774–796
18. Gibiansky LV, Sigmund O (2000) Multiphase composites with extremal bulk modulus. J Mech Phys Solids 48(3):461–498
19. Fujii D, Chen B, Kikuchi N (2001) Composite material design of two-dimensional structures using the homogenization design method. Int J Numer Methods Eng 50(9):2031–2051
20. Yin L, Ananthasuresh GK (2001) Topology optimization of compliant mechanisms with multiple materials using a peak function material interpolation scheme. Struct Multidiscip Optimiz 23(1):49–62
21. Zuo W, Saitou K (2017) Multi-material topology optimization using ordered SIMP interpolation. Struct Multidiscip Optimiz 55(2):477–491
22. Chandrasekhar A, Suresh K (2021) Multi-material topology optimization using neural networks. Comput Aided Des 136:103017
23. Cherrière T, Laurent L, Hlioui, S, Louf F, Duysinx P, Geuzaine C, Ben Ahmed H, Gabsi M, Fernández E (2022) Multi-material topology optimization using wachspress interpolations for designing a 3-phase electrical machine stator. Struct Multidiscip Optimiz 65(12):352
24. Wang MY, Wang X (2004) Color level sets: a multi-phase method for structural topology optimization with multiple materials. Comput Methods Appl Mech Eng 193(6–8):469–496
25. Liu P, Shi L, Kang Z (2020) Multi-material structural topology optimization considering material interfacial stress constraints. Comput Methods Appl Mech Eng 363:112887
26. Tavakoli R (2014) Multimaterial topology optimization by volume constrained Allen-Cahn system and regularized projected steepest descent method. Comput Methods Appl Mech Eng 276:534–565
27. Zhao L, Xu B, Han Y, Xie YM (2019) Topology optimization of dynamic stress response reliability of continuum structures involving multi-phase materials. Struct Multidiscip Optimiz 59:851–876
28. Fuchs MB, Shemesh NNY (2004) Density-based topological design of structures subjected to water pressure using a parametric loading surface. Struct Multidiscip Optimiz 28(1):11–19
29. Zhang H, Zhang X, Liu S (2008) A new boundary search scheme for topology optimization of continuum structures with design-dependent loads. Struct Multidiscip Optimiz 37(2):121–129
30. Lee E, Martins JR (2012) Structural topology optimization with design-dependent pressure loads. Comput Methods Appl Mech Eng 233:40–48
31. Ibhadode O, Zhang Z, Rahnama P, Bonakdar A, Toyserkani E (2020) Topology optimization of structures under design-dependent pressure loads by a boundary identification-load evolution (BILE) model. Struct Multidiscip Optimiz 62:1865–1883
32. Chen BC, Kikuchi N (2001) Topology optimization with design-dependent loads. Finite Elem Anal Des 37(1):57–70
33. Sigmund O, Clausen PM (2007) Topology optimization using a mixed formulation: an alternative way to solve pressure load problems. Comput Methods Appl Mech Eng 196(13–16):1874–1889

34. Zienkiewicz OC, Taylor RL (2005) The Finite Element Method for Solid and Structural Mechanics. Butterworth-Heinemann
35. Zheng B, Chang CJ, Gea HC (2009) Topology optimization with design-dependent pressure loading. Struct Multidiscip Optimiz 38(6):535–543
36. Picelli R, Neofytou A, Kim HA (2019) Topology optimization for design-dependent hydrostatic pressure loading via the level-set method. Struct Multidiscip Optimiz 60(4):1313–1326
37. Kumar P, Langelaar M (2022) Topological synthesis of fluidic pressure-actuated robust compliant mechanisms. Mech Mach Theory 174:104871
38. Kumar P, Saxena A (2022) An improved material mask overlay strategy for the desired discreteness of pressure-loaded optimized topologies. Struct Multidiscip Optimiz 65(10):304
39. Kumar P (2023) SoRoTop: a hitchhiker's guide to topology optimization MATLAB code for design-dependent pneumatic-driven soft robots. Optim Eng pp 1–35
40. Bruns TE, Tortorelli DA (2001) Topology optimization of non-linear elastic structures and compliant mechanisms. Comput Methods Appl Mech Eng 190(26–27):3443–3459
41. Svanberg K (1987) The method of moving asymptotes—a new method for structural optimization. Int J Numer Methods Eng 24(2):359–373
42. Kumar P, Pinskier J, Howard D, Langelaar M. Topology optimization of fluidic pressure-driven multi-material compliant mechanisms. In: International design engineering technical conferences and computers and information in engineering conference, vol 87363, p V008T08A038. American Society of Mechanical Engineers
43. Kumar P, Saxena A (2015) On topology optimization with embedded boundary resolution and smoothing. Struct Multidiscip Optimiz 52(6):1135–1159

Spatial Load Distribution in Composite Flapping Wing Under Small Deformation

Vivek Khare and Sudhir Kamle

Abstract Present study reports the theoretically supported wind tunnel experimental investigations on the load distribution in a flapping wing under low amplitude flapping. Wind tunnel experiments are performed at 2, 4 and 6 m/s wind speed for hummingbird inspired nanocomposite wing. The spatial deformations fields are obtained from Digital image correlation (DIC). Modified Theodorsen lift theory is implemented to obtain load distribution from the local chord deformations to obtain aerodynamic loads. Three spatial locations at 25, 50 and 75% wing-span are considered for deformation data. The theoretical formulation was used to separate out the inertial and aerodynamic components of loads. Results reveal that the mid-stroke deformation mostly contributes for the aerodynamic lift, while all other deformation instants represent the dominance of inertial loads.

Keywords Aerodynamic loads · Nanocomposite wing · Spatial deformation · Wing shape change

1 Introduction

Modern flight vehicle designers are emphasizing on the flexible, lightweight and fatigue resistant structures [1–4]. Flexible wings MAVs, drones, gossamer spacecraft and engineered metamaterials etc. are few latest advancements in line with the research efforts made in past decade [4, 5]. Flexibility induced shape change in the structures has the ability to improve the aerodynamic performance and customize the conventional composites with desired characteristics [4].

Flexible flapping wings of Natural flyers are biomimicked in unmanned aerial systems [5, 6]. They adjust their shape and size in response to the fluctuations in the wind environment and climatic changes. Mechanism of shape change during flapping

---

V. Khare (✉)
Mississippi State University, Starkville, Mississippi 39759, United States
e-mail: vivekconcept@gmail.com

S. Kamle
Indian Institute of Technology, Indore, Madhya Pradesh 453552, India

S. Sidhardh et al. (eds.), *Advances in Structural Integrity for Mechanical, Civil, and Aerospace Applications*, Lecture Notes in Mechanical Engineering,
https://doi.org/10.1007/978-981-97-6367-2_29

is influenced by the wing flexibility and vice-versa. Previous studies have reported the lift and drag distribution during flapping cycle, which do not provide significant information about the wing shape at low amplitude instantaneous deformation. Furthermore, the role of local deformation fields in load estimation is not completely understood due to theoretical paucities. Present study reports the theoretically supported experimental investigations on the load distribution in a composite flapping wing under low amplitude flapping. Modified Theodorsen lift theory is implemented to obtain load distribution from the local chord deformations. Digital image correlation (DIC)-based deformation measurement system is coupled with two component load estimation mechanism. The spatial deformations fields are obtained from DIC and fitted with the theoretical model to obtain aerodynamic loads. The theoretical formulation also differentiates in between the inertial and aerodynamic components of loads. The wind tunnel experiments are performed by syncing DIC and load estimation mechanism. Three spatial locations at 25, 50 and 75% wing-span are considered for deformation data. The experiments are performed at 2, 4 and 6 m/s wind speed for two different composite wings morphologies. Results reveal that the mid-stroke deformation mostly contributes for the aerodynamic lift, while all other locations represent the dominance of inertial loads.

2 Theoretical Load Distribution

Theodorsen and Garrick [7–9] developed formulated the lift in a pitching and plunging airfoil as mentioned in Fig. 1. The theory was later modified for the rigid wings. Shankha et al. [10] proposed the lift theory for flexible flapping wings. Similar to the rigid wings case, the lift developed in flexible flapping wings is due to both inertial and aerodynamic components. However, the flexibility attributes are more pronounced in a typical flapping cycle. It is evident from the previous studies that the magnitude of aerodynamic loads varies during upstroke and downstroke [4, 11–13]. On the other hand, identification of spatial location over the wing surface is also

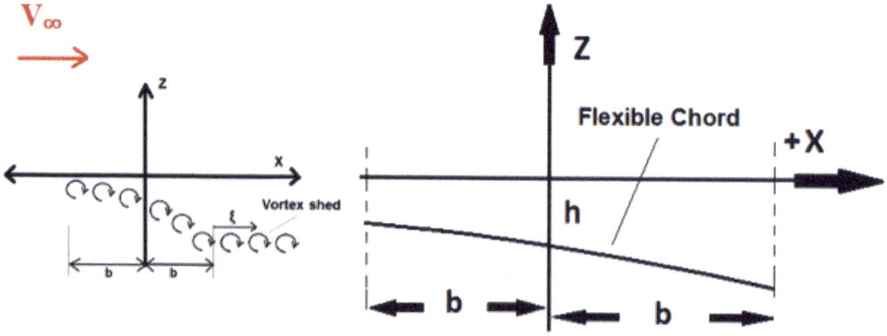

Fig. 1 Vortex shed and deformation in flexible wing chord

important to investigate. This is due to the fact that the adverse pressure gradient at the bottom surface helps the wings change their local shape and size. Such phenomenon is observed during the stroke reversal. It has been reported that downstroke produce more aerodynamic lift in bioinspired flapping wings [5, 6, 13].

Modified Theodorsen theory [7, 10] assumed out of plane chord deformation and the expression is shown in Eq. (1). The chord deformation considers heaving amplitude and flexibility parameters. Lift expression shown in Eq. (2), clearly differentiates in between the aerodynamic component of loads from the overall lift response. Another lift component is due to the inertial effects. The aerodynamic and inertial lift components are known as circulatory and non-circulatory lift respectively and given in second and first term in Eq. (2) [10, 14].

$$Z_a = -h - A_1(x-b) - A_2(x-b)^2 - A_3(x-b)^3 \quad (1)$$

$$L = \mathrm{i}\omega\pi\rho b^2\left[\dot{h} + V_\infty(A_1 + 2bA_2 + \frac{15}{4}b^2A_3)\right] + 2\pi\rho V_\infty bC(k)\left[\dot{h} + V_\infty\left(A_1 + 3bA_2 + \frac{15}{2}b^2A_3\right)\right] \quad (2)$$

where, Z_a is the out of plane chord displacement, h is heaving amplitude, L is total lift, b is semi-chord length, V_∞ is incoming wind speed, $C(k)$ is Theodorsen coefficient, A_1, A_2 and A_3 are polynomial fitting constants.

3 Experimental

Hummingbird inspired nanocomposite flapping wing was developed as reported in our previous studies [15, 16]. Nanocomposites offer excellent thermomechanical stability at extreme environment and therefore preferred for biomimicking [17, 18]. Thin nanocomposite films were developed and transformed to bioinspired wings with Carbon fiber epoxy composite as the leading edge stiffener as shown in Fig. 2.

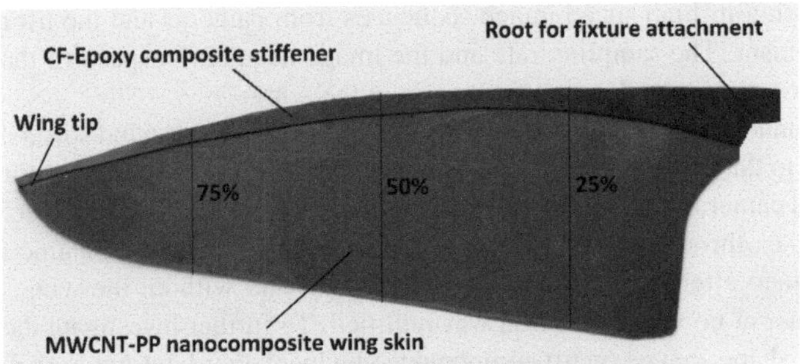

Fig. 2 Flexible nanocomposite wing with chord sections

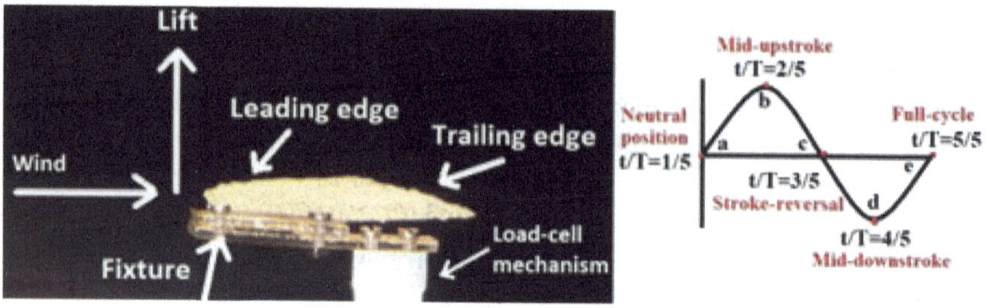

Fig. 3 **a** Flexible wing in wind tunnel test section and **b** deformation instants in normalized time scale

A Load cell mechanism is made for the estimation of lift force. A root fixture was developed to hold the wing at its root. The other end of the fixture was attached to the load cell mechanism as shown in Fig. 3a. The orientation of the fixture was measured with the angle sensor to ensure zero angle of attack against the incoming wind. The load cell mechanism with the wing attached was installed inside the wind tunnel test section. The wing was excited at 28 Hz flapping frequency which is close to the wingbeat frequency of Ruby-Throated Hummingbird [4, 13]. The excitation was given from shaker-amplifier-based system controlled using a function generator.

A stereovision-based Digital image correlation system was designed and developed for the deformation measurement in light and flexible aerospace structures, which is being utilized in the present study. The principle of stereovision setup in order to obtain both in-plane and out of plane deformation is well explained in previous studies [15, 16, 19, 20]. The stereovision system composed of two high speed cameras (Olympus, i-Speed TR, 10,000 FPS) attached to a rigid frame inside the wind tunnel test section. The DIC setup was calibrated to obtain various intrinsic and extrinsic camera characteristics which were included in the correlation program later for wing motion estimation [20]. From the global deformation field obtained from DIC, the local deformation both in spanwise and chordwise directions can be estimated.

Load cell mechanism and the DIC cameras were synchronized and configured in LabVIEW environment. The idea was to obtain both the structural deformation from the processing of high speed image sequences from cameras and the lift data from the mechanism. The sampling rate and the image acquisition speed of the cameras were chosen as per the flapping frequency of the wing.

Wind tunnel experiments were performed at 2, 4 and 6 m/s wind speed. The wing was made to flap at 28 Hz flapping frequency. Deformation field was captured using high speed cameras and the lift force was obtained from the load cell mechanism in real time, i.e., through the LabVIEW program. Inertial and aerodynamic lift forces were obtained after performing experiments with and without the wing. The load cell response at no wind condition was nullified. To further investigate the effect of local wing shape change on lift components, the local chord deformation data at five instants (refer Fig. 3b) was processed to calculate circulatory and non-circulatory lift as explained earlier. The image sequences obtained from the DIC setup were processed using an image correlation program in open source software DICe [21] to obtain out of plane wing deformation.

4 Results and Discussion

4.1 Spatial Load Estimation

Figures 4, 5 and 6 represent the spatial displacement field obtained at 25, 50 and 75% wing-span over a flapping cycle. Five consecutive deformation instants during start, mid-downstroke, downstroke reversal, mid-upstroke and full cycle are considered (refer Fig. 3). The deformation instants are shown in the normalized scale along the chord axis. These profiles represent chord deformation from the wing leading edge toward the trailing edge.

The displacements are obtained after correlating the grayscale subsets in the sequence of images. These subsets are derived along the chord section from the undeformed image. It is evident that due to wing flexibility, the chordwise deformation is substantially identified in the deformation instants. The stiff leading edge region doesn't deform much compared to the unstiffened trailing edge. The idea behind stiffening leading edge and unstiffened trailing edge is to introduce chordwise flexibility, which is mostly found in the biological flyers such as hummingbirds [4, 5, 22]. The humerus in hummingbird wings, which is the upper arm bone that joins shoulder with their elbows, can be mimicked with the leading edge stiffener.

Fruitful inferences of such chordwise flexibility are noticed at 2 and 4 m/s wind speed. These chordwise deformations are consistent both at 25 and 50% span locations. The amplitude of deformation is high at 50% location which is at the half span. Significant deformation amplitude was also observed at 75% chord section. We noticed traces of trailing edge deformation, mainly at half cycle and mid-downstroke near the wing tip, i.e., at 75% chord section. Due to these sectional chord shape attributes at different wind speeds, the net lift obtained from the conventional load cell mechanism or the sting balances in the wing tunnel would be inappropriate. It is

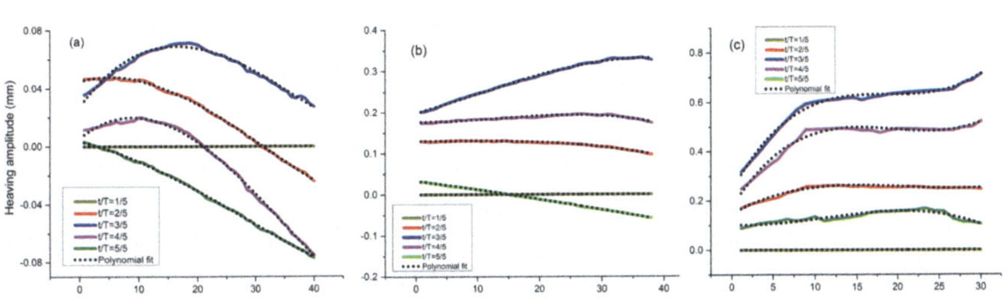

Fig. 4 Spatial chord deformation at 25, 50 and 75% span-length from the wing-root 2 m/s

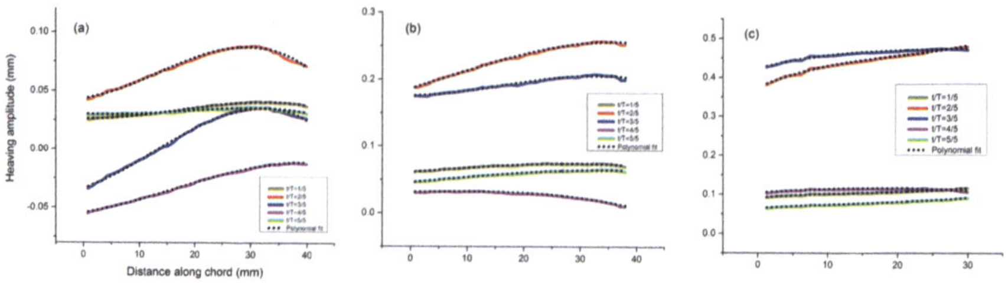

Fig. 5 Spatial chord deformation at 25, 50 and 75% span-length from the wing-root 4 m/s

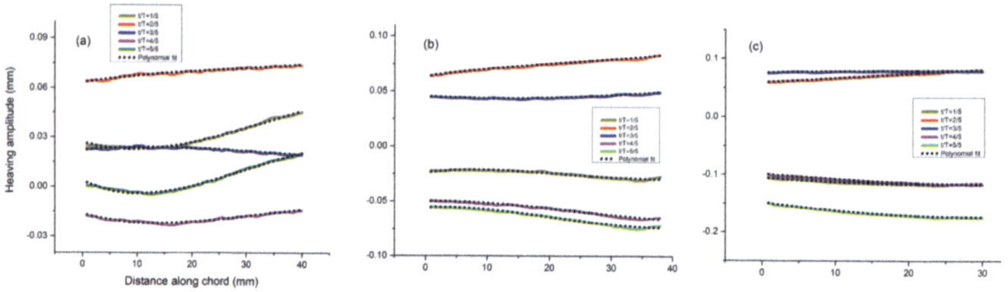

Fig. 6 Spatial chord deformation at 25, 50 and 75% span-length from the wing-root 6 m/s

important to identify the aerodynamic implications of these spatial deformations and the associated lift components. Load cell predicted aerodynamic lift and theoretically estimated sectional lift due to these deformations are discussed in next sections.

4.2 *Aerodynamic Lift*

The lift forces are obtained for every considered wind speed from the load cell mechanism. The load cell mechanism was precisely calibrated at all considered speeds and the response due to mechanism was nullified while obtained actual wing lift. Figures 7, 8 and 9 represent the cyclic lift during the flapping cycle at 2, 4 and 6 m/s wind speed. The inertial and aerodynamic lift are represented as non-circulatory and circulatory lift. The response was recorded with only stiffener of the wing and then with the whole wing. The reason behind such adjustment is that the aerodynamically effective lift is achieved due to the wing shape change and the pressure gradient at the bottom surface of the wing.

There is no actual aerodynamic lift without the wing surface and the load cell response represents overall lift. Hence, in order to obtain the inertial component of lift, the same experiment was performed with only leading edge stiffener. The only stiffener case represents the major inertial component of lift. The net aerodynamic lift is obtained after subtracting overall response from inertial component. It is to be noted that the magnitude of circulatory component of lift force was higher than the non-

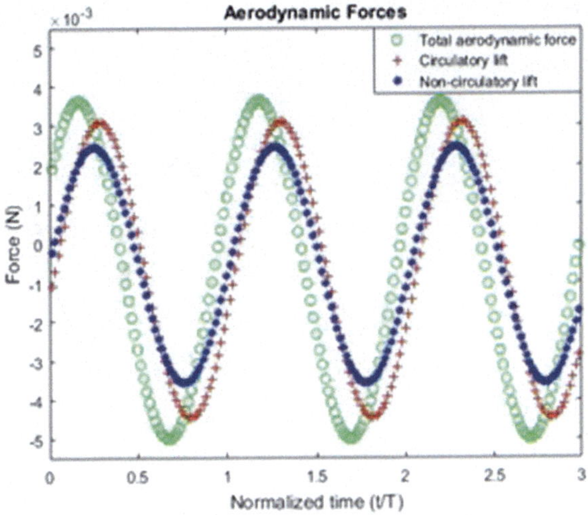

Fig. 7 Aerodynamic lift components from load cell mechanism at 2 m/s

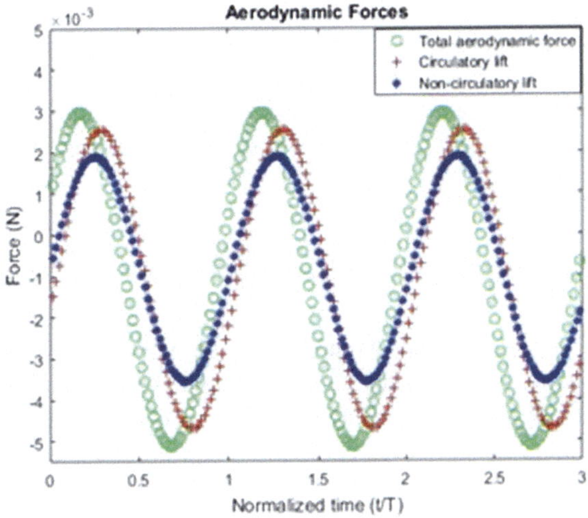

Fig. 8 Aerodynamic lift components from load cell mechanism at 4 m/s

circulatory component. The aerodynamic lift obtained here is from the experimental observation, which could also include some uncertainty such as wind instability at low speed and due to excitation mechanism.

Low amplitude flapping induced lift should be precisely investigated since the major source of lift in these wing structures is due to flexibility. The shape change in natural flyers is transient phenomenon, specifically during stroke reversal. Such factors strongly advocate the analysis of local deformation fields in the wing. Flapping wings have extended leading edge and trailing edge which sharply meet at the wing tip. The chord length generally decreases from root to tip. Hence, the deformation profile is essential to obtain and subsequently the aerodynamic lift should

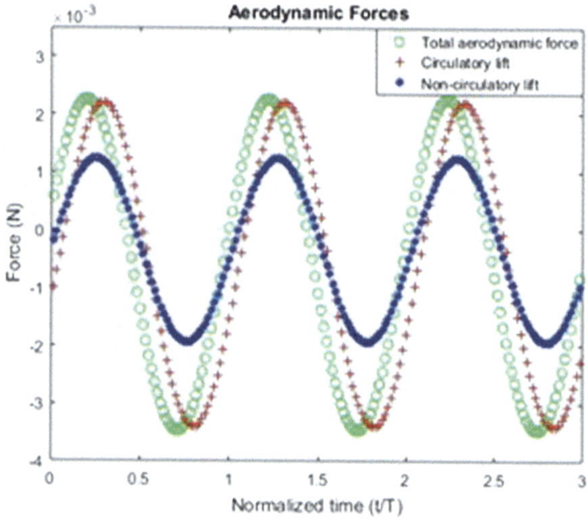

Fig. 9 Aerodynamic lift components from load cell mechanism at 6 m/s

be measured from the spatial deformation field [10, 11, 23, 24]. This also helps in differentiating between the circulatory and non-circulatory lift. The deformation profiles in Sect. 4.1 are processed with the modified theoretical lift expression shown in Eqs. (1) and (2) and discussed in the next section.

4.3 Effect of Wind Speed on Sectional Lift Distribution

Figures 10, 11 and 12 represent the circulatory and non-circulatory lift obtained at 25, 50 and 75% wing location. We included two additional wind speed of 8 and 10 m/s here for precise understanding. It is evident that the inertial forces were dominant at almost all speeds at 25% location. The wing experiences greatest acceleration at the beginning of strokes which could have increased the non-circulatory lift. However, at mid-downstroke, the circulatory lift was high. Evolution of circulatory lift was noticed over the flapping cycle at 50 and 75% wing location. This could be attributed to the chordwise shape change as depicted in Figs. 4, 5 and 6. The mid-stroke instants exhibit high circulatory lift which is evident from the amplitude of out of plane deformation from spatial fields obtained from DIC. Apart from the circulatory lift, the non-circulatory lift was also increased with the wind speed. This additional lift is due to the acceleration caused by the surrounding wind, which causes the wing to deform. A careful observation of the lift components at 75% wing location indicates that the lift was reduced near the root of the wing.

Overall, it was observed that both circulatory and non-circulatory forces are developed during flapping. Circulatory forces could be due to flexibility induced wing shape change, whereas non-circulatory forces are more pronounced due to the wind induced acceleration. High magnitude of lift components at high wind speed could

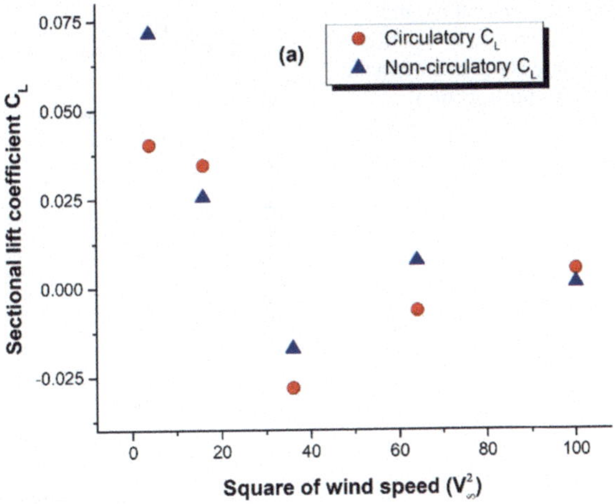

Fig. 10 Circulatory and non-circulatory forces from unsteady aerodynamic model at 25% chord section

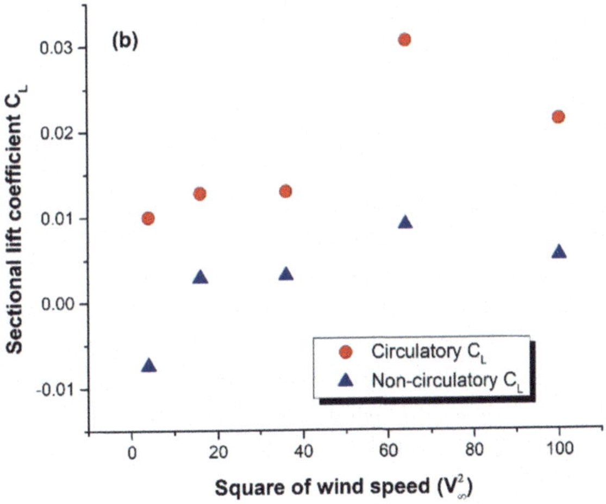

Fig. 11 Circulatory and non-circulatory forces from unsteady aerodynamic model at 50% chord section

not be considered aerodynamically effective. This is due to the fact that bird inspired wings have a limited flapping frequency range less than 80 Hz [3, 4, 22] and they take advantage of low wind speed to increase maneuverability during flapping flight. The consequences of flapping at high wind speeds are first shown in the spatial deformation fields in earlier section. It was found that the chordwise flexibility was either absent or had imposed reverse camber like effect, which is not an aerodynamically effective shape of deformation. Moreover, there are evidences of large trailing edge deformation. Such mechanisms are well reported in previous studied as flow leakage conditions.

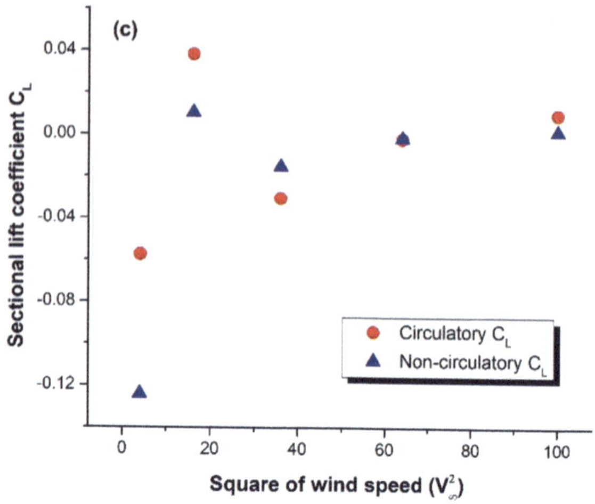

Fig. 12 Circulatory and non-circulatory forces from unsteady aerodynamic model at 75% chord section

5 Conclusion

Present study was aimed to evaluate the aerodynamic lift due to flexibility induced deformation in a nanocomposite flapping wing inspired by Ruby-Throated Hummingbird wing. A leading edge stiffened wing was tested in wind tunnel environment and the spatial deformation was studied to differentiate between non-circulatory and circulatory components of the lift. It was found that circulatory forces were dominating the overall lift during midstrokes, whereas the non-circulatory forces were also noticed at the beginning of strokes. Both lift components were more pronounced at low wind speed. High lift at high wind speeds was due to the trailing edge deformation as depicted in spatial deformation fields. The spatial field acquired from DIC results also supported the lift inferences at different spanwise locations.

Acknowledgements The authors would like to thank National Wind Tunnel Facility at Indian Institute of Technology Kanpur, for providing the test facility. Financial assistance from MHRD, New Delhi is gratefully acknowledged.

References

1. Nakata T, Liu H (2011) Aerodynamic performance of a hovering hawkmoth with flexible wings: a computational approach. Proc R Soc B 279
2. Le TQ, Truong TV, Park SH, Truong TQ, Ko JH, Park HC, Byun D (2013) Improvement of the aerodynamic performance by wing flexibility and elytra-hind wing interaction of a beetle during forward flight. J R Soc Interface 10
3. Zhao L, Huang Q, Deng X, Sane S (2009) Aerodynamic effects of flexibility in flapping wings. J R Soc Interface 44(7):485–497

4. Shyy W, Aono H, Kang CK, Liu HL (2013) An Introduction to Flapping Wing Aerodynamics, 1st edn. Cambridge University Press, New York (Chapters 3 and 4)
5. Yusoff H, Abdullah MZ, Mujeebu MA, Ahmad KA (2015) Effect of skin flexibility on aerodynamic performance of flexible skin flapping wings for micro air vehicles. Exp Tech 39:11–20
6. Kim DK, Han JH, Kwon KJ (2009) Wind-tunnel tests for a flapping wing model with a changeable camber using macro-fiber composite actuators. Smart Mater Struct 18
7. Theodorsen T (1935) General theory of aerodynamic stability and the mechanism of flutter. NACA report 496, Washington DC
8. Garrick IE (1936) Propulsion of a flapping and oscillating airfoil. National Advisory Committee for Aeronautics
9. DeLaurier JD (1993) An aerodynamic model for flapping-wing flight. Aeronaut J 97:125–130
10. Koley SS (2015) Theoretical and experimental investigation of a heaving two dimensional flexible wing. Master's thesis. Indian Institute of Technology Kanpur
11. Shyy W, Smith R (1997) A study of flexible airfoil aerodynamics with application to micro aerial vehicles. In: AIAA fluid dynamics conference. AIAA, Snowmass
12. Mazaheri K, Ebrahimi A (2010) Experimental investigation of the effect of chordwise flexibility on the aerodynamics of flapping wings in hovering flight. J Fluids Struct 26:544–558
13. Cheng B, Tobalske BW, Powers DR, Hedrick TL, Wang Y, Wethington SM, Chiu GTC, Deng X (2016) Flight mechanics and control of escape manoeuvres in hummingbirds. II. Aerodynamic force production, flight control and performance limitations. J Exp Biol 219:3532–3543
14. Khare V, Kamle S (2022) Aerostructural evaluation of bioinspired chordwise flexible flapping wing. Mech Adv Mater Struct. https://doi.org/10.1080/15376494.2022.2153187
15. Khare V, Kumar D, Kamle S (2020) Structural damage assessment of MAV flapping wings using DIC-wavelet technique. In: Non-destructive characterization and monitoring of advanced materials, aerospace, civil infrastructure, and transportation, vol 11380. SPIE, SPIE Smart Structures + Non-destructive Evaluation
16. Khare V, Kamle S (2022) Biomimicking and evaluation of dragonfly wing morphology with polypropylene nanocomposites. Acta Mech Sin 38
17. Khare V, Kamle S (2020) Effect of temperature and filler volume fraction on the creep and recovery behaviour of MWCNT–COOH–reinforced polypropylene films. Arch Appl Mech
18. Khare V, Srivastava S, Kamle S, Kamath GM (2019) Effect of filler functionalization on the thermo-mechanical behavior of polypropylene nanocomposites. Proc Struct Integrity 14:215–225
19. Sutton MA, Orteu JJ, Schreier H (2009) Image correlation for shape, motion and deformation measurements. Springer
20. Sutton MA, Yan JH, Tiwari V, Schreier HW, Orteu JJ (2008) The effect of out-of-plane motion on 2D and 3D digital image correlation measurements. Opt Laser Eng 46:746–757
21. DICe (2015) Digital image correlation, E. http://dicengine.github.io/dice
22. Combes SA, Daniel TL (2003) Flexural stiffness in insect wings. II. Spatial distribution and dynamic wing bending. J Exp Biol 206:2989–2997
23. Chen S, Li H, Guo S, Tong M, Ji B (2018) Unsteady aerodynamic model of flexible flapping wings. Aerosp Sci Technol 80:354–367
24. Walker WP, Patil MJ (2014) Unsteady aerodynamics of deformable thin airfoils. J Aircr

A CDM-Based Modelling and Prediction of Progressive Damage in Z-Pinned Unidirectional Laminated Composites

S. Borchate, A. Bagla, and P. Mohite

Abstract In recent years, the automotive and aerospace industries have accelerated interest in the use of advanced carbon fibre-reinforced composites as a primary structural material because of their great potential. However, unidirectional fibre-reinforced laminates have limitations such as low inter-laminar crack propagation resistance, low delamination resistance, low impact damage resistance, and poor post-impact damage tolerance due to the lack of through-thickness reinforcement. Z-pinning can overcome these problems as it is the only technique suitable for reinforcing prepreg laminates in the through-thickness direction. In the present study, the realistic RVE was generated from the idealization of the micro-structure. The finite element model was developed by considering a pure resin-rich region, an in-plane fibre waviness for 0.28 mm diameter, 2–5% volume fraction, and 90° inclination angle of a Z-pin. A Continuum Damage Mechanics (CDM)-based damage model was developed to predict the micro-damage initiation based on the von Mises criterion for isotropic material failure and the Tsai–Wu criterion for transversely isotropic material failure. Further, a CDM-based isotropic hardening plasticity damage model for matrix material was developed to predict damage progression based on elasticity and plasticity governing equations (yield function, equivalent plastic strain, plastic flow law, plastic dissipation). The backward Euler and Newton's methods are used to simplify the iteration algorithm. In addition to the theories and criteria that derive the damage process, the material properties/stiffness degradation method (MPDM) was carried out throughout the finite element analysis.

Keywords Z-pinned composites · Continuum Damage Mechanics · Isotropic hardening plasticity · Progressive damage

S. Borchate (✉) · A. Bagla · P. Mohite
Department of Aerospace Engineering, Indian Institute of Technology Kanpur, Kanpur, Uttar Pradesh 208016, India
e-mail: borchate@iitk.ac.in

S. Sidhardh et al. (eds.), *Advances in Structural Integrity for Mechanical, Civil, and Aerospace Applications*, Lecture Notes in Mechanical Engineering,
https://doi.org/10.1007/978-981-97-6367-2_30

1 Introduction

In recent years, the automotive and aerospace industries have accelerated interest in the use of advanced carbon fibre-reinforced composites as a primary structural material because of their great potential. These composites have high specific stiffness and strength, excellent fatigue properties, good thermal and electrical conductivity, acceptable wear, and corrosion resistance. However, unidirectional fibre-reinforced laminates have limitations such as low inter-laminar crack propagation resistance, low delamination resistance, low impact damage resistance, and poor post-impact damage tolerance due to the lack of through-thickness reinforcement [1–3]. Z-pinning can overcome these problems as it is the only technique suitable for reinforcing prepreg laminates in the through-thickness direction [4].

In Z-pinning technique, the rod or pins made of high strength and stiffness are inserted in the through-thickness direction of a composite [4]. The Z-pinning causes micro-structural damage such as fibre breakage, in-plane fibre waviness, resin-rich region, fibre distortion, fibre crimp, Z-pin offset, laminate swelling, and so on, which affects the stiffness and strength of the laminate [5, 6]. These micro-scale damages vary with the Z-pin diameter, volume fraction, pin material, inclination angle, and lamina stacking sequence, etc. However, in the literature, a limited study is available on the systematic prediction of micro-damage initiation and its progression in the Z-pinned unidirectional composites.

2 Micro-structural Study and Generation of RVE

The continuous rod with a small diameter and made up of strong material such as metals or composites are used for Z-pins. The common materials like SiC/BMI, T650/BMI, titanium, T300/Epoxy, T300/BMI, aluminium, P100/Epoxy, stainless steel, and S-Glass/Epoxy are primarily used with a rod diameter of 0.2–1.0 mm [4]. For damage prediction, the comparatively small Z-pin volume fraction (2–5%) is preferred to understand the different failure modes.

The micro-structure idealized in this study is from Refs. [4, 7] in which laminates were made using unidirectional carbon fibre/epoxy (CYCOM 970) prepreg tape with 20 plies, and the Z-pin was made of 8 mm long carbon/BMI rods. The micro-structural study was carried out by using an advanced image processing tool, and it was observed that, after the insertion of a Z-pin, there are two significant micro-structural changes, such as pure resin-rich region and fibre waviness around a Z-pin (refer Fig. 1) [4, 7–9].

The detailed study of these significant micro-structural changes is an essential task for damage prediction. The fibre waviness angle (β) was calculated from the original fibre direction, as were the resin-rich zone length (k) and fibre waviness height (h) from binary micro-graphs obtained through digital image processing.

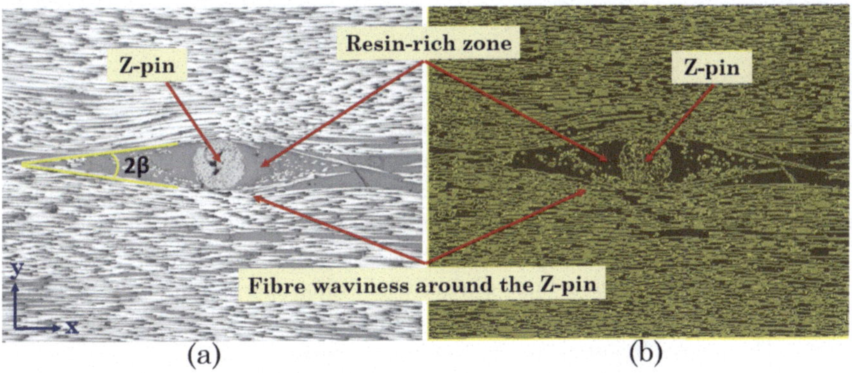

Fig. 1 **a** Fibre waviness and resin-rich zone around the Z-pin. **b** Micro-structural details using an advanced image processing tool [4, 7]

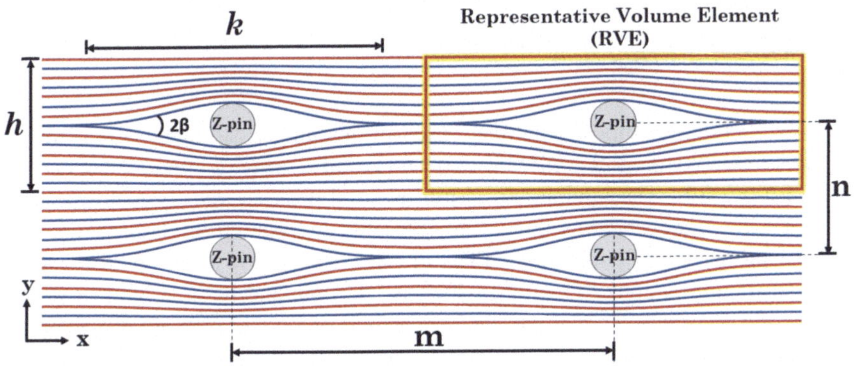

Fig. 2 Selection of Representative Volume Element (RVE) for Z-pinned unidirectional laminates and RVE dimensions

The micro-structural idealization was considered for the generation of Representative Volume Element (RVE) [10], which includes symmetrical distribution of wavy fibres and resin-rich zone, no pin offset, no fibre crimping and fibre damage, no stress concentration and manufacturing defects, and Z-pins are equally spaced and perfectly aligned. The RVE used in this study was generated from the micro-structure with a 0.28 mm Z-pin diameter (d). The other effective dimensions were characterized as, total resin-rich zone length (k) 2.0334 mm, total fibre waviness height (h) 0.7565 mm, and fibre waviness angle (β) 6.69° (refer Fig. 2). These all parameters were considered while modelling RVE in commercial ABAQUS/CAE software.

3 Modelling and FEA of RVE

For the prediction of micro-damage initiation and its progression, the RVE technique was used for modelling the complicated micro-structure of Z-pinned unidirectional composites. The three-dimensional finite element model was developed by considering in-plane fibre waviness and resin-rich zone around a Z-pin. In the modelling, fibre waviness angle (β), distorted fibre region (h), and size of resin-rich zone (k) are dependent variables on the Z-pin diameter (d). The each single wavy fibre band was modelled as a homogeneous three-dimensional orthotropic continuum. The whole RVE was divided into local regions to simplify and assign local wavy fibre orientations. The micro-structural idealization stated above was also considered while modelling RVE. The commercial standard ABAQUS/CAE tool was used to perform the finite element analysis [11, 12].

3.1 Material Properties

The unidirectional composite laminate used in this numerical analysis is made of carbon fibre/epoxy prepreg (AS4/3501-6), and Z-pin is made of impregnated carbon tows (T300/9310). The lamina and Z-pin material are transversely isotropic, and therefore, nine independent engineering constants are required to define these two material properties in commercial FE software. The matrix material in the pure resin-rich region is epoxy (3501-6) which is isotropic and defined with only two engineering constants. The longitudinal, as well as transverse strength properties in tensile, compressive, and shear are used in the CDM approach. The basic material stiffness and strength properties used throughout this study are given in Tables 1 and 2 respectively [13–15].

Table 1 Basic material stiffness properties used in the FE model

Material	E_1 (GPa)	E_2 (GPa)	E_3 (GPa)	G_{12} (GPa)	G_{13} (GPa)	G_{23} (GPa)	ν_{12}	ν_{13}	ν_{23}
Lamina (AS4/3501-6)	136.40	8.90	8.90	5.95	5.94	3.21	0.25	0.25	0.38
Z-pin (T300/9310)	144.00	7.31	7.31	4.45	4.45	2.65	0.25	0.25	0.39
Epoxy resin (3501-6)	4.44	–	–	1.65	–	–	0.34	–	–

Table 2 Basic material strength (MPa) properties used in the FE model

Description	Lamina (AS4/3501-6)	Z-pin (T300/9310)
Longitudinal tensile strength (X_T)	1950	1500
Longitudinal compressive strength (X_C)	1480	900
Transverse tensile strength (Y_T)	48	27
Transverse compressive strength (Y_C)	200	200
In-plane shear strength (S_{12})	79	80
Out-of-plane shear strength (S_{23})	55	80
Description	Epoxy resin (3501-6)	
Tensile strength (Y_{mT})	69	
Compressive strength (Y_{mC})	250	
Shear strength (S_m)	50	

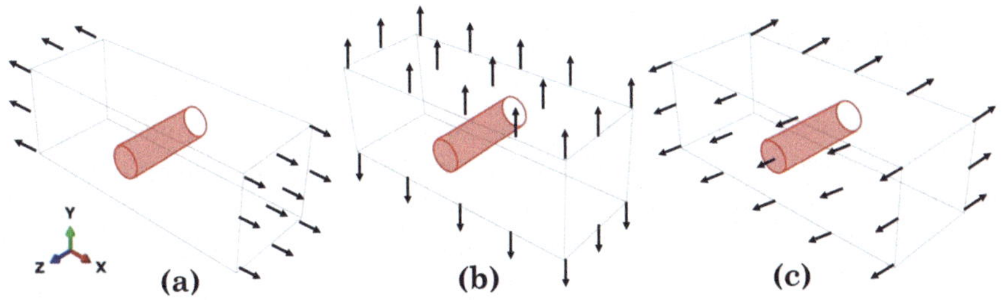

Fig. 3 Uniform displacement boundary conditions on an RVE **a** normal in x direction, **b** normal in y direction, **c** normal in z direction

3.2 *Specific Boundary Condition*

It is important to select a specific boundary condition that will give the average strain for the displacement loading condition [11]. There is no unique relation between the average strain and displacements; for the same average strain in the material, there can be several displacement combinations. The uniform displacement loading was selected as a specific boundary condition in this numerical analysis. It should be noted that the results reported in this work are based on the uniform displacement loading normal in the x direction, and displacements in the other two transverse directions (y and z) are constraints to zero (refer Fig. 3a). After performing the finite element analysis with the CDM approach, the unique nodal free body forces and spatial displacement were captured on the RVE face on which uniform displacement loading was applied. Finally, the force-time and displacement-time plots combine to generate a stress–strain curve.

4 Continuum Damage Mechanics (CDM) Approach

A Continuum Damage Mechanics based damage model was developed to predict the micro-damage initiation based on the von Mises yield criterion for isotropic material failure and Tsai–Wu criterion for transversely isotropic material failure. Further, a CDM-based isotropic hardening plasticity damage model for matrix material was developed to predict damage progression based on the elasticity and plasticity constitutive relations (yield function, effective plastic strain, plastic potential, plastic flow rule, and plastic dissipation). The backward Euler and Newton's method was used to simplify the iteration algorithm. In addition to the theories and criteria that derive the damage process, the material properties or stiffness degradation method (MPDM) was carried out throughout the finite element analysis. The relations referred in the following subsections were numerically implemented through user material subroutine UMAT.

4.1 *Damage Variable (D)*

At the mesoscale, the damage variable can be defined as the area void fraction of micro-cavities and micro-cracks, while at the micro-scale, it can be interpreted as the enlargement of micro-cavities due to the breaking of the atom bonds. For a general three-dimensional case, the effective Cauchy stress tensor σ_{ij}^* is expressed as [16, 17],

$$\sigma_{ij}^* = \frac{\sigma_{ij}}{1 - D} \tag{1}$$

4.2 *Tsai–Wu Failure Criterion*

The most familiar form of the Tsai–Wu criterion is expressed in the failure function for orthotropic materials in their material's principal axes. Here, by assuming transverse isotropy for unidirectional composites [18], we can write,

$$\begin{aligned} F = {} & F_{11}\sigma_1^2 + F_{22}\left(\sigma_2^2 + \sigma_3^2\right) + 2F_{23}\sigma_2\sigma_3 + 2F_{12}\sigma_1\left(\sigma_3 + \sigma_2\right) \\ & + F_1\sigma_1 + F_2\left(\sigma_2 + \sigma_3\right) + F_{44}\tau_{23}^2 + F_{66}\left(\tau_{13}^2 + \tau_{12}^2\right) \end{aligned} \tag{2}$$

where, F_{11}, F_{22}, F_{44}, F_{66}, F_1, and F_2 are the coefficients determined from the conventional strengths of unidirectional composites. A failure criterion is defined as the material's being safe if $F < 1$, while the critical condition for failure is predicted when $F = 1$.

4.3 von Mises Yield Criterion

The von Mises yield criterion predicts failure when the distortion energy density reaches the critical value. The von Mises stress ($V_{\mathrm{m}}^{\mathrm{stress}}$) can be calculated by using the components of stresses as [16, 17],

$$V_{\mathrm{m}}^{\mathrm{stress}} = \left[\frac{1}{2}\left\{(\sigma_1 - \sigma_2)^2 + (\sigma_2 - \sigma_3)^2 + (\sigma_3 - \sigma_1)^2\right\} + 3\left(\tau_{12}^2 + \tau_{23}^2 + \tau_{13}^2\right)\right]^{(1/2)} \tag{3}$$

If von Mises stress is greater than the matrix yield stress, then matrix damage variable $D_{\mathrm{m}} = 1$; otherwise $D_{\mathrm{m}} = 0$.

4.4 Plastic Flow Rule

The direction of plastic flow after yield is determined from the normality condition. In the case of von Mises plasticity (associated flow), the increment in the plastic strain tensor is in a direction which is normal to the yield surface.

The effective plastic strain $\varepsilon_{\mathrm{e}}^{\mathrm{p}}$ is referred as the hardening parameter for the case of strain hardening, and it is given by,

$$\varepsilon_{\mathrm{e}}^{\mathrm{p}} = \int \mathrm{d}\varepsilon_{\mathrm{e}}^{\mathrm{p}}; \quad \mathrm{d}\varepsilon_{\mathrm{e}}^{\mathrm{p}} = \sqrt{\frac{2}{3}\mathrm{d}\varepsilon_{ij}^{\mathrm{p}}\mathrm{d}\varepsilon_{ij}^{\mathrm{p}}} \tag{4}$$

where, $\mathrm{d}\varepsilon_{\mathrm{e}}^{\mathrm{p}}$ denotes the increment in effective plastic strain and $\mathrm{d}\varepsilon_{ij}^{\mathrm{p}}$ denotes plastic part of the incremental linear strain tensor.

The plastic potential for a damaged material can be expressed as [16, 17, 19, 20],

$$P = P_{\mathrm{y}} + P_{\mathrm{D}} \tag{5}$$

Here, the component P_{y} is associated with the hardening and yielding, and P_{D} is due to damage. For a material yielding according to the von Mises criterion which is expressed as,

$$P_{\mathrm{y}} = \frac{\sigma_{\mathrm{e}}}{1 - D} - \sigma_{\mathrm{y}} \tag{6}$$

where, σ_{y} is the variable yield stress calculated from the Ludwik's power law which is defined as,

$$\sigma_{\mathrm{y}} = (\sigma_{\mathrm{y}})_0 + H\left(\varepsilon_{\mathrm{e}}^{\mathrm{p}}\right)^x \tag{7}$$

where, $(\sigma_{\mathrm{y}})_0$ denotes initial yield stress, H denotes the hardening coefficient and x denotes the hardening exponent.

The plastic potential due to damage P_D for isothermal process expressed as,

$$P_D = \frac{1}{2M_0} \frac{1}{(1 - D)} (-T)^2 \tag{8}$$

where, M_0 is a material parameter and determined from experimental data.

4.5 Damage Growth Law

The dissipative part of thermodynamic force $(-T)$ corresponding to the damage rate can be given by [16, 17, 19, 20],

$$-T = \frac{\sigma_e^2}{2E(1 - D)^2} f\left(\frac{\sigma_h}{\sigma_e}\right) \tag{9}$$

Here, ratio (σ_h/σ_e) is called the triaxiality, v is the Poisson's ratio, E is the Young's modulus, σ_h denotes hydrostatic part of the Cauchy stress tensor σ_{ij}, and σ_e denotes von Mises effective stress related to the deviatoric part σ'_{ij} given by the following relation,

$$\sigma_e = \sqrt{\frac{3}{2}\sigma'_{ij}\sigma'_{ij}} \tag{10}$$

Now, if $D = 0$ in Eq. (6), then P_y reduces to the original form due to von Mises criterion. Also, P_D reduces to zero when the damage is zero.

Therefore, in incremental form, the flow rule are obtained as,

$$d\varepsilon_{ij}^{p} = \frac{3}{2} \frac{d\lambda}{(1 - D)} \frac{\sigma'_{ij}}{\sigma_e} \tag{11}$$

Then, the damage growth law becomes,

$$dD = (1 - D) \frac{\partial P_D}{\partial(-T)} d\varepsilon_e^p \tag{12}$$

where, λ denotes the plastic multiplier. The forms of P_y and P_D are not well established in the literature. The various forms of P_D proposed in the literature along with the form used in this study [20–26].

5 Results and Discussion

The FEA results for Z-pinned unidirectional composite reported here are based on RVE with a Z-pin diameter (d) of 0.28 mm and a volume fraction (V_p) 2–5%. The detailed study of micro-damage initiation and its progression is an important task for the prediction of failure modes.

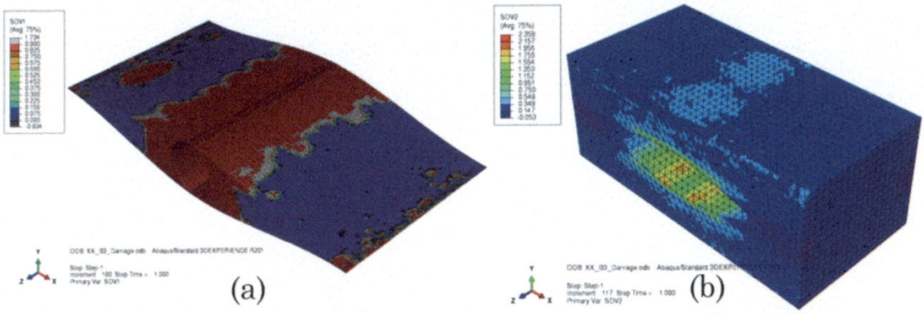

Fig. 4 The micro-damage initiation and its progression in **a** matrix region and **b** lamina region

5.1 Elastic Behaviour of Z-Pinned UD Composite

In the first part, micro-damage initiation was predicted using a damage model based on the von Mises yield criterion for isotropic material failure and Tsai–Wu criterion for transversely isotropic material failure (Fig. 4). The damage model was implemented in the user material subroutine of the ABAQUS/Standard program, UMAT. The micro-damage initiation was observed at the interface of Z-pin, pure resin-rich region, and lamina. The matrix failure was observed in the pure resin-rich region around a Z-pin and at eyelet corners. The transverse fibre splitting and in-plane shear damage at the lamina-pure resin-rich region interface due to matrix failure inside the lamina are the dominant modes of failure. The elastic response of material was studied from a stress–strain curve as shown in Fig. 5. It was observed that the Z-pinned unidirectional composite behaves linearly in the elastic region. The followings are important observations:

- RVE with Z-pin volume fraction of 2%, maximum stress is 417.02 MPa for strain 0.36%, and maximum in-plane modulus E_1 is 112.69 GPa.
- RVE with Z-pin volume fraction of 3%, maximum stress is 400.19 MPa for strain 0.37%, and maximum in-plane modulus E_1 is 105.73 GPa.
- RVE with Z-pin volume fraction of 4%, maximum stress is 443.37 MPa for strain 0.44%, and maximum in-plane modulus E_1 is 94.58 GPa.
- RVE with Z-pin volume fraction of 5%, maximum stress is 413.14 MPa for strain 0.43%, and maximum in-plane modulus E_1 is 91.13 GPa.

Also, in addition to the theories and criteria that derives the damage process, the material properties/stiffness degradation method (MPDM) was carried out throughout the finite element analysis. The degradation of in-plane Young's modulus E_1 w.r.t. strain (uniform displacement loading normal in x direction) is shown in Fig. 6. It was observed that the in-plane modulus E_1 degraded from 105.73 to 51 GPa due to micro-damage initiation and its progression in the pure resin-rich region and lamina material for Z-pin volume fraction of 3%.

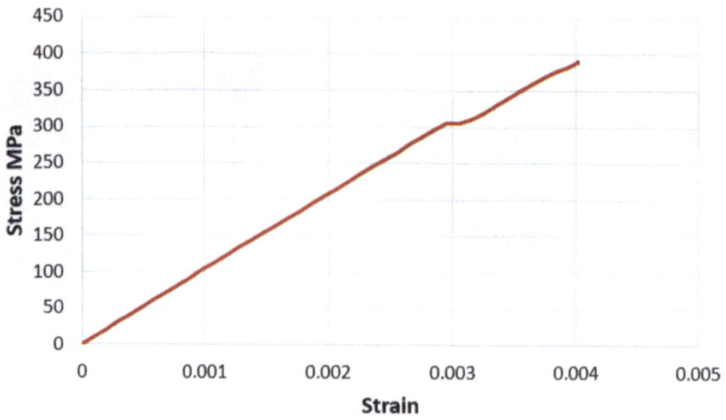

Fig. 5 Stress–strain curve of an RVE with Z-pin volume fraction 3%

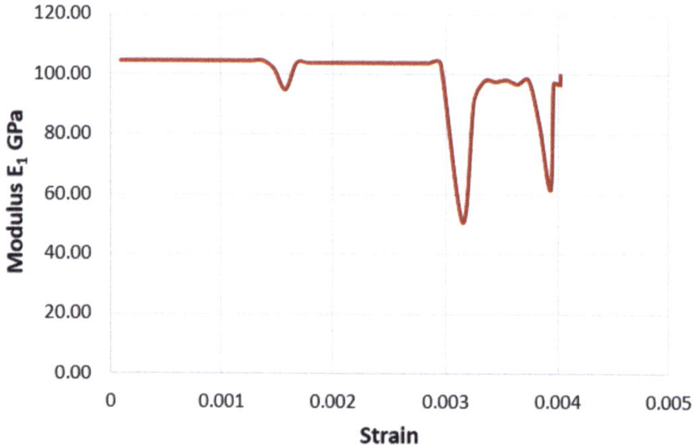

Fig. 6 Degradation of in-plane Young's modulus E_1 w.r.t. strain for Z-pin volume fraction 3%

5.2 *Elastoplastic Behaviour of Z-Pinned UD Composite*

In the second part, in addition to the damage model developed in the first part, the damage progression was predicted using a CDM-based isotropic hardening plasticity damage model for matrix material consisting of elastoplastic constitutive relations. The damage model was implemented in the user material subroutine of the ABAQUS/Standard program, UMAT. The small plasticity regions were observed in the matrix material near the interface of Z-pin, pure resin-rich region, and lamina. The elastoplastic response of material was studied from a stress–strain curve as shown in Fig. 7. It was observed that the Z-pinned unidirectional composite behaves linearly in the elastic region and a slight non-linearity in the plastic region. The followings are important observations:

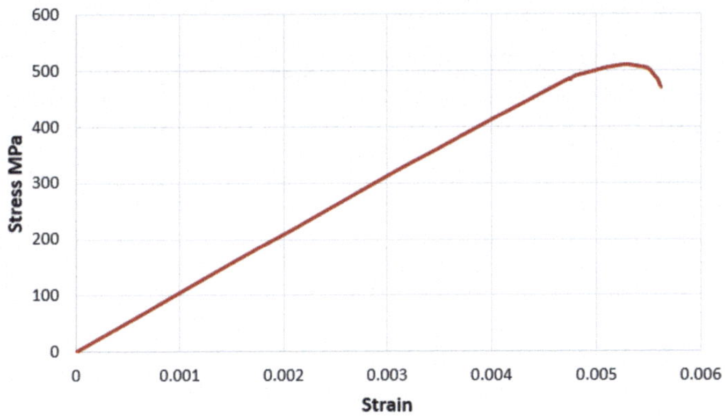

Fig. 7 Stress–strain curve of an RVE with Z-pin volume fraction 4%

- RVE with Z-pin volume fraction of 2%, maximum stress is 444.12 MPa for strain 0.41%, and maximum in-plane modulus E_1 is 117.47 GPa.
- RVE with Z-pin volume fraction of 3%, maximum stress is 420.12 MPa for strain 0.43%, and maximum in-plane modulus E_1 is 109.99 GPa.
- RVE with Z-pin volume fraction of 4%, maximum stress is 465.47 MPa for strain 0.46%, and maximum in-plane modulus E_1 is 104.77 GPa.
- RVE with Z-pin volume fraction of 5%, maximum stress is 421.14 MPa for strain 0.46%, and maximum in-plane modulus E_1 is 96.42 GPa.

Similar to first part, in addition to the theories and criteria that derives the damage process, the material properties/stiffness degradation method (MPDM) was carried out throughout the finite element analysis. The degradation of in-plane Young's modulus E_1 w.r.t. strain (uniform displacement loading normal in x direction) is shown in Fig. 8. It was observed that the in-plane modulus E_1 degraded from 104.77 to 16 GPa due to micro-damage initiation and its progression in the pure resin-rich region and lamina material for Z-pin volume fraction of 4%. This large degradation of in-plane modulus E_1 is due to a change in the material behaviour from elastic to plastic, called elastoplastic.

The micro-damage initiation and its progression near the interface of Z-pin, pure resin-rich region, and lamina is shown in Fig. 4a for matrix region and in Fig. 4b for lamina region.

6 Conclusion

The micro-structure of Z-pinned unidirectional composite laminate was examined using an advanced image processing tool. The realistic RVE was generated from the micro-structural idealization. The RVE used in this numerical analysis was generated from the micro-structure with 0.28 mm diameter and 2–5% volume frac-

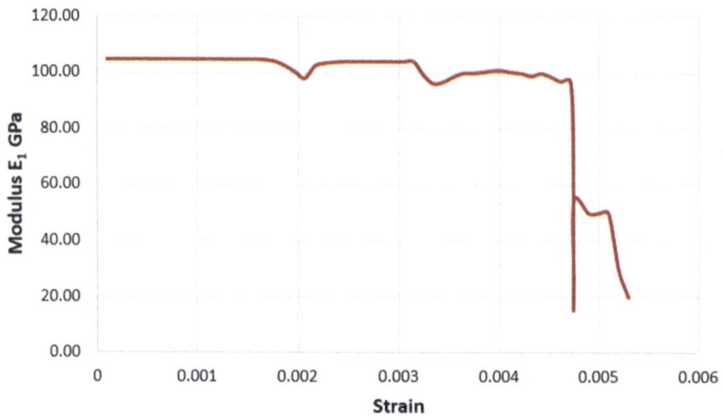

Fig. 8 Degradation of in-plane Young's modulus E_1 w.r.t. strain for Z-pin volume fraction 4%

tion of a Z-pin. The finite element 3D model was developed from this RVE by considering in-plane fibre waviness and pure resin-rich zone around a Z-pin. This finite element model was analysed to investigate the micro-damage initiation and its progression using the Continuum Damage Mechanics (CDM) based approach. The mathematical formulation was implemented in the user material subroutine of the ABAQUS/Standard program, UMAT.

The following conclusions were drawn from this study,

- The micro-damage initiation was observed at the interface of a Z-pin, pure resin-rich region, and lamina.
- The matrix failure was observed in the isolated pure resin-rich region around a Z-pin and at eyelet corners.
- The transverse fibre splitting and the in-plane shear damage at lamina-pure resin-rich region interface due to matrix failure inside lamina are the dominant modes of failure.
- The Z-pinned unidirectional composite laminate behaves linearly in the elastic region, but shows slight non-linearity in the plastic region.
- The in-plane modulus E_1 degraded in elastic behaviour also in elastoplastic behaviour of Z-pinned composite due to the micro-damage initiation and its progression at the interface of a Z-pin, pure resin-rich region, and lamina.

References

1. Freitas G, Magee C, Dardzinski P, Fusco T (1994) Fiber insertion process for improved damage tolerance in aircraft laminates. J Adv Mater 25(4):36–43
2. Ranatunga V, Clay SB (2013) Cohesive modeling of damage growth in z-pinned laminates under mode-i loading. J Compos Mater 47(26):3269–3283

3. Rugg K, Cox B, Ward K, Sherrick G (1998) Damage mechanisms for angled through-thickness rod reinforcement in carbon-epoxy laminates. Compos Part A Appl Sci Manuf 29(12):1603–1613
4. Mouritz A (2007) Review of z-pinned composite laminates. Compos Part A Appl Sci Manuf 38(12):2383–2397
5. Chang P, Mouritz A, Cox B (2007) Flexural properties of z-pinned laminates. Compos Part A Appl Sci Manuf 38(2):244–251
6. Steeves CA, Fleck NA (2006) In-plane properties of composite laminates with through-thickness pin reinforcement. Int J Solids Struct 43(10):3197–3212
7. Chang P, Mouritz A, Cox B (2006) Properties and failure mechanisms of z-pinned laminates in monotonic and cyclic tension. Compos Part A Appl Sci Manuf 37(10):1501–1513
8. Mouritz A (2007) Compression properties of z-pinned composite laminates. Compos Sci Technol 67(15–16):3110–3120
9. Gnaba I, Legrand X, Wang P, Soulat D (2019) Through-the-thickness reinforcement for composite structures: a review. J Ind Text 49(1):71–96
10. Sun C, Vaidya R (1996) Prediction of composite properties from a representative volume element. Compos Sci Technol 56(2):171–179
11. Whitcomb JD, Chapman CD, Tang X (2000) Derivation of boundary conditions for micromechanics analyses of plain and satin weave composites. J Compos Mater 34(9):724–747
12. Barrett DJ (1996) The mechanics of z-fiber reinforcement. Compos Struct 36(1–2):23–32
13. Grassi M, Zhang X, Meo M (2002) Prediction of stiffness and stresses in z-fibre reinforced composite laminates. Compos Part A Appl Sci Manuf 33(12):1653–1664
14. Cartie DD, Troulis M, Partridge IK (2006) Delamination of z-pinned carbon fibre reinforced laminates. Compos Sci Technol 66(6):855–861
15. Soden PD, Hinton MJ, Kaddour AS (1998) Lamina properties, lay-up configurations and loading conditions for a range of fibre reinforced composite laminates. Compos Sci Technol 58:1011–1022
16. Lemaitre J (1985) A continuous damage mechanics model for ductile fracture. ASME J Eng Mater Technol 107:83–89
17. Lemaitre J (2012) A course on damage mechanics. Springer
18. Li S, Sitnikova E, Liang Y, Kaddour AS (2017) The Tsai–Wu failure criterion rationalised in the context of UD composites. Compos Part A Appl Sci Manuf 102:207–217
19. Lemaitre J, Desmorat R (2006) Engineering damage mechanics: ductile, creep, fatigue and brittle failures. Springer
20. Kumar M, Dixit PM (2015) A nonlinear ductile damage growth law. Int J Damage Mech 24(7):1070–1085
21. Chandrakanth S, Pandey PC (1993) A new ductile damage evolution model. Int J Fract 60(4):73–76
22. Chandrakanth S, Pandey PC (1995) An isotropic damage model for ductile material. Eng Fract Mech 50(4):457–465
23. Chow CL, Wang J (1987) An anisotropic theory of elasticity for continuum damage mechanics. Int J Fract 33(1):3–16
24. Dhar S, Sethuraman R, Dixit PM (1996) A continuum damage mechanics model for void growth and micro crack initiation. Eng Fract Mech 53(6):917–928
25. Tai WH, Yang BX (1986) A new microvoid damage model for ductile fracture. Eng Fract Mech 25(3):377–384
26. Thakkar BK, Pandey PC (2007) A high-order isotropic continuum damage evolution model. Int J Damage Mech 16(4):403–426

Spectral Analysis of the Structural Waveguides with Defects Using the Semi-analytical Finite Element Method

Anoop Kumar Dube and S. Gopalakrishnan

Abstract The spectral analysis of the structural waveguide leads to the dispersion relation, which shows the behavior of the different wave modes with frequency. It helps in ultrasonic-guided wave-based non-destructive inspection to choose an excitation frequency based on the choice of the excitation wave mode(s). The dispersion curves for a waveguide with simple defect-free geometry work well for their inspection. However, it's important to note that the presence of defects within the structure can significantly affect the wave dispersion behavior used for diagnostics. Therefore, the use of the dispersion curves of the defect-free waveguide for the inspection of the defective ones can yield less accurate inspection results. This emphasizes the need to conduct the spectral analysis of the structural waveguide with defects and investigate the sensitivity of dispersion curves to various defect parameters such as shape, size, depth, and orientation. The dispersion curves can be obtained analytically or numerically, but they have inherent limitations. The analytical methods are limited to simple geometry; however, using the numerical methods for complex geometry will require a substantial computational cost. Therefore, this paper presents the spectral analysis of the three-dimensional structural waveguide with defects using the Semi-Analytical Finite Element Method, which uses the advantages of both analytical and numerical methods. We have performed the spectral analysis of the simple geometry without and with defects (such as cracks or delamination) of varying sizes. It is observed that the finite element discretization of the cross-section affects the number and accuracy of wave modes obtained in the dispersion curves. The validity of the SAFE framework has been initially established by comparing the dispersion curves obtained from our SAFE formulation with those generated by GUIGUW (Bocchini et al. 2011), an open-source dispersion computation software, for a defect-free structural waveguide. Subsequently, we applied our SAFE formulation to compute dispersion curves for waveguides with defects. As the defects reduce the structure's stiffness, we observed

A. K. Dube (✉) · S. Gopalakrishnan
Department of Aerospace Engineering, Indian Institute of Science, Bangalore 560012, India
e-mail: anoopdube@iisc.ac.in
URL: https://sites.google.com/view/cwml

S. Gopalakrishnan
e-mail: krishnan@iisc.ac.in

S. Sidhardh et al. (eds.), *Advances in Structural Integrity for Mechanical, Civil, and Aerospace Applications*, Lecture Notes in Mechanical Engineering,
https://doi.org/10.1007/978-981-97-6367-2_31

a significant reduction in the cutoff frequencies of the higher-order wave modes in the dispersion curves for defective waveguides compared to the defect-free waveguide.

Keywords Semi-analytical finite element method · Spectral analysis · Non-destructive testing and evaluation · Structural health monitoring · Dispersion curves · Delamination · Crack

1 Introduction

The laminated composites are widely used in aerospace structures. The composites have advantages over conventional structural materials, such as high strength-to-weight ratio, improved fatigue strength, corrosion resistance, but their complex failure mechanisms hinder their use in critical areas. The most common defects that cause failure in composites are inter-layer delamination, matrix crack, and fiber breakage. These defects may originate during manufacturing or get induced in service by impact or fatigue loading. The delamination in laminated composite generally originates from the edges or the matrix cracks due to bending or fatigue loads. The bonded lap joints [1] of the structural components are also widely used in aerospace structures. The bonded joints offer various advantages such as uniform stress distribution and lightweight; however, the presence of the micro crack, voids, aging of the bonding layer, and improper interfacial bonding limits their application as they may cause the structure to fail. Therefore, detection of the defects at the early stage is necessary to avoid any structural failure. There are various conventional Non-Destructive Inspection (NDI) techniques, such as X-ray Computed Tomography (X-ray CT), Micro Computed Tomography (Micro CT), Acoustic Emission Technique (AET), Eddy Current Testing (ECT), Thermography, and Ultrasonic Testing (UT), traditionally used in various industries, including aerospace, for inspecting structures at pre-defined intervals. These traditional NDI techniques are costly and time-consuming [2] and sometimes, they require disassembling the parts from the main structure for better accessibility. Therefore, they cannot be used for inspecting aircraft or spacecraft during service. Ultrasonic Guided Wave (UGW) based NDI (UNDI) has become popular because UGW has enhanced sensitivity to small size defects compared to vibration-based methods. Because the guided waves follow the boundary and can travel long distances with the minimal loss [2], the UGW can monitor larger and inaccessible areas to detect hidden defects [3]. As the UNDI does not necessarily require the disassembly of the component, they can be deployed on the structure either externally by using the adhesively bonded sensors and actuators or internally by embedding the sensors and actuators in the structure during manufacturing [4]. This way, the UNDI system, also known as Structural Health Monitoring (SHM) system, can monitor the structure's performance during the service.

Applying the UNDI or SHM system to the actual structure requires knowledge of the behavior of the ultrasonic waves in the structure, which is sensitive to the material and geometrical properties. The dispersion curves help to get an understanding of the behavior of the ultrasonic waves. The dispersion curves show the dependence of the wavenumber (k), phase speeds (C_p), and group speeds (C_g) with the frequency [5]. The dispersion curves help to choose the wave mode(s) and the excitation frequency with which the actuator will excite the structure for the diagnostics. Usually, the dispersion curves are calculated for the simple and defect-free structure, which works well for their UNDI. However, the presence of any defect in the structure will change the behavior of the propagating waves used for diagnostics. Therefore, the dispersion curves of the simple and defect-free structure will not give accurate inspection results for the UNDI of the complex and defective system. This emphasizes the need to obtain the dispersion curves for complex and defective structures. There are various methods to compute the dispersion curves, such as analytical [5, 6], semi-analytical [7, 8], numerical methods [9, 10], and experiments [10, 11].

The analytical methods involve two-spatial and one-temporal Fourier transform to obtain the ordinary Governing Differential Equations (GDE) of motion in the frequency-wavenumber domain from the partial GDE for a general three-dimensional (3D) solid [5, 6]. Then the harmonic solution is assumed in the wave propagation direction to obtain the dispersion relation. There are open-source software available for dispersion curve computation. Lowe et al. [12] have developed a licensed software for the dispersion curve computation, DISPERSE, which computes the analytical dispersion curves from Global Matrix Method (GMM). An open-source dispersion curve computation software developed by Dr. Armin Huber at German Aerospace Center (DLR) named Dispersion Calculator (DC) [13, 14] computes the analytical dispersion curves using Transfer Matrix Method (TMM) and Stiffness Matrix Method (SMM). A numerical method such as the Finite Element Method (FEM) discretizes the domain into a few finite numbers of small domains, and the displacement field is approximated from the spatial polynomials. After performing the discretization and FE formulation, the time-domain responses are obtained. Then the dispersion curves are obtained by performing the spatial and temporal Fourier transform on the time-domain responses [9, 10]. The analytical methods give the exact solutions to the dispersion relation; however, they can be used for uniform and simple geometry. It is difficult to obtain the dispersion curves for structures with complex geometry analytically. The numerical method such as FEM gives an approximate solution, and its accuracy depends on the discretization of the domain and the order of the approximation polynomials. FEM is useful for computing the dispersion curve for structures with complex geometry. The semi-analytical method uses the advantages of both the analytical and numerical methods. Semi-Analytical Finite Element Method (SAFEM) [7, 8] is one of the semi-analytical methods, combining the analytical and FE methods. It assumes harmonic solution in the wave propagation direction while FE solution in the perpendicular direction(s), i.e., in the cross-section, which reduces the three-dimensional wave propagation problem into one or two-dimensional problems. This way, it reduces the size of the matrices of the computational system involved compared to using complete FEM, increasing the computational efficiency. The GUIGUW [15, 16] is another open-source software for the dispersion curve computation which uses the SAFEM.

Various literature discusses the interaction of waves with the delamination in composite laminates. Most of them have modeled the delaminated parts in the defect region as separate waveguides (sub-laminates) and the rest as a single intact waveguide (base-laminate). They enforce appropriate boundary and matching conditions at the interfaces of the sub-laminates and base-laminates [17, 18]. As the contact between the delaminated surfaces produces non-linear effects, many researchers have ignored the contact or interpenetration of the delaminated surfaces under loading to avoid the complexity. Suryanarayan and Mujumdar [19] improved Wang et al. [17] solution further by imposing the constrain condition so that the delaminated layers cannot overlap and will have identical transverse displacements. Hayashi and Kawashima [20] studied the wave propagation in laminated composite with delamination using SAFEM. They observed that the lamb waves propagate individually in the two regions divided by the delamination. Therefore, the waveguides above and below the delamination behave as individual waveguides. Munian et al. [21] have studied the interaction of lamb waves with delamination in composite laminate using the time-domain spectral finite element method. They have modeled the delamination by creating separated surfaces in the delamination region. These surfaces are free to move independently with the assumption of small vibration amplitude due to waves compared to the gap created between the delamination surfaces. Therefore, the contact or interpenetration of the delamination surfaces has been ignored in their work. We have followed a similar approach as in [21] to model the delamination in the structure for computing the dispersion curve for defective structural components.

Existing literature predominantly focuses on the analysis of dispersion curves pertaining to defect-free structural waveguides. State-of-the-art UNDI techniques often employ these dispersion curves for inspecting defective waveguides, but such approaches can yield less accurate inspection results. It's important to note that the presence of defects within the structure can significantly affect the wave dispersion behavior used for diagnostics. In this study, we undertake an analysis of the wave dispersion characteristics in waveguides containing defects such as delamination or cracks and its sensitivity with respect to one of the defect parameters, i.e., the size of the defect. As the analytical dispersion analysis of the structure with defects is difficult, we have employed SAFEM. We have considered an infinite length structure with a square cross-section without and with crack or delamination cases of different sizes running throughout the length. The resultant dispersion curves offer a valuable resource for enhancing the precision of inspections conducted on defective waveguides. The paper is organized as follows. Section 2 presents the mathematical formulations for the governing equations of motion and SAFEM. The computer implementation part is discussed in Sect. 3. Then the results related to the convergence study are discussed in Sect. 4 to know the number of elements needed to perform the dispersion computation using SAFEM. The validation of the present formulation and the effect of the defects on the dispersion curves are then discussed. Finally, we conclude the paper in Sect. 5.

2 Mathematical Formulations

Here, we derive Navier's governing differential equation for a general 3D solid from Hamilton's principle. Then we perform the weak formulation of Navier's equation. We assume the y-direction to be the wave propagation direction. For the SAFEM formulation, we have taken the displacement field as the combination of the analytical and FE polynomial solutions. We have adopted a general 3D solid with xz-plane as FE discretization plane and the analytical harmonic solution in y-direction. The 3D solid waveguide shown in Fig. 1 has the dimensions, H as the height in z-direction, W as the width in x-direction, and an infinite length in y-direction. The cross-section of the solid waveguide is discretized from FEM. The defects such as cracks or delamination are introduced in the xz-plane later, which run throughout in y-direction.

2.1 *Equations of Motion*

Let's assume the time-dependent displacement field in the Cartesian coordinate system as $u = u_x(x, y, z, t)$, $v = u_y(x, y, z, t)$ and $w = u_z(x, y, z, t)$, where x, y, and z represent the Cartesian coordinates and t denotes time. Using the strain–displacement relation, the stain components and their variations are computed as,

$$\boldsymbol{\varepsilon} = \boldsymbol{D}\boldsymbol{u} \quad \text{and} \quad \delta\boldsymbol{\varepsilon} = \boldsymbol{D}\delta\boldsymbol{u} \tag{1}$$

and the stress components are computed from the generalized Hook's law for a general 3D solid as,

$$\boldsymbol{\sigma} = \boldsymbol{C}\boldsymbol{\varepsilon} \tag{2}$$

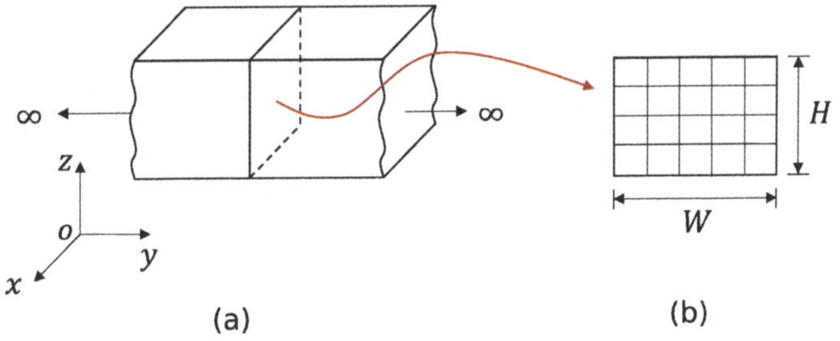

Fig. 1 **a** Structural component with the global coordinate system having finite dimensions of the cross-section in xz-plane and infinite length in y-direction. **b** The SAFEM model with FE discretization of the cross-section

where δ is the variational operator, $\boldsymbol{\varepsilon}, \boldsymbol{\sigma}$, and $\boldsymbol{u}$ are the strain vector, stress vector, and displacement field vector, respectively [22], with $\boldsymbol{\varepsilon} = \begin{bmatrix} \varepsilon_{xx} & \varepsilon_{yy} & \varepsilon_{zz} & \gamma_{yz} & \gamma_{xz} & \gamma_{xy} \end{bmatrix}^{\mathrm{T}}$; $\boldsymbol{\sigma} = \begin{bmatrix} \sigma_{xx} & \sigma_{yy} & \sigma_{zz} & \tau_{yz} & \tau_{xz} & \tau_{xy} \end{bmatrix}^{\mathrm{T}}$; $\boldsymbol{u} = \begin{bmatrix} u_x & u_y & u_z \end{bmatrix}^{\mathrm{T}}$. $\boldsymbol{D}^{\mathrm{T}} = \begin{bmatrix} \frac{\partial}{\partial x} & 0 & 0 & 0 & \frac{\partial}{\partial z} & \frac{\partial}{\partial y} \\ 0 & \frac{\partial}{\partial y} & 0 & \frac{\partial}{\partial z} & 0 & \frac{\partial}{\partial x} \\ 0 & 0 & \frac{\partial}{\partial z} & \frac{\partial}{\partial y} & \frac{\partial}{\partial x} & 0 \end{bmatrix}$; and

$$\boldsymbol{C} = \begin{bmatrix} \lambda + 2\mu & \lambda & \lambda & 0 & 0 & 0 \\ & \lambda + 2\mu & \lambda & 0 & 0 & 0 \\ & & \lambda + 2\mu & 0 & 0 & 0 \\ & & & \mu & 0 & 0 \\ & \mathrm{Sym} & & & \mu & 0 \\ & & & & & \mu \end{bmatrix}$$

are the differential and stiffness matrices, respectively. The transpose of a matrix is denoted by $(.)^{\mathrm{T}}$, and Sym represents the symmetricity of the stiffness matrix. The λ and μ are the Lame's parameters and calculated from the relation $\lambda = \frac{\nu E}{(1+\nu)(1-2\nu)}$ and $\mu = \frac{E}{2(1+\nu)}$, where E is Young's modulus and ν is the Poisson's ratio of the considered material. The governing differential equations of motion can be derived from Hamilton's principle [22], which is given, in the absence of the external forces, as,

$$\int_{t_1}^{t_2} (\delta K - \delta S)\, \mathrm{d}t = 0 \tag{3}$$

where K and S are the kinetic and strain energies, respectively, and computed from the following expression

$$K = \frac{1}{2} \int_V \left(\rho \dot{\boldsymbol{u}}^{\mathrm{T}} \dot{\boldsymbol{u}}\right) \mathrm{d}V; \qquad S = \frac{1}{2} \int_V \left(\boldsymbol{\varepsilon}^{\mathrm{T}} \boldsymbol{C} \boldsymbol{\varepsilon}\right) \mathrm{d}V \tag{4}$$

where ρ is the density of the material, and the overhead dot represents the derivative with respect to time. Substituting the expressions of the kinetic and strain energies into Hamilton's principle Eq. (3) and performing the integration by parts in the kinetic energy term for time gives,

$$\int_{t_1}^{t_2} \int_V \left(\delta \boldsymbol{\varepsilon}^{\mathrm{T}} \boldsymbol{C} \boldsymbol{\varepsilon} + \rho \delta \boldsymbol{u}^{\mathrm{T}} \ddot{\boldsymbol{u}}\right) \mathrm{d}V\, \mathrm{d}t = 0 \tag{5}$$

Substituting the strains from the strain–displacement relation Eq. (1), in Eq. (5), we obtain the following Eq. (6) for a general finite element, 'e' for which the variables are denoted with a superscript (e) as,

$$\int_{t_1}^{t_2}\int_{V}\left[\left(\boldsymbol{D}\delta\boldsymbol{u}^{(e)}\right)^{\mathrm{T}}\boldsymbol{C}\boldsymbol{D}\boldsymbol{u}^{(e)}+\frac{\partial^2}{\partial t^2}\delta\left(\boldsymbol{u}^{(e)}\right)^{\mathrm{T}}\rho\boldsymbol{u}^{(e)}\right]\mathrm{d}V\,\mathrm{d}t=0 \tag{6}$$

Equation (6) can be used for the SAFEM formulation.

2.2 Semi-analytical Finite Element Method (SAFEM) Formulation

Let's assume the displacement field solution of the form [7],

$$\boldsymbol{u}^{(e)}(x,y,z,t)=\hat{\boldsymbol{u}}^{(e)}\left(x,k_y,z,\omega\right)\mathrm{e}^{-i\left(k_y y-\omega t\right)} \tag{7}$$

where the y-direction is the wave propagation direction with k_y as the wavenumber, ω as the radial frequency, and $i=\sqrt{-1}$. The displacement field is as assumed to be the harmonic function in y-direction, and the frequency-domain displacement amplitude $\hat{\boldsymbol{u}}^{(e)}\left(x,k_y,z,\omega\right)$ is a function of space, x, and z and approximated with spatial polynomials in x and z directions, like the conventional FEM.

$$\hat{\boldsymbol{u}}^{(e)}\left(x,k_y,z,\omega\right)=\boldsymbol{N}\tilde{\boldsymbol{u}}^{(e)} \tag{8}$$

where $\boldsymbol{N}=\boldsymbol{N}\left(\xi(x,z),\eta(x,z)\right)=\begin{bmatrix} N_1 & 0 & 0 & N_2 & 0 & 0 & \cdots & N_n & 0 & 0\\ 0 & N_1 & 0 & 0 & N_2 & 0 & \cdots & 0 & N_n & 0\\ 0 & 0 & N_1 & 0 & 0 & N_2 & \cdots & 0 & 0 & N_n\end{bmatrix}$ and $\tilde{\boldsymbol{u}}=\left[\tilde{u}_{x_1}\ \tilde{u}_{y_1}\ \tilde{u}_{z_1}\ \tilde{u}_{x_2}\ \tilde{u}_{y_2}\ \tilde{u}_{z_2}\ \cdots\ \tilde{u}_{x_n}\ \tilde{u}_{y_n}\ \tilde{u}_{z_n}\right]^{\mathrm{T}}$ with n as the number of nodes in a finite element 'e'. The expressions of the shape functions for a 4-noded quadrilateral element shown in Fig. 2a are

$$N_1=\frac{1}{4}(1-\xi)(1-\eta);\qquad N_2=\frac{1}{4}(1+\xi)(1-\eta);$$
$$N_3=\frac{1}{4}(1+\xi)(1+\eta);\qquad N_4=\frac{1}{4}(1-\xi)(1+\eta) \tag{9}$$

The N_1, N_2, N_3, and N_4 are the Lagrange shape functions at the node 1, 2, 3, and 4, respectively, in the natural element shown in Fig. 2a. Figure 2 shows the mapping of a general physical element in the global coordinate system to the natural element in the local coordinate system.

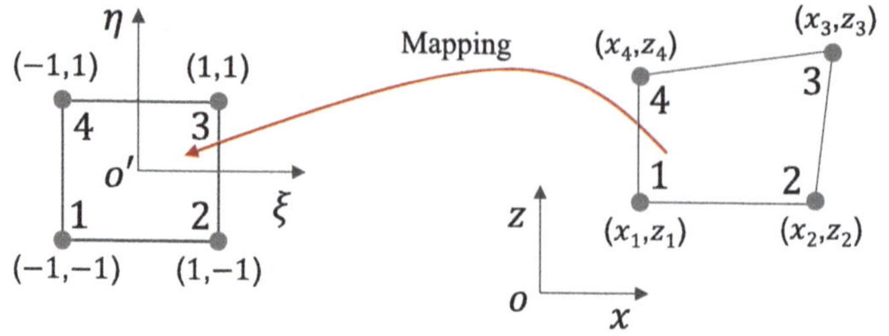

(a) Natural Element (b) Physical Element

Fig. 2 Mapping a general physical element in the global coordinate system to the natural element in the local coordinate system for FEM formulation

If the geometry is approximated with the same polynomials as the displacement field, it is known as the iso-parametric mapping [23]. By using the iso-parametric mapping, the geometry is approximated as,

$$x = x_j N_j(\xi, \eta); \qquad z = z_j N_j(\xi, \eta) \tag{10}$$

where j is a repeated index that signifies the summation over j and $j = 1, 2, 3$, and 4 for a four-noded quadrilateral element. By combining Eqs. (7) and (8), the displacement field is approximated as,

$$\boldsymbol{u}^{(e)}(x, y, z, t) = \boldsymbol{N}\mathrm{e}^{-i(k_y y - \omega t)}\tilde{\boldsymbol{u}}^{(e)} \tag{11}$$

Now substituting the displacement field expression from Eq. (11) into Eq. (6) gives,

$$\begin{aligned} \int_{t_1}^{t_2} (\delta\tilde{\boldsymbol{u}}^{(e)})^{\mathrm{T}} \int_V \Big[& \left(\boldsymbol{DN}\mathrm{e}^{-i(k_y y - \omega t)}\right)^{\mathrm{T}} \mathbf{C}\boldsymbol{DN}\mathrm{e}^{-i(k_y y - \omega t)} \\ & + \frac{\partial^2}{\partial t^2}\left(\boldsymbol{N}\mathrm{e}^{-i(k_y y - \omega t)}\right)^{\mathrm{T}} \rho\boldsymbol{N}\mathrm{e}^{-i(k_y y - \omega t)}\Big] \tilde{\mathbf{u}}^{(e)}\, \mathrm{d}V\, \mathrm{d}t = 0 \end{aligned} \tag{12}$$

Since the $(\delta\tilde{\boldsymbol{u}})^{\mathrm{T}}$ is arbitrary; using the fundamental lemma of calculus in Eq. (12), we can write,

$$\begin{aligned} \int_V \Big[& \left(\boldsymbol{DN}\mathrm{e}^{-i(k_y y - \omega t)}\right)^{\mathrm{T}} \mathbf{C}\boldsymbol{DN}\mathrm{e}^{-i(k_y y - \omega t)} \\ & + \frac{\partial^2}{\partial t^2}\left(\boldsymbol{N}\mathrm{e}^{-i(k_y y - \omega t)}\right)^{\mathrm{T}} \rho\boldsymbol{N}\mathrm{e}^{-i(k_y y - \omega t)}\Big] \tilde{\mathbf{u}}^{(e)}\, \mathrm{d}V = 0 \end{aligned} \tag{13}$$

Rewriting $\boldsymbol{D}$ as $\boldsymbol{D} = \frac{\partial}{\partial x}\boldsymbol{L}_x + \frac{\partial}{\partial y}\boldsymbol{L}_y + \frac{\partial}{\partial z}\boldsymbol{L}_z$, where

$$\boldsymbol{L}_x = \begin{bmatrix} 1\,0\,0 \\ 0\,0\,0 \\ 0\,0\,0 \\ 0\,0\,0 \\ 0\,0\,1 \\ 0\,1\,0 \end{bmatrix}; \quad \boldsymbol{L}_y = \begin{bmatrix} 0\,0\,0 \\ 0\,1\,0 \\ 0\,0\,0 \\ 0\,0\,1 \\ 0\,0\,0 \\ 1\,0\,0 \end{bmatrix}; \quad \text{and} \quad \boldsymbol{L}_z = \begin{bmatrix} 0\,0\,0 \\ 0\,0\,0 \\ 0\,0\,1 \\ 0\,1\,0 \\ 1\,0\,0 \\ 0\,0\,0 \end{bmatrix}$$

Let's simplify by substituting the expression of $\boldsymbol{D}$ in the term $\boldsymbol{DN}\mathrm{e}^{-i(k_y y-\omega t)}$ as

$$\begin{aligned} \boldsymbol{DN}\mathrm{e}^{-i(k_y y-\omega t)} &= \left(\frac{\partial}{\partial x}\boldsymbol{L}_x + \frac{\partial}{\partial y}\boldsymbol{L}_y + \frac{\partial}{\partial z}\boldsymbol{L}_z\right)\boldsymbol{N}\mathrm{e}^{-i(k_y y-\omega t)} \\ &= \left(\boldsymbol{L}_x\boldsymbol{N}_{,x} - ik_y\boldsymbol{L}_y\boldsymbol{N} + \boldsymbol{L}_z\boldsymbol{N}_{,z}\right)\mathrm{e}^{-i(k_y y-\omega t)} \end{aligned} \tag{14}$$

Substituting Eq. (14) in Eq. (13) gives the following equation,

$$\begin{aligned} \int_V \Big[&\left(\left(\boldsymbol{L}_x\boldsymbol{N}_{,x} - ik_y\boldsymbol{L}_y\boldsymbol{N} + \boldsymbol{L}_z\boldsymbol{N}_{,z}\right)\mathrm{e}^{-i(k_y y-\omega t)}\right)^{\mathrm{T}} \mathbf{C}\left(\boldsymbol{L}_x\boldsymbol{N}_{,x} - ik_y\boldsymbol{L}_y\boldsymbol{N}\right. \\ &\left.+\boldsymbol{L}_z\boldsymbol{N}_{,z}\right)\mathrm{e}^{-i(k_y y-\omega t)} + \frac{\partial^2}{\partial t^2}\left(\boldsymbol{N}\mathrm{e}^{-i(k_y y-\omega t)}\right)^{\mathrm{T}}\rho\boldsymbol{N}\mathrm{e}^{-i(k_y y-\omega t)}\Big]\tilde{\boldsymbol{u}}^{(e)}\,\mathrm{d}V = 0 \end{aligned} \tag{15}$$

Now, let's assume $\boldsymbol{B} = \boldsymbol{L}_x\boldsymbol{N}_{,x} - ik_y\boldsymbol{L}_y\boldsymbol{N} + \boldsymbol{L}_z\boldsymbol{N}_{,z} = \boldsymbol{B}_x - ik_y\boldsymbol{B}_y + \boldsymbol{B}_z$, where $\boldsymbol{B}_x = \boldsymbol{L}_x\boldsymbol{N}_{,x}$, $\boldsymbol{B}_y = \boldsymbol{L}_y\boldsymbol{N}$, and $\boldsymbol{B}_z = \boldsymbol{L}_z\boldsymbol{N}_{,z}$. $\boldsymbol{N}_{,x}$ and $\boldsymbol{N}_{,z}$ are the derivative of the shape function matrix with respect to x and z directions, respectively. Taking the transpose inside the bracket and substituting expressions of $\boldsymbol{B}$ in Eq. (15) will lead to

$$\int_V \left[\left(\boldsymbol{B}_x^{\mathrm{T}} + ik_y\boldsymbol{B}_y^{\mathrm{T}} + \boldsymbol{B}_z^{\mathrm{T}}\right)\mathbf{C}\left(\boldsymbol{B}_x - ik_y\boldsymbol{B}_y + \boldsymbol{B}_z\right) - \omega^2\boldsymbol{N}^{\mathrm{T}}\rho\boldsymbol{N}\right]\tilde{\mathbf{u}}^{(e)}\,\mathrm{d}V = 0 \tag{16}$$

Further simplifying Eq. (16) will lead to the following equation,

$$\left(k_y^2\boldsymbol{K}_2^{(e)} + ik_y\boldsymbol{K}_1^{(e)} + \boldsymbol{K}_0^{(e)} - \omega^2\boldsymbol{M}^{(e)}\right)\tilde{\boldsymbol{u}}^{(e)} = \mathbf{0} \tag{17}$$

where $\boldsymbol{K}_2^{(e)} = \int_V \boldsymbol{B}_y^{\mathrm{T}}\boldsymbol{C}\boldsymbol{B}_y\,\mathrm{d}V$; $\boldsymbol{K}_1^{(e)} = \int_V\left(-\boldsymbol{B}_x^{\mathrm{T}}\boldsymbol{C}\boldsymbol{B}_y + \boldsymbol{B}_y^{\mathrm{T}}\boldsymbol{C}\boldsymbol{B}_x + \boldsymbol{B}_y^{\mathrm{T}}\boldsymbol{C}\boldsymbol{B}_z - \boldsymbol{B}_z^{\mathrm{T}}\boldsymbol{C}\boldsymbol{B}_y\right)\mathrm{d}V$; and $\boldsymbol{K}_0^{(e)} = \int_V\left(\boldsymbol{B}_x^{\mathrm{T}}\boldsymbol{C}\boldsymbol{B}_x + \boldsymbol{B}_x^{\mathrm{T}}\boldsymbol{C}\boldsymbol{B}_z + \boldsymbol{B}_z^{\mathrm{T}}\boldsymbol{C}\boldsymbol{B}_x + \boldsymbol{B}_z^{\mathrm{T}}\boldsymbol{C}\boldsymbol{B}_z\right)\mathrm{d}V$ are the elemental stiffness matrices and $\boldsymbol{M}^{(e)} = \int_V \boldsymbol{N}^{\mathrm{T}}\rho\boldsymbol{N}\,\mathrm{d}V$ is the elemental mass matrix. These elemental stiffness and mass matrices are computed using Gauss quadrature numerical integration scheme like conventional FEM. After calculating the elemental stiffness and mass matrices, they get assembled to the respective global stiffness matrices $\boldsymbol{K}_2$, $\boldsymbol{K}_1$, and $\boldsymbol{K}_0$ and mass matrix $\boldsymbol{M}$. In the absence of external forces, the final form of Navier's equation in terms of global matrices in the frequency-wavenumber domain is obtained as,

$$\left(k_y^2\boldsymbol{K}_2 + ik_y\boldsymbol{K}_1 + \boldsymbol{K}_0 - \omega^2\boldsymbol{M}\right)\tilde{\boldsymbol{u}} = \boldsymbol{0} \tag{18}$$

For the non-trivial solution of Eq. (18), the determinant of the coefficient of $\tilde{\boldsymbol{u}}$ should be zero, which results in a Polynomial Eigenvalue Problem (PEP) with respect to wavenumber, k and standard eigenvalue problem with respect to frequency, ω. The size of the stiffness and mass matrices, equivalent to the total number of degrees of freedom, depends on the order of the shape function and the number of finite elements in the discretized region. The PEP for a known frequency, ω can be converted to the linear generalized eigenvalue problem using linearization methods (as discussed in Ref. [24]) and solved as a standard eigenvalue problem. The total number of eigenvalues and corresponding eigenvectors are twice the number of degrees of freedom for a given frequency, ω. Similarly, Eq. (18) can be solved for a range of frequencies. These eigenvalues represent the wavenumbers, $k(\omega)$, at each frequency step. In another way, Eq. (18) can be solved as a standard eigenvalue problem for known wavenumber, k and will result in eigenvalues as the frequencies, $\omega(k)$. Plotting the wavenumbers for the frequency shows the existing wave modes and their dependencies on the frequency. The phase speed (C_p) is computed using the real part of wavenumber, k_real as $C_\mathrm{p} = \frac{\omega}{k_\mathrm{real}}$. The group speed ($C_\mathrm{g}$) is computed by the method discussed by Bartoli et al. [7] by Eq. (19) as,

$$C_\mathrm{g} = \frac{\partial\omega}{\partial k_y} = \frac{\tilde{\boldsymbol{u}}_\mathrm{L}^T\left(2k_y\boldsymbol{K}_2 + i\boldsymbol{K}_1\right)\tilde{\boldsymbol{u}}_\mathrm{R}}{2\omega\tilde{\boldsymbol{u}}_\mathrm{L}^T\boldsymbol{M}\tilde{\boldsymbol{u}}_\mathrm{R}} \tag{19}$$

where $\tilde{\boldsymbol{u}}_\mathrm{L}$ and $\tilde{\boldsymbol{u}}_\mathrm{R}$ are the left and right eigenvectors respectively obtained from the solution of Eq. (18).

3 Computer Implementation

The whole formulation discussed above is implemented in MATLAB R2020b. There are different ways to solve the PEP obtained in terms of the global stiffness and mass matrices from Eq. (18). In one way, we can directly equate the determinant of the coefficient matrix of $\tilde{\boldsymbol{u}}$ to zero for the non-trivial solution of $\tilde{\boldsymbol{u}}$, which leads to the polynomial in terms of the wavenumber, k_y and frequency, ω. Computing the determinant of the coefficient matrix for a large number of degrees of freedom is very difficult. Alternatively, the PEP can also be solved as an eigenvalue problem by converting the PEP to an eigenvalue problem using various linearization techniques. We can use MATLAB inbuilt function *polyeig* [25], which uses the companion matrix linearization technique to convert the PEP into a generalized eigenvalue form.

Then it computes the eigenvalues and eigenvectors by solving them as a standard eigenvalue problem. Here, we have used the MATLAB inbuilt function *polyeig* to obtain the solution of the PEP and *eig* to solve the standard eigenvalue problem obtained from Eq. (18).

4 Results and Discussions

The geometry considered here to demonstrate the effect of various types and sizes of defects on the dispersion curve is a rectangular cross-section structural waveguide with infinite length, as shown in Fig. 3. The cross-sectional parameters of the component are height, $H = 50$ mm, width, $W = 50$ mm, and has an infinite length in y-direction. The material assumed is aluminum alloy 6060, as mentioned in the Dispersion Calculator (DC) user manual [26], with the material properties as follows, Young's modulus, $E = 70$ GPa, Poisson's ratio, $\nu = 0.33$, and density, $\rho = 2700\ \mathrm{kg/m^3}$.

The dispersion curves for the structural waveguide with horizontal crack or delamination running throughout in y-direction (see Fig. 3) of size a are computed and compared with the dispersion curve for the defect-free waveguide (see Fig. 1). The face parallel to the xz-plane is discretized using FE discretization with 4-noded quadrilateral elements. As we only discretize the cross-section (i.e., 2D geometry) of the component instead of the whole structure (i.e., 3D geometry), it results in much smaller size matrices, saving a considerable amount of computational time and cost. As discussed earlier, the cracks or delamination can be modeled in different ways. We have modeled a single crack or delamination in the structure by creating separate surfaces in the defect region [21]. These separate surfaces are created by releasing the constraints and compatibility conditions at the surface and assigning different node numbers to nodes on opposite faces of the crack or delamination region. It's important to note that because our analysis in this study operates within the linear regime and the excitation loads induce a relatively low amplitude of vibration in the structure, compared to the gap created between the delamination surfaces without interpenetration, we have neglected any interpenetration between the delamination

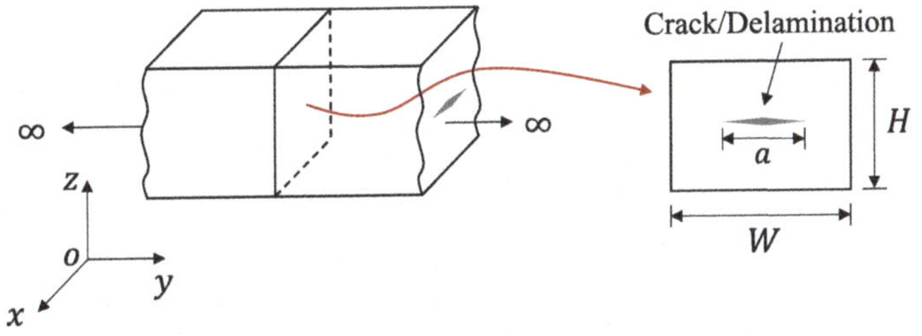

Fig. 3 Structural component with crack or delamination of size a running throughout in y-direction

or crack surfaces. Introducing interpenetration would introduce non-linear effects, which are not considered here. These cracks or delamination are assumed to be running throughout the length in y-direction, as shown in Fig. 3.

We observed that the dispersion curve obtained from SAFEM for a defect-free structural component is sensitive to the number and size of the finite elements used to discretize the cross-section. For a lower number of finite elements, they overestimate the frequencies due to the high stiffness of the discretized structure, which can be overcome by increasing the number of finite elements in the domain [8]. Therefore, before computing the dispersion curves for the structural component with defect, we perform the convergence study for the size and the number of 4-noded quadrilateral elements to be used for the accurate computation of the dispersion curves on the defect-free structural waveguide. We have used the bi-linear shape function for the geometry and the displacement approximation. We have performed the convergence study for the cutoff frequency of the shear mode (i.e., first-order anti-symmetric Lamb wave mode—A_1). Only the fundamental Lamb wave modes start as propagating waves at zero frequency in the dispersion curve. All the higher-order Lamb wave modes remain evanescent waves up to certain specific frequencies, after which they become propagating waves. Those specific frequencies are known as the cutoff frequencies. We computed the dispersion curve for defect-free structural components with the varying number of finite elements from 4 to 460, obtained the shear mode's cutoff frequency, and plotted against the respective number of finite elements. The cutoff frequency for shear mode is also calculated from the open-source dispersion curve computation software GUIGUW [15, 16]. The GUIGUW software uses constant strain triangular element mesh for dispersion computation from SAFEM and the dispersion curve is obtained for the suggested maximum edge size of the triangular element as 3.12 mm. Figure 4 shows the convergence plot of the shear mode cutoff frequency with the varying number of quadrilateral linear elements for the defect-free structural waveguide. It is observed that the shear mode cutoff frequency, 31.28 kHz for around 200 elements with 3.57 mm as the maximum edge length of the quadrilateral element, closely matches with the corresponding approximate value, 31.20 kHz, obtained from GUIGUW with less than 1% error. The converged value of the shear mode cutoff frequency from the current SAFEM formulation is obtained as 31.24 kHz for 460 finite elements, as shown in Fig. 4, with 2.50 mm maximum edge length of the quadrilateral element.

If only propagating wave modes are of interest, Eq. (18) can be solved for known real wavenumber, k_{real}, as standard eigenvalue problem for frequencies, $\omega(k_{\text{real}})$ [7]. Therefore, the solution of Eq. (18) will only represent the propagating waves. The dispersion curve for the defect-free waveguide, shown in Fig. 5, is computed for 460 number of 4-noded quadrilateral finite elements in the cross-section and is superimposed with the dispersion curve obtained from GUIGUW software for the maximum edge length of triangular element as 3.12 mm. In Fig. 5, x-axis is showing the frequency (in Hz) and the dispersion curves are plotted up to 100 kHz. Figure 5a–c shows the plots of the real part of the wavenumber, k_{real} (1/m), phase speed, C_{p}

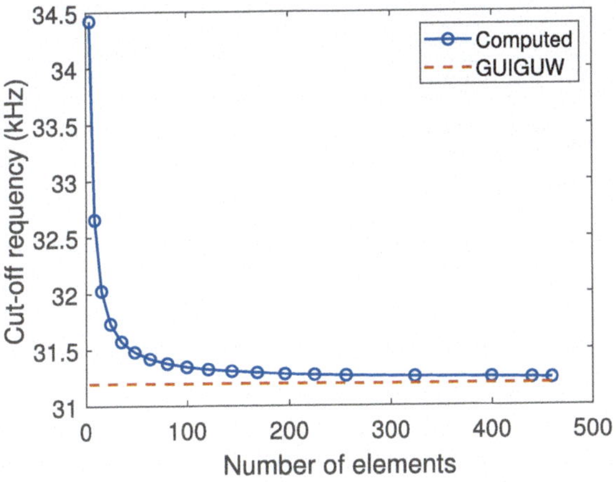

Fig. 4 The convergence plot for the shear mode (first-order anti-symmetric Lamb wave mode—A_1) cutoff frequencies obtained from the current SAFEM formulation (solid blue curve with circular markers) with the varying number of finite elements and comparing with cutoff frequency (31.20 kHz) for respective mode obtained from GUIGUW [15, 16] (dashed brown curve)

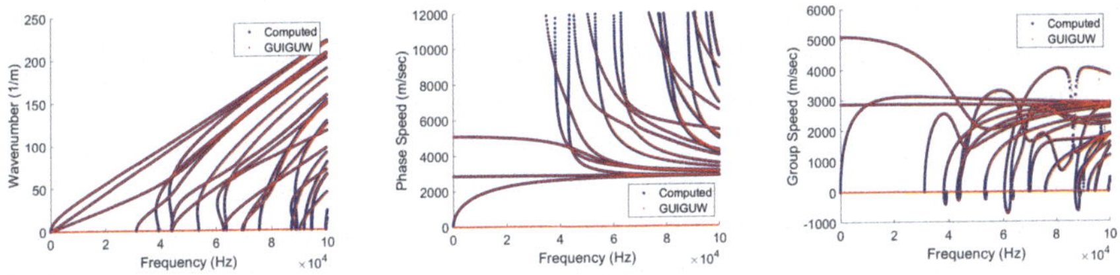

(a) k_{real} vs f (b) C_p vs f (c) C_g vs f

Fig. 5 Superposition of the dispersion curves for defect-free waveguide computed from the current SAFEM formulation (in blue color big dots), with open-source dispersion curve calculator software GUIGUW (in red color small dots)

(m/s), and group speed, C_g (m/s) respectively. All the computed dispersion curves (in blue color big dots) show a good match with the dispersion curves obtained from GUIGUW (in red color small dots).

Further, the same geometry is considered to study the effect of the defects on the dispersion curve. The crack is modeled in similar fashion as discussed earlier. The horizontal cracks or delamination of size, $a = 5$, 10, 15, and 20 mm are individually considered in the structural waveguide to study their effect on the dispersion curves. The cutoff frequency for the shear mode is again monitored in the dispersion curves obtained from the current SAFEM formulation for the waveguide with all the crack or delamination cases with the varying number of finite elements. Figure 6a shows the convergence curves for the shear mode cutoff frequency with the varying number of quadrilateral linear elements for a waveguide with the crack of sizes 5, 10, 15, and 20 mm and compared against the defect-free waveguide. The converged values

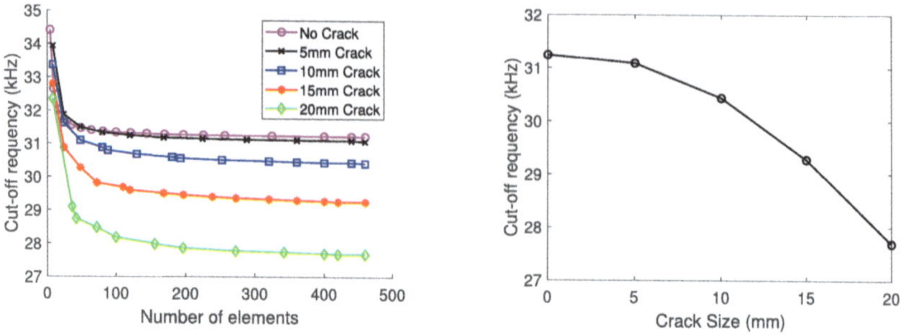

(a) Convergence plots: Cutoff frequency vs Number of finite elements

(b) Cutoff frequency vs Defect size

Fig. 6 Comparison of the shear mode (first-order anti-symmetric Lamb wave mode—A_1) cutoff frequencies obtained from the current SAFEM formulation for structural waveguide without and with crack or delamination of lengths, $a = 5$, 10, 15, and 20 mm

of the shear mode cutoff frequency for the waveguide with the considered cracks or delamination are 31.09 kHz, 30.44 kHz, 29.28 kHz, and 27.70 kHz respectively for 460 finite elements, which is shown in Fig. 6b.

Figure 6b compares the shear mode cutoff frequency for the defect-free waveguide and waveguide with cracks or delamination of various sizes. It can be seen from the figure that the presence of the defect in the structure affects the cutoff frequencies and hence the dispersion curves significantly. The reduction in the shear mode cutoff frequency is non-linear with the size of the defect, which is caused due to the reduction in the structure's stiffness in the presence of the defect. Figure 7a–c are the wavenumber, phase speed and group speed plots respectively for the structure with 20 mm crack or delamination in red color small dots compared with the defect-free structure in blue color big dots. In both the cases i.e., defect-free and with 20 mm crack or delamination, we have considered around 200 quadrilateral finite elements in the cross-section for dispersion computation. The dispersion curves are plotted up to 50 kHz frequency to clearly show the effect of the defect on the fundamental wavemodes. It can be noticed in Fig. 7 that there exists total four-fundamental modes named as the longitudinal mode (fundamental symmetric wavemode—S_0), flexural mode (fundamental anti-symmetric wavemode—A_0), shear horizontal (SH_0) and torsional (T_0) modes, which start from zero frequency. The S_0 and A_0 modes get affected due to the presence of defect; however, the SH_0 and T_0 modes almost remain unaffected.

Although these dispersion curves can be validated through both experiments and numerical simulations (using commercial Finite Element software packages), the primary focus of this study is on establishing a methodology for computing dispersion curves in complex and/or defective waveguides. The validation of these dispersion curves will be addressed in future research endeavors. Here, we have considered only

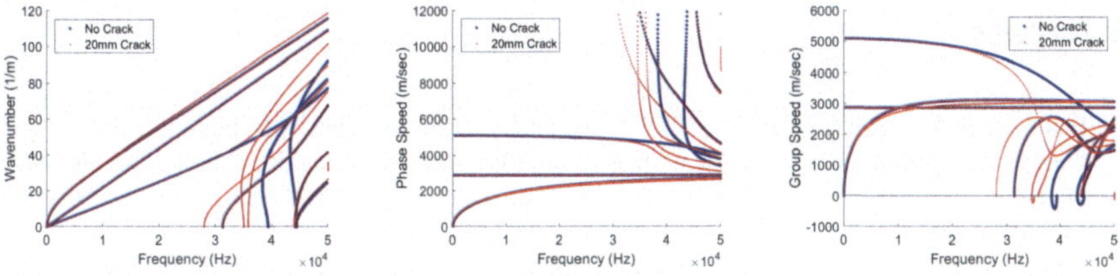

(a) k_{real} vs f (b) C_p vs f (c) C_g vs f

Fig. 7 Superposition of the dispersion curves for defect-free (in blue color big dots) and waveguide with 20 mm crack or delamination (in red color small dots) computed from the current SAFEM formulation

the horizontal mid-plane crack or delamination for the analysis. This could be further extended to analyze the structure with off-mid-plane horizontal cracks, different orientations of the cracks or delamination, multiple cracks, and their combination. This could also be used to analyze the waveguides with other kinds of defects, such as surface corrosion and the material with inclusion.

5 Conclusion

We have used SAFEM to perform the spectral analysis of the structural waveguide with and without defects. We have performed a convergence analysis for the shear mode cutoff frequency of the defect-free structure and compared it with the cutoff frequency value for the respective mode obtained from GUIGUW [16], an open-source dispersion computation software, to decide on the optimal number of finite elements to be used to discretize the cross-section of the waveguide. Our analysis indicated that approximately 200 finite elements are sufficient for the discretization of the considered geometry resulting in less than 1% error when compared with the results from GUIGUW. The validity of our SAFE framework has been established by comparing the dispersion curves obtained from our SAFE formulation with those generated by GUIGUW for a defect-free structural waveguide. Subsequently, we applied our SAFE formulation to compute dispersion curves for waveguides with defects such as cracks or delamination of different sizes to show the effect of the defect and its size on the dispersion curve. It is observed that the presence of the defect affects the shear mode cutoff frequency significantly, as shown in Fig. 6b, which is due to the reduction in the stiffness of the structure. Presently, we have only considered the horizontal mid-plane crack for our analysis. We can further extend our study to analyze the structure with off-mid-plane horizontal cracks, vertical cracks, multiple cracks, defects in different orientations, and their combinations. We can also perform the spectral analysis for the other types of defects, such as surface corrosion and the material with inclusion.

References

1. Paunikar S, Gopalakrishnan S (2022) Wave propagation in adhesively bonded metallic and composite lap joints modelled through spectrally formulated elastically coupled double beam element. Int J Mech Mater Des 18:365–393. https://doi.org/10.1007/S10999-021-09584-4/FIGURES/25
2. Cawley P, Alleyne D (1996) The use of Lamb waves for the long range inspection of large structures. Ultrasonics 34:287–290. https://doi.org/10.1016/0041-624X(96)00024-8
3. Geetha GK, Roy Mahapatra D, Gopalakrishnan S, Hanagud S (2016) Laser Doppler imaging of delamination in a composite T-joint with remotely located ultrasonic actuators. Compos Struct 147:197–210. https://doi.org/10.1016/J.COMPSTRUCT.2016.03.039
4. Mahapatra DR, Gopalakrishnan S (2004) Spectral finite element analysis of coupled wave propagation in composite beams with multiple delaminations and strip inclusions. Int J Solids Struct 41:1173–1208. https://doi.org/10.1016/J.IJSOLSTR.2003.10.018
5. Gopalakrishnan S (2016) Wave propagation in materials and structures, 1st edn. CRC Press, Boca Raton. https://doi.org/10.1201/9781315372099
6. Rose JL (2014) Ultrasonic guided waves in solid media. Cambridge University Press, Cambridge. https://doi.org/10.1017/CBO9781107273610
7. Bartoli I, Marzani A, Lanza di Scalea F, Viola E (2006) Modeling wave propagation in damped waveguides of arbitrary cross-section. J Sound Vib 295:685–707. https://doi.org/10.1016/J.JSV.2006.01.021
8. Marzani A, Viola E, Bartoli I, Lanza di Scalea F, Rizzo P (2008) A semi-analytical finite element formulation for modeling stress wave propagation in axisymmetric damped waveguides. J Sound Vib 318:488–505. https://doi.org/10.1016/J.JSV.2008.04.028
9. Munian RK, Roy Mahapatra D, Gopalakrishnan S (2020) Ultrasonic guided wave scattering due to delamination in curved composite structures. Compos Struct 239:111987. https://doi.org/10.1016/J.COMPSTRUCT.2020.111987
10. Foster EA, Bernard R, Bolton G, Jackson-Camargo JC, Gachagan A, Mohseni E, MacLeod CN (2022) Inspection of nuclear assets with limited access using feature guided waves. NDT E Int 131:102695. https://doi.org/10.1016/j.ndteint.2022.102695
11. Tian Z, Yu L, Leckey C (2015) Delamination detection and quantification on laminated composite structures with Lamb waves and wavenumber analysis. J Intell Mater Syst Struct 26:1723–1738. https://doi.org/10.1177/1045389X14557506
12. Pavlakovic B, Lowe M, Alleyne D, Cawley P (1997) Disperse: a general purpose program for creating dispersion curves. In: Review of progress in quantitative nondestructive evaluation, pp 185–192. https://doi.org/10.1007/978-1-4615-5947-4_24
13. Huber AMA, Sause MGR (2018) Classification of solutions for guided waves in anisotropic composites with large numbers of layers. J Acoust Soc Am 144:3236. https://doi.org/10.1121/1.5082299
14. Center for Lightweight-Production-Technology—The Dispersion Calculator. An open source software for calculating dispersion curves and mode shapes of guided waves. Accessed 14 July 2022. https://www.dlr.de/zlp/en/desktopdefault.aspx/tabid-14332/24874read-61142/
15. Bocchini P, Marzani A, Viola E (2011) GUIGUW. Accessed 8 Aug 2022. http://www.guiguw.com/
16. Bocchini P, Marzani A, Viola E (2011) Graphical user interface for guided acoustic waves. Am Soc Civ Eng 25(3):202–210. https://doi.org/10.1061/(ASCE)CP.1943-5487.0000081
17. Wang JTS, Liut YY, Gibby JA (1982) Vibrations of split beams. J Sound Vib 84(4):491–502. https://doi.org/10.1016/S0022-460X(82)80030-8
18. Nag A, Mahapatra DR, Gopalakrishnan S (2002) Identification of delamination in a composite beam using a damaged spectral element. Struct Health Monit 1:105–126. https://doi.org/10.1177/147592170200100107
19. Mujumdar PM, Suryanarayan S (1988) Flexural vibrations of beams with delaminations. J Sound Vib 125:441–461. https://doi.org/10.1016/0022-460X(88)90253-2

20. Hayashi T, Kawashima K (2002) Multiple reflections of Lamb waves at a delamination. Ultrasonics 40:193–197. https://doi.org/10.1016/S0041-624X(02)00136-1
21. Munian RK, Mahapatra DR, Gopalakrishnan S (2018) Lamb wave interaction with composite delamination. Compos Struct 206:484–498. https://doi.org/10.1016/J.COMPSTRUCT.2018.08.072
22. Reddy JN (2006) Theory and analysis of elastic plates and shells, 2nd edn. CRC Press, Boca Raton. https://doi.org/10.1201/9780849384165
23. Reddy JN (2018) Introduction to the finite element method, 4th edn. McGraw Hill Professional
24. Gopalakrishnan S, Chakraborty A, Roy Mahapatra D (2008) Spectral finite element method, 1st edn. Springer, London. https://doi.org/10.1007/978-1-84628-356-7
25. Polynomial eigenvalue problem—MATLAB polyeig. MathWorks India. Accessed 14 July 2022. https://in.mathworks.com/help/matlab/ref/polyeig.html
26. Huber A (2015) Numerical modeling of guided waves in anisotropic composites with application to air-coupled ultrasonic inspection. Doctoral thesis

Fracture Mechanics Based Approach for Modeling Beam–Column Connection in RCC Buildings Subjected to Seismic Loading

Praveena Rao and Hemaraju Pollayi

Abstract This paper deals with the modeling of beam–column connections in reinforced concrete (RC) buildings subjected to various seismic loadings based on a fracture mechanics approach using Python. A numerical approach framed within the fracture mechanics formulation is presented for evaluating crack initiation and propagation (fatigue life) at beam–column joints in a six-story MDOF building frame subjected to lateral cyclic loads. The performance of RC frames is significantly influenced by the response of beam–column joints, which cause partial or complete collapse of buildings during ground motion. High shear stresses are concentrated at beam–column joints; thus the present work deals with the modeling of RC beam–column connections for determining the number of loading cycles required for the nucleation of cracks leading to subsequent failure or fracture of reinforcing bars. The cross-sectional area, geometrical features, material characteristic strengths, stress–strain life parameters, and equivalent lateral forces due to earthquakes are used to determine the number of cycles for an initial crack of 1 mm. The Ramberg–Osgood relationship has been used to determine the true stress–strain curve by providing fatigue test data such as elasticity modulus (E), failure cycles, stress, and strain parameters of the material. The number of loading cycles for crack initiation and fracture is computed for joint members at every floor of the considered six-story MDOF frame building. The critical crack length and the number of cycles after which bars fractures is computed for all six floor beam–column joints with varying diameter steel bars. From the numerical model developed in Python, it was found that the cracks initiated in the joint rebars at 335.39 $\approx$ 336 cycles at the fifth floor where there are maximum lateral forces and 5030 cycles at the fourth floor. The critical crack length of different bar diameters (8–20 mm) for the 5th were found to be 0.62, 0.97, 1.37, 1.84, 2.35, 2.9, and 3.49 mm.

P. Rao (✉)
Institute of Aeronautical Engineering, Hyderabad, Telangana 500043, India
e-mail: praveenarao@iare.ac.in

H. Pollayi
GITAM (Deemed to be University), Hyderabad, Telangana 502329, India

S. Sidhardh et al. (eds.), *Advances in Structural Integrity for Mechanical, Civil, and Aerospace Applications*, Lecture Notes in Mechanical Engineering,
https://doi.org/10.1007/978-981-97-6367-2_32

Keywords Beam–column joint · Crack propagation · Cyclic loads · Fracture mechanics · Critical crack length

1 Introduction

The objective of this work is to determine the number of cycles of loading required for crack initiation and subsequent fracture/failure of beam–column joints using the fracture mechanics formulations for elastic and plastic deformations. The modeling of the reinforcing bars in beam–column joints subjected to different earthquake loading is performed using stress and strain parameters calculated by the Ramberg–Osgood relationship. Parameters for strain-life approach have been determined by the application of Coffin–Manson equation for LCF. The Ramberg–Osgood relationship has been used to determine the true stress–strain curve by providing fatigue test data such as elasticity modulus (E), failure cycles, stress, and strain parameters of the material.

The present seismic design codes for earthquake resistant buildings has provisions to withstand: (a) earthquakes of minor intensity without any damage, (b) moderate intensity earthquakes but no major damage to structural members and (c) severe earthquakes while allowing minor damages to non-structural and structural components without collapse of buildings. Several earthquakes of varying intensities occur over a couple of years that may not result in an immediate collapse or failure of structures [1]. Nevertheless, leading to damage accumulation due to repetitive low-cycle fatigue earthquake loading. The behavior of concrete due to cyclic fatigue loads is greatly influenced by the number of cycles required for failure (). When the number of loading cycles or the frequency of loading is < 10,000 cycles the structure or component is subjected to high strain, whereas for larger cycles > 10,000 results in a low-strain condition. Failure of joints due to fatigue is one of most critical aspect to be studied for the safety and serviceability of structures [2]. Typically, joint failure take place due to the crushing of concrete and yielding of steel at the plastic hinge regions [3].

There are large stresses accumulated at these joints when subjected to low cycle fatigue loads resulting in severe damages and even collapse of buildings. Lignos and Krawinkler [4] developed extensive database comprising of several experimental data collected from literature survey for structural steel components. Choi and Park [5] presented an optimal design framework with multi-objectives for enabling beam-hinging in steel frame structures.

1.1 Behavior of Joints During Earthquake

High-intensity earthquakes tend to induce large stress reversals accompanied by large strain reversals (tension and compression) in the primary steel reinforcements. It has been found in medium–high risk zones, the building frame components (flexural and compression) typically undergo cyclic strain with high amplitudes of up to 0.06 m [6]. The behavior of beam–column joints in reinforced concrete building is reported as a complex mechanism due to different loading combination such as flexure, shear and gravity as in Fig. 1. Investigations show that the failure of different members of RC frame structures is due to accumulation of large shearing stresses causing distress at joints and ultimately leading to collapse [7]. A major crack originates at the conjunction of beam and column which is the point of concentration of net stresses. The resistance of RC building frames to different load cases is significantly influenced by the behavior of beam–column joint under any loading condition [8]. The different joint details in RC frame buildings are shown in Fig. 2.

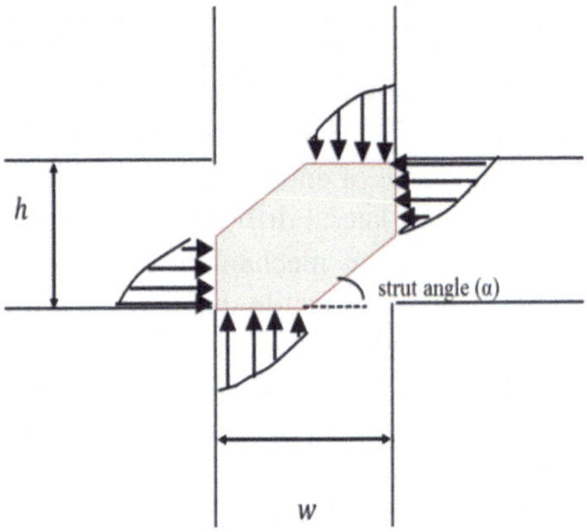

Fig. 1 Shear stresses at the beam–column connection zone (Hejazi et al.)

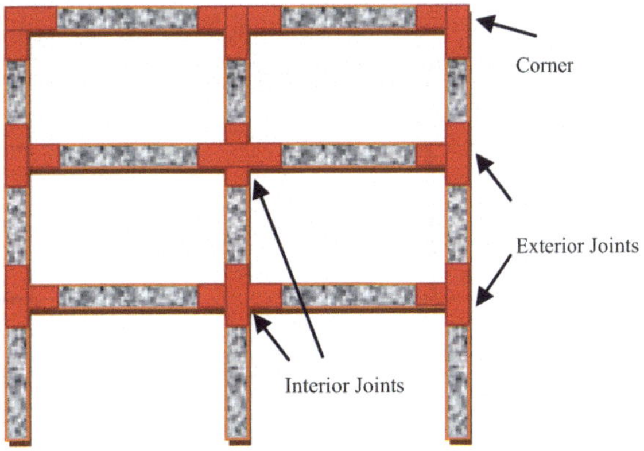

Fig. 2 Different joint details in RC frame buildings

1.2 Strong Column—Weak Beam

The current seismic design codes for RC frame buildings emphasize on providing special ductile detailing in order to avert a sudden brittle failure of joints. Thereby enabling plastic hinging and flexural failure of beams just after the joint. This is known as the 'Strong Column-Weak Beam' mechanism (SCWB) which ensures the strengthening of columns more than the beams. The fundamental objective of providing stronger columns is to prevent weak story failure of building frames and preventing severe causalities [9]. The beams are allowed to undergo plastic hinge formation before the columns, so that the building does not collapse suddenly and occupants are warned in advanced shown in Fig. 3. Material properties and yield stress of components are the two factors that largely influence the number of cycles to failure due to fatigue (Srinivasan and Pradeep 2018). This criterion states that yielding of columns should and must be prevented always. Column yielding is considered as highly undesirable for the safety of occupants as it leads to weak story and sudden collapse of entire structure. Plastic hinging at beam ends is rather desirable and a uniform lateral drift distribution can be obtained at the different story levels [10]. Column sway mechanism is highly undesirable as it results in column yielding due to inadequate ductile detailing of beam–column joints [11]. For better seismic performance, the damages are allowed to form at hinges in the beams which is characteristic behavior of SCWB. Surana et al. [12] conducted incremental dynamic analyses for different special moment-resisting frame models of low- and mid-rise RCC buildings.

All international codes follows strong column weak beam concept and have different value of β (beta) for different country codes by a formula in the form of a ratio of total flexural strengths of columns to beams framing at each joint [13].

The code provision says that the moment capacity of column should be β times stronger than beams given by Eq. (1) as follows:

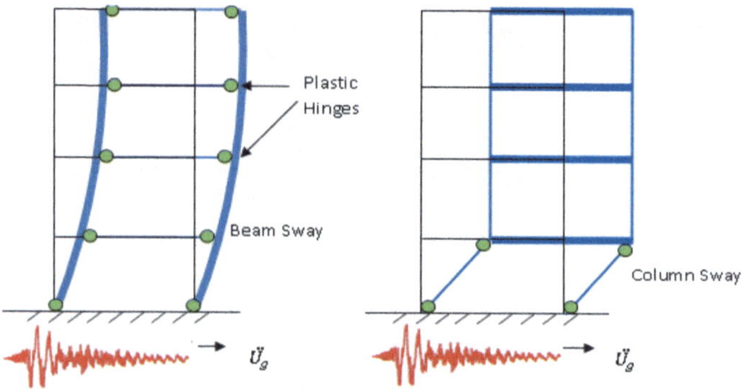

Fig. 3 Failure mechanism for **a** strong column weak beam and **b** strong beam weak column

$$\sum M_c \geq \beta \sum M_b \tag{1}$$

Values of moment capacity factor (β) for various international codes varying from 1.2 to 1.4. Indian codes prescribe a value of 1.4.

2 Fracture Mechanics Based Modeling of Beam–Column Joints

In the present work, fracture mechanics based modeling of fatigue life for beam–column joints is carried out. The mechanism of joint failure is typically due to shear typically due to concrete damage and yielding of reinforcing bars. The structural components do not collapse with concrete cracking or damage but rather after ultimate fracture of reinforcing steel, which carries most of the lateral loads and stresses. The beam–column connections in moment resisting frames undergo large moment reversals due to severe earthquake ground motion. Thereby producing much larger magnitudes of shear forces in horizontal and vertical directions at the joint sections compared to the adjoining members. The steel rebars at the joint sections coming from the beams and columns undergo tension at one boundary and compression at the other boundary due to reversals of moments.

2.1 Stages of Fracture

Crack-growth behavior of different structures signifies the physical and macroscopic fatigue failure mechanism [14]. Fatigue life of structures includes the stages of crack initiation and propagation or growth through the weaker zones in the member [15]. The crack growth process for a structural component under cyclic loads may be classified into three phases comprising initiation, propagation and fracture as shown in Fig. 4.

Fig. 4 Stages of fracture (1) crack initiation, (2) crack growth/propagation, and (3) fracture or failure

2.2 Numerical Formulation for Fracture

Fatigue is the phenomena of failure resulting from repeated loading cycles. As cyclically applied loads are not sufficiently large to result in failure within one loading cycle. A similar case is witnessed during earthquakes wherein cyclic stresses are subjected on structural element resulting in micro-cracking and cracking leading to fracture in due course of time [16]. The process of structural or elemental failure due to small number of load cycles impending fracture is called low-cycle fatigue (LCF). The steel bars provided as reinforcements are subjected to huge tensile and compressive strain cycles in the inelastic range during large earthquakes which may categorized as LCF [17–21].

The present work is based on the formulations derived from the S–N diagram using the fatigue data where the repeated cycles of stress (S) is plot as a stress ratio against the number of loading cycles (N) resulting in failure of the component. The S–N diagrams are generally represented within the Probability of Failure (PoF) formulation. The modeling of the reinforcing bars in beam–column joints subjected to different earthquake loading requires stress and strain parameters, which is calculated by the Ramberg–Osgood relationship. Parameters for strain–life approach are determined by the application of Coffin–Manson equation for LCF. The Ramberg–Osgood relationship is used to determine the true stress–strain curve by providing fatigue test data such as elasticity modulus (E), failure cycles, stress, and strain parameters of the material. The Ramberg–Osgood relationship for true stress–strain is expressed in Eq. (2) as follows:

$$\varepsilon_{\mathrm{tot}} = \underbrace{\frac{\sigma}{E}}_{\text{elastic}} + \underbrace{\left(\frac{\sigma}{K}\right)^{\frac{1}{n}}}_{\text{plastic}} \tag{2}$$

where strength coefficient and strain hardening exponent are denoted by K and n, respectively. Where the total of the true strain component $\varepsilon_{\mathrm{tot}}$ may be further decomposed as the sum of elastic strain ε_{e} and plastic strain ε_{p} as given in Eq. (3):

$$\varepsilon_{\mathrm{tot}} = \varepsilon_{\mathrm{e}} + \varepsilon_{\mathrm{p}} \tag{3}$$

The Hooke's law gives the relationship between the true stress and its elastic strain can be modeled as a line and given by Eq. (4). The plastic strain component is determined by the power–law relationship as in Eq. (5).

$$\sigma = E\varepsilon_{\mathrm{e}} \tag{4}$$

$$\sigma = K\varepsilon_{\mathrm{p}}^{n} \tag{5}$$

The values of strength coefficient (K) and strain hardening exponent (n) may be obtained by using best fit line from a log–log curve of stress versus plastic strain component as well as by the following relationship in Eq. (6):

$$K = \frac{S_y}{\varepsilon_{yp}^n} \tag{6}$$

where S_y denotes yield stress for an offset plastic strain (ε_{yp}) of 0.002.

The cycles count for growth of a crack to its final length ultimately resulting in failure (N_f) can be determined with help of Coffin–Manson relationship for different strain values as given in Eq. (7).

$$\varepsilon_{tot} = \underbrace{\frac{\sigma_f}{E}(2N_f)^b}_{\text{elastic}} + \underbrace{\varepsilon_f\left(2N_f^c\right)}_{\text{plastic}} \tag{7}$$

where

N_f number of cycles,
σ_f fatigue strength coefficient,
ε_f fatigue strain coefficient,
b elastic strain exponent, and.
c plastic strain exponent are strain life parameters obtained from Coffin–Manson equation.

3 Numerical Example

In the present work, a six-story framed building subjected to El-Centro earthquake ground motion is analyzed by response spectrum method. The strong column—weak beam mechanism is considered for the design of beam–column joints with plastic hinging of beams. The cracking of reinforcement bars at the beam–column joints is analyzed for low fatigue cycle due to earthquake ground motion based on fracture mechanics approach as discussed in the preceding section. The critical crack length and the number of cycles after which fracture of bars shall take place is determined for different diameter steel bars. The relationship of the geometrical properties and the structural properties of the bars with their fatigue life is studied.

The beam–column connections in moment resisting frames undergo large moment reversals due to severe earthquake ground motion. Thereby producing much larger magnitudes of shear forces in horizontal and vertical directions at the joint sections compared to the adjoining members. The steel rebars at the joint sections coming from the beams and columns undergo tension at one boundary and compression at the other boundary due to reversals of moments. Thus high bond strength is essential between the concrete in the joint and the main longitudinal bars in order to sustain such large magnitude forces. In this section, the mathematical modeling for fracture of reinforcing bars in the beam–column joint is performed in Python while considering rotations at beam ends due to plastic hinge formation.

The lateral forces at different floor levels are determined for the building subjected to El Centro earthquake ground motion by response spectrum method. These lateral forces are taken as the effective forces acting for the fracture modeling of bars in the beam–column joints as represented in the flowchart shown in Fig. 5. The different mechanical properties of the reinforcing HYSD bars are taken as input parameters for the fracture modeling are shown in Table 1. The depth and width of the beam column joint considered in the present work are 450 mm and 300 mm, respectively. The yield strength of rebars is considered to be 415 MPa and elastic modulus 2×10^5 MPa; whereas, values of mechanical properties such as stress concentration factor and fracture toughness are collected from different research papers.

The results of the fracture model for reinforcing bars at beam–column joints for the fifth story subjected to maximum lateral forces of the considered building frame are shown in Table 2. It may be seen that the maximum effective lateral loads are acting at the higher floors than the lower floors due to fixed cantilever characteristic of lumped mass building frame model. The number of cycles for loading up to initiation of crack was found to be $335.39 \approx 336$ cycles at the fifth floor, 5030 cycles at the

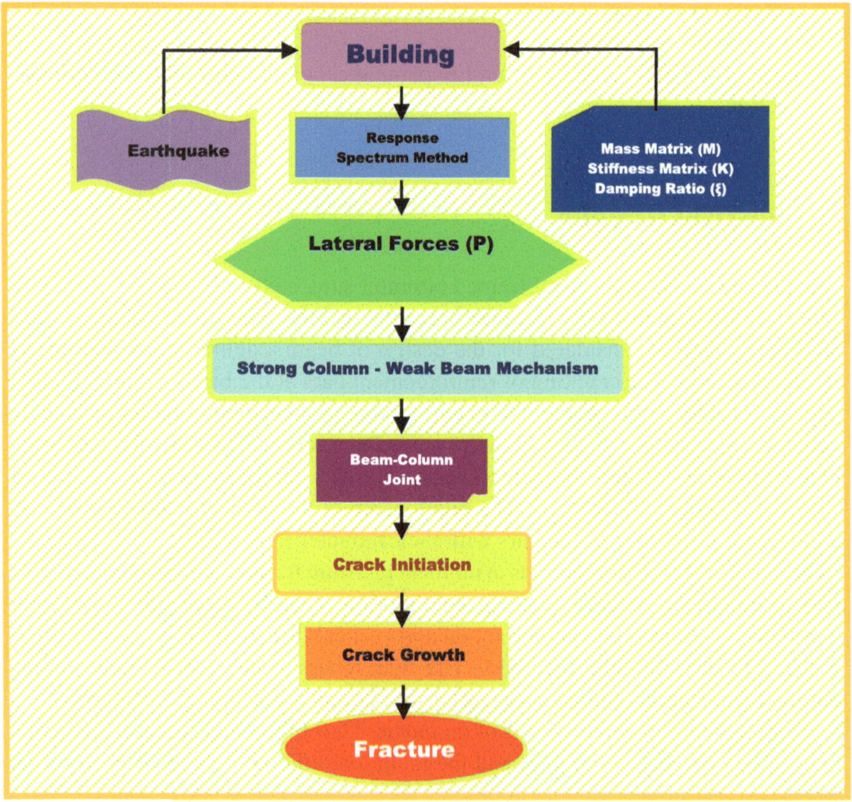

Fig. 5 Flowchart for fracture modeling in Python framework

Table 1 Input parameters used in fracture modeling framework

1	External load on material (MPa)	P
2	Area of cross section (mm^2)	A
3	Fracture toughness	Kc
4	Stress concentration factor	Kt
5	Yield strength of material (MPa)	S_y
6	Elastic modulus (MPa)	E
7	Depth of beam (mm)	d
8	width of column (mm)	w
9	Initial crack length (mm)	a

fourth floor, 70.84 × 104 cycles at third and sixth floors, 16.38 × 108 cycles at second floor and 19.40 × 1014 cycles at the first floor. The critical crack length of bars at the fifth floor were found to be 0.64 mm, 1.0 mm, 1.44 mm, 1.95 mm, 2.55 mm, 3.23 mm and 3.99 mm for bars diameters of 8 mm, 10 mm, 12 mm, 14 mm, 16 mm, 18 mm, and 20 mm, respectively, by simplified method. Similarly, the critical crack length of different bar diameters for the fifth floor were found to be 0.62, 0.97, 1.37, 1.84, 2.35, 2.9, and 3.49 mm by iterative method. Thus it may be stated that the beam–column joints at the fifth floor are exposed to maximum forces due to earthquake loads resulting in crack initiation within reinforcing bars at 336 cycles of loading and eventually leading to fracture or failure of bars after a critical crack length of 0.62–0.64 mm is reached. For the first floor, using simplified method, the crack length is 466.66 mm for 20 mm diameter reinforcing bars and minimum is 74.57 mm for 8 mm bars. It is observed that the length of crack is depending on the diameter of bars and the number of cycles for floors 1, 2, 3, and 4.

Table 2 Results from Python model for cracking of rebars in fifth floor beam–column joint subjected to maximum lateral earthquake forces

Floor No.	Lateral load P (MPa)	Diameter of bars (mm)	Number of cycles (crack initiation)	Crack propagation			
				Number of cycles		Critical crack length (mm)	
				Simplified method	Iterative method	Simplified method	Iterative method
5	0.63163	8	335.39	356	1	0.64	0.62
		10	335.39	700	1	1.0	0.97
		12	335.39	1220	9	1.44	1.37
		14	335.39	1955	29	1.95	1.84
		16	335.39	2948	65	2.55	2.35
		18	335.39	4240	123	3.23	2.9
		20	335.39	5877	210	3.99	3.49

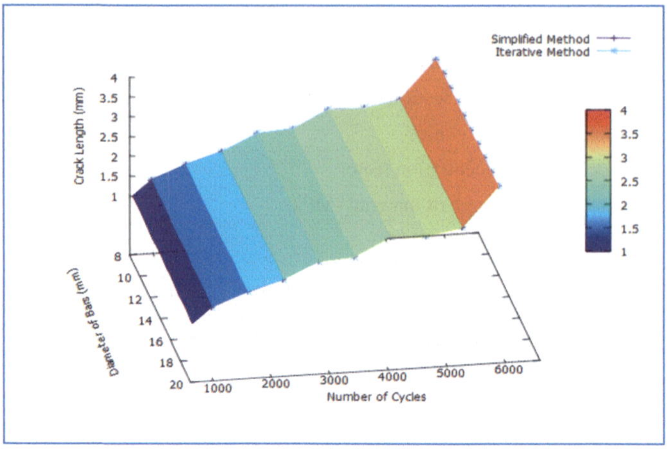

Fig. 6 Surface plot showing variation of crack length and number of cycles with different bar diameters for 5th floor

For $P = 0.63163$ MPa, and for all values of reinforcing bar diameters, crack initiated at $335.39 \approx 336$ cycles. Similarly, at the 6th floor for $P = 0.33975$ MPa, cracks in the reinforcement initiates at $708{,}415.89 \approx 708{,}416$ cycles. It was observed that the 1st floor joint rebars has maximum fatigue life and tends fail at a maximum of 71.26×10^4 cycles of repetitive loading (Fig. 6).

For the first floor, using simplified method, the crack length is 466.66 mm for 20 mm diameter reinforcing bars and minimum is 74.57 mm for 8 mm bars. It is found that the length of crack is depending on the diameter of bars and the number of cycles for floors 1, 2, 3 and 4. For $P = 0.63163$ MPa with all values of reinforcing bar diameters, crack initiate at $335.39 \approx 336$ cycles. Similarly, at the 6th floor for $P = 0.33975$ MPa, cracks in the reinforcement initiates at $708{,}415.89 \approx 708{,}416$ cycles. Figure 7a, b shows the variation of crack length with respect to the number of cycles to failure for the 20 mm diameter reinforcing bars at different floor beam–column joints. It may be observed that the 1st floor joint rebars has maximum fatigue life and tends fail at a maximum of 71.26×10^4 cycles of repetitive loading.

The number of loading cycles resulting in fracture of reinforcing bars at 2nd floor joints is 22.64×10^3cycles, 2.79×10^3 cycles for 3rd floor, 573 cycles for 4th floor, 210 cycles for 5th floor and 2.79×10^3 for the 6th floor. It may be observed that the critical crack length is greatly influenced by the lateral forces acting at each floor due to the earthquake loads. The maximum lateral forces are acting at the 5th floor and it is seen that the cracks initiate after about 335 loading cycles and consequently leads to failure at 210 cycles which is least value compared to the other floors shown in Figs. 8 (a) and (b). Thus maximum damage and shorter fatigue life may be found for reinforcing bars at beam–column joints in the 5th floor considering a six-story frame building subjected to El Centro ground motion. Cracks tend to form typically in regions with pre-existing defects or around any weaker zones.

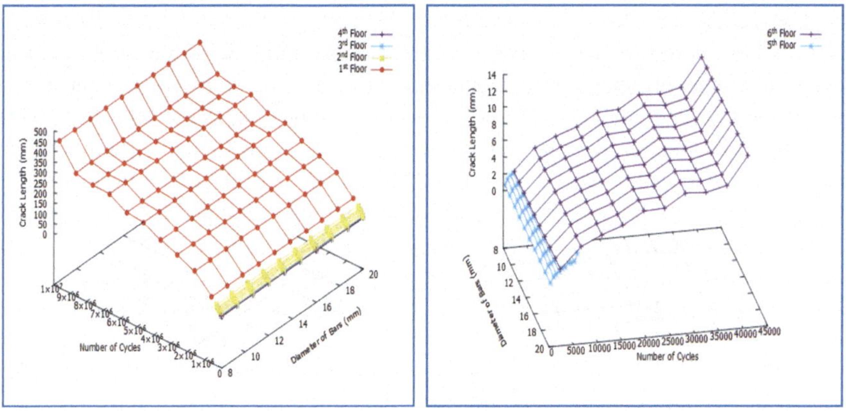

a. b.

Fig. 7 **a** Variation of crack length and number of cycles with different bar diameters for floors 1, 2, 3, and 4. **b** Variation of crack length and number of cycles with different bar diameters for floors 5 and 6

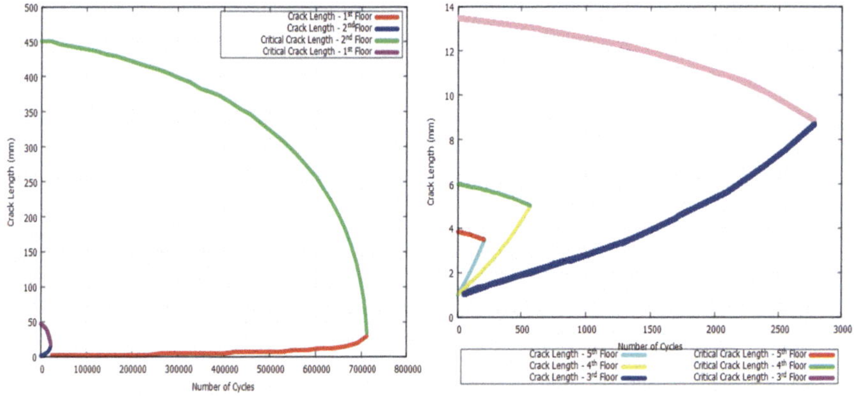

a. b.

Fig. 8 **a** Variation of crack length and critical crack length for rebars at beam–column joints in 1st and 2nd floors. **b** Variation of crack length and critical crack length for rebars at beam–column joints in 3rd or 6th, 4th and 5th floors

4 Conclusions

All engineering structures and materials are constituted with cracks or similar defects. Minor cracks exist around weaker regions where there are large stress concentrations that develop into larger cracks until failure due to fatigue. In the present work a six-story building frame subjected to El-Centro earthquake ground motion is considered and analyzed by response spectrum method. The strong column—weak beam mechanism is considered for the design of beam–column joints with plastic hinging of beams. The cracking of reinforcement bars at the beam–column joints is analyzed for low cycle fatigue based on fracture mechanics approach, as the failure of joints is governed by the behavior of reinforcing steel. The critical crack length and the number of cycles after which fracture of bars shall take place is determined for different diameter steel bars. From the numerical model developed in Python, it seen that the cracks initiated at $335.39 \approx 336$ cycles at the fifth floor, 5030 cycles at the fourth floor, 70.84×104 cycles at third and sixth floors, 16.38×108 cycles at second floor and 19.40×1014 cycles at the first floor. After the initiation of cracks at the different floor levels, the critical crack length of bars were found to be 0.64, 1.0, 1.44, 1.95, 2.55, 3.23, and 3.99 mm at the 5th floor for different bars diameters from 8 to 20 mm by simplified method. Similarly, the critical crack length of different bar diameters for the 5th were found to be 0.62, 0.97, 1.37, 1.84, 2.35, 2.9, and 3.49 mm by iterative method. Fatigue life of structures includes the stages of crack initiation and propagation or growth through the weaker zones in the member. The crack size is highly dependent on the component size in consideration and forces acting.

References

1. Shahbazpanahi S, Hejazi F, Paknahad M, Rahimipour A, Nassimi MR (2018) Modeling crack propagation in RC beam-column joints. Tech Gazette 25(4):1183–1189
2. Sabah HAH, Harba ISI (2021) A review-behavior of reinforced concrete exterior beam-column connections under cyclic loading. In: ICGE 2021—E3S web of conferences 318, 03008
3. Salgado RA, Guner S (2017) Modelling beam-column joints for progressive collapse analysis. In: Proceedings of 39th IABSE symposium, Vancouver, Canada, Engineering the Future, pp 592–599
4. Lignos DG, Krawinkler H (2011) Deterioration Modeling of steel components in support of collapse prediction of steel moment frames under earthquake loading. J Struct Eng (ASCE) 137(11):1291–1302
5. Choi SW, Park HS (2012) Multi-objective seismic design method for ensuring beam-hinging mechanism in steel frames. J Constr Steel Res 74:17–25
6. Park JW, Kim DH, Yoon SG, Lee JY (2012) Shear deterioration of reinforced concrete beam-column joints. In: Proceedings of 15th World conference on earthquake engineering, Lisbon, Portugal
7. Uma SR, Jain SK (2006) Seismic design of beam-column joints in RC moment resisting frames—review of codes. Struct Eng Mech 23(5):579–597
8. Abdelwahed BS, Kaloop MR, El-Demerdash WE (2021) Nonlinear numerical assessment of exterior beam-column connections with low-strength concrete. Buildings 11(11):562
9. Uma SR, Prasad AM (2006) seismic behaviour of beam column joints in reinforced concrete moment resisting frames—a review. Final report: A—Earthquake Codes IITK-GSDMA Project on Building Codes
10. Ning N, Qu W, Mad ZJ (2016) Design recommendations for achieving "strong column-weak beam" in RC frames. Eng Struct 126:343–352
11. Nakashima M, Sawaizumi S (2000) Column-to-beam strength ratio required for ensuring beam collapse mechanisms in earthquake responses of steel moment frames. In: 12th World conference on earthquake engineering, Auckland, New Zealand

12. Surana M, Singh Y, Lang DH (2018) Effect of strong-column weak-beam design provision on the seismic fragility of RC frame buildings. Int J Adv Struct Eng 10:131–141
13. Shamszadeh MM, Maleki S (2022) Evaluation of strong column-weak beam criterion in spliced columns of steel moment frames. Results Eng 14:100453
14. Eljufout T, Toutanji H, Al-Qaralleh M (2019) Fatigue stress-life model of RC beams based on an accelerated fatigue method. Infrastructures 4(2):16
15. Huang C, Li X, Wen M (2022) Analysis of crack formation and growth in tunnel linings using Double-K fracture criterion. Appl Sci 12(3):1064
16. Tsonos AG (2005) Cyclic load behavior of reinforced concrete beam-column sub assemblages of modern structures. Earthquake Resist Eng Struct V, WIT Trans Built Environ 81:439–449
17. Metelli G, Messali F, Beschi C, Riva P (2015) A model for beam–column corner joints of existing RC frame subjected to cyclic loading. Eng Struct 89:79–92
18. Mitra N, Mitra S, Lowes LN (2011) Probabilistic model for failure initiation of reinforced concrete interior beam–column connections subjected to seismic loading. Eng Struct 33:154–162
19. Cheung PC, Paulay T, Park R (1993) Behaviour of beam-column joints in seismically loaded RC frames. Struct Eng 71(8):129–138
20. Légeron F, Paultre P, Mazars J (2005) Damage mechanics modeling of nonlinear seismic behavior of concrete structures. J Struct Eng (ASCE) 131(6)
21. Hawileh RA, Abdalla JA, Oudah F, Abdelrahman K (2010) Low-cycle fatigue life behavior of BS 460B and BS B500B steel reinforcing bars. Fatigue Fract Eng Mater Struct 33(7):397–407

Thermomechanical Homogenization of Corrugated Core Sandwich Structure Using First Order Shear and Normal Deformation Theory

Nazim Khan and Pritam Chakraborty

Abstract The corrugated core sandwich structure is a promising candidate for the Integrated Thermal Protection System (ITPS) of Reusable Launch Vehicles (RLVs) due to its multi-functionality of load-bearing and thermal insulation capabilities. It is incorporated on the exterior of RLVs to protect the underlying structure from the severe aerodynamic and thermal loads during ascent to re-entry. ITPS can be exposed to drastically varying thermal loads during flight operations depending on its position on the vehicle, resulting in spatially varying thickness-wise temperature-dependent properties and thermal stresses. Thermomechanical analysis of RLVs components such as wing, fuselage, etc., along with actual geometry of ITPS panel makes it computationally infeasible due to their disparate length scales. This issue can be addressed by idealizing the ITPS panel as a homogeneous thick plate. In this work, a homogenization method is proposed to model the ITPS panel as a homogeneous plate to perform thermomechanical analysis. The methodology has been developed based on First Order Shear and Normal Deformation Theory (FSNDT) to incorporate the effect of out-of-plane normal strain in addition to transverse shear in the homogeneous plate model. The finite element computation of homogenized stiffness properties and sectional thermal forces and moments are presented in this paper. The efficacy of this method is verified with the comparison of the plate model and full-scale model of an ITPS panel with a thickness-wise temperature field. The comparison shows that derived plate representation is capable to encapsulate the response of the full-scale model efficiently.

Keywords First order shear and normal deformation theory · Homogenization · Corrugated core sandwich structure

---

N. Khan (✉) · P. Chakraborty
Indian Institute of Technology, Kanpur 208016, India
e-mail: nazim@iitk.ac.in

P. Chakraborty
e-mail: cpritam@iitk.ac.in

S. Sidhardh et al. (eds.), *Advances in Structural Integrity for Mechanical, Civil, and Aerospace Applications*, Lecture Notes in Mechanical Engineering,
https://doi.org/10.1007/978-981-97-6367-2_33

1 Introduction

Reusable Launch vehicles (RLVs) can significantly reduce the cost of space missions for the benefits of routine space exploration, national security, and commercial technology. Thus, their development is actively pursued by several agencies to facilitate reliable and low-cost space access [1, 2]. This can be achieved by the development of fully functional RLVs by increasing their operational life. There are various challenges associated with different technical areas for the development of RLVs. These challenges must be overcome before the development of fully operational RLVs. One such challenge is a lightweight, structurally stiff and thermally insulative external surface to protect the vehicle from severe aerodynamic heating during re-entry to maintain the aerodynamic shape under acceptable limits. Thermal insulation is a key feature that the external surface should possess for the development of RLVs. Thermal Protection Systems (TPSs) are incorporated at the exterior of the vehicle to ensure thermal insulation. These TPSs are subjected to enormous thermal loads during ascent to re-entry resulting in temperature 650 k–1700 k depending on its position on the vehicle [3]. However, these conventional TPSs have limited load-bearing capability and poor maintainability [3, 4]. These drawbacks can be addressed by the development of Integrated Thermal protection Systems (ITPSs) having both structural and thermal insulation capabilities. Therefore, these multifunctional TPSs are the potential way of weight saving to reduce the cost of space missions. Additionally, ITPSs must be designed to be robust and damage tolerant [3].

Sandwich structures are being considered for ITPSs and consist of two thin face sheets, separated by a lightweight thick core, which facilitates the high bending stiffness, enhanced energy absorbing capability, and shock resistance. The configuration of the core is an evaluation mechanism for the innovative design of sandwich structures and can be used to enhance thermal insulation performance. In the past decades, different types of core geometry of sandwich structures have been developed such as foam core, honeycomb core, truss core, corrugated core, and play an important role in the requirement of a specific application. Prior research demonstrated that the honeycomb core has better specific stiffness than foam core sandwich structure [6]. However, the interface is susceptible to debonding due to the small adhesive area between the honeycomb core and face sheets, which restricts the use of the honeycomb core in the ITPS panel. This issue can be resolved by a corrugated structure, which possesses a large interface adhesive area. Furthermore, the corrugated core structures resist bending and twisting in addition to vertical shear [7], which makes it more competent for ITPSs. The truss-core sandwich structures are lighter and better to carry compressive and bending load than corrugated core structures. However, the truss-core cannot withstand stresses generated by large thermal gradient [8]. Therefore, the use of the truss-core sandwich structure is prevented for the ITPSs design. The earlier studies also have shown that corrugated core sandwich structures are being considered for multifunctional TPS owing to both load-bearing and TPS functionality [3, 4, 8].

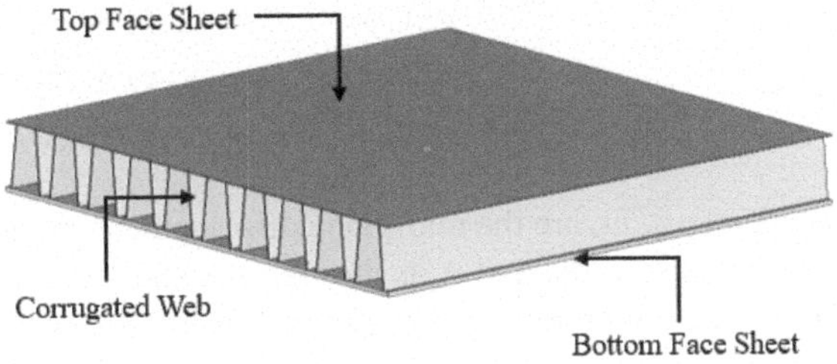

Fig. 1 Corrugated core sandwich structure [10]

The corrugated core sandwich structure is shown in Fig. 1. The empty space of the corrugated core is filled with non-loading insulating material (SAFILL) to block the heat transfer from the top face sheet (TFS) to the bottom face sheet (BFS). The function of the core is to keep the face sheets apart and stabilize the entire structure to facilitate a thick plate by resisting normal and shear deformation and curvature in a longitudinal direction [7]. The structural analysis of RLVs components such as wing, fuselage, empennage, etc., along with the actual geometry of ITPS panel is infeasible due to very small length scale of core geometry. This necessitates the evaluation of homogeneous response of the ITPS panel for computationally efficient structural analysis. The comprehensive literature review on the homogenous plate models of sandwich structure is available in [9].

In this work, the homogenization method has been addressed using First Order Shear and Normal Deformation Theory (FSNDT). The homogeneous plate representation of corrugated core sandwich structure is derived based on FSNDT to incorporate the effect of through-thickness normal strain in the plate model. This further increases the degree of freedom associated with the transverse displacement. The main objective of this work is the development of a user-defined element (UEL) to account for the normal strain across the thickness in the homogenized plate model. The homogenized stiffness properties are determined from the unit-cell analyses and assigned to the proposed plate model. The efficacy of method is verified by comparing the response of the corrugated core sandwich structure with the homogeneous plate model.

2 First Order Shear and Normal Deformation Theory

In this section, the First Order Shear and Normal Deformation Theory (FSNDT) is proposed for the mechanical behavior of the homogenized plate. The novelty of this theory is the incorporation of extensibility of the transverse normal in contrast to FSDT. This amounts for the transverse normal strain across the thickness. A linear variation of the transverse displacement is considered along the thickness direction. The displacement components at a material point (x, y, z) can be defined as

$$
\begin{aligned}
u(x, y, z) &= u_0(x, y) + z\phi_x(x, y) \\
v(x, y, z) &= v_0(x, y) + z\phi_y(x, y) \\
w(x, y, z) &= w_0(x, y) + z\phi_z(x, y)
\end{aligned}
\tag{1}
$$

where u_0, v_0, w_0 are the mid-plane displacement, ϕ_x and ϕ_y are rotation about y and x-axes respectively, and ϕ_z is through-thickness normal strain.
The strain components can be defined from Eq. (1) as follows:

$$
\begin{aligned}
\epsilon_{xx} &= \epsilon_{xx}^0 + z\kappa_x \\
\epsilon_{yy} &= \epsilon_{yy}^0 + z\kappa_y \\
\epsilon_{zz} &= \phi_z = \epsilon_{zz}^0 \\
\gamma_{xy} &= \gamma_{xy}^0 + z\kappa_{xy} \\
\gamma_{xz} &= \gamma_{xz}^0 + z\gamma_{xz}^1 \\
\gamma_{yz} &= \gamma_{yz}^0 + z\gamma_{yz}^1
\end{aligned}
\tag{2}
$$

where ϵ_{xx}, ϵ_{yy} and ϵ_{zz} are normal strain along x, y and z-directions, respectively; γ_{xy}, γ_{xz} and γ_{yz} are the engineering shear strains,

$$
\begin{aligned}
&\epsilon_{xx}^0 = \frac{\partial u_0}{\partial x}, \quad \epsilon_{yy}^0 = \frac{\partial v_0}{\partial y}, \quad \gamma_{xy}^0 = \frac{\partial u_0}{\partial y} + \frac{\partial v_0}{\partial x}, \\
&\kappa_x^0 = \frac{\partial \phi_x}{\partial x}, \quad \kappa_y^0 = \frac{\partial \phi_y}{\partial y}, \quad \kappa_{xy}^0 = \frac{\partial \phi_x}{\partial y} + \frac{\partial \phi_y}{\partial x}, \\
&\gamma_{xz}^0 = \phi_x + \frac{\partial w_0}{\partial x}, \quad \gamma_{xz}^1 = \frac{\partial \phi_z}{\partial x}, \quad \epsilon_{zz}^0 = \phi_z \\
&\gamma_{yz}^0 = \phi_y + \frac{\partial w_0}{\partial y}, \quad \gamma_{yz}^1 = \frac{\partial \phi_z}{\partial y}.
\end{aligned}
$$

where ϵ_{xx}^0 and ϵ_{yy}^0 are the in-plane normal strains; ϵ_{zz}^0 is the transverse normal strain; γ_{xy}^0 is the in-plane shear strain; γ_{xz}^0, γ_{xz}^1, γ_{yz}^0 and γ_{yz^1} are the transverse shear strains; and κ_x, κ_y and κ_{xy} are the curvatures of the mid-plane.

The stress-strain relationship is defined as

$$
\begin{Bmatrix} \sigma_{xx} \\ \sigma_{yy} \\ \sigma_{zz} \\ \sigma_{xy} \end{Bmatrix} = \begin{bmatrix} C_{11} & C_{12} & C_{13} & C_{16} \\ C_{12} & C_{22} & Q_{23} & C_{26} \\ c_{13} & C_{23} & C_{33} & C_{36} \\ C_{16} & C_{26} & C_{36} & C_{66} \end{bmatrix} \begin{Bmatrix} \epsilon_{xx} - \alpha_x \Delta T \\ \epsilon_{yy} - \alpha_y \Delta T \\ \epsilon_{zz} - \alpha_z \Delta T \\ \gamma_{xy} \end{Bmatrix}
\tag{3}
$$

and

$$
\begin{Bmatrix} \sigma_{xz} \\ \sigma_{yz} \end{Bmatrix} = \begin{bmatrix} C_{55} & C_{45} \\ C_{45} & C_{44} \end{bmatrix} \begin{Bmatrix} \gamma_{xz} \\ \gamma_{yz} \end{Bmatrix}
\tag{4}
$$

where C_{ij} are components of the stiffness matrix. ΔT represents the temperature difference and α_x, α_y and α_z are coefficients of thermal expansion along x, y and z-directions, respectively.

The stress resultant forces and moments per unit length are obtained as follows:

$$\begin{Bmatrix} N_{xx} \\ N_{yy} \\ N_{zz} \\ N_{xy} \\ M_{xx} \\ M_{yy} \\ M_{xy} \end{Bmatrix} = \begin{bmatrix} A_{11} & A_{12} & A_{13} & A_{16} & B_{11} & B_{12} & B_{16} \\ A_{12} & A_{22} & A_{23} & A_{26} & B_{12} & B_{22} & B_{26} \\ A_{13} & A_{23} & A_{33} & A_{36} & B_{13} & B_{23} & B_{36} \\ A_{16} & A_{26} & A_{36} & A_{66} & B_{16} & B_{26} & B_{66} \\ B_{11} & B_{12} & B_{13} & B_{16} & D_{11} & D_{12} & D_{16} \\ B_{12} & B_{22} & B_{23} & B_{26} & D_{12} & D_{22} & D_{26} \\ B_{16} & B_{26} & B_{36} & B_{66} & D_{16} & D_{26} & D_{66} \end{bmatrix} \begin{Bmatrix} \epsilon_{xx}^0 \\ \epsilon_{yy}^0 \\ \epsilon_{zz}^0 \\ \gamma_{xy}^0 \\ \kappa_x \\ \kappa_y \\ \kappa_{xy} \end{Bmatrix} - \begin{Bmatrix} N_{xx}^{\mathrm{T}} \\ N_{yy}^{\mathrm{T}} \\ N_{zz}^{\mathrm{T}} \\ N_{xy}^{\mathrm{T}} \\ M_{xx}^{\mathrm{T}} \\ M_{yy}^{\mathrm{T}} \\ M_{xy}^{\mathrm{T}} \end{Bmatrix}, \tag{5}$$

$$\begin{Bmatrix} Q_{xz} \\ Q_{yz} \end{Bmatrix} = \begin{bmatrix} A_{55} & A_{45} \\ A_{54} & A_{44} \end{bmatrix} \begin{Bmatrix} \gamma_{yz}^0 \\ \gamma_{xz}^0 \end{Bmatrix}, \tag{6}$$

and

$$\begin{Bmatrix} P_{xz} \\ P_{yz} \end{Bmatrix} = \begin{bmatrix} B_{55} & B_{45} \\ B_{54} & B_{44} \end{bmatrix} \begin{Bmatrix} \gamma_{yz}^1 \\ \gamma_{xz}^1 \end{Bmatrix} \tag{7}$$

where

$$([A], [B], [D]) = \int_{\frac{-h}{2}}^{\frac{h}{2}} [C] \left(1, z, z^2\right) \mathrm{d}z \tag{8}$$

and A_{ij} and B_{ij} with $(i, j = 4, 5)$ are the components of transverse shear properties. N_x, N_y and N_{xy} are the in-plane forces, N_z are the transverse force , M_x, M_y and M_{xy} are the moments and Q_{xz}, Q_{yz}, P_{xz}, P_{yz} are the transverse shear forces. N_x^T, N_y^T, N_{xy}^T and N_z^T are the thermal forces and M_x^T, M_y^T and M_{xy}^T are the thermal moments.

The elemental stiffness matrix ([K]) following the procedure given in [9] can be derived in terms of the homogenized stiffness properties ($[ABD]$ and $[D_s]$) from the following equation.

$$[K] = \int_{\Omega} \left([B_b]^T [ABD][B_b] + [B_s]^T [D_s][B_s]\right) d\Omega \tag{9}$$

The finite element formulation was implemented as a User-element (UEL) subroutine in ABAQUS [12]. The element stiffness matrix and force vector using the Gauss-Legendre quadrature rule using $2\times$ 2 quadrature rule can be determined as

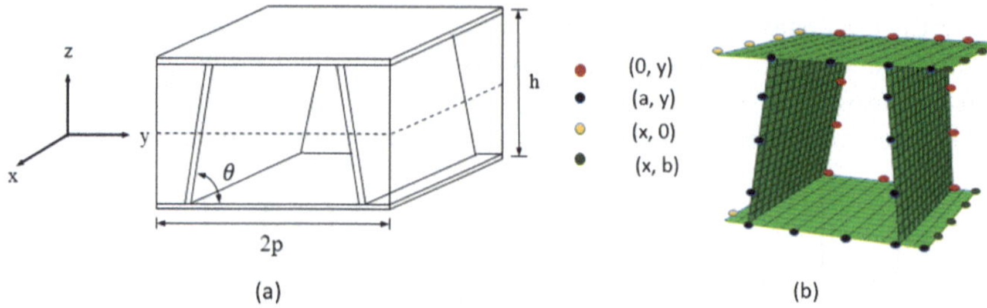

Fig. 2 **a** The periodic unit-cell of the corrugated core sandwich panel. **b** Boundary nodes on which Dirichlet boundary conditions were applied [11]

$$[K] = \sum_{i=1}^{2}\sum_{j=1}^{2}\left([B_b(\xi_i, \eta_j)]^T [ABD][B_b(\xi_i, \eta_j)] + [B_s(\xi_i, \eta_j)]^T [D_s][B_s(\xi_i, \eta_j)]\right) |J|\, W_i W_j \text{ and}$$

$$\{f_e\} = \sum_{i=1}^{2}\sum_{j=1}^{2}\{f_e(\xi_i, \eta_j)\} |J|\, W_i W_j \tag{10}$$

where (ξ_i, η_j) are Gauss points coordinates, and W_i and W_j are weight functions. The $[ABD]$, $[D_s]$, $[B_b]$ and $[B_s]$ has the same meaning as given in [9].

3 Micromechanical Analysis

The full-scale model of corrugated core sandwich structure can be idealized as 2D thick plate to perform structural analyses of ITPS panel [9, 11]. The plate representation replaces the actual geometry by incorporating the homogenized stiffness properties. This lead to reduce degrees of freedom significantly to enable thermomechanical analyses efficiently. In this work, the homogenized stiffness properties are obtained from FE unit-cell analyses. The smallest structural unit with given geometry and material properties as shown in Fig. 2a was chosen to perform these simulations. The homogenized properties are obtained by enforcing the boundary conditions on all edges intersecting with the bounding box. These boundary conditions can be obtained by the mechanical equivalence of the strains and curvatures of the unit-cell and corresponding plate element and imposed on the opposite faces of the unit-cell. The components of A-B-D matrix were obtained by enforcing linearly independent unit macroscopic strains and curvatures. For example, when $\epsilon_{x0} = 1$ is enforced to finite element model of the unit-cell (Fig. 2b) with remaining strains and curvatures equal to zero. The component of stiffness matrices were calculated from the the resultant nodal forces and moments at the restrained boundary nodes such as $A_{11} = N_x$, $A_{12} = N_y$, $A_{13} = N_z$, $B_{11} = M_x$ and $B_{12} = M_y$. Similarly, other components of stiffness matrices can be computed by imposing the rest of the macroscopic deformation.

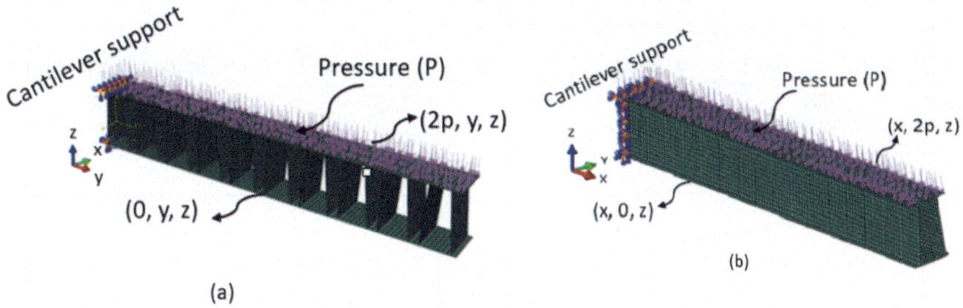

Fig. 3 FEM model of to determine **a** A_{44}, and **b** A_{55}. Faces x = 0 and x = 2p in (**a**), and faces y = 0 and y = 2p in (**b**) are subjected to states of plane strain conditions [11]

The thermal forces and moments ($\{N_T\}$ and $\{M_T\}$) shown in Eq. 5 were obtained from unit-cell analyses subjected to a thickness-wise temperature field by constraining all faces of the unit-cell along x, y and z-directions. $\{N_T\}$ and $\{M_T\}$ are equal and opposite to the resultant nodal forces and moments act on the boundary faces of the unit-cell.

For transverse shear deformation, it is assumed that the transverse shear forces are only affected by the corresponding shear strain resulting in coupling terms equal to zero. In addition to this, the components of transverse shear are considered independent of the normal transverse strain. The non-zero A_{44} and A_{55} (Eq. (6)) cannot be determined using single unit-cell analysis as it violates the conditions of anti-periodicity of moment. These constant were obtained by load-controlled cantilever beam analyses as shown in Fig. 3a, b by substituting free end transverse deflection into Eqs. (11a) and (11b) as described in [11].

$$A_{44} = \frac{4D_{22}^{*}Pa^{2}}{8wD_{22}^{*} - Pa^{4}} \quad (11a) \qquad A_{55} = \frac{4D_{11}^{*}Pa^{2}}{8wD_{11}^{*} - Pa^{4}} \quad (11b)$$

where $D_{22}^{*} = \left(D_{22} - \frac{B_{22}^{2}}{A_{22}}\right)$ and $D_{11}^{*} = \left(D_{11} - \frac{B_{11}^{2}}{A_{11}}\right)$ [4].

4 Comparison of the Full-Scale and Homogenized Plate Model

The homogenized method described in Sect. 3 has been verified for a corrugated core sandwich panel with the geometrical parameters shown in Table 1. All these geometric parameters and material of ITPS constituents (TFS, BFS, and web) were taken from the study performed by Sharma et al. [4, 5]. Titanium alloy was chosen for TFS and corrugated web due to its high service temperature, high strength, and high modulus. The material of BFS was chosen to be Beryllium alloy. The temperature-dependent Young's moduli and coefficients of thermal expansion (CTE) were linearly fitted with temperature as shown in Fig. 4. The linearly fitted properties were used

Table 1 Geometrical properties of the unit-cell [4]

Parameter	2p (mm)	h (mm)	θ (deg)	t_{TFS} (mm)	t_{BFS} (mm)	t_{WEB} (mm)
Value	50	70	85	1.2	7.49	1.63

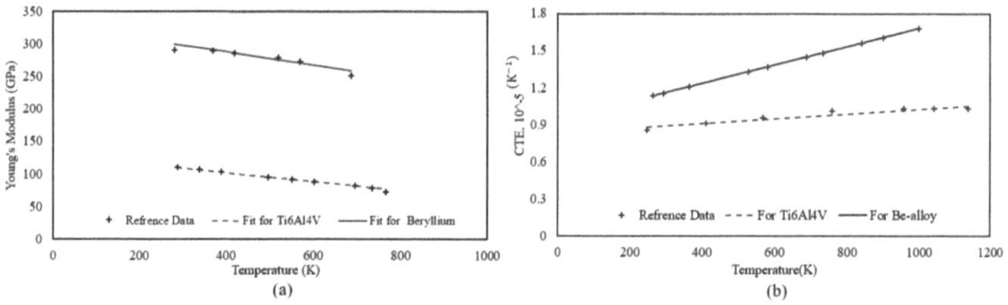

Fig. 4 Variation of **a** Elastic stiffness [10], **b** coefficient of thermal expansion with temperature

in both full-scale model and unit-cell analyses. The materials chosen for TFS, BFS and corrugated web of ITPS panel can be disparate. Hence, the temperature field throughout the thickness would be dissimilar in all three sections. The thickness of the core is very large as compared to face sheets. Additionally, the temperature gradient will be continuous due to the same material used in TFS and web. Hence, temperature distribution can be considered as a single monotonic function across the thickness of the unit-cell.

The homogenized properties are evaluated based on the method described in Sect. 3. The coupling terms (A_{16}, A_{26}, A_{23}, B_{16}, B_{26},B_{13}, B_{23}, D_{16} and D_{26}) should be non-zero due to asymmetric geometry and material properties. However, these values are very small compared to other components of the A-B-D. Therefore, these coupling terms are considered zero. The homogenized stiffness properties, and thermal forces and moments shown in Eqs. 5 and 6 subjected to thickness-wise temperature field ($T = 800$ K) are given below:

$$[ABD] = \begin{bmatrix} 2.35 & 0.15 & 0.11 & 0 & -62.10 & -3.08 & 0 \\ 0.15 & 1.97 & 0 & 0 & -3.08 & -62.10 & 0 \\ 0.11 & 0 & 0.37 & 0 & 0 & 0 & 0 \\ 0 & 0 & 0 & 0.91 & 0 & 0 & -29.45 \\ -62.10 & -3.08 & -0.24 & 0 & 2560 & 188 & 0 \\ -3.08 & -62.10 & 0 & 0 & 188 & 2414 & 0 \\ 0 & 0 & 0 & -29.45 & 0 & 0 & 1117 \end{bmatrix} \times 10^6,$$

$$\{N_T\} = \begin{Bmatrix} 20.30 \\ 17.70 \\ 2.60 \\ 0.0 \end{Bmatrix} \times 10^3, \quad \{M_T\} = \begin{Bmatrix} -570.50 \\ -570.60 \\ 0.0 \end{Bmatrix} \times 10^3 \text{ and } \quad [D_s] = \begin{bmatrix} 1.02 \times 10^5 & 0 \\ 0 & 317 \end{bmatrix}.$$

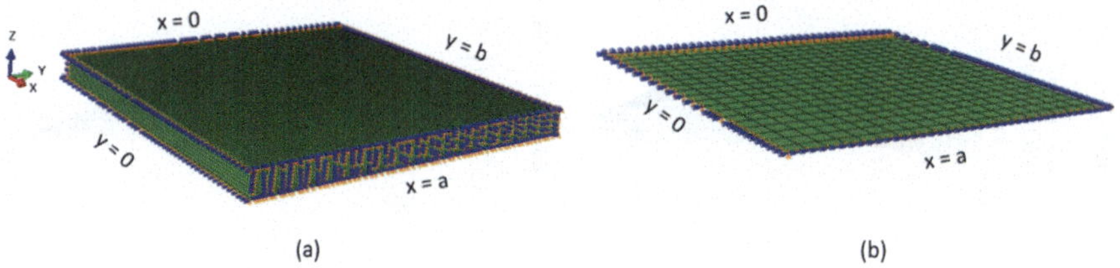

Fig. 5 Finite element model and boundary conditions for **a** full-scale model, and **b** 2D plate model

where the units of components of [A], [B] and [D] are in N/mm, N and N-mm, respectively; $\{N^T\}$ and $\{M^T\}$ are in N/mm and N, respectively; A_{44} and A_{55} are in N/mm.

Further, the response of the homogenized plate model is compared with the full-scale model of the corrugated core sandwich structure. It is assumed that symmetric load is subjected to the full-scale model of the ITPS panel. Therefore, symmetric analyses are performed with one-fourth of the ITPS panel and homogenized plate model. The dimensions of both FEM models are $a = b = 1000$ mm with 20×20 unit-cells along and across the corrugation. The directions along and across the corrugation are y and x, respectively. The ITPS panel is modeled with S4 elements using ABAQUS [12]. S4 is a general purpose four-noded quadrilateral shell element with full integration and finite strain formulation. Each node has six degrees of freedom -three displacements and three rotations with a bi-linear interpolation function. In contrast to the full-scale model, the user-defined element (UEL) is developed for the homogenized plate model based on FSNDT described in Sect. 2. The element size of 10 mm and 50 mm were considered to model the ITPS panel and 2D plate model, respectively. The finite element models of the full-scale and homogenized plate are modeled with 50702 and 441 nodes, respectively. Therefore, the full-scale model is approximately 115 times larger than the plate model. The simply supported boundary conditions were imposed along the boundary edges as: (a) $u_2 = 0$, $u_3 = 0$ and $\theta_1 = 0$ at $x = 0$, (b) $u_1 = 0$, $u_3 = 0$ and $\theta_2 = 0$ at $y = 0$ and symmetric boundary conditions were applied on symmetric edges ($x = a$ and $y = b$). In addition to this, ϕ_z is also constrained in the homogenized plate model based on FSNDT at $x = 0$ and $y = 0$. The boundary conditions on both finite element models are shown in Fig. 5.

The thermomechanical analyses were considered to verify the homogenized properties, thermal forces and moments with a through-thickness temperature field. The effect of temperature in the homogeneous plate model is incorporated through homogenized thermoelastic properties ([ABD], $[D_s]$, $\{N^T\}$ and $M^T\}$) and assigned to the plate element using UEL. The analyses were carried out to compare the response between the full-scale and homogenized plate models subjected to the uniform thickness-wise temperature field. The contours of transverse displacement (u_3) of full-scale model and plate model are shown in Fig. 6. It can be seen from the figure that the transverse displacement of the homogenized plate is in good agreement with the full-scale model of the ITPS panel. However, the local deflections are observed in the full-scale model due to the effect of the corrugated web. In contrast, these local effects are absent in the plate model.

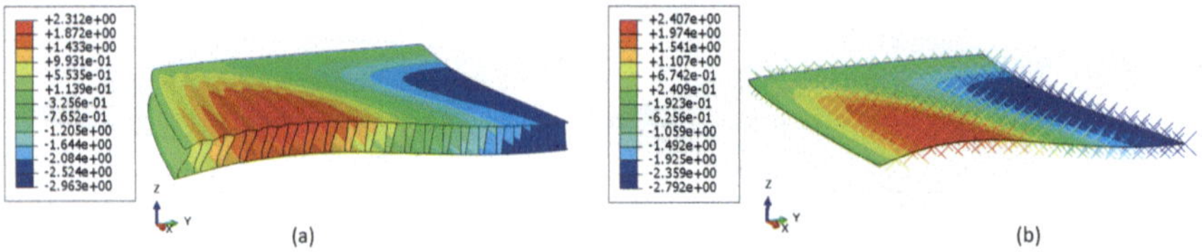

Fig. 6 Contours of the transverse displacement, u_3 (mm) of **a** full-scale model, and **b** homogenized plate subjected to thickness-wise temperature, T =800 K

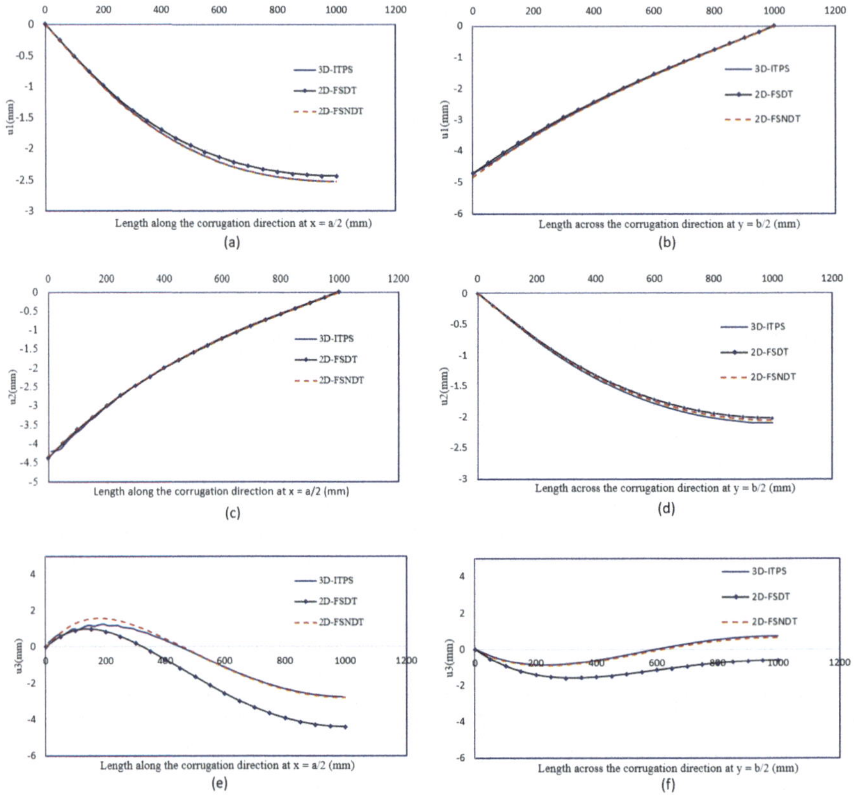

Fig. 7 Comparison of displacement components of the full-scale and the homogenized plate subjected to transverse pressure of 101 kPa at $T = 800K$: **a, b** u_1 , **c, d** u_2 and **e, f** u_3

The variation of displacement components between the full-scale and plate model subjected to thickness-wise temperature (T = 800 K) are shown in Fig. 7. The mid-plane displacement of the full-scale model is chosen for comparison with the plate model along x = a/2 and y = b/2. It can be seen from Fig. 7a–d that the variation of the in-plane displacements (u_1 and u_2) obtained from FSDT and FSNDT reasonably accurate with the full-scale model. However, the deviation can be observed in the

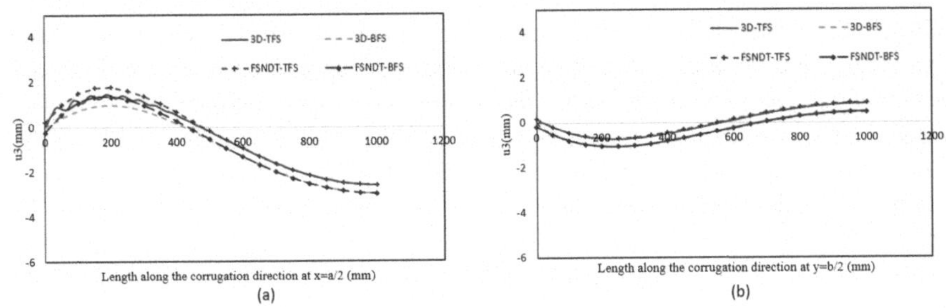

Fig. 8 The variation of the transverse displacement (u_3) of TFS and BFS along **a** $x = a/2$, **b** $y = b/2$

transverse displacement. From Fig. 7e–f, the maximum error occurs in the transverse displacement (u_3) between the FSDT plate model and full-scale model. In contrast, the homogenized plate model based on FSNDT yields a more accurate variation of u_3 compared to the FSDT plate model. Hence, the proposed plate model is competent to encapsulate the response of the detailed structure accurately. However, the deviation in u_3 can be seen along $x = a/2$ near the boundary. The error ascribed is due to the effect of the web present in the full-scale model. On the contrary, the variation of u_3 along $y = b/2$ matches well with the full-scale model. The difference in accuracy is due to the sectional geometries along these directions. The sectional geometries are identical along $y = b/2$, while geometry along $x = a/2$ are varying from each other. Thus, the plate assumption along the corrugation direction is not appropriate resulting in a large error compared to the other direction.

Additionally, the transverse displacements of the TFS and BFS are obtained from the plate solution using $w(x, y, z) = w_0(x, y) + z\phi_z(x, y)$. The variation in u_3 along x =a/2 and y =b/2 obtained from the plate model is compared with the full-scale model in Fig. 8. The comparison shows that the prediction of the homogenized plate is agreed satisfactorily with the full-scale model. Thus, the ITPS panel can be approximated as a homogenized plate derived from FSNDT.

5 Conclusion

In this work, 2D plate representation of a corrugated core sandwich structure with a thickness-wise temperature field has been derived. The First Order Shear and Normal Deformation Theory (FSNDT) is proposed to account for the through-thickness normal strain in the plate model. The temperature-dependent elastic properties and thermal expansion are considered in the homogenized plate. The effect of temperature is incorporated in the plate model through homogenized stiffness properties and thermal forces and moments. The present work is focused on the development and implementation of a user-defined element (UEL) subroutine in ABAQUS based on FSNDT. A detailed derivation of the stiffness matrix in terms of the homogenized stiffness properties is presented. The homogenized properties obtained from unit-cell and beam analyses are assigned to the plate elements. The efficacy of the homogenization method is done by comparing the response between the full-scale and homogenized plate model subjected to a thickness-wise temperature field. The comparison of displacement components shows that the results obtained from the homogenized plate are reasonably accurate with the full-scale model. The ratio of degrees of freedom between the full-scale and homogenized plate model is approximately 115. Thus, the homogenized plate model yields computationally efficient and accurate results and hence can be used in the thermomechanical analysis of the ITPS panel.

References

1. NASA, HQ.: Access to Space Study, Summary Report. NASA-TM-109693, (1994)
2. National Research Council.: Reusable launch vehicle: Technology development and test program. National Academies Press, (1996). doi: https://doi.org/10.17226/5115
3. Martinez OA, Sharma A, Sankar BV, Haftka RT, Blosser ML (2010) Thermal force and moment determination of an integrated thermal protection system. AIAA Journal 48(1):119–128. https://doi.org/10.2514/1.40678
4. Sharma, A., Sankar, B. and Haftka, R.: Homogenization of plates with microstructure and application to corrugated core sandwich panels. 51st AIAA/ASME/ASCE/AHS/ASC Structures, Structural Dynamics, and Materials Conference 18th AIAA/ASME/AHS Adaptive Structures Conference 12th 2706 (2010). doi: https://doi.org/10.2514/6.2010-2706
5. Sharma, A.: Multi-fidelity design of an integral thermal protection system for future space vehicle during re-entr. Ph.D. thesis, University of Florida (2010)
6. Yang JS, Li DL, Ma L, Zhang SQ, Schröder KU, Schmidt R (2019) Numerical static and dynamic analyses of improved equivalent models for corrugated sandwich structures. Mechanics of Advanced Materials and Structures 26(18):1556–1567. https://doi.org/10.1080/15376494.2018.1444232
7. Chang WS, Ventsel E, Krauthammer T, John J (2005) Bending behavior of corrugated-core sandwich plates. Compos Struct 70(1):81–89. https://doi.org/10.1016/j.compstruct.2004.08.014
8. Bapanapalli SK (2007) Design of an integral thermal protection system for future space vehicles. Ph.D. thesis, University of Florida
9. Khan N, Chakraborty P (2023) Direct-homogenization-based plate models of integrated thermal protection system for reusable launch vehicles. AIAA J 61(12):5242–5262. https://doi.org/10.2514/1.J063120
10. Khan N, Chakraborty P (2021) Homogenization of corrugate core sandwich structures for ITPS applications. NMIMS Eng Technol Rev 3:42–61. https://doi.org/10.2514/1.26779
11. Khan N, Chakraborty P (2021) Thermomechanical homogenization of corrugated core sandwich structures of reusable launch vehicles. AIAA J 59(10):4228–4242. https://doi.org/10.2514/1.J060246
12. Abaqus FEA (2017) Abaqus INC. Providence, Rhode Island, United States

The Effect of Hydroxyl Functionalized MWCNTs on the Interlaminar Fracture Toughness of Basalt Fiber-Reinforced Epoxy Composite

P. J. Saikia, S. Kakati, M. Kumar, and N. Muthu

Abstract With recent advances in science and technology, there is a worldwide demand for lightweight, environmentally friendly, low-cost polymer composites. Basalt fiber-reinforced composites have received a lot of interest since they are less expensive, have good chemical stability, and have better mechanical properties than glass fiber. The present study is concerned with the effect of hydroxyl functionalized multi-walled carbon nanotubes (MWCNTs) on the mode I fracture behavior of basalt fiber-reinforced epoxy composites. The composites are fabricated by conventional hand layup with vacuum bagging technique and the mode I energy released rate G_{IC} is evaluated using a double cantilever beam test with three different MWCNTs contents of 0.125, 0.25, and 0.5% per weight of epoxy/hardener. It is observed that the fracture toughness increased proportionally with the addition of MWCNTs till 0.25%. The enhancement is primarily due to improved dispersion and interfacial adhesion between the epoxy resin and the MWCNTs. However, the interfacial characteristics for 0.5% of MWCNT might be diminished due to the agglomeration of MWCNTs in the epoxy resin. The finite element modeling (cohesive zone) correlates well with the experimental results and confirms the enhancement of interlaminar fracture toughness by adding MWCNTs.

Keywords Interlaminar fracture toughness · Cohesive zone modeling · Carbon nanotube

1 Introduction

Fiber-reinforced polymer composites (FRPCs) have been widely used for decades due to their durability, high flexural strength, easiness of processing, and thermal stability. Superior specific strength-to-weight ratios of FRPCs make them a potential replacement for metals in aerospace and automobile applications. However, the weak

P. J. Saikia (✉) · S. Kakati · M. Kumar · N. Muthu
Indian Institute of Technology Guwahati, Guwahati, Assam 781039, India
e-mail: pranj176103002@iitg.ac.in

S. Sidhardh et al. (eds.), *Advances in Structural Integrity for Mechanical, Civil, and Aerospace Applications*, Lecture Notes in Mechanical Engineering,
https://doi.org/10.1007/978-981-97-6367-2_34

out-of-plane properties due to improper adhesion between fiber and the matrix often lead to fiber pullout and delamination. So, improving the interfacial adhesion to enhance the out-of-plane properties is one of the crucial research areas in present times [1, 2].

Basalt fiber is primarily used in structural and engineering applications because it possesses eco-friendly and superior mechanical properties compared to glass fiber. The cost of basalt fiber is relatively less than carbon fiber, and it is compatible with different grades of polyester and epoxy resin [3]. Therefore, numerous studies have been performed to improve the performance of basalt fiber-reinforced composites during their service [4, 5].

Numerous methods (hybridization, z-pinning, etc.) have been developed to develop a superior-quality composite for structural applications [6]. Due to the remarkable electrical, thermal, and mechanical properties of the CNTs, improving adhesion between the fiber and matrix material by adding a nano-filler (graphene and CNTs) has offered a new potential area for research. The mechanical properties of fiber-reinforced composites after incorporating different proportions of CNTs have been investigated in recent years. Bekyarova et al. [7] mentioned that the addition of CNTs (up to 25 wt%) increases the interlaminar shear strength ($\sim$ 30%) in the out-of-plane direction. Similarly, the increase in tensile strength and elastic modulus of CNT-reinforced glass fiber/epoxy composite is approximately 14% and 20%, respectively, compared to pristine glass fiber/epoxy composites. Kim et al. [3] investigated the fracture behavior of CNT-reinforced basalt fiber epoxy composites treated with acid and saline solutions. The flexural strength and fracture toughness of saline-treated composites were 10 and 40% greater than that of the multiscale composites treated with an acid solution. However, limited research work has been reported on the fracture behavior of hydroxyl functionalized multiscale basalt/epoxy composite.

Improving the mechanical properties by incorporating the CNTs in the composite matrix depends on the effectiveness of the dispersion technique [8–10]. Due to the hydrophobic nature of the CNTs, there is a possible chance of agglomeration, resulting in a reduction of the overall properties of the CNT-reinforced multiscale composite [11]. Among the various dispersion techniques available, the ultrasonication method has been used by several researchers for dispersing the CNTs in the resin [12–14]. The calendaring method has also been an effective homogenization method for dispersing large volumes of CNTs with maximum production rates [10].

Composites are prone to delamination, which can be effectively modeled using the cohesive zone model (CZM). For decades, plenty of efforts have been devoted to modeling the failure phenomenon of composites. Although Barenblatt [15] and Dugdale [16] developed the concept of the cohesive zone, other different cohesive laws (bilinear, trapezoidal, etc.) have been proposed based on the failure behavior along with the interface [17, 18]. The bilinear cohesive law has been extensively used compared to all other cohesive laws because of its simplicity and straightforward physical meaning. For instance, Giuliese et al. [19] used bilinear cohesive law to model the effect of interleaved nylon nanofiber on the global mechanical properties of carbon fiber/epoxy composite laminate. The CZM parameters were identified for double cantilever beam (DCB) and end-notched flexure (ENF) test configurations using a virtual optimization process with different nanofibers. Zhao et al. [20] have investigated the delamination response of multidirectional laminate under mode I and mixed-mode loading. A fracture toughness function has been defined based on the traction in fiber bridging, stiffness, and strength along with the interface. The proposed model has been validated by the DCB and mixed-mode bending (MMB) tests on multidirectional laminates.

In this work, DCB tests were carried out to investigate the effect of hydroxyl functionalized carbon nanotubes on the mode I fracture behavior of basalt fiber-reinforced epoxy composites. Incorporating MWCNTs increases the fracture energy at the interface, which can be easily observed by the global load–displacement response of the specimen under the experimentation. A bilinear cohesive zone model accurately modeled the delamination phenomenon along the interface. The CZM parameters were estimated for different proportions of MWCNTs.

2 Experimental Methodology

2.1 *Material*

The composite panel is made using the matrix composed of epoxy resin Epofine-1564 with hardener FINEHARD-3486–2. The matrix has a total Young's modulus of 3.42 GPa, and a Poisson's ratio of 0.3. Basalt fabric with a single fiber diameter in the range of 10–20 μm, elastic modulus = 89 GPa, and 200GSM (gram per square meter) grade is used for the composite fabrication. The hydroxyl functionalized MWCNTs (density = 0.06 g/cc, diameter = 30 nm, length = 10 μm, elastic modulus = 400 GPa is supplied by Ad nanotechnologies, India.

2.2 *Fabrication Process*

Initially, the epoxy resin part is manually mixed for 10 min with the required proportion of MWCNTs (0.125, 0.25, and 0.5% weight percentage of resin/hardener mixture). The mixture is then sonicated in a SONICS 500-watt, 20 kHz tip sonicator for 5 min for an amplitude of 40% and 1 h. 55 min for 20% amplitudes, respectively. To reduce the temperature of the mixture generated during the process of sonication, the mixture container is kept in a bowl filled with ice. Additionally, to distribute the sonication energy uniformly in the entire mixture, every 20 min, the mixture is stirred using a spoon. After sonication, the mixture is stirred mechanically at 600 rpm for 1 h. 30 min in a magnetic stirrer. During the mechanical stirring process, the temperature of the stirrer is set at 60° centigrade. The solution is allowed

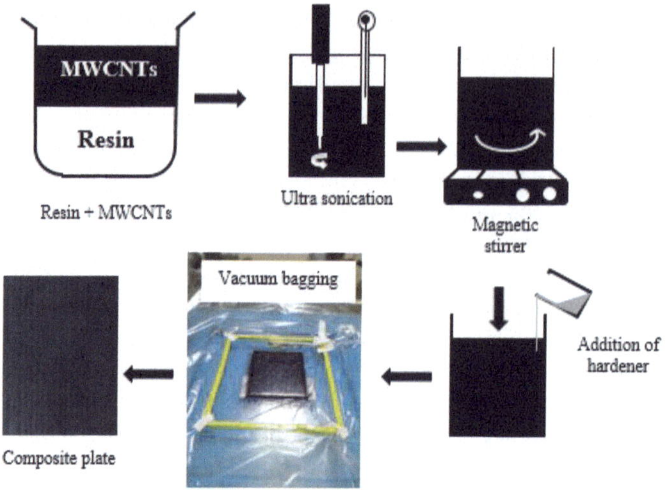

Fig. 1 Schematics of composite fabrication process

to cool to room temperature, and the hardener is added in the ratio of 30 : 100 as per the recommendation of the supplier. The entire fabrication process is shown in Fig. 1.

Because of the higher length-to-diameter ratio, the MWCNTs are segregated while passing through the multiple layers of fibers. Therefore, a conventional hand layup with the vacuum bagging process has been adapted instead of a resin transfer molding (RTM) process. Initially, the resin hardener mixture is poured into a single layer of basalt fiber with a fiber-to-matrix weight ratio of 3 : 2. The resin/hardener mixture is spread uniformly throughout a basalt fabric of dimensions $200 \times 200 \times 0.25\text{mm}^3$ using a paintbrush and hand roller. The above procedure is repeated for all 16 layers of basalt fabric. To create an initial crack for mode I delamination, a Teflon sheet with a thickness of 0.1 mm and a length of 70 mm is inserted between the middle of the laminate. The assembly is then covered with peel ply, sealed using a vacuum bag and sealant tape, and subjected to vacuum to remove the air bubble. The laminate is subjected to curing at 100 $^{\circ}$C centigrade for 24 hrs. The specimens were cut into the specified double cantilever beam (DCB) dimensions according to the standard ASTM D 5528−13.

2.3 Test Procedures

The mode I fracture test is conducted in SHIMADZU AGX-5, 100 kN load cell universal testing machine, as shown in Fig. 2. The dimensions of the specimens for DCB tests were $180 \times 21 \times 4\text{mm}^3$, with an initial crack length of 60 mm. Five specimens were tested at each condition with a loading rate of 5 mm/min to obtain reliability in the results. Figure 3 shows the average global behavior for all the variations of MWCNTs. The load linearly increases with the displacement and decreases gradually once the crack opens.

A magnified camera is used to capture the crack propagation ahead of the crack front. The strain energy released rate (SERR) for incremental crack growth is evaluated using the modified beam theory. For a DCB test configuration, SERR can be expressed as

$$G = \frac{3P\delta}{2ba}, \tag{1}$$

where P and δ are the load and displacement from global failure response; b and a are the specimen width and length of delamination, respectively.

A correction factor, Δ, has been incorporated into the crack length to nullify the effect of rotation at the delamination front. Therefore, the corrected mode I interlaminar fracture toughness can be defined as

$$G_{\text{corrected}} = \frac{3P\delta}{2b(a + \Delta)}. \tag{2}$$

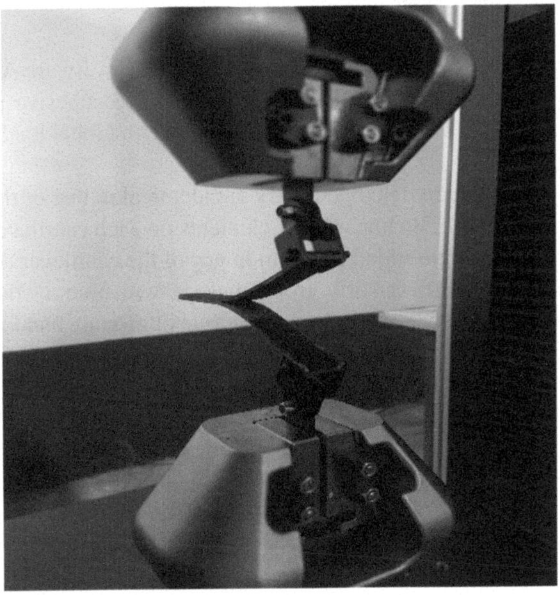

Fig. 2 Experimental setup

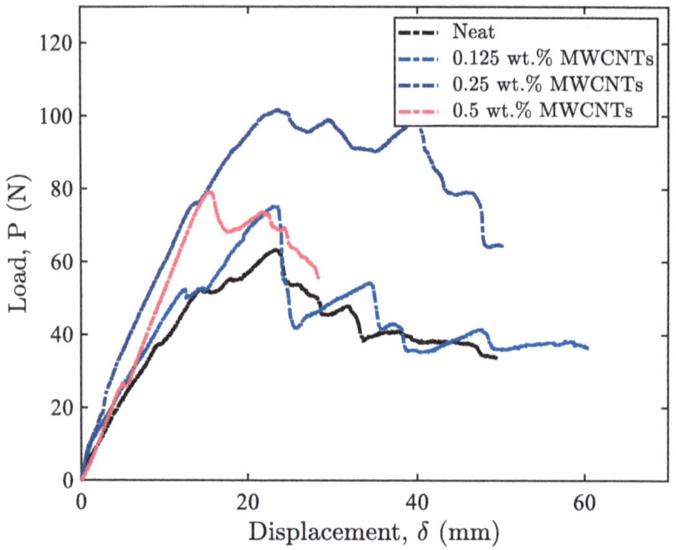

Fig. 3 Load–displacement response

3 Numerical Modeling

The specimen geometry is modeled numerically in the FE software ABAQUS 2017. The primary purpose of the modeling is to estimate the cohesive parameters for each variation of MWCNTs and to investigate its effect on global load–displacement response.

The dimensions of the modeled geometry are identical to that of the tested specimens. There were 18, 900, 3D linear brick elements on each cantilever arm with the cohesive surface behavior provided at the midplane of the cantilever across the interface. The laminate is defined as an orthotropic material with nine elastic constants. All the material properties were evaluated using the rule of mixture and EasyPBC calculator in ABAQUS. The obtained material properties with all variations of MWCNTs are shown in Table 1.

Table 1 Material properties for multiscale composites with all variations of MWCNTs

wt% MWCNTs	E_1 (GPa)	$E_2 = E_3$(GPa)	$\nu_{12} = \nu_{13}$	ν_{23}	$G_{12} = G_{13}$(GPa)	G_{23}(GPa)
Neat	54.668	14.252	0.2314	0.2119	4.912	3.449
0.125%	54.737	14.838	0.2314	0.2126	5.124	3.609
0.25%	54.805	15.399	0.2315	0.2134	5.329	3.765
0.5%	54.935	16.458	0.2315	0.2147	5.716	4.062

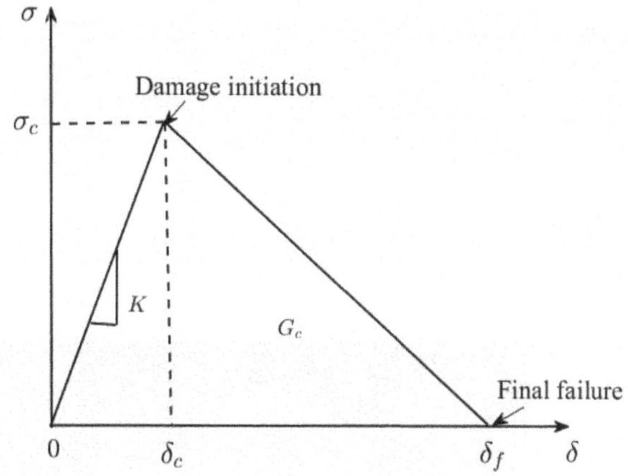

Fig. 4 Bilinear cohesive zone model

A bilinear cohesive law can effectively simulate the fracture behavior of the interface. The traction increases linearly with the separation displacement in a bilinear cohesive law until damage occurs, as shown in Fig. 4. Until the onset of damage initiation, the constitutive response at the interface can be defined as

$$\sigma = K\delta$$

where K is a non-physical parameter for the cohesive defined by the ratio between cohesive stress and the separation displacement δ.

The stiffness of the interface degraded from K to $(1 - d)K$ during the damage evolution process. The damage variable d of the cohesive element is assumed to be zero on the onset of damage initiation δ_c, and the approaches value one as it reaches the critical separation δ_f. The instantaneous cohesive stress during damage evolution can be represented as

$$\sigma = (1 - d)K\delta$$

$$d = \frac{\delta_f(\delta - \delta_c)}{\delta(\delta_f - \delta_c)}, \quad \mathrm{d} \in [0, 1]. \tag{3}$$

To simulate the load application, a vertical displacement δ is applied at the end node situated at the midplane on the upper half of the cantilever beam. Another equivalent node on the bottom part of the cantilever beam is constrained in all three directions to allow free rotation of the beam. The cohesive fracture energy G_C is initially set to calculated values of fracture energy obtained from the experiment. All other cohesive properties were iteratively calculated iteratively until the load–displacement response obtained from the numerical modeling accurately matched the experimental observations. The estimated CZM parameters for all variations of MWCNTs are shown in Table 2. Figure 5 shows the results obtained from the

Table 2 Estimated CZM parameters for all variations of MWCNTs

wt% MWCNTs	Cohesive stiffness K (MPa/mm)	Maximum cohesive strength σ_c (MPa)	Cohesive fracture energy G(N/mm)
Neat	800	5	1.10
0.125%	800	20	1.20
0.25%	1000	50	2.45
0.5%	900	32	1.28

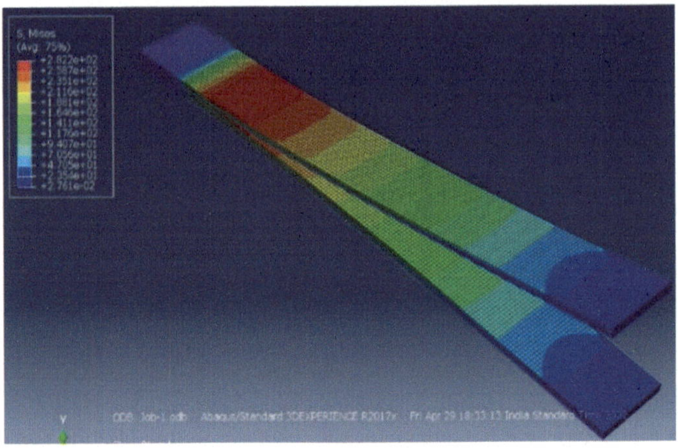

Fig. 5 Numerical results

numerical simulations. A comparison between the numerical and experimental load–displacement responses for all variations of MWCNTs is shown in Fig. 6a–d. The agreement of numerical with the experimental results is excellent, with an accuracy of more than 95%.

The numerical results demonstrate that although the FE results generally correlate well with the experimental load–displacement outcomes, certain discrepancies become apparent (shown in Fig. 6b), especially as the crack progresses. One potential explanation for these disparities lies in using a constant critical energy release rate in the numerical simulation despite fracture energy changing as delamination advances. The modified beam theory (mentioned in Eq. (2)) employed to estimate mode I fracture energy assumes a linear relationship between the cube root of compliance and crack length. This assumption is valid only when a constant fracture energy value is presumed during crack propagation. In fiber bridging cases, the experimental data points follow a curved trajectory, making a linear fitting method inadequate for determining accurate fracture energy values. Nevertheless, the FE modeling provides valuable insights into visualizing the impact of MWCNTs on interlaminar fracture toughness.

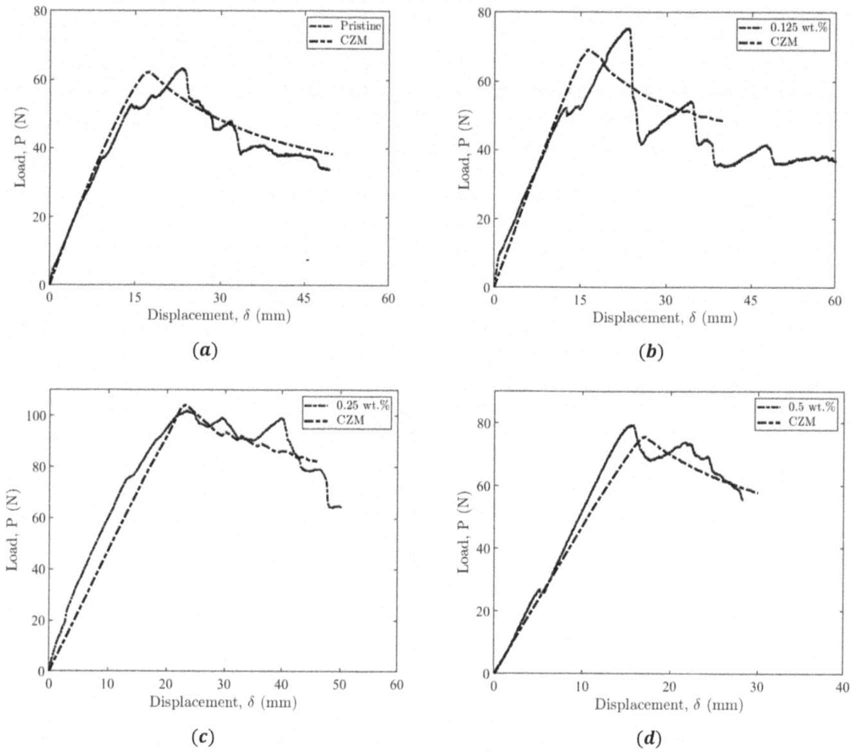

Fig. 6 Experimental and numerical load–displacement responses for **a** pristine, **b** 0.125% MWCNTs, **c** 0.25% MWCNTs, **d** 0.5% MWCNTs

4 Conclusion

Fracture tests have been carried out on basalt fiber-reinforced composite to investigate the effect of MWCNTs on its mode I fracture toughness behavior. The composites were fabricated and the mode I fracture properties were estimated with three different weight proportion of hydroxyl functionalized MWCNTs (0.125, 0.25, and 0.5%). It is observed that the load-bearing capacity and the fracture toughness of the composite coupons increase to 0.25% of MWCNTs. However, the fracture properties are reduced for 0.5% of MWCNTs due to the agglomeration of the MWCNTs in the resin mentioned in the earlier reported literature [21]. Finally, a suitable cohesive zone methodology with a bilinear cohesive law has been adopted to numerically investigate fracture toughness enhancement. The numerical results resemble the experimental observations, and the increasing values of cohesive fracture energy justify the enhancement of fracture toughness with increased proportions in MWCNTs.

References

1. Lee S-O, Choi S-H, Kwon SH, Rhee K-Y, Park S-J (2015) Modification of surface functionality of multi-walled carbon nanotubes on fracture toughness of basalt fiber-reinforced composites. Compos Part B Eng 79:47–52. https://doi.org/10.1016/j.compositesb.2015.03.077
2. Khandelwal S, Rhee KY (2019) Recent advances in basalt-fiber-reinforced composites: tailoring the fiber-matrix interface. Compos Part B Eng 192:108011. https://doi.org/10.1016/j.compositesb.2020.108011
3. Kim MT, Rhee KY, Park SJ, Hui D (2012) Effects of silane-modified carbon nanotubes on flexural and fracture behaviors of carbon nanotube-modified epoxy/basalt composites. Compos Part B Eng 43(5):2298–2302. https://doi.org/10.1016/j.compositesb.2011.12.007
4. Öztürk S (2005) The effect of fiber content on the mechanical properties of hemp and basalt fiber-reinforced phenol formaldehyde composites. J Mater Sci 40(17):4585–4592. https://doi.org/10.1007/s10853-005-1103-z
5. Liu Q, Shaw MT, Parnas RS, McDonnell A-M (2006) Investigation of basalt fiber composite mechanical properties for applications in transportation. Polym Compos 27(1):41–48
6. Kostagiannakopoulou C, Loutas TH, Sotiriadis G, Markou A, Kostopoulos V (2015) On the interlaminar fracture toughness of carbon fiber composites enhanced with graphene nano-species. Compos Sci Technol 118:217–225. https://doi.org/10.1016/j.compscitech.2015.08.017
7. Bekyarova E et al (2007) Multiscale carbon nanotube−carbon fiber reinforcement for advanced epoxy composites. Langmuir 23(7):3970–3974. https://doi.org/10.1021/la062743p
8. Gojny FH, Wichmann MHG, Köpke U, Fiedler B, Schulte K (2004) Carbon nanotube-reinforced epoxy-composites: enhanced stiffness and fracture toughness at low nanotube content. Compos Sci Technol 64(15):2363–2371. https://doi.org/10.1016/j.compscitech.2004.04.002
9. Hollertz R, Chatterjee S, Gutmann H, Geiger T, Nüesch FA, Chu BTT (2011) Improvement of toughness and electrical properties of epoxy composites with carbon nanotubes prepared by industrially relevant processes. Nanotechnology 22(12):125702. https://doi.org/10.1088/0957-4484/22/12/125702
10. Thostenson ET, Chou T-W (2006) Processing-structure-multi-functional property relationship in carbon nanotube/epoxy composites. Carbon N. Y. 44(14):3022–3029. https://doi.org/10.1016/j.carbon.2006.05.014
11. Yamamoto T et al (2008) Improved bath sonication method for dispersion of individual single-walled carbon nanotubes using new triphenylene-based surfactant. Jpn J Appl Phys 47(4):2000–2004. https://doi.org/10.1143/JJAP.47.2000
12. Miyagawa H, Mohanty AK, Drzal LT, Misra M (2004) Nanocomposites from biobased epoxy and single-wall carbon nanotubes: synthesis, and mechanical and thermophysical properties evaluation. Nanotechnology 16(1):118
13. Yu N, Zhang ZH, He SY (2008) Fracture toughness and fatigue life of MWCNT/epoxy composites. Mater Sci Eng A 494(1–2):380–384. https://doi.org/10.1016/j.msea.2008.04.051
14. Zhang X-H, Zhang Z-H, Xu W-J, Chen F-C, Deng J-R, Deng X (2008) Toughening of cycloaliphatic epoxy resin by multi-walled carbon nanotubes. J Appl Polym Sci 110(3):1351–1357
15. Barenblatt GI (1962) The mathematical theory of equilibrium cracks in brittle fracture. In: Advances in applied mechanics, vol 7. Elsevier, Amsterdam, pp 55–129
16. Dugdale DS (1960) Yielding of steel sheets containing slits. J Mech Phys Solids 8(2):100–104
17. Needleman A (1987) A continuum model for void nucleation by inclusion debonding. J Appl Mech Trans ASME 54(3):525–531. https://doi.org/10.1115/1.3173064
18. Tvergaard V, Hutchinson JW (1992) The relation between crack growth resistance and fracture process parameters in elastic-plastic solids. J Mech Phys Solids 40(6):1377–1397
19. Giuliese G, Palazzetti R, Moroni F, Zucchelli A, Pirondi A (2015) Cohesive zone modelling of delamination response of a composite laminate with interleaved nylon 6,6 nanofibers. Compos Part B Eng 78:384–392. https://doi.org/10.1016/j.compositesb.2015.03.087

20. Zhao L, Gong Y, Zhang J, Chen Y, Fei B (2014) Simulation of delamination growth in multidirectional laminates under mode I and mixed mode I/II loadings using cohesive elements. Compos Struct 116(1):509–522. https://doi.org/10.1016/j.compstruct.2014.05.042
21. Borowski E, Soliman E, Kandil U, Taha M (2015) Interlaminar fracture toughness of CFRP laminates incorporating multi-walled carbon nanotubes. Polymers (Basel) 7(6):1020–1045. https://doi.org/10.3390/polym7061020

Code Provisions to Contain Floor Flexibility in RC Buildings Under Earthquake Shaking—An Elastic Study

G. Tamizharasi and M. S. Harshit

Abstract Typically, floor diaphragms of RC buildings with a long and narrow plan or poor plan geometry, large openings, and inadequate vertical lateral load-resisting elements exhibit huge in-plane flexibility; which leads to unforeseen drift patterns and force distribution among structural elements. Building code provisions independently define rigid diaphragm by placing upper limits on (i) the plan aspect ratio, (ii) the percentage area of openings, and (iii) the relative diaphragm displacements, with additional remarks on plan geometry. Therefore, this study aims to understand the combined effects of these parameters and check the applicability of these provisions given in building codes. Thus, parameters influencing floor flexibility are identified based on relative diaphragm displacements. Equivalent Static Analysis ESA is conducted on the spectrum of rectangular building configurations consisting of RC Moment Frame and RC Structural Wall buildings having: (a) rectangular plan geometry, (b) slab with openings, (c) structural walls located at the periphery, and (d) varying height, respectively. Results agree with Linear Time History Analysis THA. Floor diaphragms of buildings with RC Frames are always rigid and those with RC structural walls (located especially at edges) show significant in-plane floor deformation under earthquake shaking. Also, defining rigid floor diaphragm solely based on plan aspect ratio, the percentage area of openings and relative diaphragm displacements may not ensure diaphragm to behave rigidly; the relative diaphragm displacements exceed the prescribed limits even if the plan aspect ratio or percentage area of opening is within the limits, especially in buildings with structural walls located at the periphery. Building code provisions (like India) must revise the criteria by satisfying all three conditions collectively. For a fixed relative diaphragm displacement, upper limits are placed on the plan aspect ratio and percentage area of the opening.

Keywords Floor diaphragm · Plan geometry · Plan aspect ratio · Percentage area of opening · Relative diaphragm displacements

---

G. Tamizharasi (✉) · M. S. Harshit
Sardar Vallabhbhai National Institute of Technology, Surat, Gujarat, India
e-mail: gtamizh@amd.svnit.ac.in

S. Sidhardh et al. (eds.), *Advances in Structural Integrity for Mechanical, Civil, and Aerospace Applications*, Lecture Notes in Mechanical Engineering,
https://doi.org/10.1007/978-981-97-6367-2_35

1 Introduction

Floor diaphragm transfers lateral forces in proportion to its lateral stiffness (i.e. rigid diaphragm) or tributary area of load distribution (i.e. flexible diaphragm). Generally, the floor slab of buildings is assumed to act as rigid diaphragm under earthquake shaking; it is almost suitable for monolithically constructed RC buildings and not always for masonry, wood or steel and pre-cast buildings (Fig. 1(a)) [26, 27, 30]. The in-plane deformation of the floor slab is ignored, and the building responses are estimated by considering only three degrees of freedom on each floor. Such assumptions may be applicable to buildings with moment frames but not for buildings that are long and narrow (e.g. long rectangular buildings), poor plan geometry in plan (e.g. buildings having shapes like L, Y, T, V, etc.) and in elevations (e.g. buildings with setbacks) and buildings with structural walls [17–19, 22]. Such buildings undergo significant in-plane deformations under earthquake shaking and cause collapse in buildings during past earthquakes (Fig. 1b).

Sometimes, in moment frame buildings with different plan geometry (like L, U etc.), the diaphragm still acts as rigid because the in-plane stiffness of the floor diaphragm is too huge in comparison with the stiffness of columns; but the assumption becomes invalid in the presence of structural walls as the diaphragm stiffness becomes insignificant in comparison with structural walls. In such conditions, in-plane deformations of the diaphragm cannot be ignored [17, 19]. Otherwise, the contribution from the relative stiffness of the diaphragm to the relative stiffness of vertical elements must be considered [21]. Also, force distribution among structural elements such as structural walls is based on relative stiffness even after defining such buildings as flexible based on building code provisions [6]. The effect becomes worse in buildings with a large plan aspect ratio. Additionally, an increase in number of stories, reduces the lateral stiffness of the building which results in a decrease in floor flexibility [19]. Besides this, based on the need and aesthetic appearance, most often buildings are provided with ducts or openings that reduce the diaphragm stiffness. It often disturbs the load path in transferring the induced inertia forces to vertical elements and causes stress concentration at a few locations of the diaphragms. Such buildings undergo huge in-plane deformation depending upon the size and location of the opening; it will increase the floor flexibility. Sometimes, it is better to use inelastic modelling to exactly capture the behaviour of such buildings [1, 6]. But building code provisions do not address these additional effects while defining the criteria for rigid diaphragm [14].

In all, parameters influencing floor diaphragm include materials and type of construction, plan geometry, plan aspect ratio (*L*/*B*, where L and B are longitudinal and lateral building plan dimensions), slenderness ratio (*H*/*B* or *H*/*L*, where *H* is the overall height of the building), presence and location of structural walls, and presence & location of openings, respectively. But, building code provisions consider a few critical parameters to classify the floor diaphragms. Usually building code provisions (such as Algeria, ASCE 7–16, Bangladesh, El Salvador, Eurocode 8, Greece, India, Iran 2800, Korea, Mexico, Pakistan, Peru, Philippines, USA, FEMA

(a)

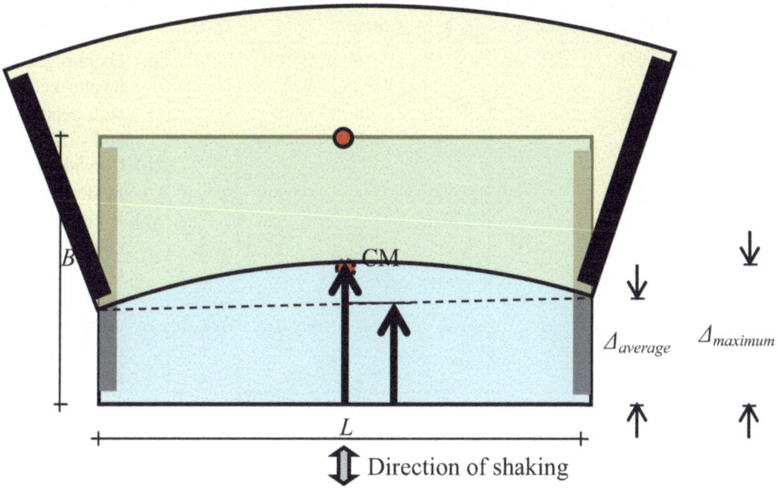

(b)

Fig. 1 **a** Poor performance of buildings from past earthquakes due to lack of diaphragm actions [17, 27, 30] and **b** behaviour of floor diaphragm under earthquake shaking

450 (i.e. [13]), UBC [2–5, 7–12, 15, 16, 20, 23–25, 28, 29]) define flexibility of floor diaphragms using (Table 1): (i) plan aspect ratio ($L/B = 2.5$–4, where L and B are dimensions of buildings along longitudinal and lateral plan direction), (ii) percentage area of openings ($\leq$ 15–50%) or area of floor opening (e.g. Turkey), and (iii) the ratio for maximum diaphragm displacement to average diaphragm displacement ($\beta = \Delta_{max}/\Delta_{avg} = 1.1$–$3$), respectively. Some building code provisions (like Chile, New Zealand, and Turkey) do not explicitly define these limits and emphasise on sufficient in-plane stiffness and strength. A few code provisions (like Chile, and Eurocode 8) directly places a restriction on plan geometry, and some (like Australia) limit the maximum diaphragm displacements using permissible deflection of structural elements. Additionally, limits are placed on height H to lateral plan dimension B of the building (i.e. *H/B*, e.g. Mexico). But defining limits on a few critical parameters may not ensure the diaphragm to behave rigid; it must consider the influence of other critical parameters.

Table 1 Building code provisions for floor diaphragm

S. no	Building code provisions	Provisions for floor diaphragm			
		L/*B*	% area of Opening	$\frac{\Delta_{max}}{\Delta_{avg}}$	Additional remarks
1	Algeria, 2003	≤ 4	< 15%	–	–
2	ASCE 7-16, 2017	≤ 3	$\leq 50\%$	≤ 3	–
3	Australia, 2007	–	–	–	Δ_{max} shall not exceed the permissible deflection of structural elements
4	Bangladesh, 2017	≤ 4	$\leq 50\%$	≤ 3	–
5	Chile, 1996	–	–	–	(a) Diaphragm should be checked for sufficient resistance and stiffness (b) Consider floor flexibility if required (c) Buildings with complex shapes (H, L, T, U, etc.) should act as one single unit; otherwise, divide them into separate unit
6	El Salvador, 1994; Korea, 2009; Pakistan, 2007; Philippines, 2010; FEMA 450; UBC, 1991	–	$\leq 50\%$	–	–
7	Eurocode 8, 2004	≤ 4	–	≤ 1.1	In C, H, I, and X plan shapes, the stiffness of the lateral branches should be comparable to that of the central part, to satisfy the rigid diaphragm condition
8	Greece, 2000	≤ 4	$\leq 30\%$	–	–
9	India, IS 1893(1):2016	≤ 4	$\leq 50\%$	≤ 1.2	–
10	Iran 2800, 2007	–	$\leq 50\%$	≤ 1.5	–
11	Mexico, 1995	< 2.5	$\leq 20\%$	–	*H*/*B* < 2.5
12	New Zealand, 1992	–	–	–	The diaphragm shall not contain abrupt variations in stiffness that influence the distribution of lateral force in the structures

(continued)

Table 1 (continued)

S. no	Building code provisions	Provisions for floor diaphragm			
		L/*B*	% area of Opening	$\frac{\Delta_{max}}{\Delta_{avg}}$	Additional remarks
13	Peru, 2016	–	≤ 50%	–	Diaphragm cross section with a net resistant area of ≥ 25% of the total cross-sectional area of the same direction computed with the total dimensions of the storey
14	Turkey, 2007	–	–	–	(a) Total area of floor opening < 1/3 gross floor area (b) Local floor opening should not hinder the smooth transfer of seismic loads to vertical elements (c) No abrupt reduction in in-plane stiffness and strength

Equivalent static analysis ESA is conducted by varying critical structural parameters using a spectrum of rectangular building configurations, consisting of RC Structural Walls and RC moment frames. The provisions given in building codes are checked for their applicability; results agree with linear time history analysis THA. Hence revisions are suggested for building code provisions using revised design limits.

2 Numerical Study

ESA of various building configurations is performed using SAP 2000 by varying critical parameters like *L*/*B*, % area of opening, height of building, presence, and location of structural walls. 5 storey building with 3 m storey height and 4 m bay length in each direction is considered; the building has fixed 3 bays in Y-direction and a varying number of bays in x-directions depending upon the aspect ratio of building. M30 grade concrete and Fe 415 steel reinforcement are considered. Sizes of beams and columns are 300 mm × 300 mm and 400 mm × 400 mm. Thickness of slab and infill walls are 150 mm and 250 mm, respectively; Unit weight of infill wall is 20 kN/m^3. The building is assumed to be in seismic zone V having hard rock type soil. Values of importance factor and response reduction factor are taken as 1 and 3. Beams and columns are modelled as line elements; structural walls, and slabs are modelled as 2D shell elements. The range of parameters and response quantities considered for present study is given in Table 2. Diaphragm displacement Δ and

Table 2 Range of parameters considered

S. No.	Parameters considered	Range of parameters
1	Plan aspect ratio, *L*/*B*	1, 2, 4, 5.33, 8, 10
2	% area of opening	2.22, 6.66, 11.11, 15.55, 20.00, 24.44, 28.88, 33,33, 37.77, 42.22, 44.44
3	Height of buildings (by varying number of storey)	3, 5, and 10 storey buildings with height 9, 15 and 30 m
4	Structural walls	With and without structural walls at periphery
5	Check for rigid floor diaphragm using IS 1893(1)	β (= $\Delta_{max}/\Delta_{avg}$) $\leq$ 1.2
6	Response quantities	Δ in mm, Δ_n and β

normalised diaphragm deformation with edge displacement Δ_n along the length of building in X-direction, and normalised displacements (expressed in terms of $\beta = \Delta_{max}/\Delta_{avg}$) are chosen as response parameters, to check the provisions given in code provisions. Δ is used to find the magnitude of deformations in the plan by varying *L*/*B* whereas Δ_n is used for finding the amplification of displacement in the plan for a fixed *L*/*B* by varying other parameters like % area of opening and height of the buildings. Also, the limit prescribed for Δ and β is $\leq$ 0.004H and $\leq$ 1.2 [IS 1893(1)] [10].

2.1 Plan Aspect Ratio

Five-storey buildings with different *L*/*B* (1, 2, 4, 5.33, 8, and 10) are considered (Table 3). All these buildings are analysed for two different cases: (i) with structural walls located at periphery of building along Y-direction and (ii) without structural walls. Following are the observations from the study (Fig. 2):

(a) In-plane deformation increases with an increase in *L*/*B*; thus, limiting *L*/*B* in building code provisions is mandatory. Otherwise, floor flexibility must be considered in design; and
(b) Buildings with structural walls located at periphery show more amplification in in-plane floor deformation than buildings without structural walls; thus, the presence of structural wall and its location highly influences the flexible behaviour of floor diaphragm. This effect must be considered while placing an upper limit on *L*/*B* in building code provisions.

Table 3 Selected building configurations with varying aspect ratio

Building configurations with different L/B having Structural Walls located at Periphery in Y-direction		
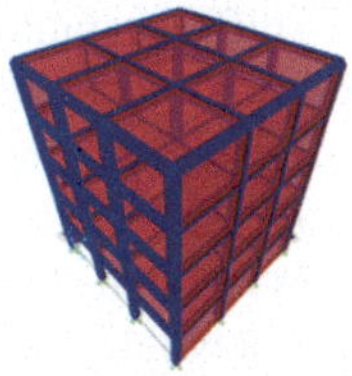 L/B = 1	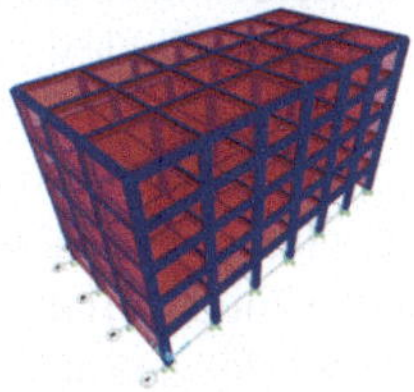 L/B = 2	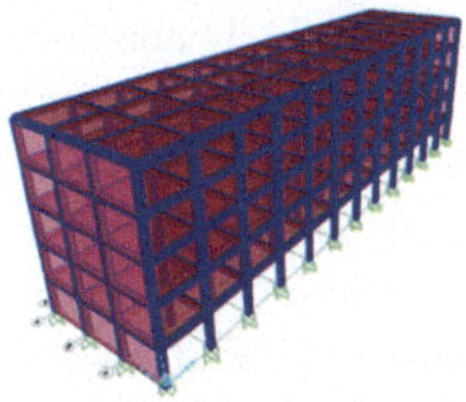 L/B = 4
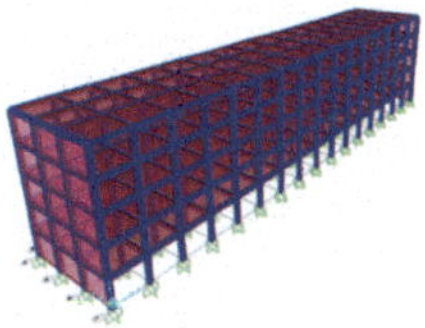 L/B = 5.33	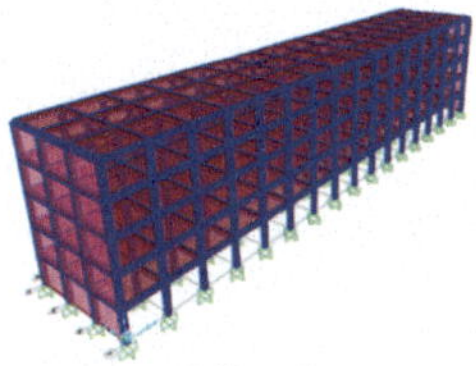 L/B = 8	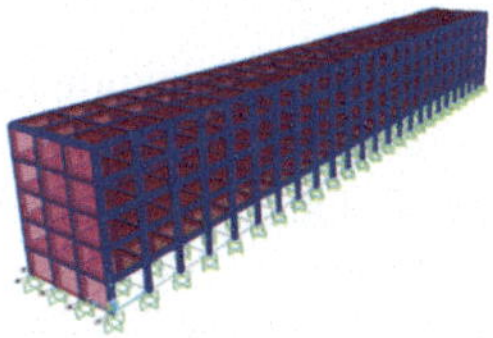 L/B = 10

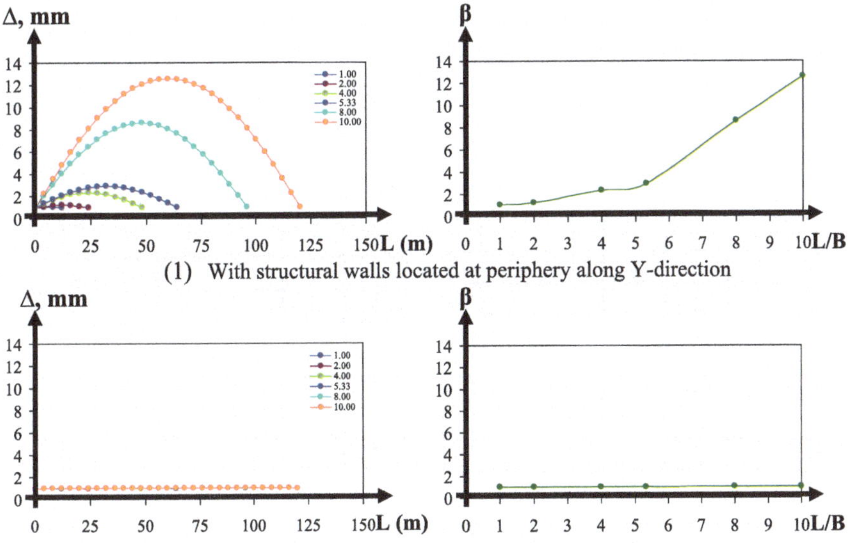

Fig. 2 Diaphragm displacement (Δ in mm) along length of building L in *X*-direction and variation of β with *L*/*B* for buildings with: **a** with structural walls and **b** without structural walls located at periphery along *Y*-direction

2.2 *Percentage Area of Opening*

Five-storey buildings having different percentage areas of openings (00.00, 08.33, 16.66, 22.22, 33.33 and 44.44) are considered (Table 4). In total, 74 different building configuration with different percentage area of openings are obtained (a few configurations may not seem to be practical, but it is included to show the maximum range of amplification); only results corresponding to buildings with $L/B = 4$ is discussed here, as the overall observations remain similar. All these buildings are analysed for two different cases: (i) with structural walls located at the periphery of building along Y-direction and (ii) without structural walls. Following are the observations made from the study (Fig. 3):

(a) In-plane deformation increases with increase in % area of openings, and hence, code provisions must have placed an upper limit on % area of opening (say 50%). Otherwise, floor flexibility due to opening must be considered in design; and

(b) Buildings with structural walls located at periphery shows more amplification in in-plane floor deformation than buildings without structural walls; thus, presence of structural wall and its location highly influences flexible behaviour of floor diaphragm. This effect must be considered while placing an upper limit of % area of openings in building code provisions.

Table 4 Selected building configurations with varying % area of openings ($L/B = 4$)

Building configurations with different % area of opening with Structural Walls located at Periphery in Y-direction	
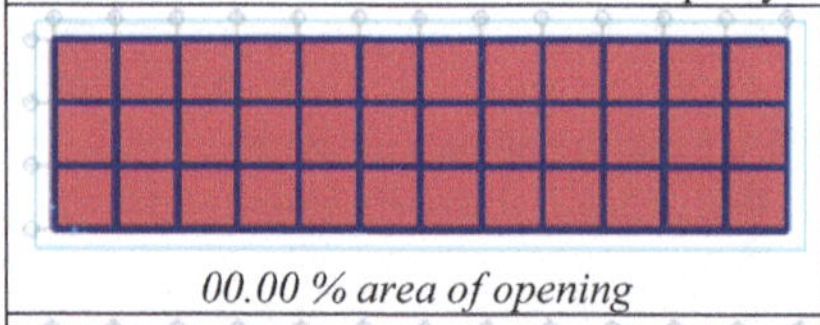 *00.00 % area of opening*	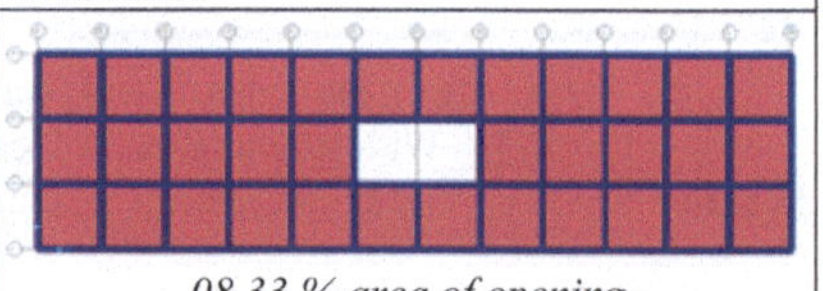 *08.33 % area of opening*
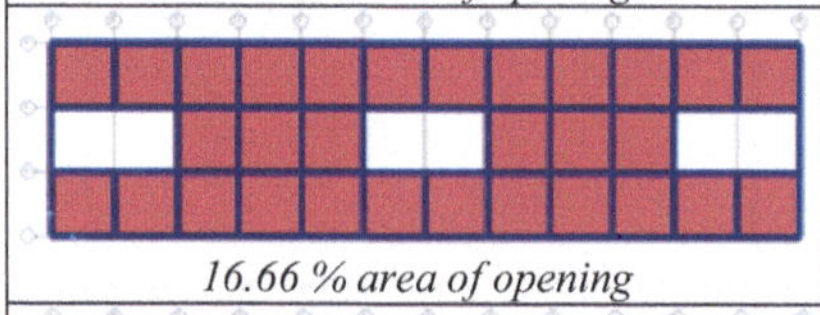 *16.66 % area of opening*	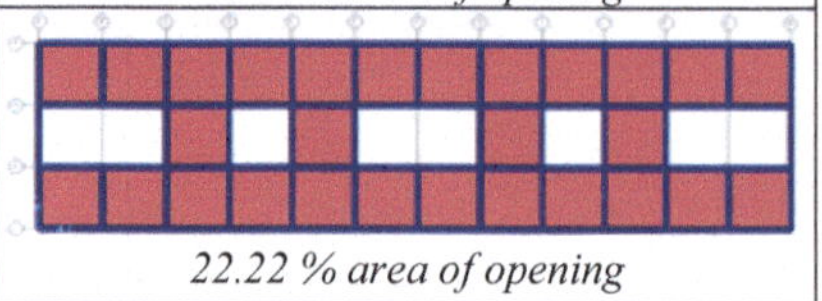 *22.22 % area of opening*
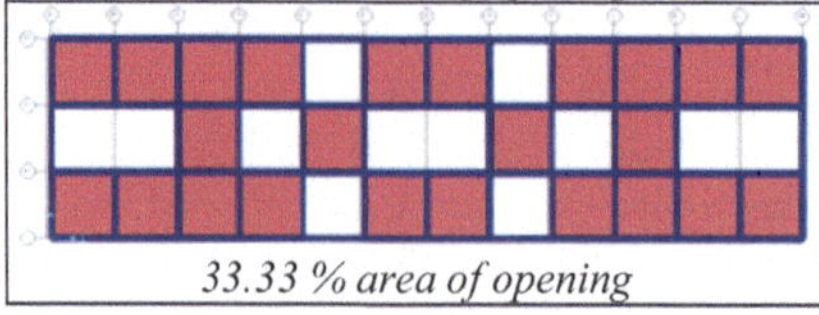 *33.33 % area of opening*	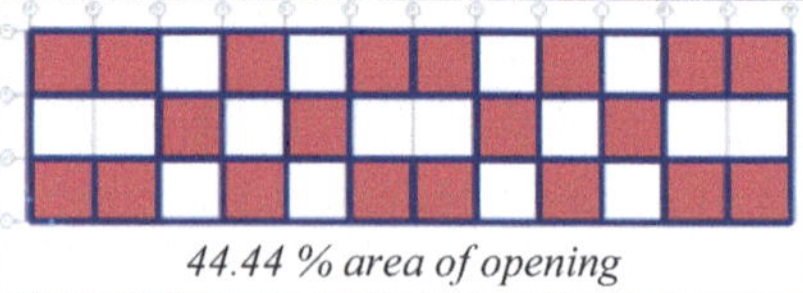 *44.44 % area of opening*

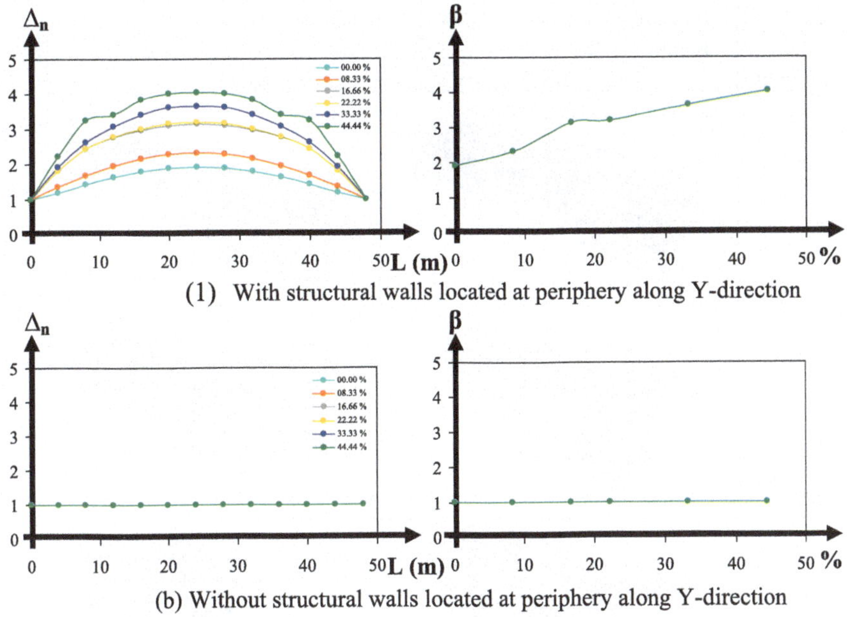

Fig. 3 Normalised diaphragm displacement (Δ_n) along length of building L in *X*-direction and variation of β with % area of opening for buildings with: **a** with structural walls and **b** without structural walls located at periphery along *Y*-direction

2.3 *Height of Building*

Twelve building configurations (having $L/B = 3$) with different heights (by varying number of stories like 3, 5, and 10) are considered (Table 5). Four different cases are studied, which include moment-resisting frame buildings:

(1) without structural wall and without opening (Case 1- Wosw + WO opening);
(2) without structural wall and with opening (Case 2- Wosw + 50% opening);
(3) with structural wall and without opening (Case 3- Wsw + WO opening); and
(4) with structural wall and with opening (Case 4- Wsw + 50% opening).

Following are the observations from the study (Figs. 4 and 5):

(a) In-plane deformation increases with a decrease in number of stories which in turn results in a decrease in the height of the building. Flexibility of building increases with height resulting in almost stiffer or rigid diaphragm; diaphragm flexibility is more pronounced in stiffer (short period) buildings; thus, the effect of height on floor diaphragms must be considered while placing limit on floor diaphragms (as like building code of Mexico); otherwise, it must be accounted for in modelling the diaphragm. Variation of normalised displacement along the number of stories evidenced a similar observation (Fig. 5); and

Table 5 Selected building configurations with varying number of stories ($L/B = 3$)

No. of stories	Building configurations with different % area of opening with Structural Walls located at Periphery in Y-direction	
	0% opening	50% opening
3	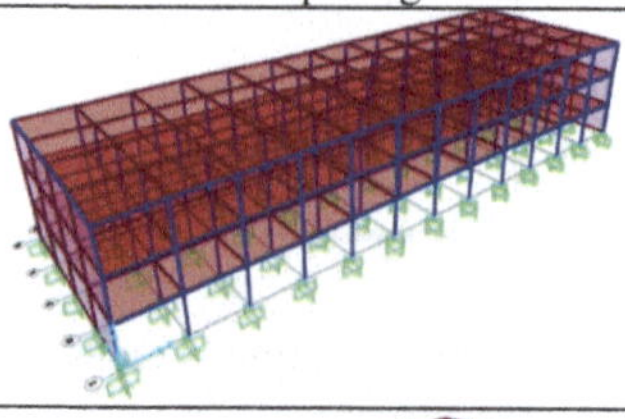	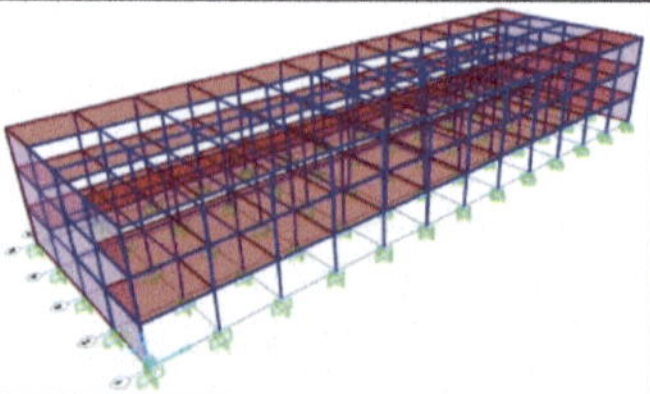
5	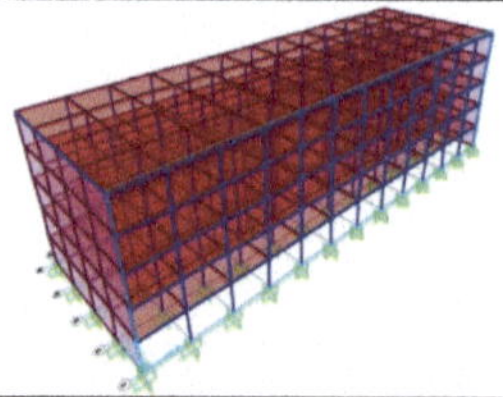	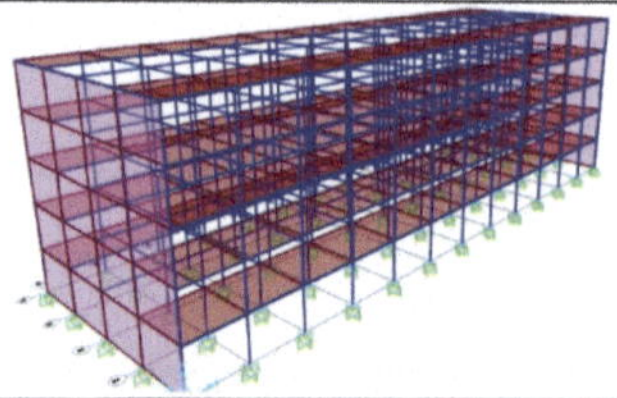
10	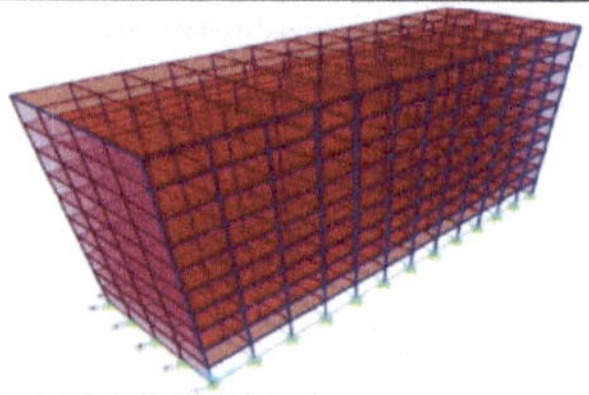	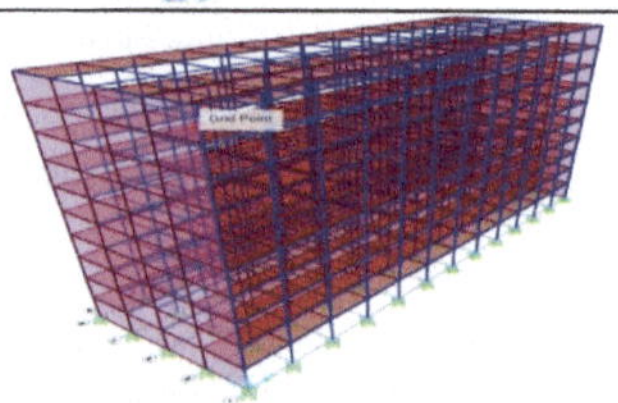

(b) Buildings with structural walls located at periphery show more amplification in in-plane floor deformation than buildings without structural walls; thus, presence of structural wall and its location highly influences flexible behaviour of floor diaphragm. The effect increases with increase in % area of opening and decrease in height of building. It is mandatory to consider the effect of structural walls while placing an upper limit on % area of openings and height of building in building code provisions.

2.4 *Presence/Location of Structural Walls*

The behaviour of floor diaphragms is greatly influenced by the presence and location of structural walls. In total, 74 building configurations are considered (including 37 buildings with structural walls and 37, without structural walls), by varying L/B and % area of openings (Table 6). Buildings with structural walls located at periphery evidenced too much of in-plane deformation due to the presence of stiffer walls at edges. Responses are completely dominated by presence and location of structural walls. The buildings become flexible even if L/B or % area of opening is within prescribed limits (say $L/B = 2$, which is < 3 and with $< 50\%$ opening). On the contrary, buildings without structural walls do not exceed the limits defined for floor

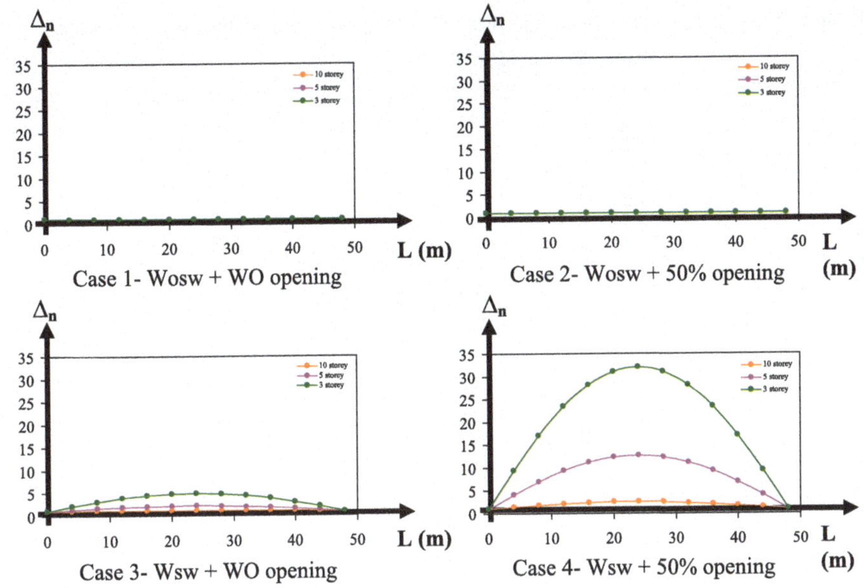

Fig. 4 Normalised diaphragm displacement (Δ_n) along length of building L in *X*-direction with $L/B = 3$, having a different number of stories: Case 1—Wosw + WO opening, Case 2—Wosw + 50% opening, Case 3—Wsw + WO opening, and Case 4—Wsw + 50% opening

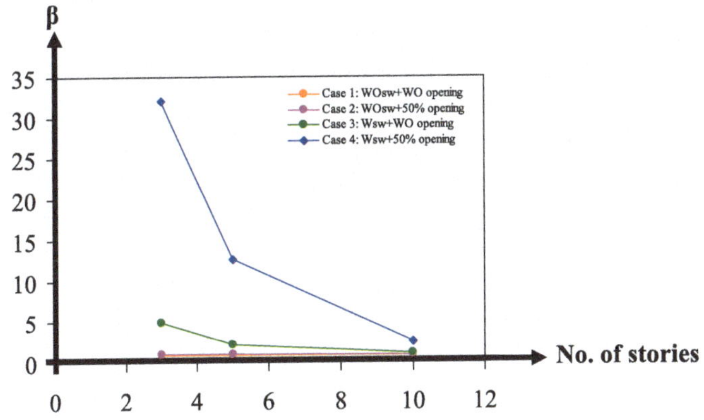

Fig. 5 Normalised displacement varying with number of stories for four different cases of buildings with $L/B = 3$

diaphragms. Hence, the influence of structural walls must be accounted for while placing upper limits on floor diaphragms; otherwise, the provisions given in building codes are applicable to buildings without structural walls only.

Table 6 Normalised diaphragm displacement of buildings with and without structural walls

S. No.	*L*/*B*	% area of opening	β	
			With structural walls	Without structural walls
1	1.00	11.11	1.05	1.00
		33.33	1.13	1.00
		44.44	0.15	1.00
		55.55	0.14	1.00
2	**2.00**	**11.11**	**1.24**	**1.00**
		22.22	1.40	1.00
		33.33	2.12	1.00
		44.44	8.56	1.02
3	4.00	5.55	2.33	1.004
		16.66	3.14	1.008
		22.22	3.20	1.007
		33.33	3.65	1.005
		44.44	4.06	1.005
4	5.33	4.16	2.91	1.01
		12.5	2.97	1.01
		20.83	3.26	1.01
		29.16	3.83	1.00
		37.5	4.11	1.00
5	8.00	2.70	8.62	1.02
		8.30	8.56	1.01
		13.80	8.63	1.01
		19.40	9.00	1.02
		25.00	10.30	1.00
		30.50	10.96	1.00
		36.11	11.44	1.00
		41.66	12.03	1.02
		44.44	12.65	1.04
6	10.00	*2.22*	12.55	1.02
		6.66	12.40	1.02
		11.11	12.33	1.01
		15.55	12.43	1.01
		20.00	12.92	1.03
		24.44	14.06	1.00
		28.88	14.60	1.00
		33.33	15.00	1.00
		37.77	15.54	1.00

(continued)

Table 6 (continued)

S. No.	*L*/*B*	% area of opening	β	
			With structural walls	Without structural walls
		42.22	16.44	1.03
		44.44	17.22	1.06

2.5 Comment on Building Code Provisions

Usually, building code provisions define limits for rigid floor diaphragms by placing upper limits on: (i) *L*/*B* (2.5–4), (ii) % area of openings (15–50%) and (iii) limiting normalised diaphragm displacements ($\beta < 1.1$–3). These limits are directly applicable to buildings without structural walls but not for buildings with structural walls especially that are located at the periphery. Also, the upper limit on β is placed to indirectly account for the same. These limits are defined independently in different building code provisions and do not consider the influence of other critical parameters. Therefore, it is recommended to consider the influence of all critical parameters while placing upper limit for rigid floor diaphragms. As per IS 1893(1), considering influence of critical parameters, upper limit shall be revised as $L/B = 2$, % area of opening $\leq 10\%$ for β to be within 1.2 (Table 6); it may be revised further based on inelastic analysis. Otherwise, as a conservative estimate, the provisions given in IS 1893(1) shall be applied collectively where β shall be limited within 1.2 for $L/B = 3$, % area of opening $\leq 50\%$. Thus, all three limits must be satisfied together. Following are the overall summary of the present study:

(a) Limits defined for *L*/*B*, % area of opening and β must consider one to one correlation between each other, especially in buildings with structural walls; otherwise, additional walls must be included in the middle bays to control the effect of flexible floor diaphragms;
(b) An upper limit on the height of the building $H/B \leq 2.5$ (as like Mexico) shall be imposed on building code provisions; also, such building is shorter and are affected by the flexible behaviour of floor diaphragms; these provisions shall be applicable to such buildings only; and
(c) Linear time history analysis is conducted on four cases of buildings discussed in Sect. 2.3. 7 different ground motions are considered (namely, Cape Mendocino—Fortuna, Cape Mendocino—Eureka, Coalinga—Parkfield (which includes Cholame 3W, Fault Zone 10, Fault Zone 3, and Gold Hill 3E), Imperial Valley—Plaster City) and linear PGA scaling corresponding to building natural period is adopted for similar demand considered to ESA. Average value of maximum β is estimated using the 7 ground motions. Results from linear time history analysis THA (considering) evidence almost all the observations from ESA (Fig. 6); Case 4 shows drastic variations due to the presence of structural walls. Hence, it is mandatory to revise the design limits given in building code provisions considering influence of all critical parameters.

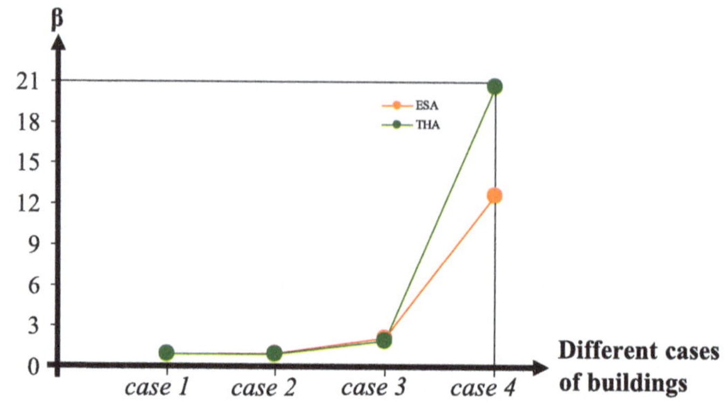

Fig. 6 Comparison of normalised displacement for four different cases of 5 storey buildings with $L/B = 3$ using ESA and THA

3 Conclusions

Floor diaphragms act rigid if the diaphragm displacement is almost uniform and force distribution among structural elements is based on its relative stiffnesses. But in-plane deformations increase in buildings due to various critical parameters like geometry (*L*/*B* and shape), % area of openings, material used, and type of construction adopted (like monolithic or pre-cast), presence/location of structural wall, and height of building. Usually, it is not possible to completely ignore the in-plane deformation of floor diaphragms, hence building codes additionally placed limits on normalised diaphragm displacements β along with *L*/*B* and % area of openings. But limits defined in building code provisions are applicable to buildings without structural walls. ESA is conducted on spectrum of rectangular building configurations with and without structural walls by varying critical parameters like *L/B*, % area of openings (and its position) and height of buildings (by varying number of stories); results agree with linear THA. Influence of all these parameters is studied using β and revised design limits are proposed considering combined effect of all these parameters. As per IS 1893(1), for $\beta \leq 1.2$, *L*/*B* and % area of opening in the building should be restricted to within 2 and 10%. The criteria for rigid floor diaphragms should be clearly re-defined in code provisions by satisfying proposed limits of all three parameters together. It is also applicable to shorter buildings and hence an upper limit of *H/B* should be included in code provisions (as like Mexico, $H/B \leq 2.5$). The proposed limits shall be revised considering various plan geometry and based on nonlinear time history analysis. Alternately, at present using IS 1893(1), for buildings with any complex geometry, additional structural walls shall be added to ensure that $\beta \leq 1.2$, $L/B \leq 3$, and % area of opening $\leq 50\%$, respectively.

References

1. Al Harash MT (2008) Inelastic seismic response of reinforced concrete buildings with floor diaphragm openings. In: Proceedings of the 15th World conference on earthquake engineering, Beijing, China
2. Algerian Earthquake Resistant Regulations. R P A 99/Version (2003)
3. ASCE/SEI 7-16 (2012) Minimum design loads for buildings and other structures. American Society of Civil Engineers, Reston, Virginia
4. Association of Structural Engineers of the Philippines (2010) National structural code of the Philippines, NSCP C101-10. Philippines
5. Australian Standards (2007) Seismic design actions, Part 4: Earthquake actions in Australia, AS1170.4. Australia
6. Bahar S, Benanane A, Belarbi A (2019) The influence of deformability of horizontal diaphragms in the distribution of seismic loads to bracing elements in rectangular buildings. J Mater Eng Struct JMES 6(1):105–118
7. Bangladesh National Building Code (2017)
8. Building and Housing Research Center (2007) Iranian code of practice for seismic resistant design of buildings, Standard No. 2800. Third edition, Iran
9. Building Code of Pakistan (2007) Seismic provision, SP-2007. Pakistan
10. Bureau of Indian Standards (2016) Indian standard criteria for earthquake resistant design of structures, IS 1893(1). New Delhi, India
11. Disaster and Emergency Management Directorate (2007) Regulations for buildings in disaster regions. Turkey
12. Eurocode 8 (2004) European standard for design of structures for earthquake resistance, EN 1998-1. The European Union
13. Federal Emergency Management Agency (FEMA 450–1) (2003) NEHRP recommended provisions for seismic regulations for new buildings and other structures. Building Seismic Safety Council, Washington, DC
14. Fleischman RB, Farrow KT (2001) Dynamic behaviour of perimeter lateral-system structures with flexible diaphragms. Earthquake Eng Struct Dynam 30(5):745–763. https://doi.org/10.1002/eqe.36
15. EAK2000 (2000) Greek code for seismic resistant structures
16. International Council of Building Officials (1997) Uniform building code (UBC). Whittier, CA, USA
17. Ju SH, Lin MC (1999) Comparison of building analyses assuming rigid or flexible floors. J Struct Eng (ASCE) 125(1):25–31. https://doi.org/10.1061/(ASCE)0733-9445(1999)125:1(25)
18. Kunnath SK, Panahshahi N, Reinhorn AM (1991) Seismic response of RC buildings with inelastic floor diaphragms. J Struct Eng (ASCE) 117(4):1218–1237. https://doi.org/10.1061/(ASCE)0733-9445(1991)117:4(1218)
19. Moeini M, Rafezy B, Howson WP (2008) Investigation into the floor diaphragm flexibility in rectangular reinforced concrete buildings and error formula. In: Proceedings of the 14th world conference on earthquake engineering, Beijing, China
20. Official Chilean Code (1996) Earthquake resistant design of buildings, NCH433. National Institute of Standardization, INN, Chile
21. Roskelley BA (2010) An investigation of the influence of diaphragm flexibility on building design through a comparison of forced vibration testing and computational analysis. Master of Science Dissertation submitted to California Polytechnic State University
22. Saffarini HS, Qudaimat MM (1992) In-plane floor deformations in RC structures. J Struct Eng (ASCE) 118(11):3089–3102. https://doi.org/10.1061/(ASCE)0733-9445(1992)118:11(3089)
23. Seismic Code Evaluation. El Salvador (1997)
24. Seismic Design Code. Korea (2009)
25. Standards Association of New Zealand (1992) Code of practice for general structural design and design loadings for buildings, NZS 4203. New Zealand

26. Su RK, Lee YY, Lee CL, Ho JCM (2011) Typical collapse modes of confined masonry buildings under strong earthquake loads. Open Constr Build Technol J 5(1):50–60
27. Mejía LG (2017) AZ of seismic failures. Procedia Eng 210:433–440
28. Mexico City Building Code (1995) Complementary technical norms on earthquake resistant design. Department of Federal District, Mexico
29. National Building Code (2016) Earthquake resistant design. Technical Standard of Buildings, Ministry of Housing, Construction and Sanitation, Peru
30. Vlachakis G, Vlachaki E, Lourenço PB (2020) Learning from failure: damage and failure of masonry structures, after the 2017 Lesvos earthquake (Greece). Eng Fail Anal 117:104803. https://doi.org/10.1016/j.engfailanal.2020.104803

On the Satisfaction of Natural and Essential Boundary Conditions for Bending in Nanobeams Within the Framework of Eringen's Nonlocal Elasticity Theory

Gaurab Kumar Khanra, I. R. Praveen Krishna, and P. Raveendranath

Abstract This work highlights the existing contradictions in satisfying boundary conditions while modeling nanobeams in cantilevered configurations. Various analytical and numerical models can analyze it. Depending on the configuration, these models behave differently. Hence, it is difficult for an engineer to select the method to design the component/structure. Cantilevered nanobeams under point and uniformly distributed load (UDL) are investigated. Different methods' results are contrasted. Inconsistencies found while satisfying essential and natural boundary criteria are examined. The nonlocal moment-curvature relationship and its effect on the solution are also discussed. All fundamental quantities are compared. While some methods anticipate cantilever hardening, others predict softening. The results show technique flaws. A unified and consistent approach to problem-solving is needed.

Keywords Analytical and numerical solutions · Eringen's nonlocal elasticity · Essential and natural boundary conditions · Euler-Bernoulli nanobeams

1 Introduction

The discovery of Carbon Nanotubes (CNTs) by Iijima et al. [1] of NEC corporation in the early 1990s marked the start of nanotechnology. Geim and Novoselov's discovery of graphene [2] at Manchester University has sparked interest in theoretical, computational, and experimental study. Nanomaterials outperform conventional materials in thermal, mechanical, electrical, magnetic, and electronic characteristics. Thus, they offer great potential for building Nano Electro-Mechanical Systems (NEMS), nanobots for precision medicine administration, nano-electronics for wearable computing, and more. These devices are made from nanostructures like CNTs or graphene sheets.

G. K. Khanra (✉) · I. R. Praveen Krishna · P. Raveendranath
Indian Institute of Space Science & Technology, Valiamala, Thiruvananthapuram, Kerala 695547, India
e-mail: gaurabkumar.17@res.iist.ac.in

S. Sidhardh et al. (eds.), *Advances in Structural Integrity for Mechanical, Civil, and Aerospace Applications*, Lecture Notes in Mechanical Engineering,
https://doi.org/10.1007/978-981-97-6367-2_36

In micro/nanodevices, size effects due to material microstructure cannot be neglected [3]. Molecular Dynamics simulations can predict micro/nanomaterial system behavior. However, modeling actual devices is computationally expensive. Continuum-based theories offer an alternative engineering solution strategy. Effective mathematical modeling of micro/nanostructures necessitates size-dependent continuum theories. Accurate and efficient micro/nanostructure mathematical models reduce resource-intensive investigations. Some size-dependent continuum theories are Couple Stress Theory (CST), Strain Gradient Theory (SGT), Eringen's nonlocal elasticity theory, Multiscale models, and Hybrid models. The nonlocal theory, developed by Kroner [4] and revised by Eringen [5], was widely accepted by engineers due to its simplicity.

Peddieson et al.[6] developed a nonlocal Euler-Bernoulli beam formulation based on Eringen's theory, solving a cantilever beam with sinusoidal, uniformly distributed, and point loads. They found that MEMS devices had negligible nonlocal effects, whereas NEMS devices could have them. Wang and Liew[7] reached a similar conclusion. A study by Wang et al. [8] used virtual work to construct the governing equations for the nonlocal Timoshenko beam and solve beam problems with various boundary conditions. Challamel and Wang [9] identified an inconsistency in Euler-Bernoulli cantilever nanobeams with point loads. This situation with Eringen's differential nonlocal model is known as the "cantilever beam paradox" problem. Eringen's theory was used to produce finite element formulations for nonlocal elastic Euler-Bernoulli beam and Kirchhoff plate by Phadikar and Pradhan [10], which displayed a similar paradoxical phenomenon as it resulted in an element stiffness matrix devoid of the nonlocal parameter. Nguyen et al. [11] developed a mixed finite element approach for nonlocal Euler-Bernoulli beams with varying end conditions. The stiffness matrix naturally contains the nonlocal parameter, resolving the cantilever beam paradox. Thus, mixed finite element formulations offer the benefit of including size effects in the stiffness matrix, unlike the conventional finite element formulations.

Fernandez-Saez et al. [12] presented nonlocal formulations that do not have contradictions like the nonlocal differential model. In their work, solutions to Fredholm's integro-differential equations were obtained. These solutions often lack uniqueness and are ill-posed. Tuna and Kirca [13] solved the nonlocal Euler-Bernoulli beam for various boundary conditions and load scenarios using Eringen's nonlocal integral formulation. They converted the integral formulation to a Volterra integral equation of the first kind to prevent the ill-posedness of the problem. Analytical solutions are found by solving the resulting equations with Laplace transforms. However, Romano and Barretta [14] noted that Tuna and Kirca's solution fails to meet the Dirac-delta property for boundary conditions and concentrated loads.

Most literature only discusses nanobeam deformation. Slopes and stress resultants are essential engineering parameters. The present work examines the satisfaction of boundary conditions (both essential and natural) and the application of concentrated loads by different solution methodologies available in the literature. A mixed finite element formulation based on Eringen's differential nonlocal model by Nguyen et al. [11] is employed as a reference solution.

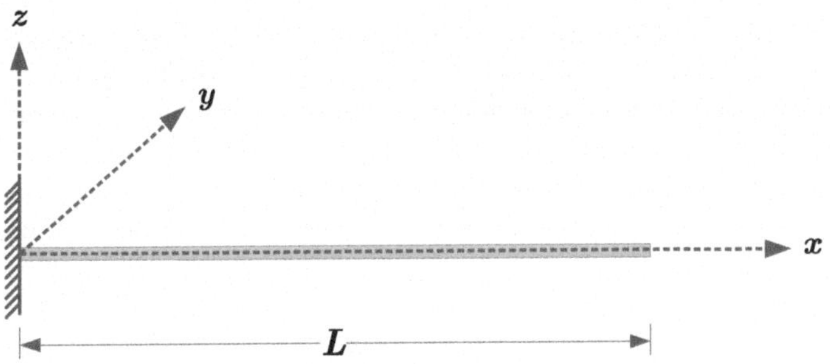

Fig. 1 Schematic diagram of an Euler-Bernoulli nanobeam

Nonlocal Euler-Bernoulli beam formulations based on Eringen's differential model are discussed in this work. In Sect. 2, Eringen's integral and differential nonlocal models' governing integro-differential equations are briefly reviewed. Finally, the section discusses the relationship between the two models. This study presents various solution approaches in Sect. 3. The analytical solutions by Wang [7], Kitipornchai [8] and Tuna [13] and numerical solutions by Pradhan [10] and Nguyen [11] are presented concisely. The cantilever beam paradox and similar anomalies are examined in Sect. 4, focusing on three specific cases, viz:

- Cantilever nanobeam subjected to point load at the tip.
- Cantilever nanobeam subjected to point load at an arbitrary location.
- Cantilever nanobeam subjected to UDL throughout it's span.

This study focuses on beam deflections, slopes, and bending moments, which comprise the kinetic and kinematic quantities of the beam problem (including essential and natural boundary conditions). Analytical and numerical methods given in [7, 8, 10, 11, 13] were used to obtain the presented results. The solution approaches are based on Erigen's differential nonlocal model, except for Tuna and Kirca's analytical solutions [13], which are developed from Eringen's integral nonlocal model.

2 Overview of Eringen's Nonlocal Theory

Numerous literature [3, 5, 12, 13, 16] explores Eringen's nonlocal integral and differential theory. An overview is provided here for completeness (Fig. 1).

Eringen's nonlocal elasticity theory accounts for "size-effects" through long-range inter-atomic interactions. Thus, stress at a specific point is a function of strains at all points in the domain of interest, as shown below :

$$\sigma_{ij} = \int_V K(|x - \bar{x}|, e_0 a)\sigma_{ij}^c(\bar{x})dV(\bar{x}) \tag{1}$$

Where, σ_{ij}^{c} is the classical stress tensor at a point, $K(|x-\bar{x}|, e_0 a)$ is the kernel function to account for the "size-effects", $|x-\bar{x}|$ is the Euclidean norm, e_0 is the nonlocal material constant, a is the nonlocal characteristic length. Also, the classical stress tensor $\sigma_{ij}^{c}(x)$ is related to the strain tensor ϵ_{kl} at a point through the 4th order elasticity tensor C_{ijkl}, as per the generalized Hooke's law $\sigma_{ij}^{c} = C_{ijkl}\epsilon_{kl}$, V is the volume of the computational domain. For a nonlocal 1-D Euler-Bernoulli beam. Equation 1 can be written as:

$$\sigma_x = \int_0^L K(|x-\bar{x}|, e_0 a) E \epsilon_x(\bar{x}) d\bar{x} \tag{2}$$

In Eq. 2, σ_x denotes nanobeam bending stress, E is the material's modulus of elasticity, and $\epsilon_x(\bar{x})$ is the fiber's axial bending strain. The following integral formulation of Eringen's equations can be obtained by minimizing total potential energy with the Euler-Lagrange formulation. The governing integro-differential equation is as follows [12]:

$$\frac{\mathrm{d}^2}{\mathrm{d}x^2}\left[EI\int_0^L K(|x-\bar{x}|, e_0 a)\frac{\mathrm{d}^2 w(\bar{x})}{\mathrm{d}\bar{x}^2}\, d\bar{x}\right] - q(x) = 0 \; in \; \Omega \in [0, L] \tag{3}$$

In which $\frac{d^2 w(\bar{x})}{d\bar{x}^2}$ represents the curvature of the nanobeam. I is the 2nd area moment of inertia of the nanobeam cross-section and $q(x)$ is the intensity of the applied distributed load. Note that Eq. 3 represents a Fredholm integro-differential equation of the first kind [17, 18].

To convert Eq. 1 into its differential form, Eringen [16] assumed the kernel function $K(|x-\bar{x}|, e_0 a)$, a two-dimensional nonlocal modulus defined by Eq. (3.6) in Eringen [16], where K_0 is the modified Bessel function. The nonlocal modulus function is a δ sequence, resulting in the Dirac-delta function when $\tau \to 0$, transforming a nonlocal elasticity problem into a classical one. Additionally, $K(|x-\bar{x}|, e_0 a)$ was assumed as a Green's function of a linear differential operator, transforming the kernel function into a Dirac-delta measure.

Henceforth, after a series of mathematical operations on Eq. 1 [3, 16] and subsequent application of D'Alembert's principle, an equivalent differential form can be obtained as:

$$EI\frac{\mathrm{d}^4 w}{\mathrm{d}x^4} + (e_0 a)^2\frac{\mathrm{d}^2 q(x)}{\mathrm{d}x^2} = q(x) \; in \; \Omega \in [0, L] \tag{4}$$

3 Essential and Natural Boundary Conditions

Integral and differential models can solve nanobeam bending problems with different boundary and load cases. One must note that the two models define natural boundary conditions differently.

The differential nonlocal model uses a bi-exponential kernel function: $K(|x - \bar{x}|, e_0 a) = \frac{1}{2e_0 a} e^{-\frac{|x-\bar{x}|}{e_0 a}}$ (According to literature [12, 13, 19]), the bending moment expression is:

$$M(x) = \int_0^L \frac{1}{2e_0 a} e^{-\frac{|x-\bar{x}|}{e_0 a}} \circledast M^E(\bar{x})\, d\bar{x} \tag{5}$$

With $M^E(\bar{x})$ being the equilibrium bending moment as per the local or classical elasticity theory. Using the above convolution integral definition and substitution of the expression for the moment-curvature relation we have :

$$M(x) = \int_A \sigma_x(\bar{x}) z \,\mathrm{d}A = -EI \int_0^L K(|x - \bar{x}|, e_0 a) \frac{\mathrm{d}^2 w(\bar{x})}{\mathrm{d}\bar{x}^2}\, \mathrm{d}\bar{x} \tag{6}$$

In the integral formulation the above moment equation Eq. 6 has to be modified at the boundary as follows:

$$M'(0) = \frac{M(0)}{e_0 a} \tag{7}$$

$$M'(L) = -\frac{M(L)}{e_0 a} \tag{8}$$

This modification is known as the constitutive boundary conditions in literature [19], and present a contradictory need for the bending fields to concurrently fulfill the criteria of the constitutive boundary conditions and the equilibrium equations. This results in non unique solutions for nanobeam elastostatic problems.

4 Existing Solution Methods

4.1 Analytical Solutions

Various available solution methods are listed in Table 1. Multiple authors have described beam mechanics using different variables and parameters. However, standard notations as described are used for all the cases. However, standard notations as described are used for all the cases.

Table 1 Expressions concerned with the bending analysis of nonlocal Euler-Bernoulli nanobeams subjected to various loading and boundary conditions (For expressions of the variables α, β, γ & ϕ refer Appendix C)

Case	Tuna and Kirca [13]	Wang et al. [7]
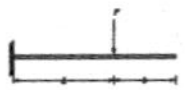	$u_z(x) = \begin{cases} \frac{Px(6(e_0a)((e_0a)+L)+x(3L-x))}{6EI} & in\ 0 < x < L \\ \frac{P(6x(e_0a)(L+(e_0aL))-L^2(L-3x))}{6EI} & in\ L < x < (L+b) \end{cases}$	$u_z(x) = -\frac{P((L-x)H(x-L)((L-x)^2-6(e_0a)^2)+x^2(x-3L))}{6EI}$
	$\theta(x) = \begin{cases} \frac{P(2(e_0a)((e_0a)+L)+x(2L-x))}{2EI} & in\ 0 < x < L \\ \frac{P(L^2+2(e_0a)(L+(e_0a)))}{2EI} & in\ L < x < (L+b) \end{cases}$	$\theta(x) = -\frac{P(\delta(x-L)(6(x-L)(e_0a)^2+(L-x)^3)-3H(x-L)((L-x)^2-2(e_0a)^2)+3x(x-2L))}{6EI}$
	$M(x) = \begin{cases} \frac{P}{2}(L(\alpha+\gamma)+\frac{1}{e_0a}(\beta+\phi)) & in\ 0 < x < L \\ 0 & in\ L < x < (L+b) \end{cases}$	$M(x) = 2P(e_0a)^2\delta(L-x) - PH(L-x)^2 + \frac{P(L-x)\delta'(L-x)(6(e_0a)^2+2Lx-x^2-L^2)}{6}$
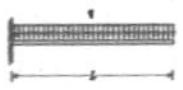	$u_z(x) = \frac{qx(12(e_0a)((e_0a)(2L-x)+L^2)+x(6L^2-4Lx+x^2))}{24EI}$	$u_z(x) = \frac{qx^2(-12(e_0a)^2+6L^2-4Lx+x^2)}{24EI}$
	$\theta(x) = \frac{q(3(e_0a)(2(e_0a)(L-x)+L^2)+x(3L^2-3Lx+x^2))}{6EI}$	$\theta(x) = \frac{qx(x^2-3Lx+3L^2-6(e_0a)^2)}{6EI}$
	$M(x) = \frac{q}{4}((2(e_0a)^2 + L^2 + 2(e_0a)L)e^{-\frac{x}{e_0a}} + 2(e_0a)^2e^{\frac{x-L}{e_0a}} + 2(-2(e_0a)^2 + L^2 + 2Lx - x^2))$	$M(x) = \frac{q}{4}\left(L(2e_0a + L)e^{-\frac{x}{e_0a}} - 2(L-x)^2\right)$

Equation 5 was used to calculate analytical solutions by Tuna and Kirca [13]. As stated, these are Fredholm integro-differential equations of the first kind. Thus, the existence of a solution is not guaranteed. If a solution exists, it may not be unique [17, 18]. After converting Eq. 5 into Volterra integro-differential equations of the first kind, Tuna and Kirca obtained a well-posed problem [13]. This was done by some rearrangements in the limits of integration. Subsequently, using Laplace transforms, the Volterra integro-differential equation was solved. According to Eringen's integral nonlocal model, the nonlocal bending field across the nanobeam is calculated using the convolution integral definition of Romano and Barretta [19]:

5 Numerical Solutions

5.1 Finite Element Solutions

Galerkin's weighted residual method is applied on Eringen's differential nonlocal model Eqn.4 based on the work of Phadikar and Pradhan [10]. Cubic hermite polynomials are used as shape functions leading to:

$$\left[K^{FEM}\right]\{w\} - \{Q_{dist}^{NL}\} = \{Q_{dist}\} + \{P\} \tag{9}$$

Here, $\left[\boldsymbol{K}^{FEM}\right]$ is the element level stiffness matrix, $\{\boldsymbol{w}\}$ is the vector of nodal displacements and rotations, $\{\boldsymbol{Q}_{dist}\}$ is the vector of distributed loads as per classical theory, $\{\boldsymbol{Q}_{dist}^{NL}\}$ is the vector of distributed loads associated with the nonlocal parameter, $\{\boldsymbol{P}\}$ is the point or concentrated load vector. After imposing the appropriate boundary conditions, Eqn.9 can be solved for the nodal deflection and rotations. Note that the nonlocal parameter appears only on the load vector $\{\boldsymbol{Q}_{dist}^{NL}\}$.

5.2 Mixed Finite Element Solutions

A mixed finite element approach was utilized to solve Eq. 4 by Nguyen et al. [11]. Here, the equilibrium equations and the nonlocal Euler-Bernoulli moment-curvature relations are discretized using Galerkin's weighted residual method with linear Lagrange shape function to obtain a set of algebraic equations.

$$\left[\boldsymbol{K}^{MFEM}\right]\begin{Bmatrix}\{\boldsymbol{w}\}\\ \{\boldsymbol{M}\}\end{Bmatrix} = \begin{Bmatrix}\{\boldsymbol{Q}_{dist}\}\\ \{\boldsymbol{\theta}\}\end{Bmatrix} + \begin{Bmatrix}\{\boldsymbol{P}\}\\ \frac{(e_0a)^2}{EI}\{\boldsymbol{P}\}\end{Bmatrix} \tag{10}$$

Here, $\left[\boldsymbol{K}^{MFEM}\right]$ is the element level stiffness matrix, $\{\boldsymbol{w}\}$ is the vector of nodal displacements, interpolated as $\{\boldsymbol{w}\} = \left[\left(1-\frac{x}{L}\right)\ \frac{x}{L}\right]\begin{Bmatrix}w_1\\ w_2\end{Bmatrix}$, $\{\boldsymbol{M}\}$ represents the vector

Table 2 Nanobeam material and geometric properties

$L\ (nm)$	h (nm)	b (nm)	E (nN/nm^2)	$P\ (nN)$	$q\ (nN/nm)$	$e_0a\ (nm)$
10	**1**	**1**	**5500**	$\mathbf{10^{-3}}$	$\mathbf{10^{-4}}$	**0.5, 2**

of nodal moments, also interpolated as $\{\boldsymbol{M}\} = \left[\left(1-\frac{x}{L}\right)\ \frac{x}{L}\right]\begin{Bmatrix} M_1 \\ M_2 \end{Bmatrix}$, $\{Q_{dist}\}$ is the vector of distributed loads as per classical theory, $\{\boldsymbol{\theta}\}$ is the vector of section rotations, $\{\boldsymbol{P}\}$ is the load vector corresponding to point or concentrated loading. The parameter $\frac{(e_0a)^2}{EI}\{\boldsymbol{P}\}$ is the effective section rotation caused by the nonlocal parameter due to the action of the concentrated load. Here, the dependency of element stiffness matrix $\left[\boldsymbol{K}^{MFEM}\right]$, on the nonlocal parameter (e_0a) unlike the classical finite element method is worth noting.

6 Results and Discussion

The above described analytical and numerical methods were applied on different beam models to understand the suitability of the methods in satisfying the essential and natural boundary conditions. The physical quantities such as displacement, slope, and bending moments obtained for the different models are compared against each other. Table 2 summarizes the material and geometric properties of the beam models [11].

6.1 Tip Loaded Nonlocal Euler-Bernoulli Cantilever Beam.

Challamel and Wang [9] identified a flaw in Eringen's differential nonlocal constitutive model, resulting in the "cantilever beam paradox" in literature. This occurs because Eringen's differential nonlocal model omits the 2nd term in the moment-curvature relationship of the Euler-Bernoulli cantilever nanobeam, for the moment calculation as seen in Eq. 11.

$$M(x) - (e_0a)^2 M''(x) = -EI\frac{\mathrm{d}^2 w}{\mathrm{d}x^2} \tag{11}$$

The analytical solutions by [7, 8]) likewise ignore this additional term in the moment-curvature relationship. This yields deflection solutions devoid of the nonlocal parameters and resembles the classical Euler-Bernoulli beam. As the element stiffness matrix lacks the nonlocal parameter, classical FEM nanobeam bending solutions are also independent of it. However, MFEM solutions accommodate the complete moment expression by embedding the nonlocal parameter in the element stiffness matrix (see Appendix B.1). A detailed explanation can be found in references [10, 11]. Analytical solutions of clamped-clamped nanobeams exposed to UDL show similar behavior. For clarity, the deflections are normalized against the tip deflection of a local (classical) cantilever beam. Since the EB formulation does not account for shear deformation, thin and thick beams behave similarly for the same nonlocal parameter. The classical, FEM and Wang analytical solutions are identical because they lack the nonlocal parameter. MFEM solutions soften due to the nonlocal parameter. Tuna and Kirca [13] found higher beam deflections under the same loading condition, indicating a more significant softening effect than the MFEM solution. All formulations meet the fixed-end essential displacement boundary condition. Slopes are analyzed in Fig. 2b to assess essential boundary conditions fulfillment. The solution by Tuna and Kirca [13] fails to meet the critical slope boundary condition (null rotation at fixity), as seen in the figure. Every alternative formulation meets the zero slope at the fixity criterion. Figure 2c shows the bending moment distribution in a nonlocal cantilever beam with a concentrated load at its tip. MFEM accurately captures the analytical bending moment distribution solutions. The bending moment distribution estimated by Tuna and Kirca [13] differs significantly from the analytical solution due to the solution's non-satisfaction with the Dirac-delta property. The nonlocal EB formulation also shows that the moment-curvature relationship is independent of the aspect ratio. When shear deformation effects are included, the aspect ratio may affect the solution, which warrants the extension of the current MFEM formulation to the Timoshenko beam theory.

Next, we analyze the impact of nonlocal parameters in the EB formulation, which alters the moment-curvature relationship in Eq. 11. Figure 2d shows that MFEM, Tuna, and Kirca [13] predict softening behavior with a rise in the nonlocal parameter e_0a. Nonlocal parameters insensitivity in other solution methods results in "cantilever paradox" due to the missing term in the moment-curvature relationship (Eq. 11). The deflections and elastic curvature of a tip-loaded nonlocal EB cantilever beam are compared in Fig. 2e, f with two nonlocal parameter values ($e_0a = 0.5$ and 2.0 nm). Both MFEM and Tuna and Kirca [13] exhibit softening behavior when nonlocal parameter increases. As in the previous instance, aspect ratio does not affect the normalized deflection curve. For smaller nonlocal parameter values, Tuna and Kirca's elastic curvature calculated as per [13] matches the linear elastic curvature, except at the ends. However, for higher values of nonlocal parameter, it deviates grossly from the true curvature, making the method inconsistent in capturing the exact behavior of the nonlocal EB beam.

6.2 A Nonlocal Euler-Bernoulli Cantilever Beam Subjected to Point Load at an Arbitrary Location

This particular case is of special interest due to the strange behavior of the beam deflection predicted by both analytical solution of Wang and MFEM solution of Nguyen et al. [11]. In their work, it is suggested to introduce scale factors for moments

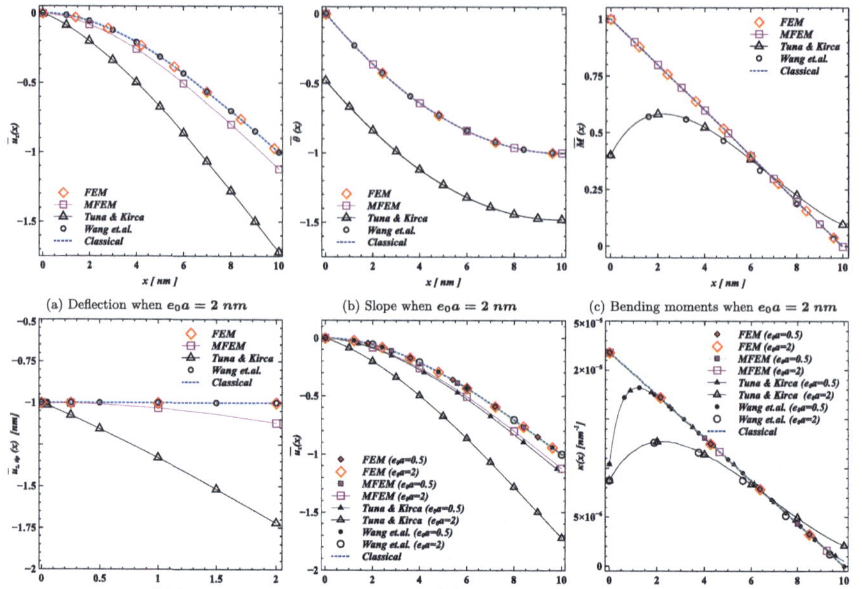

(a) Deflection when $e_0a = 2\ nm$ (b) Slope when $e_0a = 2\ nm$ (c) Bending moments when $e_0a = 2\ nm$

(d) Tip-deflections Vs e_0a (e) Deflections for 2 e_0a values (f) Elastic curvatures for 2 e_0a values

Fig. 2 Non-dimensionalization factor used for deflections is $\frac{3EI}{PL^3}$, for slopes is $\frac{2EI}{PL^2}$ and for bending moments is $\frac{1}{PL}$

to obtain more realistic deflections. In the present work, such scale factors are not included, as the focus of the study is on the satisfaction of the kinetic and kinematic boundary conditions. In nano cantilever beams, subjected to point load at arbitrary location, one can observe a change in the direction of slope after the point of load application.

This phenomenon can be observed in Fig. 3, for Wang's solution as well as for MFEM. The other solutions do not exhibit this behavior. However, this behavior is primarily a function of the nonlocal parameter. With a reduced nonlocal parameter ($e_0a = 0.5\ nm$), it can be observed that deflection becomes more close to the classical beam deflection as shown in Fig. 3c. This is because the effect of the second term in the left-hand side of the moment-curvature relationship (refer Eqn. 11) becomes less significant, with the reduction of the nonlocal parameter, thus making the equation more closer to the classical one. Further, to understand this change in direction of slope at the point of application of load, slopes are evaluated and shown in figures Fig. 4a, b. It can be seen that the MFEM satisfies the zero slope condition at the fixed boundary, whereas the Tuna and Kirca solution fails to satisfy the same. Again, the MFEM and Wang solutions predict a jump in the slopes at the point of load application, whereas all the other methods predict a C^0 continuity in slope. From the figure, it is evident that the magnitude of this jump depends on the nonlocal parameter while keeping all other parameters constant. The bending moment and shear force

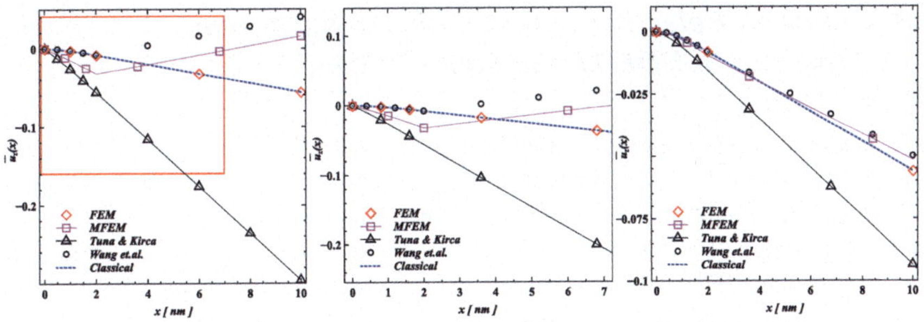

(a) Beam deflections (b) Deflections near $x = 2\ nm$ (c) Deflections for $e_0a = 0.5\ nm$

Fig. 3 **a** Non-dimensionalized beam deflections when point load is applied at $x = 2$ nm **b** enlarged figure corresponding to the highlighted portion in (**a**) when $e_0a = 2$ nm

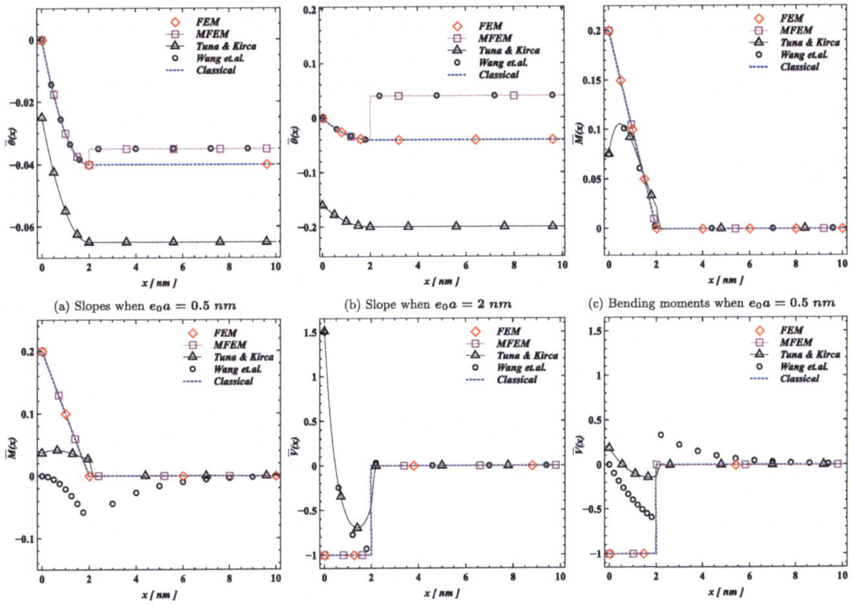

(a) Slopes when $e_0a = 0.5\ nm$ (b) Slope when $e_0a = 2\ nm$ (c) Bending moments when $e_0a = 0.5\ nm$

(d) Bending moments when $e_0a = 2\ nm$ (e) Shear forces when $e_0a = 0.5\ nm$ (f) Shear forces when $e_0a = 2\ nm$

Fig. 4 Non-dimensionalization factor used for deflections is $\frac{3EI}{PL^3}$, for slopes is $\frac{2EI}{PL^2}$, for bending moments is $\frac{1}{PL}$ and for shear forces is $\frac{1}{P}$

are evaluated over the beam as shown in the Fig. 4c–f. Shear force and bending moment distributions evaluated by Tuna and Kirca's [13] analytical solution exhibit an unacceptable dependency on nonlocal parameter leading to erroneous results in the region close to the fixed boundary. All other solution methods are able capture them accurately.

6.3 Nonlocal Euler-Bernoulli Cantilever Nanobeam Subjected to UDL Throughout It's Span

This section compares cantilever nanobeam UDL deflections, slopes, and curvature responses predicted by different solution approaches. As in previous cases, the aspect ratio affects the nanobeam response, but the non-dimensional response remains the same. Referring to Fig. 5a for the non-dimensionalized deflection of the cantilever nanobeam under UDL. All formulations except classical theory account for the non-local parameter. In FEM, the nonlocal parameter only affects the force vector as a correction term (see Appendix 7). Wang's analytical solution has a comparable impact since the moment-curvature relationship does not depend on the nonlocal parameter. This reduces net force, stiffening the solution relative to the classical solution. The moment-curvature relation has the nonlocal correction term in MFEM and Tuna and Kirca's solutions. Adding nonlocal parameters in the stiffness matrix and force vector of the MFEM formulation (see Appendix B) softens the nanobeam deflection. The slopes evaluated using various approaches are compared in Fig. 5b. As in the preceding sections, Tuna and Kirca's slopes have a nonzero value at the fixed end, making the results kinematically inadmissible. The classical solution predicts a maximum slope at the free end of the cantilever beam, while all other approaches pre-

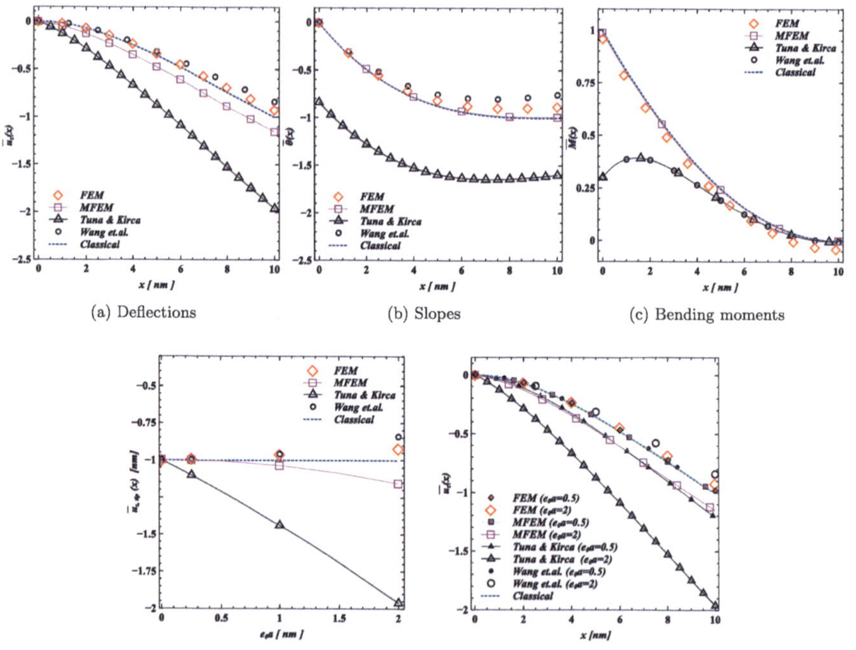

(a) Deflections (b) Slopes (c) Bending moments

(d) Tip deflections vs. e_0a (e) Deflections for 2 e_0a values

Fig. 5 .

dict a slope at some distance from the free end. Figure 5c shows the beam's bending moment distribution. FEM-calculated bending moments differ slightly from actual values across beams, including endpoints. Tuna and Kirca's approach predicts bending moments incorrectly near the fixed boundary but overpredicts at the free end. All other methods reliably calculate free end moments to zero.

7 Conclusions

Most of the available literature on nanobeam analysis focuses only on their deformation behavior. However, a reliable computational model for the structural design nanobeams demands accurate evaluation of derived quantities (like slopes and stress resultants) and the deformation. This work objectively evaluates analytical and numerical methods in the available literature for their ability to satisfy essential and natural boundary conditions. The current analysis chooses three key cantilevered boundary beam configurations. The inconsistencies in satisfying essential and natural boundary conditions in some analytical and numerical formulations are thoroughly examined. This is perhaps the first time all fundamental quantities, not just beam deflection, are compared between approaches. The effect of nonlocal parameters on beam kinematics is studied. Some approaches based on Eringen's nonlocal elasticity theory, which gives sensible deformations, fail to meet boundary conditions. The vast disparities in nature and accuracy of the various solution methods demand further research to develop more versatile methods.

Appendix Stiffness and load matrices (FEM)

$$\left[K^{FEM}\right] = \frac{EI}{L^3}\begin{bmatrix} 12 & 6L & -12 & 6L \\ 6L & 4L^2 & -6L & 2L^2 \\ -12 & -6L & 12 & -6L \\ 6L & 2L^2 & -6L & 4L^2 \end{bmatrix} \qquad \{Q_{dist}\} = \begin{Bmatrix} \frac{qL}{2} \\ \frac{qL^2}{12} \\ \frac{qL}{2} \\ \frac{-qL^2}{12} \end{Bmatrix} \tag{A.12}$$

$$\{Q^{NL}_{dist}\} = (e_0 a)^2 \begin{Bmatrix} 0 \\ q \\ 0 \\ -q \end{Bmatrix} \qquad \{P\} = \begin{Bmatrix} -Q_1 \\ M_1 \\ Q_2 \\ -M_2 \end{Bmatrix} \tag{A.13}$$

B Stiffness and load matrices (MFEM)

$$\left[\boldsymbol{K}^{MFEM}\right] = \begin{bmatrix} 0 & 0 & \frac{1}{L} & -\frac{1}{L} \\ 0 & 0 & -\frac{1}{L} & \frac{1}{L} \\ \frac{1}{L} & -\frac{1}{L} & -\frac{(e_0a)^2}{EIL} - \frac{L}{3EI} & \frac{(e_0a)^2}{EIL} - \frac{L}{6EI} \\ -\frac{1}{L} & \frac{1}{L} & \frac{(e_0a)^2}{EIL} - \frac{L}{6EI} & -\frac{(e_0a)^2}{EIL} - \frac{L}{3EI} \end{bmatrix} \quad \{\boldsymbol{Q}_{dist}\} = \begin{Bmatrix} \int_{x_e}^{x_{e+1}} qN_1\, dx \\ \int_{x_e}^{x_{e+1}} qN_2\, dx \end{Bmatrix} \tag{B.1}$$

$$\{\boldsymbol{\theta}\} = \begin{Bmatrix} \theta_1 \\ \theta_2 \end{Bmatrix} \qquad \{\boldsymbol{P}\} = \begin{Bmatrix} Q_1 \\ Q_2 \end{Bmatrix} \qquad \frac{(e_0a)^2}{EI}\{\boldsymbol{P}\} = \begin{Bmatrix} \frac{(e_0a)^2}{EI} Q_1 \\ \frac{(e_0a)^2}{EI} Q_2 \end{Bmatrix} \tag{B.2}$$

C Expressions for α, β, γ and ϕ in Table 1

$$\begin{aligned} \alpha &= e^{-\frac{x}{e_0a}} - 1 \\ \beta &= (e_0a)^2 e^{-\frac{x}{e_0a}} - (e_0a)((e_0a) - x) \\ \gamma &= e^{-\frac{L-x}{e_0a}} - 1 \\ \phi &= (e_0a)((e_0a) + x) - (e_0a)e^{-\frac{L-x}{e_0a}}(L + (e_0a)) \end{aligned} \tag{C.1}$$

References

1. Iijima S, Brabec C, Maiti A, Bernholc J (1996) Structural flexibility of carbon nanotubes. J Chem Phys 104:2089–2092. https://doi.org/10.1063/1.470966
2. Snehashish C, Laxmi B (2016) Static and dynamic problems of nanobeams and nanoplates. World Scientific, Rourkela, Odisha. https://doi.org/10.1142/10137
3. Karlicic D, Murmu T, Adhikari S, McCarthy M (2015) Non-local structural mechanics. John Wiley & Sons, Hoboken, New Jersey
4. Kroner E (1967) Elasticity theory of materials with long range cohesive forces. Int J Solids Struct 3:731–742. https://doi.org/10.1016/0020-7683(67)90049-2
5. Eringen AC, Edelen D (1972) On nonlocal elasticity. Int J Eng Sci 10:233–248. https://doi.org/10.1016/0020-7225(72)90039-0
6. Peddieson J, Buchanan GR, McNitt RP (2003) Application of nonlocal continuum models to nanotechnology. Int J Eng Sci 41:305–312. https://doi.org/10.1016/S0020-7225(02)00210-0
7. Wang Q, Liew KM (2007) Application of nonlocal continuum models to static analysis of micro-and nano-structures. Phys Lett A 363:236–242. https://doi.org/10.1016/j.physleta.2006.10.093

8. Wang CM, Kitipornchai S, Lim CW, Eisenberger M (2008) Beam bending solutions based on nonlocal Timoshenko beam theory. J Eng Mech 134:475–481. https://doi.org/10.1061/(ASCE)0733-9399(2008)134:6(475)
9. Challamel N, Wang CM (2008) The small length scale effect for a non-local cantilever beam: a paradox solved. Nanotechnology 19:345703. https://doi.org/10.1088/0957-4484/19/34/345703
10. Phadikar JK, Pradhan SC (2010) Variational formulation and finite element analysis for nonlocal elastic nanobeams and nanoplates. Comput Mater Sci 49:492–499. https://doi.org/10.1016/j.commatsci.2010.05.040
11. Nguyen NT, Kim NI, Lee J (2015) Mixed finite element analysis of nonlocal Euler-Bernoulli nanobeams. Finite Elements Anal Des 106:65–72. https://doi.org/10.1016/j.finel.2015.07.012
12. Fernandez Saez J, Zaera R, Loya A, Reddy J (2016) Bending of Euler-Bernoulli beams using Eringen's integral formulation: a paradox resolved. Int J Eng Sci 99:107–116. https://doi.org/10.1016/j.ijengsci.2015.10.013
13. Tuna M, Kirca M (2016) Exact solution of Eringen's nonlocal integral model for bending of Euler-Bernoulli and Timoshenko beams. Int J Eng Sci 105:80–92. https://doi.org/10.1016/j.ijengsci.2016.05.001
14. Romano G, Barretta R (2016) Comment on the paper "Exact solution of Eringen's nonlocal integral model for bending of Euler-Bernoulli and Timoshenko beams" by Meral Tuna & Mesut Kirca. Int J Eng Sci 100:240–242. https://doi.org/10.1016/j.ijengsci.2016.09.009
15. Tuna M, Kirca M (2017) "Respond to the comment letter by Romano and Barretta on the paper"Exact solution of Eringen's nonlocal integral model for bending of Euler-Bernoulli and Timoshenko beams". Int J Eng Sci 116:141–144. https://doi.org/10.1016/j.ijengsci.2017.02.009
16. Eringen AC (1983) On differential equations of nonlocal elasticity and solutions of screw dislocation and surface waves. J Appl Phys 54:4703–4710. https://doi.org/10.1063/1.332803
17. Wazwaz A (2011) Linear and nonlinear integral equations. Springer, Heidelberg
18. Wazwaz AM (2015) First course in integral equations. World Scientific Publishing Company, Singapore
19. Romano G, Barretta R, Diaco M, de Sciarra F (2017) Constitutive boundary conditions and paradoxes in nonlocal elastic nanobeams. Int J Mech Sci 121:151–156. https://doi.org/10.1016/j.ijmecsci.2016.10.036
20. Polyanin A, Manzhirov A (2008) Handbook of integral equations. Chapman and Hall/CRC, Boca Raton

Thermally Induced Concrete Properties at Elevated Temperature

Ananda Mitra, Shiwanand R. Suryawanshi, and Banti A. Gedam

Abstract The fire-damaged properties of concrete are one of the concerns for the structural application that may affect structural stability and integrity against collapse. Nowadays, infrastructure industries prefer to use high-strength concrete (HSC) or high-performance concrete (HPC) over normal strength concrete (NSC) due to structural sustainability and serviceability aspects, and such HSC/HPC is observed to be highly susceptible to temperature than NSC. Knowledge of thermal-induced properties, i.e., conductivity, specific heat, and diffusivity, is essential for evaluating heat transfer phenomenon that not only helps in the selection of appropriate materials but also helps to obtain HSC/HPC structural member' desired fire resistance rating and behaviour. Hence, a comprehensive set of experimentally measured data of thermally induced properties of HSC/HPC have been collected from different literature, and this reveals that the type of aggregates, cementitious content with water, and the chemico-physico-hygro-thermo process of the mix design can significantly influence thermally induced properties. Also, it is observed that their behavioural response to temperature and relationship with each other is still not very well established. Therefore, this paper aims to review and discuss the influence of elevated temperature on the type of concrete properties, i.e., mechanical, chemical, and physical properties, and its suitability for structural application.

Keywords Thermal-induced properties · Heat transfer · High-strength concrete · High-performance concrete · Thermal degradation

A. Mitra (✉) · S. R. Suryawanshi · B. A. Gedam
Sardar Vallabhbhai National Institute of Technology, Surat, India
e-mail: anandamitra27@gmail.com

S. R. Suryawanshi
e-mail: ssr@amd.svnit.ac.in

B. A. Gedam
e-mail: bantiagedam@ced.svnit.ac.in

S. Sidhardh et al. (eds.), *Advances in Structural Integrity for Mechanical, Civil, and Aerospace Applications*, Lecture Notes in Mechanical Engineering,
https://doi.org/10.1007/978-981-97-6367-2_37

1 Introduction

High-intensity fire load in accidental conditions is a severe hazard for high-rise buildings. In such circumstances, fire resistance evaluation of structural members for their stability and integrity against failure is most important. However, safety and evacuation measures are acknowledged in the national building code as the active and passive measures system [1, 2], but when these measures fail, only structural stability and integrity are the last options for safe evacuation. Many research studies elucidated by experimental studies show that, due to the attribution of severe damage in standard and non-standard fire loading, the structural members cannot maintain their stability and integrity against collapse, even though structural members' fire resistance measures are followed from codes and standard practices [3]. Basically, existing codes and standard practices, i.e., NBC 2016, IS 456, EC II, and ASCE acknowledging prescriptive approaches of fire-resisting using only adequate concrete cover provision for reinforced concrete members, and any detailed advanced analysis/design methodology is not adopted nor proposed [1, 2, 4–6]. Hence, a performance-based fire-safe design solution and analysis technique for fire resistance evaluation of reinforced concrete (RC) members at exposure to fire load are more necessary than any prescriptive approach.

The effect on the thermal properties of HSC and HPC varies from that of NSC and introduces mechanical properties variation with temperature. A highly dense microstructure makes the heat transfer rate more rapid in RC members. Therefore, a consideration of appropriate thermally induced concrete properties, i.e., specific heat, conductivity, and diffusivity, are the prime requirements for heat transfer evaluation. This helps further evaluation to finalise behaviour and fire resistance rating of RC members at fire load considering mechanical properties of concrete, i.e., stress–strain relation, compressive strength, tensile strength, thermal expansion, and mass loss [7]. Hence, this paper aims at material properties data compilation from different literature and discusses the present fire resistance problem for RC member design and analysis.

2 Research Significance

Concrete is a widely used construction material due to its durability and strength. However, when exposed to high temperatures, its physical and chemical properties can be significantly altered. Therefore, it is essential to research the behaviour of concrete when subjected to thermal loads. The study of thermally induced concrete properties at elevated temperatures can provide essential insights into the material's behaviour, enabling the development of more resilient and durable structures that can withstand the effects of fire, explosions, and other extreme events. Through in-depth research, we can identify the factors that influence the behaviour of concrete under high temperatures, such as the composition of the material, the heating rate, and the cooling method. This knowledge can be applied to improve the design and construction of concrete structures, making them safer and more resilient.

3 Heat Transfer Analysis

Heat transfer analysis is most important for structural member fire resistance evaluation to finalise mechanical properties and the critical behaviour at exposure to the fire load. In heat transfer evaluation, three basic mechanisms are involved, i.e., conduction, convection, and radiation [8]. When the RC member solid homogeneous body comes in contact with fire load, the inherent heat transfer as a gradient of temperature at section is due to the conduction mechanism that depends on the thermally induced concrete properties, i.e., conductivity, specific heat, and diffusivity concerning temperature. However, the exposed surfaces as boundary of RC member are subjected to convection by the atmospheric temperature of hot gases and due to the emission of fire load emitting radiation. To evaluate the heat transfer rate in RC members, Fourier's law may be used to calculate the heat flow per unit area by conduction [9], as shown in Eq. 1.

$$q = -k\left(i\frac{\partial T}{\partial x} + j\frac{\partial T}{\partial y} + k\frac{\partial T}{\partial z}\right) = -k\nabla T. \tag{1}$$

For one-dimensional and two-dimensional heat flow, the equation is simplified and shown in Eqs. 2 and 3, respectively:

$$q = -k\frac{\partial T}{\partial x} \tag{2}$$

$$q = -k\left(i\frac{\partial T}{\partial x} + j\frac{\partial T}{\partial y}\right). \tag{3}$$

The amount of heat flow is given by:

$$q = -kA\frac{\partial T}{\partial x}. \tag{4}$$

However, the conduction mechanism for transfer analysis is responsible, but heat transfer also varies based on the type of fire load with time and the exposure boundary surface of RC members. Figure 1a shows one-dimensional heat transfer variation concerning the depth in distance X from the fire exposure surface temperature T. From computational evaluation, illustrated heat transfer concerning different depths

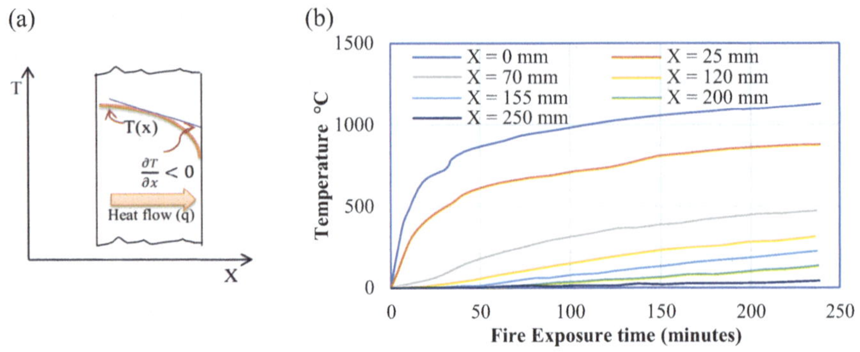

Fig. 1 **a** One-dimensional heat transfers due to the conduction mechanism [10] and **b** heat transfer evaluation with respect to depth RC member [11]

is shown in Fig. 1b. This reveals that the shallow depth faces a high rate of heat transfer at a high-intensity fire load. However, the depth increases heat transfer rate is low, and some platen-type variation at 100–150 °C can be observed. This is due to the thermo-hygro process, in which concrete moisture changes its physical form from liquid to gas, and as a result, temperature variation in the range of 100ºC–150ºC can be observed.

4 Chemico-physico-hygro-thermo Properties of Concrete

When concrete is subjected to high temperatures, introduces physiochemical changes in the inherent concrete binding matrix, and affects the bonding between the aggregates and cement paste matrix, this results in the degradation of concrete properties, depending on the heating rate and applied load [10]. In fact, such variation directly affects the thermal and mechanical properties of concrete. Therefore, the chemico-physico-hygro-thermo properties are discussed in this section.

4.1 Cement Binding Matrix

The behaviour of cement binding matrix paste at high temperatures depends on porosity, water-to-cement ratio, CaO, SiO_2, $Ca(OH)_2$, degree of hydration, and moisture content. As the cement binding matrix temperature reaches up to 110°C, available free moisture content evaporates and introduces disintegration of ettringite and gypsum [11]. This is an initial deterioration process in which dehydration of the chemical paste causes microstructural changes in the cement binding matrix.

At a temperature of 120–500 °C, C-S–H dehydrates to larnite, resulting in the commencement loss of cement paste. This introduces micro–macrocrack formation and easy movement of moisture-based evaporated steam to condense in the porous section, resulting in high pore pressure development in a dense matrix that causes concrete spalling and release of microtension for transient creep development [12].

Tobermorite undergoes reorientation and breakdown around 200–300 °C. In contrast, calcium hydroxide ($Ca(OH)_2$) decomposes into calcium oxide (CaO) and water (H_2O) at temperatures between 450 and 550 °C, resulting in enhanced shrinkage and microcracking in the cement binding matrix.

Calcium carbonate ($CaCO_3$) is decarburised when it decomposes into calcium oxide (CaO) and carbon dioxide (CO_2) at temperatures beyond 650°C. At around 700 °C, C-S-H decomposes into Beta-dicalcium silicate β-C_2S, βCS, and H_2O, shrinking the cement paste and losing some of its strength.

From 20 to 300 °C, there is no substantial increase in porosity, while porosity and micro-cracks grow dramatically over 300 °C. Another abrupt rise in porosity occurs above 900 °C. The temperature considerably influences non-hydrated cement clinkers contained in cement paste. This non-hydrated cement clinker expands while the hydrated cement pastes contracts. Differential thermal expansion ultimately leads to microcracking [10].

4.2 *Aggregates*

The type of aggregate influences the heat conductivity, stability, and change in concrete's residual strength. Due to high temperatures, aggregates undergo thermal expansion and various physicochemical changes, resulting in microcracking and even melting at 1000 °C. When exposed to high temperatures, these effects depend on the aggregates' thermal stability. In general, two types of aggregates are used: siliceous and carbonate aggregates.

The primary component of siliceous aggregates is SiO_2; therefore, sand also fits into this group. At 600 °C, there is a transition from α to β of the crystalline quartz to siliceous aggregates, leading to a volumetric expansion of 5.7%. Even though this transition is reversible, it undergoes volumetric cracking due to the expansion process [13]. $CaCO_3$ is the primary component of carbonate aggregates, which become unstable at 650 °C due to the breakdown of calcium carbonate ($CaCO_3$) into calcium oxide (CaO) and carbon dioxide (CO_2). This process reaches its zenith at 800 °C and ends at around 900 °C. This transition is an endothermic process; as a result, it absorbs a substantial quantity of heat, and the heat flow through the concrete slows down.

When concrete made with carbonate aggregate is subjected to temperatures ranging from 600 to 800 °C, it demonstrates a higher specific heat capacity. This is due to an endothermic reaction that occurs during the decomposition of dolomite. This reaction absorbs a significant amount of energy (heat/temperature), resulting in concrete made by carbonate aggregate having more fire resistance than siliceous aggregate [7].

4.3 Transition Zone

The interface between the aggregates and the cement paste is weak because of inadequate bonding surfaces, large voids, and weak C-S-H crystal [14]. This interface is around 50μm thick and is known as the transition zone [15]. When the concrete is heated, the aggregates expand and the cement paste contracts due to dehydration. The transition zone is the first-place cracking appears due to uneven expansion and shrinkage. This eventually can result in the spalling of concrete. Figure 2 illustrates the thermal incompatibility that exists between aggregates and cement paste.

5 Thermal Properties

Code and standard practices, i.e., Eurocode 2 EN 1992–2-2 [4] and ASCE manual No. 78 [5], acknowledge concrete's thermally induced properties. However, these properties primarily focused on NSC, and a lack of HSC/HPC can be observed. In fact, such properties used without understating the heat transfer mechanism may serve as erroneous information to finalise the behaviour of RC members at standard and non-standard fire loads.

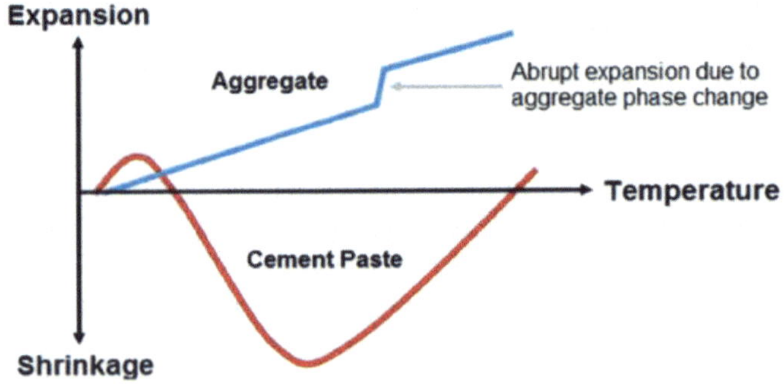

Fig. 2 Incompatibility between cement paste and aggregates when subjected to high temperature

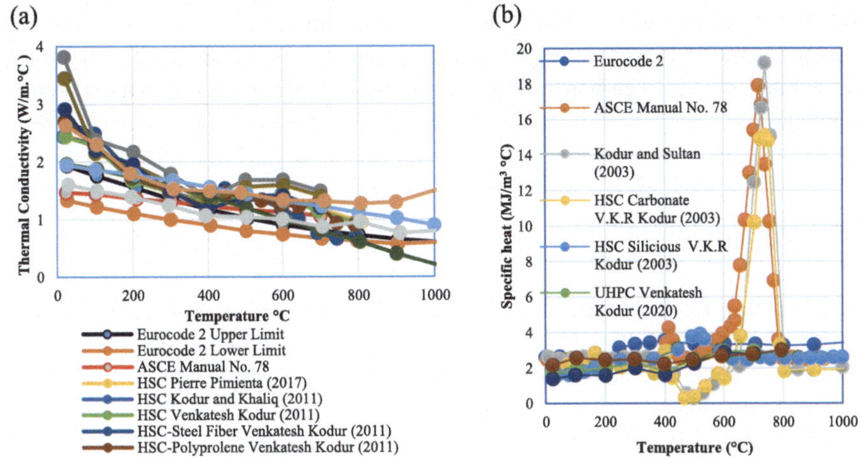

Fig. 3 Variation of thermal properties for NSC, HSC, HPC, and UHPC: **a** conductivity and **b** specific heat-carrying capacity

5.1 Thermal Conductivity

Figure 3a illustrates a compiled database of thermal conductivity from different literature. Eurocode 2 and ASCE provide a limiting value and preliminary proposed for NSC, whereas HSC and Ultra-HPC (UHPC) thermal conductivity values are higher than for NSC [4, 5]. This is because of the various material factors, i.e., moisture content, type of aggregate, water-to-cement ratio, and density. The thermal conductivity ranges 1.94–2.91 W/m °C observed for HSC and UHPC, and the addition of steel fibres and polypropylene fibres has increased by 3.44 – 3.81 W/m °C in ambient conditions. However, concerning increasing temperature, thermal conductivity for all types of concrete was reduced due to the decomposition of cement binding matrix $Ca(OH)_2$ into CaO and H_2O, and evaporation of moisture content. After 400 °C, a minor rise in thermal conductivity for HSC/UHPC was observed due to the thermo-hygral process. Also, the performance-wise thermal conductivity of siliceous aggregate concrete is higher than that of carbonate aggregate concrete due to the more crystalline nature of aggregate.

5.2 Specific Heat Capacity

Figure 3b illustrates a compiled database of the specific heat-carrying capacity of NSC, HSC, and UHPC [4, 5, 16]. However, any significant variation of specific heat-carrying capacity of siliceous type aggregate up to 1000°C was not observed but the use of carbonate aggregate in concrete improved heat-carrying capacity from 400 to 800 °C. In which, maximum specific heat reached 15.09–19.17 MJ/m^3 K at

700–750 °C have been observed. This is due to the attribution of dissociation of the dolomite in the carbonate aggregate. Therefore, to improve the fire resistance rating RC member, the use of carbonate aggregate seems useful technique than the siliceous aggregate.

5.3 *Diffusivity*

As seen in Fig. 1b, the heat transfer rate concerning time is reduced when the heating temperature increases. This is because of the disintegration of the concrete binding matrix with the temperature that introduced micro–macrolevel crack formation. In computation/mathematical numerical approaches to heat transfer evaluation, such damage is considered a diffusivity, and it is related to conductivity, specific heat, and density of concrete. Figure 4a shows the thermally induced diffusivity of concrete concerning temperature is gradually reducing, and it can be evaluated by using Eq. 5 [17].

$$\alpha = \frac{k}{\rho c_p}, \quad (5)$$

where C_p is the material's specific heat, k is thermal conductivity, and ρ is density.

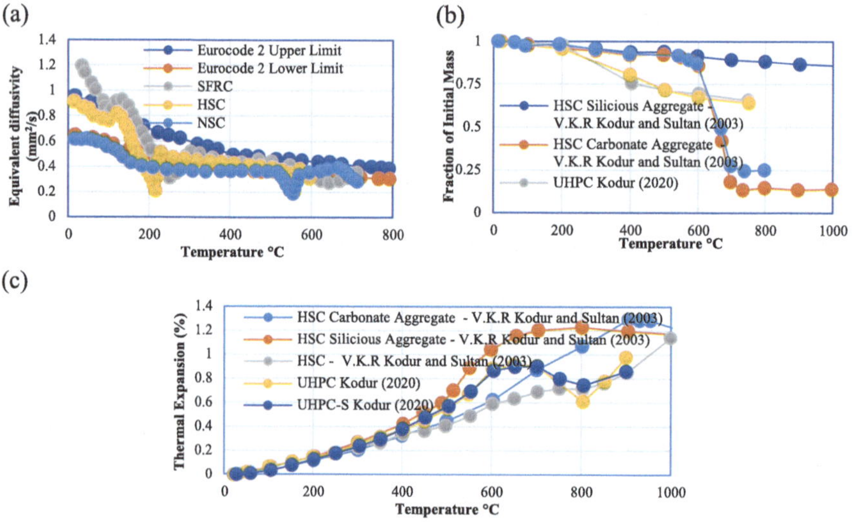

Fig. 4 Thermal properties variation for different concretes: **a** thermal diffusivity, **b** mass loss comparison, and **c** temperature-dependent thermal expansion

5.4 *Mass Loss*

High elevated temperatures also affect the cement binding matrix and introduce chemical combined hydrating moisture losses that gradually weaken the paste and paste-aggregate bond. Figure 4b shows a different concrete mass loss due to the drying effect at elevated temperatures [7]. At above 600 °C, the transition of $CaCO_3$ into CaO and CO_2 causes a significant amount of mass loss in carbonate aggregate [7]. Mass loss of NSC is less than compared to HSC/UHPC [18]. This is due to the low water-to-binder content with high supplementary cementitious materials in HSC/UHPC, in which chemical decomposition is higher than NSC. Significant mass loss occurs due to evaporation of free saturated moisture, which is trapped in the pores of the concrete at temperatures of 100–300 °C. At 500 °C, rapid evaporation of the water left over from decomposition and breakdown of $Ca(OH)_2$ can be observed. The total H_2O evaporation in concrete mass loss can be observed at 500–750 °C. HSC comprising siliceous aggregates, showing mass loss is less than carbonate aggregate HSC.

5.5 *Thermal Expansion*

Thermal strain variation of concrete depends on moisture content, binder type, and aggregate type [19]. Figure 4c shows the thermal expansion for various types of concrete. Linear thermal expansion behaviour of concretes up to 500 °C due to drying shrinkage of the concrete binding matrix has been observed. There is a high-intensity thermal expansion between 600 and 800 °C due to the breakdown of the cement paste and the loss of chemically bonded water observed. Decarbonisation of limestone-based aggregate concrete' thermal expansion increases as the temperature exceeds 800 °C [16]. Dehydration of calcium hydroxide above 800 °C, the highest thermal expansion occurs in this temperature range. The aggregate type also significantly influences the thermal expansion of concrete. Thermal expansion of siliceous aggregate HSC rises to 800 °C and then stabilises.

6 Mechanical Properties of Concrete at Elevated Temperature

In to obtain structural members' behaviour at elevated temperatures in computational, numerical, analysis, and structural member design approaches, mechanical properties play significant role.

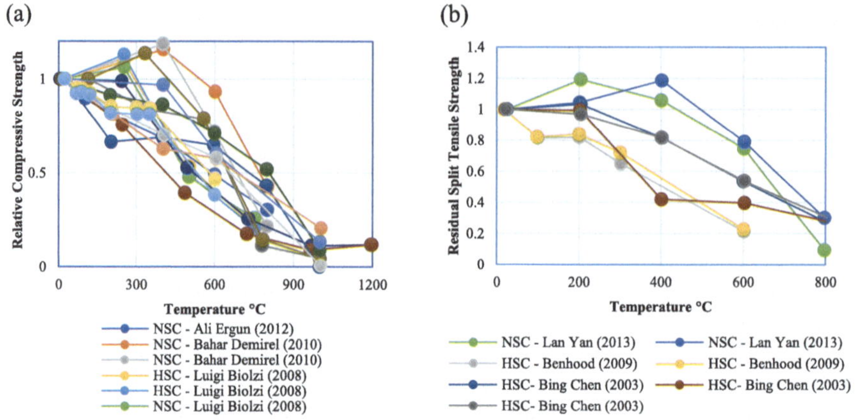

Fig. 5 Variation in relative strength of NSC and HSC with temperature: **a** compressive and **b** split tensile

6.1 Compressive Strength

A designed flexural member load-carrying capacity is associated with compression stress above a neutral axis of a member, and their failure depends on ultimate compressive strength carrying capacity. Under fire load, the designed load-carrying capacity diminishes and depends on the temperate-dependent compressive strength of concrete. Therefore, the temperature-dependent compressive strength of NSC and HSC for RC members' behaviour evaluation is most important. Figure 5a shows compressive strength variation of NSC and HSC with respect to temperature. It is observed that the NSC strength reduction factor is lower than that of HSC [20–23]. However, in a few cases, the strength of concrete due to the hydration process at 200–400 °C increases and then decreases at a very high rate. The strength of NSC/HSC is almost 50% of the designed strength at 600 °C and does not seem suitable for structural application.

6.2 Tensile Strength

Fire-induced spalling of highly durable concrete is a common problem. This directly changes the transmission mechanism of the internal force of RC members and jeopardises load-carrying capacity. Therefore, to mitigate the spalling of RC members at elevated temperatures, the tensile strength limit of concrete plays a significant role. Spalling of concrete 190–250 °C and explosive spalling in the range 280–422 °C can be observed due to internal vapour pressure. Once the tensile permissible limit of concrete exceeds the spalling can be observed, it depends on the durability of the concrete. As seen in Fig. 5b, it is observed that the NSC split tensile strength

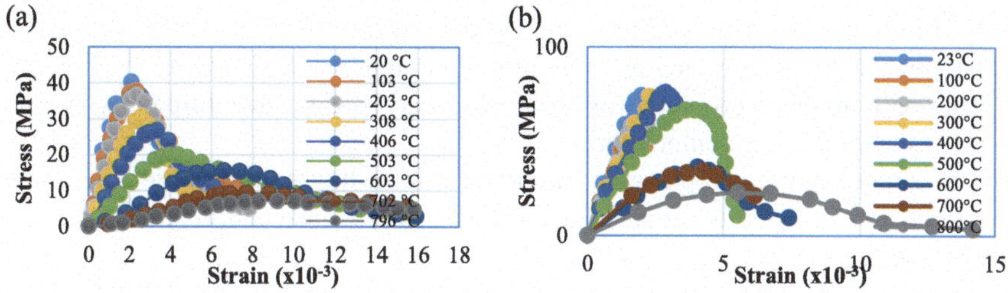

Fig. 6 **a** Residual stress–strain curves of NSC with temperature and **b** residual stress–strain curves of HSC with temperature [7]

improves with respect to temperature up to 400 °C and then reduced significantly, whereas the split tensile strength of HSC compared to NSC is more severe and reduces after 200 °C [24–26]. This implies that the low split tensile strength of HSC is more severe in spalling compared to NSC.

6.3 Stress–Strain Relation

Designed RC members' fire resistance evaluation with respect to fire load depends on stress–strain design curve consideration [3]. In which, ultimate loading strain and failure strain of concrete are most important to finalise the fibre stress carrying capacity of the RC member. As seen in Fig. 6a, b, the stress–strain relation with respect to temperature is varying for both NSC and HSC. Nevertheless, the ultimate failure strains do not show much deviation for both concrete, but the failure strain of HSC is more severe than that of NSC.

This result is caused by different temperatures due to the varying properties of their constituent materials. The main factors that contribute to this difference include the type and amount of cementitious materials used, the water-to-cement ratio, and the aggregate characteristics. NSC exhibits a higher strain capacity at low temperatures than HSC seen in Fig. 6a, b, which has a lower ductility and higher stiffness. This is because NSC has a higher cementitious content and a more porous microstructure, which allows for greater internal deformations. In contrast, HSC has a denser microstructure and lower porosity, resulting in a stiffer and more brittle material.

7 Conclusion

The following conclusions were drawn from this compiled database and comparative study for thermally induced concrete properties at elevated temperatures:

- Existing code and standard practices do not consider HSC and HPC properties variation with temperature and only limit their applicability to NSC.
- Thermally induced properties of NSC, HSC, and HPC behave differently in each other under elevated temperatures.
- The type of aggregate plays an important role in thermal properties, in which carbonate aggregate is suitable for improving fire resistance rating compared to siliceous aggregate.
- NSC, HSC, and HPC strength properties are not suitable for structural application after 600 °C.
- The properties of concrete significantly diminish in elevated temperatures due to the decomposition and demineralisation of the cement binding matrix under the chemico-physico-hygro-thermo process.

8 Future Research Work

In order to improve serviceability and sustainability after fire damage, it is important to conduct future research on the thermally induced properties of concrete at elevated temperatures. This research should investigate how different factors, such as the type of aggregate, cementitious material, and curing conditions, affect the behaviour of different types of concrete under high-temperature conditions. Additionally, this research can explore different methods for improving the fire resistance of concrete structures, including the use of fire retardants and incorporating fibres or other additives into the concrete mixture [27]. By gaining a better understanding of the thermal changes in concrete properties, this research can contribute to the development of more robust and resilient concrete structures that can withstand high-temperature exposure to ensure the safety of occupants and protect property.

Acknowledgements Dr. Banti A. Gedam, Associate Professor (Grade-I) in the Department of Civil Engineering at SVNIT Surat, would like to express their sincere gratitude and appreciation to the Science and Engineering Research Board (SERB), India, for funding their research work under Project File No: EEQ/2023/000130.

CRediT Authorship Contribution Statement

Ananda Mitra: Investigation, Methodology, Conceptualization, Data curation, Formal analysis, Writing.

Shiwanand R. Suryavanshi: Supervision, Conceptualization.

Banti A. Gedam: Funding acquisition, Investigation, Methodology, Project administration, Supervision, Conceptualization, Data curation, Formal analysis, Writing—review and editing.

Declaration of Competing Interest The authors declare no known competing financial interest or personal relationship that could have appeared to influence the work reported in this paper.

References

1. NBC, National Building Code Volume 1, Bur. INDIAN Stand. 1 (2016)
2. NBC, National Building Code of India, 2016 Volume 2, Bur. INDIAN Stand. 2 (2016)
3. Gedam BA (2021) Fire resistance design method for reinforced concrete beams to evaluate fire-resistance rating. Structures 33:855–877. https://doi.org/10.1016/j.istruc.2021.04.046
4. Eurocode 2: Design of concrete structures—Part 1–2: general rules—structural fire design, Eur. Union. 2 (2008)
5. ASCE (1992) Structural fire protection american society of civil engineers structural fire protection
6. Plain and Reinforced Concrete—Code of Practice IS: 456, Bur. INDIAN Stand. (2021)
7. Kodur V (2014) Properties of concrete at elevated temperatures. ISRN Civ Eng 2014. https://doi.org/10.1155/2014/468510
8. Incropera F, Dewitt D (1985) Introduction to heat transfer. https://inis.iaea.org/search/search.aspx?orig_q=RN:19023113
9. Cannon JR, Browder FE (1984) The one-dimensional heat equation. Cambridge University Press. https://doi.org/10.1017/CBO9781139086967
10. Khoury GA (2000) Effect of fire on concrete and concrete structures. Prog Struct Eng Mater 2:429–447. https://doi.org/10.1002/pse.51
11. Hager I (2013) Behaviour of cement concrete at high temperature. Bull Polish Acad Sci Tech Sci 61:145–154. https://doi.org/10.2478/bpasts-2013-0013
12. International Federation for Structural Concrete, fib Bulletin 38: Fire design of concrete structures (2007), pp 91–92
13. Annerel E, Taerwe L (2009) Revealing the temperature history in concrete after fire exposure by microscopic analysis. Cem Concr Res 39:1239–1249. https://doi.org/10.1016/j.cemconres.2009.08.017
14. Mehta PK, Monteiro PJM (2019) Concreto: Microestrutura, Propriedades e Materiais, 2ed. IBRACON, São Paulo, 782p
15. Paulon V, Kirchheim AP (n.d.) Nanoestrutura e Microestru- tura do Concreto Endurecido
16. Kodur V, Khaliq W (2011) Effect of temperature on thermal properties of different types of high-strength concrete. J Mater Civ Eng 23:793–801. https://doi.org/10.1061/(asce)mt.1943-5533.0000225
17. Felicetti R (2007) Assessment of the equivalent thermal diffusivity for fire analysis of concrete structures. In: Proceedings of Fib Task Group 4.3 Workshop Fire Design of Concrete Structures, Coimbra, pp 8–9
18. Kore Sudarshan D, Vyas AK (2019) Impact of fire on mechanical properties of concrete containing marble waste. J King Saud Univ Eng Sci 31:42–51. https://doi.org/10.1016/j.jksues.2017.03.007
19. Khoury GA (2008) Fire and concrete. Encontro Nac Betão Estrutural. https://doi.org/10.1016/0008-8846(88)90088-9
20. Ergün A, Kürklü G, Serhat Başpnar M, Mansour MY (2013) The effect of cement dosage on mechanical properties of concrete exposed to high temperatures. Fire Saf J 55:160–167. https://doi.org/10.1016/j.firesaf.2012.10.016
21. Biolzi L, Cattaneo S, Rosati G (2008) Evaluating residual properties of thermally damaged concrete. Cem Concr Compos 30:907–916. https://doi.org/10.1016/j.cemconcomp.2008.09.005
22. Husem M (2006) The effects of high temperature on compressive and flexural strengths of ordinary and high-performance concrete. Fire Saf J 41:155–163. https://doi.org/10.1016/j.firesaf.2005.12.002
23. Noumowe AN, Clastres P, Debicki G, Costaz JL (1996) Transient heating effect on high strength concrete. Nucl Eng Des 166:99–108. https://doi.org/10.1016/0029-5493(96)01235-6
24. Yan L, Zhang J, Xing YM, Li JJ (2013) High-temperature mechanical properties and microscopic analysis of nano-silica steel fibre RC. Mag Concr Res 65:1472–1479. https://doi.org/10.1680/macr.13.00143

25. Behnood A, Ghandehari M (2009) Comparison of compressive and splitting tensile strength of high-strength concrete with and without polypropylene fibers heated to high temperatures. Fire Saf J 44:1015–1022. https://doi.org/10.1016/j.firesaf.2009.07.001
26. Siddique R, Kaur D (2012) Properties of concrete containing ground granulated blast furnace slag (GGBFS) at elevated temperatures. J Adv Res 3:45–51. https://doi.org/10.1016/j.jare.2011.03.004
27. Gedam BA (2023) Spalling mitigation techniques for high-performance concrete at elevated temperature. Mater Today Proc. https://doi.org/10.1016/j.matpr.2023.03.556

Strength and Deformation Assessment of a Tall RC-MRF Designed by Force-Based Design Method and Performance-Based Plastic Design Method

Rohit Vyas and Anoop I. Shirkol

Abstract Past seismic events have shown that high-rise Reinforced Concrete Moment Resisting Frames (RC-MRFs) designed according to current building codes often experience significant inelastic deformations. Predicting and controlling these deformations has proven challenging, as there is no precise method available. This study aims to introduce a performance-based plastic design (PBPD) approach for analyzing and designing high-rise RC-MRF frames while accounting for the inelastic behavior of the structure. In this study, a 24-story RC-MRF frame is selected for analysis and design, utilizing both the force-based design (FBD) method and the PBPD method for seismic loads. The PBPD approach considers three performance levels: immediate occupancy (IO), life safety (LS), and collapse prevention (CP). Indian standard code guidelines and principles are applied in the analysis and design of the frame. The performance assessment of the frame is conducted through nonlinear static pushover analysis. The results obtained indicate that the PBPD-LS frame outperforms the other frames. It is noteworthy that design sections in the PBPD-LS frame are found to be the most efficient, and the displacements remain within defined limits. On the other hand, the PBPD-IO frame is found to be highly uneconomical among the frames. Additionally, secondary moment effects (P-delta) appear to dominate in the PBPD-CP frame. Based on these observations, it can be concluded that, among all the frames, the PBPD-LS frame meets the desired PBPD outcomes and performs exceptionally well within the specified limits.

Keywords RC-MRF · Performance level · Nonlinear static pushover analysis · Force-based design (FBD) method · Performance-based plastic design (PBPD)

R. Vyas (✉) · A. I. Shirkol
Malaviya National Institute of Technology, Jaipur 302017, India
e-mail: 2020RCE9515@mnit.ac.in

S. Sidhardh et al. (eds.), *Advances in Structural Integrity for Mechanical, Civil, and Aerospace Applications*, Lecture Notes in Mechanical Engineering,
https://doi.org/10.1007/978-981-97-6367-2_38

Abbreviations

A_h	Design horizontal earthquake acceleration coefficient
S_a	Spectral acceleration coefficient
S_d	Design spectral displacement
T	Fundamental natural period
g	Gravitational acceleration
h	Height
h_n	Height of top floor
I	Importance factor
Q_i	Lateral load at ith level
R	Response reduction factor
R_μ	Ductility reduction factor
R_R	Redundancy factor adopted as unity
V	Shear at interior column
V'	Shear at exterior column
V_o	Base shear at performance point
w_n	Lumped weight of top floor
Z	Earthquake zone factor
β_i	Shear distribution factor for PBPD
γ	Energy modification factor
V_B	Design base shear
W	Lumped weight of structure
w_i	Lumped weight at ith level
μ	Ductility factor.
$\theta_{\max}$	Selected target drift.
θ_p	In-elastic drift
$\boldsymbol{\theta_y}$	Selected yield drift.
M	Mass of system
M_{pc}	Plastic moment
M_u	Beam moment capacity
$C'V_i$	Lateral load distribution factor PBPD

1 Introduction

The Reinforced Concrete Moment Resisting Frame (RC-MRF) consists of horizontal beams, slabs, and vertical columns designed in accordance with standard codes [11, 12]. These standards dictate the proportions and detailing of concrete and reinforcement to achieve the required strength and desired ductility during seismic events. However, observations from past seismic events have revealed that the current code practices often fall short in delivering satisfactory performance when the structure exhibits nonlinear or inelastic behavior under strong ground motions. This can result in displacements and drift exceeding defined limits, and in extreme cases, the risk of complete collapse. To address these challenges, this study introduces the performance-based plastic design (PBPD) method, well-known for its capability to incorporate nonlinear structural behavior. The PBPD method streamlines the iterative process compared to standard design methods because it predefines the target drift [1]. It employs the energy balance concept, where the principle of virtual work is used to analyze the structure. The work required to push the structure to the preselected target drift is considered as the energy demand of an equivalent single degree of freedom (SDOF) system, from which the base shear is calculated [2]. Initially designed for different types of steel frames [2], the PBPD method has been successfully adapted for use in reinforced concrete structures [3]. Studies [4, 5] attempting PBPD, based on the guidelines in [10], have reported satisfactory results. The selection of the target drift is based on assumptions made in [6], considering the possibility of a ground motion event with a high intensity relative to the past. This may be an earthquake hazard level corresponding to a Basic Safety Earthquake-2 (BSE-2) with a 2% chance of occurrence within the next fifty years. To validate the method, a 24-story RC-MRF structure is analyzed and designed using both the traditional Indian standard (IS) code method, also known as the force-based design method (FBD), and the PBPD method, considering all three performance levels (immediate occupancy—IO, life safety—LS, and collapse prevention—CP) as per [6]. The guidelines from [10] are applied to calculate the design base shear and other design parameters. Subsequently, a seismic performance comparison is conducted to assess the strength and deformation characteristics of the frame design by the PBPD and FBD methods, as described in [6]. This comprehensive analysis aims to demonstrate the effectiveness of the PBPD method over the FBD approach.

2 Study Frame

The study focuses on a 24-story Reinforced Concrete Moment Resisting Frame (RC-MRF) structure, as depicted in Fig. 1. The schematic design procedure for the RC-MRF is illustrated in Fig. 2. Given the frame's symmetry along both axes, it can be simplified into an equivalent 2D frame for analysis. Details about the frame are provided in Table 1.

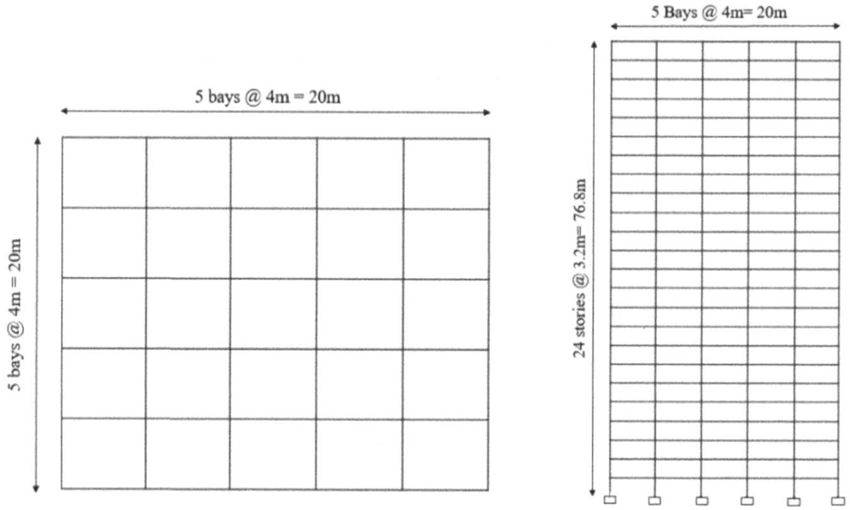

Fig. 1 Plan and elevation of 24-story study frame

3 Earthquake Forces

3.1 Earthquake Forces and Its Distribution in FBD Frame

Initially, the equivalent static load method is employed to calculate the base shear and the analytical distribution of lateral loads across the height of the structure. Subsequently, response spectrum analysis is conducted using ETABS [18], applying the complete quadratic combination method as outlined in [10]. The equations used for the determination of the design seismic horizontal coefficient (A_h) are as follows:

$$A_\text{h} = \frac{\left(\frac{Z}{2}\right)\left(\frac{S_\text{a}}{g}\right)}{\left(\frac{R}{I}\right)} \tag{1}$$

$$V_\text{B} = A_\text{h} \cdot W \tag{2}$$

$$Q_i = \frac{W_i h_i^2}{\sum W_i h_i^2}. \tag{3}$$

The response acceleration coefficient (S_a/g) for medium soil type is detailed in Table 1, enabling the determination of the spectral acceleration coefficient for the force-based design (FBD) frame. In Earthquake Zone 5, it is considered that the vertical seismic effect is two-thirds of the horizontal seismic effect. After taking into account all the factors outlined in [10], the total base shear (V_B) is calculated to

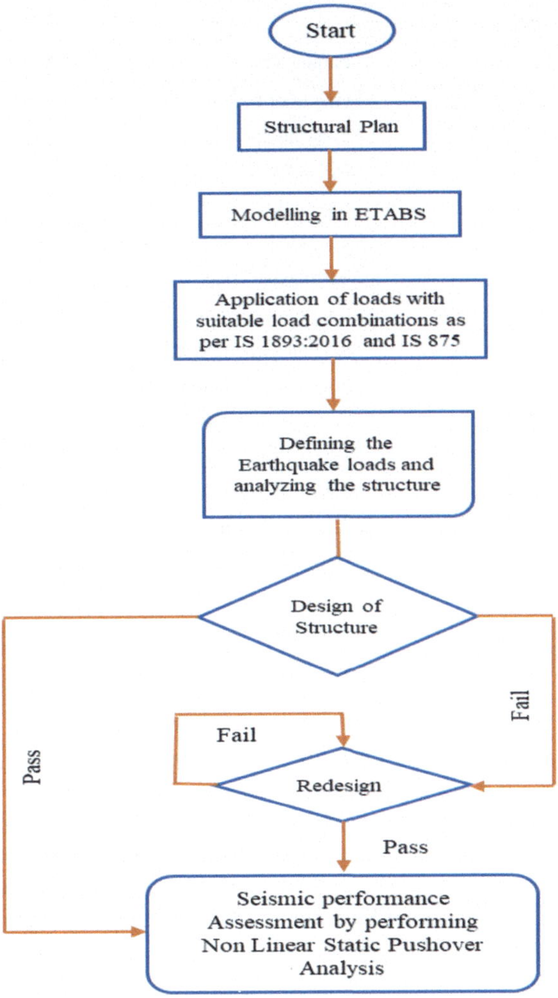

Fig. 2 Schematic view analyzing RC-MRF

be 287.93 kN. Due to the influence of the response reduction factor (*R*), the base shear demand is found to be lower, making the structure appear more flexible. The distribution of shear across the stories is determined according to Eq. 3, as presented in Table 2.

3.2 Earthquake Forces and Its Distribution in PBPD Frame

To calculate lateral loads and base shear using the PBPD method, we consider a target drift value of 0.01 for immediate occupancy (IO), 0.02 for life safety (LS), and

Table 1 Frame descriptions

S. No.	Parameters	Value
1	Height of building	76.8 m
2	Story height	3.2 m
3	No. of bays	5 bays in both direction
4	Bay width	5 m
5	Live load	3.0 kN/m^2 at typical floor
6	Earthquake zone	Zone-V As per [12]
7	Importance factor	1
8	Response reduction factor (R)	5
9	Spectral acceleration (sa/g)	0.69
10	Fundamental natural period (T)	1.94 s
11	Type of soil	Type II, medium as per [12]
12	Size of beam	300 × 400 mm
13	Size of column	600 × 600 mm
14	Slab thickness	150 mm
15	Lumped weight (W) of building	11,442.65 kN
16	Concrete grade	M30
17	Steel grade	HYSD bars (Fe-415)

0.04 for collapse prevention (CP) conditions, which we denote as PBPD-IO, PBPD-LS, and PBPD-CP frames, respectively. A yield drift value of 0.004 is adopted. As columns are treated as non-yielding members, we define a single hinge at the base of each column and two hinges for the beams at both ends. The base reactions for the study frames are computed using the principle of virtual work for an Elasto-Plastic Single Degree of Freedom (SDOF) system. Subsequently, lateral force distribution is determined following the recommendations provided in [7].

$$V_{\mathrm{B}} = \left(\frac{-A_{\mathrm{h}} + \sqrt{A_{\mathrm{h}}^2 + 4\gamma S_{\mathrm{a}}^2}}{2}\right)W \tag{4}$$

$$\gamma = \frac{2\mu - 1}{R_{\mu}^2} \tag{5}$$

$$A_{\mathrm{h}} = \left(\sum_{i=1}^{n}(\beta_i - \beta_{i+1})h_i\right)\left(\frac{w_n h_n}{\sum_{i=1}^{n} w_i h_i}\right)^{0.75T^{-0.2}}\left(\frac{\theta_{\mathrm{p}} 8\pi^2}{T^2 g}\right) \tag{6}$$

$$\beta_i = \left(\frac{\sum_{i=1}^{n} w_i h_i}{w_n h_n}\right)^{0.75T^{-0.2}} \tag{7}$$

Table 2 Lateral load distribution for FBD frames

Floor level	h_i (m)	W_i (kN)	$W_i h_i^2$	$Q_i = \frac{W_i h_i^2}{\sum W_i h_i^2}$	Lateral load distribution (kN)
24	76.8	371.413	2,190,680.06	0.0932	26.836
23	73.6	481.358	2,607,497.84	0.1109	31.942
22	70.4	481.358	2,385,688.01	0.1015	29.225
21	67.2	481.358	2,173,736.39	0.0925	26.629
20	64	481.358	1,971,642.98	0.0839	24.153
19	60.8	481.358	1,779,407.79	0.0757	21.798
18	57.6	481.358	1,597,030.82	0.0679	19.564
17	54.4	481.358	1,424,512.05	0.0606	17.45
16	51.2	481.358	1,261,851.51	0.0537	15.458
15	48	481.358	1,109,049.18	0.0472	13.586
14	44.8	481.358	966,105.061	0.0411	11.835
13	41.6	481.36	833,019.16	0.0354	10.205
12	38.4	481.36	709,791.474	0.0302	8.6951
11	35.2	481.36	596,422.002	0.0254	7.3063
10	32	481.36	492,910.746	0.021	6.0382
9	28.8	481.36	399,257.704	0.017	4.891
8	25.6	481.36	315,462.877	0.0134	3.8645
7	22.4	481.36	241,526.265	0.0103	2.9587
6	19.2	481.36	177,447.868	0.0075	2.1738
5	16	481.36	123,227.686	0.0052	1.5096
4	12.8	481.36	78,865.7193	0.0034	0.9661
3	9.6	481.36	44,361.9671	0.0019	0.5434
2	6.4	481.36	19,716.4298	0.0008	0.2415
1	3.2	481.36	4929.10746	0.0002	0.0604
Total		11,442.65	23,504,141		287.93

$$Q_i = \left[(\beta_i - \beta_{i+1})\left(\frac{w_n h_n}{\sum_{i=1}^{n} w_i h_i}\right)^{0.75T^{-0.2}}\right] V_{\mathrm{B}}. \tag{8}$$

The spectral acceleration coefficient (S_{a}) for the PBPD frame is determined using the spectrum described for the force-based design (FBD) method, as outlined in [6] and illustrated in Fig. 3a. In the case of PBPD frames, the ductility reduction factor R_μ is considered the same as the ductility ratio (μ) since the effective time period of the selected frame exceeds 0.55 s [6]. In the elasto-plastic system, energy is taken as an input unit, but due to nonlinearity, this energy requires modification. Therefore, an energy modification factor (γ) is introduced, as per [4]. The temporal variation of the energy modification factor is depicted in Fig. 3b. The P-delta effect is accounted for when calculating the base shear (V_B), as detailed in Table 3, in the PBPD method [4]. The lateral load distribution for PBPD frames is provided in Table 4 and illustrated in Fig. 4.

According to Fig. 4, the distribution curves for lateral forces and base shear for both FBD and PBPD-LS exhibit nearly identical values. This similarity can be attributed to the direct reduction of the base shear in the FBD method, which is reduced to one-fifth due to the response reduction factor (R) set at 5. Similarly, the PBPD-LS frame is designed with a ductility ratio (μ) of 5, resulting in a similar one-fifth reduction in the base shear coefficient. In contrast, the PBPD-CP frame is designed for the

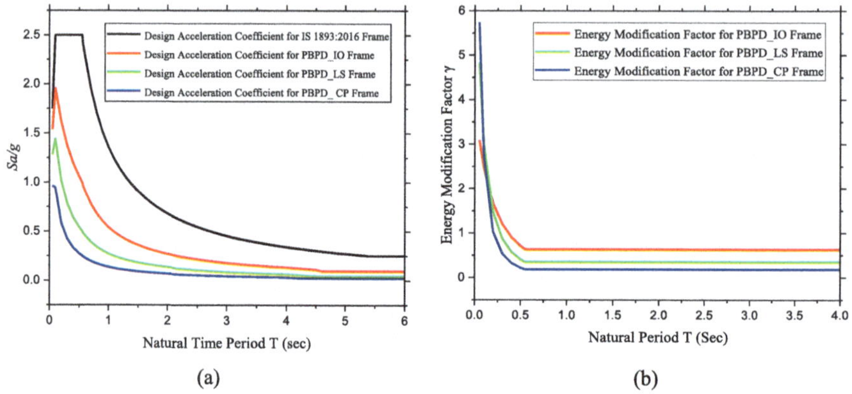

Fig. 3 **a** Response acceleration coefficient curve for all frames. **b** Energy modification factor curve for PBPD frames [7]

Table 3 Design parameters for PBPD frames

Parameters	PBPD-IO	PBPD-LS	PBPD-CP
W (kN)	11,442.650	11,442.650	11,442.649
T (s)	1.946	1.946	1.946
θ_y	0.004	0.004	0.004
θ_{max}	0.01	0.02	0.04
θ_p	0.006	0.016	0.036
μ (θ_{max}/θ_y)	2.5	5	10
S_a/g	0.280	0.140	0.0698
R_μ	2.5	5	10
γ	0.64	0.36	0.19
A_h	1.1323	1.959	4.409
$V_{B/W}$	0.043	0.0035	0.00021
V_B (kN)	487.234	41.004	2.409
$V_{B\,(p\text{-delta})}$	114.426	228.852	457.706
Total V_B (kN)	601.660	269.857	460.115

Table 4 Lateral load distribution for PBPD frames

Floor level	h_i (m)	$\beta_i - \beta_{i+1}$	$C'V_i$	Q_i (PBPD-IO)	Q_i (PBPD-LS)	Q_i (PBPD-CP)
24	76.8	1	0.163	97.845	43.886	74.826
23	73.6	0.699	0.114	68.396	30.677	52.306
22	70.4	0.547	0.089	53.53	24.009	40.937
21	67.2	0.463	0.075	45.331	20.332	34.666
20	64	0.405	0.066	39.667	17.792	30.335
19	60.8	0.361	0.059	35.321	15.842	27.011
18	57.6	0.325	0.053	31.772	14.25	24.297
17	54.4	0.294	0.048	28.752	12.896	21.988
16	51.2	0.267	0.043	26.106	11.709	19.965
15	48	0.243	0.039	23.737	10.646	18.152
14	44.8	0.221	0.036	21.578	9.6781	16.501
13	41.6	0.200	0.032	19.584	8.7838	14.977
12	38.4	0.181	0.030	17.722	7.9486	13.553
11	35.2	0.163	0.027	15.966	7.1611	12.21
10	32	0.146	0.024	14.297	6.4126	10.934
9	28.8	0.130	0.021	12.7	5.6962	9.7123
8	25.6	0.114	0.019	11.162	5.0064	8.536
7	22.4	0.099	0.016	9.6726	4.3384	7.397
6	19.2	0.084	0.014	8.2233	3.6883	6.2887
5	16	0.0695	0.011	6.8065	3.0529	5.2052
4	12.8	0.0553	0.009	5.4157	2.429	4.1416
3	9.6	0.0413	0.006	4.0449	1.8142	3.0933
2	6.4	0.0275	0.004	2.6886	1.2059	2.0561
1	3.2	0.0137	0.002	1.342	0.6019	1.0262
Total				601.7	269.9	460.11

highest ductility ratio value of 10, leading to the lowest base shear value among the frames. On the other hand, PBPD-IO is designed with a ductility ratio of 2.5, resulting in the largest base shear value when compared to the other frames. To account for the P-delta effect, PBPD frames incorporate the multiplication of respective drift limits by the lumped weight. This approach proves beneficial in compensating for the reduction of base shear due to the ductility ratio (μ).

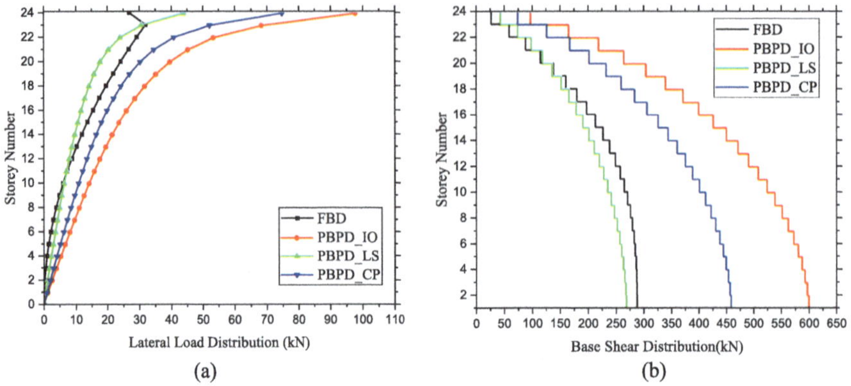

Fig. 4 **a** Lateral load. **b** Story shear distribution curve for study frame

4 Design of Sections

The selected frames were subjected to lateral loads at all the joints across all bays, with gravity loads acting in the vertical direction. For the force-based design (FBD) frame, base shear, beam, and column moments were calculated. In the performance-based plastic design (PBPD) frames, beam and column moments were analytically calculated based on defined yield mechanisms and plastic theory. Columns were considered non-yielding members, while beams were treated as yielding members to ensure that failure would occur in beams first. The principle of virtual work was applied to compute beam and column moments. Notably, the column moments in PBPD frames were found to be higher than the beam moments, indicating a 'weak beam-strong column' condition, as justified here. Additionally, the moment distribution curves for all PBPD frames exhibited a more uniform distribution when compared to the FBD frames. Subsequently, the sections were designed in accordance with the provisions outlined in [11, 12], and efforts were made to adopt optimum sections for all frames. The PBPD-IO frame required the heaviest section due to the higher section modulus requirement. Details of moment distribution and section design are presented in Figs. 5 and 6, as well as Tables 2 and 3.

5 Seismic Performance Evaluation

The seismic performance of the study frame is assessed through a nonlinear pushover analysis, which takes into account lateral load distribution and higher mode effects as recommended in [8]. The performance point is determined using the procedure outlined in [13]. Lateral and gravitational loads are applied to the study frame as specified in [6]. To evaluate the strength and deformation of the study frames, plastic hinges are strategically placed. For beam and column sections, one hinge is positioned

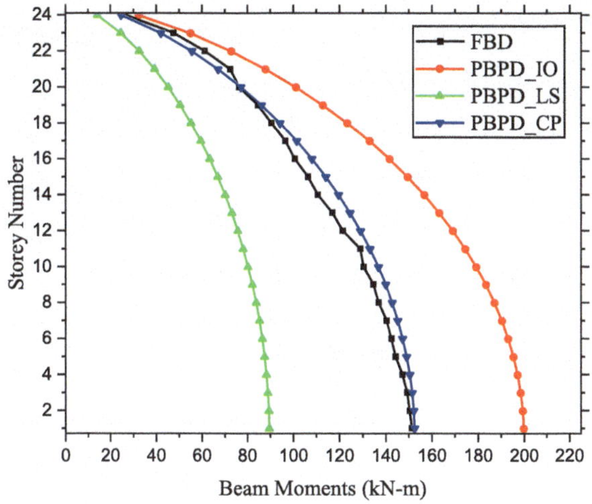

Fig. 5 Beam moment curves

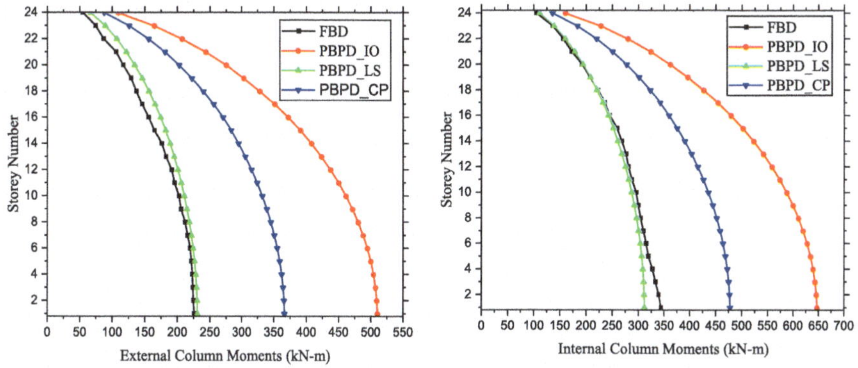

Fig. 6 External and internal moment distribution curve

at a distance of 0.1 times the length from one end and 0.9 times the length from the other end. The performance point of the study frames is identified by examining the base shear-displacement curve, as depicted in Fig. 7, and obtained through pushover analysis as described in [15]. The base shear values obtained at the performance point of the pushover curve are 1040.43 kN, 573.20 kN, and 574.15 kN for the PBPD-IO, PBPD-LS, and PBPD-CP frames, respectively. These values significantly exceed the base shear obtained through the PBPD procedure, indicating that the frames designed using the PBPD method are much stiffer than initially expected (Tables 5, 6 and 7).

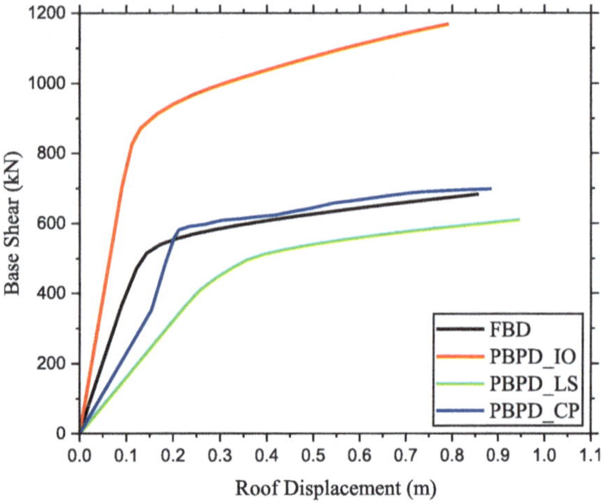

Fig. 7 Pushover curve for the study frames

Table 5 Section details of beams for FBD and PBPD frames

Floor level	FBD frames		PBPD-IO		PBPD-LS		PBPD-CP	
	Beam size (mm)	A_{st} (mm^2)	Beam size (mm)	A_{st} (mm^2)	Beam size (mm)	A_{st} (mm^2)	Beam size (mm)	A_{st} (mm^2)
21–24	300 × 400	1256.63	350 × 550	1900.66	230 × 400	2280.80	250 × 450	1570.79
11–20	300 × 400	1570.79	350 × 550	1900.66	230 × 400	1900.66	250 × 450	1900.66
1–10	300 × 400	1884.96	400 × 600	2280.80	300 × 400	1884.96	300 × 500	1884.96

Note 10 mm ϕ @ 150 mm c/c shear reinforcement is provided in all beams

Table 6 Section details of exterior columns for FBD and PBPD frame

Floor level	FBD frames		PBPD-IO		PBPD-LS		PBPD-CP	
	Column size (mm)	P_t (%)	Column size (mm)	P_t (%)	Column size (mm)	P_t (%)	Column size (mm)	P_t (%)
21–24	600 × 600	0.8	550 × 550	1.05	500 × 500	0.8	550 × 550	0.8
11–20	600 × 600	2.61	750 × 750	1.22	550 × 550	2.01	600 × 600	1.05
1–10	600 × 600	5.77	800 × 800	1.32	600 × 600	2.75	700 × 700	2.03

Note 10 mm ϕ @ 170 mm pitch is provided for columns

5.1 Assessment of Strength

The strength assessment of all the study frames is conducted using the actual response reduction factor (R). This factor describes the structure's energy dissipation capacity

Table 7 Section details of interior columns for FBD and PBPD frames

Floor level	FBD frames		PBPD-IO		PBPD-LS		PBPD-CP	
	Column size (mm)	P_t (%)	Column size (mm)	P_t (%)	Column size (mm)	P_t (%)	Column size (mm)	P_t (%)
21–24	600 × 600	0.8	550 × 500	0.95	500 × 500	0.8	550 × 550	0.8
11–20	600 × 600	2.72	750 × 750	3.01	550 × 550	2.01	600 × 600	1.59
1–10	600 × 600	3.08	800 × 800	5.36	600 × 600	2.72	700 × 700	4.43

Note 10 mm ϕ @ 160 mm pitch is provided for columns

due to inelastic behavior during an earthquake. The method for assessing strength using R is detailed in [13], where the factor is calculated by multiplying three factors: R_s, R_μ, and R_R, representing the strength factor, ductility factor, and redundancy factor, respectively The strength factor, known as R_s, is calculated by dividing the performance point base shear (V_o) by the design seismic base shear (V_B). The ductility reduction factor (R_μ) is calculated by comparing the maximum displacement to the yield displacement, following the method proposed in [9]. In this assessment, the redundancy factor (R_R) is considered to be unity. The calculations for the R factor are presented in Table 8.

$$R = R_s R_\mu R_R \tag{9}$$

The results above indicate that all frames have an R-value greater than 1, signifying that each frame possesses some degree of reserve strength. Notably, the PBPD-IO frame exhibits the highest reserve strength, as it is designed for a target drift of only 1% and features the heaviest section among all conditions. However, this design makes it less cost-effective. Conversely, the PBPD-CP frame, designed for a maximum target drift of 4%, possesses the least reserve strength. It's worth noting that the PBPD-LS frame, designed for a 2% drift, has an R-value closest to that of the FBD frames. This suggests that, among all the PBPD conditions, the life safety (LS) condition offers superior seismic performance.

Table 8 Calculation of response reduction factor (R) for FBD and PBPD frames

Frame type	V_o (kN)	V_B (kN)	R_s	R_μ	R_R	R
FBD	619.54	287.9	2.152	2.37	1	5.10
PBPD_IO	1040.43	601.66	1.729	5.65	1	9.81
PBPD_LS	573.20	269.86	2.124	2.32	1	4.92
PBPD_CP	574.15	460.11	1.248	2.76	1	3.44

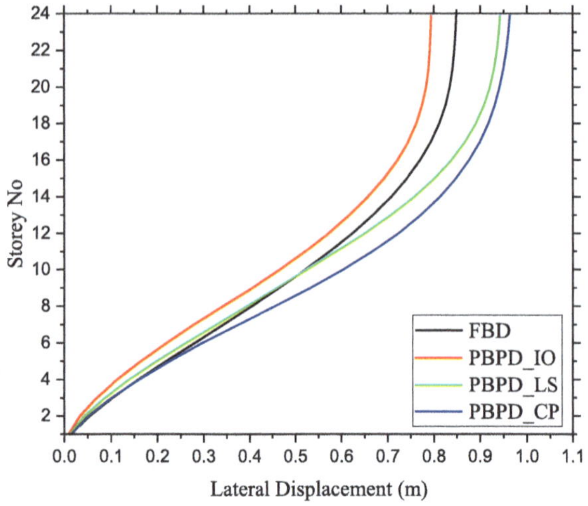

Fig. 8 Maximum lateral displacement curve for the study frame

5.2 *Deformation Assessment*

The deformation of the selected study frames is represented by the maximum lateral displacement curve obtained after the pushover analysis, as depicted in Fig. 8. It is evident that the PBPD-CP frame experiences the maximum lateral displacement of 0.9 m, in line with its design for a maximum inelastic drift of 3.6%. In contrast, the PBPD-IO frame exhibits the least displacement at 0.76 m, as it is designed for a 0.6% inelastic drift. Interestingly, both the FBD and PBPD-LS frames display similar maximum displacements, measuring 0.83 m and 0.88 m, respectively. However, it is important to note that the PBPD-LS frame is designed for a 1.6% inelastic drift, whereas the FBD frame is designed only up to the elastic limit. This suggests that, among all the frames, the PBPD-LS frame exhibits the most favorable seismic performance.

6 Conclusion

A 24-storied RC-MRF frame was analyzed and designed by FBD and PBPD methods. In FBD method, all parameters were calculated analytically and validated by preparing mathematical model. The P-delta effect was considered as a secondary effect in FBD method, whereas parameters of PBPD frames with all three conditions were analyzed as per [6]. The P-delta effect was considered at the initial stage of analysis. Later to design the frames, guidelines of [11, 12] were adopted from which the following observations were carried out.

- The PBPD-IO frame is the most uneconomical section among all PBPD conditions because it is designed for maximum base shear of 601.7 kN and has heaviest sections.
- The PPBD-IO and PBPD-CP frames are designed for 25 and 14.3% higher section modulus than FBD frame while PBPD-LS frame is designed for almost 1.8% lesser section modulus than FBD frame which means that PBPD-LS is the most economical section in nature.
- On an average moment, capacity of column is 3.3 times of beam moments in PBPD method which means that PBPD method justifies weak beam strong column concept.
- As discussed earlier, the strength and deformation characteristics of PBPD-LS frame are within defining limits which means that PBPD-LS has better seismic performance.

The above-mentioned results clearly represent that the frame designed with suitable target drift in PBPD method shows better seismic performance but there is a need to investigate the effect of pinched hysteresis behavior in nonlinear region. Similarly, PBPD method is suitable for 2D frames. The effect of adjacent column on seismic performance is still unidentified. So a further research is required in PBPD method of performance-based analysis.

References

1. Goel SC et al (2009) Performance-based plastic design (PBPD) method for earthquake-resistant structures: an overview. Struct Design Tall Spec Build 19(1–2):115–137. https://doi.org/10.1002/tal.547
2. Goel SC, Chao SH (2008) Performance-based plastic design: earthquake-resistant steel structures. International Code Council
3. Liao WC et al (2012) Performance-based plastic design and energy-based evaluation of seismic resistant RC moment frame. J Mar Sci Technol 20(3):9
4. Qammer SS et al (2019) Displacement-Based design of RC frames using design spectra of Indian Code and its seismic Performance evaluation. J Inst Eng (India) Ser A 100(3):367–379. https://doi.org/10.1007/s40030-019-00373-z.
5. Qammer SS et al (2018) Seismic performance evaluation of RC frames designed by PBPD method attuned with indian code of practice. In: 16th Symposium on earthquake engineering. Indian Society of Earthquake Technology, Roorkee, p 12
6. Dalal SP et al (2021) Strength, deformation and fragility assessment of reinforced concrete moment resisting frame designed by force based design and the performance based plastic design method for seismic loads. Structures 29:1154–1164. https://doi.org/10.1016/j.istruc.2020.11.029
7. Chao S-H et al (2007) A seismic design lateral force distribution based on inelastic state of structures. Earthq Spectra 23(3):547–569. https://doi.org/10.1193/1.2753549
8. Chopra AK et al (2001) A modal pushover analysis procedure for estimating seismic demands for buildings. Earthq Eng Struct Dynam 31(3):561–582. https://doi.org/10.1002/eqe.144
9. Newmark NM, Hall WJ (1982) Engineering monographs on earthquake criteria, structural design, and strong motion records. Earthquake Engineering Research Institute, Berkeley
10. Indian Standard, I. S. (2016) Criteria for earthquake resistance design of structures. Part-I. Bureau of Indian Standards, New Delhi, India

11. IS 456:2000. Plain and reinforced code of practice (Fourth Revision). Bureau of Indian standard, New Delhi
12. BIS, IS 13920 ductile detailing of reinforced concrete structures subjected to seismic forces–code of practice. Bureau of Indian Standards, New Delhi (India)
13. Biradar BB et al (2022) Comparative study and performance evaluation of steel moment resisting frames design with: force-based design and performance-based plastic design. Structures 43:696–709. https://doi.org/10.1016/j.istruc.2022.07.001
14. American Society of Civil Engineers (2017) Seismic evaluation and retrofit of existing buildings. American Society of Civil Engineers
15. Rojahn C et al (1995) ATC 19 structural response modification factors. Applied Technology Council, Redwood City, CA, USA
16. FEMA 356, FEDERAL EMERGENCY (2000) Prestandard and commentary for the seismic rehabilitation of buildings. Federal Emergency Management Agency, Washington, DC, USA
17. Computers and Structures (2016) Software verification examples. CSI, America
18. ETABS Version 20 (2020) Structural and earthquake engineering software, computers and structures, America

Establishing Process-Structure Linkages Using Generative Adversarial Networks

Mohammad Safiuddin, CH. Likith Reddy, Ganesh Vasantada, C. H. J. N. S. Harsha, and S. Gangolu

Abstract The microstructure of material strongly influences its mechanical properties and the microstructure itself is influenced by the processing conditions. Thus, establishing a Process-Structure-Property relationship is a crucial task in material design and is of interest in many engineering applications. We develop a GAN (Generative Adversarial Network) to synthesize microstructures based on given processing conditions. This approach is devoid of feature engineering, needs little domain awareness, and can be applied to a wide variety of material systems. Results show that our GAN model can produce high-fidelity multiphase microstructures that have a good correlation with the given processing conditions.

Keywords Generative adversarial network · Conditional image synthesis · Processing-structure relationship · Deep learning · Microstructure

1 Introduction

A significant emphasis in the design of advanced material systems is the delineation of process-structure-property relationships [1]. Although analytical and statistical methods for the design of some materials have been successfully used [2], the underlying assumptions of homogeneity restrict their ability to generalize to other material systems. As solutions to these challenges, machine learning and data-driven approaches have piqued interest in the field of material science. Microstructure reconstruction is important in computational material design because it provides an effective method for understanding the high-dimensional microstructure space. Deep learning has been used in the past to model material properties from microstructures [3] and to synthesize microstructures with desirable properties [4, 5].

M. Safiuddin (✉) · CH. Likith Reddy · G. Vasantada · C. H. J. N. S. Harsha
National Institute of Technology Calicut, Kozhikode, Kerala, India
e-mail: srinu@iitk.ac.in

S. Gangolu
Department of Materials Science and Engineering, Indian Institute Of Technology Kanpur, Kanpur, India

S. Sidhardh et al. (eds.), *Advances in Structural Integrity for Mechanical, Civil, and Aerospace Applications*, Lecture Notes in Mechanical Engineering,
https://doi.org/10.1007/978-981-97-6367-2_39

Deep learning-based microstructure reconstruction relies heavily on generative models like variational autoencoders (VAE) [6] and generative adversarial networks (GAN) [7]. VAE was used by Cang et al. [4] to create two-phase microstructures, demonstrating the ability of convolutional networks to model material characteristics from microstructures. To synthesize microstructures, Yang et al. [5] used deep convolutional GAN and transfer learning to enhance structure-property predictions.

The aforementioned line of work does not account for two aspects of the problem at hand. Previous research has concentrated on two-phase microstructures, while multiphase microstructures are present in many material systems. Characterization and generation of multiphase microstructures have been studied rarely, owing to the difficulty of determining the higher order correlation of multiphase materials. Also, recent research has not considered the impact of processing conditions on the microstructure. This is a critical component of material design since the interest is to not only design an ideal microstructure but also to figure out how to produce it.

We overcome these two problems by creating a GAN architecture capable of synthesizing multiphase alloy microstructures from processing conditions and illustrate this approach by using the Ultra High Carbon Steel Database [8]. This is the first attempt, to our knowledge, to represent the processing-structure relationship as a multi-conditional image synthesis problem.

2 Methodology

2.1 The Ultra High Carbon Steel Database (UHCSDB)

The UHCS database contains 961 microstructures produced by scanning electron microscopy (SEM) of samples with similar compositions but different processing conditions. These steels have a carbon content of 1–2.1% per cent, and the steel grade used in this dataset is Adamite grade steel (AS100). Due to the increased carbon concentration of these alloys in comparison to standard steels, pro eutectoid cementite (Fe_3C) produces a carbide network coupled with the grain boundaries of the high-temperature austenite process and is a distinguishing microstructure feature of these alloys. These brittle, powerful carbides contribute to the well-known strength and wear resistance of ultra-high carbon steel [9]. Various microconstituents found in the dataset are shown in Fig. 1.

Out of the 961 images present, 598 images have metadata on processing conditions in the form of three parameters, i.e. annealing temperature, annealing time, and the cooling method. It is evident from Fig. 2 that spheroidite is the major micro constituent in many classes. This imbalance poses an issue to the GAN which can be solved by eliminating samples from the majority class (under-sampling) or adding more examples from the minority class (over-sampling). Over-sampling can cause overfitting while under-sampling can cause a loss of information. Therefore, when sampling from an unbalanced dataset, we rebalance the class distributions and mit-

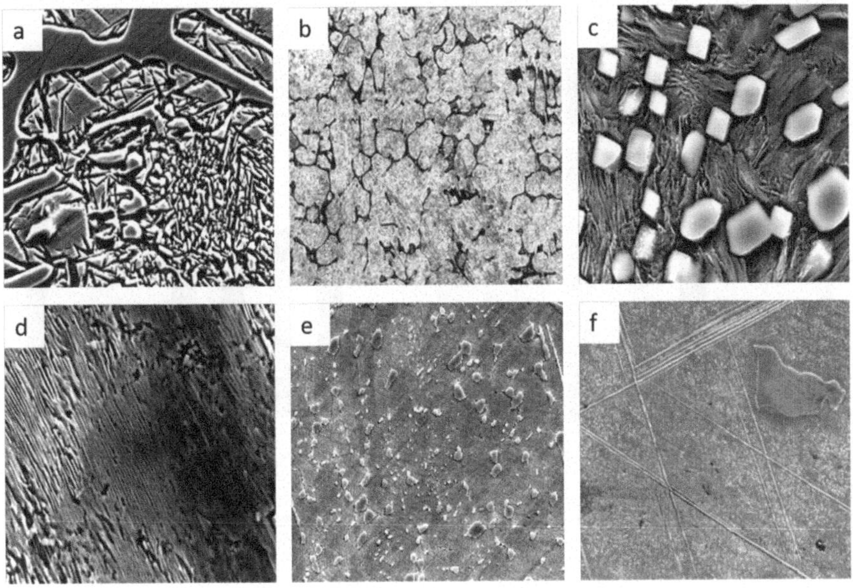

Fig. 1 Primary microstructure constituents in the UHCS dataset: **a** Pearlite, **b** Cementite Network, **c** Spheroidite, **d** Pearlite + Spheroidite, **e** Widmanstätten, **f** Martensite

igate overfitting using data augmentations. Owing to the small size of the dataset, using a deep learning model like GAN effectively is a challenging task. We can overcome this to an extent using DiffAugment [10] with Colour, Translation, and Cutout policy.

2.2 *Learning Framework*

To learn underlying latent distributions from high-dimensional data, such as images, GANs formulate the learning problem as a two-player zero-sum game between a generator which tries to produce synthetic images indistinguishable from real ones, and a discriminator which tries to differentiate if an image is real or has been synthesized. Conditional generative models also offer an effective way to better manage features in synthesized samples by imposing additional structure onto the GAN latent space. Conditional GAN [11] suggests an adaptation of this technique by supplying side details (class labels) to both the generator and discriminator. This increases the consistency and variety of the synthesized images. Following Goodfellow et al.'s [7] introduction of GAN, several improvements have been proposed to promote consistent training and rapid convergence. An overview of the GAN framework is shown in Fig. 3.

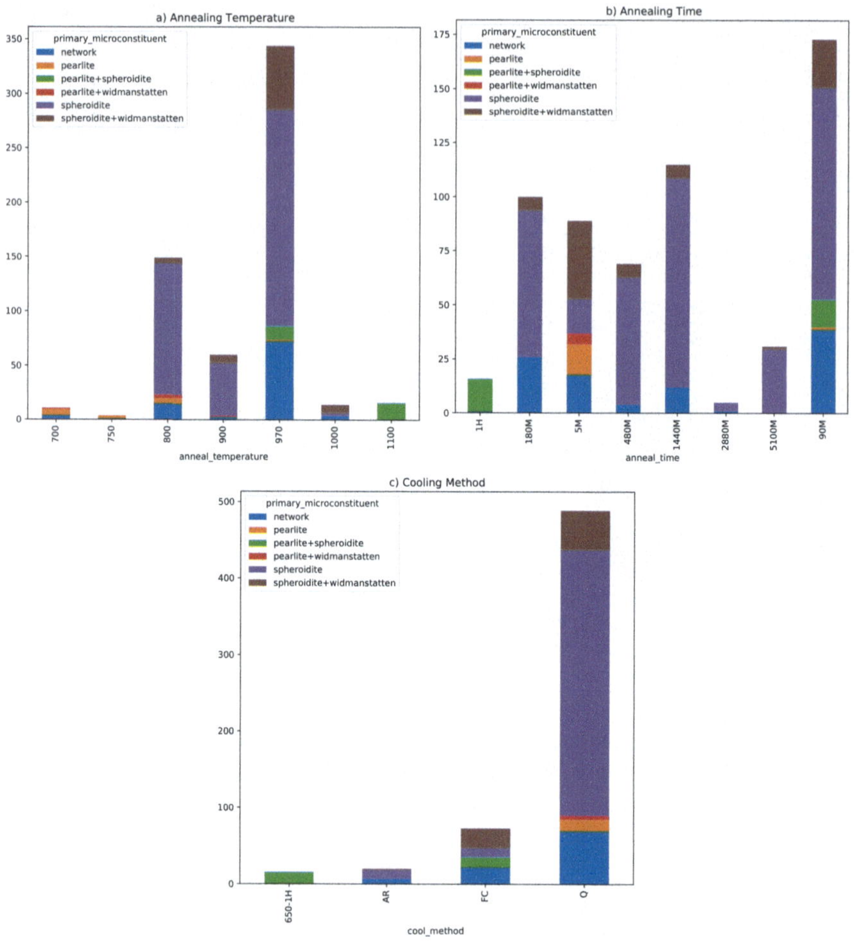

Fig. 2 Stacked bar plots for process parameters. The imbalance in the data set can be seen

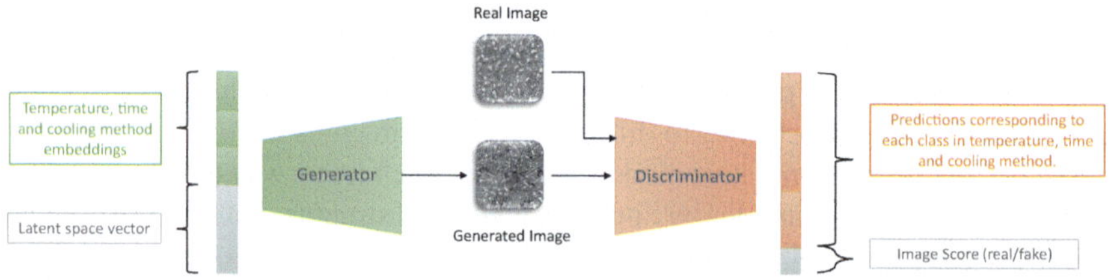

Fig. 3 Overview of the GAN framework used to synthesize microstructures based on processing conditions

Architecture BigGAN [12] improves efficiency by introducing general architectural improvements for better scalability and by modifying a regularization scheme to improve conditioning. This current work builds on the BigGAN-deep architecture to create microstructures from specified processing parameter values. The generator's input mechanism was modified to accommodate three conditional inputs as opposed to just one. The discriminator was modified to output a 21-dimensional vector (refer to Loss function) as output instead of a scalar. The aforementioned modifications are described in more detail in Sect. 6.

Loss function We use Omni loss [13] instead of hinge-loss used in the BigGAN paper [12]. Loss function can be elucidated as follows.

Let $\hat{y}$ and y denote a classifier output and the corresponding multi-label vector respectively. Assume you have K positive labels and L negative labels. Then $\hat{y}$ is a score vector with $K + L$ dimensions. The omni-loss is defined as

$$\mathcal{L}_{\text{omni}}(\hat{\boldsymbol{y}}, \boldsymbol{y}) = \log\left(1 + \sum_{i\in\mathbb{I}_{\text{neg}}} e^{\hat{y}_i}\right) + \log\left(1 + \sum_{j\in\mathbb{I}_{\text{pos}}} e^{-\hat{y}_j}\right) \tag{1}$$

$\mathbb{I}_{\text{neg}}$ contains indexes of negative scores (i.e. $|\mathbb{I}_{\text{neg}}| = L$), and $\mathbb{I}_{\text{pos}}$ contains indexes of positive scores (i.e. $|\mathbb{I}_{\text{pos}}| = K$). $\hat{y}_k$ represents the element k of vector $\hat{y}$. Using Eq. 1 we can define the loss for both the generator and discriminator.

The discriminator loss is composed of two parts, one for x_{real} (from the training set) and another for x_{fake} (from the generator). The multi-label vector associated with x_{real} is given by

$$\boldsymbol{y}_{\text{real}} = [\underbrace{-1,\ldots,1_{\text{gt}_{\text{temp}}},\ldots,-1}_{C_{\text{temp}}},\underbrace{-1,\ldots,1_{\text{gt}_{\text{time}}},\ldots,-1}_{C_{\text{time}}},\underbrace{-1,\ldots,1_{\text{gt}_{\text{cool}}},\ldots,-1}_{C_{\text{cool}}},\underbrace{1_{\text{real}},-1}_{2}] \tag{2}$$

$C_{\text{temp}}, C_{\text{time}}, C_{\text{cool}}$ are the number of classes in anneal temperature, anneal time, and cooling method, respectively. The dimension of $\boldsymbol{y}_{\text{real}}$ is given by $C = C_{\text{temp}} + C_{\text{time}} + C_{\text{cool}} + 2 = 7 + 8 + 4 + 2 = 21$. The value of $1_{\text{gt}_{\text{temp}}}$, $1_{\text{gt}_{\text{time}}}$ and $1_{\text{gt}_{\text{cool}}}$ is 1 if the index in the vector is equal to the ground truth label of x_{real}, else -1. 1 is used to denote that the corresponding score belongs to the positive set and -1 is used for the negative set. The multi-label vector for x_{fake} is also a 21-dimensional vector and is given by

$$\boldsymbol{y}_{\text{fake}} = [\underbrace{-1,\ldots,-1,\ldots,-1}_{C_{\text{temp}}},\underbrace{-1,\ldots,-1,\ldots,-1}_{C_{\text{time}}},\underbrace{-1,\ldots,-1,\ldots,-1}_{C_{\text{cool}}},\underbrace{-1,1_{\text{fake}_2}}] \tag{3}$$

From Eqs 1, 2, 3, we can define discriminator loss as follows

$$\mathcal{L}_D = \mathbb{E}_{\boldsymbol{x}_{\text{real}}\sim p_{\text{d}}}\left[\mathcal{L}_{\text{omni}}\left(D(\boldsymbol{x}_{\text{real}}), \boldsymbol{y}_{\text{real}}\right)\right] + \mathbb{E}_{\boldsymbol{x}_{\text{fake}}\sim p_{\text{g}}}\left[\mathcal{L}_{\text{omni}}\left(D(\boldsymbol{x}_{\text{fake}}), \boldsymbol{y}_{\text{fake}}\right)\right] \tag{4}$$

The generator intends to deceive the discriminator into thinking the samples are real. As a result, the multi-label vector is

$$y_{\text{fake}}^{(G)} = [\underbrace{-1, \ldots, 1_{G_{\text{temp}}}, \ldots, -1}_{C_{\text{temp}}}, \underbrace{-1, \ldots, 1_{G_{\text{time}}}, \ldots, -1}_{C_{\text{time}}}, \underbrace{-1, \ldots, 1_{G_{\text{cool}}}, \ldots, -1}_{C_{\text{cool}}}, \underbrace{1_{\text{real}}, -1}_{2}] \tag{5}$$

The value of $1_{G_{\text{temp}}}$, $1_{G_{\text{time}}}$ and $1_{G_{\text{cool}}}$ is 1 if the index in the vector is equal to the label used by the generator to generate x_{fake}, else -1. The generator loss is given by

$$\mathcal{L}_G = \mathbb{E}_{\boldsymbol{x}_{\text{fake}} \sim p_g}\left[\mathcal{L}_{\text{omni}}\left(D(\boldsymbol{x}_{\text{fake}}), \boldsymbol{y}_{\text{fake}}^{(G)}\right)\right] \tag{6}$$

Training We use Adam optimizer [14] with a learning rate of 0.0002, $\beta_1 = 0.9$ and $\beta_2 = 0.999$ for both the generator and discriminator. Additionally, the weight decay of the generator and discriminator is set to 0.001 and 0.0005, respectively. Model is trained for 6000 epochs with a batch size of 96 and a $D : G$ ratio of $4 : 1$. The implementation was done using Pytorch [15] and it was trained on the free GPU available on Kaggle.[1]

3 Results

After adequate training, the GAN model's generator component generates microstructures that visually mimic the intricacy of the original microstructures with surprising realism. While the GAN-generated microstructures and the original data appear to be extremely comparable, a closer examination reveals that the original microstructures are more variable than the synthetic microstructures. This is consistent with observations demonstrating that GANs produce less variance than the data used to train them [16]; they learn the average structure more quickly than the outliers. Nonetheless, the created microstructures resemble those in the training set. Also, It is evident from Fig. 4 that there is a strong correlation between the processing parameters and the generated microstructures. The characteristics of the generated microstructures that are worth noting are described in Discussion.

Notation for processing conditions Let us consider: 800-5-Q, where the anneal temperature is 800 °C, anneal time is 5 min and the cooling method is Quenching. Cooling methods present in the dataset are Quenching, Air Cooling, Furnace Cooling, and constant heating at 650 °C for 1 h which are represented as Q, AR, FC, 650-1H, respectively.

[1] www.kaggle.com.

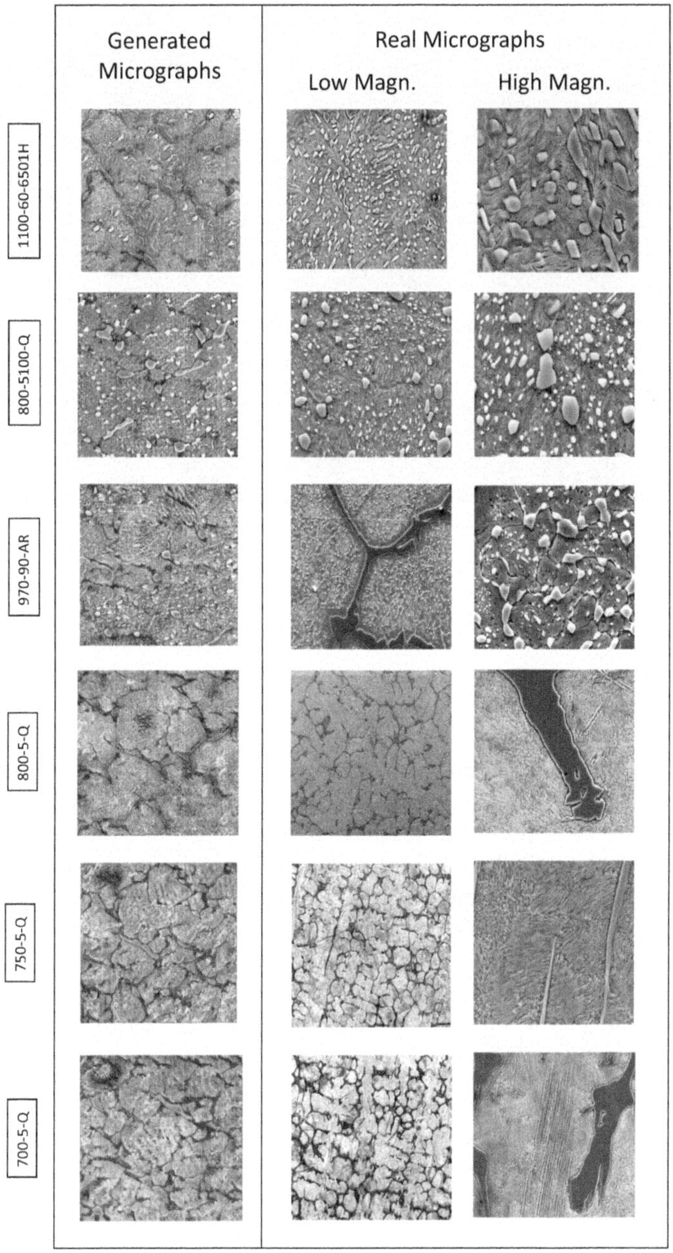

Fig. 4 A visual comparison of generated microstructures with the real ones in the dataset

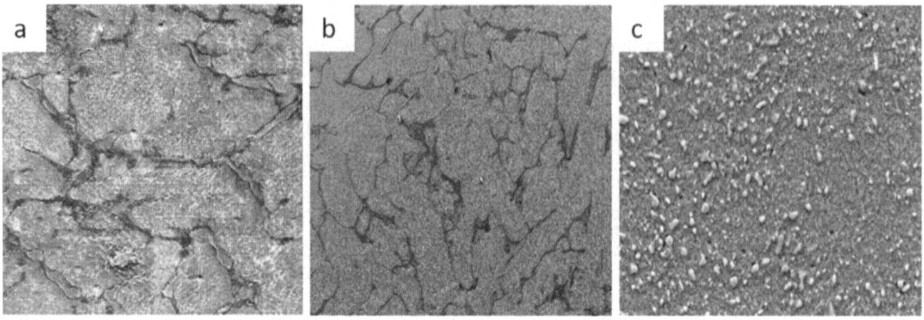

Fig. 5 Microstructures corresponding to 800-1440-Q: **a** Generated, **b** Real (Magn. 98×), **c** Real (Magn. 1964×).

Fig. 6 Generated microstructures corresponding to 1000-X-Q, where X is increased from 5 to 5100 min.

4 Discussion

Effect of Magnification The training data consists of images with varying magnifications. The current GAN model peculiarly adapts to this. The relative scale of a certain feature concerning other features in a generated microstructure is similar to the microstructures in the dataset. Let us consider the case represented in Fig. 5. Spheroidite grains can be seen in Fig. 5c. However, there is a network structure in Fig. 5b and the spheroidite is not visible. The generated image has a level of magnification where both spheroidite and network can be seen. The size of spheroidite grains is much smaller compared to the ones in Fig. 5c and the network web is less spread out compared to Fig. 5b. This adaptability is seen across other processing parameters as well (see Fig. 4).

Microstructure Evolution To understand if the model has learned any physically accurate patterns based on the changes in the processing conditions, let us consider the case in Fig. 6. It can be observed that, as the annealing time increases, the size of the sphere keeps on increasing by combining with nearby spheres. This phenomenon is in line with the work of Hecht et al.[17] though not completely consistent from 60 to 90 M.

5 Conclusion

We used a modified version of the BigGAN-deep architecture with a reframed Omni loss to generate microstructures under specific conditions. We demonstrate the fidelity of the generated microstructures and their correlation with the processing methods used. The GAN model, in our opinion, offers the following advantages:

- The GAN framework can be used to model any type of material system.
- The generator once trained can generate an arbitrarily large number of unique microstructures.
- Although the GAN model needs a substantial amount of computing power to train, it is quite efficient when it comes to generating synthetic microstructures.

Despite being trained on an extremely small dataset (598 images), our GAN model offers decent fidelity. We believe that the variation and quality of the images can be improved by using a large dataset. Also, the trained model can be used in transfer learning for other steel microstructure datasets.

This GAN framework can be included in a larger framework to establish Process-Structure-Property (PSP) linkages. The generated microstructures can be added to an already available dataset to improve performance for the classification of micro-constituents [18].

6 Data Availability

The data used in this work was publicly made available by DeCost et al. [8] here. The pre-trained model and code used in this work can be found here.

Appendix Architectural Details

In this work, we aim to establish a processing-structure relationship via multi-conditional image generation. BigGAN is more intuitive for multi-conditional scenarios since it provides class conditional information at multiple levels of the generator as opposed to just providing it at the initial level which is the case for typical GAN architectures. This is illustrated in Fig. 7.

As mentioned earlier, we modified the BigGAN-deep to be better suited for the task at hand. We use three embeddings for annealing temperature, annealing time, and cooling method as opposed to just one. We are not using a projection-style generator. We observed that the model collapsed fairly quickly with a default latent dimension of 128. Therefore, we increase the latent space dimension to 384 to avoid collapse

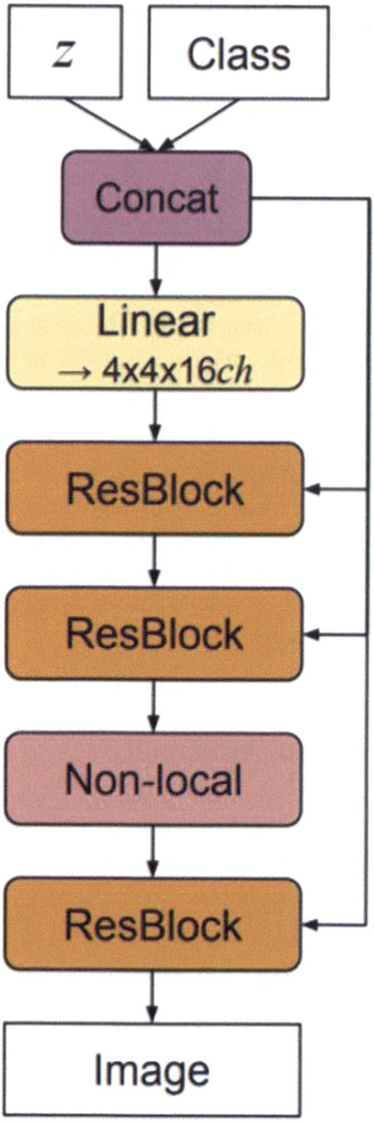

Fig. 7 Layout of the architecture for BigGAN-deep generator [12]

as increasing the latent space dimension gives the model more degrees of freedom to fit the individual samples, thereby increasing the quality of generated images. The aforementioned modifications can be observed in the Figs. 8 and 9.

$z \in R^{384} \sim \mathcal{N}(0, I)$ Embed $(y_{temp}) \in R^{128}$ Embed $(y_{time}) \in R^{128}$ Embed $(y_{cool}) \in R^{128}$
Linear $(384 + (128 \times 3)) \rightarrow 4 \times 4 \times 16ch$
ResBlock $16ch \rightarrow 16ch$
ResBlock up $16ch \rightarrow 16ch$
ResBlock $16ch \rightarrow 16ch$
ResBlock up $16ch \rightarrow 8ch$
ResBlock $8ch \rightarrow 8ch$
ResBlock up $8ch \rightarrow 8ch$
ResBlock 8ch $\rightarrow 8ch$
ResBlock up $8ch \rightarrow 4ch$
Non-Local Block (64×64)
ResBlock $4ch \rightarrow 4ch$
ResBlock up $4ch \rightarrow 2ch$
ResBlock $2ch \rightarrow 2ch$
ResBlock up $2ch \rightarrow ch$
BN, ReLU, 3×3 Conv $ch \rightarrow 1$
Tanh

Generator

Image $x \in R^{256 \times 256 \times 1}$
3×3 Conv $1 \rightarrow ch$
ResBlock down $ch \rightarrow 2ch$
ResBlock $2ch \rightarrow 2ch$
ResBlock down $2ch \rightarrow 4ch$
ResBlock $4ch \rightarrow 4ch$
Non-Local Block (64×64)
ResBlock down $4ch \rightarrow 8ch$
ResBlock $8ch \rightarrow 8ch$
ResBlock down $8ch \rightarrow 8ch$
ResBlock $8ch \rightarrow 8ch$
ResBlock down $8ch \rightarrow 16ch$
ResBlock $16ch \rightarrow 16ch$
ResBlock down $16ch \rightarrow 16ch$
ResBlock $16ch \rightarrow 16ch$
ReLU, Global sum pooling
$linear \rightarrow 21$

Discriminator

Fig. 8 Our architecture for 256 × 256 images. ch = 64 [12]

$z \in R^{128} \sim \mathcal{N}(0, I)$ Embed $(y) \in R^{128}$
Linear $(128 + 128) \rightarrow 4 \times 4 \times 16ch$
ResBlock $16ch \rightarrow 16ch$
ResBlock up $16ch \rightarrow 16ch$
ResBlock $16ch \rightarrow 16ch$
ResBlock up $16ch \rightarrow 8ch$
ResBlock $8ch \rightarrow 8ch$
ResBlock up $8ch \rightarrow 8ch$
ResBlock 8ch $\rightarrow 8ch$
ResBlock up $8ch \rightarrow 4ch$
Non-Local Block (64×64)
ResBlock $4ch \rightarrow 4ch$
ResBlock up $4ch \rightarrow 2ch$
ResBlock $2ch \rightarrow 2ch$
ResBlock up $2ch \rightarrow ch$
BN, ReLU, 3×3 Conv $ch \rightarrow 3$
Tanh

Generator

Image $x \in R^{256 \times 256 \times 3}$
3×3 Conv $1 \rightarrow ch$
ResBlock down $ch \rightarrow 2ch$
ResBlock $2ch \rightarrow 2ch$
ResBlock down $2ch \rightarrow 4ch$
ResBlock $4ch \rightarrow 4ch$
Non-Local Block (64×64)
ResBlock down $4ch \rightarrow 8ch$
ResBlock $8ch \rightarrow 8ch$
ResBlock down $8ch \rightarrow 8ch$
ResBlock $8ch \rightarrow 8ch$
ResBlock down $8ch \rightarrow 16ch$
ResBlock $16ch \rightarrow 16ch$
ResBlock down $16ch \rightarrow 16ch$
ResBlock $16ch \rightarrow 16ch$
ReLU, Global sum pooling
Embed$(y) \cdot h +$ $(linear \rightarrow 1)$

Discriminator

Fig. 9 Original BigGAN-deep architecture for 256 × 256 images. ch = 64

References

1. Olson GB (1997) Computational design of hierarchically structured materials. Science 277(5330):1237–1242. arXiv:https://science.sciencemag.org/content/277/5330/1237.full.pdf. https://doi.org/10.1126/science.277.5330.1237. https://science.sciencemag.org/content/277/5330/1237
2. Fullwood DT, Niezgoda SR, Adams BL, Kalidindi SR (2010) Microstructure sensitive design for performance optimization. Progr Mater Sci 55(6):477–562 https://doi.org/10.1016/j.pmatsci.2009.08.002. https://www.sciencedirect.com/science/article/pii/S0079642509000760
3. Cecen A, Dai H, Yabansu YC, Kalidindi SR, Song L (2018) Material structure-property linkages using three-dimensional convolutional neural networks. Acta Materialia 146:76–84. https://doi.org/10.1016/j.actamat.2017.11.053
4. Cang R, Li H, Yao H, Jiao Y, Ren Y (2018) Improving direct physical properties prediction of heterogeneous materials from imaging data via convolutional neural network and a morphology-aware generative model. Comput Mater Sci 150:212–221. https://doi.org/10.1016/j.commatsci.2018.03.074. https://www.sciencedirect.com/science/article/pii/S0927025618302337
5. Yang Z, Li X, Catherine Brinson L, Choudhary AN, Chen W, Agrawal A. Microstructural materials design via deep adversarial learning methodology. J Mech Des 140(11). https://doi.org/10.1115/1.4041371
6. Kingma DP, Welling M. An introduction to variational autoencoders. CoRR abs/1906.02691. arXiv:1906.02691
7. Goodfellow IJ, Pouget-Abadie J, Mirza M, Xu B, Warde-Farley D, Ozair S, Courville A, Bengio Y. Generative adversarial networks, NIPS'14: proceedings of the 27th international conference on neural information processing systems, vol 2. arXiv:1406.2661
8. DeCost B, Hecht M, Francis T, Webler B, Picard Y, Holm E (2017) Uhcsdb: ultrahigh carbon steel micrograph database: tools for exploring large heterogeneous microstructure datasets. In: Integrating materials and manufacturing innovation, vol 6. https://doi.org/10.1007/s40192-017-0097-0
9. DeCost B, Francis T, Holm E (2017) Exploring the microstructure manifold: image texture representations applied to ultrahigh carbon steel microstructures. Acta Materialia 133. https://doi.org/10.1016/j.actamat.2017.05.014
10. Zhao S, Liu Z, Lin J, Zhu J, Han S (2020) Differentiable augmentation for data-efficient GAN training. CoRR abs/2006.10738. arXiv:2006.10738
11. Mirza M, Osindero S (2014) Conditional generative adversarial nets. CoRR abs/1411.1784. arXiv:1411.1784
12. Brock A, Donahue J, Simonyan K (2018) Large scale GAN training for high fidelity natural image synthesis. CoRR abs/1809.11096. arXiv:1809.11096
13. Zhou P, Xie L, Ni B, Tian Q (2021) Omni-gan: on the secrets of cgans and beyond. CoRR abs/2011.13074. arXiv:2011.13074
14. Kingma D, Ba J (2015) Adam: a method for stochastic optimization. In: International conference on learning representations. arXiv:1412.6980
15. Paszke A, Gross S, Massa F, Lerer A, Bradbury J, Chanan G, Killeen T, Lin Z, Gimelshein N, Antiga L, Desmaison A, Kopf A, Yang E, DeVito Z, Raison M, Tejani A, Chilamkurthy S, Steiner B, Fang L, Bai J, Chintala S (2019) Pytorch: an imperative style, high-performance deep learning library, pp 8024–8035. arXiv:1912.01703
16. Salimans T, Goodfellow IJ, Zaremba W, Cheung V, Radford A, Chen X (2016) Improved techniques for training gans. CoRR abs/1606.03498. arXiv:1606.03498
17. Hecht MD (2017) Effects of heat treatments and compositional modification on carbide network and matrix microstructure in ultrahigh carbon steels. https://doi.org/10.1184/R1/6716156.v1. https://kilthub.cmu.edu/articles/thesis/Effects_of_Heat_Treatments_and_Compositional_Modification_on_Carbide_Network_and_Matrix_Microstructure_in_Ultrahigh_Carbon_Steels/6716156

18. Azimi SM, Britz D, Engstler M, Fritz M, Mücklich F (2018) Advanced steel microstructural classification by deep learning methods. https://doi.org/10.1038/s41598-018-20037-5

An Integrated Approach for Damage Detection, Localization and Quantification in Beam Like Structure Using Vibration Measurements

Revanth Dugalam and Guru Prakash

Abstract Cracks are the most common types of damage in beams that need to be detected early so that remedial measures can be taken on time otherwise it leads to complete structural failure. Hence, the damage detection in beam like element is important for structural integrity, long service life, and operational cost minimization. In the last few years, several damage detection and quantification methods have been developed using vibration measurements. However, most of them fail to detect the damage, quantify its magnitude and localize it accurately. In this study, an integrated approach has been proposed that will address these issues in a single framework. The proposed approach utilizes the concept of modal analysis to detect any damage in beams. Both the numerical and experimental analyses were used to determine the modal frequencies and results were compared. Moreover, the proposed approach has been utilized to localize and quantify the damage in a steel beam. The developed integrated approach demonstrated superior results on a test specimen of a cantilever beam and successfully approximated the location and magnitude of the damage.

Keywords Cantilever beam · Damage detection · Modal analysis · Natural frequencies

1 Introduction

Structural damage detection (SDD) has a paramount value in civil engineering community. SDD can be defined as the early detection of an anomaly in a structure that adversely affects the structure's current or future performance. Vibration based damage detection is an area of significant research activity recently because of its advantage of using dynamic characteristics of the structure [1]. Hence in this study vibra-

R. Dugalam · G. Prakash (✉)
Department of Civil Engineering, IIT Indore, Indore, MP 453552, India
e-mail: guruprakash@iiti.ac.in

R. Dugalam
e-mail: phd2101104004@iiti.ac.in

S. Sidhardh et al. (eds.), *Advances in Structural Integrity for Mechanical, Civil, and Aerospace Applications*, Lecture Notes in Mechanical Engineering,
https://doi.org/10.1007/978-981-97-6367-2_40

tion based damage detection technique has been adopted by utilizing the dynamic property of a structure. A cantilever beam with damage has been considered and a damage detection algorithm has been developed using vibration measurements.

Cantilever beams are most commonly used in the engineering structures such as trusses and the construction of bridges, towers and buildings. Due to some suddenly occurring and uneven operational loads, damage in the form of cracks occur in these beams. These cracks may grow and coalesce affects the performance of the entire beam or sometimes may leads to failure. Hence, this study focuses on the early detection and quantification of damage in the cantilever beams, which may help to increase the service-life of the beam. Most of the researchers utilized different dynamic characteristics to achieve this, for example Alvandi et al. [2–4] a few mode shapes and natural frequencies to identify and localize the damage in beam and plate structures. The authors applied some noisy signals to the dynamic properties in order to create real environment and found using strain energy damage detection was efficient.

Sivasubramanian et al. [5] presented a graphical method and conducted numerical studies to detect and quantify the damage and later generalized it for all the beam structures. Fan et al. [6] performed a review and comparative study on vibration based damage detection techniques on beams and plate like structures. The authors classified damage detection methods based on vibration features, natural frequency-based methods, mode shape-based methods, curvature mode shape-based methods and methods using both mode shapes and frequencies, and their advantages and disadvantages were discussed. However, not so much work has been done on cantilever beams and most of the researchers focused on numerical studies for damage detection instead of experimental studies.

Ahmad et al. [7] conducted vibration analysis on cantilever beams both analytically and experimentally. The authors estimated the first four natural frequencies by applying Fourier transform (FFT) for the vibration measurements. Rucevskis et al. [8, 9] proposed a damage detection algorithm without using healthy structure data and localized the damage using the mode shapes and natural frequencies of damaged structures. The authors introduced the smoothing polynomial concept, which seems to be interesting because it doesn't requires baseline data. The authors also proposed an algorithm for multiple damages in beam structures without using baseline data. Kannappan [10] conducted studies numerically, experimentally and formulated damage indexes using natural frequencies, mode shape and strain energy. Gunawan et al. [11] proposed a new damage indicator and sensitively detected the crack in the beams. The authors proposed the damage indicator in the time and frequency domains derived from Euler Bernoulli's beam theory.

In studies where natural frequency served as the parameter for damage detection, researchers often formulated damage indexes involving extensive mathematical calculations and intricate solutions. This paper makes the following main contributions:

(i) A robust method has been proposed that eliminates the need for complex calculations and can be easily understood and implemented.
(ii) The proposed approach has been validated through both experimental and numerical simulations, and the results have been compared.

Fig. 1 Freely vibrating system

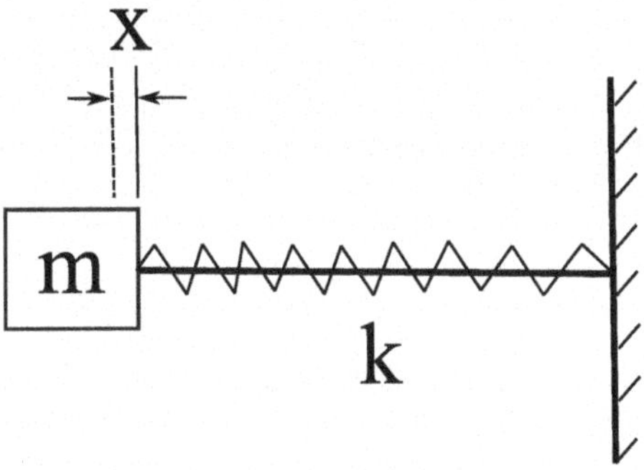

(iii) An attempt has been made using the first mode modal frequencies for the localization and quantification of damage in beam structures, which demonstrated satisfactory results.

These contributions collectively highlight the simplicity, effectiveness and practical applicability of the proposed methodology in the context of damage detection in beam structures.

2 Damage Detection Method

The idea of the proposed work is to utilize frequency relationship with the stiffness of the structure. Structural damage is considered to be a result of a change in the stiffness matrix only, i.e., there is no change in the mass matrix [12].

Consider a free vibration system Fig. 1 with mass m and stiffness k with a displacement x, the force $m\ddot{x}$ exerted by the mass on the spring is equal and opposite to the force kx applied by the spring on the mass:

$$m\ddot{x} + kx = 0 \tag{1}$$

The solution to the above equation is given by

$$x = A \sin\sqrt{\frac{k}{m}}t + B \cos\sqrt{\frac{k}{m}}t \tag{2}$$

where $\sqrt{\frac{k}{m}}$ is the angular natural frequency which is denoted by ω_n

$$\omega_n = \sqrt{\frac{k}{m}} \tag{3}$$

The sinusoidal oscillation of the mass repeats continuously, and the time interval to complete one cycle is the period (T):

$$T = \frac{2\pi}{\omega_n} \tag{4}$$

The reciprocal of the time period is the natural frequency (f_n), from Eqs. 3 and 4 :

$$f_n = \frac{\omega_n}{2\pi} = \frac{1}{2\pi}\sqrt{\frac{k}{m}} \tag{5}$$

From Eq. 5, it is clear that frequency is directly proportional to stiffness, it represents that if there is a change in stiffness, there will be a change in frequency. By using this relationship damage detection algorithm has been developed. To detect any damage that occurs along the length of the beam, a single framework has been formed using ratios of damaged frequencies to undamaged frequencies, which here we called normalized frequencies. Let ω_o are the undamaged frequencies and ω_1 are the damaged frequencies for first mode of the beam structures, now the normalized frequencies can be determined as (Fig. 2)

$$\text{Normalized frequencies} = \frac{\omega_1}{\omega_o} \tag{6}$$

3 Numerical Study

The proposed method has been validated by numerical modeling of cantilever beam of dimensions of length $l = 900$ mm and depth $d = 12$ mm shown in Fig. 3 made of steel with material properties- young's modulus of elasticity $E = 210$ GPa, poisson's ratio $\nu = 0.27$. For finding frequencies of a cantilever beam analytically, required to solve the determinant of a eigen value matrix, may sometimes be larger to solve. Numerical modeling helps us to solve determinant of larger matrices with less effort and without mistakes. Hence, numerical modeling has been adopted in this study for the proposed method (Fig. 4).

The finite element software $ANSYS$ has been adopted for finding the frequencies of the cantilever beam. The very first step after creating geometry was meshing of the cantilever beam. The formation of mesh plays an important role on the frequencies. Hence, mesh convergence study has been conducted by using the root means square (RMS) error of frequencies. We started with a coarser mesh with the certain number of elements and found frequencies of the first mode by doubling the number of elements and again found the frequencies of first mode. Now, the RMS percentage error of both the frequencies has been calculated. The process has been continued till the percentage of error is minimized (Table 1). At 52,098 elements, the error was found to be 0.0005% with a mesh size of 1.4 mm (Fig. 5). The finer mesh has been used at the location of the crack for better results.

The geometry has been fixed at one end and free at the other end, and modal analysis has been conducted to extract the frequencies of the first three modes for the cantilever beam. Subsequently, a rectangular crack (Fig. 6) has been introduced

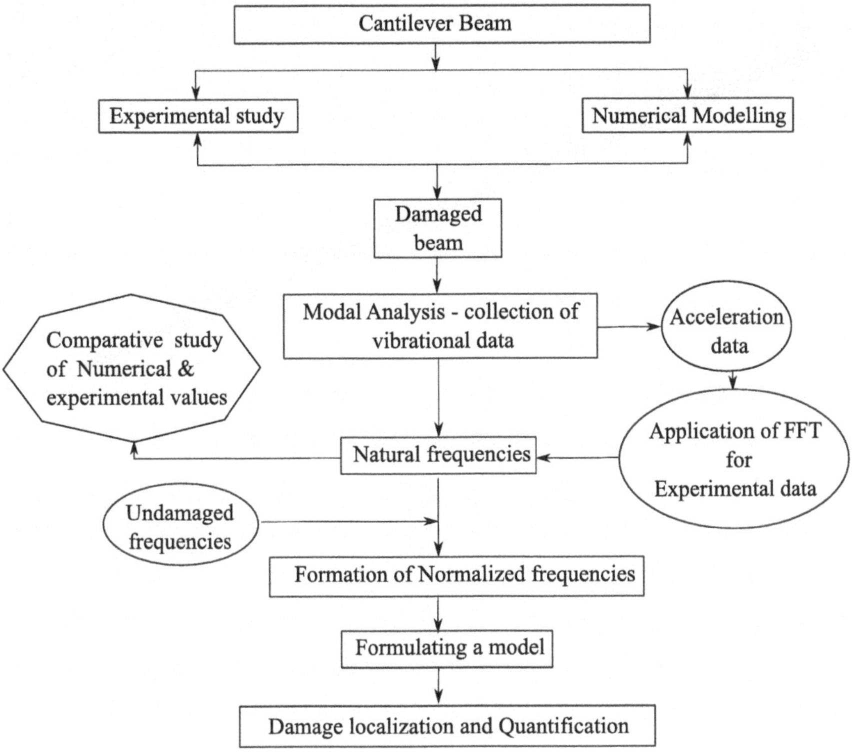

Fig. 2 Flowchart showing the steps involved in the proposed methodology

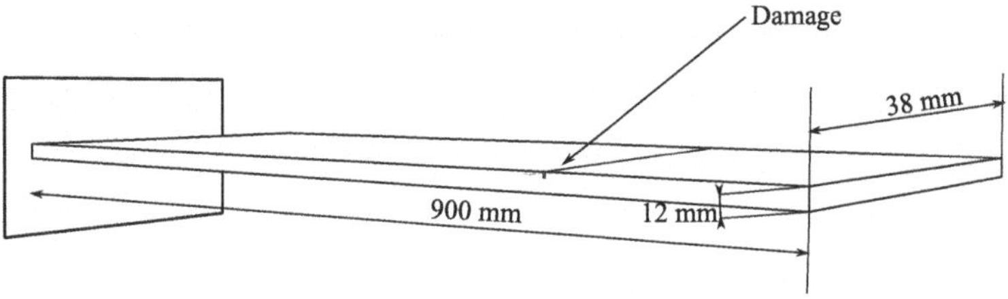

Fig. 3 Cantilever beam

to the geometry at a location 0.1 times the length of the beam (l) from the fixed end, with a crack depth of 0.3 times the depth of the beam (d) and an approximate crack width of 1 mm. Since, rectangular cracks are most common types of cracks in beam structures due to suddenly occurring loads, fatigue, etc. Frequencies of the cracked beam have been extracted. The same procedure has been repeated at the initial location with an increased crack depth of 0.4d and then with 0.5d at the same location. Similarly, cracks have been introduced at locations 0.2l, 0.3l up to 0.9l, and frequencies have been calculated. A total of 27 simulations have been done, and

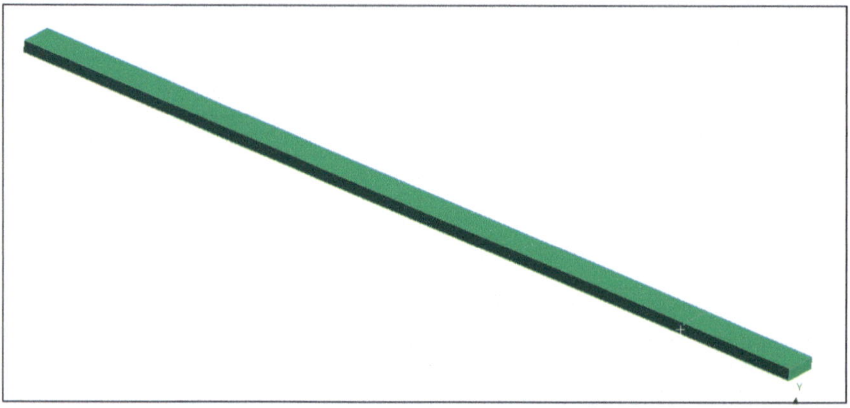

Fig. 4 Numerically modeled cantilever beam (900 mm × 38 mm × 12 mm)

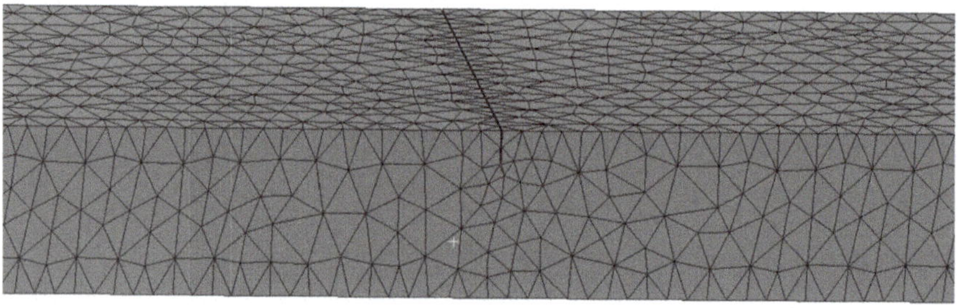

Fig. 5 Closeup view of meshing

Table 1 Mesh convergence by RMS percentage error

Number of elements	% of RMS error
92	0.29
183	0.199
308	0.349
672	0.099
1150	0.049
2142	0.0249
4392	0.0249
11528	0.0049
52098	0.0005

frequencies have been calculated. The normalized frequencies, obtained by taking the ratio of damaged frequencies to undamaged frequencies, have been presented in Sect. 5.

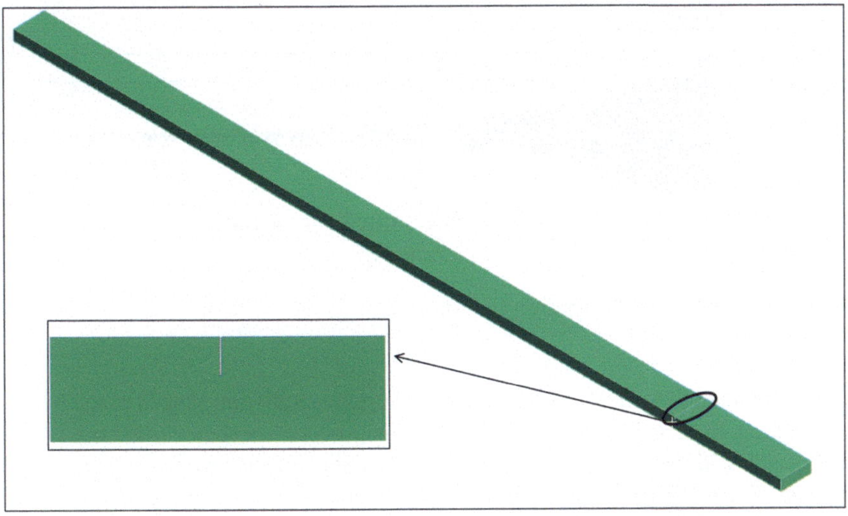

Fig. 6 Cantilever beam with rectangular crack at a distance of 810 mm from fixed end

Fig. 7 Inducing crack on the beam

4 Experimental Study

The proposed method has been validated using an experimental study to resemble real-life structural conditions. Experimental data has been obtained in the time domain and subsequently converted to the frequency domain. To capture the frequencies of the cantilever beam, a data acquisition system, specifically the MX840B, has been employed. This system features 8 universal inputs and supports over 15 transducer technologies. For the time domain data collection, an impact hammer and accelerometer have been utilized. The impact hammer, of type 8207 with a soft tip, has been employed to excite the beam, while an accelerometer sensor of type 4534-B-002 has been mounted on the beam to record the resulting vibration data. To induce a crack in the steel beam, an angle grinder equipped with a 4-inch diameter metal cutting blade (see Fig. 7) has been used.

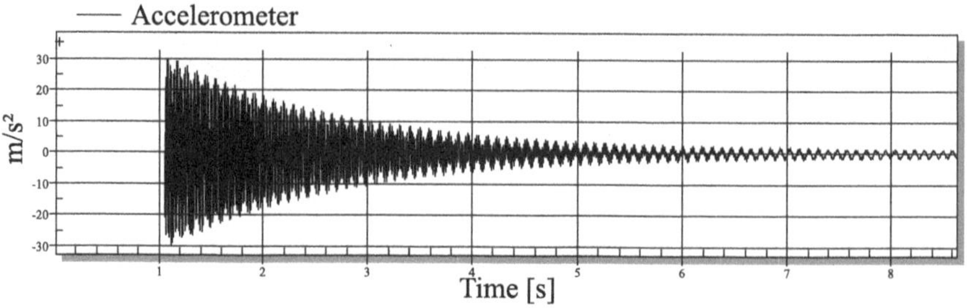

Fig. 8 Time domain signal of undamaged beam

As explained in Sect. 3, cracks have been systematically introduced to the beams one after another, as illustrated in Fig. 7b. This process has been carried out for all nine beams with varying depths of cracks. The data collected by the accelerometer has been recorded using the catmaneasy software with a sampling frequency of 1200 Hz. At this point, the data has existed in the time domain, representing time versus acceleration (Fig. 8).

The data should be in the frequency domain so that the proposed methodology would be able to apply. Hence an algorithm named Fast Fourier transform (FFT) has been used in order to get the frequencies. It is an algorithm that converts a signal from a time domain to the frequency domain. By using this algorithm, all the signals of the beams obtained from the accelerometer sensor were converted to frequency domains. For the application of FFT, computational software MATLAB has been used. For the locations of $0.2l$, $0.4l$ for all the depths of cracks, the converted signals have been presented in Fig. 9.

Similarly, frequencies have been found for all the remaining crack locations. Healthy beam frequencies of the first three modes have also been found. By taking the ratio of damaged and undamaged frequencies, normalized frequencies were found and presented in the next section (Table 2).

5 Comparative Study

The natural frequencies of both damaged and undamaged cantilever beams have been extracted using both numerical and experimental approaches. Given the significance of validating experimental results and the possibility of manual or instrumental errors, this study prioritized numerical analysis. After calculating the normalized frequencies through both approaches, a comparison was conducted by presenting the percentage error in tabular form and using a plot (see Fig. 10).

At different locations of the beam with different depths of cracks, the frequencies were changed. There was a maximum error of 2.9% at $0.3l$ length for $0.4d$ depth crack. A minimum error of 0.03% at $0.9l$ length for $0.3d$ depth crack. It was observed that more differences with experimental and numerical approaches has been

Table 2 Mode1 Normalized frequencies in hertz for different samples of the damaged cantilever beam

	Method	0.1*l*	0.2*l*	0.3*l*	0.4*l*	0.5*l*	0.6*l*	0.7*l*	0.8*l*	0.9*l*
0.3*d*	ANSYS	0.978755	0.984845	0.989873	0.993838	0.996742	0.998583	0.999645	1.000212	1.000495
	Experimental	0.961352	0.967052	0.969885	0.978017	0.981997	0.989215	0.991017	0.998938	1.000187
	Error (%)	1.7	1.8	1.9	1.6	1.5	0.9	0.86	0.12	0.03
0.4*d*	ANSYS	0.959847	0.971106	0.980525	0.988032	0.993555	0.997167	0.999221	1.000212	1.000637
	Experimental	0.939278	0.945050	0.951836	0.967033	0.972647	0.979050	0.986413	0.997836	0.998544
	Error (%)	2.06	2.6	2.9	2.09	2.09	1.81	1.28	0.24	0.21
0.5*d*	ANSYS	0.929537	0.948587	0.948587	0.978259	0.988102	0.994688	0.998442	1.000141	1.000778
	Experimental	0.928593	0.931863	0.942487	0.951476	0.970566	0.972566	0.981574	0.989567	0.999512
	Error (%)	0.09	1.67	0.6	2.7	1.75	2.2	1.12	1.05	0.12

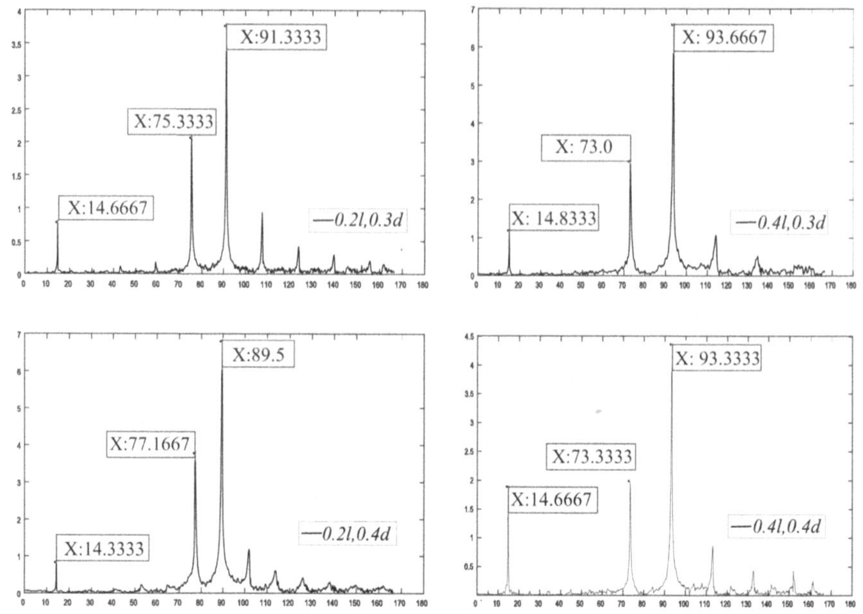

Fig. 9 Frequency domain signals of 0.2l, 0.4l location cracks

happened for 0.4*d* crack, it may be because of either manual indentation of crack or beam specimens having some improper dimensions at some locations. However, because of having less percentage of error, this study has proceeded further with the application of the proposed methodology using the normalized frequencies of first mode. Using the proposed method graphs have been plotted and these graphs approximately localized and quantified the crack which was presented in Sect. 6.

6 Results and Discussions

The graphs have been plotted in terms of crack locations (Fig. 11) and crack depths (Fig. 12). It is clear that at the same location as the depth of crack increases, the frequency values have been reduced. As the crack moved away from the fixed end, the frequency values increased and the normalized frequency was almost near 1 or more than 1. Combined the different depths of cracks for different locations by taking the average of the normalized frequencies and plotted in the form of crack locations *vs* normalized frequencies for localization and crack depth *vs* time (Fig. 13) for quantification of crack. Using these plots by knowing the normalized frequencies of any cantilever beam with dimensions used in this study, one can easily localize and quantify the crack.

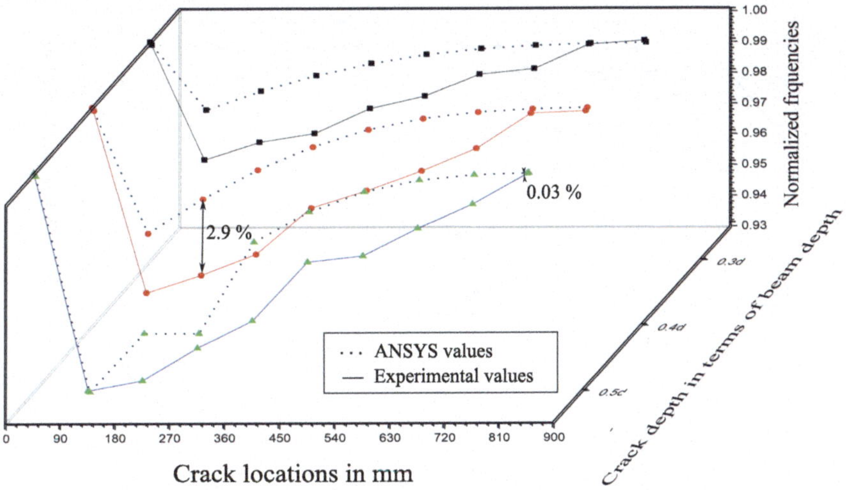

Fig. 10 Comparison of numerical and experimental values showing maximum and minimum errors

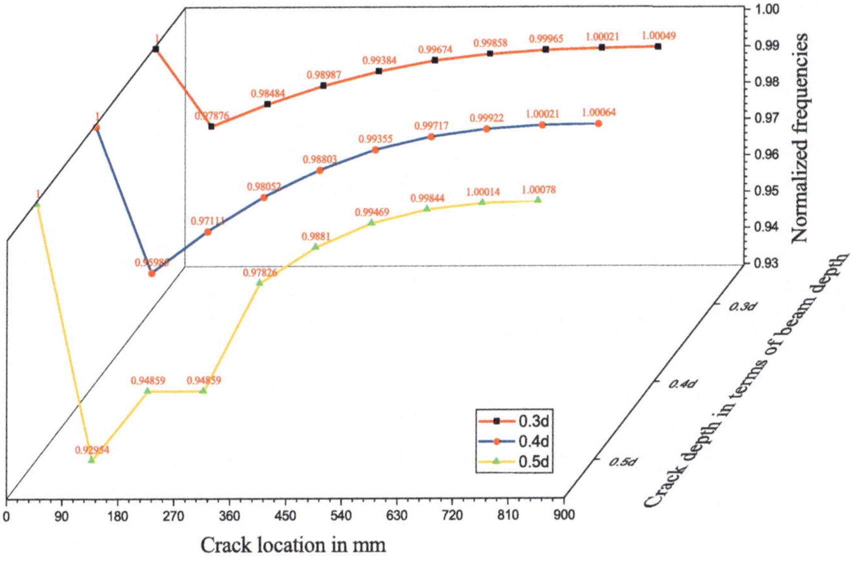

Fig. 11 Mode1 normalized frequencies in terms of crack locations

A cantilever test specimen of dimensions as per present study with the known location of the crack was taken, extracted frequencies of that specimen and tried for localization and quantification of crack using above graphs. The proposed plots were able to find the location and depth of the crack approximately.

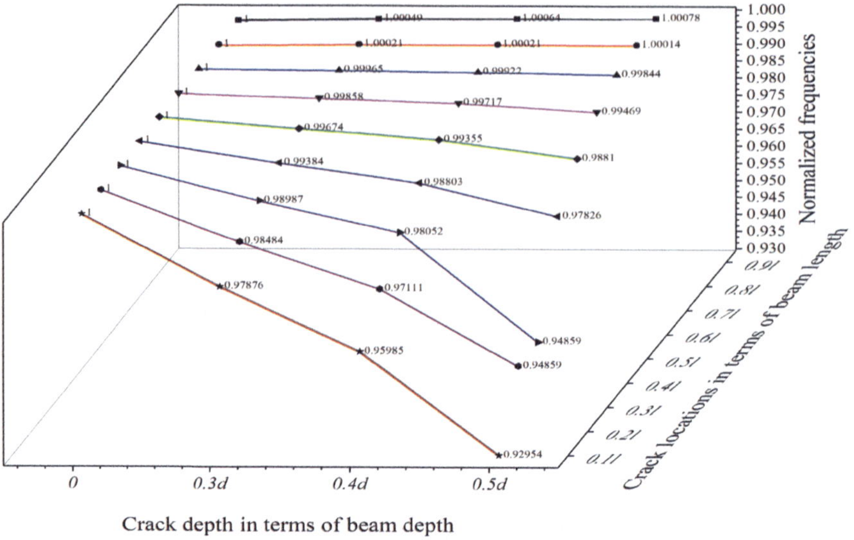

Fig. 12 Mode1 normalized frequencies in terms of crack depth

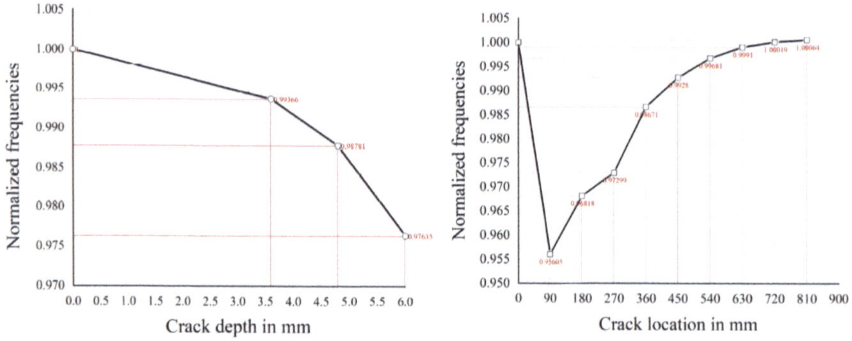

Fig. 13 Damage localization and quantification plots

7 Conclusion and Future Direction

In the present study, a robust method has been proposed for localizing and quantifying the damage. The main conclusions of this paper are as follows: (i) Numerical simulations and experimental methods were employed to extract frequencies from beam structures. The frequency values obtained through these two approaches demonstrated a high level of consistency, with a maximum error observed at 2.9%. This close agreement of results validates our approach's accuracy and reliability, reinforcing our methodology's robustness in beam structures. (ii) This study focused on utilizing only the first mode of frequencies to localize and quantify damage in beams.

The outcomes of our investigation revealed that this first modal frequencies proved to be highly effective, yielding optimal results in the context of damage detection. This approach not only simplifies the analysis but also demonstrates its potential in structural health monitoring. However, rigorous research is necessary to apply this method to real-life structures. This study could be further extended by incorporating different modes of modal frequencies for damage detection. A primary limitation of this study is the utilization of healthy structural data for damage detection, which may not be feasible for all real-life structures, especially those constructed many years ago. Therefore, it would be more relevant to isolate the location and extent of damage based on vibration measurements from unhealthy structures.

Acknowledgements We are thankful to Indian Institute of Technology Indore for providing financial support and encouragement for the present study.

References

1. Fugate ML, Sohn H, Farrar CR (2001) Vibration-based damage detection using statistical process control. Mech Syst Signal process 15(4):707–721
2. Alvandi A, Cremona C (2006) Assessment of vibration-based damage identification techniques. J Sound Vib 292(1–2):179–202
3. Choi S, Park S, Park N-H, Stubbs N (2006) Improved fault quantification for a plate structure. J Sound Vib 297(3–5):865–879
4. Perera R, Ruiz A, Manzano C (2007) An evolutionary multiobjective framework for structural damage localization and quantification. Eng Struct 29(10):2540–2550
5. Sivasubramanian K, Umesha PK (2008) A graphical method of damage detection and quantification for structures
6. Fan W, Qiao P (2011) Vibration-based damage identification methods: a review and comparative study. Struct Health Monit 10(1):83–111
7. Ahmad I, Siddiqui MR, Khan M, Iqbal N, Rizwan M, Ahmad SM (2016) Vibration analysis of a multi degree of freedom cantilever beam using theoretical and computational tools Solid-Electrolyte-Interface Fracture View project Vibration Analysis View project Vibration analysis of a multi degree of freedom cantilever beam using theoretical and computational tools. Technical report
8. Rucevskis S, Janeliukstis R, Akishin P, Chate A (2016) Mode shape-based damage detection in plate structure without baseline data. Struct Control Health Monit 23(9):1180–1193
9. Ručevskis S, Wesolowski M, Chate A (2009) Vibration-based damage detection in a beam structure with multiple damage locations. Aviation 13(3):61–71
10. Scientia TE, Kannappan L (2008) Damage detection in structures using natural frequency measurements. Technical report
11. Gunawan FE, Nhan TH, Asrol M, Kanto Y, Kamil I et al (2021) A new damage index for structural health monitoring: a comparison of time and frequency domains. Proc Comput Sci 179:930–935
12. Bicanic N, Chen H-P (1997) Damage identification in framed structures using natural frequencies

Extended Finite Element Method for Sub-interface Crack in Limited Permeable Piezoelectric Bi-materials

J. Jena, I. V. Singh, and V. Gaur

Abstract This work investigates the numerical simulation of the sub-interface crack in the limited permeable piezoelectric bi-material. The mechanical and dielectric equilibrium is satisfied by incorporating electrostatic tractions on the crack surface. A crack opening model is used to incorporate the crack face tractions in the framework of the extended finite element method (XFEM). The level-set method is used in association with XFEM to track the discontinuity in the discretized domain. The sixfold electro-mechanical enrichment functions are adopted to get the effect of arbitrary poling direction in the context of XFEM. The electric displacement intensity factor (EDIF) is obtained using the electro-mechanical interaction integral in conjunction with asymptotic crack tip variables. The effect of poling direction and permittivity of the medium are studied to understand the behavior of EDIF.

Keywords Piezoelectric material · Limited permeable boundary condition · XFEM · EDIF

J. Jena (✉) · I. V. Singh · V. Gaur
Department of Mechanical and Industrial Engineering, Indian Institute of Technology Roorkee, Roorkee, Uttarakhand 247667, India
e-mail: j_jena@me.iitr.ac.in

I. V. Singh
e-mail: ivsingh@me.iitr.ac.in

V. Gaur
e-mail: vidit.gaur@me.iitr.ac.in

S. Sidhardh et al. (eds.), *Advances in Structural Integrity for Mechanical, Civil, and Aerospace Applications*, Lecture Notes in Mechanical Engineering,
https://doi.org/10.1007/978-981-97-6367-2_41

1 Introduction

In the last few years, smart/intelligent materials [barium titanate ($BaTiO_3$) or lead-zirconate-titanate (PZT)] have received more attention due to their inherent use in modern technical areas in sensors, actuators, or transducers [1, 2]. However, piezoelectric materials are mostly brittle and have low fracture toughness. Hence, the study of fracture mechanics is very essential to understand the strength, reliability, and lifetime behavior of defects (cracks) in the piezoelectric material.

Concerning the fracture mechanics in the linear piezoelectric solid, the boundary condition (BC) across the crack cavity plays an essential role in accurately estimating fracture parameters. Several researchers have worked on crack face BCs (impermeable, permeable, and limited permeable BC) in the piezoelectric material [3–7]; they found that the limited permeable BC is more accurate as the dielectric property of the crack interior is considered into account. Hao and Shen [6] initially proposed this limited permeable 'or' semipermeable 'or' exact BC considering electric displacement on the crack surface. After a series of examinations over this limited permeable BC, it is concluded that the electrostatic tractions (both electrical and mechanical tractions) must be considered on the crack surface to satisfy the mechanical and dielectric equilibrium at the crack surfaces [8, 9]. It is also found that [9], with an increase in electric load, the fracture parameters (stress intensity factor (SIF) and EDIF) are mostly affected due to the loading of electrostatic tractions on either side of the crack cavity. Apart from that, the cracks in the bi-material domain are increasing much attention in both isotropic [10–13] and in piezoelectric material [14, 15]. Hence, studying the effect of electrostatic tractions in the sub-interface crack could be more useful in estimating the fracture parameters for piezoelectric bi-material used in several applications in micro-electro-mechanical-system (MEMS) and nano-electro-mechanical-system (NEMS).

In general, several numerical tools such as the finite element method (FEM) [16], polygonal finite element method (PFEM) [17, 18], boundary element method (BEM) [19], extended isogeometric analysis (XIGA) [20], extended finite element method (XFEM) [21], and scaled boundary FEM (SBFEM) [22, 23] are used to solve the fracture mechanics problems in many engineering applications considering both conventional and smart material. However, the FEM has been mostly used numerical tool to study the cracked piezoelectric material in the last few years. However, in the context of FEM, there is a remeshing issue during crack propagation in the finite element mesh. To overcome this issue, the XFEM was developed to treat several discontinuities (cracks, holes, inclusions, etc.) without changing the mesh structure [24–28]. As per the literature study, it is found that most of the fracture study is based on impermeable BC in piezoelectric material in the framework of XFEM [29–31]. However, few authors have examined the limited permeable BC to understand the effect of crack face loading in 2D piezoelectric material using XFEM [32, 33]. Jena et al. [32, 33] found that both isotropic fourfold and electro-mechanical sixfold enrichment functions can be used without loss of accuracy in finding the EDIF in the limited permeable piezoelectric material. Finally, they concluded that the crack opening model in the framework XFEM is not limited to any crack configurations, crack face electrostatic tractions, and far-field loading in the limited permeable piezoelectric material.

The above studies show the effect of electrostatic tractions on the fracture parameter of limited permeable piezoelectric material based on analytical/numerical solutions. As per the author's knowledge, no such study is available specifically for the sub-interface crack in a limited permeable piezoelectric material using FEM/XFEM. Hence in this study, the crack opening model is used to simulate the sub-interface crack in a limited permeable piezoelectric bi-material in the framework of XFEM. In addition to that, the electro-mechanical interaction integral technique is used to evaluate EDIF. The effect of polarization angle and permittivity of the crack medium are studied to understand the behavior of EDIF for the sub-interface crack in the limited permeable piezoelectric bi-material.

2 Basic Equations for Linear Piezoelectric Solid

The governing equations and BCs for the piezoelectric media under in-plane electro-mechanical loading are discussed in the following section.

2.1 *Governing Equations*

The governing equations for the linear piezoelectric solid in the absence of body forces and surface charges are expressed as follows:

- Constitutive relations

$$\sigma_{ij} = C_{ijkm}\varepsilon_{km} - e_{mij}E_m, \quad D_i = e_{ikm}\varepsilon_{km} + \kappa_{im}E_m \tag{1}$$

- Kinematic equations

$$\varepsilon_{ij} = \frac{1}{2}\left(u_{i,j} + u_{j,i}\right), \quad E_l = -\varphi_{,l} \tag{2}$$

- Equilibrium equations

$$\sigma_{ij,j} = 0, \quad D_{i,i} = 0 \tag{3}$$

where, σ_{ij}, D_i, ε_{ij}, and E_m are representing components of stress tensors, electric displacement vectors, strain tensors, and electric field vectors; C_{ijkm}, κ_{im}, and e_{ikm} denote the elastic stiffness tensor, dielectric constant, and piezoelectric constants, respectively; u_i and φ are symbolized as displacement vector and electric potential.

2.2 Boundary Conditions

The boundary Γ of an arbitrary piezoelectric domain Ω is partitioned as $\Gamma = \Gamma_{\tilde{f}} \cup \Gamma_{\tilde{m}} = \Gamma_{\tilde{q}} \cup \Gamma_{\tilde{e}}$ (see Fig. 1), where the traction $\tilde{f}$ and the surface charge $\tilde{q}$ are imposed on the outer boundary as,

$$\sigma_{ij} n_j = \tilde{f} \quad \text{on}\, \Gamma_{\tilde{f}}, \quad D_i n_i = -\tilde{q} \quad \text{on}\, \Gamma_{\tilde{q}} \tag{4}$$

where n_j is the unit normal on the boundary Γ. Next, the primary variables (u_i) and (φ) are prescribed as:

$$u_i = \tilde{u}_i \quad \text{on}\, \Gamma_{\tilde{m}}, \quad \varphi = \tilde{\varphi} \quad \text{on}\, \Gamma_{\tilde{e}} \tag{5}$$

The crack domain Γ_c is split as Γ_c^+ and Γ_c^- to represent the top and bottom boundary of the crack surfaces. The crack face tractions t_j (Coulomb traction) and surface charge ω_s can be prescribed as:

$$\sigma_{ij} n_j = t_j \quad \text{on}\, \Gamma_c, \quad D_i n_i = -\omega_s \quad \text{on}\, \Gamma_c \tag{6}$$

3 The Classical XFEM Formulation for Piezoelectric Material

XFEM was initially developed for elastic media to characterize the local feature by introducing the enrichment functions into the standard displacement-based approximation. The XFEM has also been extended to be used in piezoelectric material to

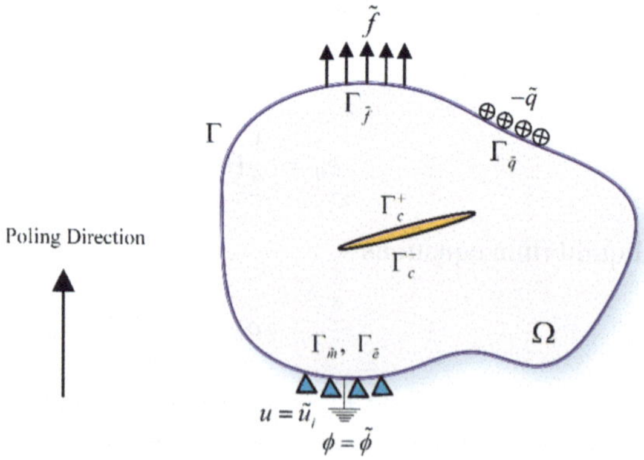

Fig. 1 Schematic of limited permeable piezoelectric domain prescribed with natural and essential boundary conditions

capture the electro-mechanical field at the crack tip [21]. In this section, the formulation of XFEM with enrichment strategy is discussed for cracked piezoelectric material.

3.1 Crack Description with Electro-mechanical Enrichment Functions

In the XFEM framework, the resulting displacement $\mathbf{u}^h(\mathbf{x})$ and electric potential field $\varphi^h(\mathbf{x})$ can be obtained by adding enrichment approximations to the standard finite element approximation. This resulting field in the XFEM framework can be expressed as:

$$\mathbf{u}^h(\mathbf{x}) = \underbrace{\sum_{i\in\tilde{n}} N_i(\mathbf{x})\mathbf{u}_i}_{\text{Standard FE Approximation}} + \underbrace{\sum_{j\in\tilde{n}^{sp}} N_j(\mathbf{x})\left[H\left(\overline{f}^h(\mathbf{x})\right) - H\left(\overline{f}_j\right)\right]\overline{\mathbf{a}}_j}_{\text{Enrichment Approximation for crack}}$$
$$\underbrace{\sum_{k\in\tilde{n}^{tp}} N_s(\mathbf{x}) \sum_{\alpha}\left[\psi^{\alpha}\left(r,\theta,\overline{\mu}_{\alpha}^{(\mathrm{re})},\overline{\mu}_{\alpha}^{(\mathrm{im})}\right) - \psi^{\alpha}\left(\mathbf{x}_k,\theta,\overline{\mu}_{\alpha}^{(\mathrm{re})},\overline{\mu}_{\alpha}^{(\mathrm{im})}\right)\right]\overline{\mathbf{b}}_s^{\alpha}}_{\text{Enrichment Approximation for crack}} \tag{7}$$

$$\varphi^h(\mathbf{x}) = \underbrace{\sum_{i\in\tilde{n}} N_i(\mathbf{x})\varphi_i}_{\text{Standard FE Approximation}} + \underbrace{\sum_{j\in\tilde{n}^{sp}} N_j(\mathbf{x})\left[H\left(\overline{f}^h(\mathbf{x})\right) - H\left(\overline{f}_j\right)\right]\overline{\mathbf{c}}_j}_{\text{Enrichment approximation for Crack}}$$
$$+\underbrace{\sum_{k\in\tilde{n}^{tp}} N_k(\mathbf{x}) \sum_{\alpha}\left[\psi^{\alpha}\left(r,\theta,\overline{\mu}_{\alpha}^{(\mathrm{re})},\overline{\mu}_{\alpha}^{(\mathrm{im})}\right) - \psi^{\alpha}\left(\mathbf{x}_k,\theta,\overline{\mu}_{\alpha}^{(\mathrm{re})},\overline{\mu}_{\alpha}^{(\mathrm{im})}\right)\right]\overline{\mathbf{d}}_k^{\alpha}}_{\text{Enrichment approximation for Crack}} \tag{8}$$

where $\tilde{n}$, $\tilde{n}^{\mathrm{sp}}$ and $\tilde{n}^{\mathrm{tp}}$ are used to represent nodes of the standard, split, and tip element (see Fig. 2), respectively: The $H\left(\overline{f}\right)$ and $\psi^{\alpha}\left(r,\theta,\overline{\mu}_{\alpha}^{(\mathrm{re})},\overline{\mu}_{\alpha}^{(\mathrm{im})}\right)$ denote Heaviside and electro-mechanical crack tip enrichment functions; the $H\left(\overline{f}\right)$ is defined by

$$H\left(\overline{f}\right) = \begin{cases} 1 & \text{if } \overline{f} \geq 0 \\ -1 & \text{otherwise} \end{cases} \tag{9}$$

The electro-mechanical enrichment function $\psi^{\alpha}\left(r,\theta,\overline{\mu}_{\alpha}^{(\mathrm{re})},\overline{\mu}_{\alpha}^{(\mathrm{im})}\right)$ is given by

$$\psi^{\alpha}\left(r,\theta,\overline{\mu}_{\alpha}^{(\mathrm{re})},\overline{\mu}_{\alpha}^{(\mathrm{im})}\right) = \left\{\begin{array}{c} \sqrt{r}\tilde{g}_1(\theta), \sqrt{r}\tilde{g}_2(\theta), \sqrt{r}\tilde{g}_3(\theta), \\ \sqrt{r}\tilde{g}_4(\theta), \sqrt{r}\tilde{g}_5(\theta), \sqrt{r}\tilde{g}_6(\theta) \end{array}\right\} \tag{10}$$

In Eq. (10), the function $\tilde{g}_{\tilde{m}}(\theta)$ can be written as:

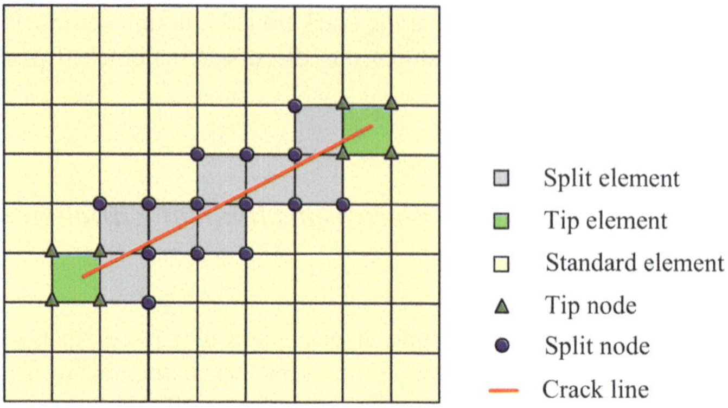

Fig. 2 Structured mesh with the schematic representation of enriched node across the crack containing element

$$g_{\tilde{m}}(\theta) = \begin{cases} \rho_{\tilde{m}}\left(\tilde{\omega}(\theta,\tilde{\phi}), \overline{\mu}_{\tilde{m}}^{(\mathrm{re})}, \overline{\mu}_{\tilde{m}}^{(\mathrm{im})}\right)\cos\left(\frac{\chi_{\tilde{m}}(\tilde{\omega}(\theta,\tilde{\phi}),\overline{\mu}_{\tilde{m}}^{(\mathrm{re})},\overline{\mu}_{\tilde{m}}^{(\mathrm{im})})}{2}\right) & \text{if } \overline{\mu}_{\tilde{m}}^{(\mathrm{im})} > 0 \\ \rho_{\tilde{m}}\left(\tilde{\omega}(\theta,\tilde{\phi}), \overline{\mu}_{\tilde{m}}^{(\mathrm{re})}, \overline{\mu}_{\tilde{m}}^{(\mathrm{im})}\right)\sin\left(\frac{\chi_{\tilde{m}}(\tilde{\omega}(\theta,\tilde{\phi}),\overline{\mu}_{\tilde{m}}^{(\mathrm{re})},\overline{\mu}_{\tilde{m}}^{(\mathrm{im})})}{2}\right) & \text{if } \overline{\mu}_{m}^{(\mathrm{im})} \leq 0 \end{cases} \tag{11}$$

where the arbitrary poling direction can be assigned by $\tilde{\omega} = \theta - \tilde{\phi}$ with rotation of the material axis at an angle $\tilde{\phi}$, whereas $\overline{\mu}_m^{(\mathrm{re})}$ and $\overline{\mu}_m^{(\mathrm{im})}$ are representing on real and imaginary parts of the complex number $\overline{\mu}_m$. The complex number $\overline{\mu}_m$ mainly depends on the reduced material constant and is evaluated from the characteristic equation (see Béchet et al. [34] for more detail). The modified angle χ_m and ρ_m is expressed as:

$$\chi_m(\tilde{\omega}(\theta,\tilde{\phi}), \overline{\mu}_m^{(\mathrm{re})}, \overline{\mu}_m^{(\mathrm{im})}) = \frac{\pi}{2} + \pi\,\mathrm{int}\left(\frac{\tilde{\omega}}{\pi}\right) - \arctan\left(\frac{\cos\left(\theta - \pi\,\mathrm{int}\left(\frac{\tilde{\omega}}{\pi}\right)\right) + \overline{\mu}_m^{(\mathrm{re})}\sin\left(\tilde{\omega}(\theta,\tilde{\phi}) - \pi\,\mathrm{int}\left(\frac{\tilde{\omega}}{\pi}\right)\right)}{\left|\overline{\mu}_m^{(\mathrm{im})}\right|\sin\left(\tilde{\omega} - \pi\,\mathrm{int}\left(\frac{\tilde{\omega}}{\pi}\right)\right)}\right) \tag{12}$$

$$\rho_{\tilde{m}}(\tilde{\omega}(\theta,\tilde{\phi}), \overline{\mu}_{\tilde{m}}^{(\mathrm{re})}, \overline{\mu}_{\tilde{m}}^{(\mathrm{im})}) = \frac{1}{\sqrt{2}}\sqrt[4]{|\overline{\mu}_{\tilde{m}}|^2 + \overline{\mu}_{\tilde{m}}^{(\mathrm{re})}\sin(2\tilde{\omega}) - \left(|\overline{\mu}_{\tilde{m}}|^2 - 1\right)\cos(2\tilde{\omega})} \tag{13}$$

4 Interaction Integral for Limited Permeable Piezoelectric Material

In this section, the formulation of interaction integral and extraction of EDIF from interaction integral are discussed for limited permeable piezoelectric media.

4.1 J-Integral

The electro-mechanical J-integral for limited permeable piezoelectric solids can be written as [33]:

$$J = \int_{\Gamma} \left(H\delta_{1j} - \sigma_{ij}u_{i,1} - D_j\varphi_{,1}\right) n_j \mathrm{d}\Gamma + \int_{\Gamma^+ + \Gamma^-} \left(\omega_s\varphi_{,1} - t_i u_{i,1}\right)\mathrm{d}\Gamma \tag{14}$$

where $H = (\sigma_{ij}\varepsilon_{ij} - D_jE_j)/2$ is the electric enthalpy density function and n_j is the unit normal vector which is outward direction from the boundary of the J-domain (see Fig. 3). Next, the electro-mechanical J-integral using two independent equilibrium states (actual $(\sigma_{ij}^{ac}, D_j^{ac}, u_i^{ac}, \varphi^{ac})$ and the auxiliary state $(\sigma_{ij}^{ax}, D_j^{ax}, u_i^{ax}, \varphi^{ax})$) can be written as:

$$J^{\mathrm{Sum}} = \int_{A} \left((\sigma_{ij}^{ac} + \sigma_{ij}^{ax})(u_{i,1}^{ac} + u_{i,1}^{ax}) + (D_j^{ac} + D_j^{ax})(\varphi_{,1}^{ac} + \varphi_{,1}^{ax}) - H^{(ac,ax)}\delta_{1j}\right) q_{,j}\mathrm{d}A$$
$$+ \int_{\Gamma^+ + \Gamma^-} \left(\omega_s^{ac}\varphi_{,1}^{ax} + \omega_s^{ax}\varphi_{,1}^{ac} - t_i^{ac}u_{i,1}^{ax} - t_i^{ax}u_{i,1}^{ac}\right) q\mathrm{d}\Gamma \tag{15}$$

where q is the plateau function having unity value on Γ_ε and zero on the boundary of Γ as shown in Fig. 3. The interaction integral $I^{(Ac,Ax)}$ for two states can be extracted from Eq. (15) and written as [33, 35].

$$I^{(ac,ax)} = \int_{A} \left(\sigma_{ij}^{ac}u_{i,1}^{ax} + D_j^{ac}\varphi_{,1}^{ax} + \sigma_{ij}^{ax}u_{i,1}^{ac} + D_j^{ax}\varphi_{,1}^{ac} - H^{(ac,ax)}\delta_{1j}\right) q_{,j}\mathrm{d}A$$
$$+ \int_{\Gamma^+ + \Gamma^-} \left(\omega_s^{ac}\varphi_{,1}^{ax} - t_i^{ac}u_{i,1}^{ax}\right) q\mathrm{d}\Gamma \tag{16}$$

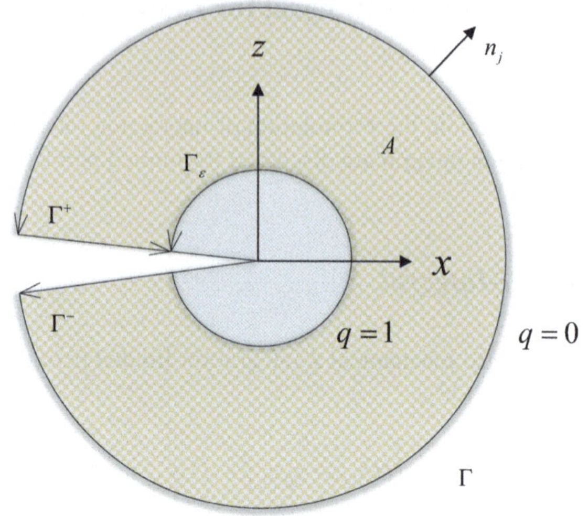

Fig. 3 Schematic representation of contour path to obtain the J-integral

4.2 Extraction of SIFs and EDIF from the Interaction Integral

The electro-mechanical J-integral under mixed-mode loading conditions is equal to the energy release rate, hence J can be written as [36],

$$J = \frac{1}{2} K^T Y K \tag{17}$$

where $K = \{K_{II}\ K_I\ K_{III}\ K_{IV}\}^T$ which contains the vector of all the fracture modes and Y is the generalized Irwin matrix (depends on the piezoelectric constants) and is represented by $\tilde{Y}_{MN} = -\mathrm{Im}\{A_{M\xi} N_{\xi N}\}$

Therefore, the interaction integral based on Eq. (17) can be given as [36],

$$\begin{aligned} I^{(ac,ax)} &= K_{II}^{ac} K_{II}^{ax} Y_{11} + K_I^{ac} K_I^{ax} Y_{22} + K_{IV}^{ac} K_{IV}^{ax} Y_{44} + (K_I^{ac} K_{II}^{ax} + K_{II}^{ac} K_I^{ax}) Y_{12} \\ &+ (K_{II}^{ac} K_{IV}^{ax} + K_{IV}^{ac} K_{II}^{ax}) Y_{14} + (K_I^{ac} K_{IV}^{ax} + K_{IV}^{ac} K_I^{ax}) Y_{24} \end{aligned} \tag{18}$$

Next, the SIF and EDIF can be extracted from the interaction integral in choosing the auxiliary state, e.g., if the auxiliary state for mode-I fracture is to be chosen as $K_1^{ax} = 1$; $K_{II}^{ax} = K_{IV}^{ax} = 0$. Then, Eq. (18) reduces to,

$$I^{(ac,I)} = K_I^{ac} Y_{22} + K_{II}^{ac} Y_{12} + K_{IV}^{ac} Y_{24} \tag{19}$$

Similarly, for *mode*-II and *mode*-IV fracture, Eq. (18) reduces to,

$$I^{(ac,II)} = K_I^{ac} Y_{12} + K_{II}^{ac} Y_{11} + K_{IV}^{ac} Y_{14} \tag{20}$$

$$I^{(ac,IV)} = K_I^{ac} Y_{24} + K_{II}^{ac} Y_{14} + K_{IV}^{ac} Y_{44} \tag{21}$$

5 Introduction to Electric Crack Face BC and Procedure for the Numerical Implementation

This section discusses the several types of crack face BCs and XFEM-based numerical implementation for the sub-interface crack in the limited permeable piezoelectric material.

5.1 Crack Face BCs

In general, three types of BCs exist for the piezoelectric media as follows:

1. **Impermeable BC**: In this BC, the electric field does not pass through the crack interior of the specimen because this BC ignores the presence of the dielectric medium inside the crack cavity (see Fig. 4a). This impermeable BC can be expressed as,

$$\Delta\varphi = \varphi^+ - \varphi^- \neq 0, \quad D_2^+ = D_2^- = 0 \tag{22}$$

   where the superscripts '+' and '−' denote the top and bottom sides of the crack surface.
2. **Permeable BC**: This BC is represented by equal potential on either side of the crack surface. Also, the electric field is fully penetrated through the crack interior of the specimen (see Fig. 4b). This BC can be expressed as:

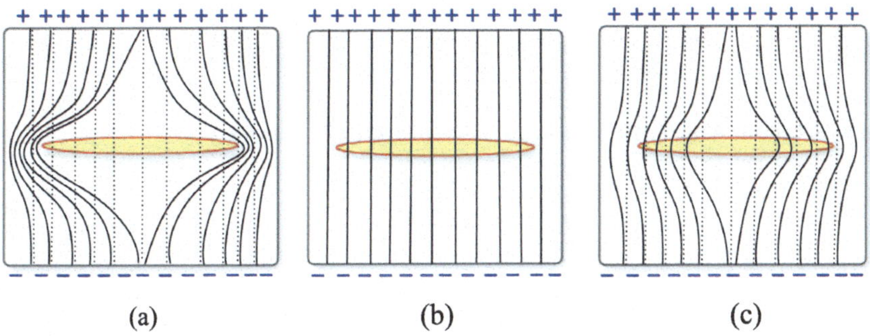

Fig. 4 Schematic of the BCs in piezoelectric material: **a** impermeable BC, **b** permeable BC, and **c** semipermeable BC

$$\varphi^+ = \varphi^-, \quad D_2^+ = D_2^- \neq 0 \tag{23}$$

3. **Limited Permeable BC**: In the limited permeable BC, the electric field is partially passed through on either side of the crack cavity of the specimen, as depicted in Fig. 4c. Initially, this BC was proposed by Hao and Shen [6] to incorporate electric tractions D_2^c on the crack surface by considering the finite value of permittivity inside the crack cavity. Then this model is extended by Ricoeur and Kuna [9] to make the thermodynamic consistent by incorporating coulomb forces (t_2^c) with electric traction (D_2^c) on the crack surface as depicted in Fig. 5. Mathematically, this BC can be represented as,

$$D_2^+ = D_2^- = D_2^c, \quad D_2^c = -\kappa_c \frac{\varphi^+ - \varphi^-}{u_2^+ - u_2^-} \tag{24}$$

$$t_2^+ = t_2^- = t_2^c, \quad t_2^c = \frac{1}{2}\frac{(D_2^c)^2}{\kappa_C} = \frac{1}{2}E_2^c D_2^c \tag{25}$$

where κ_c is the dielectric constant of the crack medium, κ_c is defined as $\kappa_c = \kappa_r \kappa_0$ where κ_r is relative permittivity and κ_0 is the permittivity of the vacuum.

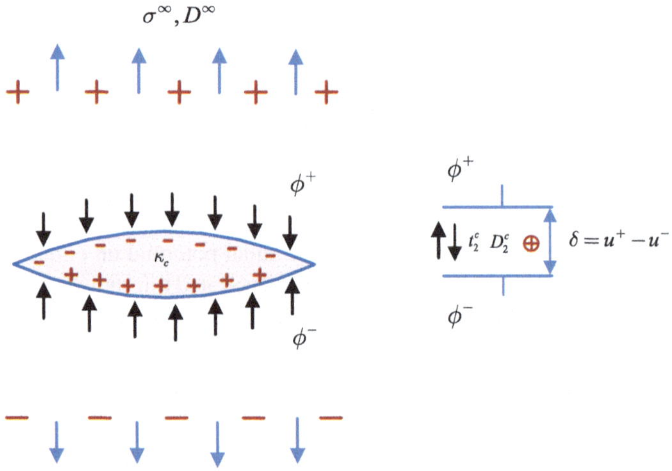

Fig. 5 Schematic of limited permeable BC in the piezoelectric material

5.2 *Numerical Implementation of Limited Permeable Sub-interface Crack in Piezoelectric Material*

This section describes the implementation of limited permeable BC in the piezoelectric material. Basically, the limited permeable BC implementation is not possible using the closed crack concept, as there is no assumption of a dielectric medium inside the crack cavity. So, the opened crack concept is used to implement this limited permeable BC in the piezoelectric material, as depicted in Fig. 5. Initially, the impermeable BC is used to open the crack cavity with the application of external loading. Then the electrostatic tractions D_2^c and t_2^c are applied on the crack surface based on Eqs. (24) and (25), respectively. Next, the primary variables are calculated based on the deformed configuration of the domain after the application of electrostatic tractions. Then the EDIF is extracted using the interaction integral approach (see in Sect. 4). Next, the above procedure is followed by updating the primary variables at each load step throughout the simulation.

6 Numerical Results and Discussions

In Sect. 6, one numerical example is solved by considering the electrostatic tractions on the crack surface of the limited permeable piezoelectric material. In this example, some basic assumptions are taken for the present study as follows:

- Isoperimetric Q4 element is taken to discretize the specimen.
- All the units are taken as SI.
- Plane strain assumption is taken.
- Complex sixfold enrichment functions are used in each example.
- The PZT-4 and PZT-5H material constant is taken (see Table 1) for the present study.

Table 1 Piezoelectric material properties for PZT-4 and PZT-5H [37]

	PZT-4	PZT-5H
Elastic constants ($\times 10^9$ Pa)	$C_{11} = 13.9, C_{44} = 2.56$ $C_{12} = 7.78, C_{13} = 7.43$ $C_{33} = 11.3$	$C_{11} = 12.6, C_{44} = 3.53$ $C_{12} = 5.5, C_{13} = 5.3$ $C_{33} = 11.7$
Piezoelectric constants (C/m^2)	$e_{31} = -6.98$ $e_{33} = 13.84$ $e_{15} = 13.44$	$e_{31} = -6.5$ $e_{33} = 23.3$ $e_{15} = 17$
Dielectric permittivity ($\times 10^{-9}$ C/(Vm))	$\kappa_{11} = 6$ $\kappa_{33} = 5.47$	$\kappa_{11} = 15.1$ $\kappa_{33} = 13$

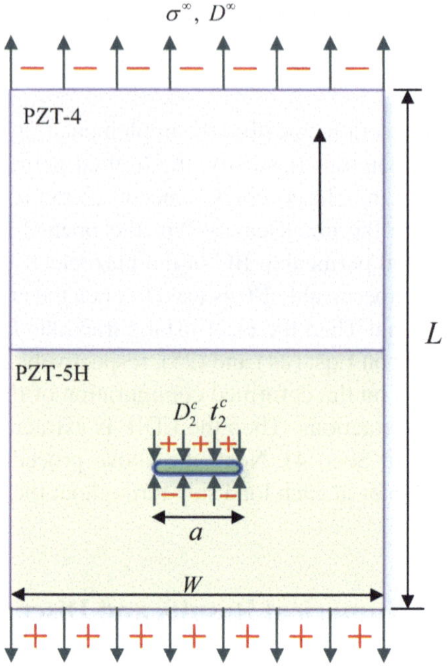

Fig. 6 Limited permeable piezoelectric bi-material with a sub-interface crack subjected to electrostatic tractions

6.1 A Limited Permeable Sub-interface Crack in a Piezoelectric Bi-material

In this example, we consider a sub-interface crack of length a is taken for the investigation of a limited permeable piezoelectric bi-material (see Fig. 6). The geometric data of the specimen used as length $L = 0.04$, width $W = 0.02$ and crack length $a = 0.002$. This piezoelectric domain is discretized with the number of nodes 80 in the x-direction and 160 in the z-direction. In this example, the upper and lower half of the specimen is assigned with the piezoelectric material constant PZT-4 and PZT-5H, respectively. The specimen is loaded with far-field stress $\sigma^{\infty} = 15$, and incremental electric displacement D^{∞} varies 0.005–0.02. The top and bottom surface side of the crack is subjected to electrostatic tractions D_2^c and t_2^c which is considered based on Eqs. (24) and (25), respectively. This section examines the influence of poling direction on EDIF by a change in poling angle from $\tilde{\phi} = 0$ to $\pi/6$. It is found that there is a significant difference in the result of EDIF due to a change in poling angle with the increase in electric displacement D^{∞}, as depicted in Fig. 7. Also, the effect of dielectric medium on EDIF is investigated, and it is found that the result of EDIF increases with decrease in relative permittivity of the cracked medium (see Fig. 8). The stress σ_{zz} and electric displacement D_z contour plots obtained from the present example are shown in Fig. 9.

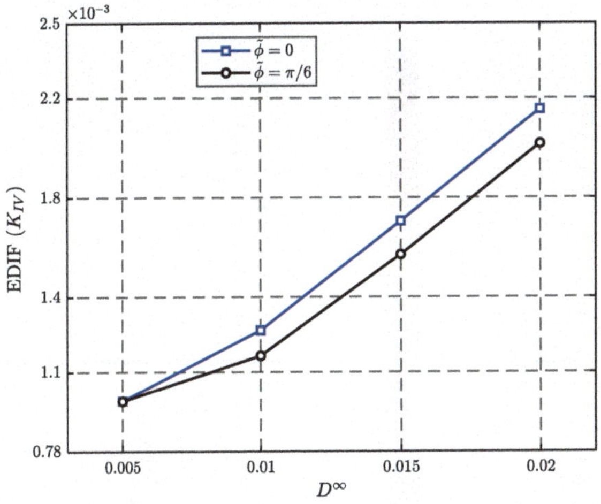

Fig. 7 Comparison of the result of EDIF K_{IV} at a different poling angle in limited permeable piezoelectric bi-material with sub-interface crack subjected to electrostatic tractions

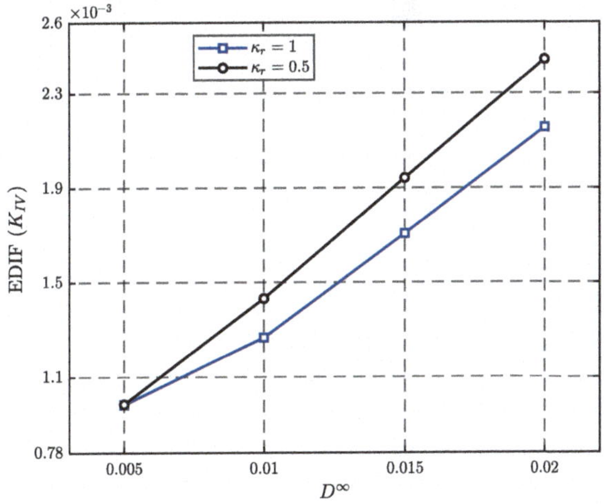

Fig. 8 Comparison of the result of EDIF K_{IV} concerning the change in relative permittivity of the medium inside the crack cavity of piezoelectric bi-material

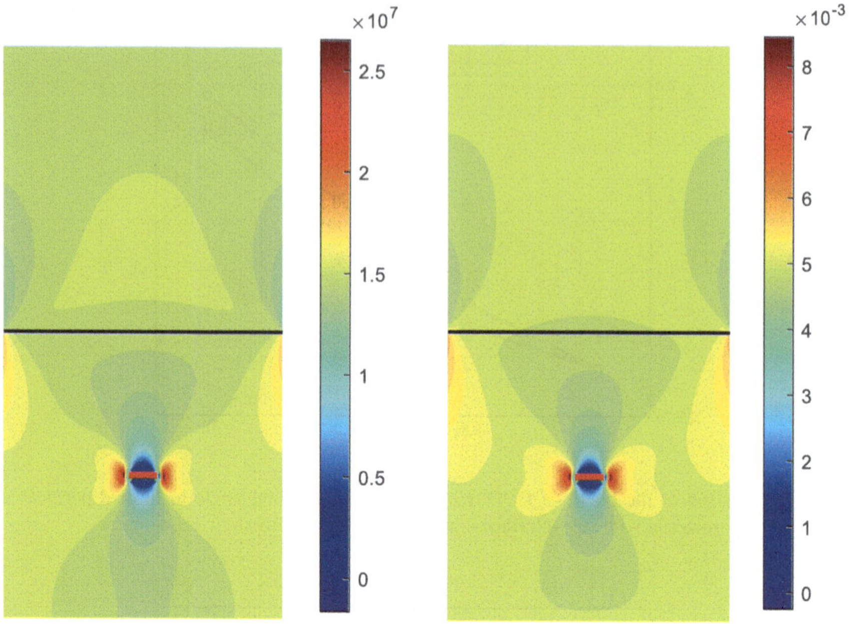

(a) (b)

Fig. 9 Contour plots of **a** stress σ_{zz} and **b** electric displacement D_z considering a sub-interface crack (crack length $a = 0.002$) in a limited permeable piezoelectric bi-material subjected to the far-field load of $\sigma^{\infty} = 15$ and $D^{\infty} = 0.015$

7 Conclusion

This work presents a sub-interface crack analysis in the 2D limited permeable piezoelectric bi-material using XFEM. The electro-mechanical interaction integral technique is utilized to evaluate the EDIF for limited permeable piezoelectric material. As per the above snumerical study, the following conclusions are drawn,

- The sixfold electro-mechanical enrichment functions are successfully implemented for the sub-interface crack to obtain the result of EDIF.
- It is found that there is a significant difference in the result of EDIF with incremental electric loading due to a change in poling angle and relative permittivity of the medium.

References

1. Narita F, Fox M (2018) A review on piezoelectric, magnetostrictive, and magnetoelectric materials and device technologies for energy harvesting applications. Adv Eng Mater 20:1700743. https://doi.org/10.1002/adem.201700743
2. Sharma SK, Kumar S, Kumar R (2022) Parametric analysis of hybrid tribo-piezoelectric energy harvester. In: Mechanics based design of structures and machines, pp 1–14. https://doi.org/10.1080/15397734.2022.2041436
3. McMeeking RM (1999) Crack tip energy release rate for a piezoelectric compact tension specimen. Eng Fract Mech 64:217–244. https://doi.org/10.1016/S0013-7944(99)00068-5
4. McMeeking RM (2001) Towards a fracture mechanics for brittle piezoelectric and dielectric materials. Int J Fract 108:25–41
5. McMeeking RM (2004) The energy release rate for a Griffith crack in a piezoelectric material. Eng Fract Mech 71:1149–1163. https://doi.org/10.1016/S0013-7944(03)00135-8
6. Tian-Hu H, Zi-Yuan S (1994) A new electric boundary condition of electric fracture mechanics and its applications. Eng Fract Mech 47:793–802. https://doi.org/10.1016/0013-7944(94)90059-0
7. Wippler K, Ricoeur A, Kuna M (2004) Towards the computation of electrically permeable cracks in piezoelectrics. Eng Fract Mech 71:2567–2587. https://doi.org/10.1016/j.engfracmech.2004.03.003
8. Ricoeur A, Kuna M (2009) Electrostatic tractions at dielectric interfaces and their implication for crack boundary conditions. Mech Res Commun 36:330–335. https://doi.org/10.1016/j.mechrescom.2008.09.009
9. Ricoeur A, Kuna M (2009) Electrostatic tractions at crack faces and their influence on the fracture mechanics of piezoelectrics. Int J Fract 157:3. https://doi.org/10.1007/s10704-009-9321-z
10. Yang M, Kim KS (1993) The behavior of subinterface cracks with crack-face contact. Eng Fract Mech 44:155–165. https://doi.org/10.1016/0013-7944(93)90088-A
11. Lu H, Lardner TJ (1992) Mechanics of subinterface cracks in layered material. Int J Solids Struct 29:669–688. https://doi.org/10.1016/0020-7683(92)90120-I
12. Venkatesha KS, Ramamurthy TS, Dattaguru B (1998) A study of the behaviour of subinterface cracks in bimaterial plates. Eng Fract Mech 59:241–252. https://doi.org/10.1016/S0013-7944(97)00129-X
13. Yao XF, Yeh HY, Xu W (2005) Dynamic initiation and propagation behaviour of sub-interfacial cracks in PMMA/aluminium bi-material system. Fatigue Fract Eng Mater Struct 28(12):1191–1198. https://doi.org/10.1111/j.1460-2695.2005.00960.x
14. Beom HG, Jeong KM, Kim YH (2003) Intensity factors for subinterface cracks in dissimilar anisotropic piezoelectric media. Arch Appl Mech 73:184–198. https://doi.org/10.1007/s00419-003-0280-y
15. Yang PS, Liou JY, Sung JC (2008) Subinterface crack in an anisotropic piezoelectric bimaterial. Int J Solids Struct 45:4990–5014. https://doi.org/10.1016/j.ijsolstr.2008.05.001
16. Kuna M (1998) Finite element analyses of crack problems in piezoelectric structures. Comput Mater Sci 13:67–80. https://doi.org/10.1016/S0927-0256(98)00047-0
17. Khoei AR, Yasbolaghi R, Biabanaki SOR (2015) A polygonal finite element method for modeling crack propagation with minimum remeshing. Int J Fract 194:123–148. https://doi.org/10.1007/s10704-015-0044-z
18. Sharma D, Singh IV, Kumar J (2022) A microstructure based elasto-plastic polygonal FEM and CDM approach to evaluate LCF life in titanium alloys. Int J Mech Sci 107356. https://doi.org/10.1016/j.ijmecsci.2022.107356
19. Groh U, Kuna M (2005) Efficient boundary element analysis of cracks in 2D piezoelectric structures. Int J Solids Struct 42:2399–2416. https://doi.org/10.1016/j.ijsolstr.2004.09.023
20. Singh SK, Singh IV, Mishra BK, Bhardwaj G (2019) Analysis of cracked functionally graded material plates using XIGA based on generalized higher-order shear deformation theory. Compos Struct 225:111038. https://doi.org/10.1016/j.compstruct.2019.111038

21. Chadaram S, Yadav SK (2021) Three-dimensional thermal fracture analysis of piezoelectric material by extended finite element methods. Eng Fract Mech 256:107981. https://doi.org/10.1016/j.engfracmech.2021.107981
22. Nguyen DTD, Javidan F, Attar M, Natarajan S, Yang Z, Ooi EH, Song C, Ooi ET (2022) Fracture analysis of cracked magneto-electro-elastic functionally graded materials using scaled boundary finite element method. Theoret Appl Fract Mech 103228. https://doi.org/10.1016/j.tafmec.2021.103228
23. Li C, Tong L (2015) 2D fracture analysis of magnetoelectroelastic composites by the SBFEM. Compos Struct 132:984–994. https://doi.org/10.1016/j.compstruct.2015.07.015
24. Kumar M, Singh IV (2020) Numerical investigation of creep crack growth in plastically graded materials using $C(t)$ and XFEM. Eng Fract Mech 226:106820. https://doi.org/10.1016/j.engfracmech.2019.106820
25. Pandey VB, Singh IV, Mishra BK (2021) A Strain-based continuum damage model for low cycle fatigue under different strain ratios. Eng Fract Mech 242:107479. https://doi.org/10.1016/j.engfracmech.2020.107479
26. Jha A, Singh IV, Mishra BK, Singh R, Singh RN (2021) Numerical study of coupled elasto-plastic hydrogen diffusion at crack tip using XFEM. Recent advances in computational mechanics and simulations. Springer, Singapore, pp 177–186
27. Sarkar S, Singh IV, Mishra BK (2021) A simplified continuous–discontinuous approach to fracture based on decoupled localizing gradient damage method. Comput Methods Appl Mech Eng 383:113893. https://doi.org/10.1016/j.cma.2021.113893
28. Duhan N, Patil RU, Mishra BK, Singh IV, Pak YE (2022) Nonlinear thermo-elastic analysis of edge dislocations with internal heat generation in semiconductor materials. Mech Mater 169:104322. https://doi.org/10.1016/j.mechmat.2022.104322
29. Béchet E, Scherzer M, Kuna M (2008) Fracture of piezoelectric materials with the X-FEM. Eur J Comput Mech Rev Eur Mécan Numér 17:637–649. https://doi.org/10.3166/remn.17.637-649
30. Bhargava RR, Sharma K (2011) A study of finite size effects on cracked 2-D piezoelectric media using extended finite element method. Comput Mater Sci 50:1834–1845. https://doi.org/10.1016/j.commatsci.2011.01.026
31. Bhargava RR, Sharma K (2012) X-FEM simulation for two-unequal-collinear cracks in 2-D finite piezoelectric specimen. Int J Mech Mater Des 8:129–148. https://doi.org/10.1007/s10999-012-9182-x
32. Jena J, Singh SK, Gaur V, Singh IV, Natarajan S (2021) A new framework based on XFEM for cracked semipermeable piezoelectric material. Eng Fract Mech 253:107874. https://doi.org/10.1016/j.engfracmech.2021.107874
33. Jena J, Singh SK, Gaur V, Singh IV, Natarajan S (2022) A new framework based on XFEM to study the role of electrostatic tractions in semipermeable piezoelectric material. Eng Fract Mech 266:108398. https://doi.org/10.1016/j.engfracmech.2022.108398
34. Béchet E, Scherzer M, Kuna M (2009) Application of the X-FEM to the fracture of piezoelectric materials. Int J Numer Meth Eng 77(11):1535–1565. https://doi.org/10.1002/nme.2455
35. Kuna M (2010) Fracture mechanics of piezoelectric materials—where are we right now? Eng Fract Mech 77:309–326. https://doi.org/10.1016/j.engfracmech.2009.03.016
36. Rao BN, Kuna M (2008) Interaction integrals for fracture analysis of functionally graded piezoelectric materials. Int J Solids Struct 45:5237–5257. https://doi.org/10.1016/j.ijsolstr.2008.05.020
37. Xu XL, Rajapakse RKND (2001) On a plane crack in piezoelectric solids. Int J Solids Struct 38:7643–7658. https://doi.org/10.1016/S0020-7683(01)00029-4

Residual Stress Prediction for Butt-Welded Plate Joint with Unequal Plate Thickness Using Artificial Neural Networks

Sandipan Baruah, Subrato Sarkar, and Indra Vir Singh

Abstract The present work describes the usefulness of artificial neural networks (ANN) for predicting residual stresses in a butt-welded plate joint. The motivation for the present work is driven by the recently emerging usage of machine learning techniques to reduce computational cost of finite element (FE) analysis. A sample specimen with unequal plate thickness, often used in aircraft panels and shipyards, is investigated here, in order to consider the effects of plate thickness variation in the prediction process. Datasets for building ANNs are generated from sequentially coupled thermal and elasto-plastic FE analysis of the welded specimens using varying inputs like length, width, thickness of plates, weld power, and welding speed. The longitudinal and transverse residual stresses in the domain are obtained as outputs for every set of inputs. ANN models comprising of different activation functions and hidden layers are trained and tested using inputs of three-dimensional coordinate points of the specimens along with the inputs–outputs of FE simulations. The ANN models are then used to predict residual stresses using new input data, and their validity is investigated through FE simulations. It was observed that predictions of ANN models agree well with the simulations, thereby establishing the ANNs as powerful alternatives to FE analysis of welding. For the welding process and material investigated in the present work, the ANN scheme can be used to predict the residual stresses at a considerably faster speed than the conventional FE computations for varying weld-process parameters and geometric parameters.

S. Baruah (✉) · I. V. Singh (✉)
Department of Mechanical and Industrial Engineering, Indian Institute of Technology Roorkee, Uttarakhand 247667, India
e-mail: s_baruah@me.iitr.ac.in

I. V. Singh
e-mail: ivsingh@me.iitr.ac.in

S. Sarkar
Department of Mechanical, Aerospace, and Nuclear Engineering, Rensselaer Polytechnic Institute, Troy, NY 12180, USA
e-mail: sarkas6@rpi.edu

S. Sidhardh et al. (eds.), *Advances in Structural Integrity for Mechanical, Civil, and Aerospace Applications*, Lecture Notes in Mechanical Engineering,
https://doi.org/10.1007/978-981-97-6367-2_42

Keywords Neural networks · Welding · Residual stress · Thermo-elasto-plastic FE analysis

1 Introduction

In manufacturing domains, welding is one of the most popular methods used for producing leak-proof and durable joints in metallic components. It consumes lesser time as compared to other joining processes and has widespread applications in industries concerning automobiles, shipbuilding, railways, civil constructions, etc. However, a major challenge in the engineering service of welded components is the presence of residual stresses in them. Due to severely high temperatures and thermal gradients developed during welding, there is an evolution of thermal stresses in the welded component which remain in tensile or compressive form even after the component fully cools down to room temperature. The tensile residual stresses induce undesirable effects on the service life of the component, especially during fatigue loading or in corrosive environments. Therefore, it is necessary to estimate the residual stress distribution in a welded structure, prior to the start of its intended service.

Some commonly used experimental techniques for determining the residual stresses are hole-drilling method, X-ray diffraction, and neutron diffraction. However, due to their complexities, risks, and costs, experimental methods are mostly replaced by numerical techniques, with the most popular technique being the thermo-elastic-plastic finite element (FE) analysis of welding. In this regard, early works like the sequentially performed thermal and mechanical FE analysis of welding by [1] are notable. To simulate the heat source in arc welding, [2] proposed a circular flux model, whereas [3] later developed a more realistic double-ellipsoidal model. The FE analysis was extended to pipe welds by [4]. The nonlinear analysis involved in estimating welding residual stresses was described by [5]. Various other researchers have also investigated single-pass as well as multi-pass FE analysis of welding for different materials and specimens [6–13].

However, the nonlinear FE analysis of welding, with a fine mesh near the weld zone, requires significantly high computation time and cost. As such, supervised machine learning tools like artificial neural networks (ANNs) and support vector machines (SVMs) have received significant attention as alternatives to FE simulations. Baruah and Singh [14] developed a back propagation ANN model for predicting residual stresses by varying weld-process parameters and plate thickness. [15, 16] used results of FE analysis to train ANNs and predicted welding-induced distortions at selected locations on a specimen. Other similar works include residual stress or deformation prediction for girth welded pipes [17], T-joint fillet welds [18, 19], and dissimilar metallic plate welds [20]. Recently, [21] used different types of machine learning models combined with genetic algorithms to minimise the peak tensile residual stresses.

Despite several works related to the application of ANNs for welding analysis, there is a scarce amount of research concerning the development of ANN models which can predict residual stresses by incorporating the variations of both geometric parameters and weld-process parameters. Moreover, most of the works in literature are concerned with predicting residual stresses only at specific locations in the specimen domain, without the capability to predict at any desired geometric location. The validity of ANN predictions also requires further investigation. Based on these research gaps, an effort has been made in the present work to address some of the concerns. The research highlights of the present work are:

- A framework is developed for predicting welding residual stresses through ANNs for varying geometric parameters such as length, width, and thickness of plates and weld-process parameters such as weld power, welding speed.
- A plate joint of unequal thickness is analysed in order to incorporate any inherent effect of thickness variation on the residual stresses developed due to welding.
- The ANN framework is used for prediction of residual stresses at various random locations on the specimen. The validity of the predictions is established by comparing the results with FE simulations.
- Different forms of ANN incorporating different hidden layers and activation functions are investigated, and thereby, the effectiveness of ANN as an alternate tool to FE analysis for fast prediction of welding residual stress has been highlighted.

This paper is divided into five sections. Section 1 briefly introduces the importance of investigating residual stresses in welding, including literature survey and the present research highlights. Section 2 describes the theoretical background comprising of FE analyses of welding and the ingredients of ANNs. The details related to the specimen weld problem investigated in the present work are provided in Sect. 3. The obtained results and their discussion are presented in Sect. 4. Finally, Section 5 highlights the major conclusions drawn from the present work.

2 Theoretical Background

The analysis of welding, investigated in the present study, includes FE analyses and development of an ANN model for predicting residual stresses in the welded specimen. The theoretical aspects behind them are described as follows.

2.1 Finite Element Analysis of Welding

For investigating the residual stresses, the FE analyses consist of a thermal simulation followed by an elasto-plastic mechanical simulation as described in [21]. The thermal simulation is transient and nonlinear and requires the inputs of weld voltage, current, arc efficiency, and welding speed, along with the temperature-dependent thermal conductivity, specific heat, density, and the constant latent heat of the material. The temperature distribution in the components is obtained at discrete times during the welding and its subsequent cooling. The temperature history is provided as an input to an elasto-plastic nonlinear mechanical simulation, along with temperature-dependent mechanical properties. The incremental stresses are computed for the entire heating and cooling regimes. The unrelaxed stresses obtained at the end of weld cooling are considered as the residual stresses remaining in the component.

For performing the thermal simulation, the heat input of the arc is modelled using Goldak's double-ellipsoidal model. The volumetric heat distribution from the arc is mostly concentrated near the arc-centre (O) in the form of two ellipsoids in the front and rear as described in Fig. 1 as per [3]. The heat distribution for the Goldak's model is given as:

$$q = \frac{6\sqrt{3} f_j \eta\, V\, I}{a\, b\, c_j\, \pi \sqrt{\pi}} \exp\left(-3\left(\frac{x^2}{a^2} + \frac{y^2}{b^2} + \frac{(z + wt)^2}{c_j^2}\right)\right) \quad ; \quad \begin{array}{l} j = 1 \text{ (front ellipsoid)} \\ j = 2 \text{ (rear ellipsoid)} \end{array} \tag{1}$$

In Eq. (1), V is the weld voltage, I is the weld current, η is the arc efficiency, w is the weld speed, and f_1, f_2 are distribution factors. The governing equation for energy balance in transient thermal analysis incorporating phase transformation is taken from [21] as:

$$\nabla \cdot (k_c \cdot \nabla T) + q - \rho c_s \frac{\partial T}{\partial t} - \rho L \frac{\partial f_{\text{pc}}}{\partial T} \frac{\partial T}{\partial t} = 0. \tag{2}$$

The quantities incorporated in Eq. (2) for computation of temperature (T) are thermal conductivity (k_c), heat input (q), density (ρ), specific heat (c_s), latent heat (L), and material-specific phase-change function (f_{pc}). The ambient temperature is used as an initial condition ($T = T_\infty$). The boundary condition is heat convection applied on the free surfaces (n_s), given in terms of the heat transfer coefficient (h) by Eq. (3) as,

$$-(k_c \cdot \nabla T) \cdot n_s = h(T - T_\infty). \tag{3}$$

The FE model for nonlinear, transient thermal analysis of welding is given in time-incremental and iterative form to solve for the time–temperature distribution is taken from [21] as:

$$\begin{aligned} &\left(\mathbf{K}_c(T)^{(i)}_{t+\Delta t} + \mathbf{K}_h + \Delta t^{-1}\left[\mathbf{C}_h(T)^{(i)}_{t+\Delta t} + \mathbf{C}_{\text{pc}}(T)^{(i)}_{t+\Delta t}\right]\right)\delta\mathbf{T}_{t+\Delta t} \\ &= \mathbf{Q}_{t+\Delta t} + \mathbf{S} - \left(\mathbf{K}_c(T)^{(i)}_{t+\Delta t} + \mathbf{K}_h\right)\mathbf{T}^{(i)}_{t+\Delta t} \\ &\quad - \Delta t^{-1}\left\{\mathbf{C}_h(T)^{(i)}_{t+\Delta t}\left(\mathbf{T}^{(i)}_{t+\Delta t} - \mathbf{T}_t\right) - \left(\mathbf{L}_{\text{pc}}(T)^{(i)}_{t+\Delta t} - \mathbf{L}_{\text{pc}}(T)_t\right)\right\} \end{aligned} \tag{4}$$

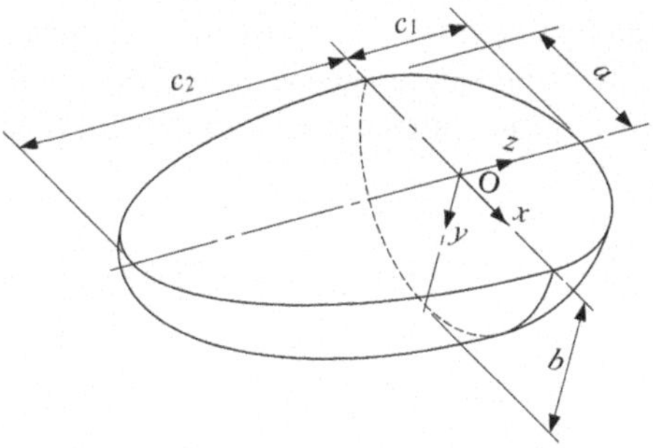

Fig. 1 Representation of double-ellipsoid heating-zone model developed by [3]

$$\mathbf{T}_{t+\Delta t}^{(i+1)} = \mathbf{T}_{t+\Delta t}^{(i)} + \delta \mathbf{T}_{t+\Delta t}. \tag{5}$$

Equation (4) incorporates the thermal conductivity matrix ($\mathbf{K}_c$), thermal convective matrix ($\mathbf{K}_h$), heat capacity matrix ($\mathbf{C}_h$), phase-change capacity matrix ($\mathbf{C}_{\text{pc}}$) and rate vector ($\mathbf{L}_{\text{pc}}$), input heat load vector (**Q**), and convective boundary flux vector (**S**). The computation of these matrices and vectors is provided in [21]. The temperature field is updated using Eq. (5) until some convergence is reached. After the completion of the thermal simulation, the mechanical elasto-plastic simulation for stress field is performed using the thermal loads. The governing equation for mechanical analysis is taken from [21] as shown in Eq. (6). The 3-2-1 principle of constraining is used as a boundary condition to prevent rigid-body motion. An incremental strain approach as given in Eq. (7) is followed as in order to solve the mechanical stress field

$$\nabla \cdot (\mathrm{d}\sigma) = 0 \tag{6}$$

$$\mathbf{d}\varepsilon = \mathbf{d}\varepsilon^{\mathbf{el}} + \mathbf{d}\varepsilon^{\mathbf{pl}} + \mathbf{d}\varepsilon^{\mathbf{th}}. \tag{7}$$

Equation (7) incorporates incremental forms of total strain ($\mathbf{d\varepsilon}$), elastic strain ($\mathbf{d\varepsilon^{el}}$), plastic strain ($\mathbf{d\varepsilon^{pl}}$), and thermal strain ($\mathbf{d\varepsilon^{th}}$). In order to consider the plastic flow, von-Mises yield function is used and the constitutive relation between incremental strain and incremental stress is given as,

$$\mathbf{d}\sigma = \mathbf{D}^e\left(\mathbf{d}\varepsilon - \mathbf{d}\varepsilon^{\text{th}}\right) - \frac{9}{4\sigma_{\text{eff}}^2(3G+H)}\left(\mathbf{D^e s s}^T\mathbf{D}^{eT}\left(\mathbf{d}\varepsilon - \mathbf{d}\varepsilon^{\text{th}}\right) + \mathbf{D}^e\,\mathbf{s}\frac{\partial F}{\partial T}\Delta T\right). \tag{8}$$

Equation (8) is fully applicable for the plastic regime. For the elastic regime, the second term of Eq. (8) is zeroed out. $\mathbf{D^e}$ is the elastic modulus matrix which includes temperature-dependent Young's modulus and Poisson's ratio. **s** is the deviatoric stress tensor, σ_{eff} is the effective von-Mises stress, G is the modulus of rigidity, H is the hardening modulus, and F is the temperature-dependent yield function. The hardening modulus and yield function are evaluated respectively as $H = \partial\sigma_y(T)/\partial\varepsilon_{\text{eq}}^{\text{pl}}(T)$

and $F = \sigma_{\text{eff}}(T) - \sigma_y(T)$ where σ_y is the temperature-dependent yield stress and $\varepsilon_{\text{eq}}^{\text{pl}}$ is the temperature-dependent equivalent plastic strain. The incremental and iterative FE form to compute the displacement field (u) and stress field (σ) are taken from [21] as:

$$\left(\mathbf{K}_m(\sigma)_{t+\Delta t}^{(i)}\right)\delta\mathbf{u}_{t+\Delta t} = \int_{\Omega} \mathbf{B}^T\left(\mathbf{D}(\sigma)_{t+\Delta t}^{(i)}\right)\alpha(T_{t+\Delta t})\Delta\mathbf{T}\mathrm{d}\Omega$$

$$+\frac{9}{4}\int_{\Omega} \frac{\mathbf{B}^T\mathbf{D}_{t+\Delta t}^{e}\mathbf{s}_{t+\Delta t}^{(i)}}{\left(\sigma_{eff}^{2}(3G+H)\right)_{t+\Delta t}^{(i)}}\left(\frac{\partial F}{\partial T}\Delta T\right)_{t+\Delta t}\mathrm{d}\Omega - \left(\mathbf{K}_m(\sigma)_{t+\Delta t}^{(i)}\right)\left(\mathbf{u}_{t+\Delta t}^{(i)} - \mathbf{u}_t\right) \tag{9}$$

$$\mathbf{u}_{t+\Delta t}^{(i+1)} = \mathbf{u}_{t+\Delta t}^{(i)} + \delta\mathbf{u}_{t+\Delta t}. \tag{10}$$

Equation (9) incorporates the mechanical stiffness matrix ($\mathbf{K}_m$), elasto-plastic modulus matrix ($\mathbf{D}$), temperature-dependent coefficient of thermal expansion (α), and all other quantities involved in Eq. (8). The computation proceeds through the stress-return mapping using the effective stress function (ESF) algorithm of [22], which has also been described in [21]. Finally, Eq. (10) is used to update the displacement field until some convergence is reached. The strains and stresses are also computed in every iteration. The weld deposition along the weld path is modelled using the element activation and deactivation scheme [21], where the elements which have not been formed or have been melted are deactivated by providing a severely reduced value of stiffness, thermal conductivity, and heat capacity. As welding progresses, the elements along the weld path are successively activated by restoring them with their original properties.

The FE procedure described in this section can be used to generate input–output datasets for training and testing an ANN model. The inputs would comprise of some of the physically admissible weld-process parameters and some geometrical dimensions of the specimen. The outputs would be the longitudinal and transverse residual stress at the point of query. In this regard, a brief description of ANN models is provided in the following section.

2.2 Artificial Neural Networks (ANNs) for Estimating Residual Stresses

Artificial neural networks (ANNs) have significant importance in computational engineering domains. They can assist in avoiding simulations which are repetitive with respect to similar type of inputs and thereby save huge computation time. The basic components of an ANN are described below.

(a) **Neuron**: A neuron is basically a cell or node which consists of information such as an input feature, a processed or activated data point or an output.
(b) **Layer**: A layer is basically a group of neurons, which intakes information from a preceding group of neurons and processes the information for a succeeding set of neurons.
(c) **Weights and biases**: These are used to form linear combinations of the input features or processed features.
(d) **Activation functions**: These are used to modify the linear combinations in order to incorporate the nonlinearities which are inherent in the data.

For the input layer, the values at the neurons are the values of the input features for a training example ($\mathbf{x}^{(l)}$). The values at the neurons of any layer (l) are modified to new values ($\mathbf{j}^{(l)}$) as:

$$\mathbf{j}^{(l)} = \mathbf{W}^{(l)}\mathbf{x}^{(l)} + \mathbf{b}^{(l)}. \tag{11}$$

In Eq. (11), $\mathbf{W}^{(l)}$ is the weight matrix for mapping from layer l to $l + 1$, and $\mathbf{b}^{(l)}$ is the bias in layer l. An activation function (h) processes the modified values as:

$$\boldsymbol{\psi}^{(l+1)} = h(\mathbf{j}^{(l)}) \tag{12}$$

In most cases, sigmoid or ReLU activation functions are used, which are given as:

$$h(z) = \begin{cases} (1 + \exp(-z))^{-1} & , \text{ sigmoid} \\ \max(0, z) & , \text{ ReLU} \end{cases}. \tag{13}$$

The above process of determining the activated values sequentially from the first hidden layer to the last (output) layer is known as forward propagation. Subsequently, the errors accumulated in each layer are computed as:

$$\mathbf{e}^{(l)} = \begin{cases} \boldsymbol{\psi}^{(l)} - \mathbf{y} & , \text{ if } l \text{ is the last layer} \\ \mathbf{W}^{T}\mathbf{e}^{(l+1)} * \boldsymbol{\psi}'^{(l)} & , \text{ for other layers except input layer} \end{cases}. \tag{14}$$

Equation (14) is computed sequentially from the last layer to the first hidden layer. The symbol (*) in the second part of Eq. (14) indicates piecewise product of vectors. This process is known as backward propagation. Both forward and backward propagation are carried out for all training examples (n), and the total gradient ($\mathbf{R}$) of the cost incurred in the neural network at any layer (l) is obtained as,

$$\mathbf{R}^{(l)} = \frac{1}{n}\sum_{i=1}^{n}\left(\mathbf{e}^{(l)}\left(\boldsymbol{\psi}^{(l)}\right)^{T}\right)_{n}. \tag{15}$$

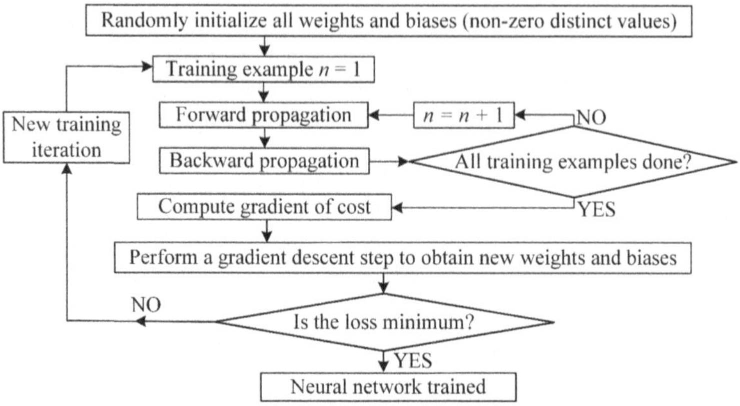

Fig. 2 Flowchart for training an ANN

This cost is then minimised using optimisation algorithms such as gradient descent. The overall algorithm for ANN is summarised in the flowchart shown in Fig. 2.

3 Summary of the Present Work Scheme

The FE method described in Sect. 2.1 is used for input–output data generation. The accuracy of the FE model is ensured through validation with experimental estimation of residual stresses available in literature. A part of the generated data is used train the ANN model described in Sect. 2.2. The other part of the data is used to test the ANN model for its effectiveness. The developed ANN model can be used to predict the longitudinal and transverse residual stresses for new input data, by fully avoiding the time-consuming FE analysis. In order to ensure validity of the ANN model, predictions made by the ANN are cross-checked for a few data points with FE simulation results. The summary of the entire process is described in Fig. 3.

4 Details of Weld Problem with Unequal Plate Thickness

In order to implement the procedure of residual stress prediction described in Sects. 2.1–2.3, a sample weld problem is taken up for study. The specimen description is taken from the work of [23], as shown in Fig. 4 which consists of welding two plates of stainless steel (SS) grade 304 having widths of 150 mm, lengths of 400 mm along which they are welded, and thickness of each plate being 2.5 and 4 mm. The specimen of unequal plate thickness is taken up for analysis, in order to incorporate the variation of thicknesses as input parameters for the present study of ANN model.

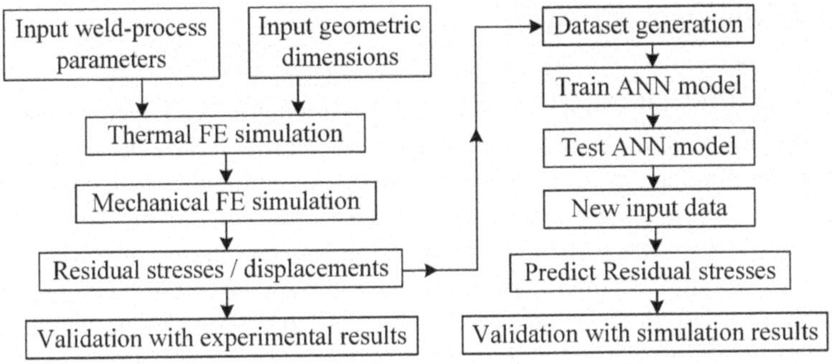

Fig. 3 Flowchart for overall procedure of predicting residual stresses using ANN

Fig. 4 Specimen geometry for welding analysis [23]

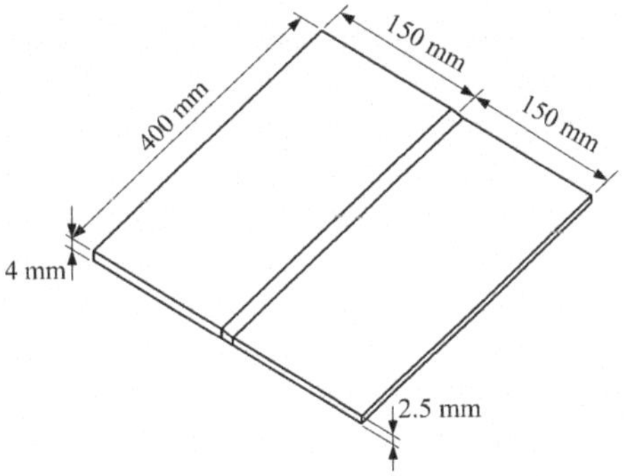

The temperature-dependent thermal and mechanical properties of SS304, phase-transition parameters, and weld-process parameters for the simulated sample problem (Fig. 4) are directly taken from [23] and are not reported here. The incorporation of latent heat into the numerical model is performed using a nonlinear phase-change model for SS304, which has been described in [21].

5 Results and Discussion

For the weld problem described in Sect. 3, the implementation of the FE analysis and ANN are both performed using in-house codes. The results obtained are discussed as follows:

5.1 FE Simulation Results

For performing the FE simulation, 8-noded isoparametric brick elements are used for meshing the specimen. The welding arc passes through the centreline of plates along the longitudinal direction, thereby causing severe thermal gradients around the centreline. Therefore, an adequately refined mesh is provided in this central region. In order to estimate the sufficiency of the refinement, a mesh convergence study is performed on the specimen. As shown in Fig. 5a–c, a coarse mesh, a medium mesh, and a fine mesh are constructed by varying the characteristic length (l_e) of the elements near the weld central region. The thermal FE simulation is performed on the specimen using the Goldak's parameters stated in [21]. The time-temperature distribution is determined from the FE simulation for the entire specimen, using all three mesh sizes.

Experimental measurements of temperature were performed by [23] at some selected locations on the specimen by fixing thermocouples during the welding process. In the present work, the numerically obtained temperature profiles are verified with the experimental results of [23] at two points having distances of 10 mm on either side of the weld centreline at the mid-region of the top surface. The obtained temperature plots are shown in Fig. 6. It can be observed from the plots that the temperature at the points of investigation has a steep rise when the moving heat source passes by their vicinity, thereby obeying the physical process of heat flow. Subsequently, the temperature starts to drop gradually, owing to the convective heat transfer occurring from the surfaces of the specimen. Further, from the comparison of Fig. 6a, b, it is observed that the peak temperature at the thicker region of the plate is higher than the thinner region. This can be attributed to the higher amount heat transferring through the thicker region of the specimen, which has also been ascertained by [23].

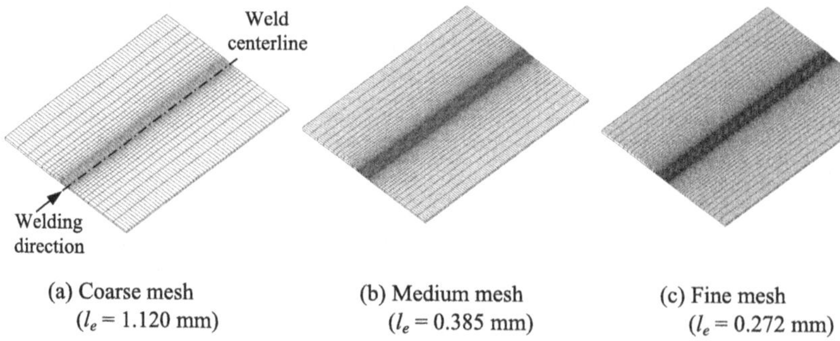

Fig. 5 Mesh sizes used for convergence study of FE thermal simulation

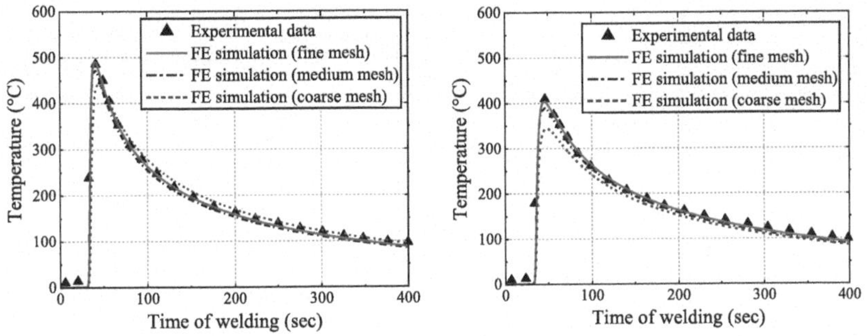

(a) Time-temperature plot at 10 mm from the weld centerline for the 2.5 mm thick region

(b) Time-temperature plot at 10 mm from the weld centerline for the 4 mm thick region

Fig. 6 Comparison of the temperature profiles between experimental data [23] and FE simulations of present work using different mesh sizes

In Fig. 6, the FE simulations for each mesh appear to follow the experimental trend of temperature evolution. In order to quantify the convergence of the mesh, the error (δ) in the simulation using any particular mesh is computed as per Eq. (16). Here, m is the number of times the temperature is measured experimentally during welding, T_{exp} are the values of measured temperatures, and T_{fe} are the values of simulated temperatures.

$$\delta = \left[\frac{1}{m} \sum_{i=1}^{m} \left(\frac{T_{\text{exp}} - T_{\text{fe}}}{T_{\text{exp}}} \right) \right] \times 100. \quad (16)$$

The percentage errors obtained for each mesh using Eq. (16) are plotted in Fig. 7. It can be observed from the plot that the error between the experimental measurements and FE simulations initially decreases with increase of mesh density (represented by $1/l_e$) but tends to become constant for the medium and fine meshes. Therefore, the fine mesh used in the present work is considered to have sufficient refinement in order to accurately capture the high thermal gradients occurring during the welding process. This mesh is represented in Fig. 5c and is used further for the mechanical FE analysis of estimating the residual stresses in the domain.

The residual distortions and residual displacements are computed through the mechanical analysis by maintaining one-to-one correspondence with thermal simulation. This is ensured by using the same number of time increments and time steps in both thermal and the mechanical FE analyses. The experimental measurements of the vertical displacements along the transverse axis at the upper surface of the specimen were provided by [23]. Therefore, the values of vertical displacement computed by the FE simulation using the fine mesh are compared with it and plotted as shown in Fig. 8. It can be observed from the plots that the displacements at the 4-mm-thick portion as well as the 2.5-mm-thick portion of the specimen are agreeable with the experimental measurements. The 2.5-mm-thick portion has a lower

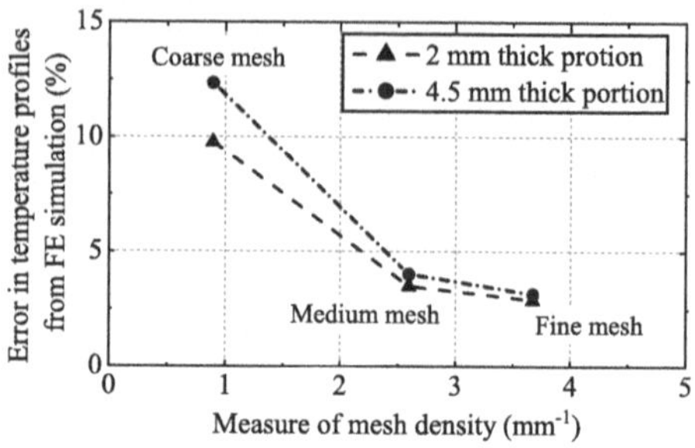

Fig. 7 Variation of error between experimental and simulated temperature profiles with change of mesh density ($1/l_e$)

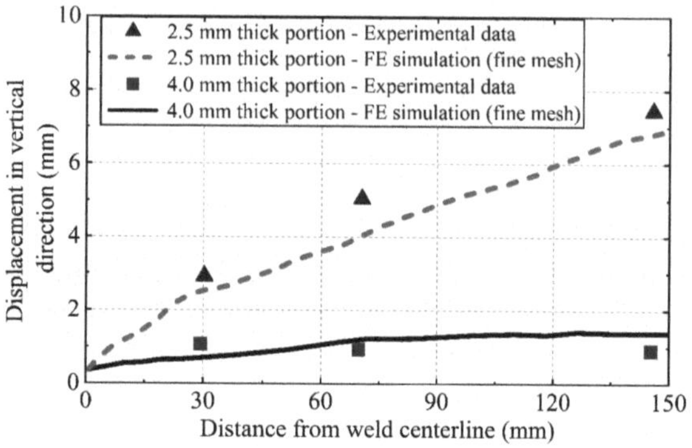

Fig. 8 Comparison of the vertical displacement at the top surface of the welded specimen between experimental data [23] and FE simulation using fine mesh (present work)

stiffness than 4-mm-thick portion, and therefore, it undergoes a higher amount of vertical displacement. Moreover, the unconstrained sides of the specimen undergo higher displacement as compared to the inner regions.

The longitudinal residual stresses (σ_1) and transverse residual stresses (σ_2) corresponding to the residual displacements are shown as contour plots in Fig. 9a, b. It can be observed that the residual stresses are tensile in nature near the weld centreline, with longitudinal stresses being more severe than transverse stresses. This is mostly due to the longitudinally unrelaxed heat-affected zone (HAZ) near the weld bead. The material ahead gets compressed by the expanding HAZ, thereby developing compressive residual longitudinal stresses. The bulk of the specimen towards the edges, further away from the weld centreline, undergoes relatively free expansions and contractions and hence develops very less stress.

5.2 *ANN Model Development and Results*

The results of FE analysis presented in Sect. 4.1 correspond to the input weld conditions and input geometrical dimensions of [23]. In order to generate data to train an ANN, the input conditions and the geometrical dimensions are varied, and the values of longitudinal and transverse residual stresses are obtained at coordinate locations

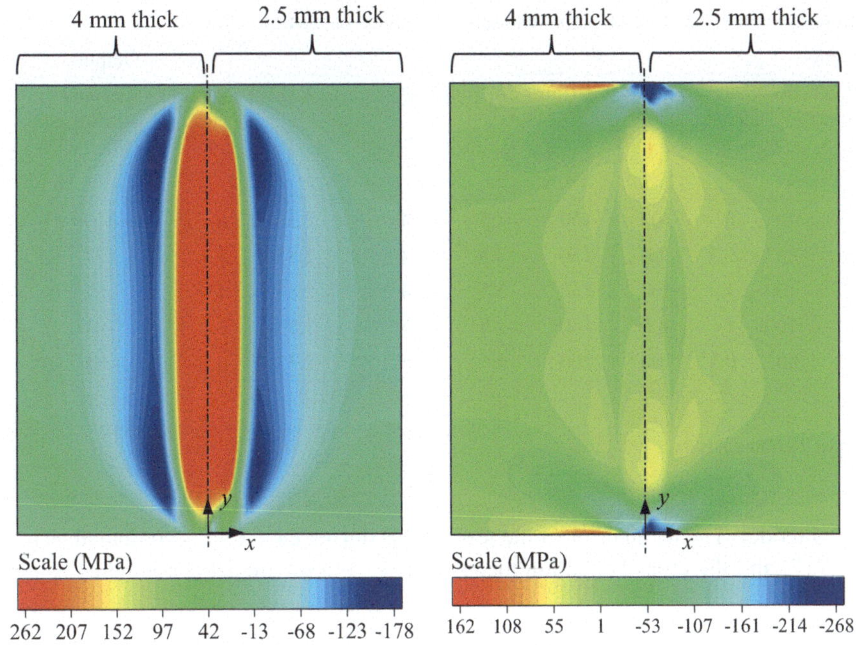

(a) Longitudinal residual stress (σ_1) (b) Transverse residual stress (σ_2)

Fig. 9 Residual stress distribution on the upper surface of the welded specimen

(*x*, *y*, *z*) of the integration points in the FE mesh. Therefore, a set of input–output data is obtained to develop an ANN model. The variations of the input parameters are listed in Table 1. A total of 216 combinations of input parameters are analysed to obtain the residual stresses. A total of 100 datasets based on coordinate locations are selected from the results of each simulation, thereby accumulating a total of 21,600 examples for developing an ANN model. A sample list of some example data is shown in Table 2.

The training set for ANN comprised of 80% of the examples, while the testing set comprised of 20% of the examples. A fivefold cross-validation is utilised by shuffling the training and testing sets, in order to ensure consistent accuracy of the

Table 1 Variations of input parameters for welding analysis

Specimen length (*l*) (mm)	Specimen width (*b*) (mm)	Plate thickness (h_1) (mm) (1st plate)	Plate thickness (h_2) (mm) (2nd plate)	Welding power (P = *VI*) (kW)	Welding speed (*w*) (mm/s)
300	150	2.5	4	1.83	4
400	200	3.5	5	2.33	5
500				2.83	6

Table 2 Sample list of input–output datasets for training and testing ANN

Inputs									Outputs	
Point coordinates (mm)			l (mm)	b (mm)	h_1 (mm)	h_2 (mm)	P (kW)	w (mm)	Residual stresses (MPa)	
x	y	z							σ_1	σ_2
25	40	0	300	150	2.5	4	2.33	6	210	51
35	80	0.5	300	150	2.5	4	2.83	5	177	12
40	100	1	300	150	3.5	4	2.83	4	165	19
80	100	1	400	150	2.5	5	1.83	6	– 104	7
80	80	0.5	400	200	2.5	5	2.33	6	– 85	6
⋮	⋮	⋮	⋮	⋮	⋮	⋮	⋮	⋮	⋮	⋮
Upto 21,600 datasets										

ANN model. The mean error of the test set (e_t) during each fold is computed as per Eq. (17) and the cross-validation mean error (e_{cv}) in the model is computed as per Eq. (18).

$$e_t = \frac{1}{2n_t}\left(\sum_{i=1}^{n_t}(\sigma_l^{(i)} - y_l^{(i)})^2 + \sum_{i=1}^{n_t}(\sigma_t^{(i)} - y_t^{(i)})^2\right) \tag{17}$$

$$e_{cv} = \frac{1}{n_{cv}}\sum_{i=1}^{n_{cv}}(e_t)_i. \tag{18}$$

In Eq. (17), n_t denotes the number of test examples, $\sigma_l^{(i)}$ and $\sigma_t^{(i)}$ denotes the computed longitudinal and transverse residual stresses, $y_l^{(i)}$ and $y_t^{(i)}$ denotes the predicted longitudinal and transverse residual stresses. In Eq. (18), $n_{\text{cv}} = 5$ for a five-fold cross-validation strategy. The cross-validation errors are computed for various ANN models having different numbers of neurons, hidden layers, and activation functions. The results are plotted in Fig. 13. It can be observed from Fig. 13 that the errors on the test-data are within 8% for all the trained ANN models. Moreover, the test-data error becomes almost constant after training using three hidden layers for sigmoid as well as ReLU-activated models. The error decreases slightly when the number of neurons in each hidden layer is increased from 15 to 25. However, it was felt that further increase in the number of neurons or hidden layers would increase the training time of the ANN without causing any significant reduction in the test-data error. Therefore, the sigmoid-activated ANN model with 3 hidden layers and 25 neurons per hidden layer is selected for predicting outputs for previously unseen inputs. As such, datasets with new inputs were fed into the trained and tested ANN model, and predictions were obtained as listed in Table 3.

Table 3 New input parameters and prediction of residual stress by ANN

Data no	New input parameters[a]									Predictions[a]	
	x	y	z	l	b	h_1	h_2	P	s_w	σ_1	σ_2
1	0	175	0	350	175	3	5	2	5	269	66
2	20	250	0	350	175	3	5	2	5	228	37
3	40	300	1	350	175	3	5	2	5	134	10
4	0	250	0.5	550	225	2.5	3.5	2.5	5.5	272	89
5	10	300	0.5	550	225	2.5	3.5	2.5	5.5	238	45
6	20	525	1	550	225	2.5	3.5	2.5	5.5	197	– 80

[a]units are same as Table 2

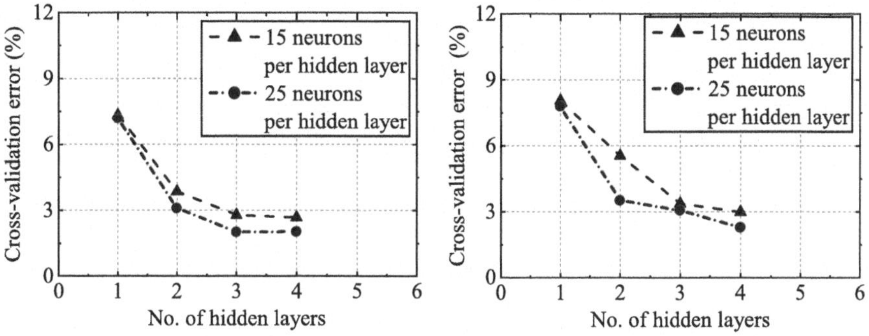

(a) Plot for sigmoid activated ANN (b) Plot for ReLU activated

Fig. 10 Plots showing the variation of ANN test-error with varying numbers of hidden layers, neurons per hidden layer, and activation functions

For evaluating the performance of the ANN model, the inputs in Table 3 are utilised in new FE simulations to compute the longitudinal and transverse residual stresses. The percentage variation between the ANN-predicted residual stresses and the FE-computed residual stresses is shown in Fig. 11. It can be observed that for all new data points, the ANN-predicted values are within 6% of the FE-simulated values (Fig. 10).

6 Conclusions

The present work has firstly described a procedure using finite element (FE) simulations and artificial neural networks (ANNs) for predicting the residual stresses in the component. The major conclusions which drawn from the present work are:

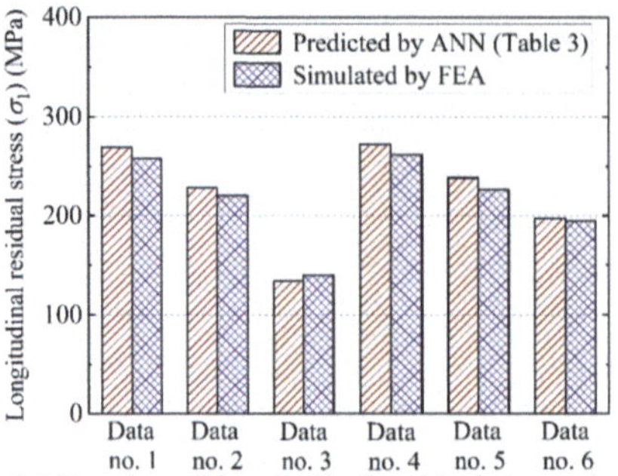

(a) Bar plots of longitudinal residual stress at specific data-points

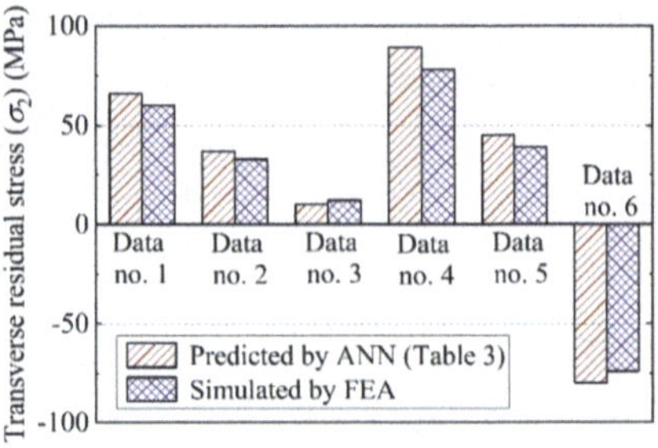

(b) Bar plots of transverse residual stress at specific data-points

Fig. 11 Comparison of the magnitudes of residual stresses predicted using a trained ANN with the magnitudes computed using FEA simulation

- FE simulations act as a suitable tool for setting up a framework for ANN models. Various parameters of the welding process can be varied to obtain residual stresses, thereby generating a set of inputs and outputs to train the ANN.
- The variation of geometric dimensions, including length, width, and thickness of both plates, can be suitably incorporated in training the ANN model. Therefore, the trained and tested models can accommodate specimen size effects in predicting the residual stresses.
- The residual stresses can be predicted by ANN at any desired location in the domain of the welded plate and are not restricted to any specific or predefined location as described in various literatures.
- The ANN model predicts the residual stresses for new data with close agreement (within 6%) of the FE simulations. Therefore, for any plate specimen of the same material with different dimensions and weld conditions, the welding residual stresses can be directly predicted by the ANN. Therefore, experimental measurements and FE simulations for new inputs can be fully avoided, which would be beneficial in saving experimentation costs and computational time.

Acknowledgements The present work was carried out as a part of the doctoral research work of Mr. Sandipan Baruah under the fellowship of the Ministry of Education, Government of India. We gratefully acknowledge the computational support provided by the Department of Mechanical and Industrial Engineering, Indian Institute of Technology (IIT), Roorkee.

References

1. Ueda Y, Yamakawa T (1973) Analysis of thermal elastic-plastic behavior of metals during welding by finite element method. J Jpn Welding Soc 42(6):567–577. https://doi.org/10.2207/qjjws1943.42.567

2. Friedman E (1975) Thermomechanical analysis of the welding process using the finite element method. J Pressure Vessel Technol 97(3):206–213. https://doi.org/10.1115/1.3454296
3. Goldak J, Chakravarti A, Bibby M (1984) A new finite element model for welding heat sources. Metall Trans B 15(2):299–305. https://doi.org/10.1007/BF02667333
4. Karlsson RI, Josefson BL (1990) Three-dimensional finite element analysis of temperatures and stresses in a single-pass butt-welded pipe. J Pressure Vessel Technol 112(1):76–84. https://doi.org/10.1115/1.2928591
5. Murthy YVLN, Rao GV, Iyer PK (1996) Numerical simulation of welding and quenching processes using transient thermal and thermo-elasto-plastic formulations. Comput Struct 60(1):131–154. https://doi.org/10.1016/0045-7949(95)00359-2
6. Lindgren LE, Runnemalm H, Näsström MO (1999) Simulation of multipass welding of a thick plate. Int J Numer Meth Eng 44(9):1301–1316. https://doi.org/10.1002/(SICI)1097-0207(19990330)44:9%3c1301::AID-NME479%3e3.0.CO;2-K
7. Berglund D, Lindgren LE, Lundbäck A (2001). Three-dimensional finite element simulation of laser welded stainless steel plate. In: Proceedings of the seventh international conference on numerical methods in industrial forming processes, Toyohashi, Japan, pp 1119–1124
8. Deng D, Liang W, Murakawa H (2007) Determination of welding deformation in fillet-welded joint by means of numerical simulation and comparison with experimental measurements. J Mater Process Technol 183(2–3):219–225. https://doi.org/10.1016/j.jmatprotec.2006.10.013
9. Barsoum Z (2008) Residual stress analysis and fatigue of multi-pass welded tubular structures. Eng Fail Anal 15(7):863–874. https://doi.org/10.1016/j.engfailanal.2007.11.016
10. Seyyedian CM, Haghpanahi M, Sedighi M (2010) Investigation of the effect of clamping on residual stresses and distortions in buttwelded plates. Trans B Mech Eng 17(5):387–394. http://scientiairanica.sharif.edu/article_3338_dd1fd1dfa45c135bf89a5997af6c6894.pdf
11. Perić M, Tonković Z, Rodić A, Surjak M, Garašić I, Boras I, Švaić S (2014) Numerical analysis and experimental investigation of welding residual stresses and distortions in a T-joint fillet weld. Mater Des 53:1052–1063. https://doi.org/10.1016/j.matdes.2013.08.011
12. Almeida DF, Martins RF, Cardoso JB (2017) Numerical simulation of residual stresses induced by TIG butt-welding of thin plates made of AISI 316L stainless steel. Proc Struct Integrity 5:633–639. https://doi.org/10.1016/j.prostr.2017.07.032
13. Ahmadzadeh M, Fard AH, Saranjam B, Salimi HR (2012) Prediction of residual stresses in gas arc welding by back propagation neural network. NDT E Int 52:136–143. https://doi.org/10.1016/j.ndteint.2012.07.009
14. Baruah S, Singh IV (2023) A framework based on nonlinear FE simulations and artificial neural networks for estimating the thermal profile in arc welding. Finite Elem Anal Des 226:104024. https://doi.org/10.1016/j.finel.2023.104024
15. Choobi MS, Haghpanahi M, Sedighi M (2012) Prediction of welding-induced angular distortions in thin butt-welded plates using artificial neural networks. Comput Mater Sci 62:152–159. https://doi.org/10.1016/j.commatsci.2012.05.032
16. Narayanareddy VV, Chandrasekhar N, Vasudevan M, Muthukumaran S, Vasantharaja P (2016) Numerical simulation and artificial neural network modeling for predicting welding-induced distortion in butt-welded 304L stainless steel plates. Metall Mater Trans B 47(1):702–713. https://doi.org/10.1007/s11663-015-0468-x
17. Tian L, Luo Y (2020) A study on the prediction of inherent deformation in fillet-welded joint using support vector machine and genetic optimization algorithm. J Intell Manuf 31(3):575–596. https://doi.org/10.1007/s10845-019-01469-w
18. Mathew J, Moat RJ, Paddea S, Fitzpatrick ME, Bouchard PJ (2017) Prediction of residual stresses in girth welded pipes using an artificial neural network approach. Int J Press Vessels Pip 150:89–95. https://doi.org/10.1016/j.ijpvp.2017.01.002
19. Tian L, Luo Y, Wang Y, Wu X (2014) Prediction of transverse and angular distortions of gas tungsten arc bead-on-plate welding using artificial neural network. Mater Des 1980–2015(54):458–472. https://doi.org/10.1016/j.matdes.2013.08.082
20. Liu F, Tao C, Dong Z, Jiang K, Zhou S, Zhang Z, Shen C (2021) Prediction of welding residual stress and deformation in electro-gas welding using artificial neural network. Mater Today Commun 29. https://doi.org/10.1016/j.mtcomm.2021.102786

21. Baruah S, Sarkar S, Singh IV, Mishra BK (2022) A computational framework based on FEA, ML and GA for estimation of welding residual stresses. In: Finite elements in analysis and design, vol 205. https://doi.org/10.1016/j.finel.2022.103753
22. Kojić M, Bathe KJ (1987) The 'effective-stress-function' algorithm for thermo-elasto-plasticity and creep. Int J Numer Meth Eng 24(8):1509–1532. https://doi.org/10.1002/nme.1620240808
23. Hashemzadeh M, Chen BQ, Soares CG (2015) Numerical and experimental study on butt weld with dissimilar thickness of thin stainless steel plate. Int J Adv Manuf Technol 78(1):319–330. https://doi.org/10.1007/s00170-014-6597-6.pdf

Blended Binder Geopolymer Concrete: The-Current-State-of-Art

K. Bavithra Devi, M. Dinesh Kumar, and C. Umarani

Abstract Ordinary Portland cement is a versatile material in the construction industry. The production of cement is identified to be the major greenhouse emitter, ejecting about 5–7% of total carbon dioxide (CO_2) released into the environment. Cement industry analysis report—2022 advocates, 'in 2021, the post-COVID-19 pandemic, the hopeful prospect about the growth in Infrastructure and real estate sector is probably increasing the demand for cement, which is expected to rise 8 million tons per annum additionally in the production of cement.' Addressing the above issue, Geopolymers are identified to be an environmentally friendly construction material that alternates ordinary Portland cement. Geopolymers are produced by alkali activation of alumino-silicates, obtained from sources rich in silica and alumina. Geopolymer concrete utilizes supplementary cementing materials (SCM) both natural pozzolans like metakaolin, calcined shale, volcanic tuffs, and industrial by-products such as Fly ash (FA), ground granulated blast furnace slag (GGBS), rice husk ash (RHA), and silica fume (SF) as the sources. Geopolymers have undergone notable development in the research and application field from the day the material stepped into the construction industry. This paper aims to comprehend the different factors such as characteristics of raw materials, the need for blending and proportion of blending, the effect of alkali activator solution, and the effects of proportionately blending raw materials. Furthermore, the review advocates that more research should be done on both material and structural levels to enhance Geopolymer, a potential material in the construction industry.

Keywords Alkali activator solution (AAS) · Fly ash (FA) · Geopolymer (GP) · Ground granulated blast furnace slag (GGBS)

K. Bavithra Devi (✉) · M. Dinesh Kumar · C. Umarani
Division of Structural Engineering, Department of Civil Engineering, College of Engineering-Guindy, Anna University Chennai, Tamil Nadu, Chennai 600025, India
e-mail: kbavithradevi@gmail.com

C. Umarani
e-mail: umarani@annauniv.edu

S. Sidhardh et al. (eds.), *Advances in Structural Integrity for Mechanical, Civil, and Aerospace Applications*, Lecture Notes in Mechanical Engineering,
https://doi.org/10.1007/978-981-97-6367-2_43

1 Introduction

Conventional concrete is a versatile material, with a wide range of mechanical properties, durability, long-life performance, and cost-effectiveness. In contrast, it is a known fact that the cement production process consumes approximately 94.76×10^6 Joules/ton of energy [1], which results from about 5–7% of the total carbon dioxide released into the environment.

Along the side, the production of cement consumes large quantities of raw materials causing over-exploitation of available raw materials. Since properties, lime is the main source of ordinary Portland cement (OPC), it is foreseen that an acute shortage of material may come after 25 years, the World Earth Summit warns, the cement industry to supplement a greener alternative binder for ordinary Portland cement with an expedient material possessing similar or better structural, mechanical, and durability properties [2].

Geopolymers patented by Professor. Joseph Davidovits (1978) show a promising approach in the construction industry as a supplement to ordinary Portland cement. Joseph Davidovits coined the term Geopolymer, to represent a broad range of materials characterized by a chain or network of inorganic molecules and also highlighted the possibility of Geopolymer used by Egyptians in the construction of pyramids [1]. Geopolymers have undergone notable development in the research and application field from the day the material stepped into the construction industry.

This paper aims to comprehend the different factors such as characteristics of raw materials, the need for proportionately blending raw materials and its effect in producing blended Geopolymer concrete, the effect of alkali activator solution, water-binder ratio, etc. These are considered to be the important factors to enhance the structural, mechanical, and durability properties of Geopolymer concrete both in fresh and hardened states.

2 Geopolymer and Geopolymerisation

Geopolymerisation is an integrated process that involves leaching, diffusion, reorientation, condensation, and polymerization. The resulting 3D network structure rearranges and reorganizes creating a wider network, which possesses the binding property. Thus, the highly alkaline environment enhances the polycondensation of silica and alumina sources resulting in the binding compound, Geopolymer concrete[3, 4].

Davidovits formulated the term 'Geopolymer' for a binder material acquired from alkali activation of metakaolin and suggested the name 'Polysialates' with the following empiric formulae [4].

$$\mathrm{Mn}\{-(\mathrm{SiO_2})z-\mathrm{AlO_2}\}n, w\mathrm{H_2O}. \quad (1)$$

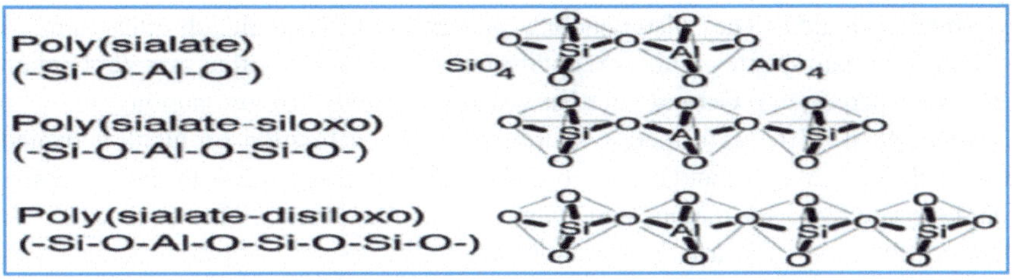

Fig. 1 Chemical structure of polysialate [5]

Geopolymers are chains or networks of mineral molecules linked with co-valent bonds. Davidovits introduces the sialate nomenclature (Fig. 1) to describe the aluminosilicate structures. Sialate is an abbreviation for silicon-oxo-aluminate. The sialate network consists of aluminum and silica tetrahedrally interlinked alternately by sharing all the oxygen atoms [5, 6]. This nomenclature helps in describing the composition of Geopolymers according to their Si/Al ratio, with a ratio of 1.0 being a poly(sialate), 2.0 being a poly(sialate-siloxo), and 3.0 a poly (sialate-di siloxo). In this structure of Geopolymers, the silicon can bond directly with each other (Si–Si) or be linked through 'bridging' oxygen (Si–O-Si) to form independent polymer chains. While aluminum atoms can only replace the silicon atoms in Si–O-Si polymer chains to form Si–O-Al instead.

Puertas et al. [7] studied the consequence of NaOH molar concentration, curing temperature, and Fly ash to slag ratios on the strength and hydration products. The mixture of 50% Fly ash and 50% slag activated with 10 M NaOH and cured at 25 °C, at 28 days of the reaction, develop compressive strengths of about 50 MPa. The complete reaction of slag results in mainly C–S–H gel, with high amounts of tetra-coordinated Al in its structure and Na ions in the interlayer spaces. Though Fly ash is partially dissolved, no hydrated alkaline alumino-silicates with three-dimensional structural characteristics of alkaline activation of Fly ash are reported.

Wang and Scrivener [8] investigated to determine the hydration process of alkali-activated slag cement. The conclusion given by them was regardless of the activator used, the main hydration product is calcium silicate hydrate with a low C–S ratio and varying degree of crystallinity. It is to be noted that hydrates of zeolite groups were not formed.

Lecomte et al. [9] worked on a comparative study about the composition, crystallographic and microstructure by considering four types of cementitious material samples such as ordinary Portland cement (PC), an alkali-activated slag (AAS), a Geopolymer (GP) and a sample containing both alkali-activated slag phase and Geopolymer phase (GP-AAS). In X-ray diffraction, the PC sample contains C–S–H gel phase, ettringite, Portlandite, and quartz, but no crystalline or semi-crystalline phase is formed in GP, GP-AAS, and AAS samples. Compared to other samples, Geopolymer appears to be a composite composed of GGBFS particles surrounded by a binding matrix. They also conclude that except GP sample, all other samples contain carbonate species. Thus, except the GP sample, all the other samples contain carbonate species, and Geopolymer has greater resistance to carbonation.

Ismail et al. [10] performed synchrotron X-ray diffraction on alkali-activated ground blast furnace slag and Fly ash-based Geopolymers to explore the phase formation, followed by SEM imaging and compressive strength tests. They concluded that the calcium class-C Fly ash does not dissolve in the activator solution as readily as the calcium in slag. This seems to increase the final mechanical strength because the calcium is readily available for C–S–H. It was reported, that the addition of water glass to the NaOH activator does not create any new phase in class-C Fly ash Geopolymer, although it essentially enhanced the compressive strength as well as the rate of strength development.

The microstructural evolution of alkali-activated binders based on blast furnace slag, Fly ash, and their blends for the initial six months of sealed curing was assessed by Rangan et al. [11]. They figured out that the nature of the gel in alkali-activated slag-to-Fly ash blended binders is strongly dependent on the slag-to-Fly ash ratio. This study strongly witnessed different mechanisms of calcium release from slag upon the activation in the presence of aluminum silicate precursor. In silicate-activated slag, a structure dominated by C–A–S–H type gel was noticed, while silicate-activated Fly ash is dominated by N–A–S–H Geopolymer gel.

2.1 Geopolymer Concrete

Extensive research on Fly ash-based Geopolymer concrete was carried out by Rangan [12] along with his colleagues and students. They strongly concluded that the low calcium Fly ash-based Geopolymer concrete has excellent compressive strength and is suitable for structural application. In addition, the factors that influence the properties of fresh concrete and hardened concrete were identified. Technically, they suggested that the design provision of current [13] standards and codes can be used to design reinforced low calcium Fly ash-based Geopolymer concrete structural members. Additionally, they concluded that heat-cured low calcium Fly ash-based Geopolymer concrete shows excellent resistance to sulfate attack, good acid resistance, low creep, and very little drying shrinkage. Furthermore, several economic benefits of Geopolymer concrete were also reported.

Wallah and Rangan [12] conducted an elaborate work on the long-term properties of low calcium Fly ash-based Geopolymer concrete, like creep, drying shrinkage, sulfate resistance, and sulfuric acid resistance. They concluded that heat-cured Fly ash undergoes very little drying shrinkage and low creep. It also has excellent resistance to sulfate attacks.

Whereas Sumajouw et al. [14] conducted experimental and analytical studies on the behavior and strength of reinforced Fly ash-based Geopolymer concrete beams and columns. The studies insist that class F Fly ash requires an alkaline solution of high concentration and heat curing for the activation, and materials with the addition of silicates. For activation purposes, a concentration of 8–10 molar hydroxide is often required at a lower concentration. The behavior and elastic properties of hardened Geopolymer concrete were found to be similar to that of Portland cement concrete, using the design standards and codes. Overall, they concluded that low calcium Fly ash-based Geopolymer concrete could be used to replace portland cement.

Yip et al. [15] investigated seven different calcium silicate materials to find out the role of calcium in Geopolymerisation. From the results, they encountered that the coexistence of Geopolymeric gel and C–S–H gel provides satisfactory mechanical properties of the matrices synthesized at low alkalinities. At high alkalinity, Geopolymeric gel appears to be dominant, and thus, calcium has a lesser role in affecting the end product.

With the addition of slag, an increase in the hardening rate was observed in work done by Pulligula and Mondal [16] which concludes that calcium that dissolves from ground granulated blast slag (GGBS) significantly influences both early and later age properties. Due to the availability of free calcium ions, Fly ash dissolution is prolonged, and enhancement in Geopolymer gel formation was found. Nevertheless, the Si to Al ratio is necessary for C–S–H gel formation. The compressive strength of Geopolymer increased with an increase in slag content. The rate of increase in strength is more effective after 20% slag addition in Geopolymer. The strength increment with time is mainly due to C–S–H cementations [17].

Dinesh Kumar and Umarani [18] experimentally investigated the effect of alkali activator for the different alkali activator ratios and with 2.5 M and 5 M NaOH molarity for varying amounts of Fly ash and GGBS and concluded that consistency and setting time decreases with increased in the molarity of NaOH.

Mechanical properties are found to improve when adding calcium compounds such as a Fly ash substitute for the ambient temperature cured samples, whereas decreasing properties have been noted for the 70 °C cured samples [19]. The coexistence of Geopolymeric gel and calcium silicate hydrate (C–S–H) gel was revealed [16, 20]. The formation of CSH gel together with the Geopolymeric gel occurs only at low alkalinity. In the presence of high concentrations of NaOH (greater than 7.5 Molar), the Geopolymeric gel is the predominant phase and traces of calcium precipitates have been found.

A comprehensive overview of the various works carried out on using different additives in the AAS system on the effect on workability was done by Rashad [21, 22]. The findings were that sodium lignosulfonate or sulfonated naphthalene-based superplasticizers failed to improve the workability, whereas improved with the help of polycarboxylate and vinyl copolymer-based superplasticizers.

Lloyd et al. [23] did the studies based on the microstructure of inorganic polymer cement. He reported that the binder is predominantly an aluminosilicate gel, charge-balanced by alkali metal cations, the calcium becomes relatively well dispersed throughout the gel, which was supplied by slag particles.

The microstructure of the binder resulting from hydroxide activation of Fly ash was found less uniform than that which forms in a corresponding silicate-activated system, and this can be justified in terms of a newly developed explanation for the differences in reaction mechanisms between these two systems. In hydroxide activation, the newly formed gel phase nucleates and grows outwards from the ash particle surfaces. In a silicate-activated system, the high silica concentration enables a more homogeneous gelation process to occur throughout the inter-particle volume. Hence, it can be understood to some extent concerning the effects of adding calcium-rich compounds to Fly ash during alkali activation.

An extensive experiment on low calcium Fly ash-based Geopolymer concrete by Noushini et al. [24] has concluded that the existing stress–strain models such as Collins and Mitchell's do not accurately predict the actual stress–strain behavior of low calcium Fly ash-based Geopolymer concrete. They have also proposed a model with a parameter for curing conditions.

Ganesan et al. [25] have attempted studies to determine the stress vs. strain behavior of confined Geopolymer concrete. They have concluded that Mander's model for stress vs. strain behavior for Portland cement concrete can be adopted for Geopolymer concrete with modification in the curve-fitting factor.

The alkali-activated slag performs better in durability properties such as water sorptivity, chloride, and carbonation resistance than portland cement and blended concretes [26, 27]. The performance of alkali-activated Fly ash in terms of weight loss, compressive strength, variations in volume, presence of the products of degradation, and microstructural changes perform better than Portland cement concrete [27]. Zuhua et al. [28] have studied the role of water in Geopolymeric reactions and found out that an optimum percentage of non-evaporable water is necessary for strength characteristics.

3 Strength Aspect of Geopolymer Concrete

Factually, the incorporation of steel fibers increases the tensile and flexural strength of concrete and improves the energy absorption capacity, so as of ductility of concrete. There can be a considerable difference between a Geopolymer binder and an ordinary Portland cement concrete OPC binder in terms of chemical reactions and microstructure developed. Furthermore, only a few literatures are available on the fracture behavior of Geopolymer concrete.

Pan et al. [29] have concluding that Geopolymer Concrete (GC) is more brittle than OPC and the tensile strength of Geopolymer concrete is more, fracture energy and elastic modulus are lower.

Accordingly, Sarker et al. [30] have also reported that the fracture energy of heat-cured Fly ash-based GC is similar to OPC with the higher tensile, and bond strength and possesses more brittle behavior of GC.

The results of the work done by Nath and Sarker [31] on fracture properties of GGBS—Fly ash blended Geopolymer concrete cured in ambient temperature reported the following findings. The fracture properties are influenced by the mixture compositions and water content. Geopolymer concrete with slag and Fly ash blend exhibited more ductility than low calcium Fly ash-based Geopolymer concrete with heat curing.

An improvement in tension stiffening and cracking characteristics of GC with the addition of steel fibers was reported by Ganesan et al. [25]. They have also proposed a method for predicting the width of cracks in reinforced GC elements under uniaxial tension. Rao et al. [32] have disclosed the fact that a 25% more characteristic strength is required for Fly ash and GGBS-based Geopolymer concrete. However, there are no published data on fracture properties of ambient air-cured Fly ash and GGBS blended steel fiber-reinforced Geopolymer concrete, where a hybrid microstructure is formed.

On the contrary, [17, 29] there is a necessity to develop a large database on material properties of Geopolymer concrete such as elastic modulus, stress–strain behavior, and fracture parameters to arrive at design parameters.

Ng et al. [33]based on SEM images, proposed that binders mixed with a lower mass ratio of SS/SH from 2.0 to 2.5 tend to react more efficiently due to the high surface area and bind without being blocked by the presence of extra water. It shows that a higher mass ratio of sodium silicate resulted in unreacted Fly ash particles in the case of 100% Fly ash. It is understood that too much AAS in the mix hampered the process of Geopolymerisation by limiting the interaction between the binder and AAS.

Pratyush Kumar et al. [34] studied the experimental investigation on GGBS, Metakaolin—80–20%, 50–50%, and 20–80%, a 10 M solution of sodium hydroxide with sodium silicate as an alkaline activator solution. The samples characterized and the result shows that FTIR peaks at around 972 cm^{-1} indicate some degree of Geopolymerisation, representing the presence of stretching vibrations of Al-O and Si–O. The 1448 cm^{-1} indicates the presence of a slight amount of carbonation as it represents the reaction of atmospheric carbon dioxide with the Geopolymer matrix. The peaks at around 1000 cm^{-1} indicate the degree of Geopolymerisation that has taken place.

4 Structural Behavior of Geopolymer Concrete

An extensive study on low calcium Fly ash-based doubly reinforced Geopolymer concrete beams and columns was carried out by Sumajouw et al. [14]. Twelve beams of 3.3-m length with 0.2 × 0.3-m cross-sections were tested. The beams were subjected to four-point bending with the mid-one-third span in pure flexure. They have concluded the following:

- The crack patterns observed for reinforced Geopolymer concrete beams were similar to those reported in the literature for reinforced Portland cement concrete beams.
- The ductility of reinforced Geopolymer concrete beams increased as the tensile reinforcement ratio decreased.
- Service load deflections of test beams were correlated with values calculated using the serviceability provisions of Australian draft code AS 3600.

Datteraya et al. [35] investigated ambient temperature cured reinforced Geopolymer concrete beams. Six Geopolymer concrete mixes were formulated with varying ratios of Fly ash and GGBS, and eighteen beams were tested. The Geopolymer concrete mixes yielded a strength in the range of 17–63 MPa. The reinforcements were designed considering a balanced section for the expected characteristic strength. It was observed that deflections at different stages, cracking load, service load, and peak load were higher for Geopolymer concrete beams. The ductility factor was comparable to that of OPC beams. The studies showed that the conventional reinforced concrete theory could be used for reinforced Geopolymer concrete flexural beams for computation of moment capacity, deflection, and crack width within reasonable limits.

Yong Sing et al. [36] Fly ash (FA) studied ladle furnace slag (LFS), sodium silicate, and sodium hydroxide solution were mixed to prepare the alkali activator. The NaOH molarity of 6, 8, 10, 12, 14 M, alkali activator (Na_2SiO_3/ NaOH) mass ratio—2.0, 2.5, 3.0, 3.5, 4.0, 4.5 was used. An increase in the molarity of NaOH produced a denser Geopolymer with lower porosity and water absorption. In this context, an increase in the molarity of the alkali solution facilitated the formation of a Geopolymer matrix due to the improved dissolution rate of Fly ash particles. The high Na_2SiO_3/ NaOH ratio promoted the formation of mobile precursors, which improved the reaction kinetics by increasing the Geopolymerisation extent and thus enhanced the flexural strength and stiffness of thin Geopolymers. An adverse effect was observed in Geopolymer activated with 14 M owing to the disruption of the Geopolymerisation process due to the presence of excessive OH ions in the Geopolymer matrix, causing more pores, unreacted Fly ash, and subsequently lower flexural strength. The coexistence of C–S–H gels with C–A–S–H gels in the Geopolymer matrix helped to achieve the highest flexural strength of Geopolymer as observed.

The shear tests were conducted by Ng et al. [37] on five series of 250 mm deep by 120 mm wide beams spanning 2250 mm without web reinforcement (stirrups) but with end-hooked and straight steel fibers in various fiber volumetric dosages, ranging from 0 to 1.5%. The test results show that the shear strength increases significantly as the fiber content increases and improvement in the cracking behavior is achieved with the addition of fibers.

The standard RILEM pull-out test was carried out by Castel and Foster [38] to determine the Geopolymer concrete bond on both deformed and smooth reinforcing steel bars. These tests aim to assess the development of the bond strength from 24 h to 28 days after casting, with different heat curing conditions. The 28-day bond strength and the overall bond stress–slip behavior of the Geopolymer concrete were similar to those previously reported for OPC-based concretes. Heat curing enhanced the early bond strength.

The work done by Ganesan et al. [39] reports that the addition of steel fibers to GPC significantly enhances the bond strength of Geopolymer concrete with reinforcement. Pull-out tests under monotonic loading were carried out on high-yield strength deformed bars of 10, 12, and 16 mm diameter and M 40 grade concrete. Different volume fractions (0.25, 0.50, 0.75, and 1%) of crimped steel fibers were used.

Furthermore, studies conducted by Vijai et al. [40] dealt with fiber-reinforced Geopolymer concrete composite beams. Three types of fibers such as steel, polypropylene, and glass fibers in different volume fractions, were incorporated. The ultimate load-carrying capacity is more significant with an increase in Steel fiber content. Whereas ultimate load-carrying capacity is marginal with propylene fiber and not so significant in the case of glass fiber. Similarly, reinforcing steel fibers to Geopolymer concrete increases the ductility compared to propylene and Glass fiber.

5 Geopolymer Concrete Applications

Geopolymer and the concept of Geopolymerisation are well accepted by many scientists [5, 9]. In the recent past, many studies on Geopolymer cement and Geopolymer concrete have been carried out, recent research on Geopolymer concrete includes the development of ultra-high-performance Geopolymer concrete [41]. Also, researchers have reported on ultra-high-performance concrete incorporating copper slag as fine aggregate [42], etc. Although Geopolymer concrete is a compelling alternative for sustainable material, the present application on buildings and other structural components is limited due to the lack of structural design and codes [18].

However, a few modifications are needed to incorporate the difference in fresh property behavior such as viscosity and properties controlling setting time. Some of the many such applications are featured in this section, and Fig. 2a displays the first building constructed in the world with structural Geopolymer concrete with three suspended floors made from structural Geopolymer concrete [44]. Fig. 2b represents the Demo platform laid with Geopolymer concrete at CSIR-SERC.

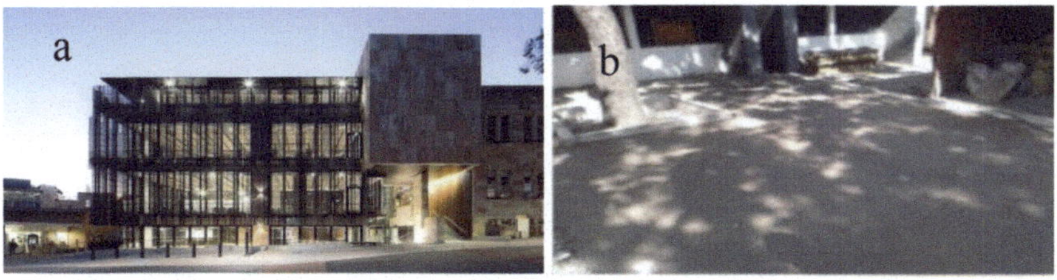

Fig. 2 **a** Queen's Land University GCI Building. *Source* www.geopolymer.org/news/worlds-first-public-building-with-structural-geoopolymer-concrete/. **b** Demo platform laid with Geopolymer concrete in CSIR-SERC. *Source* https://serc.res.in/pilot-cum-demonstration-study-use-fly-ash-geopolymer-concrete-pavement-construction-ntp-unchahar

6 Recent Advancements in Geopolymer

A growing pattern is evidenced in favor of the usability, durability, and marketability of Geopolymer concrete under service conditions in civil infrastructure applications. The practical implementation of the Geopolymer technology done worldwide is noticeable. The application of Geopolymers has been developed since 1979 in a vast range, and nowadays, India is working a lot in the field of Geopolymer concrete.

The highly alkaline environment enhances the polycondensation of silica and alumina sources resulting in the binding compound, Geopolymer concrete. Hence, generally, these materials are acid-resistant.

Australia's first greenfield airport constructed for 40,000 m^3 is a potential development of the Geopolymer concrete synthesized by blending Fly ash and slag. The massive quantity of Geopolymer concrete required for which a twin mix batch was (Fig. 3) established in the site for producing Geopolymer concrete appears to be a challenging task. Unlike ordinary concrete Geopolymer needs provision for alkali activators, which are added to raw materials as solutions [43].

The Brisbane well camp airport is the Australia's first greenfield airport developed by Wagner's Earth friendly. This airport is considered as the largest application of the newly developed Geopolymer concrete. The successful functioning of the runway of the Brisbane well camp airport, Australia is shown in Fig. 4 [43]. In recent days, Geopolymer cements have been produced and more researchers are growing in the 3D printing technique using Geopolymer. The world's first Geopolymer 3D printed house was constructed by Renca which is shown in Fig. 5. This was considered a huge milestone in the progress of sustainable materials [44].

Furthermore, proper standards, guidelines, and codal provisions are required to streamline the production of the material and its application. The growing trend in the field of research to mitigate the difficulties in arriving at the mix design ratio and perplexity in the curing process of concrete can be observed in recent times.

Fig. 3 Twin mobile wet mix batch plant. *Source* Glasby et al. [43]

Fig. 4 Brisbane well camp airport [43]

Fig. 5 3D printed house using Geopolymer [44]. *Source* Glasby et al. [43]. https://www.renca.org/blog

7 Conclusion

The following are the conclusions drawn from the extensive literature review carried out on blended binder Geopolymer concrete.

1. Class F Fly ash was an acceptable material for producing Geopolymer concrete compared to class-C Fly ash.
2. Calcium-rich class-C Fly ash does not dissolve in the activator solution as readily as the calcium present in slag.
3. The fresh Fly ash-based Geopolymer is easily handled for up to 120 min without any sign of setting.
4. The requirement of high-temperature curing and the need for calcium-containing material to enhance the reactivity of Fly ash particles are the challenges experienced by low calcium class F Fly ash-based Geopolymer.

5. Addressing the above issue, experimental investigations are carried out by proportionately blending ground granulated blast slag with Fly ash-based Geopolymer concrete.
6. Blended binder Geopolymer show improvement in their mechanical properties when cured at ambient temperature (23–47 °C)
7. The coexistence of the C–S–H gel together with the Geopolymeric gel occurs only at low alkalinity.
8. Fresh concrete suffers for workability due to flash setting, while in hardened concrete mechanical properties are found to be satisfactory.
9. The 28-day bond strength and the overall bond stress–slip behavior of the Geopolymer concrete were found similar to that of ordinary Portland cement concrete.
10. Due to the lack of proper mix design and codal provisions, the present application of Geopolymer concrete on buildings and other structural components is limited.

References

1. Jindal BB (2018) Feasibility study of ambient cured geopolymer concrete—a review. Adv Concr Constr 6(4):387–405. https://doi.org/10.12989/acc.2018.6.4.387
2. Sambucci M, Sibai A, Valente M (2021) Recent advances in geopolymer technology. A potential eco-friendly recent advances in geopolymer technology. A potential eco-friendly solution in the construction materials industry: a review. https://doi.org/10.3390/jcs5040109
3. Davidovits J (1991) Geopolymers—inorganic polymeric new materials. J Therm Anal 37(8):1633–1656. https://doi.org/10.1007/BF01912193
4. Davidovits J (1979) Synthesis of new high temperature geo-polymers for reinforced plastics/composites. SPE PACTEC 79:151–154
5. Davidovits J. Properties of geopolymer cements. [Online]. Available www.geopolymer.org
6. Davidovits J (2017) Geopolymers: ceramic-like inorganic polymers. J Ceram Sci Technol 8(3):335–350. https://doi.org/10.4416/JCST2017-00038
7. Puertas F, Martínez-Ramírez S, Alonso S, Vázquez T (2000) Alkali-activated fly ash/slag cements. Strength behaviour and hydration products. Cem Concr Res 30(10):1625–1632. https://doi.org/10.1016/S0008-8846(00)00298-2
8. Wang S-D, Scrivener KL (1995) Hydration products of alkali activated slag cement. Cem End Concr Res 25(3):561–571
9. Lecomte I, Henrist C, Liégeois M, Maseri F, Rulmont A, Cloots R (2006) (Micro)-structural comparison between geopolymers, alkali-activated slag cement and Portland cement. J Eur Ceram Soc 26(16):3789–3797. https://doi.org/10.1016/j.jeurceramsoc.2005.12.021
10. Ismail I, Bernal SA, Provis JL, San Nicolas R, Hamdan S, Van Deventer JSJ (2014) Modification of phase evolution in alkali-activated blast furnace slag by the incorporation of fly ash. Cem Concr Compos 45:125–135. https://doi.org/10.1016/j.cemconcomp.2013.09.006
11. Rangan BV, Rangan BV, Fieaust P, Faci C (2008) Fly ash-based geopolymer concrete
12. Wallah SE, Rangan BV (2006) Low-calcium fly ash-based geopolymer concrete: long-term properties
13. Lloyd NA, Rangan BV (2010) Geopolymer concrete with fly ash [Online]. Available http://www.claisse.info/Proceedings.htm
14. Sumajouw MDJ, Rangan BV (2006) Low-calcium fly ash-based geopolymer concrete : reinforced beams and columns. Faculty of Engineering, Curtin University of Technology

15. Yip CK, Lukey GC, Provis JL, van Deventer JSJ (2008) Effect of calcium silicate sources on geopolymerisation. Cem Concr Res 38(4):554–564. https://doi.org/10.1016/j.cemconres.2007.11.001
16. Puligilla S, Mondal P (2013) Role of slag in microstructural development and hardening of fly ash-slag geopolymer. Cem Concr Res 43(1):70–80. https://doi.org/10.1016/j.cemconres.2012.10.004
17. Nath SK, Kumar S (2013) Influence of iron making slags on strength and microstructure of fly ash geopolymer. Constr Build Mater 38:924–930. https://doi.org/10.1016/j.conbuildmat.2012.09.070
18. Dineshkumar M, Umarani C (2020) Effect of alkali activator on the standard consistency and setting times of fly ash and GGBS-based sustainable geopolymer pastes. Adv Civ Eng 2020:1–10. https://doi.org/10.1155/2020/2593207
19. Temuujin J, van Riessen A, Williams R (2009) Influence of calcium compounds on the mechanical properties of fly ash geopolymer pastes. J Hazard Mater 167(1–3):82–88. https://doi.org/10.1016/j.jhazmat.2008.12.121
20. Yip CK, Lukey GC, Van Deventer JSJ (2005) The coexistence of geopolymeric gel and calcium silicate hydrate at the early stage of alkaline activation. Cem Concr Res 35(9):1688–1697. https://doi.org/10.1016/j.cemconres.2004.10.042
21. Rashad AM (2013) A comprehensive overview about the influence of different additives on the properties of alkali-activated slag—a guide for civil engineer. Constr Build Mater 47(2013):29–55. https://doi.org/10.1016/j.conbuildmat.2013.04.011
22. Rashad AM (2014) A comprehensive overview about the influence of different admixtures and additives on the properties of alkali-activated fly ash. Mater Des 53:1005–1025. https://doi.org/10.1016/j.matdes.2013.07.074
23. Lloyd RR, Provis JL, Van Deventer JSJ (2009) Microscopy and microanalysis of inorganic polymer cements. 2: the gel binder. J Mater Sci 44(2):620–631. https://doi.org/10.1007/s10853-008-3078-z
24. Noushini A, Aslani F, Castel A, Gilbert RI, Uy B, Foster S (2016) Compressive stress-strain model for low-calcium fly ash-based geopolymer and heat-cured Portland cement concrete. Cem Concr Compos 73:136–146. https://doi.org/10.1016/j.cemconcomp.2016.07.004
25. Ganesan N, Indira PV, Santhakumar A (2014) Influence of steel fibres on tension stiffening and cracking of reinforced geopolymer concrete. Mag Concr Res 66(6):268–276. https://doi.org/10.1680/macr.13.00273
26. Law DW, Adam AA, Molyneaux TK, Patnaikuni I (2012) Durability assessment of alkali activated slag (AAS) concrete. Mater Struct Constr 45(9):1425–1437. https://doi.org/10.1617/s11527-012-9842-1
27. Duxson P, Fernández-Jiménez A, Provis JL, Lukey GC, Palomo A, Van Deventer JSJ (2007) Geopolymer technology: the current state of the art. J Mater Sci 42(9):2917–2933. https://doi.org/10.1007/s10853-006-0637-z
28. Zuhua Z, Xiao Y, Huajun Z, Yue C (2009) Role of water in the synthesis of calcined kaolin-based geopolymer. Appl Clay Sci 43(2):218–223. https://doi.org/10.1016/j.clay.2008.09.003
29. Pan Z, Sanjayan JG, Rangan BV (2011) Fracture properties of geopolymer paste and concrete. Mag Concr Res 63(10):763–771. https://doi.org/10.1680/macr.2011.63.10.763
30. Sarker PK, Haque R, Ramgolam KV (2013) Fracture behaviour of heat cured fly ash based geopolymer concrete. J Mater 44:580–586. https://doi.org/10.1016/j.matdes.2012.08.005
31. Nath P, Sarker PK (2017) Fracture properties of GGBFS-blended fly ash geopolymer concrete cured in ambient temperature. Mater Struct Constr 50(1). https://doi.org/10.1617/s11527-016-0893-6
32. Gunneswara Rao TD, Alfrite P, Mallikarjuna Rao G, Andal M (2015) Fracture parameters of fly ash and GGBS based geopolymer concrete. Appl Mech Mater 764–765:1090–1094. https://doi.org/10.4028/www.scientific.net/amm.764-765.1090
33. Ng C, Alengaram UJ, Wong LS, Mo KH, Jumaat MZ, Ramesh S (2018) A review on microstructural study and compressive strength of geopolymer mortar, paste and concrete. Constr Build Mater 186:550–576. https://doi.org/10.1016/j.conbuildmat.2018.07.075

34. Pratyush Kumar SA (2018) Study of mechanical and microstructural properties of geopolymer concrete with GGBS and Metakaolin. In: ICCMMEMS_2018 Study, pp 28127–28135
35. Dattatreya JK et al (2011) Flexural behaviour of reinforced Geopolymer concrete beams with GGBS and metakaoline. Int J Civ Struct Eng 2(1):138–159
36. Yong-Sing N et al (2021) Evaluation of flexural properties and characterisation of v10-mm thin geopolymer based on fly ash and ladle furnace slag. J. Mater. Res. Technol. 15:163–176. https://doi.org/10.1016/j.jmrt.2021.08.016
37. Ng TS, Amin A, Foster SJ (2013) The behaviour of steel-fibrereinforced geopolymer concrete beams in shear. Mag Concr Res 65(5):308–318. https://doi.org/10.1680/macr.12.00081
38. Castel A, Foster SJ (2015) Bond strength between blended slag and Class F fly ash geopolymer concrete with steel reinforcement. Cem Concr Res 72:48–53. https://doi.org/10.1016/j.cemconres.2015.02.016
39. Ganesan N, Indira PV, Santhakumar A (2015) Bond behaviour of reinforcing bars embedded in steel fibre reinforced geopolymer concrete. Mag Concr Res 67(1):9–16. https://doi.org/10.1680/macr.14.00125
40. Vijai K, Kumutha R, Vishnuram BG (2015) Flexural behaviour of fibre reinforced geopolymer concrete composite beams. Comput Concr 15(3):437–459. https://doi.org/10.12989/cac.2015.15.3.437
41. Ambily PS, Ravisankar K, Umarani C, Dattatreya JK, Iyer NR (2014) Development of ultra-high-performance geopolymer concrete. Mag Concr Res 66(2):82–89. https://doi.org/10.1680/macr.13.00057
42. Ambily PS, Umarani C, Ravisankar K, Prem PR, Bharatkumar BH, Iyer NR (2015) Studies on ultra high performance concrete incorporating copper slag as fine aggregate. Constr Build Mater 77:233–240. https://doi.org/10.1016/j.conbuildmat.2014.12.092
43. Glasby T, Day J, Genrich R, Aldred J (2015) Gp-airport. In: Concrete 2015 Conference, vol 11, no. 1, pp 1–9
44. Reggiani A (2020) State of construction 3D printing today [Online]. Available www.renca.org

Websites

45. https://www.geopolymer.org/news/worlds-first-public-building-with-structural-geoopolymer-concrete/
46. https://serc.res.in/pilot-cum-demonstration-study-use-fly-ash-geopolymer-concrete-pavement-construction-ntp-unchahar

Effect of H-Shaped Plan Geometry on Lateral-Torsional Response of Buildings Under Earthquake Shaking—An Elastic Study

G. Tamizharasi, Abhishek Mandala, Dheeraj Gashikanti, and Parth Suthar

Abstract Torsional effects are prominent in buildings having poor: (i) plan geometry in plan and elevation, (ii) distribution of vertical structural elements, and (iii) distribution of mass, respectively. It causes torsional eccentricity and torsional flexibility (whose fundamental mode is torsional) in buildings. Hence, building code provisions like India design for torsion using torsional eccentricity and define torsional irregularity using ratio of edge element displacements and natural period ratios (which defines *torsional flexibility*). Additionally, limits are placed on plan geometry using projection plan dimensions or area, and aspect ratio. Effect of plan shape on lateral-torsional response of building is studied on different H-shaped plan geometries using equivalent static approach (ESA) and response spectrum approach (RSA), based on local or global response parameters, like displacement and rotation about the centre of mass CM, force distribution among structural elements, base shear and torsional moment, and torsional irregularity criteria. Comparison of normalized responses of rectangular and H-shaped buildings showed that unsymmetrical buildings show more translation in ESA and more rotation in RSA, but the overall variation is marginal, i.e. 1.1–1.3 irrespective of level of torsional eccentricity. Alternately, the normalized force demand imposed on each structural element based on ESA is more than RSA. Hence, it is required to provide a dynamic amplification factor (DAF) for torsion in both ESA and RSA. Also, it is preferred to avoid re-entrant corners in H-shaped buildings by dividing them into individual blocks. Amplification in normalized shear

G. Tamizharasi (✉) · A. Mandala · D. Gashikanti · P. Suthar
Sardar Vallabhbhai National Institute of Technology, Surat, Gujarat 395007, India
e-mail: gtamizh@amd.svnit.ac.in

A. Mandala
e-mail: u18ce062@ced.svnit.ac.in

D. Gashikanti
e-mail: u18ce060@ced.svnit.ac.in

P. Suthar
e-mail: u18ce098@ced.svnit.ac.in

S. Sidhardh et al. (eds.), *Advances in Structural Integrity for Mechanical, Civil, and Aerospace Applications*, Lecture Notes in Mechanical Engineering,
https://doi.org/10.1007/978-981-97-6367-2_44

forces of structural elements of H-shaped buildings and their corresponding individual blocks varies from 1.1 to 3.1; thus, it is preferred to limit building projections. Otherwise, design the structural elements by providing with adequate stiffness, strength, and deformability to ensure better performance.

Keywords Plan geometry · Torsional eccentricity · Torsional flexibility · Re-entrant corners

1 Introduction

Usually, buildings with larger plan aspect ratio and poor plan geometry (like L, T, + , U, V, Y, and H) evidence poor performance under earthquake shaking. Damage or collapse in such building is mainly due to lateral-torsional response (e.g. complete collapse of a wing in the open ground storey of H-shaped building due to shear failure in columns at ground storey during *2001 Bhuj Earthquake*, Fig. 1a [12]), flexibility in floor diaphragm, presence of re-entrant corners (inducing local stress concentration), non-parallel systems, etc. Each wing of these buildings tends to vibrate separately, causing serious problems at the common core region, leading to potential collapse (Fig. 1b) [11, 19]. Additionally, the lack of lateral load-resisting elements and their distribution cause stiffness asymmetry and torsional flexibility (when the first mode of building becomes torsional mode) in buildings, which amplifies torsional effects [21]. Hence, it is required to provide almost uniform distribution of lateral load-resistant elements, especially along the periphery to ensure symmetry and avoid torsional flexibility [5]. Also, varying plan geometry or asymmetry in buildings highly influences their natural period [1, 2]; hence, code provisions must provide additional considerations for unsymmetrical buildings [14].

Fig. 1 **a** Wing failure of H-shaped building during 2001 Bhuj Earthquake and **b** local stress concentration in corner during Qeshm Earthquake [12, 19]

Simple analytical methods used for regular rectangular buildings shall be used in the estimation of a few critical parameters influencing torsional effects [23]. Torsional effects in buildings usually depend on structural parameters and ground motion characteristics. Structural parameters include torsional eccentricity, natural period ratio, translational natural period, and damping. But additional parameters are required to consider the effect of plan geometry; it includes: (i) plan shape, projection ratio, projection direction [13, 20]. Similarly, response parameters studied include overall displacement, inter-storey drift, diaphragm rotation, normalized base shear, overturning moment, torsional irregularity ratio, and variation in natural periods [1, 2, 13, 14]. Additionally, such buildings require the need for checking torsional irregularity, diaphragm flexibility, re-entrant corners, and vertical setbacks [20]. Inelastic behaviour of these buildings can be obtained using nonlinear static analysis. The capacity curve and the damage pattern evidenced the performance of building failed in past earthquakes (e.g. damage of an L-shaped building during 2010 Chile Earthquake, due to unsymmetrical location of structural walls) [15]. Mostly, damage in such buildings shall be reduced by providing structural walls and its orientation [5]. Thus, studies require clarity on structural parameters and response quantities influencing torsional effects in buildings with irregular plan geometry.

Usually building code provisions account for torsional effects considering torsional eccentricity, torsional flexibility, and ratio of edge element displacements (e.g. Eurocode 8, ASCE 7-16, NBCC, IS 1893(1), Iran, Turkey, Peru). A few code provisions do not explicitly specify guidelines for buildings with irregular plan geometry and provide limits for: (i) re-entrant corners using limits for projection length or projection area and (ii) floor flexibility using plan aspect ratio, percentage area of openings, and ratio of maximum to average of edge displacements (e.g. Eurocode 8, ASCE 7-16, Bangladesh, IS 1893(1), Iran, Turkey, Peru). And some code provisions suggest designing buildings with irregular plan geometry as monolithic construction if it is modelled appropriately in the structural analysis program considering all sources of floor flexibility (e.g. Chile); otherwise, the projections must have comparable area in comparison with central region (e.g. Eurocode 8). But the effect of plan geometry in inducing lateral-torsional response in buildings is not addressed explicitly.

A set of parametric studies is required to identify the parameters influencing torsional effects in buildings with irregular plan geometry. Elastic behaviour of H-shaped building configuration is studied using equivalent static analysis (ESA) and response spectrum analysis (RSA). A spectrum of configuration is obtained by varying the stiffness distribution. Using local and global response parameters, the effect of plan shape is studied, and suggestions are provided for the building code provisions. The study is limited to elastic behaviour and requires inelastic analysis to validate the responses.

2 Numerical Study

Ten-storey H-shaped building configurations with different stiffness distributions are considered, namely (i) BC1, symmetrical and torsionally stiff, (ii) BC2, symmetrical and torsionally flexible, (iii) BC3, unsymmetrical in one direction and torsionally stiff, (iv) BC4, unsymmetrical in one direction and torsionally flexible, (v) BC5, unsymmetrical in two directions and torsionally stiff, and (vi) BC6, unsymmetrical in two directions and torsionally flexible, respectively (Fig. 2). Building modal analysis (BMA), ESA, and RSA of these building configurations are performed using ETABS software. All building has 3 m storey height and 4 m bay length in each direction. M30 grade concrete and Fe 415 steel reinforcement are considered. The size of beams is taken as 300 mm × 300 mm; whereas overall size of columns is 400 mm × 400 mm, and at a few locations, it is increased to 800 mm × 800 mm (marked in red and yellow). Thicknesses of slab and infill walls are 150 mm and 250 mm, respectively; unit weight of infill wall is 20 kN/m^3. Floor finish at all floors and roofs is taken as 1 and 1.5 kN/m^2, whereas, live load at all floor levels and roofs is taken as 3 and 1.5 kN/m^2. The building is assumed to be in seismic zone V having hard rock type soil. Values of the importance factor and response reduction factor are taken as 1 and 5. One-dimensional line elements are used for modelling beams and columns, but structural walls and slabs use 2D shell elements. All buildings have rigid diaphragms with uniform mass distribution. The effect of diaphragm flexibility, non-uniform mass distributions, and uni- or bi-directional or skewed directional shaking are not considered.

The critical structural parameters like natural period ratio τ and torsional eccentricity ratio e_K/B are estimated using BMA and ESA (Table 2). Using the fundamental natural period for translation and torsion, τ is estimated as the ratio of torsional to translational natural periods. If $\tau > 1$, then the building is said to be *torsionally flexible* and for $\tau < 1$, *torsionally stiff* [22, 23]. Alternately, torsional eccentricity can be directly obtained from the locations of centre of mass and centre of resistance obtained from ETABS. For the selected building configurations, torsional eccentricity or torsional flexibility is obtained by increasing the stiffness of structural elements around 1.56, 1.67, 2.22, 1.94, and 2.47 times of BC1 for building configurations BC2, BC3, BC4, BC5, and BC6, respectively. Uncoupled translational T and torsional T_θ natural periods, and their corresponding percentage of mass participation (PMP) in translation (PMP_X or PMP_Y) and rotation (PMP_θ) of different configurations are obtained from BMA (Table 1).

2.1 *Effect of Plan Geometry*

The effect of plan geometry is studied based on (i) displacement and rotation about CM, (ii) checking the criteria for torsional irregularity as per IS 1893(1), (iii) overall base shear and torsional moment, (iv) force distribution among structural elements

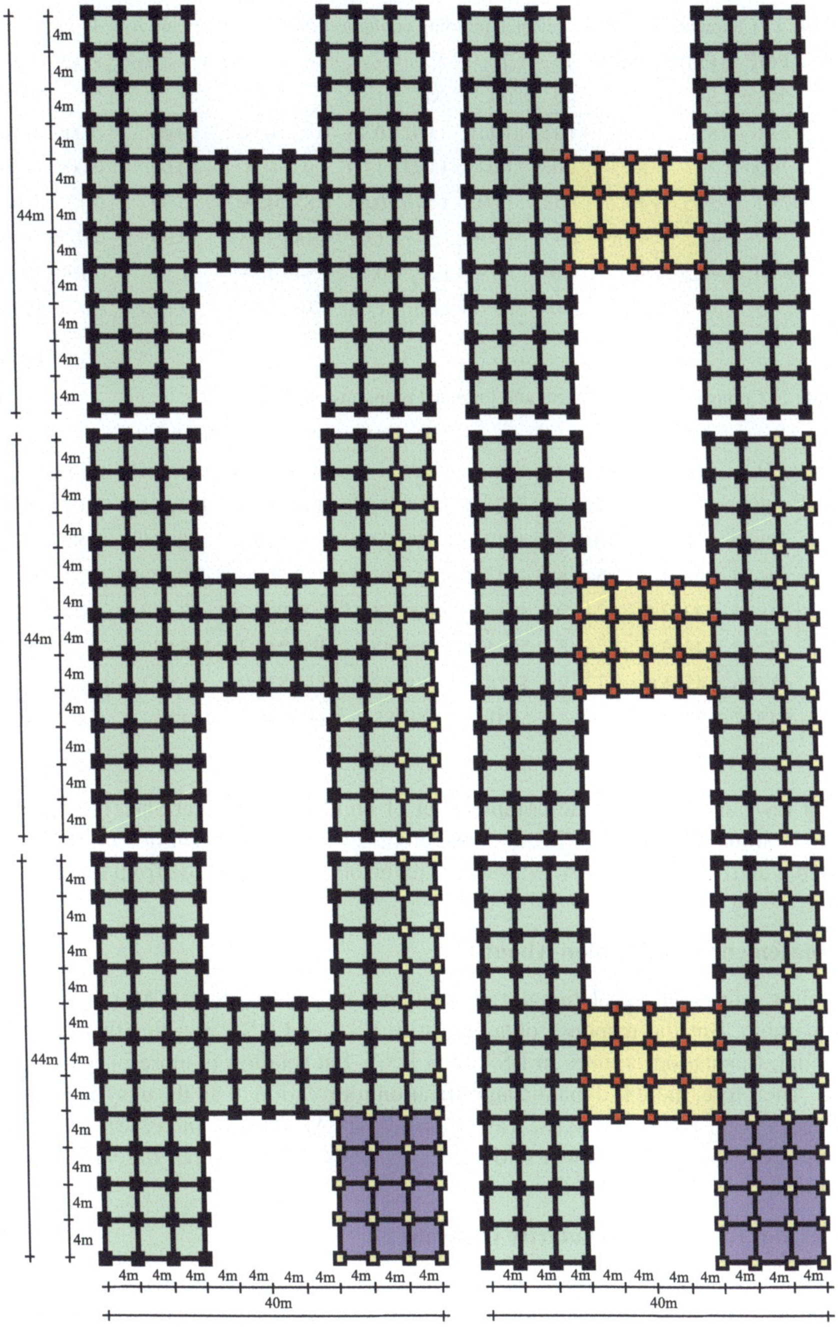

Fig. 2 Set of building configurations obtained by varying stiffness distribution of vertical structural elements

Table 1 Characteristics of the selected building configurations; $L = 40$ m and $B = 44$ m

BC	PMP_X %	PMP_Y %	PMP_θ %	T_X (s)	T_Y (s)	T_θ (s)	τ_X	τ_Y	e_{KX} (m)	e_{KY} (m)	e_{KX}/L	e_{KY}/B
1	81.6	82.1	81.8	1.154	1.116	1.092	0.95	0.98	0.00	0.00	0.000	0.000
2	79.2	79.6	81.3	1.068	1.043	1.073	1.00	1.03	0.00	0.00	0.000	0.000
3	78.7	51.5	49.3	1.561	1.602	1.370	0.88	0.85	4.09	0.00	0.102	0.000
4	77.7	40.2	41.5	1.456	1.336	1.517	1.04	1.14	2.88	0.00	0.072	0.000
5	68.0	40.7	46.6	1.505	1.580	1.322	0.88	0.84	4.18	1.12	0.105	0.025
6	71.6	38.4	40.2	1.421	1.297	1.501	1.06	1.16	3.07	0.76	0.077	0.017

Table 2 Comparison of base shear and torsional moments of different BCs

BC	Base shear (kN)				Base rotational moment (kN m)			
	EQ_X	RS_X	EQ_Y	RS_Y	EQ_X	RS_X	EQ_Y	RS_Y
0	5207.1	5207.1	5221.7	5221.7	114,572.5	114,581.4	104,425.1	104,420.9
1	3887.5	3887.5	4019.5	4019.5	85,524.0	85,519.3	80,380.4	80,373.5
2	4441.1	4441.1	4543.8	4543.8	97,701.9	97,707.1	90,868.2	90,891.3
3	4357.5	4357.5	4245.0	4245.0	95,865.7	95,865.7	87,939.8	101,745.2
4	4829.2	4829.2	5264.8	5264.8	106,242.2	106,242.2	108,947.3	135,575.1
5	4595.6	4595.6	4378.9	4378.9	100,053.5	111,521.0	91,372.6	110,687.5
6	5020.7	5020.7	5501.4	5501.4	109,388.8	118,098.0	114,585.0	145,318.8

using ESA and RSA for the complete set of building configurations. The natural period estimated from ETABS is considered; the response quantities of RSA are scaled with reference to ESA. Following are the observations made from the present study.

Displacement and Rotation About CM

Absolute values of translational and rotational responses about CM may not be comparable. But the response obtained from ESA and RSA clarifies that in each building, translation is more in ESA than RSA; but rotation is more in RSA than ESA. Therefore, there is dynamic amplification for rotation of all the unsymmetrical buildings (such as BC3, BC4, BC5, and BC6) about CM. With the increase in torsional eccentricity, the rotation about CM increases (Figs. 3 and 4), but the variation is marginal.

Check for Torsional Irregularity Criteria

IS 1893(1) defines a building to be torsionally irregular: (i) when $\Delta_{max}/\Delta_{min} > 1.5$, where Δ_{max} and Δ_{min} are the maximum and minimum displacements at extreme edges, and (ii) when $\tau > 1$. BC1, BC3, and BC5 are found to be torsionally regular (with maximum value of $\Delta_{max}/\Delta_{min}$ ranging from 1.00, 1.41, 1.38 using ESA and RSA); but BC2, BC4, and BC6 are torsionally irregular (since $\tau > 1$, with maximum

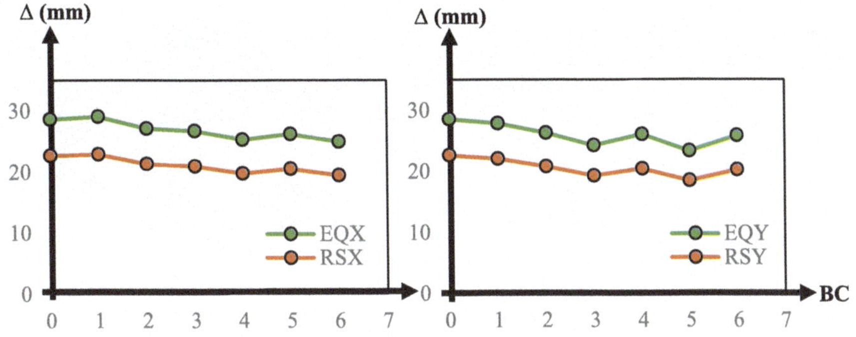

Fig. 3 Translational displacements of various building configurations

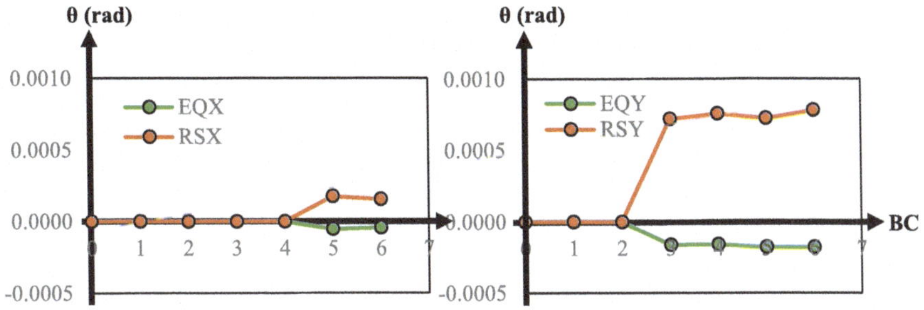

Fig. 4 Rotational response of various building configurations

value of $\Delta_{max}/\Delta_{min}$ ranging from 1.00, 1.28, 1.32). With the increase in eccentricity, relative edge displacements increased, but it is within the prescribed limits; even then buildings become irregular because of torsional flexibility (e.g. BC2, BC4, and BC6).

Base Shear and Torsional Moment

Along with BC1–BC6, a rectangular building configuration (named as BC0) with plan dimension 40 m × 44 m is considered; it has 10 bays in *X*-direction and 11 bays in *Y*-direction with same beam and column (in black) dimensions as like BC1. Based on ESA and RSA, appropriately scaled base shear and base rotational moments are estimated. Rotational moments increased for RSA than ESA with same base shear for unsymmetrical buildings (Table 2), as like rotational displacement; it is prominent in torsionally flexible buildings. Thus, observed amplification in torsional moment is only marginal and may be due to contribution from other modes. The dynamic amplification of response quantities (e.g. rotation about CM and torsional moments) is estimated to be ~ 1.0 to ~ 1.3, respectively. The variation is marginal with different levels of structural eccentricity; hence, it is mandatory to consider dynamic amplification factor for both ESA and RSA in building code provisions.

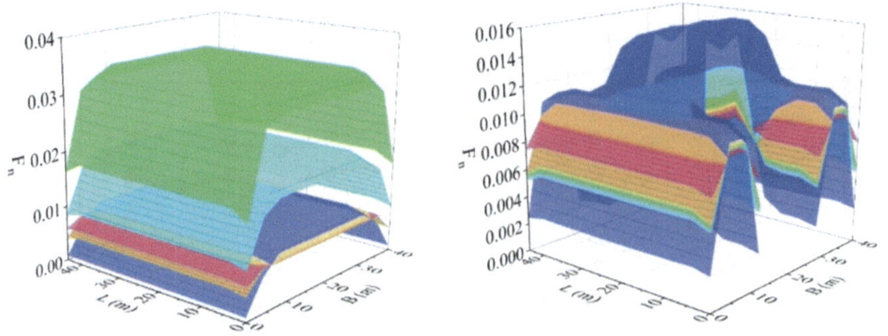

Fig. 5 Comparison of force distribution among structural element of: **a** rectangular (BC0) and **b** H-shaped building configurations (BC1) from stories 1 to 10

Force Distribution Among Structural Elements

Lateral-torsional response in building induces unequal force demand in vertical structural elements. This induced forced demand on structural elements depends on their location, lateral stiffness, and overall plan geometry. Therefore, the force distribution among structural elements of H-shaped buildings (BC1–BC6) is compared with rectangular plan geometry (BC0) (Figs. 5, 6, and 7). In each building (BC0–BC6), values of base shear and storey shear are estimated considering the software calculated natural periods, and the lateral force demand in structural elements are normalized using corresponding storey shear at each storey.

Variation of Force Distribution Along Height of Building

The distribution of forces is almost uniform near intermediate stories (say 3, 4, 5, 6, and 7, respectively) and varies at the extreme locations like ground storey and rooftop due to variations in boundary conditions. Hence, the force distribution at storey 5 is considered for comparison with other building configurations. By changing the shape of building from rectangle to H-shape, the interior elements draw more force demand in comparison with exterior elements (Fig. 5); also evidences more damage under earthquake shaking in past earthquakes.

Comparison of Force Distribution Among Different Building Configurations

EQ_X and EQ_Y represent the response obtained using ESA along *X*- and *Y*-directions, whereas RS_X and RS_Y, the response obtained using RSA along *X*- and *Y*-directions. Interior elements draw more forces in comparison with exterior elements having similar dimensions of beams and columns (BC0 and BC1). By changing the building configuration from torsionally stiff to torsionally flexible (by increasing column dimensions at the middle portion, BC2), the middle elements draw more forces than exterior elements at the periphery. These peripheral elements may demand more displacements and lead to collapse under earthquake shaking due to lack of sufficient deformation capacity (e.g. based on a comparison of the response of BC0, BC1,

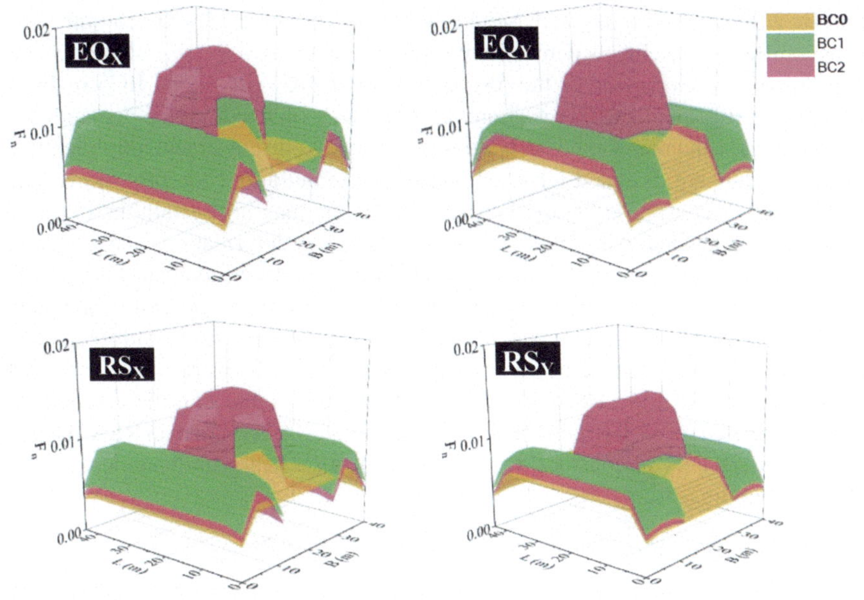

Fig. 6 Comparison of force distribution among structural element of: **a** rectangular (BC0) and **b** H-shaped building configuration (BC1 and BC2)

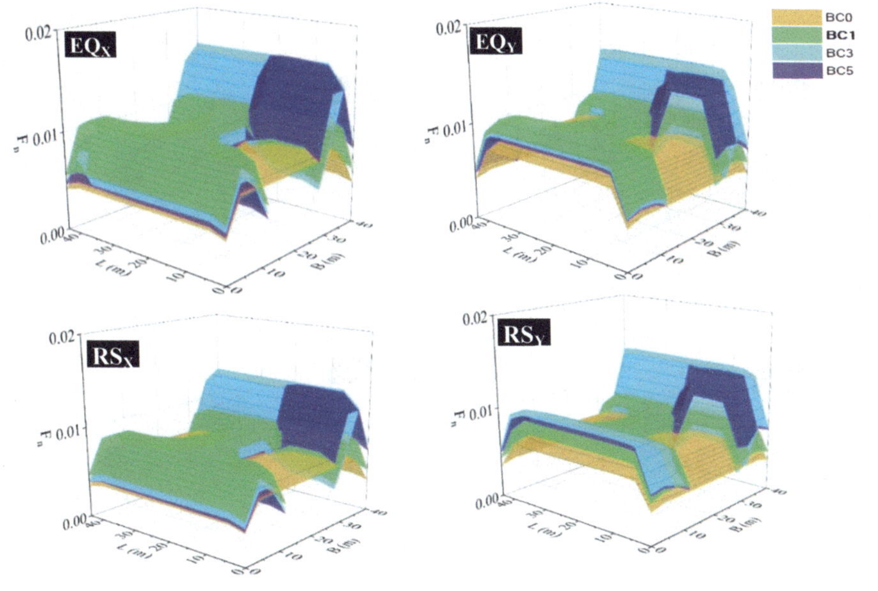

Fig. 7 Comparison of force distribution among building configurations BC0, BC1, BC3, and BC5 using ESA and RSA along *X*- and *Y*-directions

and BC2, Fig. 6). With the increase in eccentricity (by increasing the column stiffness), stiffer side elements demand more lateral forces than flexible side elements (e.g. based on comparison of the response of BC0, BC1, BC3, and BC5 or BC0, BC1, BC2, BC4, and BC6, Figs. 7 and 8). On the flexible side, force demands are almost smaller than in symmetric buildings. But due to displacement amplification, these elements are designed for amplified forces in building code provisions (e.g. IS 1893(1)). Increasing strength of structural element may not sustain the imposed displacement demand; it is required to provide sufficient deformation capacity along with strength. Thus, structural elements should be provided with adequate lateral stiffness, strength, and deformation capacity. It shall be ensured by providing adequate width and depth of cross sections, longitudinal, and transverse steel reinforcements. The observations remain same in ESA and RSA. Though overall base torsional moment estimated using RSA is more than ESA, it is found that the normalized demand imposed on each structural element based on ESA is more than RSA. Therefore, it is mandatory to consider dynamic amplification factor in dynamic analysis (RSA) of building.

Simplifying H-Shaped Buildings into Simple Rectangular Blocks

Presence of re-entrant corners in buildings demands huge force and displacement demand in structural elements. Therefore, it is better to avoid re-entrant corners in buildings. Hence, buildings shall be split into simple rectangular blocks, provided with sufficient gaps specified in IS 1893(1). Therefore, H-shaped building is divided

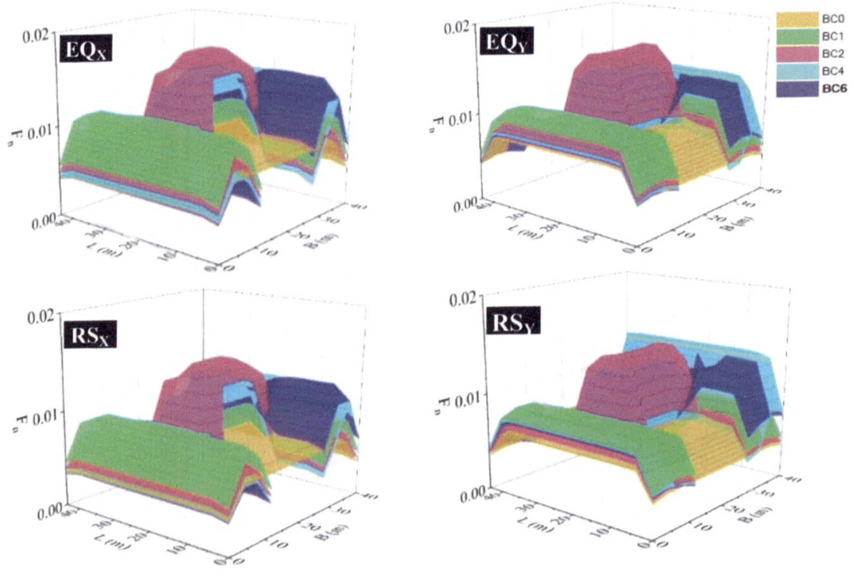

Fig. 8 Comparison of force distribution among building configurations BC0, BC1, BC2, BC4, and BC6 using ESA and RSA along *X*- and *Y*-directions

into simple rectangular blocks. The separation distance is calculated using response reduction factor and displacement at centre of mass CM of each building block using IS 1893(1). Critical load cases (like EQ_X or EQ_Y) governing the CM displacements are chosen. Finally, the normalized shear force distribution (calculated at the fifth storey) between H-shaped buildings with individual blocks of the building is compared for all configurations separately. Buildings with individual blocks are provided with required separation gap. Normalized forces are highly amplified in H-shaped buildings than the individual blocks (Figs. 9 and 10, for simplicity, only BC 3 and BC4 responses are given here as similar observations are noticed in other building configurations). In the end, the amplification of normalized shear forces of H-shaped with individual blocks is calculated (Table 3) and the amplification in normalized shear force is increased from 1.1 to 3.1. Thus, it is better to construct buildings with irregular plan geometry into respective individual rectangular blocks by providing an adequate separation gap. Also, amplification in shear force among structural elements located at the periphery of the building is marginal, but these elements demand more displacements. Hence, it should be provided with adequate stiffness, strength, and deformation capacity. Sometimes, while changing a building into individual blocks, some building may become torsionally irregular; such buildings may be converted into torsionally regular and then designed accordingly.

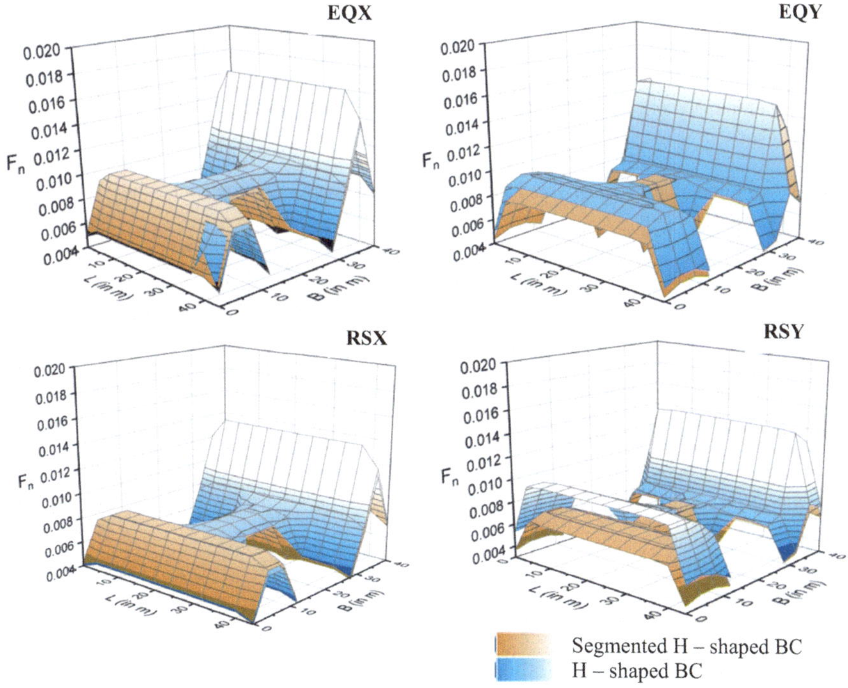

Fig. 9 Comparison of responses between BC3 and individual blocks of BC3

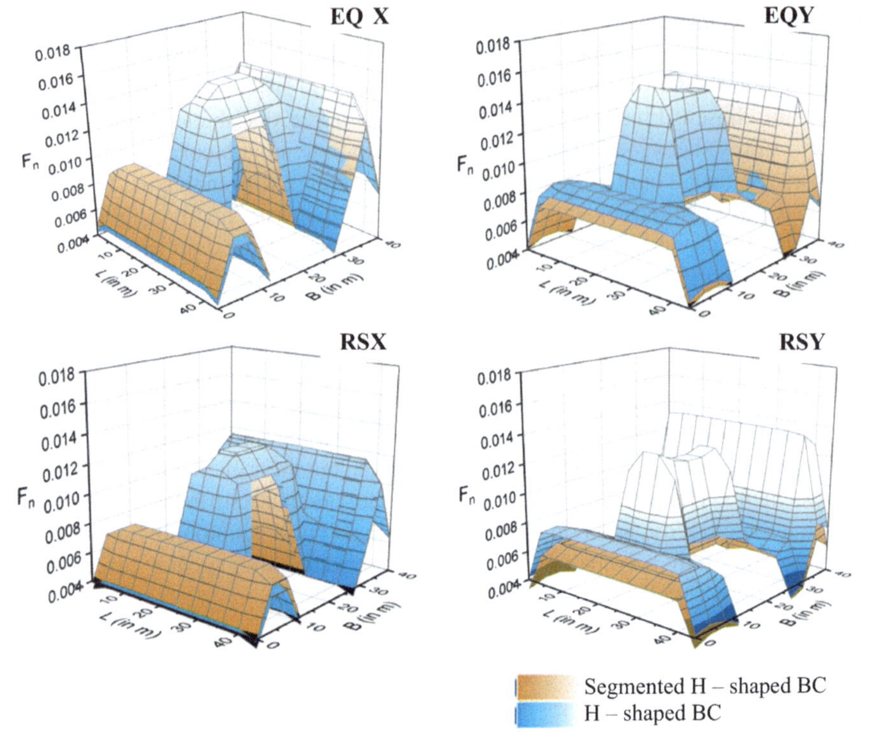

Fig. 10 Comparison of responses between BC4 and individual blocks of BC4

Table 3 Amplification of shear force in H-shaped buildings in comparison with building with individual blocks

BC	EQ_X			RS_X			EQ_Y			RS_Y		
	Left	Centre	Right	Left	Centre	Right	Left	Centre	Right	Left	Centre	Right
1	1.7	1.8	1.7	1.7	1.8	1.7	1.2	1.1	1.2	1.2	1.2	1.2
2	2.8	2.0	2.8	2.9	2.1	2.9	3.0	3.1	3.1	2.1	1.3	2.1
3	1.8	1.9	1.9	1.7	1.8	1.9	1.1	1.1	1.2	1.2	1.2	1.1
4	2.9	2.0	3.1	2.8	2.0	3.1	2.0	1.3	1.9	2.1	1.2	2.0
5	1.6	1.7	1.7	1.5	1.6	1.0	1.2	1.1	1.1	1.3	1.1	1.1
6	2.7	1.9	1.8	2.6	1.9	1.9	2.1	1.2	1.2	2.2	1.3	1.3

3 Conclusions

Torsional effects of different H-shaped building geometry are studied by varying two critical parameters namely torsional eccentricity and natural period ratio τ. Simple approaches used for regular rectangular buildings shall be applied to these buildings, and structural parameters are calculated using BMA and ESA. The effect of plan geometry on inducing lateral-torsional response is studied using ESA and RSA on the spectrum of building configurations. The key response parameters include displacement and rotation about CM, checking for torsional irregularity, base shear and torsional moment, and force distribution among structural elements. The following are the salient conclusions of the study:

(1) Unsymmetrical buildings show more translation displacements based on ESA than RSA, whereas the same show more rotational demands due to RSA than ESA. Also, the variation of rotational displacement and torsional moment is only marginal ~ 1.0–1.3 for buildings with different levels of torsional eccentricity, due to contribution from other modes. Hence, it is required to provide a dynamic amplification factor for torsion in ESA and RSA.
(2) Based on normalized force distribution among structural elements, it is found that elements located at central portion or on stiffer side of the building attract more forces, whereas elements located at periphery or on flexible side of the building demand more deformations. Hence, the structural elements should be provided with sufficient stiffness, strength, and deformability to sustain the torsional effects.
(3) The buildings considered satisfy the criteria for torsional regularity, but they are irregular in plan due to the re-entrant corner. Also, elements located near the re-entrant corners are vulnerable to earthquake shaking. Hence, it is desired to divide the building into simpler units to reduce the demands induced on critical structural elements. The amplification in normalized shear force between H-shaped buildings and individual building blocks is found to be 1.1–3.0. These elements shall be designed for amplified demand (or) reduce the demand by dividing it into individual blocks. But always elements located on periphery are highly vulnerable demanding large displacements and hence require to be detailed properly to sustain additional deformation.

Plan geometry highly influences the earthquake behaviour of buildings. Building with poor plan geometry influences torsion, re-entrant corners, and diaphragm flexibility in buildings. Also, provisions given in building codes must consider the cumulative effect of plan geometry rather than simply limiting the plan aspect ratio and projection of dimensions. This study is restricted to the effect of plan geometry on inducing torsional effects with rigid diaphragm and uniform mass distribution based on elastic analysis. It is required to validate the responses using inelastic analysis and to propose revised guidelines considering the combined effects due to torsion, re-entrant corner, and diaphragm flexibility.

References

1. Abdel Raheem SE, Ahmed MM, Ahmed MM, Abdel-Shafy AG (2018) Evaluation of plan configuration irregularity effects on seismic response demands of L-shaped MRF buildings. Bull Earthq Eng 16(9):3845–3869. https://doi.org/10.1007/s10518-018-0319-7
2. Ahmed MM, Abdel Raheem SE, Ahmed MM, Abdel Shafy AG (2016) Irregularity effects on the seismic performance of L-shaped multi-story buildings. J Eng Sci 44(5):513–536. https://doi.org/10.21608/JESAUN.2016.111440
3. ASCE/SEI 7-16 (2012) Minimum design loads for buildings and other structures. American Society of Civil Engineers. Reston, Virginia
4. Bangladesh National Building Code (2017)
5. Botis MF, Cerbu C (2020) A method for reducing of the overall torsion for reinforced concrete multi-storey irregular structures. Appl Sci 10(16), 5555, pp 1–25. https://doi.org/10.3390/app10165555
6. Building and Housing Research Center (2007) Iranian code of practice for seismic resistant design of buildings, standard no. 2800, 3rd edn. Iran
7. Bureau of Indian Standards (2016) Indian standard criteria for earthquake resistant design of structures, IS 1893(1). New Delhi, India
8. Computers and Structures Inc. (CSI) (2022) ETABS, building analysis and design. Version 20. USA
9. Disaster and Emergency Management Directorate (2007) Regulations for buildings in disaster regions. Turkey
10. Eurocode 8 (2004) European standard for design of structures for earthquake resistance, EN 1998-1. The European Union
11. Handbook on seismic retrofit of buildings (2007) Central Public Works Department and Indian Building Congress in association with Indian Institute of Technology Madras, pp 1–617
12. Humar JM, Lau D, Pierre JR (2001) Performance of buildings during the 2001 Bhuj earthquake. Can J Civ Eng 28(6):979–991. https://doi.org/10.1139/l01-070
13. Inan T, Korkmaz K, Çağatay İH (2012) An investigation on plan geometries of RC buildings: with or without projections in plan. Comput Concrete 9(6):439–455. https://doi.org/10.12989/cac.2012.9.6.439
14. Khanal B, Chaulagain H (2020) Seismic elastic performance of L-shaped building frames through plan irregularities. Structures 27:22–36. https://doi.org/10.1016/j.istruc.2020.05.017
15. Matsui T, Kawai N, Kumagai S (2014) Seismic performance of damaged buildings in 2010 Chile earthquake offshore Maule. In: Proceedings of the 10th national conference on earthquake engineering. Anchorage, Alaska
16. National Building Code (2016) Earthquake resistant design. Technical Standard of Buildings, Ministry of Housing, Construction and Sanitation, Peru
17. National Research Council (2010) National Building Code of Canada (NBCC), Canada
18. Official Chilean Code (1996) Earthquake resistant design of buildings, NCH433. National Institute of Standardization, INN, Chile
19. Omidvar B, Sadeghian A, Nojavan M (2014) The role of architectural considerations in seismic performance of buildings: a case study in Iran. J Civ Eng Archit 8(3):268–275. https://doi.org/10.17265/1934-7359/2014.03.002
20. Ravikumar CM, Babu Narayan KS, Sujith BV, Venkat Reddy D (2012) Effect of irregular configurations on seismic vulnerability of RC buildings. Archit Res 2(3):20–26. https://doi.org/10.5923/j.arch.20120203.01
21. Rodgers JE (2012) Why schools are vulnerable to earthquakes. Report by a nonprofit working towards global earthquake safety. Geohazards International, pp 1–27

22. Tamizharasi G, Prasad AM, Murty CVR (2020) A simple method to identify torsional flexibility in buildings without performing detailed structural analysis. In: Proceedings of the 17th world conference on earthquake engineering (17 WCEE), paper no: C003403, Sendai, Japan
23. Tamizharasi G, Prasad AM, Murty CVR (2021) Lateral-torsional seismic behaviour of plan unsymmetric buildings. Earthq Struct Techno Press 20(3):239–260. https://doi.org/10.12989/eas.2021.20.3.239

Discrimination of Acoustic Emission Signals Generated from Different Sources Using Support Vector Machine

Aishwarya Banerjee and Arpita Mukherjee

Abstract In the health monitoring of steel structures, the acoustic emission (AE) testing method is the most eminent method as it can identify the micro-level cracks in inaccessible places. AE waves are a quick release of energy inside any material that can be recorded using piezoelectric sensors. It is furthermore possible while capturing the AE signals that the sensors can record the noise signals which can result in an error. So the true information about the structure is restrained. Hence it is important to discriminate the actual AE signal from the other signals captured by the sensors. The employment of artificial intelligence will reduce human involvement and increase the accuracy of any AE testing-related applications such as damage detection or source localization in any structure. Therefore, this paper proposes a discriminating method of separating different signals captured by the sensors from the actual AE crack signals with the help of the support vector machine (SVM) algorithm. The AE crack data are collected on a 1.5 mm thick steel plate, which implements an R6α sensor using a pencil lead break-up test. Other noisy signals, i.e., rubbing and impact signals are generated by rubbing a small metal block and dropping it on the steel plate, respectively. The generated AE signals, sensed by the sensors are amplified since AE waves release a very small amount of energy, and then they are sent to the data acquisition system for further processing.

Keywords Acoustic emission signal · Signal discrimination · Signal processing · Support vector machine

A. Banerjee · A. Mukherjee
Academy of Scientific and Innovative Research, Ghaziabad 201002, India

A. Banerjee · A. Mukherjee (✉)
CSIR-Central Mechanical Engineering Research Institute, Durgapur, India
e-mail: a_mukherjee@cmeri.res.in

S. Sidhardh et al. (eds.), *Advances in Structural Integrity for Mechanical, Civil, and Aerospace Applications*, Lecture Notes in Mechanical Engineering,
https://doi.org/10.1007/978-981-97-6367-2_45

1 Introduction

Nowadays, the acoustic emission (AE) technique is extensively used in non-destructive type (NDT) structural health monitoring (SHM) systems. In this process, a transient elastic wave is created inside the material due to the release of localized stress energy caused by sources like friction, cavitation, impact, active corrosion, and leakage. This method is fast, robust, and reproducible; also gives more dynamic range as well as fidelity. Various AE tools are designed for analyzing the emissions generated inside the material during failure in case of unattended monitoring. It is also important to discriminate the AE signal generated from the actual crack signal among different types of noisy AE signals generated on the structure due to various activities, e.g., vibration, impact, rubbing, etc. The commonly used AE source discrimination methods are reviewed by [1] who give attention to the SHM applications. Machine learning (ML) algorithms proved to be successful tools for actual damage classification. The AE signal analysis combined with the ML algorithms for fault diagnosis and monitoring of different structures is reviewed by [2]. The performance of different ML algorithms like artificial neural networks, fuzzy logic, genetic algorithms (GA), and support vector machines (SVM). Among all the existing ML algorithms the SVM algorithm shows more efficient results for the classification of AE signals. The SVM is generally used for the classification of two classes', i.e., binary classification. Therefore, Xu [3] proposes a method for optimization of the SVM algorithm for multiclass classification and applies this method to the classification of AE signals. One more method is proposed by [4] for the classification of AE waves and the noise present in the signal based on the application of damage detection in oil storage tanks using the SVM algorithm. The radial basis function (RBF) kernel is used for damage classification at the simulation level. This algorithm is compared with the performance of a GA and the achieved finding indicates that SVM is better than GA in regard with performance. A grinding wheel wear monitoring system was developed by [5]. This development depends on the discrete wavelet decomposition and the SVM algorithm for the application of AE signal classification. Another similar type of work for the classification of different types of wear categories with the help of various ML algorithms using the AE data is done by [6]. They conducted experiments on a pin-on-disc tribometer with a digital holographic microscope using alumina and steel balls against steel discs. All the used ML algorithm's performances are shown in terms of accuracy. Their observation discloses that the SVM algorithm has the highest rate of accuracy for classification. Another automated fault detection method is proposed by [7] that is based on the SVM algorithm and AE parameters. A hybrid ML approach is proposed by [8] combining the wavelet packet transform, fast Fourier transform (FFT), and linear support vector classifier. A linear polarization resistance test is performed in the laboratory. A carbon-steel sample is used as the specimen. Three different types of corrosion severity stages are classified by applying the SVM algorithm for the classification of multiple classes. The advantages of AE along with the ML algorithms for damage detection in carbon fiber reinforced polymer cable are demonstrated by [9]. A hybrid classification method is proposed by combining the particle swarm optimization algorithm with the SVM algorithm to analyze the performance of AE wave characteristic parameters for damage degree judgment. The accuracy of this method is found near 77%. The performance of AE characteristic parameters for crack mode classification in concretes is observed by [10]. Fault recognition and classification of centrifugal pumps used in oil and gas industries is done by [11] in the laboratory. SVM along with multilayer perception is used for the classification of faults. The analysis is done on the available real-time historical data.

It is observed from the literature survey that the SVM was used earlier for various applications. Still, SVM has not been applied yet in the classification of AE signals related to the health monitoring of steel structures to the best of our knowledge. The health monitoring of steel structures has numerous challenges and requirements. Challenges include SHM of lightweight steel materials in aerospace applications, long-term monitoring, and environmental factors for SHM of steel bridges. These structures are also vulnerable to sources of noise. Hence, the present work focuses on discriminating different AE signals caused by micro-level cracks (AE crack signal), and by any other activity (rubbing and impact signal), using SVM because of its flexibility. The discrimination is important as it assists in distinguishing the actual crack signal in the presence of the noisy signal.

Additionally, the present study is different in several aspects from the existing literature that is reviewed. This paper includes a detailed process of experimental data collection, feature extraction, and signal discrimination. The authors use various signal processing techniques like parametric analysis, FFT, and wavelet transform (WT) to extract features from AE signals in the time, frequency, and time–frequency domains, respectively. These features are crucial for the classification task. The unique contribution of the article is the multiclass classification and optimization of the SVM algorithm for identifying the actual crack in a noisy situation on the steel structure. Focus is also given to selecting the most suitable AE characteristic parameters among the other characteristic parameters for feature extraction by comparing their performance in terms of accuracy. The methodology used in this presented work can benefit the steel industries as well as reduce the failure accidents of steel bridges.

In this study, a steel plate is used for the discrimination of three different types of acoustic emission signals generated by separate experimentation. Three different types of tests are rubbing rest, impact test, and pencil lead break-up test. After the collection of AE signals, their characteristic features are extracted using various signal processing tools. Therefore, the extracted AE data is used to discriminate different signals utilizing a supervised machine learning SVM algorithm. The theoretical framework of this study is discussed in Sect. 2. The data collection process is elaborated in Sect. 3. Feature extraction using different signal processing tools is presented in Sect. 2.2 and signal discrimination is examined in Sect. 4.2 before concluding the paper in Sect. 5.

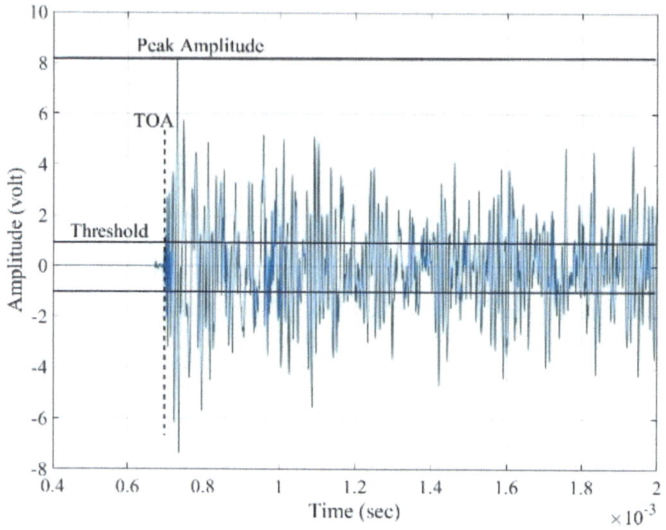

Fig. 1 Characteristic parameters of AE wave

2 Theoretical Frameworks

2.1 *Acoustic Emission Testing*

Acoustic emission (AE) [12] waves are generated by the rapid release of energy inside a material such as crack inception. It is a highly delicate method. It has the ability to identify the position of damage immediately. This is a passive technique, therefore, an external supply of energy is not needed. In this technique, the generated waves are recorded by the sensors planted on the surface of the material and they are analyzed to get the position of occurrence of the source. The origin of AE energy is the stress inside the material. AE waves have many important characteristic parameters that help to find the origin of the source. In this paper, such properties of three types of AE waves such as impact signal, rubbing signal, and pencil lead break-up test signal are studied. Some of the properties are presented in Fig. 1.

2.2 *AE Signal Processing Tools*

The main aim of this study is to observe the possible classification of three different types of AE signals. The main basis of classification is the characteristic features of the AE waves. Hence various signal processing tools are used to extract these features. The recorded AE signals have been analyzed in the frequency domain, time domain, and time–frequency domain. The following features are used for the AE signal in different domains of analysis.

Time Domain: Various time-domain features of the AE waves are extracted. These features are amplitude in volts, time of arrival of the dominant frequency at the sensor in microseconds, the energy content of the signal in volt square, and counts [13]. Amplitude is the maximum displacement moved by a point on a wave measured from its equilibrium position. The count is the number of times the signal crosses the threshold value within the duration. The threshold is a boundary beyond which an entirely different state of case exists. It can be the minimum or maximum value established for a parameter to serve as a benchmark to compare or guide the need for a complete review of the system. The duration is the difference between the time when the signal has arrived and the time taken by the signal to reach below a threshold value. The time of arrival (TOA) is the time when the first threshold is crossed. The whole energy of a continuous-time signal $x(t)$ is presented as

$$E_g = \int_{-\infty}^{\infty} x(t)^2 \mathrm{d}t. \tag{1}$$

Some of these features are indicated in Fig. 1.

Frequency Domain: The fast Fourier transform (FFT) is used primarily to determine the dominant frequency components. It is an algorithm to compute the discrete Fourier transform of a sequence or its inverse. It converts a signal from the time domain to a representation in the frequency domain. But it has the disadvantage of losing the information of time therefore the wavelet transform is used to get both the time and frequency information simultaneously [14].

Time–Frequency Domain: After determining the dominant frequency, the time of arrival of the signal at the sensor at the particular dominant frequency is calculated with the help of wavelet transform (WT). A wavelet is a kind of oscillation similar to a wave with a start from zero, then increasing and decreasing again to zero. It is a mathematical means of performing signal analysis when signal frequency varies over time. For certain types of signals and images, wavelet analysis gives more precise information about signal data than any other technique. By performing this transformation both the time and frequency data can be analyzed simultaneously. The visual representation of WT is known as a scalogram, which consists of the time, scale, and co-efficient value axes, similar to the spectrogram [15].

2.3 Support Vector Machine for Multiclass Classification

A support vector machine (SVM) is an essential supervised machine learning algorithm used for the classification of unknown data. Classification of two classes and

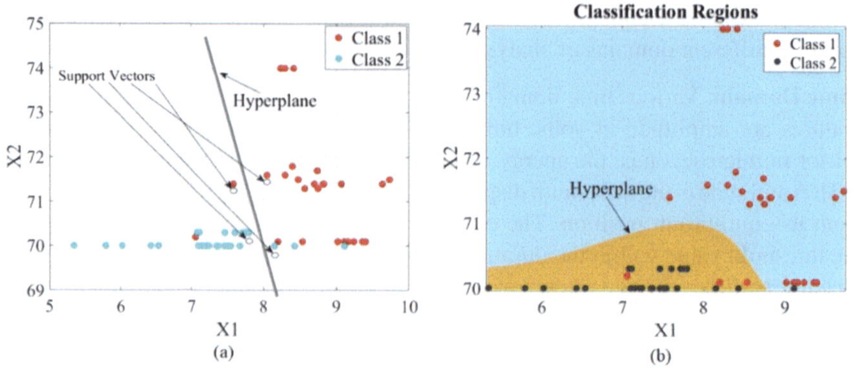

Fig. 2 Classification of two classes; **a** linear classification and **b** polynomial classification using RBF kernel

more than two are addressed as binary and multiclass classification, respectively. It creates a hyperplane between any two classes to separate them from each other during the training process [7]. The shape of the hyperplane can be linear or nonlinear depending on the data points. Figure 2 shows the classification of two classes with both the hyperplanes. The data points that are close to the hyperplane are referred to as the support vectors as shown in Fig. 2a. Natively, SVM can classify two classes at a time. More precisely it supports binary classification that separates the data into two classes. In multiclass classification, all the classes are separated with the help of several binary classifications added together. The data points are separated depending on a kernel function. Some of the used kernel functions are Gaussian, linear, polynomial, radial basis function (RBF), and sigmoid. In this study, the RBF kernel is chosen as it helps to find the nonlinear lines of separation of three classes. The RBF for two classes can be expressed mathematically as follows [16, 17]:

$$f(x_1, x_2) = \mathrm{e}^{(-\|(x_1 - x_2)\|^2)/2\sigma^2}, \tag{2}$$

where σ is the variance and x_1 and x_2 are two different sets of input parameters.

3 Data Collection

3.1 Experimental Arrangement

The data acquisition system and specimen used for this experiment are shown in Fig. 3a, b, respectively. The used specimen is a steel plate having a length of 1500 mm, breadth of 400 mm, and thickness of 1.5 mm. A R6D AE sensor [18] is mounted on the steel plate using hot glue at a certain position to capture the signals. A 2/

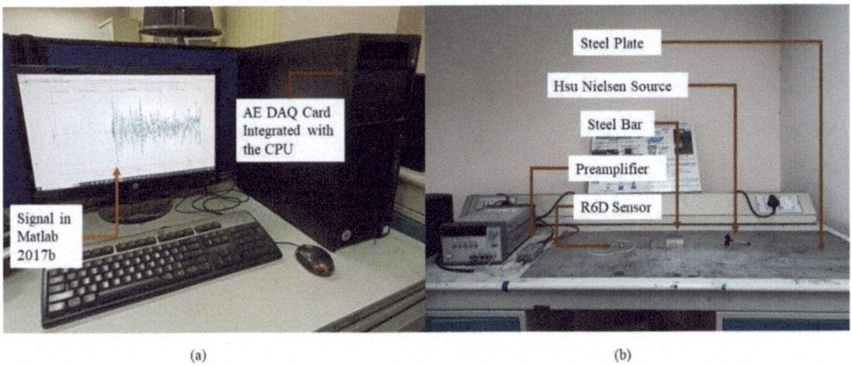

Fig. 3 Experimental setup; **a** AE data acquisition system and **b** experimental specimen, apparatus along with sensor and preamplifier

4/6 C Mistras preamplifier is used to amplify the low amplitude AE voltage signal by selecting a 40 dB gain with 10 kHz–2 MHz bandwidth. The frequency range of AE waves of crack signals generally lies between 10 kHz and 2 MHz [19] on metallic structures. After capturing and amplifying signals, they are sent to the data acquisition (DAQ) system for storing and further processing. The four-channel DAQ card is designed by Advantech [20] and is integrated with the central processing unit (CPU). A 1 MHz sampling rate is applied for the digitalization of signals. The DAQ system is shown in Fig. 3a.

3.2 Experiments Performed

Three different types of tests are performed to achieve various AE waves. The performed tests are the rubbing, impact, and PLB. A solid steel bar is rubbed over the steel plate to generate the rubbing signal in the rubbing test. These signals can be considered as the simulated source of friction between the actual structure and any other structure. Similarly, in the impact test, a solid steel ball is dropped over the steel plate from a certain height to generate the impact signal which can be considered the simulated source of any sudden hit on the actual structure. Finally, in the pencil lead break-up (PLB) test the pencil leads are broken at an angle of 45° from the plate, also known as the Hsu-Nielsen source as shown in Fig. 3b. 0.5 mm HB pencil leads are utilized to perform this test. The PLB-generated signals are referred to as the PLB signals that are considered the simulated source of the actual crack [21]. The PLB tests are performed at specific distances from the sensor. Excluding the PLB signals, the other two are noisy signals. The time-domain response of three different signals is presented in Fig. 4.

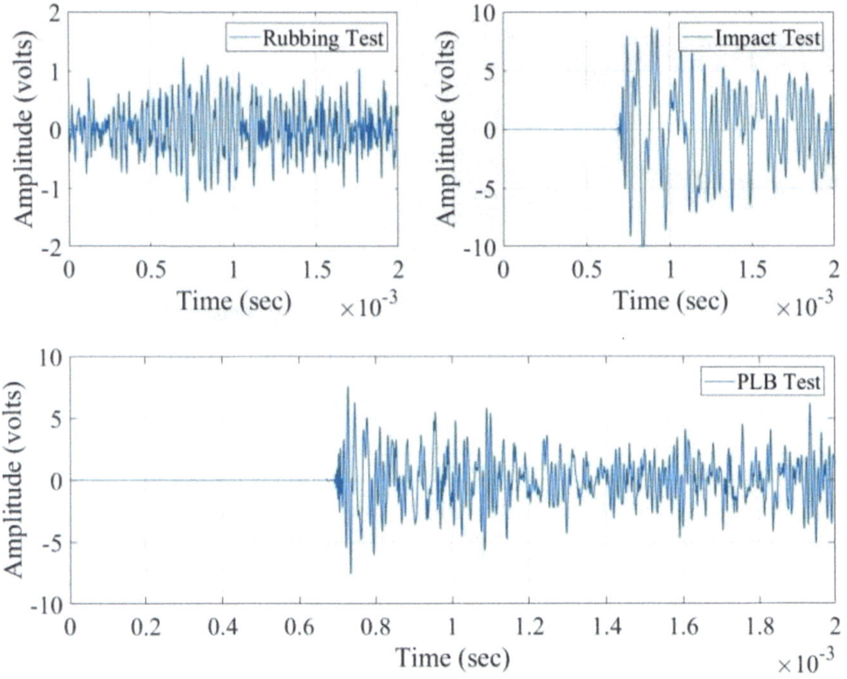

Fig. 4 Signals of different tests performed

4 Results and Discussions

4.1 Analysis of the Extracted Features of AE Wave

The features of AE waves are extracted into the domains such as frequency domain, time domain, and time–frequency domain. A tree consisting of all these features is shown in Fig. 5.

The abovementioned features are extracted with the help of MATLAB 2017b [22]. The AE threshold voltage for the three different tests is the same and it is taken as 1 V. The total number of data points used for each AE waveform of duration 2 ms is 20,000. The acquired dominant frequencies are shown in Fig. 6 for the three different signals. It is found to be 27.5 kHz for all types of signals according to Table 1. Also, the presented plots of dominant frequency components show that all the frequencies are approximately similar for three different signals, therefore this feature is not used for the discrimination of different signals. However, it is useful for determining the time of arrival of the signal to the sensor.

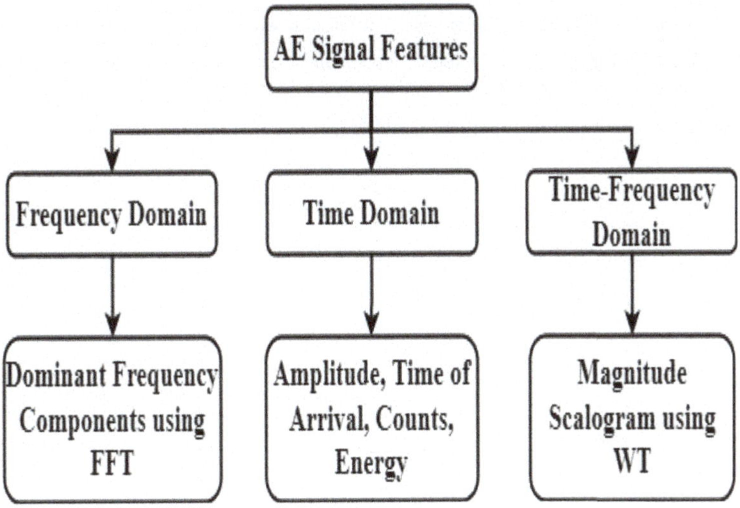

Fig. 5 Extracted features in various domains

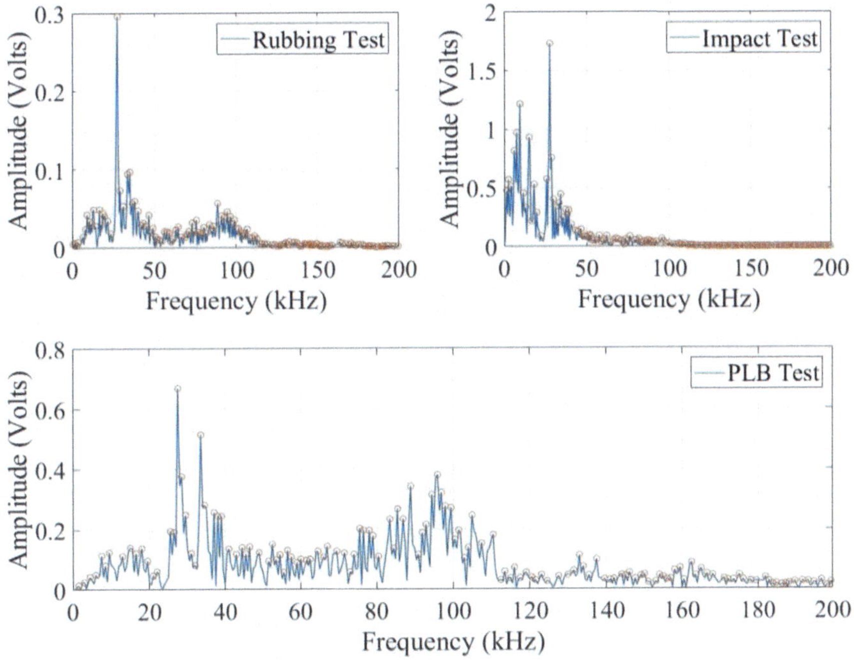

Fig. 6 Determination of frequency components using FFT

Table 1 Time-domain features of AE waves of different tests

Tests performed	Amplitude (V)	Count	Dominant frequency (kHz)	Time of arrival (μs)	Energy (V^2)
Rubbing test	1.23	78	27.5	20.5	3223
	0.93	80	27	1.5	2249
	0.94	70	27	6	1730
	0.97	76	27	90	1534
	0.97	76	27	6.5	2300
	1.01	79	27	1.5	2060
	0.93	90	27	1.1	1224
	0.98	78	27	6.3	2245
	1	73	27	5	2680
Impact test	7.58	23	27.5	714	105,227
	8.2	22	28	701	112,295
	7.06	19	9.5	702	95,338
	8.48	24	9.5	715	121,819
	8.4	22	27	718	128,854
	8.3	22	27	716	117,457
	8.05	21	15.5	716	97,040
	8.54	30	27	701	168,539
	9.11	23	15.5	701	135,304
	8.82	23	27.5	714	138,009
PLB test	8.15	107	96	700	46,369
	5.34	86	27.5	700	28,887
	7.51	100	27.5	700	45,216
	8.43	97	27.5	700	55,213
	7.21	95	27.5	700	38,397
	7.6	101	27.5	703	50,628
	7.54	90	27.5	700	57,233
	6.43	87	27.5	700	40,367
	7.24	99	27.5	700	58,699
	7.1	96	27.5	700	51,076

The time-domain features are extracted using the wavelet transform. Such four important features along with the dominant frequency component are presented in Table 1. The scalogram of different test signals is presented in Fig. 7. The wavelet function used for the evaluation of the scalogram is "Morlet". It helps to analyze the time and frequency features of the AE wave together. The energy region is also visible.

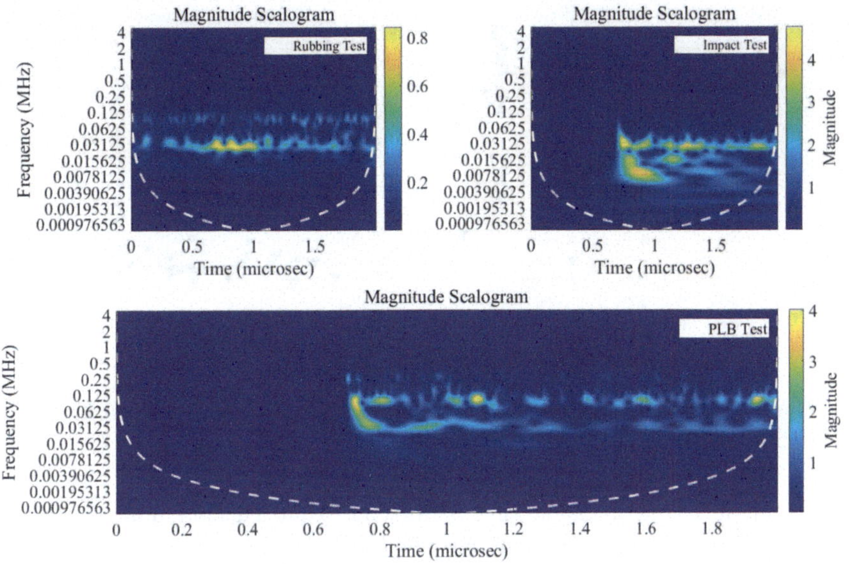

Fig. 7 Determination of magnitude scalogram using WT

4.2 Discrimination of AE Signals with SVM

Here, a supervised machine learning algorithm, the support vector machine (SVM) algorithm is used for the classification of three different types of AE signal generation sources. SVMs are generally used for the classification of two classes, but multiclass classification is also performed. In this study, the defined three classes are rubbing signal, impact signal, and PLB signal. Two time-domain features are used at a time for classification. It is observed that count-energy and amplitude-time of arrival provide the most suitable classification results. The classification regions using count-energy and amplitude-time of arrival are shown in Fig. 8, respectively. It is noted that both Fig. 8a, b are nonlinear separable SVM cases.

In this study, the hyperparameters are selected very carefully. The tuning of hyper-parameters gamma and sigma for an SVM algorithm using the radial basis function (RBF) kernel is very important to enhance performance. The shape of the decision boundary is controlled by the tuning parameter gamma (γ). The smaller value of this parameter results in the better smoothness of the boundary. Simultaneously, another tuning parameter used in this kernel function is sigma (σ). It also helps to give a smoother shape to the boundaries between the classes. Both parameters are inversely related to each other. The value of σ is determined by cross-validation. Here, the K-fold cross-validation technique is used, where the value of K is 10. The prepared dataset is divided into three parts. 60% of the total dataset is considered the training set, 20% is the validation set, and the rest 20% is the testing set. After validation, the value of σ is selected as 0.1.

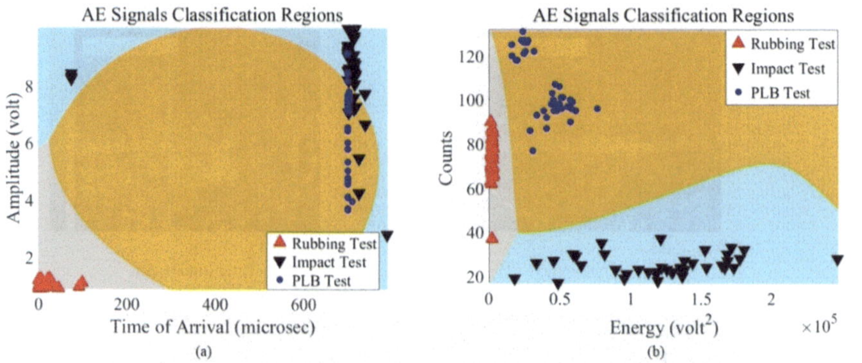

Fig. 8 Multiclass classification uses the SVM with features **a** count-energy and **b** amplitude-time of arrival

The multiclass classification using SVM is applied to a dataset of three different types of AE wave classes. The dataset is prepared in such a way that no data imbalance can be present there. The number of data points for each parameter of the corresponding class (AE wave) is the same.

The total process of data collection, feature extraction, and discrimination of three different types of signals are illustrated by a flowchart shown in Fig. 9.

4.3 Performance Evaluation of the SVM Model

The performance of the SVM model is evaluated with the help of different performance evaluation metrics. The performance of the SVM model based on the different AE parameters is visible from the figures, but evaluation of performance metrics is also necessary. The ability of the model to classify accurately can be observed from the receiver operating characteristics curves (ROC) [23]. It is plotted between the true-positive rate and the false-positive rate. The ROC of the model is shown in Fig. 10. This figure shows that for three different classes, the area under the curve (AUC) is significantly closer to 1. The closer the AUC value, the better the models' performance [24]. The value of the AUC for the class impact test is 0.9487, for the class PLB test it is 0.9737, and for another class rubbing test AUC is 0.974. The average performance of the whole model is found to be 96.537%.

The performance of the proposed SVM model is evaluated depending on the confusion matrix created from the dataset. The confusion matrix is tabulated in Table 2.

The cells of the matrix are having the true-positive (TP), true-negative (TN), false-positive (FP), and false-negative (FN) values concerning the class PLB test. The important performance evaluation parameters are precision, accuracy, recall, and *F*1 score [11]. Precision is determined by the ratio of predicted TP value and TP

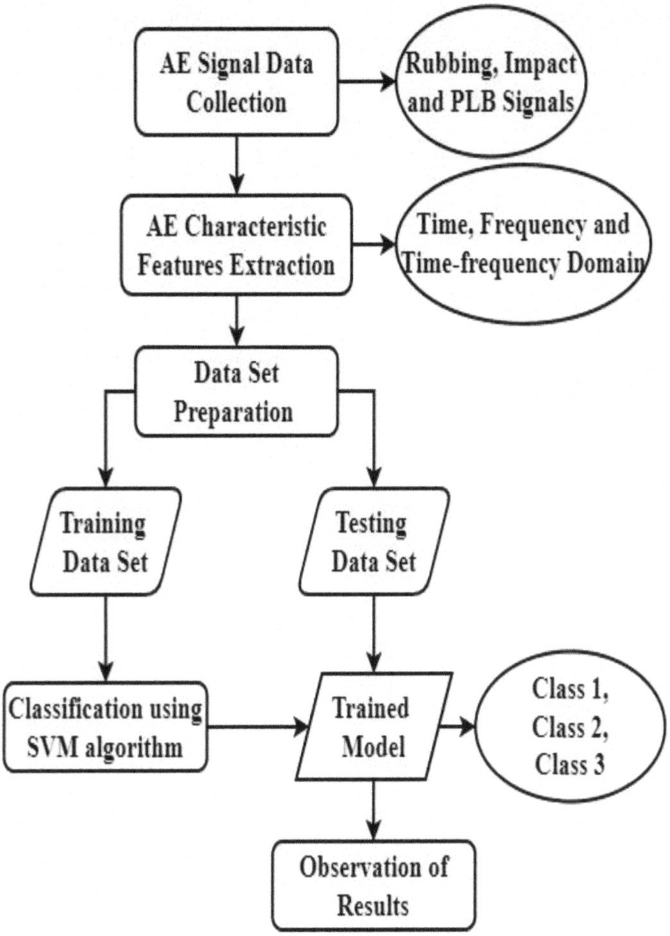

Fig. 9 Flowchart of feature extraction and signals discrimination

as well as FP value of the concerning class as shown in Eq. (3). Precision depends on the number of truly predicted data.

$$\text{Precision} = \frac{\text{TP}}{\text{TP} + \text{FP}} \tag{3}$$

Accuracy is determined using Eq. (4). It depends on all the values present in the confusion matrix.

$$\text{Accuracy} = \frac{\text{TP} + \text{TN}}{\text{TP} + \text{FP} + \text{FN} + \text{TN}} \tag{4}$$

Fig. 10 ROC of the model

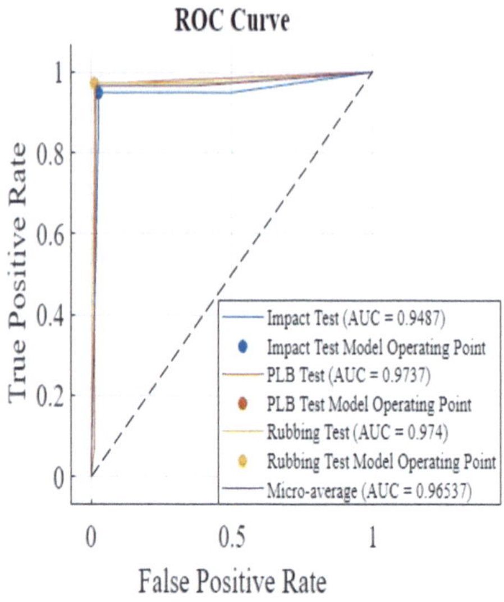

Table 2 Confusion matrix

	Classes	Predicted values		
		PLB test	Rubbing test	Impact test
Actual values	PLB test (*P*)	TP—*P*	FN1—*P*	FN2—*P*
	Rubbing test (*R*)	TP—*P*	TP—*R* or TN1—*P*	TN2—*P*
	Impact test (*I*)	FP1—*P*	TN3—*P*	TP—*I* or TN4—*P*

Recall is determined by the ratio of predicted TP value of the concerning class and addition of TP value of the concerning class as well as FN value of another class as shown in Eq. (5). The recall is also addressed as the true-positive rate.

$$\text{Recall} = \frac{\text{TP}}{\text{TP} + \text{FN}} \tag{5}$$

$F1$ score is the harmonic mean of the values of precision and true-positive rate. It is given in Eq. (6).

$$F1\text{-Score} = \frac{2 \times (\text{Precision} \times \text{Recall})}{(\text{Precision} + \text{Recall})} \tag{6}$$

The performance analysis of the proposed SVM model is presented in Table 3.

It is observed from Table 3 that the accuracy of the model to predict the PLB test is 94.87%. The obtained percentages of these specifications such as precision, recall, and $F1$ score are 92.3%, 92.3%, and 92.29%, respectively. Therefore, the performance of the model is found acceptable for the classification of different types

Table 3 Performance metrics of the SVM model based on its confusion matrix

Specifications	Values (%)
Input dataset	117 × 4
Training dataset	90 × 4
Testing dataset	27 × 4
Model	Multiclass RBF SVM
Library	SVM
Environment	MATLAB 2017b
Precision	92.3
Accuracy	94.87
Recall	92.3
*F*1 score	92.29

of AE signals and automatic prediction of the actual crack signal or PLB signal among the other signals present such as rubbing and impact signals.

5 Conclusion

In this study, automatic discrimination of various AE signals utilizing the support vector machine (SVM) algorithm is presented. The classification is done based on the extracted AE signal parameters. A steel plate is used as the experimental specimen. Three different signals are generated by rubbing a steel bar over the plate, dropping the ball, and breaking a pencil lead on the plate. After that, the features of these AE waves are extracted using several signal processing tools such as fast Fourier transform, wavelet transform, etc. Therefore, the SVM model is applied for the classification of signals. The outcome shows that SVM can classify the signals successfully based on many of the characteristic parameters of the AE waves. Depending upon the ROC of the model and the performance evaluation specification it can be concluded that the proposed method is potent for the prediction of crack signals in the presence of other signals. The accuracy of the model is found to be 94.87%.

References

1. Kaphle M, Tan ACC, Thambiratnam DP, Chan THT (2012) Effective discrimination of acoustic emission source signals for structural health monitoring. Adv Struct Eng 15(5):707–716. https://doi.org/10.1260/1369-4332.15.5.707
2. Hassan Ali Y, Abd Rahman R, Raja Hamzah RI (2014) Acoustic emission signal analysis and artificial intelligence techniques in machine condition monitoring and fault diagnosis: a review. J Teknol (Sci Eng) 69(2):121–126. https://doi.org/10.11113/jt.v69.3121

3. Xu P, Chan AK (2003) Support vector machines for multi-class signal classification with unbalanced samples. In: Proceedings of the international joint conference on neural networks, vol 2, pp 1116–1119. https://doi.org/10.1109/ijcnn.2003.1223847
4. Zhao J, Wang K, Guo Y (2010) Acoustic emission signals classification based on support vector machine. In: ICCET 2010—2010 international conference on computer engineering and technology, proceedings, vol 6, no 5, pp 1027–1032. https://doi.org/10.1109/ICCET.2010.5486240
5. Yang Z, Yu Z (2012) Grinding wheel wear monitoring based on wavelet analysis and support vector machine. Int J Adv Manuf Technol 62(1–4):107–121. https://doi.org/10.1007/s00170-011-3797-1
6. Deshpande P, Pandiyan V, Meylan B, Wasmer K (2021) Acoustic emission and machine learning based classification of wear generated using a pin-on-disc tribometer equipped with a digital holographic microscope. Wear 476:203622. https://doi.org/10.1016/j.wear.2021.203622
7. Ali SM, Hui KH, Hee LM, Leong MS (2018) Automated valve fault detection based on acoustic emission parameters and support vector machine. Alex Eng J 57(1):491–498. https://doi.org/10.1016/j.aej.2016.12.010
8. May Z, Alam MK, Nayan NA, Rahman NAA, Mahmud MS (2021) Acoustic emission corrosion feature extraction and severity prediction using hybrid wavelet packet transform and linear support vector classifier. PLoS ONE 16:1–23. https://doi.org/10.1371/journal.pone.0261040
9. Xu J, Liu X, Han Q, Wang W (2021) A particle swarm optimization–support vector machine hybrid system with acoustic emission on damage degree judgment of carbon fiber reinforced polymer cables. Struct Health Monit 20(4):1551–1562. https://doi.org/10.1177/1475921720922824
10. Neha Vivek A (2017) Crack classification of concrete under uniaxial compression using acoustic emission monitoring technique. Int J Eng Res Technol V6(03):137–141. https://doi.org/10.17577/ijertv6is030177
11. Orrù PF, Zoccheddu A, Sassu L, Mattia C, Cozza R, Arena S (2020) Machine learning approach using MLP and SVM algorithms for the fault prediction of a centrifugal pump in the oil and gas industry. Sustainability (Switzerland) 12(11). https://doi.org/10.3390/su12114776
12. Rao AK (1990) Acoustic emission and signal analysis. Defence Sci J 40(1):55–70. https://doi.org/10.14429/dsj.40.4450
13. Sause MGR (2016) In situ monitoring of fiber-reinforced composites. In: Springer series in materials science, vol 242
14. Harčarik T, Bocko J, Masláková K (2012) Frequency analysis of acoustic signal using the fast Fourier transformation in MATLAB. Procedia Eng 48:199–204
15. Hanteh M, Rezaifar O, Gholhaki M (2021) Selecting the appropriate wavelet function in the damage detection of precast full panel building based on experimental results and wavelet analysis. J Civ Struct Health Monit 11(4):1013–1036. https://doi.org/10.1007/s13349-021-00497-6
16. Cho BH, Yu H, Lee J, Chee YJ, Kim IY (2008) Nonlinear support vector machine visualization for risk factor analysis using nomograms and localized radial basis function kernels. https://doi.org/10.1109/TITB.2007.902300
17. Soman K, Sathiya A, Suganthi N (2015) Classification of stress of automobile drivers using radial basis function kernel support vector machine. In: 2014 international conference on information communication and embedded systems, ICICES 2014. https://doi.org/10.1109/ICICES.2014.7034000
18. Mistras Group (2013) R6D-AST sensor. Mistras Group, Inc.
19. Marfo A, Luo Y, Zhong-an C (2013) Quantitative acoustic emission fatigue crack characterization in structural steel and weld. Adv Civ Eng 2013(6)
20. Advantech Co. (2015) PCI-E-1800 series. Instrumental DAQ, Inc.

21. Bruzelius K, Mba D (2004) An initial investigation on the potential applicability of acoustic emission to rail track fault detection. NDT E Int 37(7):507–516. https://doi.org/10.1016/j.ndteint.2004.02.001
22. https://in.mathworks.com/matlabcentral/answers/498405-matlab-2017b
23. Rakotomamonjy A (2004) Optimizing area under ROC curve with SVMs, pp 71–80
24. Zhang XL, Jiang C (2007) Improved SVM for learning multi-class domains with ROC evaluation. In: Proceedings of the sixth international conference on machine learning and cybernetics, ICMLC 2007, pp 2891–2896. https://doi.org/10.1109/ICMLC.2007.4370641

Numerical Analysis of Miniature Disk Bend Specimens Under Creep Condition

Ritesh Gupta, Awanish Kumar Mishra, Krishna Kumar, and Abhishek Tiwari

Abstract Since it only requires a small surveillance sample, small punch creep test (SPCT) analysis has been used to estimate the lifespan of power stations and nuclear reactors. As the specifications in uniaxial creep test (UCT) and SPCT differ, numerous correlations have been proposed for calculation of conventional tensile characteristics for elastic plastic behavior of metallic materials. Due to the high operating temperatures in nuclear power plants and highly developed ultra-super critical plants, many of the components experience creep deformation. In order to determine how long a component will last, it is crucial to understand the correlations between creep properties calculated through SPCT. Thin disk specimens subjected to SPCT can be thought of as a helpful method for identifying the creep characteristics of components exposed to high temperatures. The challenge is to estimate the conventional bulk creep properties from a small punch sample. Finite element analysis (FEA) of SPCT of a metallic material can help in estimation of bulk properties and therefore FEA of a metallic material having $E/Y_S = 277$ with a power law creep behavior is carried out using ABAQUS software. This particular paper focuses on the influence of specimen thickness on the site of maximum stress, sometimes known as the weakest section, when the punch travel time was varied.

Keywords Creep · SPCT · Power law creep · FEA

R. Gupta · K. Kumar
Department of Metallurgical and Materials Engineering, Malaviya National Institute of Technology Jaipur, Jawahar Lal Nehru Marg, Malviya Nagar, Jaipur, Rajasthan 302017, India

A. K. Mishra · A. Tiwari (✉)
Fracture Mechanics Lab, Department of Metallurgical and Materials Engineering, Indian Institute of Technology Ropar, Nangal Road, Hussainpur, Rupnagar, Punjab 140001, India
e-mail: abhishek.tiwari@iitrpr.ac.in

S. Sidhardh et al. (eds.), *Advances in Structural Integrity for Mechanical, Civil, and Aerospace Applications*, Lecture Notes in Mechanical Engineering,
https://doi.org/10.1007/978-981-97-6367-2_46

1 Introduction

Many components employed in traditional and nuclear reactors, thermal power plants, chemical plants, and aero-engines operate at elevated temperatures, which can lead to time-dependent creep deformation [1]. These occurrences increase the risk of premature component failure [2]. To effectively quantify the mechanical and creep properties of materials with samples as thin as 0.1–0.5 mm, the SPCT can be utilized. The SPCT minimizes material removal and avoids the need for large quantity of materials for design and safety of nuclear and thermal power plants. It is considered a non-destructive method for assessing the properties of in-service components without compromising their structural integrity, as it utilizes small test specimens [3, 4]. Small material samples have been predominantly employed in determining the creep parameters of three types of specimens: small tensile creep specimens [5], impression creep specimens [6] and small punch test specimens [7, 8].

The small punch test (SPT) functions based on a displacement-controlled mechanism. In this setup, the punch is consistently driven through the specimen at a constant velocity by the crosshead. The resulting force is then monitored over time using a load cell, which measures the force exerted on the underside of the sample in the direction of punch contact. Using the SPT, it is possible to estimate parameters such as yield stress and even model a complete stress–strain curve [9–14]. However, the SPT technique has faced challenges in establishing a reliable correlation with ultimate tensile strength (UTS) due to conflicting associations based on the maximum load point (P_{max}) of the load–displacement curve. Unlike the uniaxial tensile test, P_{max} does not coincide with a necking situation. To address this discrepancy, Kumar et al. [15] sought to identify the necking zone preceding P_{max} through a combination of experimental methods and FEA. Their findings suggest that the load linked with a 0.48 mm displacement closely aligns with the necking zone, offering a more suitable fit for UTS correlation. This observation holds true for materials such as 20MnMoNi55, CrMoV ferritic steel, and SS304LN. These insights highlight the potential of the small punch technique as a methodology worthy of further exploration to enhance its application in structure-critical design. However, it is worth noting that the literature does not provide logical reasoning as to why the maximum thinning should occur at a constant distance. The thinning of the material is expected to be dependent on the dimensions of the specimens and the amount of deformation, which is the primary focus of this research work.

The small punch creep test, conversely, is a force-controlled test where a constant force is applied to press the punch against the specimen, and the resulting deflection or movement is monitored over time (t). Typically, there is no need to clamp the specimen with any load by the dies holding it in place, a practice known as holding without load. However, it is important to note that applying a clamping force could alter the specimen's stress response. To facilitate testing under controlled conditions, thermocouples are often strategically positioned throughout the setup, which is usually housed in a temperature- and atmosphere-controlled chamber. This setup allows for testing at elevated temperatures or in controlled environments, such as

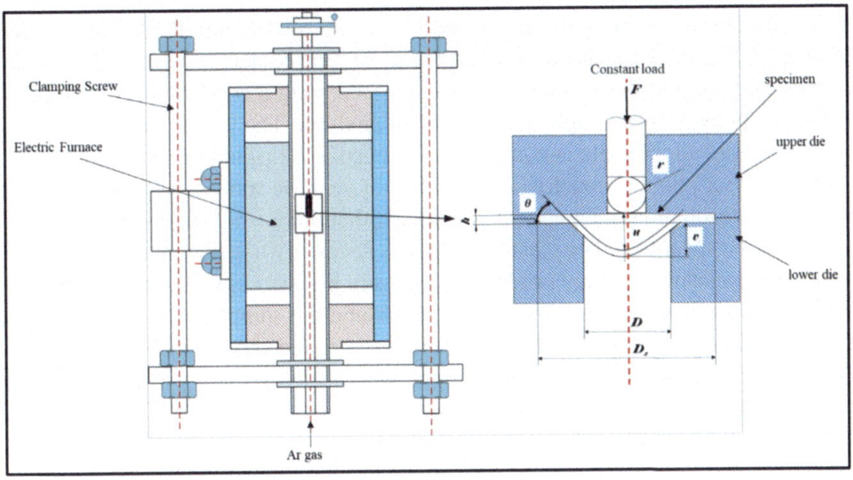

Fig. 1 Detailed experimental conditions for SPCT

conducting tests at high temperatures in an atmosphere of argon (Ar) or another inert gas to minimize oxidation effects. The SPCT setup is illustrated schematically in Fig. 1. This test is inherently complex and involves geometrical, material and contact nonlinearity.

Recent studies have highlighted the influence of excessive plasticity during loading on the creep behavior observed in SPCT [16–18]. Efforts to enhance the interpretation techniques for SPCT have been underway, with various conferences and symposia convened to facilitate the exchange of ideas and experiences aimed at addressing these issues [19–21]. This research presents the outcomes of comprehensive FEA conducted on SPCT.

2 Finite Element Analysis (FEA)

The ABAQUS software provides a versatile and effective method for adding constitutive equations to the program library [21]. The relationship between load and deflection displacement was determined through a time-dependent numerical simulation of a SPCT using the ABAQUS software. Since the SPCT apparatus, specimen, deformation and damage evolution all exhibit axisymmetric shapes, a two-dimensional FEM model, similar to the one depicted in Fig. 2a, is sufficient to simulate the SPCT at a computationally reasonable cost. Simply supported and symmetric constraints are applied to the respective edges of this specimen model in accordance with the displacement boundary conditions of the actual experimental settings. Figure 2a depicts the geometrical model for the SPCT, consisting of upper and lower dies, a punch made of ceramic, and a specimen. The specimen, a thin, round disk with a radius of 2.5 mm and a punch radius of 1.25 mm, may vary in thickness depending on the circumstances. Both the upper and lower dies are 0.1 mm thick, with a fillet radius of 0.01 mm on the upper die. Surface-to-surface contact is described between the dies exterior and the specimen's upper surface, with a coefficient of friction of 0.25 and a penalty form employed for contact analysis. The upper die is restrained, and the bottom support die is completely fastened at its rigid body standard point. Additionally, a punch, modeled as a rigid body, can be moved vertically. For all simulations throughout the paper, it is assumed that there is no friction between the punch and the specimen. The plasticity phase is necessary to generate a considerable stress, as the elastic step may produce excessively high stress within the interacting areas due to the interaction between the punch and the sample, as well as the dies and the sample. The specific chemical composition and other details of the material are undisclosed due to research confidentiality. However, it is presumed that the specimen substance follows Norton's power creep law. The Norton law constitutive equation is as follows,

$$\dot{\varepsilon} = A\sigma^n t^m \tag{1}$$

In the equation provided, '*A*' represents the coefficient, whose value is kept constant to accommodate the variability in material properties, while '*n*' and '*m*' are the indices for the power law. The value of '*m*' is set to zero because the initial stage creep is not considered in this study. Specifically, this research adopts $A = 3.31\text{e}{-21}$ with units of $\text{MPa}^{-6.943}\ \text{h}^{-1}$.

The simulation of the SPCT axisymmetric design utilized 625 elements with an element size of 0.02. These simulations were conducted under a load of 200 N for each of the three distinct thicknesses. In Fig. 2b, the deformation of the SPCT sample and the distribution of corresponding stress are depicted. Interestingly, the maximum stress location, also known as the weakest location, was not precisely at the center

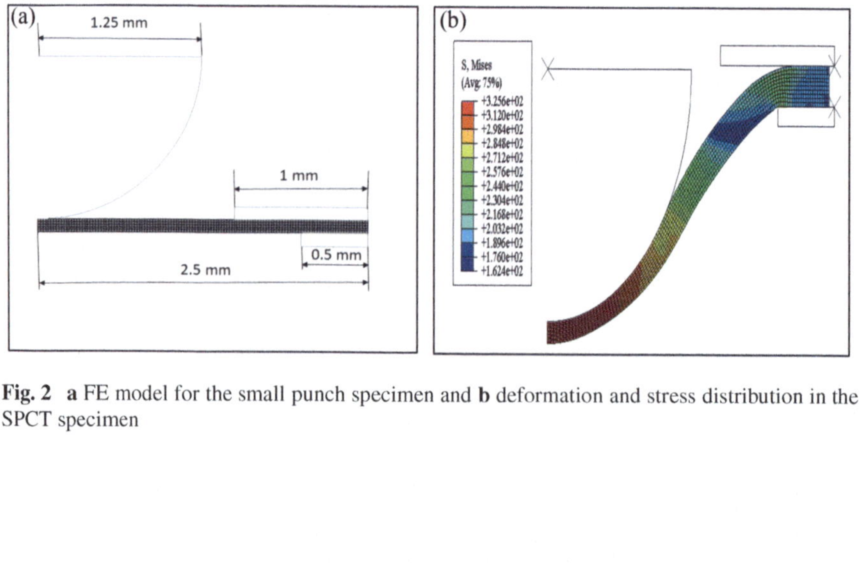

Fig. 2 **a** FE model for the small punch specimen and **b** deformation and stress distribution in the SPCT specimen

of the specimen but was instead offset by a certain amount. Ultimately, the specimen ruptured at this weakest region, where the thickness gradually decreased over time. This observation highlights the importance of considering the nonuniformity of stress distribution in SPCT simulations and its impact on specimen failure.

3 Results

The thickness of the model was varied throughout the course of the analyses, ranging from 0.2 to 0.3 mm to 0.4 mm, while the other parameters remained same. Figures 3, 4 and 5 display contour plots of the necking phenomenon for varied punch travel times (0, 50, 100, 200 and 300 h) in specimens of varying thickness.

Figure 6a–c illustrates the displacement of the disk under the punch for three distinct thicknesses across various phases of punch travel. These figures reveal that during the initial phases of punch travel, the disk undergoes more deformation through bending than stretching. However, stretching becomes more prominent as the punch travel progresses. Consequently, as deformation increases with punch travel time, there is a noticeable reduction in the disk's thickness along the radial direction. This gradual thinning of the disk eventually leads to significant thinning under further punch travel, culminating in failure along the thinnest region. It is important to note that at this stage, a considerable portion of the material beneath the punch undergoes significant flexibility. This observation underscores the complex deformation behavior experienced by the disk during the SPCT and highlights the critical role of understanding material behavior under such conditions.

To obtain the displacement of the top and bottom surfaces of the specimen along the radial direction for various punch travel times, upper and lower paths are generated. The difference between these paths is then plotted against radial displacement from the disk's center to determine the shortest distance between the top surface node and the bottom surface node. This process allows for identifying the thinnest region of the disk at a specific punch travel time, indicated by the lowest distance on such a graph. Figures 3, 4 and 5 illustrate the change in nominal thickness on the disk for three distinct thicknesses using various punch travel times. Figure 6a–c is aimed at identifying the location of the most thinned segment with punch travel time, revealing the radial location of the minimum disk thickness as it shifts away from the specimen's center with increasing punch travel time. Additionally, it should be noted that the minimum disk thickness increases as punch travel time increases. Furthermore, as depicted in Figs. 6a and 7, the graph pattern for a 0.2 mm thickness suggests two thinnest spots on the disk for a particular punch travel time. Figure 7 demonstrates the shift in the radial location of the most thinned region for a certain punch travel time. This figure helps determine how thickness affects the location of the minimum disk thickness radially, indicating that the specimen's necking is moving toward the center.

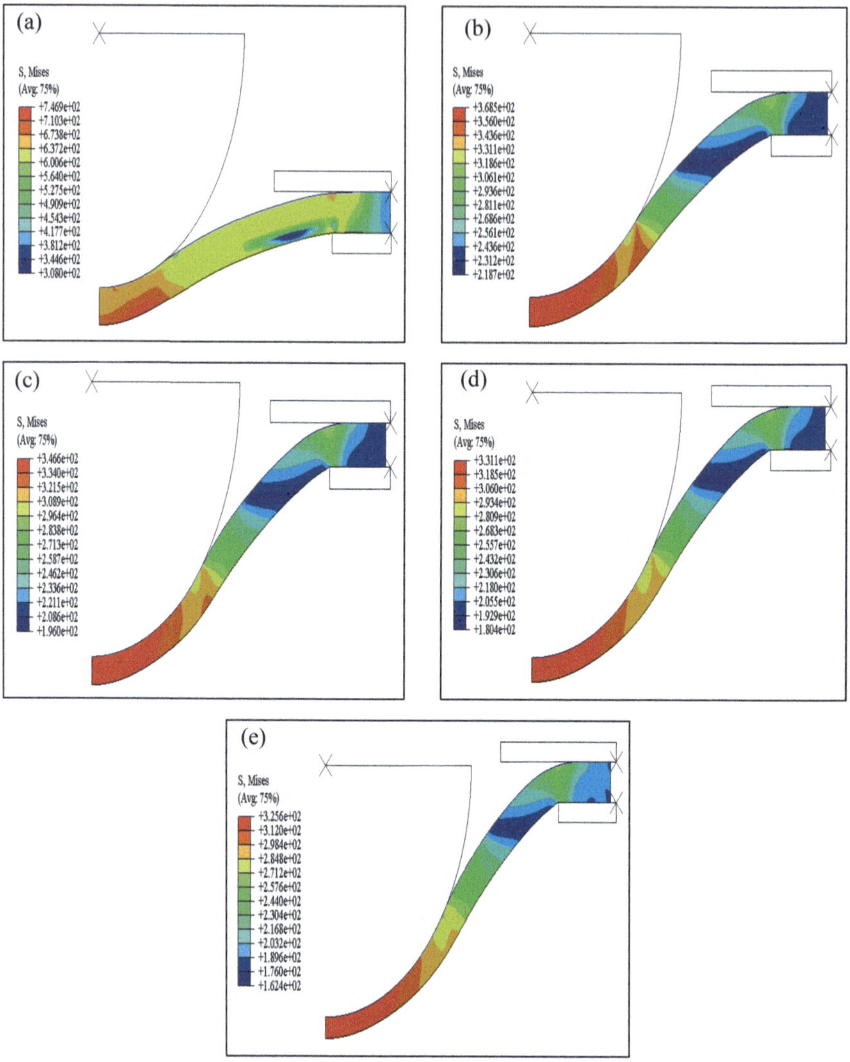

Fig. 3 Contour plots of 0.2 mm thickness for von-Mises total stress on FE model for punch travel time of **a** 0 h, **b** 50 h, **c** 100 h, **d** 200 h and **e** 300 h

4 Conclusion

Finite element analysis of the SPCT was conducted to investigate the influence of specimen thickness on the location of maximum stress under different punch travel times. Despite consistent test setup and conditions, the maximum stress location for a specific disk thickness was found to deviate from the center of the specimen for a given punch travel time. Over time, the thickness at this weakest point gradually

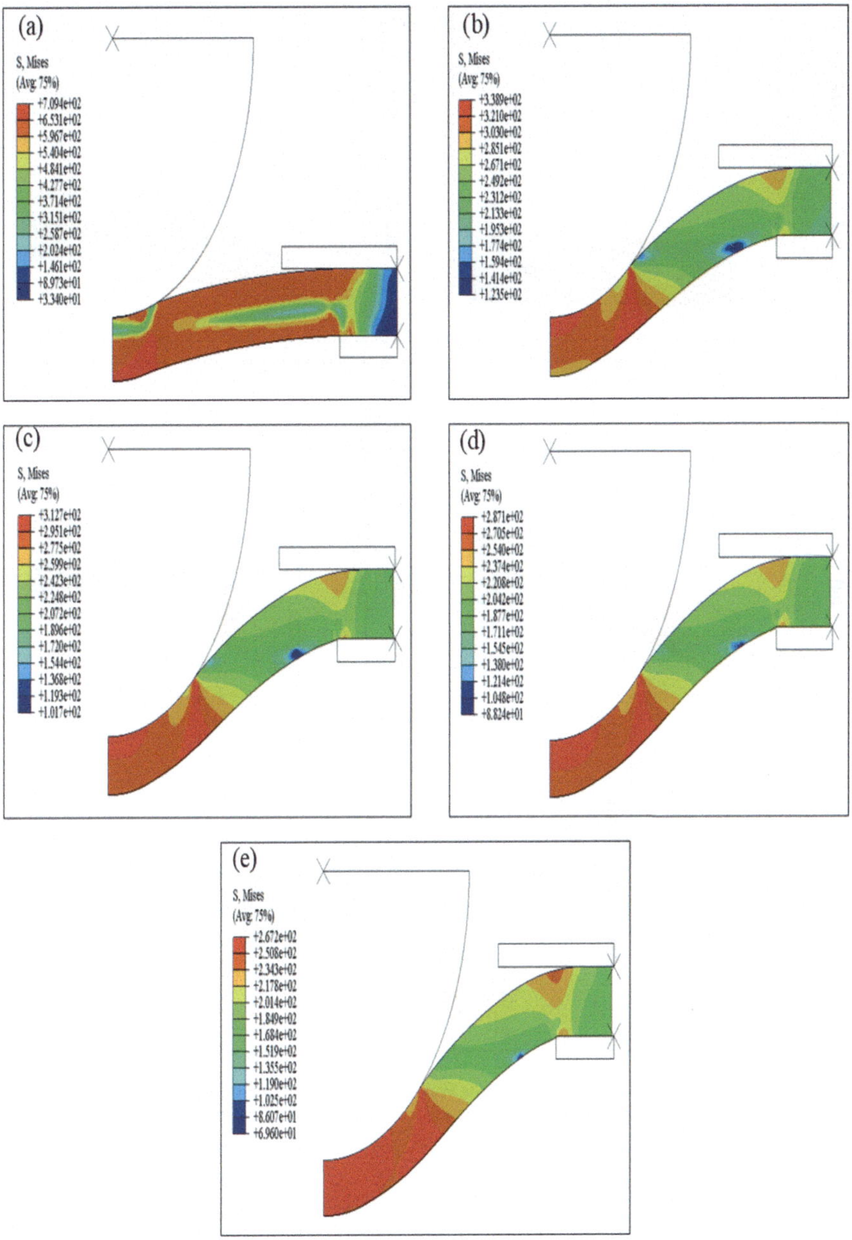

Fig. 4 Contour plots of 0.3 mm thickness for von-Mises total stress on FE model for punch travel time of **a** 0 h, **b** 50 h, **c** 100 h, **d** 200 h and **e** 300 h

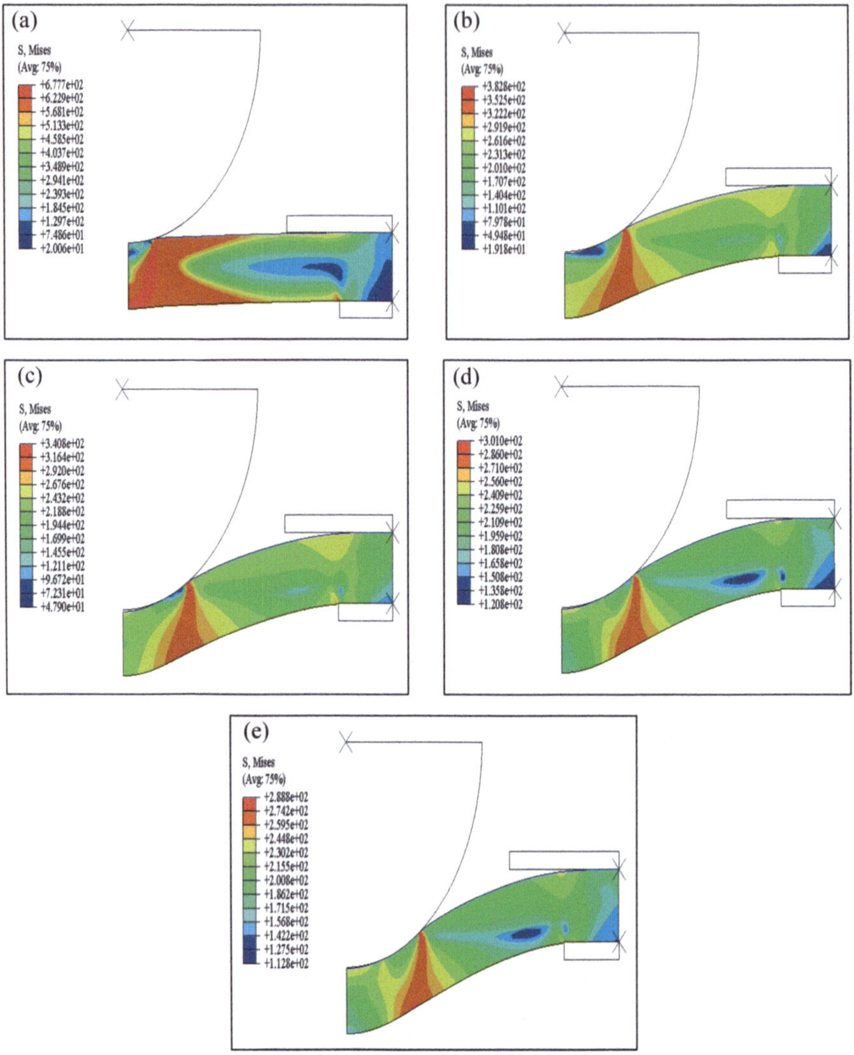

Fig. 5 Contour plots of 0.4 mm thickness for von-Mises total stress on FE model for punch travel time of **a** 0 h, **b** 50 h, **c** 100 h, **d** 200 h and **e** 300 h

decreased, leading to specimen rupture. However, upon analysis of varying thicknesses at the same punch travel time, it was observed that the radial location of the most thinned section shifted toward the center of the specimen. This observation indicates a dynamic change in the specimen's behavior with respect to thickness and punch travel time, highlighting the complexity of the SPCT and its sensitivity to experimental parameters.

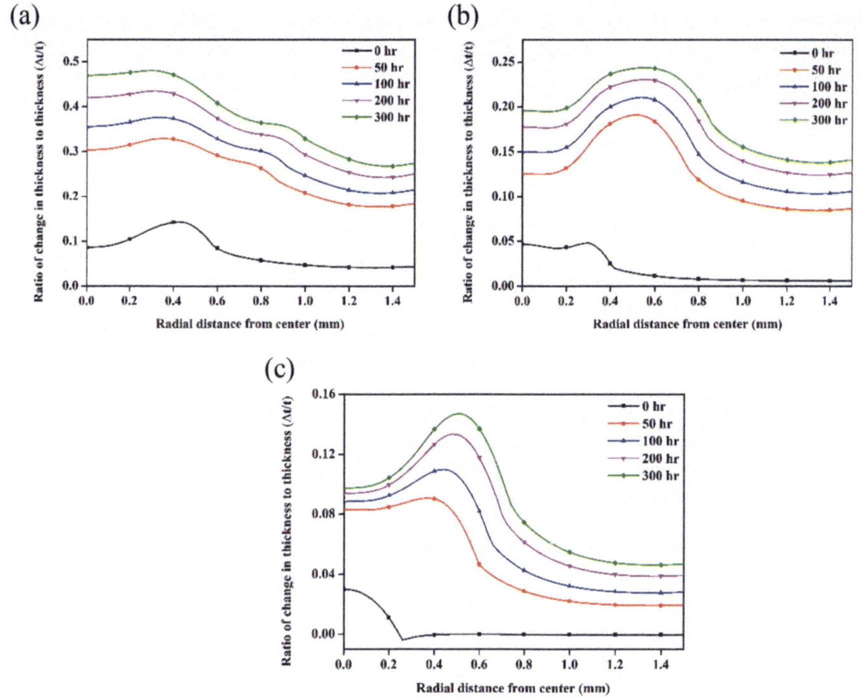

Fig. 6 Plots of thickness of disk at various radial locations for different punch travel time for **a** 0.2 mm, **b** 0.3 mm and **c** 0.4 mm disk thickness

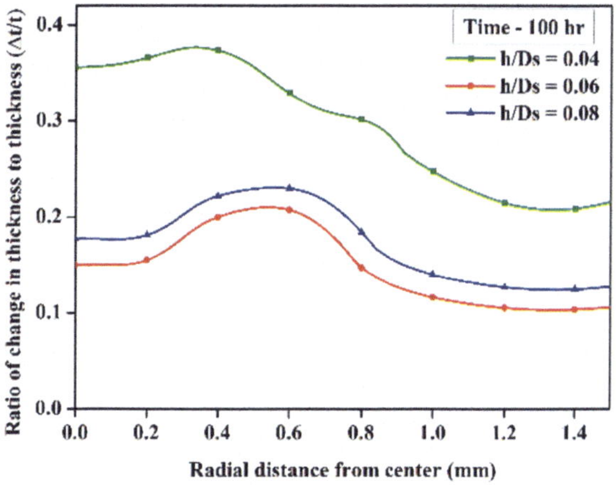

Fig. 7 Plot of thickness of disk at various radial locations for 100 h for different h/Ds ratio

Acknowledgements The authors acknowledge the support from Indian Institute of Technology, Ropar and Malaviya National Institute of Technology, Jaipur for the computational and infrastructural support.

References

1. Penny RK, Marriott DL (1995) Design for creep. Springer Netherlands, Dordrecht
2. Viswanathan R, Gandy D, Coleman K (2005) Advances in materials technology for fossil power plants: proceedings from the fourth international conference, October 25–28, 2004, Hilton Head Island, South Carolina. In: International conference on advances in materials technology for fossil power plants. ASM International
3. Sun W, Wen W, Lu J, Becker AA (2018) Determining full stage creep properties from miniature specimen creep test. Int J Mater Metall Eng 12(10):551–556
4. Hyde TH, Sun W, Becker AA, Williams JA (2004) Creep properties and failure assessment of new and fully repaired P91 pipe welds at 923 K. Proc Inst Mech Eng Part L J Mater Des Appl 218(3):211–222
5. Hyde TH, Sun W, Williams JA (2007) Requirements for and use of miniature test specimens to provide mechanical and creep properties of materials: a review. Int Mater Rev 52:213–255
6. Hyde TH, Sun W, Becker AA (1996) Analysis of the impression creep test method using a rectangular indenter for determining the creep properties in welds. Int J Mech Sci 38(10):1089–1102
7. Yu B, Han W, Tong Z, Geng D, Wang C, Zhao Y, Yang W (2019) Application of small specimen test technique to evaluate creep behavior of austenitic stainless steel. Materials 12(16):2541
8. Hyde TH, Sun W (2009) A novel, high-sensitivity, small specimen creep test. J Strain Anal Eng Des 44:171–185
9. Pan W, Boyle J, Mackenzie D (2009) Simulation of the small punch creep test with consideration of variation of material properties. In: 17th UK conference on computational mechanics. ACME UK, pp 1–4
10. Brown V (2020) Finite element modelling of the small punch test for structure critical design
11. Chen J, Ma YW, Yoon KB (2010) Finite element study for determination of material's creep parameters from small punch test. J Mech Sci Technol 24:1195–1201
12. Zhou Z, Zheng Y, Ling X, Hu R, Zhou J (2010) A study on influence factors of small punch creep test by experimental investigation and finite element analysis. Mater Sci Eng A 527:2784–2789
13. Bruchhausen M, Altstadt E, Austin T, Dymacek P, Holmström S, Jeffs S, Lacalle R, Lancaster R, Matocha K, Petzova J. European standard on small punch testing of metallic materials. In: Pressure vessels and piping conference (57908, V01AT01A065). American Society of Mechanical Engineers
14. Dymáček P, Dobeš F, Li Y, Holmström S, Stevens P (2018) Determination of creep properties from small punch test using experimental correlation. Ubiquity Proc 1(S1)
15. Kumar K (2017) Evaluation of mechanical properties of structural materials using miniature tensile and small punch test techniques
16. Milička K, Dobeš F (2006) Small punch testing of P91 steel. Int J Press Vessel Pip 83(9):625–634. https://doi.org/10.1016/j.ijpvp.2006.07.009
17. Dymáček P, Dobeš F, Li Y, Holmström S, Stevens P. Determination of creep properties from small punch test using experimental correlation
18. Allen C, Coules H, Truman C (2022) Investigation into the effects of prior plasticity on creep accumulation in 316H stainless steel, p 16. https://doi.org/10.3390/psf2022004016
19. Kumar K, Pooleery A, Madhusoodanan K, Singh RN, Chatterjee A, Dutta BK, Sinha RK (2016) Optimisation of thickness of miniature tensile specimens for evaluation of mechanical properties. Mater Sci Eng A 675:32–43

20. Hyde TH, Stoyanov M, Sun W, Hyde CJ (2010) On the interpretation of results from small punch creep tests. J Strain Anal Eng Des 45:141–164
21. Manahan MP, Argon AS, Harling OK (1981) The development of a miniaturized disk bend test for the determination of postirradiation mechanical properties. J Nucl Mater 104:1545–1550

Analyzing the Saint Venant End Effects in Cellular Lattice Structures

Paul Sudhahar, Nuli Sidhardha, Tirumala Rao Koka, and Perumal Rakkiyagounder

Abstract Lattice structures find applications in many structural areas where the weight to strength ratio needs to be minimum, for example: in space and aerospace applications. With the proper choice of lattice and selection of their geometric features functionally graded structures are designed. Lattice structures are non-homogeneous periodic structures at macroscopic level and the distribution of load, stress and strain will not be as uniform as in a continuous and homogeneous solid metal structures and components. For example, a plate made of lattice structure with face sheets will exhibit a different distribution of the stress and strain compared to a solid plate. This poses a different challenge and opportunity to study and understand the end effects of structural components made with lattice structures. In this article Saint Venant's end effect of a lattice plate is analyzed using finite element analysis. End effects of a solid plate are analyzed and considered as base line. The end effect of a plate made with simple lattice structures like square, triangular and honeycomb are analyzed and its relationship with volume fraction of the lattice is studied.

Keywords Lattice structures · End effect · Saint Venant effect · Finite element analysis · Volume fraction

---

P. Sudhahar (✉) · N. Sidhardha · T. R. Koka · P. Rakkiyagounder
Honeywell Technology Solutions Lab Pvt Ltd., Bangalore, Karnataka 560103, India
e-mail: Paul.Sudhahar@Honeywell.com

N. Sidhardha
e-mail: SGRSSIDHARDHA.NULI@Honeywell.com

T. R. Koka
e-mail: TirumalaRao.Koka@Honeywell.com

P. Rakkiyagounder
e-mail: Perumal.Rakkiyagounder@honeywell.com

S. Sidhardh et al. (eds.), *Advances in Structural Integrity for Mechanical, Civil, and Aerospace Applications*, Lecture Notes in Mechanical Engineering,
https://doi.org/10.1007/978-981-97-6367-2_47

1 Introduction

In structural mechanics and design, real boundary conditions and loadings are complicated to model them mathematically and need several approximations and assumptions. So, all the designs tend to strengthen the locations where either boundary conditions are applied, or loads are applied. Stress distributions near the point of load application, and near weak boundary conditions are going to be complex to model them accurately. Saint–Venant's principle allows the designers to simplify this for the ease of calculations. The principle roughly states that when a load applied on one end of a long plate, strain produced at the loaded end is much larger than the strain/stress seen by the plate at some distance from the load application point. After distance '*D*' from edge of application, the stress distribution becomes uniform across the cross section and becomes insensitive to the kind of load. Characteristic length *D* represents the axial distance from the edge of load application, beyond which these end effects cease to exist. This characteristic length depends on the material and the thickness of the plate. In case of lattice structure, characteristic length varies based on the unit cell shape and volume fraction of the unit cell. This information is very useful in designing the lattice-based plates and components to decide upon the design parameters accordingly (Fig. 1).

The current work tries to establish the end effect in several lattice-based designs, namely square, triangular and hexagonal by the extensive use of finite element analysis. The characteristics length for these two-dimensional lattice shapes was established as a function of the characteristic length of a solid plate and the effect on their volume fractions. This will guide the lattice-based designs particularly at the end zones.

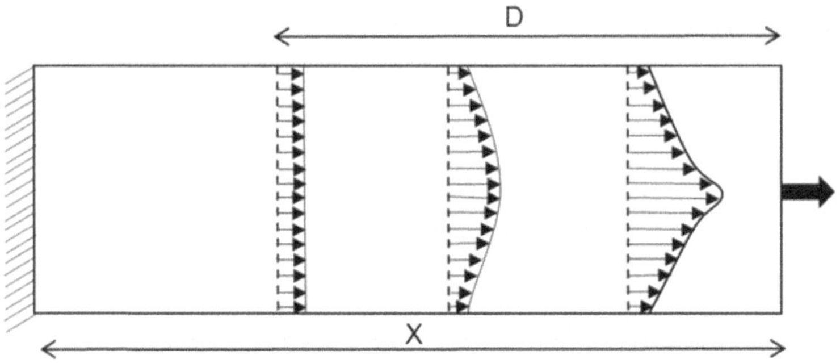

Fig. 1 Variation of stress distribution due to the application of point load

2 Literature Survey

Saint Venant principle can be stated as if the point force acting on a surface of an elastic body are replaced by another equivalent distributed force acting on the same portion of the surface, this redistribution of loading produces considerable differences in the stresses locally but has a negligible effect on the stresses at distances from the application of load. In UTM test the placement of strain gauge is attributed to the Saint Venant principle, Saint Venant's principle, named after Adhémar Jean Claude Barré de Saint Venant, a French elasticity theorist. This effect is verified using commercial multiphysics solver on a plate by Fung et al. [1]. In this study author considered three types of loadings on a plate with a hole. The stress distribution at the hole is not affected with type of loading. In the next case, hole is moved towards the loading edge. This changed the stress distribution significantly. And the stress distribution is affected by the type of loading. It is also observed that characteristic length that gives the distance where stress fields converge is also affected. Borrelli et al. [2] analyzed the Saint Venant end effects for plane deformations of a piezoelectric semi-infinite strip. For this study author formulated airy-stress function and an induction function. Later these equations are coupled and formulated into an Eigen value problem. The real part of that gives the rate of decay of end effects for piezo electric materials. Using these relations author found that the decay lengths of piezo electric materials are larger than that of isotropic elastic materials. For a piezoelectric material decay lengths may go up to five times the considered strip width.

Jog and Mokashi [3] applied Saint Venant principle to torsion and bending problems. To formulate torsion problem author used wrapping function over Prandtl function. Prandtl function is quite cumbersome to implement in terms of finite element analysis. Numerical implementation of a bending problem is based on linearized elasticity theory. Linear elasticity is a well-known approach and can be easily implemented numerically to find out deformations in solid objects and stress based on the applied boundary conditions and loads. Author developed formulations for bending and torsion which are valid for compound, multiply connected and anisotropic bars. These formulations are also verified on coarse mesh and for thin-walled cross sections. He et al. [4] applied Saint Venant principle on a cylindrical wave guide under transient loading. Author experimentally and numerically investigated axial decay of end effects. Author introduced dynamic Saint Venant end effects associated with the dynamic loading, end effects in dynamic load wave propagation. In this study he tried to validate Saint Venant principle in dynamic loading conditions and compared surface strains experimentally and numerically which validates the dynamic Saint Venant principle. Nakamura and Lakes [5] studied Saint Venant end effects for micropolar elastic solids. Author used a finite element approach called MIRACS which considers degree of freedom of the micropolar solid. He compared decay rates between micropolar solid and classical solid. The results in this study are completely dependent on the geometry. Stress decay in a 2D strip with a point load at one end is dependent on micropolar characteristic length. For the strip geometry a Cosserat solid exhibits slower stress decay than a classical solid. El Fatmi and Zenzri [6] discussed a finite element computation technique and applied this to laminated structures. Structural beam rigidities and couplings, warpings and 3D stresses are extracted using this methodology and compared these results with the available results. In this study author applied this principle to simple laminated structure. Author faced challenges to apply this principle to the complex geometries. He suggested to utilize exact beam theory in such situations which can be systematically achieved. Liu and Taciroglu [7] introduced a method to quantify Saint Venant principle for laminated piezoelectric circular cylinders. To define decay length, an algebraic Eigen equation is constructed. Author states that inverse of the real parts of constructed Eigen value gives information about decay lengths. The author analyzed and quantified end effects for laminated piezoelectric cylinders. These results are also compared with isotropic solid cylinder. To get the results from laminated piezoelectric cylinders, author introduced a semi-analytical method which calculates decay rates of electric and mechanical fields for a transversely isotropic piezoelectric cylinder. Xue and Liu [8] studied decay of Saint Venant end effects of piezoelectric-piezomagnetic sandwich structures. Piezoelectric-piezomagnetic structures are the structures where each piezomagnetic structure is embedded between piezoelectric structures or vice versa. Author found out characteristic equations for the structures and derived Eigen values for both piezoelectric and piezomagnetic structures. Later decay lengths for piezoelectric and piezomagnetic structures are obtained. Author derived decay lengths for both sandwiched configurations.

There were several studies in estimating the equivalent/effective properties of cellular materials. Ongaro reviewed on estimation of the effective properties of cellular materials by means of analytical approach [9]. The end effects are rather complicated to analytically estimate for cellular structures and to date, there is no attempt to identify decay length for lattice structures. This study will fill the gap in literature and gives decay lengths for different lattice structures.

3 Problem Statement/Numerical Simulation

Quantify the characteristic length/edge effect of a lattice plate with respect to a solid plate of similar dimension and how this effect varies with respect to the lattice's volume fraction. To analyze this problem statement, a two-dimensional long solid plate with a width of 80 mm is considered as a baseline. In this study, total four different unit cell constructions are considered. They are, namely triangular, square, hexagonal in both the directions called type-A and type-B. Type-A hexagonal cell is aligned horizontally to the next hexagonal cell, sharing a vertical edge and type-B hexagonal cell is aligned horizontally to the next hexagonal cell by sharing a vertex. Figure 2 shows the different lattices considered for the analysis.

Solid plate is modelled with plane stress elements. One end of the solid plate is fixed, and the other end of the solid plate is loaded with a point load at the midpoint of the width. The lattice plates are modelled as beams with a circular cross section. The cross section is varied from the lowest possible radius additively printable and to

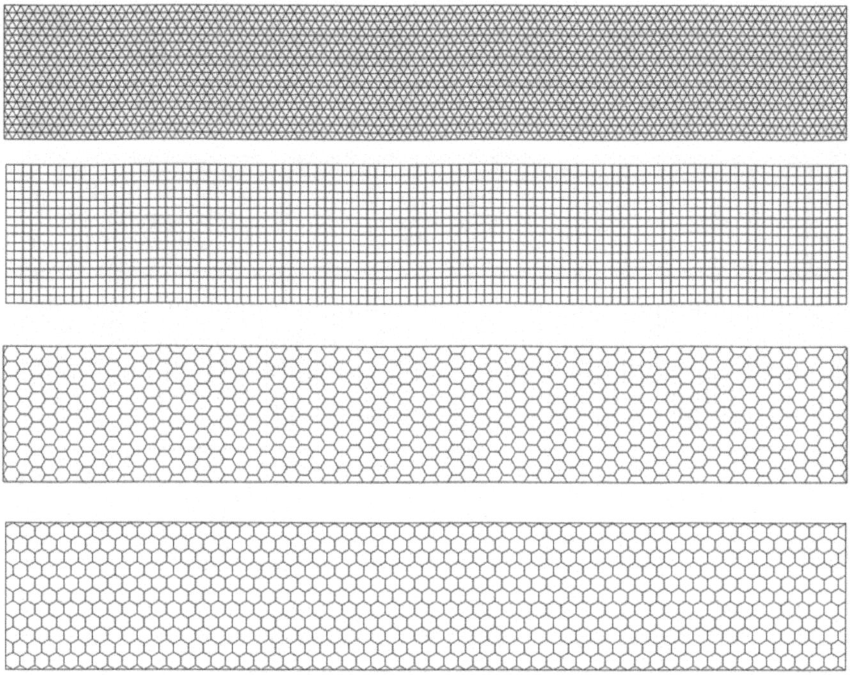

Fig. 2 Triangular, square, hex type-A and hex type-B lattices used for numerical

a maximum radius where the volume fraction becomes close to unity. Table 1 gives the diameter of the beams and the corresponding volume fraction of the lattice plate. For triangular lattice, beam diameter of 1.5 mm and above is resulting in volume fraction of more than one and so only first three diameters were considered for the study using triangular lattice. The boundary of the lattice structure is closed with similar beam to avoid any open unit cells. One end of the lattice plate is fixed and a point load is applied at the other end of the lattice plate at the mid width of the plate. Linear static analysis is conducted, and the stress patterns are observed in the solid and lattice plates. Later post processing was carried out to estimate the characteristic length of each lattice at different volume fraction at which the stress distribution across the width becomes uniform (Fig. 3).

Table 1 Lattice structures beam diameter and corresponding volume fractions

Lattice\dia. (mm)	0.5	0.75	1.0	1.5	2
Triangular	0.35	0.52	0.69	*1.04*	~
Square	0.21	0.31	0.41	0.62	0.83
Hex type-A	0.12	0.19	0.25	0.37	0.50
Hex type-B	0.12	0.19	0.25	0.37	0.50

Fig. 3 Solid, triangular, square, hex type-A and hex type-B lattices under point load at one end

4 Results and Discussion

For the generalized applicability of this study, the stress and the characteristic lengths are normalized with maximum stress of the overall lattice and the characteristic length of the solid plate, respectively. From the analysis conducted, the effect of volume fraction on the characteristic length for different lattice structures and the effect of lattice type on the characteristic length for a similar volume fraction are studied. Plots in the later sections between normalized length and normalized stress are drawn in such a way that, the point of load application is at $x = 0$ and the normalized stress is plotted as we move away from the load application. All these plots start at a high normalized stress value (near load application point) and as it moves away from the load application point the max. normalized stress reduces and stabilizes beyond certain normalized length indicating that the end effects are non-existent beyond that point. Figures 4, 5, 6 and 7 show the normalized stress versus normalized length for different volume fraction of lattices for triangular, square, hex type-A and hex type-B, respectively.

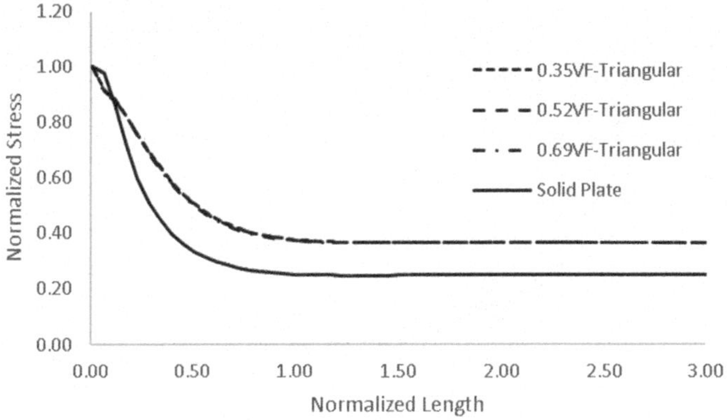

Fig. 4 Peak stresses across different sections of triangular lattice

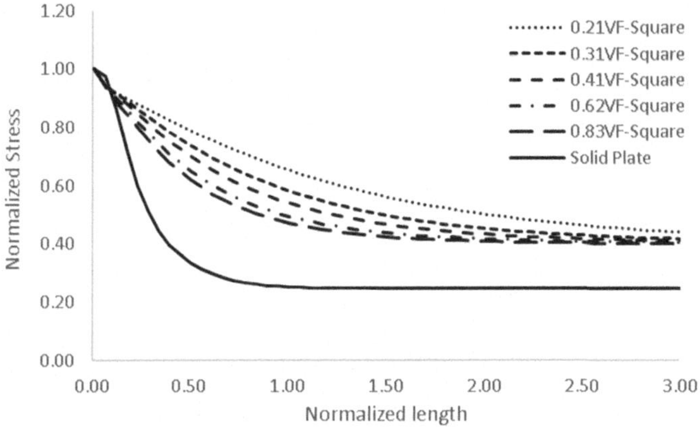

Fig. 5 Peak stresses across different sections of hex type-A lattice

For triangular lattice plate the effect of volume fraction on the characteristic length is insignificant. Compared to solid plate the characteristic length is 1.11. For square lattice, effect of volume fraction is significant and the characteristic length is varying from 3.89 to 1.67 for volume fractions of 21–83%. Both triangular and square lattices show higher stress across far away section when compared with a solid plate. For hexagonal lattices, we can see that the stress distribution across section far away from the end effected zone is lower than the solid plate. For hex type-A, this stress increases as the volume fraction increases and is below solid plate till 25% of volume fraction. Beyond that the stress is more than solid plate. For hex type-B hex, the stress is always less than solid plate beyond the end effected zone. Characteristic length for hex type-A is almost similar to triangular mesh and for type-B it is higher than triangular mesh and lower than square mesh. The relation between the volume fraction and the characteristic length is shown in Fig. 7. For reduced end effects either triangular

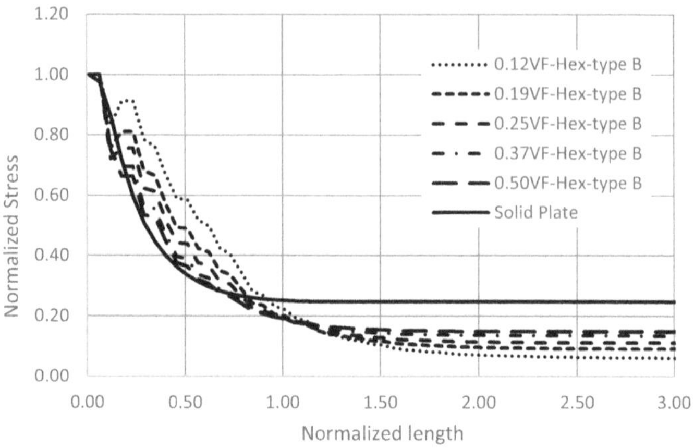

Fig. 6 Peak stresses across different sections of hex type-B lattice

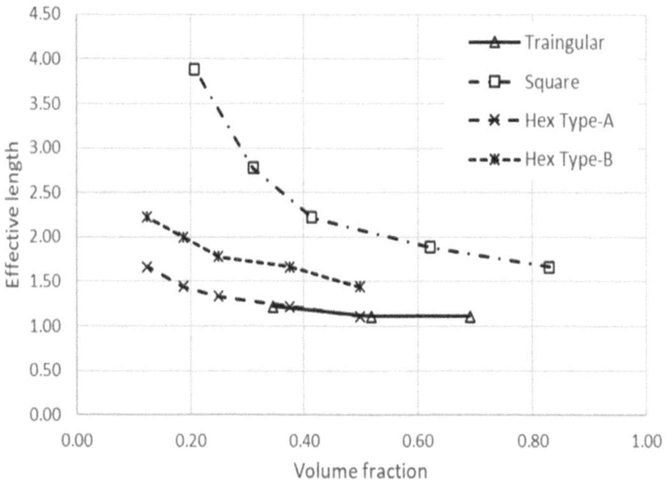

Fig. 7 Variation of characteristic lengths for lattices w.r.t. volume fractions

lattice or hexagonal lattice type-A are best suited. For reduced stress propagation hexagonal lattice type-B is best suited from the list of lattices under study. Table 2 shows the quantitative values of normalized characteristic length of different lattices at different volume fractions.

Another outcome of this study is how the normalized stress changes, compared to a solid plate and other lattices structures at similar volume fraction. Figure 8 shows the normalized stress versus normalized length for all the lattices at same volume fraction. We can see that the square lattice has a higher characteristic length and higher stress. Triangular lattice mesh and hexagonal type-A mesh comes close to solid plate for characteristic length, but the normalized stress is higher than the solid

Table 2 Volume fraction versus normalized characteristic length for different lattices

V.F. →	0.12	0.19	0.21	0.25	0.31	0.35	0.37	0.41	0.5	0.52	0.62	0.69	0.83
Triangular						1.11				1.11		1.11	
Square			3.89		2.78			2.22			1.89		1.67
Hex-A	1.67	1.44		1.33			1.22		1.11				
Hex-B	2.22	2		1.78			1.67		1.44				

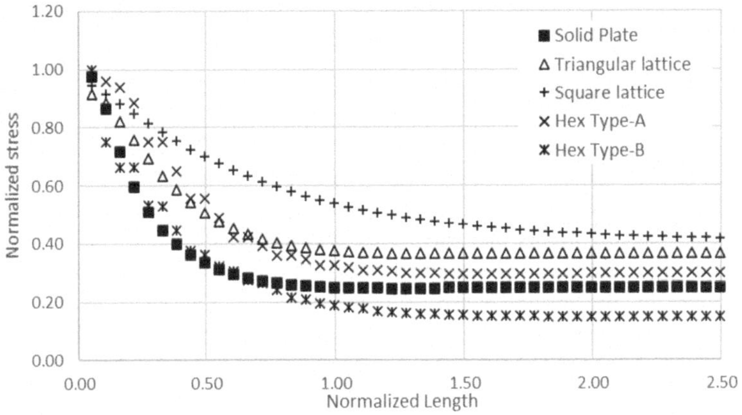

Fig. 8 Variation of peak stresses across different sections for lattices of similar volume fractions

plate which means for the same loading the max stress seen by the triangular and hexagonal type-A lattice are higher than the solid plate, as volume fraction is less than one. But hexagonal type-B lattice is having less normalized stress than the solid plate despite volume fraction less than one. But the normalized length is higher than the triangular and hex type-A lattices and solid plate.

5 Conclusions

The aim of this study is to understand the end effects on various lattice types in two-dimensional space, so that while designing the lattice-based components these end effects are taken care and doesn't affect the functionality of the component. In this study we conducted numerical simulations on four different lattices, namely triangular, square, hexagonal type-A and hexagonal type-B. From this study, it is understood that the stress transmitted by the hexagonal type-B is lower than that of solid plate but has a higher characteristic length. Triangular mesh and hexagonal mesh type-A are having characteristic lengths close to solid plate at higher volume

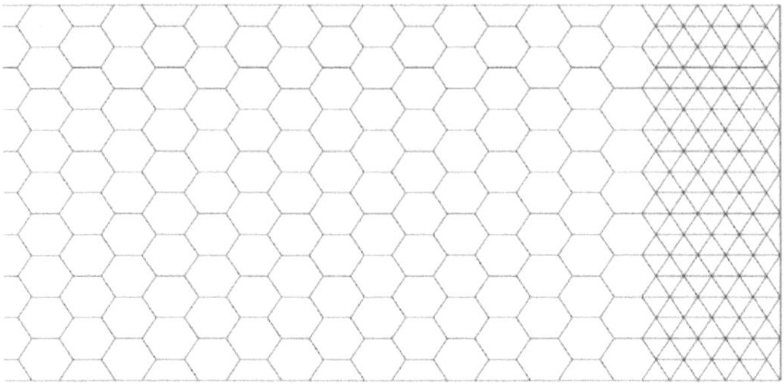

Fig. 9 Hybrid lattice deign with hex and triangular lattices

fractions. The normalized stress in both triangular and hexagonal type-A lattice are higher than the solid plate. For square lattice, both the characteristic length and normalized stress are higher than any lattice and solid plate.

This gives designer several options to design a lattice-based design which reduces the end effect and the normalized stress by hybrid lattice. One such method is, as the lattice design approaches the edge or a corner of the component, the volume fraction can be increased gradually by keeping the same lattice structure. This will result in mitigating the higher characteristic length at the edges than that of solid plate. One more option is to have a hybrid design of lattice. For example, a component with hexagonal type-B, when it comes to the edge can be transformed to triangular lattice, thus reducing the end effect as shown in Fig. 9.

The current studies need to be extended for several other lattices in both two and three dimensions. As the analytical representation of these effects are going to be complicated, numerical investigations will give us a faster insight into these aspects. If the lattice structure under study is of auxetic nature and if the final application is high strain rate like impact loads or a quasi-static loading will also determine if these studies need to be conducted at unit cell level and/or actual component level.

Acknowledgements This research is supported by Honeywell Technology Solutions Pvt. Ltd., India. We thank our leaders Parasuraman Ramakrishnan, Sr. Director, Engines and Power Systems, Venkat Rao Ganji, Sr. Engineering Manager, Engines and Power Systems and Venkata Neeladri Raju Kucharlapati, Sr. Engineering Manager, Engines, and power systems, who provided insights and expertise that greatly assisted this research.

References

1. Fung YC, Tong P, Chen X (2017) Classical and computational solid mechanics, vol 2. World Scientific Publishing Company
2. Borrelli A, Horgan CO, Patria MC (2001) Saint-Venant end effects in anti-plane shear for classes of linear piezoelectric materials. J Elasticity Phys Sci Solids 64(2):217–236

3. Karp B, Durban D (2011) Saint-Venant's principle in dynamics of structures. Appl Mech Rev 64(2)
4. He L, Ma GW, Karp B, Li QM (2015) Investigation of dynamic Saint-Venant's principle in a cylindrical waveguide—experimental and numerical results. Exp Mech 55(3):623–634
5. Nakamura S, Lakes RS (1995) Finite element analysis of Saint-Venant end effects in micropolar elastic solids. Eng Comput
6. El Fatmi R, Zenzri H (2002) On the structural behavior and the Saint Venant solution in the exact beam theory: application to laminated composite beams. Comput Struct 80(16–17):1441–1456
7. Liu CW, Taciroglu E (2007) Numerical analysis of end effects in laminated piezoelectric circular cylinders. Comput Methods Appl Mech Eng 196(17–20):2173–2186
8. Xue Y, Liu J (2010) Decay rate of Saint-Venant end effects for plane deformations of piezoelectric-piezomagnetic sandwich structures. Acta Mech Solida Sin 23(5):407–419
9. Ongaro F (2018) Estimation of the effective properties of two-dimensional cellular materials: a review. Theor Appl Mech Lett 8(4):209–230

Modeling and Diagnosis of High Contact Ratio Gear Faults

Rajeev Kumar, Chintamani Mishra, and Ranjan Kumar Mitra

Abstract Vibration characteristics of a machine change with the operating conditions (speed, load, fault, and etc.). Gears are widely used as power transmission device due to its flexibility to operate at wide range of speed and load. Any faults in the gear can lead to complete failure or even seizure of machine. Therefore, predictive maintenance strategy is adopted where the monitoring of condition of gear is done by analyzing the signal (vibration, acoustics, current, and etc.) of gear. Modeling of a physical system is economical due to the cost involves in experiments. Moreover, the models are scalable and several types of faults can be introduced into it. In the present study, a gear kinematics based 9-degrees of freedom (DOFs) model is considered to find the response of the gear in healthy as well as in faulty (broken or may be missing tooth) conditions. A higher contact ratio spur gear system is considered for analysis. The model consists of a gear-pinion pair, a loader and a driver. The equations of motion of planer block diagram model of gear-pinion system are solved in Matlab-Simulink environment to study the response of the healthy and faulty gear. This study can be extended for different gear systems to find out the fault symptoms under missing/broken tooth.

Keywords Modeling and simulation · Vibration analysis · Condition monitoring · Rotor dynamics · Single stage spur gear

1 Introduction

This Gearbox is widely used in industry machineries, automobile applications etc. for transmitting power and/or motion. Different types of faults like spalling, fatigue, crack, pitting, and broken tooth are generally observed in gears. These faults may cause catastrophic failure of the gear system. To avoid such failure, condition-based monitoring method can be used. Out of many condition-based monitoring techniques, vibration-based condition monitoring is adopted in the present study. In this tech-

R. Kumar · C. Mishra (✉) · R. K. Mitra
NIT Durgapur, Durgapur, India
e-mail: chintamani.mishra@me.nitdgp.ac.in

S. Sidhardh et al. (eds.), *Advances in Structural Integrity for Mechanical, Civil, and Aerospace Applications*, Lecture Notes in Mechanical Engineering,
https://doi.org/10.1007/978-981-97-6367-2_48

nique, the vibration characteristics of the system have the key role for identifying the faults in the system. To study these vibration characteristics, modeling of the gear system is an economical solution. A mathematical model of gear dynamics for healthy condition was proposed in [1]. A comprehensive review on modeling of one-stage spur gear was proposed in [2], where the modeling of gear with 6-DOF, 8-DOF and 12-DOF is discussed in details. The research papers [3–6], discuss the gear system modeling with faults in gear. In these reviews, how the time varying meshing stiffness (TVMS) varies with the gear faults and the variation of this TVMS leads to the change in dynamic response of system is also discussed. Many research papers use finite element method (FEM) for calculation of TVMS [3, 4]. As the finite element method is a complex process and has lower computational efficiency, some research papers focus on analytical method for the calculation of the TVMS [7–9]. The analytical method to calculate TVMS of a higher contact ratio gear system is discussed in [10]. The calculation of dynamic friction forces in spur gears using traction and frictional force is reported in [11].

The analytical method is used in gear dynamic model in some research papers [5, 6]. Analytical method also has some limitations as compared to the finite element method, so any of these two methods can be used for calculation of the TVMS. There are many work reported where vibration characteristics of different gear faults are analyzed. The faults like tooth root crack and its propagation is modeled in [3, 4, 12]. The faults like tooth spalling and tooth breakage is modeled in [5]. The diagnosis of tooth pitting is modeled in [6]. But very less research has been done on the vibration characteristics of fault such as missing tooth/broken tooth. So in the present study, vibration characteristics of the gear fault like missing tooth/broken tooth in the spur gear system of higher contact ratio (C.R. > 2) is analyzed. There are different signal processing methods proposed in many different research papers for the vibration characteristic. Some important signal processing methods are spectral kurtosis [3, 4], wavelet analysis [13–15], root mean square (RMS) value analysis [16], empirical model decomposition (EDM) method [17], and autoregressive (AR) model [18]. In the present work, the TVMS is calculated using analytical method proposed by Chen et al. [4] for both healthy as well as faulty conditions and by using this TVMS value in single stage spur gear system vibration response was computed for the spur gear system of higher contact ratio.

2 Equivalent Stiffness

2.1 Healthy Gear

The main cause of excitation for the spur gear system is the equivalent stiffness, which varies with the movement of contact point along the line of action. In the

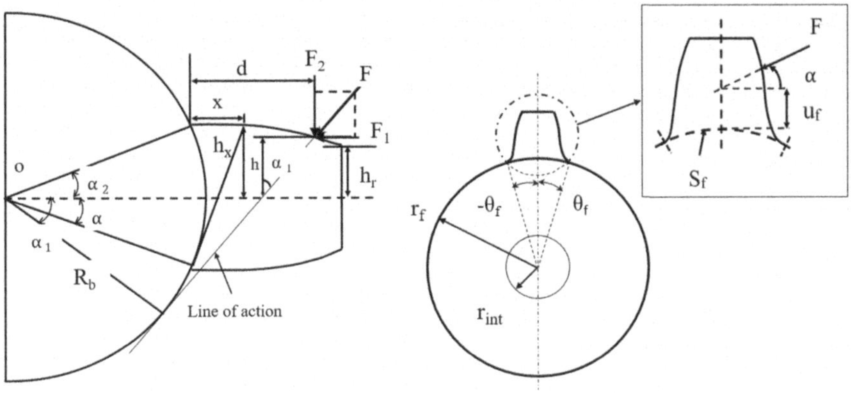

Fig. 1 **a** Healthy tooth model. **b** Geometric parameter of the fillet foundation deflection

present work, an analytical model is used for the calculation of TVMS. The potential energy method proposed by Chen et al. [4], is used for the calculation of gear mesh stiffness. In this method, gear tooth is considered as non-uniform cantilever beam and the potential energies stored due to the bending, shearing, and axial compressive deformation is calculated. The mathematical models for involvement of Hertzian, bending, and axial stiffness components are given in the work by Yang et al. [19] whereas involvement of shear stiffness is proposed in the work by Tian et al. [20] and the stiffness related to fillet foundation deflection is discussed by Sainsot et al. [21]. In the present work, the Hertzian stiffness (k_{h}), bending stiffness (k_{b}), shear stiffness (k_{s}), and the axial compressive stiffness (k_{a}) are calculated as per Eqs. (1)–(4), respectively [9].

$$k_{\mathrm{h}} = \frac{\pi E L}{4\left(1-\nu^2\right)} \tag{1}$$

$$\frac{1}{k_{\mathrm{b}}} = \int\limits_{-\alpha_1}^{\alpha_2} \frac{3\{1+\cos\alpha_1\left[(\alpha_2-\alpha)\sin\alpha-\cos\alpha\right]\}^2(\alpha_2-\alpha)\cos\alpha}{2EL\left[\sin\alpha+(\alpha_2-\alpha)\cos\alpha\right]^3}\mathrm{d}\alpha \tag{2}$$

$$\frac{1}{k_{\mathrm{s}}} = \int\limits_{-\alpha_1}^{\alpha_2} \frac{1.2\,(1+\nu)\,(\alpha_2-\alpha)\cos\alpha\,(\cos\alpha_1)^2}{EL\left[\sin\alpha+(\alpha_2-\alpha)\cos\alpha\right]}\mathrm{d}\alpha \tag{3}$$

$$\frac{1}{k_{\mathrm{a}}} = \int\limits_{-\alpha_1}^{\alpha_2} \frac{(\alpha_2-\alpha)\cos\alpha\,(\sin\alpha_1)^2}{2EL\left[\sin\alpha+(\alpha_2-\alpha)\cos\alpha\right]}\mathrm{d}\alpha \tag{4}$$

where, 'E' is the elastic modulus of rigidity for gear material, L is the width of tooth and α_1 and α_2 are the angles presented in Fig. 1.

Table 1 Values of coefficient used in Eq. (6) [5]

	$A_i \times 10^5$	$B_i \times 10^3$	$C_i \times 10^4$	$D_i \times 10^3$	$E_i \times 10^2$	F_i
$L^*(h_{f_i}, \theta_f)$	− 5.574	1.9986	− 2.3015	4.7702	2.71	6.8045
$M^*(h_{f_i}, \theta_f)$	60.111	28.100	− 83.431	− 9.9256	16.24	0.9086
$P^*(h_{f_i}, \theta_f)$	− 50.952	185.50	0.0538	53.3	28.95	0.9236
$Q^*(h_{f_i}, \theta_f)$	− 6.2042	9.0889	− 4.0964	7.8297	− 14.72	0.6904

The gear fillet foundation stiffness (k_f) is calculated using the formulation proposed by Sainsot et al. [21].

$$\frac{1}{k_f} = \frac{(\cos \alpha_m)^2 \left\{L^* (u_f/s_f)^2 + M^* (u_f/s_f) + P^* \left(1 + Q^* (\tan \alpha_m)^2\right)\right\}}{EL} \quad (5)$$

where, α_m is pressure angle of the system and 'L' is the width of tooth. The coefficients L^*, M^*, P^* and Q^* are calculated using equation:

$$X_i^* (h_{fi}, \theta_f) = \frac{A_i}{\theta_f^2} + B_i h_{fi}^2 + \frac{C_i h_{fi}}{\theta_f} + \frac{D_i}{\theta_f} + E_i h_{fi} + F_i \quad (6)$$

where, X_i^* denotes the coefficients L^*, M^*, P^* and Q^*. The other terms such as h_f, r_{int}, r_f and $h_{fi} = r_f/r_{int}$, are the geometric parameter as described in Fig. 1, the values of coefficient used in Eq. (6) are given in Table 1. Thus, the total TVMS for a single-tooth-pair in contact can be expressed as

$$K_m = \frac{1}{\frac{1}{k_h} + \frac{1}{k_{b1}} + \frac{1}{k_{s1}} + \frac{1}{k_{a1}} + \frac{1}{k_{f1}} + \frac{1}{k_{b2}} + \frac{1}{k_{s2}} + \frac{1}{k_{a2}} + \frac{1}{k_{f2}}} \quad (7)$$

where, k_f, k_a, k_s, k_b and k_h represent the fillet foundation deflection mesh stiffness, axial, shear, bending compressive, and Hertzian stiffness. Subscript 1 and 2 used in the expressions denotes pinion and gear respectively.

For more than one tooth in contact the TVMS is the sum the stiffness of each pair, and can be expressed as

$$K_T = \sum_{i=1}^{n} \frac{1}{\frac{1}{k_h} + \frac{1}{k_{b1i}} + \frac{1}{k_{s1i}} + \frac{1}{k_{a1i}} + \frac{1}{k_{f1i}} + \frac{1}{k_{b2i}} + \frac{1}{k_{s2i}} + \frac{1}{k_{a2i}} + \frac{1}{k_{f2i}}} \quad (8)$$

where, n is the number of tooth pair in meshing.

In the present study, a contact ratio of 2.1667 is considered, which indicates the average numbers of tooth pairs in contact between the gears. For contact ratio of 2.1667 there are three numbers of pairs in mesh for a time period of 0.1167% of the gear meshing period (T_e) and two numbers of pairs of tooth is in mesh for the rest of

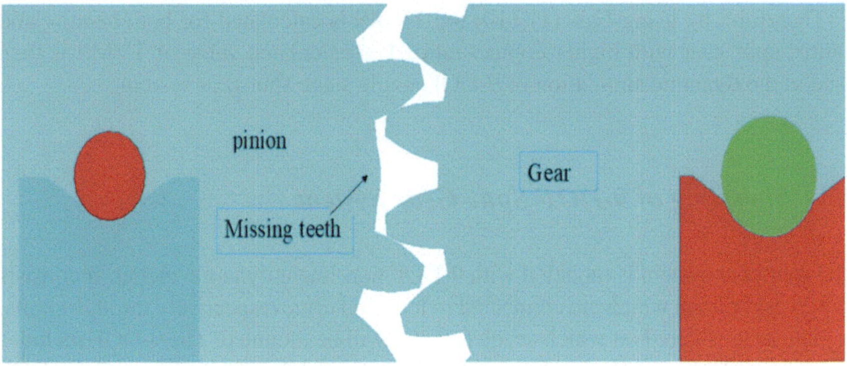

Fig. 2 Meshing zone with missing tooth on pinion

gear meshing period. This fluctuation of numbers of pairs of tooth in meshing is the gear system's main source of excitation.

For a gear which revolves at a speed of N rpm with Z numbers of teeth, the gear meshing period (in s) of the gear and gear mesh frequency (GMF) is defined [5]:

$$T_{\mathrm{m}} = \frac{60}{N * Z} \quad \text{and} \quad f_{\mathrm{m}} = \frac{N * Z}{60} \tag{9}$$

2.2 *Faulty Gear*

In the present study, the gear fault like complete broken/ missing tooth on the pinion is considered. When this completely broken tooth comes in the meshing zone, the corresponding tooth of meeting gear will not make any contact with it as shown in Fig. 2. Therefore, the mesh stiffness corresponding to this tooth will not contribute to the total TVMS. If, the ith tooth on the pinion is completely broken, the corresponding gear mesh stiffness $K_i(t)$ will be zero. Therefore, when a broken tooth comes to the contact zone, the time varying stiffness can be expressed as

$$k_{\mathrm{m}}(t) = \begin{cases} k_{i-2}(t) + k_{i-1}(t) + k_i(t), & \text{for } i * t_{\mathrm{m}} \le t < i * (\mathrm{c.r} - 2) * t_{\mathrm{m}} \\ k_{i-1}(t) + k_i(t), & \text{for } i * (\mathrm{c.r} - 2) * t_{\mathrm{m}} \le t < (i+1) * t_{\mathrm{m}} \\ k_{i-1}(t) + k_i(t) + k_{i+1}(t), & \text{for } (i+1) * t_{\mathrm{m}} \le t < (i+1) * (\mathrm{c.r} - 2) * t_{\mathrm{m}} \\ k_i(t) + k_{i+1}(t), & \text{for } (i+2) * t_{\mathrm{m}} \le t < (i+2) * (\mathrm{c.r} - 2) * t_{\mathrm{m}} \\ k_i(t) + k_{i+1}(t) + k_{i+2}(t), & \text{for } (i+2) * t_{\mathrm{m}} \le t < (i+2) * (\mathrm{c.r} - 2) * t_{\mathrm{m}} \end{cases} \tag{10}$$

Therefore, by using Eqs. (1)–(10) the TVMS is calculated for both healthy and faulted spur gear with higher contact ratio. The calculated value of TVMS is then used in the dynamic simulation of 9-DOF single stage spur gear system.

2.3 Modeling of 9-DOF Spur Gear System

The spur gear system is modeled with 9-DOF non-linear dynamic model. It consists of gear and pinion which are connected to load and drive respectively through shafts. A plate is considered on which accelerometer will be mounted. The 9-DOF includes 4-rotational DOF (one each of pinion, gear, load, and drive) and 5-transverse DOF (one each of pinion, gear, bearing foundations, and plate). The suitable springs and dampers are considered in this model and the diagram of the gear dynamic system is shown in Fig. 3. The equations of motion of 9-DOF spur gear system can be written as [22]:

$$I_{\mathrm{d}}\ddot{\theta}_{\mathrm{d}} + c_{\mathrm{t1}}\left(\dot{\theta}_{\mathrm{d}} - \dot{\theta}_{1}\right) + k_{\mathrm{t1}}\left(\theta_{\mathrm{d}} - \theta_{1}\right) = T_{\mathrm{d}} \tag{11}$$

$$I_{1}\ddot{\theta}_{1} + c_{\mathrm{t1}}\left(\dot{\theta}_{1} - \dot{\theta}_{\mathrm{d}}\right) + k_{\mathrm{t1}}\left(\theta_{1} - \theta_{\mathrm{d}}\right) = -NR_{\mathrm{b1}} \tag{12}$$

$$I_{2}\ddot{\theta}_{2} + c_{\mathrm{t2}}\left(\dot{\theta}_{2} - \dot{\theta}_{\mathrm{l}}\right) + k_{\mathrm{t2}}\left(\theta_{2} - \theta_{\mathrm{l}}\right) = NR_{\mathrm{b2}} \tag{13}$$

$$I_{\mathrm{l}}\ddot{\theta}_{\mathrm{l}} + c_{\mathrm{t2}}\left(\dot{\theta}_{\mathrm{l}} - \dot{\theta}_{2}\right) + k_{\mathrm{t2}}\left(\theta_{\mathrm{l}} - \theta_{2}\right) = -T_{\mathrm{l}} \tag{14}$$

$$m_{1}\ddot{y}_{1} + c_{1}\left(\dot{y}_{1} - \dot{y}_{3}\right) + k_{1}\left(y_{1} - y_{3}\right) = N \tag{15}$$

$$m_{2}\ddot{y}_{2} + c_{2}\left(\dot{y}_{2} - \dot{y}_{4}\right) + k_{2}\left(y_{2} - y_{4}\right) = -N \tag{16}$$

$$\begin{aligned} m_{3}\ddot{y}_{3} + c_{34}\left(\dot{y}_{3} - \dot{y}_{4}\right) + k_{34}\left(y_{3} - y_{4}\right) + c_{1}\left(\dot{y}_{3} - \dot{y}_{1}\right) + k_{1}\left(y_{3} - y_{1}\right) \\ + c_{\mathrm{p}}\left(\dot{y}_{3} - \dot{y}_{\mathrm{p}}\right) + k_{\mathrm{p}}\left(y_{3} - y_{\mathrm{p}}\right) + c_{3}\dot{y}_{3} + k_{3}y_{3} = 0 \end{aligned} \tag{17}$$

$$\begin{aligned} m_{4}\ddot{y}_{4} + c_{34}\left(\dot{y}_{4} - \dot{y}_{3}\right) + k_{34}\left(y_{4} - y_{3}\right) + c_{2}\left(\dot{y}_{4} - \dot{y}_{2}\right) \\ + k_{4}\left(y_{4} - y_{2}\right) + c_{4}\left(\dot{y}_{4}\right) + k_{4}y_{4} = 0 \end{aligned} \tag{18}$$

$$m_{\mathrm{p}}\ddot{y}_{\mathrm{p}} + c_{\mathrm{p}}\left(\dot{y}_{\mathrm{p}} - \dot{y}_{3}\right) + k_{\mathrm{p}}\left(y_{\mathrm{p}} - y_{3}\right) = 0 \tag{19}$$

where N is the dynamic meshing force given by

$$N = k_{\mathrm{m}}\left(R_{\mathrm{b1}}\theta_{1} - R_{\mathrm{b2}}\theta_{2} - y_{1} + y_{2}\right) + c_{\mathrm{m}}\left(R_{\mathrm{b1}}\dot{\theta}_{1} - R_{\mathrm{b2}}\dot{\theta}_{2} - \dot{y}_{1} + \dot{y}_{2}\right) \tag{20}$$

where I_{d}, I_{l}, I_{1} and I_{2} are the inertia of drive, load, pinion and gear respectively. m_{1}, m_{2}, m_{3}, m_{4} and m_{p} are the masses of pinion, gear, foundation, bearings and the plate respectively. k_{1} and k_{2} are the transverse stiffness c_{1} and c_{2} are the transverse damping

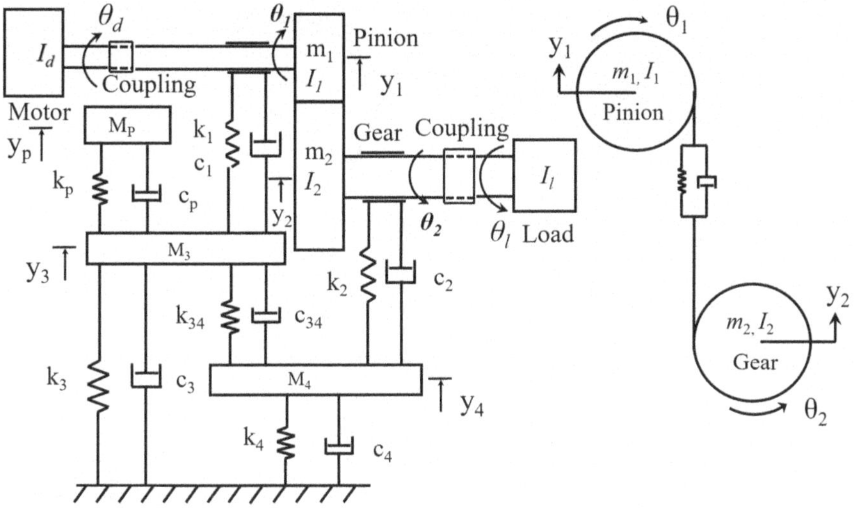

Fig. 3 Nine DOF spur gear system

of each shafts. The bearing foundations m_3 and m_4 are connected to each other by spring with stiffness k_{34} and stiffness c_{34}. The pinion and gear bearing foundations are connected to the main base by springs with stiffness k_3, k_4 and dampers with damping c_3, c_4 respectively. The plate is connected with pinion bearing foundation by a spring with stiffness k_p and a damper with damping c_p. $k_{\mathrm{t}1}$ and $k_{\mathrm{t}2}$ are torsional stiffness and $c_{\mathrm{t}1}$ and $c_{\mathrm{t}2}$ torsional damping between the shafts and load and again in between the shaft and drive. In this model, the transverse vibration of the gears and all masses are taken along the gear line of action as presented in Fig. 3.

2.4 Modeling of High Contact Ratio Gear System

A higher contact ratio spur gear system can be designed using many methods. In the present study, the method proposed by Buckingham for the instrument gearing is adopted [23] and the Buckingham dynamic load carrying capacity of pinion is calculated using Eq. (21):

$$F_\mathrm{d} = F_\mathrm{t} + \frac{0.05 * V_\mathrm{m}\,(bc + F_\mathrm{t})}{0.05 * V_\mathrm{m} + (bc + F_\mathrm{t})^{0.5}} \tag{21}$$

where V_m denotes pitch line speed, F_t denotes transmitting force, b and c are the constant.

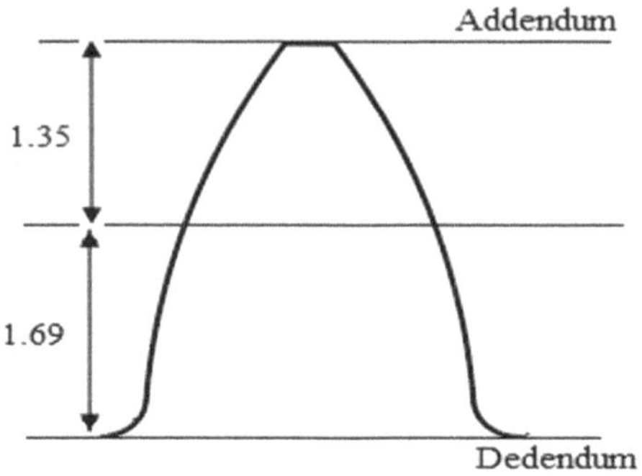

Fig. 4 Buckingham 1.35 addendum spur gear tooth

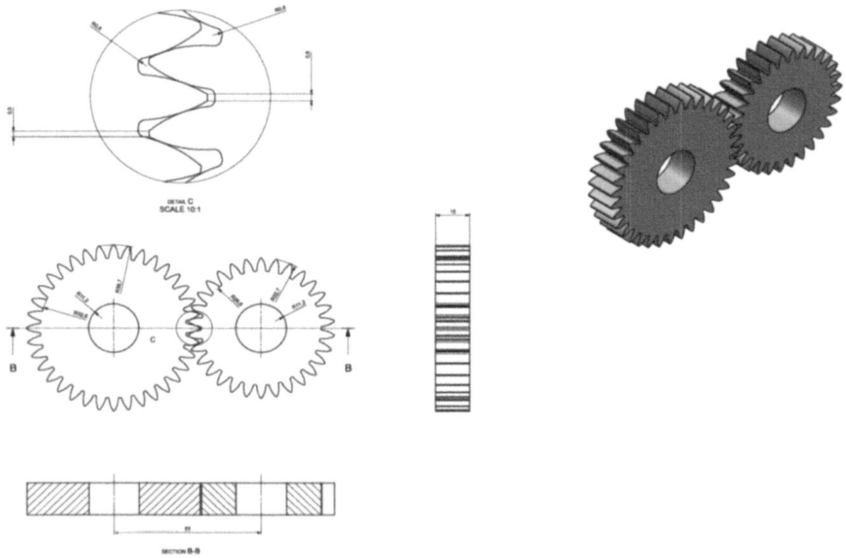

Fig. 5 Final assembly model of higher contact ratio spur gear system

In this method a 1.35 addendum and 1.69 dedendum coefficient, 15 mm teeth width, module of 2 mm and 20° normal pressure angle is chosen for the modeling of higher contact ratio spur gear system (Fig. 4). The criteria to avoid interface between the matting gears and the criteria for the top-land (top-land should be greater than or equals to 0.25/NDP) of the tooth tip are taken care during the design for negating the banding failure. In the present work, the spur gears are modeled using CATIA software and the final assembly is carried out in the SIEMENS NX software. The final assembly model of the higher contact ratio spur gear system is represented in Fig. 5. The specification of higher contact ratio spur gear system is listed in Table 2.

Table 2 Specification of HCR spur gear system

Parameter	Pinion	Gear
Number of teeth	30	36
Mass moment of inertia ($kg\ m^2$)	2×10^{-4}	4.14×10^{-4}
Mass (kg)	0.4439	0.6393
Speed (rpm)	2400	2000

Table 3 Inertia, stiffness and damping parameters of 9-DOF model

Inertia parameters	Stiffness parameters	Damping parameters
$m_1 = 0.4439$ (kg)	$k_1 = 2.5 \times 10^7$ (N/m)	$c_1 = 43.5$ (N s/m)
$m_2 = 0.6393$ (kg)	$k_2 = 2.5 \times 10^7$ (N/m)	$c_2 = 54.37$ (N s/m)
$m_3 = 0.33$ (kg)	$k_3 = 7.8 \times 10^8$ (N/m)	$c_3 = 321$ (N s/m)
$m_4 = 0.55$ (kg)	$k_4 = 7.8 \times 10^8$ (N/m)	$c_4 = 414$ (N s/m)
$m_p = 0.23$ (kg)	$k_{t1} = 276$ (N/m)	$c_{34} = 297$ (N s/m)
$I_1 = 2 \times 10^{-4}$ (kg m^2)	$k_{t2} = 248$ (N/m)	$c_{t1} = 0.0459$ (N s/m)
$I_2 = 4.14 \times 10^{-4}$ (kg m^2)	$k_{34} = 10.7 \times 10^8$ (N/m)	$c_{t2} = 0.0611$ (N s/m)
$I_d = I_l = 4.1 \times 10^{-3}$ (kg m^2)	$K_p = 1 \times 10^8$ (N/m)	$c_p = 74$ (N s/m)

3 Simulation Setup

The equations of the motions (Eqs. 11–19) are solved in MATLAB-Simulink environment using ODE 45 solver with a relative tolerance of 1×10^{-4}. The parameters of nine-DOF dynamic model are listed in Table 3 and the corresponding MATLAB-Simulink block diagram model of this system is shown in Fig. 6. A MATLAB function block is used to calculate the dynamic forces N and the corresponding moment M_i acting on the dynamic system. There are 9 inputs given to this block such as linear displacements (y_1, y_2), linear velocities ($\dot{y}_1$, $\dot{y}_2$), angular displacements (θ_1, θ_2), angular velocities ($\dot{\theta}_1$, $\dot{\theta}_2$), and the simulation time. Respective MATLAB functions are used for the lower contact ratio and higher contact ratio system and calculation of the TVMS for both healthy and faulted conditions are included in the code of MATLAB function.

4 Results and Discussion

The dynamic models of healthy as well as faulty gears are used to study the vibration response of the gear under constant speed and constant load conditions. The TVMS is calculated by analytical method and incorporated in the equations of motion of single stage spur gear system. A constant driving torque (T_d) of 10 Nm is applied to

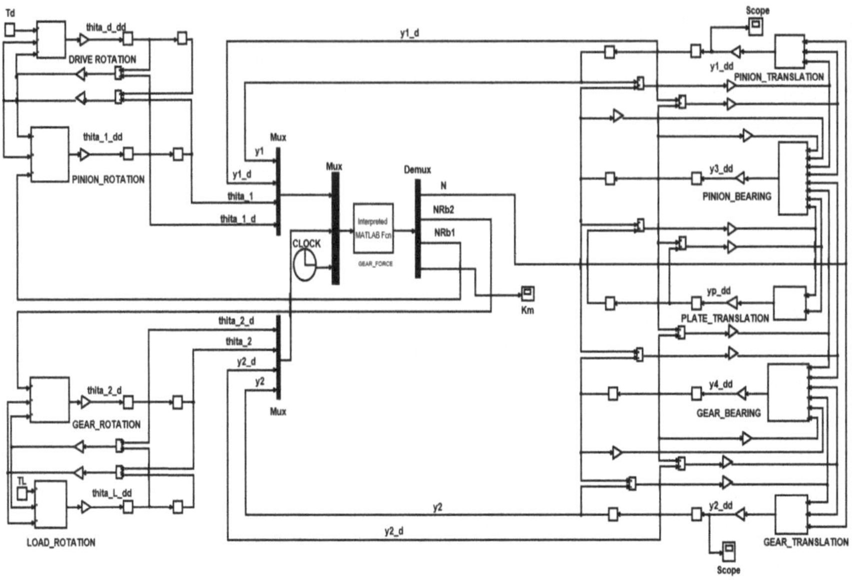

Fig. 6 MATLAB-Simulink model for 9-DOF spur gear system

the system and the load torque kept constant and set at (T_d * S.R), where S.R is the speed ratio. A comparative study of the analytical result of the TVMS of the spur gear system for both healthy and faulted condition is presented in Fig. 7. It can be observed that when the faulted tooth came in to meshing zone the TVMS reduced significantly due to loss and gain of tooth in contact. The TVMS of faulted condition, deviated from the healthy condition for a time period of c.r * T_m as the missing tooth region is present in the meshing zone during this period.

4.1 9-DOF Spur Gear System (Healthy)

The vibration acceleration response of pinion and gear is considered for the analysis. The healthy condition vibration acceleration in short window (for 0.05 s) and the corresponding envelope spectrum of both pinion and gear are presented in Figs. 8 and 9, respectively.

The symbol A_p and A_g are the acceleration signal of pinion and gear respectively and $A_p^e\ (j\omega)$ and $A_g^e\ (j\omega)$ are the envelope spectrum of the corresponding signal. In the above figures the amplitude of vibration is more in the pinion compared to the gear as the pinion drives the gear in case of simple spur gear train. Mesh frequency and its harmonics is exposed in the envelope spectrum of both pinion as well as gear.

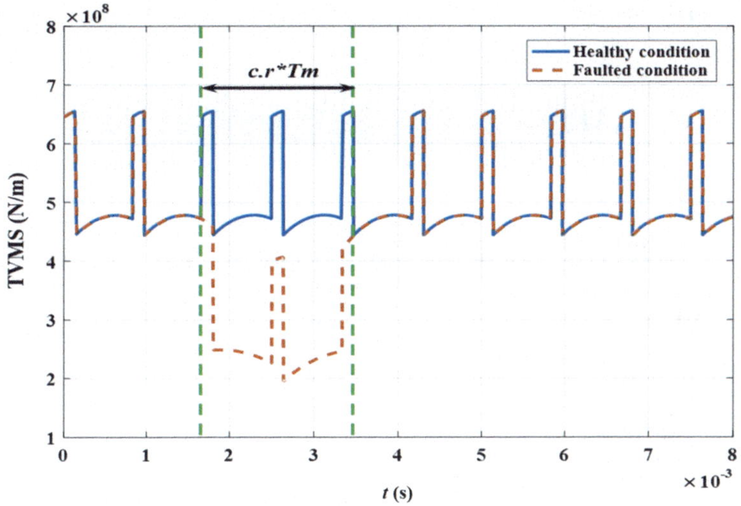

Fig. 7 Comparison of TVMS of healthy and faulted condition spur gear systems

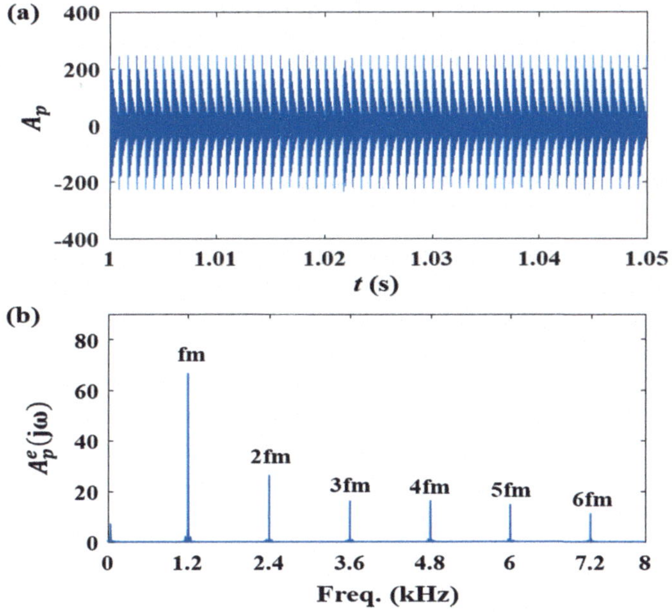

Fig. 8 **a** Acceleration signal of pinion and **b** its envelope spectrum of 9-DOF healthy simple spur gear system

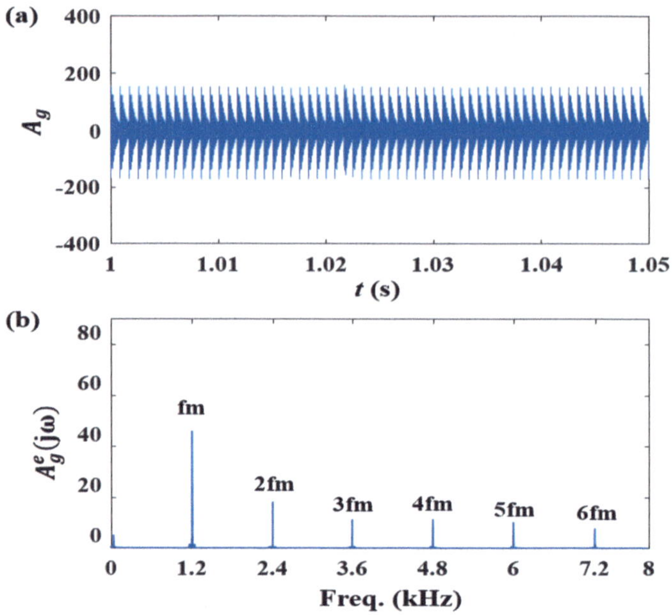

Fig. 9 **a** Acceleration signal of gear and **b** its envelope spectrum of 9-DOF healthy simple spur gear system

4.2 9-DOF Spur Gear System (Faulty)

In faulty condition vibration acceleration and the corresponding envelope spectrum of both pinion and gear are presented in Figs. 10 and 11. In those figures, it can be observed that significant peaks present in the time domain vibration signal of faulted system in both pinion and gear as compare to the healthy system. Due to the presents of missing tooth region in the meshing zone a larger amount of impulsive dynamic force is induced in the system and it results in high magnitude peaks in the time domain signal. The span between peaks is the pinion rotation period T_{rp}, as the missing tooth is on the pinion. The span between the consecutive higher peaks gives an indicator of the time period from the loss and gain of contact of the teeth. The span between the other peaks in the time domain gives the time for meshing of one teeth to the other. The envelope spectrum of corresponding signal of both pinion and gear in healthy condition is characterized with the gear mesh frequency (f_{m}) and its harmonics ($n * f_{\mathrm{m}}$). In envelope spectrum of the faulty system, amplitude modulation occurs in both pinion and gear due to the presence of high magnitude peaks in the vibration signal. Because of the amplitude modulation, higher magnitude sidebands are formed around gear mesh frequencies and its harmonics. The sidebands are characterized as ($i * f_{\mathrm{m}} \pm j * f_{\mathrm{rp}}$). Where i, j are integers and f_{rp} is the pinion rotational frequency. Characterization of sidebands with f_{rp} indicates the source of the fault which is the pinion. In the envelope spectrum it can be observed that the

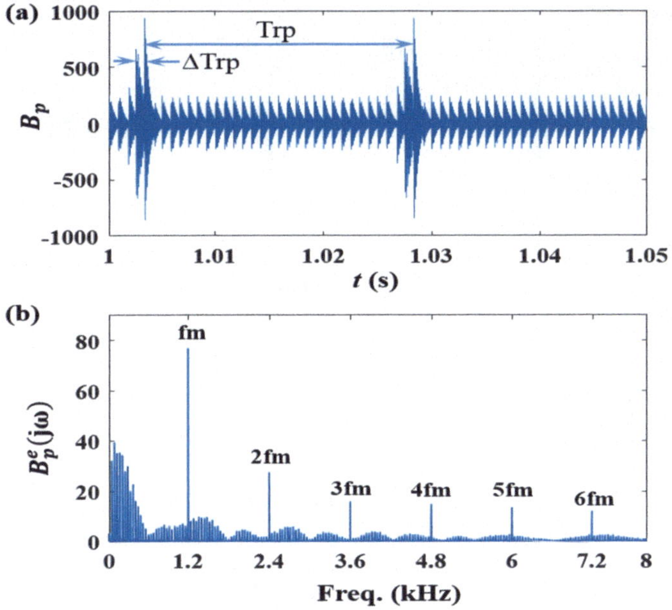

Fig. 10 **a** Acceleration signal of pinion and **b** its envelope spectrum of 9-DOF faulty simple spur gear system

amplitude of peaks at gear mesh frequency and its harmonics are more for the signal of pinion (Fig. 10b) as compared to that of the gear (Fig. 11b). The amplitude of peaks of the side bands are also more for pinion signal as compared to that of gear signals. In the above observations, the following inferences can be drawn:

- Whenever a gear teeth encounters any defects in its path it can excite the total systems as well as the surroundings. We can take the response at any parts of the gear systems.
- The response of the faulty parts of the simple gear train gives a better result as compared to the healthy parts.
- By monitoring the signal in time or frequency domain we can find out the health state of the machine.
- A proper maintenance strategy can be adopted by monitoring the signal.

5 Conclusions

In the present work, TVMS of spur gear system is calculated by analytical method. Gear fault like missing tooth is modeled in higher contact ratio single stage spur gear system and a dynamic-based model (9-DOF) is considered for the analysis of

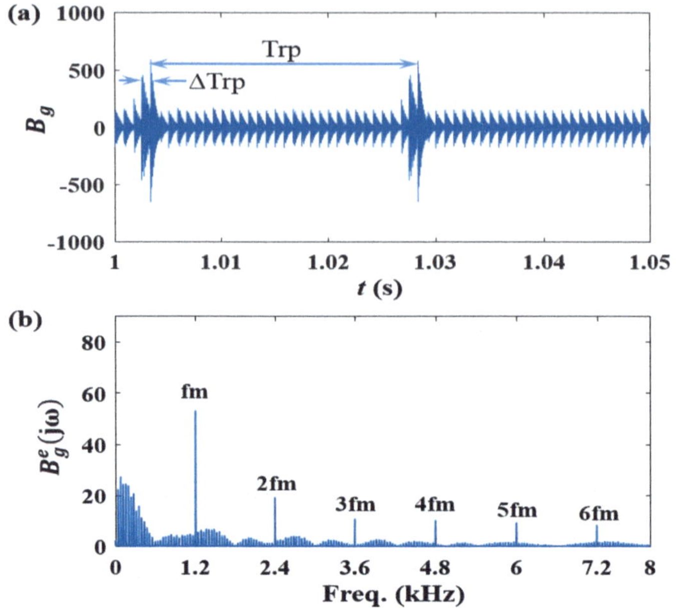

Fig. 11 **a** Acceleration signal of gear and **b** its envelope spectrum of 9-DOF faulty simple spur gear system

the vibration response of the systems. MATLAB-Simulink environment are used to solve the equations of motion of dynamic model using ODE45 solver. The obtained vibration responses for both healthy and faulted condition are analyzed in time and frequency domain. In time domain there are higher magnitudes peaks are observed as compare to the healthy condition, which indicates the impact of induced impulsive force generated due to the presents of fault in the system. The time intervals of the peaks are found as the pinion rotation time period, which specifies the source of fault. Similarly in the envelope spectrum, higher amplitude of GMF and its harmonics along with sidebands are observed in faulted condition. The sidebands are characterized with pinion rotational frequency which specifies the source of the fault. Thus, it can be inferred that in real time case of situation such types of vibration characteristics are expected in both healthy and faulty gear system.

References

1. Ozguven HN, Houser DR (1988) Dynamic analysis of high speed gears by using loaded static transmission error. J Sound Vib 125:71–83
2. Mohammed OD, Rantatalo M, Aidanpaa OJ (2015) Dynamic modelling of a one-stage spur gear system and vibration-based tooth crack detection analysis. Mech Syst Signal Process 54–55:293–305

3. Haword I, Jia S, Wang J (2001) The dynamic modeling of a spur gear in mesh including friction and a crack. Mech Syst Signal Process 15:831–853
4. Chen C, Shao Y (2011) Dynamic simulation of spur gear with tooth root crack propagating along tooth width and crack depth. Eng Fail Anal 18:2149–2164
5. Chaari F, Baccar W, Abbes MS, Haddar M (2008) Effect of spalling or tooth breakage on gear mesh stiffness and dynamic response of a one-stage spur gear transmission. Eur J Mech A/Solids 27:691–705
6. Luo Y, Baddour N, Liang M (2019) Dynamical modelling and experimental validation for tooth pitting and spalling in spur gears. Mech Syst Signal Process 119:155–181
7. Saxena A, Parey A, Chouksey M (2016) Time varying mesh stiffness calculation of spur gear pair considering sliding friction and spalling defects. Eng Fail Anal 70:200–211
8. Chen Z, Zhai W, Shao Y, Wang K, Sun G (2016) Analytical model for mesh stiffness calculation of spur gear pair with non-uniformly distributed tooth root crack. Eng Fail Anal 66:502–514
9. Saxena A, Chouksey M, Parey A (2017) Effect of mesh stiffness of healthy and cracked gear tooth on model and frequency response characteristics of geared rotor system. Mech Mach Theory 107:261–273
10. Wang Q, Zhao B, Fu Y, Kong X, Ma H (2018) An improved time-varying mesh stiffness model for helical gear pairs considering axial mesh force component. Mech Syst Signal Process 106:413–429
11. He S, Cho S, Singh R (2008) Prediction of dynamic friction forces in spur gears using alternate sliding friction formulations. J Sound Vib 309:843–851
12. Dogan O, Karpat F (2019) Crack detection for spur gears with asymmetric teeth based on the dynamic transmission error. Mech Mach Theory 133:417–431
13. Zheng H, Li Z, Chen X (2002) Gear fault diagnosis based on continuous wavelet transforms. Mech Syst Signal Process 16:447–457
14. Staszewski WJ, Tomlinson GR (1994) Application of the wavelet transform to fault detection in a spur gear. Mech Syst Signal Process 8:289–307
15. Rajeswari C, Sathiyabhama B, Devendiran S, Manivannan K (2014) A gear fault identification using wavelet, rough set based GA, ANN and C4.5 algorithm. Procedia Eng 97:1831–1841
16. Igba J, Alemzadeh K, Durugbo C, Eiriksson ET (2016) Analysing RMS and peak values of vibration signals for conditional monitoring of wind turbine gearboxes. Renew Energy 91:90–106
17. Li B, Zhang X, Wu J (2017) New procedure for gear fault detection and diagnosis using instantaneous angular speed. Mech Syst Signal Process 85:45–428
18. Wang W, Wong AK (2002) Autoregressive model-based gear fault diagnosis. J Vib Acoust 124:172–179
19. Yang DCH, Lin JY (1987) Hertzian damping, tooth friction and bending elasticity in gear impact dynamics. J Mech Trans Autom Des 109:189–196
20. Tian X (2005) Dynamic simulation for system response of gearbox including localized gear faults. University of Alberta, Canada, Library and Archives
21. Sainsot P, Velex P, Duverger O (2004) Contribution of gear body to tooth deflections—a new bi-dimensional analytical formula. J Mech Des 126:748–752
22. Omar FK, Moustaf KAF, Emam S (2011) Mathematical modeling of gearbox including defects with experimental verification. J Vib Control 18:1310–1321
23. Buckingham E (1988) Analytical mechanics of gears. Buckingham Associates, Inc., Springfield, Vermont in Association with Dover Publications, Inc., New York

Micromechanical Analysis of Carbon/Carbon Composites with Pore Characterization

O. S. Vishnu, Jhon Paul, and G. S. Pavan

Abstract In this study, the elastic properties of carbon/carbon (C/C) composite are computed by incorporating the presence of pores. Microstructure of C/C composites is analysed using scanning electron microscope (SEM) images. The pores inside the C/C composite are characterized based on area, shape and dimensions. Thirty-six SEM images are analysed. Based on this analysis, a three-dimensional RVE of C/C composite is constructed. The number of pores, their size and spatial distribution in the RVE is given by the analysis. Carbon fibres inside the representative volume element (RVE) are generated using the random sequential adsorption algorithm (RSA). Once the model is generated, periodic boundary conditions are imposed on the RVE model using Python script in Abaqus CAE. Effective elastic properties of C/C composites are computed using the finite element analysis (FEA)-based homogenization method. The effect of pore size distribution on the elastic properties of C/C composite can be understood from this study.

Keywords C/C composites · FEA-based homogenization · Pores · RVE

O. S. Vishnu · G. S. Pavan (✉)
Department of Civil Engineering, National Institute of Technology Karnataka Surathkal, Mangaluru, Karnataka 575025, India
e-mail: pavan.gs@nitk.edu.in

O. S. Vishnu
e-mail: vishnuos39@gmail.com

J. Paul
Anisotropic Private Limited, Hyderabad, Telangana, India

S. Sidhardh et al. (eds.), *Advances in Structural Integrity for Mechanical, Civil, and Aerospace Applications*, Lecture Notes in Mechanical Engineering,
https://doi.org/10.1007/978-981-97-6367-2_49

1 Introduction

Composites are materials consist of two or more materials which are distinct and identifiable in practice. A composite consists of fibres or reinforcements and a matrix which holds the fibres together. When both the fibres and matrix are of carbon origin, it is termed as carbon/carbon composites (C/C composites). Due to its excellent mechanical properties, low density and high thermal conductivity, C/C composites are widely employed in aerospace applications [1–3].

In general, C/C composites consist of carbon fibres and carbon matrix. But depending upon the manufacturing processes, the microstructure of the composite becomes complex. The microstructure of C/C composite is examined with the help of scanning electron microscope (SEM). It is observed that the carbon fibres embedded in the matrix are randomly distributed. There are pores or voids present in the matrix which could affect the effective elastic properties of the C/C composite [4]. When such complexity is present in the microstructure, analytical methods such as Mori Tanaka are proven to be less effective. Experimental approaches are not economic and even need complex testing facilities.

The complexity of the microstructure is better characterized by micromechanical methods such as FEA-based homogenization method. FEA-based homogenization techniques are widely used to evaluate the elastic properties of C/C composites [5, 6]. The most important step in FEA-based homogenization is to model a representative volume element (RVE) or a repeating unit cell (RUC). An RVE should be capable of incorporating all the important features present in the microstructure. For accurate prediction of effective elastic properties, the RVE must be constructed by incorporating all the important details in the microstructure [7]. Inside the RVE, the fibres are inserted by various methods such as Monte Carlo scheme (MC) [8] and random sequential adsorption algorithm (RSA) [9]. In MC scheme, all the fibres are inserted inside the empty RVE, and the RVE is squeezed to avoid existing overlapping between the fibres. But in RSA method, each fibre is inserted consecutively without overlapping with the existing fibres. The pores or voids present in the microstructure are observed to be dominant in C/C composites, and hence, they should be properly accounted. The shapes and geometry of the pores also play a significant role in the effective elastic properties of C/C composite. In general, pores are modelled as polygons for 2D analysis and as ellipsoids in 3D RVEs [10, 11]. The effective elastic properties of C/C composites are reducing considerably with an increase in porosity. It is observed that the pores with different sizes and shapes are distributed randomly in the carbon matrix of the C/C composite.

In this study, the effective elastic properties of C/C composites are evaluated using FEA-based homogenization. The details of the microstructure of C/C composite are obtained from SEM images. An RVE is constructed, and the fibres are inserted using RSA algorithm. The details about the sizes and shapes of pores are accurately accounted and inserted as ellipsoids inside the RVE. The numerical results obtained are compared with the values obtained from Mori Tanaka method [12] and also an RVE comprising of single fibre and matrix.

Table 1 Elastic properties of pyrolytic carbon matrix and carbon fibre

	Matrix	Carbon fibre
E_{11} (GPa)	30	233.13
E_{22} (GPa)	–	23.31
γ_{12}	0.28	0.2
G_{12} (GPa)	–	8.97
G_{23} (GPa)	–	8.23

2 Material Specification

2.1 Elastic Modulus

The elastic modulus of the carbon fibre and carbon matrix is obtained using manufacturer's data as shown in Table 1. It is assumed that carbon matrix exhibits isotropic behaviour and carbon fibre is transversely isotropic.

2.2 SEM Image Details

Images are taken with the help of a scanning electron microscope (SEM) to better understand the complex microstructure of C/C composite. Images of higher magnification from the SEM are used to determine the spatial arrangement of fibres and the diameter of carbon fibres. Carbon fibre has a diameter of 8.2 μm. Lower magnification images are used to understand the microstructure details of the C/C composite and compute porosity and overall fibre volume fraction. The porosity (pore volume fraction) and fibre volume fraction are found to be 8% and 42%, respectively.

3 RVE Generation

The elastic properties of C/C composite are obtained using a representative volume element (RVE). An RVE should be capable of mimicking the key features inside a composite. The important details of the microstructure cannot be included if the RVE is too small. When the RVE is bigger in size, the modelling and simulations become more complex. To avoid these issues, the RVE size is fixed as 50 times the radius of carbon fibres. So, in this paper, a cubical RVE with edge length of 205 μm is considered. An empty cuboid ($L \times L \times L$) is generated initially.

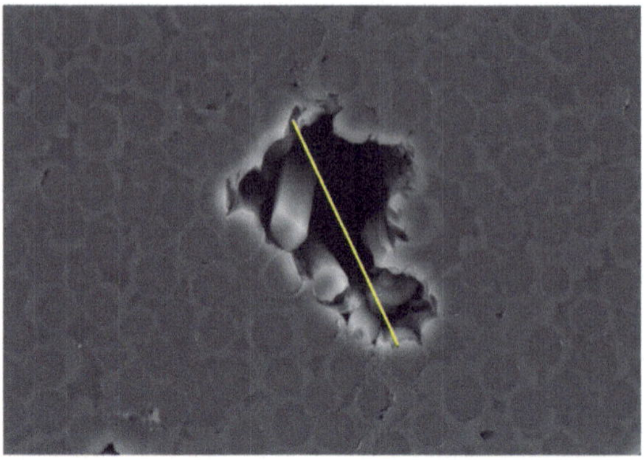

Fig. 1 C/C composite with pore showing l_{max}

3.1 *Pore Characterization and Modelling*

The porosity of the C/C composite is measured using SEM images. The areas of the pores inside the C/C composites are measured using image processing techniques in MATLAB and with the help of ImageJ software. The pores in the C/C composite are arbitrarily shaped and distributed at random. In the RVE model, the arbitrarily-shaped pores are assumed to be ellipsoids. The longest distance (l_{max}) inside the pore contour is measured as shown in Fig. 1 and is considered as the major axis length ($2a$) of an ellipse. Area of the pore is also computed. From the given value of areas and major axis length, the minor axis length is obtained using Eqs. (1) and (2).

$$a = \frac{l_{max}}{2} \tag{1}$$

$$b_1 = b_2 = \frac{2A_{pore}}{\pi l_{max}} \tag{2}$$

where b_1, b_2 are the semi major axis of the ellipsoid.

Using values of major axis length and minor axes lengths, an ellipsoid is generated. The ellipsoids are inserted inside the RVE at random positions without intersecting with each other as shown in Fig. 2.

3.2 *Fibre Distribution and Modelling*

The spatial distribution of fibres is found to be random in the C/C composites. So random sequential adsorption algorithm (RSA) is used to generate the fibres inside the RVE [13]. The procedure adopted to implement RSA algorithm in present study is as follows:

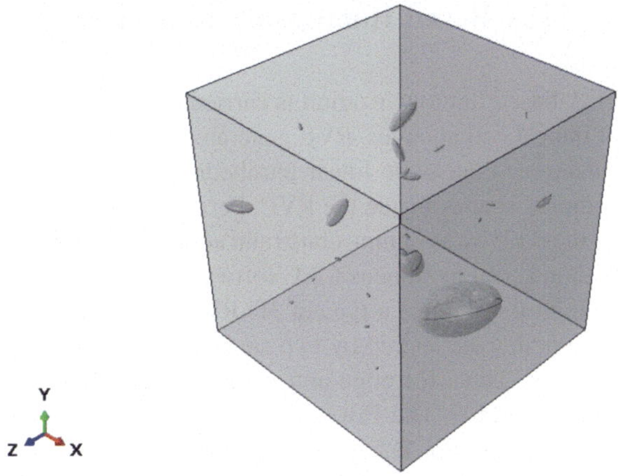

Fig. 2 3D view of RVE with pores

Step 1. Inside the 0.98L window of the RVE, a single fibre is inserted randomly. If the fibre is intersecting with the existing pores, it is deleted and regenerated.

Step 2. After successfully inserting the first fibre, the second fibre is generated. Similar to the first step, if it is intersecting with the existing pores or with the first fibre, it is deleted and regenerated.

Step 3. Step 2 is carried out continuously till the fibre volume fraction of the RVE reaches the desired value.

The minimum distance between fibres to fibres, pores to pores and fibres to pores is kept as a constant value based on the observations from SEM image data. The 3D RVE with pores and fibres is shown in Fig. 3.

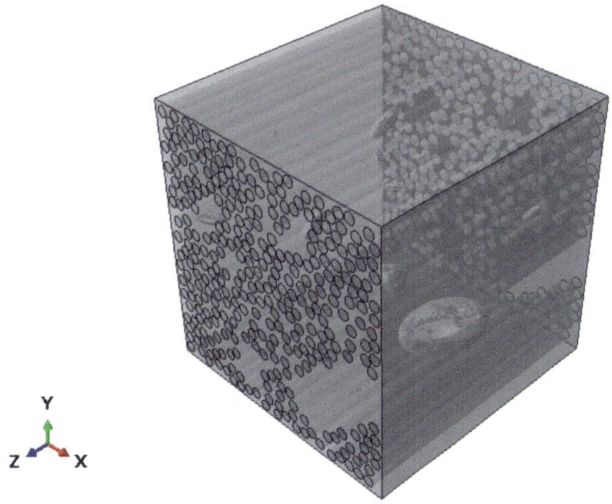

Fig. 3 3D RVE with pores and fibres

4 FEA-Based Homogenization Method

FEA-based homogenization is carried out to analyse the effective elastic behaviour of the C/C composites. RVE generated in Sect. 3 is adopted for FEA-based homogenization. Four-node linear tetrahedron elements (C3D4) are used because of the complex shapes inside the RVE. Periodic boundary conditions (PBCs) are imposed on the RVE by assigning constraint equations to pairs of nodes, edges and faces using python script in Abaqus CAE software.

Figure 4 depicts a typical 3D RVE with marked faces, edges and vertices. The edge nodes are shared by two surfaces, while the vertex nodes are shared by three surfaces. Thus, for edge and vertex nodes, the equation for surface nodes can be simplified [14]. The PBCs are formulated for face (Eqs. 3, 4, 5), edge (Eq. 6) and vertex nodes (Eq. 7), as shown below [15].

$$\begin{aligned} (u|_{x=L} - u|_{x=0})|_{y,z} &= L\varepsilon_{11}^{0} \\ (\upsilon|_{x=L} - \upsilon|_{x=0})|_{y,z} &= 0 \\ (\omega_{x=L} - \omega|_{x=0})|_{y,z} &= 0 \end{aligned} \tag{3}$$

abbreviated as $U_A - U_B = U_{AB}$

$$\begin{aligned} \left(u|_{y=B} - u|_{y=0}\right)|_{x,z} &= B\varepsilon_{12}^{0} \\ \left(\upsilon|_{y=B} - \upsilon|_{y=0}\right)|_{x,z} &= B\varepsilon_{22}^{0} \\ \left(\omega_{y=B} - \omega|_{y=0}\right)|_{x,z} &= 0 \end{aligned} \tag{4}$$

abbreviated as $U_C - U_D = U_{CD}$

$$\begin{aligned} (u|_{z=D} - u|_{z=0})|_{x,y} &= W\varepsilon_{13}^{0} \\ (\upsilon|_{z=D} - \upsilon|_{z=0})|_{x,y} &= W\varepsilon_{23}^{0} \\ (\omega_{z=D} - \omega|_{z=0})|_{x,y} &= W\varepsilon_{33}^{0} \end{aligned} \tag{5}$$

abbreviated as $U_E - U_F = U_{EF}$

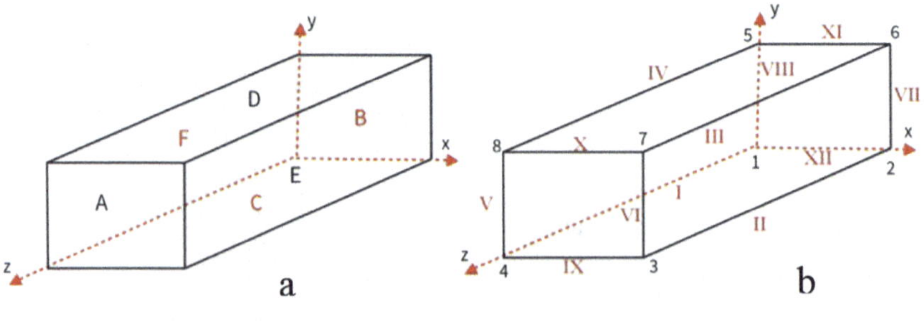

Fig. 4 Cubic unit cell for application of PBCs. **a** Edges and vertices identified and **b** faces identified

where *L*, *B* and *W* are the edge lengths of the RVE. Terms ε_{11}^0, ε_{12}^0, ε_{22}^0, ε_{13}^0, ε_{23}^0 and ε_{33}^0 are the far-field strains applied to the RVE. PBCs are implemented in the same way for all faces, with six different macroscopic strain components. Equations (3), (4) and (5) are simplified and expressed for edge and vertex and are given in Eqs. (6) and (7).

$$\left.\begin{aligned} U_{\text{II}} - U_{\text{I}} &= U_{AB} \\ U_{\text{III}} - U_{\text{I}} &= U_{AB} + U_{CD} \\ U_{\text{IV}} - U_{\text{I}} &= U_{CD} \\ \\ U_{\text{VI}} - U_{\text{V}} &= U_{AB} \\ U_{\text{VII}} - U_{\text{V}} &= U_{AB} + U_{EF} \\ U_{\text{VIII}} - U_{\text{V}} &= U_{EF} \\ U_{\text{X}} - U_{\text{IX}} &= U_{CD} \\ U_{\text{XI}} - U_{\text{IX}} &= U_{CD} + U_{EF} \\ U_{\text{XII}} - U_{\text{IX}} &= U_{EF} \end{aligned}\right\} \text{Edge Equations} \tag{6}$$

$$\left.\begin{aligned} U_2 - U_1 &= U_{AB} \\ U_3 - U_1 &= U_{AB} + U_{EF} \\ U_4 - U_1 &= U_{EF} \\ U_5 - U_1 &= U_{CD} \\ U_6 - U_1 &= U_{AB} + U_{CD} \\ U_7 - U_1 &= U_{EF} + U_{AB} + U_{CD} \\ U_8 - U_1 &= U_{CD} + U_{EF} \end{aligned}\right\} \text{Vertex Equations} \tag{7}$$

The next step is to apply the different far-field strains to generate the effective stiffness matrix of the composite. Six different load cases of far-field strains are applied to the RVE and are given by Eq. (8)

$$\begin{pmatrix} \varepsilon_{11} \\ \varepsilon_{22} \\ \varepsilon_{33} \\ \gamma_{12} \\ \gamma_{13} \\ \gamma_{23} \end{pmatrix} = \begin{pmatrix} 1 \\ 0 \\ 0 \\ 0 \\ 0 \\ 0 \end{pmatrix} \tag{8a}$$

$$\begin{pmatrix} \varepsilon_{11} \\ \varepsilon_{22} \\ \varepsilon_{33} \\ \gamma_{12} \\ \gamma_{13} \\ \gamma_{23} \end{pmatrix} = \begin{pmatrix} 0 \\ 1 \\ 0 \\ 0 \\ 0 \\ 0 \end{pmatrix} \tag{8b}$$

$$\begin{pmatrix} \varepsilon_{11} \\ \varepsilon_{22} \\ \varepsilon_{33} \\ \gamma_{12} \\ \gamma_{13} \\ \gamma_{23} \end{pmatrix} = \begin{pmatrix} 0 \\ 0 \\ 1 \\ 0 \\ 0 \\ 0 \end{pmatrix} \tag{8c}$$

$$\begin{pmatrix} \varepsilon_{11} \\ \varepsilon_{22} \\ \varepsilon_{33} \\ \gamma_{12} \\ \gamma_{13} \\ \gamma_{23} \end{pmatrix} = \begin{pmatrix} 0 \\ 0 \\ 0 \\ 1 \\ 0 \\ 0 \end{pmatrix} \tag{8d}$$

$$\begin{pmatrix} \varepsilon_{11} \\ \varepsilon_{22} \\ \varepsilon_{33} \\ \gamma_{12} \\ \gamma_{13} \\ \gamma_{23} \end{pmatrix} = \begin{pmatrix} 0 \\ 0 \\ 0 \\ 0 \\ 1 \\ 0 \end{pmatrix} \tag{8e}$$

$$\begin{pmatrix} \varepsilon_{11} \\ \varepsilon_{22} \\ \varepsilon_{33} \\ \gamma_{12} \\ \gamma_{13} \\ \gamma_{23} \end{pmatrix} = \begin{pmatrix} 0 \\ 0 \\ 0 \\ 0 \\ 0 \\ 1 \end{pmatrix} \tag{8f}$$

For each case of the far-field strain applied to the RVE, volume averaged stress components σ_{ij}^* are computed using Eq. (9).

$$\sigma_{ij}^* = \frac{1}{V} \int\limits_V \sigma_{ij} \mathrm{d}V. \tag{9}$$

For each case of far-field strain applied to RVE, six values of volume averaged strain components can be computed. In total, thirty-six values of stress components are obtained. These thirty-six values make up the coefficients of stiffness matrix $[C]$ which relate stress and strain components as shown in Eq. (10).

$$\sigma_{ij}^{*} = C_{ijkl}\varepsilon_{kl}, \tag{10}$$

where C_{ijkl} is the effective stiffness matrix of the composite.

$$\begin{pmatrix} \sigma_{11} \\ \sigma_{22} \\ \sigma_{33} \\ \tau_{12} \\ \tau_{13} \\ \tau_{23} \end{pmatrix} = \begin{bmatrix} C_{11} & C_{12} & C_{13} & C_{14} & C_{15} & C_{16} \\ C_{21} & C_{22} & C_{23} & C_{24} & C_{25} & C_{26} \\ C_{31} & C_{32} & C_{33} & C_{34} & C_{35} & C_{36} \\ C_{41} & C_{42} & C_{43} & C_{44} & C_{45} & C_{46} \\ C_{51} & C_{52} & C_{53} & C_{54} & C_{55} & C_{56} \\ C_{61} & C_{62} & C_{63} & C_{64} & C_{65} & C_{66} \end{bmatrix} \begin{pmatrix} \varepsilon_{11} \\ \varepsilon_{22} \\ \varepsilon_{33} \\ \gamma_{12} \\ \gamma_{13} \\ \gamma_{23} \end{pmatrix} \tag{11}$$

Corresponding compliance matrix $[S]$ is obtained by inverting the effective stiffness matrix $[C]$. The engineering properties of the C/C composite are evaluated from the coefficients of the compliance matrix $[S]$ using Eq. (12).

$$\begin{array}{lll} E_{11} = 1/S_{11} & E_{22} = 1/S_{22} & E_{33} = 1/S_{33} \\ \nu_{12} = -S_{12}/S_{11} & \nu_{23} = -S_{23}/S_{22} & \nu_{13} = -S_{13}/S_{11} \\ G_{12} = 1/S_{44} & G_{13} = 1/S_{55} & G_{23} = 1/S_{66} \end{array} \tag{12}$$

5 Results and Discussions

In addition to the RVE with multiple fibres, an RVE with single carbon fibre and matrix is modelled in Abaqus CAE. The diameter of the single carbon fibre in this RVE is kept as 150 μm, so that the fibre volume fraction is matching with that of RVE with multiple fibres.

For far-field strain given by Eq. (8a), the stress fields for RVE with multiple fibres and single fibre are shown in Fig. 5.

The effective elastic properties of the C/C composite with multiple carbon fibres are compared with the values obtained from Mori Tanaka method and RVE with single fibre and shown in Table 2. Mori Tanaka method is one of the oldest methods used for calculating the effective elastic properties of unidirectional composite. The RVE models with single fibre are used in many studies over the last decade by assuming a periodical distribution of fibres inside the composite [14].

In longitudinal direction (E_{11}), the modulus of elasticity obtained using Mori Tanaka method is matching with the values of the RVE with single fibre. RVE with multiple fibres shows lower modulus of elasticity than the RVE with single fibre. In the case of RVE with single fibre and Mori Tanaka method, it is evident that the interaction between fibres and the effect of pores is not considered. Comparing to the

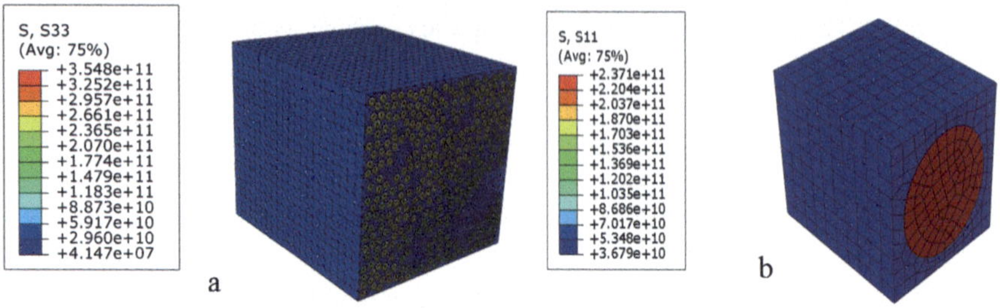

Fig. 5 Stress contours of RVE with: **a** multiple fibres, **b** single fibre

Table 2 Effective elastic properties of C/C composite

	FEA-based homogenization ($V_f = 42\%$)		Mori Tanaka method porosity = 0
	Single fibre porosity = 0	Multiple fibres porosity = 8%[a]	
E_{11} (GPa)	114.3	103.749	115.36
E_{22} (GPa)	27.87	27.8	46.45
γ_{12}	0.248	0.2519	0.248
G_{12} (GPa)	10.49	10.47	10.48
G_{23} (GPa)	10.02	10.16	10.06

[a] Porosity obtained from 2D SEM images

Mori Tanaka method, the effect of pores and the interaction between multiple fibres reduces the longitudinal elastic modulus (E_{11}) value by 11%. In this study, the edge length of the cubical RVE is equal to 50 times the radius of fibre. Hence, to achieve a fibre volume fraction of 42%, a total of 334 fibres are inserted inside the RVE. For the RVE with single fibre, a single fibre is placed at the centre of the RVE. Hence, the stress distribution inside the two RVE models varies significantly for the longitudinal loading case. The carbon fibre has very high longitudinal elastic modulus compared to the carbon matrix which will also contribute to the longitudinal elastic properties (E_{11}) of C/C composite.

In the case of elastic modulus in transverse direction (E_{22}), Mori Tanaka values do not compare well with the values obtained from FEA-based homogenization. The effect of multiple fibres in E_{22} is considerably low. However, RVE with multiple fibre and single fibre exhibited similar values of E_{22}.

It is found that the individual elastic properties of matrix and fibre play an important role in the final elastic properties of C/C composite. In the longitudinal direction, the modulus of elasticity (E_{11}) of carbon fibre is 233.13 GPa which is comparatively higher than the modulus of elasticity of matrix, 30 GPa. So, the effective elastic properties in longitudinal direction (E_{11}) are predominantly depends on the individual fibre properties. Hence, the distribution of fibres and the size of fibres affect the final elastic properties in longitudinal direction of C/C composite. But in transverse direction, the modulus of elasticity of fibre is equal to 23.11 GPa which is lower than the modulus of carbon matrix. So, changing the fibre arrangement does not influence the effective elastic property of C/C composite in transverse direction.

6 Concluding Remarks

In this study, 3D RVE of C/C composite was modelled with multiple fibres and pores. The pores were modelled as ellipsoids in the RVE. The volume fraction of fibres was 42%. The elastic properties of C/C composite were computed using FEA based homogenization techniques. The modulus of elasticity in longitudinal direction (E_{11}) was compared with Mori Tanaka method and RVE with single fibre. The values obtained by FEA-based homogenization for RVE with multiple fibres were 11% lower than the values given by Mori Tanaka method and RVE with single fibre. Presence of multiple fibres, their interaction and pores, whose effect was captured by FEA-based homogenization. The reduction in the modulus of elasticity value can be attributed to these effects. Because of the higher value of modulus of elasticity of carbon fibres in longitudinal direction, it has a predominant effect on the modulus of elasticity of C/C composites in longitudinal direction. Since the modulus of elasticity of carbon fibres is comparable to that of matrix, the effective elastic modulus of C/C composite along transverse direction is comparable with its constituents. The present model can be used to analyse structural elements consisting of C/C composites.

The modelling and evaluation of the complex geometry of pores and their distribution inside the C/C composite are the future scope of this study.

References

1. Martin JW (2006) Composite materials. Mater Eng 185–215. https://doi.org/10.1533/9781845691608.2.185
2. Park SJ, Seo MK (2011) Types of composites. Interface Sci Technol 18:501–629. https://doi.org/10.1016/B978-0-12-375049-5.00007-4
3. Ogin SL, Brøndsted P, Zangenberg J (2016) Composite materials: constituents, architecture, and generic damage. In: Modeling damage, fatigue and failure of composite materials, pp 3–23. https://doi.org/10.1016/B978-1-78242-286-0.00001-7
4. Tsukrov I, Novak J (2002) Effective elastic properties of solids with defects of irregular shapes. Int J Solids Struct 39:1539–1555. https://doi.org/10.1016/S0020-7683(01)00285-2
5. Rao MV, Mahajan P, Mittal RK (2008) Effect of architecture on mechanical properties of carbon/carbon composites. Compos Struct 83:131–142. https://doi.org/10.1016/J.COMPSTRUCT.2007.04.003
6. Sharma R, Mahajan P, Mittal RK (2013) Elastic modulus of 3D carbon/carbon composite using image-based finite element simulations and experiments. Compos Struct 98:69–78. https://doi.org/10.1016/J.COMPSTRUCT.2012.11.019
7. Drago A, Pindera MJ (2007) Micro-macromechanical analysis of heterogeneous materials: macroscopically homogeneous vs periodic microstructures. Compos Sci Technol 67:1243–1263. https://doi.org/10.1016/J.COMPSCITECH.2006.02.031

8. Gusev A, Heggli M, Lusti HR, Hine PJ (2002) Orientation averaging for stiffness and thermal expansion of short fiber composites. Adv Eng Mater 4:931–933. https://doi.org/10.1002/adem.200290008
9. Kari S, Berger H, Gabbert U (2007) Numerical evaluation of effective material properties of randomly distributed short cylindrical fibre composites. Comput Mater Sci 39:198–204. https://doi.org/10.1016/J.COMMATSCI.2006.02.024
10. Piat R, Böhlke T, Dietrich S, Gebert JM, Wanner A (2009) Modeling of effective elastic properties of carbon/carbon laminates. In: ICCM international conference on composite materials
11. Qi L, Chao X, Tian W, Ma W, Li H (2018) Numerical study of the effects of irregular pores on transverse mechanical properties of unidirectional composites. Compos Sci Technol 159:142–151. https://doi.org/10.1016/J.COMPSCITECH.2018.02.020
12. Liu L, Huang Z (2014) A note on Mori-Tanaka's method. Acta Mech Solida Sin 27:234–244. https://doi.org/10.1016/S0894-9166(14)60033-1
13. Böhm HJ, Eckschlager A, Han W (2002) Multi-inclusion unit cell models for metal matrix composites with randomly oriented discontinuous reinforcements. Comput Mater Sci 25:42–53. https://doi.org/10.1016/S0927-0256(02)00248-3
14. Sudhir A, Nagaraja AM, Gururaja S (2014) Effective mechanical properties of carbon-carbon composites. https://doi.org/10.1115/IMECE2014-36583
15. Li S (2001) General unit cells for micromechanical analyses of unidirectional composites. Compos Part A Appl Sci Manuf 32:815–826. https://doi.org/10.1016/S1359-835X(00)00182-2

The manufacturer's authorised representative in the EU is Springer Nature Customer Service Centre GmbH, Europaplatz 3, 69115 Heidelberg, Germany. If you have any concerns regarding our products, please contact ProductSafety@springernature.com

Printed and bound by CPI Group (UK) Ltd, Croydon, CR0 4YY
01/10/2026
02225911-0019